Oil and Gas Well Engineering Measurement and Control

Editors

Tianshou Ma
Yuqiang Xu

MDPI

Basel • Beijing • Wuhan • Barcelona • Belgrade • Novi Sad • Cluj • Manchester

Editors
Tianshou Ma
Petroleum Engineering School
Southwest Petroleum University
Chengdu
China

Yuqiang Xu
Department of Petroleum Engineering
China University of Petroleum (East China)
Qingdao
China

Editorial Office
MDPI
St. Alban-Anlage 66
4052 Basel, Switzerland

This is a reprint of articles from the Special Issue published online in the open access journal *Processes* (ISSN 2227-9717) (available at: www.mdpi.com/journal/processes/special_issues/Oil_Gas_Well_Engineering_Measurement_Control).

For citation purposes, cite each article independently as indicated on the article page online and as indicated below:

Lastname, A.A.; Lastname, B.B. Article Title. *Journal Name* **Year**, *Volume Number*, Page Range.

ISBN 978-3-7258-1306-3 (Hbk)
ISBN 978-3-7258-1305-6 (PDF)
doi.org/10.3390/books978-3-7258-1305-6

Contents

About the Editors

Tianshou Ma

Dr. Tianshou Ma is currently a professor in the Petroleum Engineering School at Southwest Petroleum University (SWPU). He received his bachelor's degree in petroleum engineering from SWPU in 2009 and his master's and Ph.D. degrees in oil and gas well engineering from SWPU in 2012 and 2015, respectively. His research interests focused on petroleum engineering, petroleum-related rock mechanics, and measurement while drilling (MWD), including rock mechanics characterization, wellbore stability analysis, lost circulation prevention, wellbore integrity, new MWD tools, and advanced drilling technology for safe and efficient drilling and exploitation of complex hydrocarbon reservoirs. He has published 90+ peer-reviewed journal articles, 20+ peer-reviewed international conference papers, and received 30+ registered patents. He has served as a co-chair or organizing committee member of academic conferences more than 10 times and is a member of the ARMA Future Leader Program. He is currently an associate editor of the *Journal of Petroleum Exploration and Production Technology* and *Frontiers in Earth Science* on Geohazards and Georisks section and an editorial board member of *Geomechanics* and *Geophysics* for *Geo-Energy* and *Geo-Resources, Geofluids, Advances in Civil Engineering, Advances in Mechanical Engineering, Shock and Vibration, Natural Gas Industry*, and *Drilling & Production Technology*.

Yuqiang Xu

Dr. Xu Yuqiang is currently a professor at the China University of Petroleum (East China), a Mount Taishan scholar, the winner of the Shandong Excellent Youth Fund, and the chief editor of Trends in Petroleum Engineering. His main research direction is deep-dwelling risk diagnosis and intelligent decision-making. He has hosted three projects funded by the National Natural Science Foundation of China, participated in more than 10 national-level projects, authorized more than 20 invention patents, and published more than 50 high-level academic papers. He received seven provincial and ministerial-level scientific research awards, including the second prize for scientific and technological progress in Shandong Province (ranked first) and the first prize for innovation in the China Invention Association (ranked first).

Preface

Oil and gas wells represent a unique channel in regard to oil and gas exploration and production. The measurement, control, and optimization of oil and gas wells are crucial for ensuring the safety, efficiency, and cost-effectiveness of drilling, logging, completion, stimulation, and production operations. Indeed, the application and development of information theory, cybernetics, system theory, and optimization theory have led to significant advancements in well-drilling, logging, completion, stimulation, and production operations, with more scientific approaches being utilized to help improve these areas of oil and gas production. Furthermore, numerous advanced technologies have become significantly involved in the fields of measurement, control, and optimization of wells. The accelerated exploration and development of deep-water, deep-formation, and unconventional oil and gas resources has led to the continued challenges that are frequently faced in the realm of oil and gas well engineering currently, such as high risks, low efficiency, and high costs. These challenges are further compounded by complicated surface conditions, complex geological conditions, high temperature and pressure (HTHP), strong vibrations, strong corrosion, multi-physics, and multi-phase effects. Due to the rapid development and progress of measurement and control technologies, big data, machine learning (ML), and artificial intelligence (AI), the abovementioned issues are expected to become even more challenging, with interventions being required in order to ensure the safety, high efficiency, and low costs of drilling, logging, completion, and stimulation moving forward.

This Special Issue aimed to collect research papers and review articles encompassing the latest advancements in measurement, control, and optimization technologies related to oil and gas well drilling, logging, completion, and stimulation.

Tianshou Ma and Yuqiang Xu
Editors

processes

Editorial

Oil and Gas Well Engineering Measurement and Control

Tianshou Ma [1,*] and Yuqiang Xu [2]

[1] State Key Laboratory of Oil and Gas Reservoir Geology and Exploitation, Southwest Petroleum University, Chengdu 610500, China
[2] School of Petroleum Engineering, China University of Petroleum (East China), Qingdao 266580, China; xuyuqiang@upc.edu.cn
* Correspondence: matianshou@126.com

Oil and gas wells represent a unique channel in regard to oil and gas exploration and production [1,2]. The measurement, control, and optimization of oil and gas wells are crucial for ensuring the safety, efficiency, and cost-effectiveness of drilling, logging, completion, stimulation, and production operations [1–4]. Indeed, the application and development of information theory, cybernetics, system theory, and optimization theory have led to significant advancements in well-drilling, logging, completion, stimulation, and production operations, with more scientific approaches being utilized to help improve these areas of oil and gas production [1–4]. Furthermore, numerous advanced technologies have becoming significantly involved in the fields of well-measurement, control, and optimization. The accelerated exploration and development of deep-water, deep-formation, and unconventional oil and gas resources has led to the continued challenges that are frequently faced in the realm of oil and gas well engineering currently, such as high risks, low efficiency, and high costs [5–8]. These challenges are further compounded by the complicated surface conditions [9], complex geological conditions [10], high temperature and pressure (HTHP) [11], strong vibrations [12], strong corrosions [13], multi-physics [14], and multi-phase effects [15]. Due to the rapid development and progress of measurement and control technologies, big data, machine learning (ML), and artificial intelligence (AI) [16–18], the abovementioned issues are expected to become even more challenging, with interventions being required in order to ensure the safety, high efficiency, and low costs of drilling, logging, completion, and stimulation moving forward.

This Special Issue, "Oil and Gas Well Engineering Measurement and Control" (https://www.mdpi.com/journal/processes/special_issues/Oil_Gas_Well_Engineering_Measurement_Control), showcases the latest advancements in the measurement, control, and optimization technologies related to oil and gas well drilling, logging, completion, and stimulation, featuring a total 33 original articles authored by 167 scientists from China, Egypt, Germany, Iraq, Kuwait, Russia, Saudi Arabia, Slovakia, Thailand, and the USA.

Citation: Ma, T.; Xu, Y. Oil and Gas Well Engineering Measurement and Control. *Processes* **2024**, *12*, 1034. https://doi.org/10.3390/pr12051034

Received: 8 May 2024
Accepted: 17 May 2024
Published: 20 May 2024

1. Advances in Oil and Gas Well Design and Drilling

Contribution 1 proposed a site selection model for multi-well pads of shale gas development in mountainous areas that incorporates both surface pad and underground wellbore construction costs, and it is based on digital elevation model (DEM) data. The optimal site selection can help to minimize the environmental damage and total costs of shale gas well construction.

Contribution 2 developed an optimization method for the lateral length of shale-gas horizontal wells, integrating geological, engineering, and economic factors. The economic lateral length was determined through a net present value model, and comprehensive assessments were conducted to identify the cost-optimal lateral lengths of shale gas horizontal wells in the Changning Block.

A mechanical specific energy (MSE)-based approach was formulated in Contribution 3, with the aim of detecting deep formation pore pressure. This new technique involved correcting downhole torque on bit (TOB), weight on bit (WOB), and revolution per minute

(RPM) data in order to obtain a more accurate MSE value under the conditions of compound drilling. This approach is anticipated to overcome the limitations of the compaction theory, creating a significant advantage over traditional pore pressure detection methods.

A cutting-edge method for detecting gas kicks using modern drilling technology and sophisticated software was detailed in Contribution 4. An investigation into the most appropriate procedure for gas well control, utilizing both the "driller's method" and the "wait and weight method", was conducted, and the correctness of the procedure in regard to gas well control was verified using the reaction to a gas kick from a well drilled in Hungary.

Contribution 5 involved the establishment of an experimental device for examining the friction between planes throughout periodic changes in normal force. The results of this experiment were used to modify the amplitude distribution model, featuring an average error between experiment and calculation of 3.28%. It is anticipated that this innovative concept will be implemented to reduce drill-string drag in drilling engineering.

The authors of Contribution 6 designed a new downhole robot actuator that incorporates a planetary roller screw mechanism (PRSM). The mechanical behaviors, contact load distribution, and fatigue life of this PRSM in the downhole robot system were investigated. It is believed that the work of this contribution will establish a solution to the challenges of high-accuracy and long-life transmission in downhole robot systems.

Contribution 7 discussed a discrete element model that could simulate the cutting of pebbled sandstone via a polycrystalline diamond compact (PDC) cutter, and the influence of composite impact load on the cutting performance and vibration behavior was also simulated and examined. The results revealed that the composite impactor can markedly enhance the rate of penetration (ROP), diminish vibration, and also safeguard drill bits and measurement while drilling (MWD) instruments in gravel-bearing formations.

Contribution 8 hypothesized a thermal-hydro-mechanical coupling model capable of investigating the bottomhole stress and rock breakage mechanism involved in deep-well drilling. The influence of formation properties and wellbore conditions on the bottomhole stress was also simulated and evaluated. In future, this detailed article should assist in comprehending the bottomhole stress and rock breakage mechanisms involved in the in deep-well drilling process.

Contribution 9 proposed a surface axial vibration technology to reduce the drag force of the drill-string in horizontal wells. This would consequently enhance the WOB transfer efficiency, the ROP, and the extended-reach limit of horizontal wells. The results demonstrated that the amplitude and frequency of the exciting force are the primary factors influencing the efficacy of surface vibration.

A stick–slip vibration model of the drill-string based on the multidimensional torsional vibration model was put contemplated in Contribution 10, and the impact of WOB, rotary torque, and rotary speed on stick–slip vibration was also emulated and discussed within the article. Finally, the drilling parameters were optimized for three different types of formations, namely those that were soft, medium-hard, and hard.

Contribution 11 conducted a comprehensive evaluation of global well-killing technology while also developing a formula for the blocking compositions that prevent the fluid loss during well-drilling operations. The physico-chemical, rheological, and filtration properties were subjected to rigorous testing, and the efficacy of the formula in complex conditions was validated. The solution provides a valuable means for enhancing fluid-loss control in carbonate reservoirs.

2. Advances in Oil and Gas Well Cement and Completion

Contribution 12 presented a potential breakthrough pressure model for the cement matrix and interfacial transition zone in underground gas-storage wells. The proposed model was verified by using a breakthrough-pressure-testing device and program. This work can be employed to assess the sealing capacity and sustained casing pressure of the cement–casing system in underground gas-storage wells.

Contribution 13 put forward a novel elastoplastic model for the cement sheath under acid fracturing conditions. The 3D yield state of cement and temperature stress were taken into account. The generation and evaluation mechanisms of the microannulus were analyzed and clarified. This study can provide theoretical guidance for the preventing of cement sealing failure and sustained casing pressure.

Contribution 14 provided a comprehensive overview of the characteristics of casing deformation observed in shale gas wells in the Sichuan Basin. The primary factors that exert a significant influence on casing deformation in shale gas wells were identified. To prevent casing deformation, it is recommended to optimize well trajectory, improve casing strength and cementing quality, or optimize fracturing operations.

Contribution 15 proposed an improved wellbore temperature and pressure model in offshore HTHP production wells. The implementation of the gas–liquid two-phase separated method has resulted in a notable improvement in the accuracy of the calculations in comparison to that of the traditional model. This model can be utilized to facilitate a comprehensive safety assessment of offshore HTHP production wells.

Contribution 16 hypothesized an elastic-plastic assessment model of the gas sealability of a sphere-type premium connection. The influence of make-up torque on gas sealability and contact pressure was investigated, and the additional make-up torque exerted a profound influence on gas sealability. This method can be utilized to optimize the structural design and technical parameters.

Contribution 17 presented a novel 3D stress analytical solution for the casing-cement-formation system in an inclined well. A comparative analysis of the previous models indicates a tendency to overestimate the absolute values of stress components and the potential for failure of the casing and cement in both 2D and 3D scenarios. This novel solution provides a foundation for benchmarking numerical simulation and a rapid assessment of wellbore integrity.

Contribution 18 proposed a transient surge pressure model for casing running. A multi-density slurry column structure integrated with an accurate wellbore pressure calculation and exerting annular back pressure was utilized to address the issue of casing running in a narrow safety mud-window formation. This approach is capable of achieving the desired outcomes of leak-proofing and pressure stabilization.

Contribution 19 developed a numerical temperature and pressure prediction model for steam injection in shale oil wells. The utilization of this model allows for the optimization of completion methods and injection-production parameters, thereby reducing the incidence of casing damage and enabling the extension of the casing's operational life.

3. Advances in Oil and Gas Well Production and Monitoring

Contribution 20 proposed a novel water holdup measurement approach, which can be utilized for monitoring the oil–water two-phase production profile in horizontal wells. The test prototype was used for dynamic testing on the horizontal well simulation facility. This novel approach is anticipated to address the challenges associated with water holdup profile measurements in horizontal wells.

Contribution 21 introduced an accurate detection system for the scale thickness in three-phase flow inside the oil pipes based on the attenuation of gamma rays. This detection system consists of a dual-energy gamma source and a sodium iodide detector. The system is capable of accurately detecting the scale value with the use of a single detector, thereby providing a solution to the problem of sediment scale settlement.

Contribution 22 conducted a computational fluid dynamic (CFD) simulation to investigate the impacts of fluid state, pressure distribution, sand volume, and particle sizes on the erosion rate of the electric choke valve in HTHP gas wells. This study can be utilized to optimize the layout of the electric choke valve, thereby reducing the associated costs and the number of required switching wells.

Contribution 23 conducted a CFD simulation to investigate the methane movement in the goaf drained by a large-diameter drilling machine under "U"-shaped ventilation.

The ideal spacing between large-diameter boreholes and the optimal distance between the borehole and the upper corner were deliberated.

Contribution 24 saw the proposal of a dynamic model of sealing surfaces for the premium connection, and the hysteresis features and energy dissipation mechanism of sealing surfaces were analyzed. The results demonstrated the effectiveness of energy dissipation theory in the analysis of the sealing performance of the premium connection, which contrasts with the traditional static contact analysis.

Contribution 25 developed a novel perforation-creating approach for thorough hydro jetting fracturing. This approach allows for the precise control of the perforation position, angle, and length in true triaxial hydraulic fracturing specimens. The true triaxial acid fracturing experiments were conducted. The perforation created by the novel approach exhibited a high degree of similarity to that observed in situ. The acid pretreatment was found to be an effective method of dissolving minerals, resulting in a notable reduction in breakdown pressure of 7.7 MPa.

Contribution 26 developed a dynamic simulation system for HTHP ignition. The system is designed to facilitate real-time monitoring and recording of oxidation parameters during the combustion process of crude oil. The experiments were performed to determine the ignition point of crude oil under a range of pressure conditions, heating rates, oil–water ratios, and gas injection rates. This system can provide essential data for the implementation of the project and numerical simulation.

4. Application of ML and AI in Oil and Gas Well Engineering

A new approach involving the combining of supervised and unsupervised ML was postulated by Contribution 27. This method is aimed at joint predicting produced water and natural gas associated with oil production from unconventional reservoirs. The approach has been demonstrated to achieve an accuracy of 91% in joint prediction. Such model-derived outlooks can assist operators in formulating management or remedial solutions.

Contribution 28 put forward a comprehensive artificial neural network (ANN) model for the prediction of formation permeability. The ANN model was trained using approximately 500 core data points collected from the Western Desert and Gulf regions of Egypt. The results demonstrate that the ANN model is capable of accurately forecasting core permeability with a high degree of precision, achieving a 98% accuracy rate.

Contribution 29 presented a finite element model (FEM) for shale gas production with free and adsorbed gas in both the matrix and fractures, and an ANN model was developed to predict the gas rate and inflow performance in shale reservoirs. The proposed ANN model exhibits considerable robustness and predictive capability in relation to gas rate prediction in shale reservoirs.

Contribution 30 developed a multiscale generative adversarial network-based image inpainting method for the treatment of formation micro imager (FMI) images. The residual blocks were incorporated into the U-Net network with the objective of enhancing the quality of the filled logging images. In contrast to the majority of existing filling algorithms, the proposed method demonstrated superior performance when applied to images of complex lithology.

Contribution 31 proposed a novel pattern-recognition-based approach to the early detection of kick in offshore drilling. This approach integrated data filtering, pattern recognition, Bayesian framework, and multiphase flow models. The results demonstrated that the proposed approach exhibited a high precision in monitoring early kick while maintaining a low false positive rate.

Contribution 32 established an integrated framework combining the sparrow search algorithm (SSA) with the long short-term memory (LSTM) neural network. The integration of data-driven techniques and mechanistic models has resulted in an improvement in the precision of forecasting formation fluid-loss pressure, as well as the generation of valuable insights for the prevention of lost circulation during drilling.

Finally, Contribution 33 proposed a lost circulation prediction model combining the improved whale optimization algorithm (WOA) and bidirectional long short-term memory (BiLSTM) algorithm. In comparison to the LSTM, the BiLSTM, and WOA-BiLSTM models, the improved WOA-BiLSTM model demonstrated superior performance in terms of lost circulation prediction, as proposed in this study. This was evidenced by a 22.3%, 18.7%, and 4.9% higher prediction accuracy, respectively.

Author Contributions: Conceptualization, T.M. and Y.X.; investigation, Y.X.; writing—original draft preparation, T.M.; writing—review and editing, T.M. and Y.X. All authors have read and agreed to the published version of the manuscript.

Conflicts of Interest: The authors declare no conflicts of interest.

Abbreviation

AI	artificial intelligence
ANN	artificial neural network
BiLSTM	bidirectional long short-term memory
CFD	computational fluid dynamic
DEM	digital elevation model
FMI	formation micro imager
FEM	finite element method
HPHT	high-temperature high-pressure
LSTM	long short-term memory
ML	machine learning
MSE	mechanical specific energy
MWD	measurement while drilling
PDC	polycrystalline diamond compact
PRSM	planetary roller screw mechanism
ROP	rate of penetration
RPM	revolution per minute
SSA	sparrow search algorithm
TOB	torque on bit
WOB	weight on bit
WOA	whale optimization algorithm

List of Contributions

1. Zhang, J.; Hu, N.; Li, W. Rapid site selection of shale gas multi-well pad drilling based on digital elevation model. *Processes* **2022**, *10*, 854. https://doi.org/10.3390/pr10050854.
2. Zhu, J.; He, S.; Lin, L. Optimization of the lateral length of shale-gas horizontal wells based on geology-engineering-economy integration. *Processes* **2023**, *11*, 249. https://doi.org/10.3390/pr11010249.
3. Yin, H.; Cui, H.; Gao, J. Research on pore pressure detection while drilling based on mechanical specific energy. *Processes* **2022**, *10*, 1481. https://doi.org/10.3390/pr10081481.
4. Huszar, T.; Wittenberger, G.; Skvarekova, E. Warning signs of high-pressure formations of abnormal contour pressures when drilling for oil and natural gas. *Processes* **2022**, *10*, 1106. https://doi.org/10.3390/pr10061106.
5. Liu, Y.; Liu, Y.; Guan, Z.; Niu, Y. Calculation model of the effect of periodic change of normal force on sliding friction characteristics between the planes. *Processes* **2022**, *10*, 1138. https://doi.org/10.3390/pr10061138.
6. Dong, X.; Zhu, H.; Liu, Q.; Wang, Q.; Wang, X. Design of downhole robot actuator system and mechanical behavior analyses of the PRSM by considering elastic errors and radial loads. *Processes* **2022**, *10*, 1520. https://doi.org/10.3390/pr10081520.
7. Zhang, H.; Ni, H.; Yang, H.; Fu, L.; Wang, Y.; Liu, S.; Huang, B.; Wang, Z.; Chen, G. Numerical simulation and field test research on vibration reduction of PDC cutting pebbled sandstone under composite impact load. *Processes* **2023**, *11*, 671. https://doi.org/10.3390/pr11030671.

8. Yang, B.; Xu, H. Analysis of bottomhole rock stress in deep-well drilling considering thermal-hydro-mechanical coupling. *Processes* **2023**, *11*, 683. https://doi.org/10.3390/pr11030683.
9. Long, Y.; Wang, X.; Wang, P.; Zhang, F. A method of reducing friction and improving the penetration rate by safely vibrating the drill-string at surface. *Processes* **2023**, *11*, 1242. https://doi.org/10.3390/pr11041242.
10. Wang, C.; Chen, W.; Wu, Z.; Li, J.; Liu, G. Stick–slip characteristics of drill strings and the related drilling parameters optimization. *Processes* **2023**, *11*, 2783. https://doi.org/10.3390/pr11092783.
11. Islamov, S.; Islamov, R.; Shelukhov, G.; Sharifov, A.; Sultanbekov, R.; Ismakov, R.; Agliullin, A.; Ganiev, R. Fluid-loss control technology: from laboratory to well field. *Processes* **2024**, *12*, 114. https://doi.org/10.3390/pr12010114.
12. Yang, Y.; Li, L.; Yu, W.; Zhou, Y.; Zhu, K.; Yuan, B. The application of breakthrough pressure in the evaluation of the sealing ability of cement–casing interface and cement matrix in underground gas-storage wells. *Processes* **2022**, *10*, 620. https://doi.org/10.3390/pr10040620.
13. Su, D.; Wu, X.; Li, Z.; Huang, S.; Li, J.; Sun, J.; Zheng, G. Theoretical analysis of the micro annulus of an oil-well cement sheath formed via cooling under acid-fracturing conditions. *Processes* **2022**, *10*, 966. https://doi.org/10.3390/pr10050966.
14. Xu, B.; Yang, S.; Yuan, B.; Ma, L.; Wang, L. Mechanism analysis and potential solutions for casing deformation of shale gas fracturing wells in Sichuan Basin. *Processes* **2022**, *10*, 1711. https://doi.org/10.3390/pr10091711.
15. Jing, J.; Shan, H.; Zhu, X.; Huangpu, Y.; Tian, Y. Wellbore temperature and pressure calculation of offshore gas well based on gas-liquid separated flow model. *Processes* **2022**, *10*, 2043. https://doi.org/10.3390/pr10102043.
16. Yang, B.; Xu, H.; Xiang, S.; Zhang, Z.; Su, K.; Yang, Y. Effects of make-up torque on the sealability of sphere-type premium connection for tubing and casing strings. *Processes* **2023**, *11*, 256. https://doi.org/10.3390/pr11010256.
17. Wang, X.; Jiang, T.; Zhang, Y.; Zhou, J.; Xiao, H.; Li, W. A three-dimensional analytical solution of stress field in casing-cement-stratum system considering initial stress state. *Processes* **2023**, *11*, 1164. https://doi.org/10.3390/pr11041164.
18. Mei, Y.; Yang, H.; Zhang, Z.; Ji, M. Research on managed-pressure running casing in oil and gas wells with the negative pressure window. *Processes* **2023**, *11*, 2210. https://doi.org/10.3390/pr11072210.
19. Yu, X.; Cen, X.; Kan, C.; Hu, Y.; Yang, Y.; Tao, S.; Chen, X.; Chen, X.; Hu, Z. Numerical simulation analysis of wellbore integrity and casing damage in high-temperature injection and production of shale oil. *Processes* **2023**, *11*, 3053. https://doi.org/10.3390/pr11113053.
20. Wang, Y.; Han, J.; Hao, Z.; Zhou, L.; Wang, X.; Shao, M. A new method for measuring water holdup of oil-water two-phase flow in horizontal wells. *Processes* **2022**, *10*, 848. https://doi.org/10.3390/pr10050848.
21. Mayet, A.; Chen, T.; Alizadeh, S.; Al-Qahtani, A.; Alanazi, A.; Ghamry, N.; Alhashim, H.; Eftekhari-Zadeh, E. Optimizing the gamma ray-based detection system to measure the scale thickness in three-phase flow through oil and petrochemical pipelines in view of stratified regime. *Processes* **2022**, *10*, 1866. https://doi.org/10.3390/pr10091866.
22. Guo, L.; Wang, Y.; Xu, X.; Gao, H.; Yang, H.; Han, G. Study on the erosion of choke valves in high-pressure, high-temperature gas wells. *Processes* **2022**, *10*, 2139. https://doi.org/10.3390/pr10102139.
23. Lei, Y. Movement law of methane drained by large-diameter borehole drilling machine in the goaf. *Processes* **2022**, *10*, 1669. https://doi.org/10.3390/pr10091669.
24. Yu, Y.; Cao, Y.; Qu, Z.; Dou, Y.; Wang, Z. Finite-element analysis on energy dissipation and sealability of premium connections under dynamic loads. *Processes* **2023**, *11*, 1927. https://doi.org/10.3390/pr11071927.

25. Jia, W.; Mou, J.; Wang, G.; Li, X.; Wang, X.; Ma, X. A new experimental method for acid pretreatment in perforated horizontal wells: a case study of Mahu conglomerate reservoir. *Processes* **2023**, *11*, 3353. https://doi.org/10.3390/pr11123353.
26. Yin, Y.; Chen, X.; Yu, X.; Liu, D.; Chen, C.; Zhou, X.; Li, X.; Zhang, L.; Kan, C. Experimental tests on in situ combustion using dynamic ignition simulation system in high-temperature and high-pressure conditions. *Processes* **2024**, *12*, 52. https://doi.org/10.3390/pr12010052
27. Vikara, D.; Khanna, V. Application of a deep learning network for joint prediction of associated fluid production in unconventional hydrocarbon development. *Processes* **2022**, *10*, 740. https://doi.org/10.3390/pr10040740.
28. Abdel Azim, R.; Aljehani, A. Neural network model for permeability prediction from reservoir well logs. *Processes* **2022**, *10*, 2587. https://doi.org/10.3390/pr10122587.
29. Abdel Azim, R.; Aljehani, A. Finite element and neural network models to forecast gas well inflow performance of shale reservoirs. *Processes* **2022**, *10*, 2602. https://doi.org/10.3390/pr10122602.
30. Sun, Q.; Su, N.; Gong, F.; Du, Q. Blank strip filling for logging electrical imaging based on multiscale generative adversarial network. *Processes* **2023**, *11*, 1709. https://doi.org/10.3390/pr11061709.
31. Xu, Y.; Yang, J.; Hu, Z.; Xu, D.; Li, L.; Fu, C. A novel pattern recognition based kick detection method for offshore drilling gas kick and overflow diagnosis. *Processes* **2023**, *11*, 1997. https://doi.org/10.3390/pr11071997.
32. Chen, D.; He, B.; Wang, Y.; Han, C.; Wang, Y.; Xu, Y. Prediction of leakage pressure during a drilling process based on SSA-LSTM. *Processes* **2023**, *11*, 2608. https://doi.org/10.3390/pr11092608.
33. Liu, X.; Jia, W.; Li, Z.; Wang, C.; Guan, F.; Chen, K.; Jia, L. Prediction of lost circulation in southwest Chinese oil fields applying improved WOA-BiLSTM. *Processes* **2023**, *11*, 2763. https://doi.org/10.3390/pr11092763.

References

1. Ma, T.; Chen, P.; Zhao, J. Overview on vertical and directional drilling technologies for the exploration and exploitation of deep petroleum resources. *Geomech. Geophys. Geo-Energy Geo-Resour.* **2016**, *2*, 365–395. [CrossRef]
2. Ma, T.; Liu, J.; Fu, J.; Wu, B. Drilling and completion technologies of coalbed methane exploitation: An overview. *Int. J. Coal Sci. Technol.* **2022**, *9*, 68. [CrossRef]
3. Lai, J.; Wang, G.; Fan, Q.; Zhao, F.; Zhao, X.; Li, Y.; Zhao, Y.; Peng, X. Toward the scientific interpretation of geophysical well logs: Typical misunderstandings and countermeasures. *Surv. Geophys.* **2022**, *44*, 463–494. [CrossRef]
4. Li, L.; Tan, J.; Wood, D.A.; Zhao, Z.; Becker, D.; Lyu, Q.; Shu, B.; Chen, H. A review of the current status of induced seismicity monitoring for hydraulic fracturing in unconventional tight oil and gas reservoirs. *Fuel* **2019**, *242*, 195–210. [CrossRef]
5. Wei, Y.; Feng, Y.; Tan, Z.; Yang, T.; Yan, S.; Li, X.; Deng, J. Simultaneously improving ROP and maintaining wellbore stability in shale gas well: A case study of Luzhou shale gas reservoirs. *Rock Mech. Bull.* **2023**, 100124, *in press*. [CrossRef]
6. Xie, H.P.; Liu, T.; Gao, M.Z.; Chen, L.; Zhou, H.W.; Ju, Y.; Gao, F.; Peng, X.B.; Li, X.J.; Peng, R.D.; et al. Research on in-situ condition preserved coring and testing systems. *Pet. Sci.* **2021**, *18*, 1840–1859. [CrossRef]
7. Yang, B.; Wang, H.Z.; Li, G.S.; Wang, B.; Chang, L.; Tian, G.H.; Zhao, G.M.; Zheng, Y. Fundamental study and utilization on supercritical CO_2 fracturing developing unconventional resources: Current status, challenge and future perspectives. *Pet. Sci.* **2022**, *19*, 2757–2780. [CrossRef]
8. Gao, D. Some research advances in well engineering technology for unconventional hydrocarbon. *Nat. Gas Ind. B* **2022**, *9*, 41–50. [CrossRef]
9. Zhang, J.; Hu, N.; Li, W. Rapid site selection of shale gas multi-well pad drilling based on digital elevation model. *Processes* **2022**, *10*, 854. [CrossRef]
10. Yin, H.; Cui, H.; Gao, J. Research on pore pressure detection while drilling based on mechanical specific energy. *Processes* **2022**, *10*, 1481. [CrossRef]
11. Yang, W.; Wang, B.; Yao, J.; Ranjith, P.G.; Zhang, X. Experimental study on the physical and mechanical properties of carbonatite rocks under high confining pressure after thermal treatment. *Deep. Undergr. Sci. Eng.* **2024**. [CrossRef]
12. Yuan, L.; Jiangang, S.; Minghu, N.; Chi, P.; Yingjie, W. Drill string dynamic characteristics simulation for the ultra-deep well drilling on the south margins of Junggar Basin. *Petroleum* **2023**, *9*, 205–213. [CrossRef]
13. Zhao, Y.L.; Ye, F.X.; Zhang, G.; Yao, J.; Liu, Y.F.; Dong, S.G. Investigation of erosion-corrosion behavior of Q235B steel in liquid-solid flows. *Pet. Sci.* **2022**, *19*, 2358–2373. [CrossRef]

14. Feng, D.; Chen, Z.; Wu, K.; Li, J.; Dong, X.; Peng, Y.; Jia, X.; Li, X.; Wang, D. A comprehensive review on the flow behaviour in shale gas reservoirs: Multi-scale, multi-phase, and multi-physics. *Can. J. Chem. Eng.* **2022**, *100*, 3084–3122. [CrossRef]
15. Jing, J.; Shan, H.; Zhu, X.; Huangpu, Y.; Tian, Y. Wellbore temperature and pressure calculation of offshore gas well based on gas-liquid separated flow model. *Processes* **2022**, *10*, 2043. [CrossRef]
16. Mohammadpoor, M.; Torabi, F. Big Data analytics in oil and gas industry: An emerging trend. *Petroleum* **2020**, *6*, 321–328. [CrossRef]
17. Mirza, M.A.; Ghoroori, M.; Chen, Z. Intelligent petroleum engineering. *Engineering* **2022**, *18*, 27–32. [CrossRef]
18. Li, G.; Song, X.; Tian, S.; Zhu, Z. Intelligent drilling and completion: A review. *Engineering* **2022**, *18*, 33–48. [CrossRef]

processes MDPI

Article

Rapid Site Selection of Shale Gas Multi-Well Pad Drilling Based on Digital Elevation Model

Jijun Zhang, Nan Hu * and Wenjing Li

School of Economics and Management, Southwest Petroleum University, Chengdu 610500, China; 198931010017@swpu.edu.cn (J.Z.); 201031010092@swpu.edu.cn (W.L.)
* Correspondence: hunan@swpu.edu.cn

Abstract: Drilling and completion platform construction is a fundamental part of oil and gas reservoir development, and the location of construction directly affects the whole process of shale gas drilling and development. Due to the complex surface conditions and fragile ecological environment in mountainous areas, having an appropriate platform location can significantly reduce shale gas development and environmental costs. The DEM (digital elevation model) includes geographic elevation, surface complexity, land use type, and other data, so it can be used for rapid site selection for shale gas multi-well pad drilling. In this study, first, research results related to drilling platform site selection were analyzed and summarized, and then a platform site selection method aiming to minimize the total well construction cost was developed. Second, the well construction costs were decomposed into the surface construction costs and the underground construction costs, and the site selection model with the lowest total multi-well pad construction costs was established. Third, ground feature data obtained from DEM (digital elevation model) processing were substituted into the site selection model and solved using the genetic clustering algorithm. Finally, two practical cases were used to verify the research method developed in this study. The results show that the platform site selection results can be used to not only guide the formulation of development plans, but also to reduce the scope of the field investigation in the process of site selection, reduce the intensity of field work, and improve the work efficiency.

Keywords: shale gas; multi-well pad drilling; platform location; genetic clustering algorithm

Citation: Zhang, J.; Hu, N.; Li, W. Rapid Site Selection of Shale Gas Multi-Well Pad Drilling Based on Digital Elevation Model. *Processes* **2022**, *10*, 854. https://doi.org/10.3390/pr10050854

Academic Editors: Tianshou Ma and Yuqiang Xu

Received: 4 March 2022
Accepted: 21 April 2022
Published: 26 April 2022

1. Introduction

The success of the "shale gas revolution" in the United States and the rapid growth of shale gas production in China have proven that the multi-well development mode has a significant impact on the improvement of shale gas development efficiency [1,2]. The shale gas production cycle has a direct impact on commercial development benefits. Shortening well construction time, accelerating production, and improving development efficiency are the only ways to realize commercial development, resulting in higher requirements for shale gas development engineering technology. The location of a multi-well drilling platform for shale gas directly affects the total drilling footage, which, in turn, affects the total cost of drilling and completion. In addition, the location of the platform has a great impact on the cost of the preliminary infrastructure construction, the difficulty of drilling and completing wells, and the implementation of stimulation measures for shale gas development. Under the current trend of the integration of geology and engineering in shale gas development, the determination of the platform and well location in pre-drilling engineering designs should first meet the geology and engineering requirements. When the complex surface shape does not meet the construction requirements, the wellhead position should be adjusted within the surface range permitted by the underground geological targets [3]. The marine shale development area in South China is considered to be the most promising shale gas block in existence [4–6]. However, the complex surface conditions in the area impose constraints on shale gas well placement, heavy cargo transportation, water and electricity supplies, and environmental protection, increasing the costs and technical difficulty of shale gas development in the area. Therefore, scientific site selection for multi-well pad drilling is an important prerequisite for the efficient development of shale gas.

Devine and Lesso [7] established an optimal location planning model for offshore platforms with the objective of minimizing drilling costs. Later, Frair and Devine [8] extended the above model, taking into account factors such as the platform construction cost and output at each stage, and optimizing the model with the goal of maximizing the net present value. Costa [9], Dogru [10], Grimmett [11], Garcia-Diaz JG [12], and others also conducted further studies on this problem. Rodrigues H W L [13], Walesca [14], and Mohammed Almedallah [15] et al. established a location optimization model for offshore drilling platforms and proposed a linear programming model that minimizes the overall development costs of a given oilfield. It addresses the number, location, and capacity of production platforms; the number and location of wells; where manifolds must be installed; the interconnections between platforms, manifolds, and wells; and whether each part of the well should be vertical or horizontal. Zeeshan Tariq [16–20], Mohamed Mahmoud [21], Mohammad Rasheed Khan [22], and Ah-med Sadeed [23] used machine learning, particle swarm optimization, and other methods to optimize drilling and completion plans and injection–production plans in real time, achieving results that significantly improved their economic indicators.

Ge Yunhua et al. [24] established an optimization model aimed at minimizing the sum of the investment in oilfield surface engineering construction and oil and gas well construction, studying the internal rules concerning the size, number, and position of platforms and the investment in oilfield surface engineering construction and oil and gas well construction under the technical condition of horizontal well clusters. Yan Tie [25] investigated well targets, such as the objective function of minimizing the sum of the horizontal projection distance using projection disjoint or fewer intersections as constraint conditions. They constructed a well cluster distribution model using the ant colony algorithm, which can be used to quickly identify the optimal distribution plan for the wellhead; however, they did not consider that the difference between the actual well track and the projection may lead to errors in the model. Liu Zhen et al. [26] used graph theory and the weighted center problem of network analysis to develop an optimal selection model for choosing the location of the central platform for offshore oil fields. Shi Yucai [27] considered the drilling target distribution and the minimum target horizontal displacement sum of squares as the optimization targets, basing their construction on the number of platforms. Regarding subordinate well relations, they used a combination of several different ocean drilling platform location optimization model scenarios. Then, they used the dynamic clustering analysis method for the evaluation.

According to the distribution of geological targets, Li Wenfei et al. [28] established a mathematical programming model for the location optimization of clustered well drilling platforms, solving the model with the genetic algorithm. Based on the minimum sum of total well depth method, Zhang Yuchen et al. [29] comprehensively considered the influences of the different drilling costs of different well types on the platform location and optimized the platform location. Wang Zhiyue et al. [30] used the drilling learning curve of the "well factory" to establish and solve a position optimization model for a horizontal well platform under the well factory mode. Huo Hongbo et al. [31] established position optimization technology based on an economic evaluation of an integrated exploration and development platform in the Bohai Sea through a comparison between the new production platform based on the exploration of the well's location and the optimization scheme without considering the location of the exploration well platform. This was combined with an investigation of the influence of the location of the exploration well's building platform on the drilling difficulty.

Although there have been many achievements regarding the optimization of drilling sites or platform positions, all of the developed methods have some limitations: (1) The common site selection method uses topographic and geological mapping combined with a field investigation to identify the appropriate location for a well drilling site. However, the traditional method mainly depends on the person's subjective judgment, so it is difficult to consider the complex location factors and their influences, and China's enriched areas of shale gas resources located in places such as forests or mountains cannot rely on having artificial universal coverage advantageous drilling areas in field investigations, which eventually leads to the development of a scheme to meet the demands of drilling and the subsequent project. (2) All of the research methods fail to take into account the restrictions of the surface conditions in terms of their effects on platform site selection, which has a significant impact on the construction of shale gas drilling pads (from the aspects of construction cost and environmental protection). Based on previous research results, this study analyzed the influence of surface conditions on platform site selection and developed a multi-well drilling pad site selection method based on regional digital elevation data to produce site selection results that are more suitable for the efficient development of shale gas in areas with complex surface conditions.

2. Problem Statement and Formulation

Of the total reservoir development costs, drilling and fracturing costs account for a large proportion. Since the fracturing costs are not affected by the location of the platform, the impact of the fracturing costs is not considered when the optimization goal is to minimize the well construction costs. After determination of the reservoir development plan, the more drilling and total footage of the drilling pad there is, the more land area is required, the greater the surface construction costs for the drilling pad and for moving and installing the rig are, and the greater the total well construction costs of the platform are. Therefore, for a certain number of drilling pads, the total cost of well construction should be the lowest possible to achieve the maximum development benefit. On the basis of geological modeling, the locations and number of target points are known, and the optimization objective is to minimize the total well construction costs of the horizontal pad. A multi-objective optimization model was established for the location of the shale gas platform to optimize the location of the multi-well drilling pad.

Based on the known coordinates of the underground well pattern, the surface morphological characteristics, the land use, and various cost standards, investment in shale gas development and construction can be divided into two aspects. On one hand, there is the investment in drilling pad construction, including the construction costs of the platform or well site and the cost of moving or installing drilling rigs. On the other hand, there are the costs related to underground well construction, mainly drilling and completion costs. Thus, the total investment in shale gas multi-well drilling pad construction is

$$C = CG + CD \tag{1}$$

where C is the total investment in platform development and construction (in 10^4 \$), CG is the ground construction investment of the platform (in 10^4 \$), and CD is the investment in the underground well construction of the platform (in 10^4 \$).

2.1. Ground Construction Cost Model

The amount of land required for shale gas development depends on the density of the well cluster, the size of the well cluster, the number of wells in each well cluster, and the specific conditions of the shale reservoir being developed. Shale gas development typically uses the factory drilling model with multiple horizontal wells on a single tablecloth. This significantly reduces the land use demand, especially in mountainous areas of southern China and in places with complex surface conditions. The area of the multi-well drilling pad for shale gas is determined by the number of wells on the platform, the well layout, and the rig configuration. The platform wellhead arrangement can be divided into single layout and double layout wells. Drilling operations can be completed by single or double drilling rigs. Commonly used drilling rig models can be divided into "ZJ50" and "ZJ70" drilling rigs. The "ZJ50" drilling rig is suitable for drilling operations with depths of less than 5000 m, and the "ZJ70" drilling rig is used for drilling operations with depths of more than 5000 m. In addition, the "ZJ70" drilling rig is equipped with more equipment than the "ZJ50" rig, and the cost of rig installation and relocation is also more than for the "ZJ50" rig. Therefore, it is necessary to select a suitable type of rig to complete the drilling project at the lowest possible cost.

In addition, the area of the well site should also meet the requirements of drilling and fracturing operations. According to the demands of the well site for shale gas development and construction, a calculation model for the drilling platform area was established:

$$S = [X_R + (N/A - 1) \times 5] \times Y_A \tag{2}$$

where S is the area of the platform well site (in m^2); X_R is the length of the well site which drilling with different rig, such as a single drilling rig (X_R = 65 m), double "ZJ50" drilling rig (X_R = 105 m), or another drilling rig combination (X_R = 115 m); and Y_A is the width of the well site which set with different layout wells. For single-layout wells, this is Y_A = 50 m; for double-layout wells, it is Y_A = 80 m. N is the number of platform cloth wells and A is the platform wellhead arrangement, where a single well has a value of A = 1 and a double well has a value of A = 2.

The Digital Elevation Model (*DEM*) is a digital simulation of a topographic surface or a digital representation of a topographic surface form developed with limited topographic elevation. The *DEM* is equally spaced in the horizontal and vertical directions, and the plane coordinates of the grid dot are hidden in the column number, which is usually stored in the matrix structure; that is, the elevation values of the grid cells are recorded one-by-one according to the row (or column). When adopting the *DEM* for site selection optimization, the accuracy of the *DEM* in the study area is ε (in m). Then, the regional *DEM* can be divided into grids with a size of $\varepsilon \times \varepsilon$, and platforms with different well layouts and rig configurations can be transformed into a region composed of $X/\varepsilon \times Y/\varepsilon$ *DEM* grids.

The costs involved in well site construction include the land requisition costs, well site leveling costs, and infrastructure construction costs. The land requisition fee is determined by the local government according to the area occupied and the type of land requisitioned. It can be expressed as

$$C_{zd} = S \times C_{zds} \times K \tag{3}$$

where C_{zd} is the platform land acquisition cost (in 10^4 \$), C_{zds} is the unit price of the land acquisition cost (in 10^4 \$/m^2), and K is the land type coefficient occupied by the platform.

The cost of well site leveling is determined by the level and area of land occupied by the well site: the smaller the elevation difference is, the flatter the land will be, and the less basic engineering needed to level the well site. Therefore, in terms of site selection, a location with the flattest surface possible, a gentle slope, and a small degree of cutting should be chosen. *DEM* raster data are used to extract the raster average elevation difference to represent the land flatness, which can be specifically expressed as

$$C_{pz} = S \times \sum_{i=1}^{Ng} G_i \times C_{pzs}. \tag{4}$$

where C_{pz} is the cost of well site leveling engineering (in 10^4 \$), C_{pzs} is the unit price for well site leveling in (10^4 \$/m^3), N_g is the number of grids that make up the platform, and G_i is the elevation difference in grid i (in m). G_i is calculated using the average elevation of a single grid minus the average height of all grids in the well site. It is a single grid elevation difference that levels a well site to the same elevation throughout the grid. Above average heights can be extracted from the *DEM* data. The quantity of the well site can be determined from the composition of the well site elevation difference for all grids multiplied by the area of the corresponding grid, where positive values indicate excavation and negative values indicate filling, both of which are included in the engineered area.

Multi-well drilling pad infrastructure engineering mainly involves leveling the well site, tamping the foundation, laying the ground of the well site, and building the foundation conditions before drilling. Thus, the cost of the multi-well drilling pad infrastructure project is mainly determined by the area of the well site. It can be expressed as

$$C_{pc} = S \times C_{pcs} \tag{5}$$

where C_{pc} is the cost of infrastructure construction projects (in 10^4 \$) and C_{pcs} is the unit price of the infrastructure project cost (in 10^4 \$/m^2).

Multi-well pad drilling for shale gas can significantly reduce rig moving and installation costs. In general, the fewer platforms there are in the block, the less the rig moves, and the lower the rig moving cost is. The moving and installing costs of the platform rig are determined by the number of wells on the platform, the method of well placement, and the configuration of the rig. They can be expressed as

$$C_m = C_{m1} + C_{m2} \times (N-1)/A + C_{m3}/A \tag{6}$$

where C_m is the cost of moving and installing the rig (in 10^4 \$), C_{m1} is the initial installation cost of the rig (in 10^4 \$), C_{m2} is the cost of translating the well within the same row (in 10^4 \$ per unit of time), and C_{m3} is the cost of moving the drill well between rows (in 10^4 \$ per unit of time).

Based on the above analysis, the surface construction cost of a multi-well drilling pad can be expressed as

$$CG = C_{zd} + C_{pz} + C_{pc} + C_{pr} + C_m \tag{7}$$

2.2. Drilling and Completion Cost Model

The costs of drilling and completion are determined by the well structure, borehole trajectory, drilling construction procedure, and drilling and completion technology. When the drilling and completion plan for the block have been determined, the main factors affecting the drilling and completion costs are the shape of the well track and the length of footage. Once the reservoir engineer has determined the well location, the target position, vertical depth, and horizontal section length of the horizontal well are all fixed. When the platform position changes, the wellbore trajectory and well depth change with the wellhead position, resulting in changes in the drilling and completion costs.

The well trajectory design generally follows the following principles: (1) ensuring that the purpose of drilling directional wells is achieved; (2) consideration of the ground conditions; (3) correct selection of the deflection point, borehole curvature, and maximum deviation angle; and (4) a profile design that is conducive to safe and fast drilling whilst also reducing drilling costs [32]. Under the premise of satisfying the drilling purpose, the vertical section should be kept as long as possible, and in order to adjust the vertical depth at the end of the deflection section and improve the hit target rate, we adopted the three-dimensional profile of the "vertical section + increasing deviation section + steady deviation section + torsional azimuth section + horizontal section" to determine the well trajectory. When calculating the trajectory parameters of a 3D horizontal well, it is necessary to first provide some trajectory parameters and then determine the rest of the trajectory parameters and the total footage.

In Figure 1, O is the wellhead position, K_a is the deviation building point, K_b is the initial point in the stabilized section, K_c is the end point in the stabilized section, T_A is the first target in the horizontal section, and T_B is the end target in the horizontal section. R_1 and R_2 are the curvature radii of the deflection section and the torsional azimuth section, respectively. L_1, L_2, L_3, and L_h are the length of the building section, the length of the steady slope section, the length of the torsion bearing section, and the length of the horizontal section, respectively.

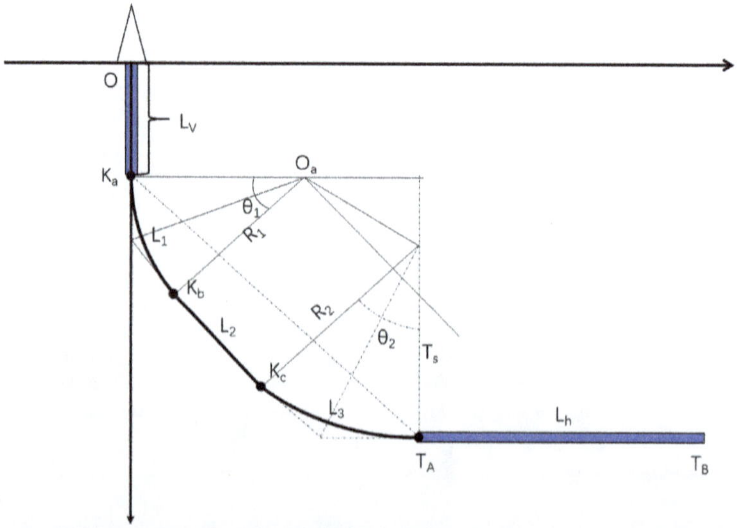

Figure 1. Drilling section design (projection of the plumb plane).

The known conditions include the T_A coordinates of the first target in the horizontal section (X_{ti},Y_{ti},Z_{ti}), the K_a coordinates of the deflection point (X_{ki},Y_{ki},Z_{ki}), the well inclination angle α_3, the azimuth angle φ_3, and the deflection slopes K_1 and K_2 of the deflection section and torsional azimuth section. According to the analytical method used in [33,34], the building section length L_1, the steady section length L_2, the torsional azimuth section length L_3, as well as the well inclination angle α_2 and the azimuth angle φ_2 in the steady section were obtained through the following steps.

The solution to the 3D borehole trajectory design model was transformed into three dimensional nonlinear equations:

$$\begin{cases} 2(T_s + L_2)L_m + 2(T_t + L_2)L_n + 2(1 - \cos\theta)L_m L_n = H^2 - L_2{}^2 \\ K_1{}^2 L_m{}^2[T_s + (1 - \cos\theta)L_n + L_2] = 2L_m + (1 + \cos\theta)L_n + L_2 - T_s \\ K_2{}^2 L_n{}^2[T_t + (1 - \cos\theta)L_m + L_2] = 2L_n + (1 + \cos\theta)L_m + L_2 - T_t \end{cases} \tag{8}$$

where H is the straight distance between the building angle and the first target (in m); T_s and T_t are the projection lengths of the straight segment $K_A T_A$ on the direction vector of the skew point and the horizontal segment, respectively (in m); and θ is the angle between the borehole direction vector at the deflection point and the borehole direction vector in the horizontal section.

$$H = \sqrt{(x_t - x_k)^2 + (y_t - y_k)^2 + (z_t - z_k)^2} \tag{9}$$

$$T_s = z_t - z_k \tag{10}$$

$$T_t = [(x_t - x_k)\cos\varphi_3 + (y_t - y_k)\sin\varphi_3]\sin\alpha_3 + (z_t - z_k)\cos\alpha_3 \tag{11}$$

Suppose that $\begin{cases} x = \frac{L_m}{H} \\ y = \frac{L_n}{H} \\ z = \frac{L_2}{H} \end{cases}$, and $x, y, z \in (0,1)$; meanwhile, $\begin{cases} t_1 = \frac{T_s}{H} \\ t_2 = \frac{T_t}{H} \end{cases}$, $\begin{cases} k_1 = \frac{K_1}{H} \\ k_2 = \frac{K_2}{H} \end{cases}$, $\begin{cases} a = 1 - \cos\theta \\ b = 1 + \cos\theta \end{cases}$. Thus, the ternary nonlinear equations can be transformed into

$$\begin{cases} k_1{}^2 x^2 (t_1 + ay + z) = 2x + by + z - t_1 \\ k_2{}^2 y^2 (t_2 + ax + z) = bx + 2y + z - t_2 \\ 2(t_1 + z)x + 2(t_2 + z)y + 2axy = 1 - z^2 \end{cases} \tag{12}$$

The nonlinear equations can be quickly solved by the quasi-analytical method, and the design parameters of borehole trajectory were obtained. The values of the central angles θ_1 and θ_2 of the two arc segments can be calculated using the following formula:

$$\cos\theta_1 = \cos\alpha_2 \tag{13}$$

$$\cos\theta_2 = \cos\alpha_2\cos\alpha_3 + \sin\alpha_2\sin\alpha_3\cos(\varphi_3 - \varphi_3) \tag{14}$$

After solving the track design parameters, the calculation expression of the total track footage of a single well hole can be obtained as follows:

$$L_i = L_v + L_1 + L_2 + L_3 + L_H \tag{15}$$

In summary, the total well depth of the horizontal well on the platform is

$$L_T = \sum_{i=1}^{N} L_i. \tag{16}$$

The drilling and completion cost is a function of the well depth and the drilling and completion cost, expressed as

$$CD = \sum_{i=1}^{N}(L_{vi} \times P_{vs} + L_{ki} \times P_{ks} + L_{hi} \times P_{hs}). \tag{17}$$

where P_{vs} is the unit price of drilling and completion in the vertical section (in 10^4 \$/m), P_{ks} is the unit price of drilling and completion in the deflecting section (in 10^4 \$/m), and P_{hs} is the unit price of drilling and completion in the horizontal segment (in 10^4 \$/m).

2.3. Constraint Conditions

In order to avoid fishhook boreholes and reduce the drilling difficulty, the center of the platform should be located in the polygon area composed of all targets, that is, the boundary of the area where all well targets are selected for the platform:

$$\begin{cases} x_{min} < X < x_{max} \\ y_{min} < Y < y_{max} \end{cases} \tag{18}$$

According to the requirements of oil drilling HSE, the selected platform site should be no less than 500 m away from the plane of residence:

$$\sqrt{\left(X \pm \frac{X_R}{2} - x_h\right)^2 + \left(Y \pm \frac{Y_A}{2} - y_h\right)^2} \geq 500 \tag{19}$$

where (x_h, y_h) are the grid coordinates of human settlements (in m).

The selected position of the platform should not occupy agricultural land or a river water source. Based on the *DEM* grid data of land use in the block, unsuitable site selection areas such as agricultural land and river water source areas were removed, and the feasible region R for the platform's location was determined:

$$(X, Y) \in R \tag{20}$$

In order to obtain the range of R, different types of *DEM* data (such as elevation data, land use type data, etc.) are used for a Boolean operation, and the intersection is obtained as the area in line with the site selection requirements.

3. The Method and Process Used to Solve the Optimization Model

Because the subordination relationship between the platform and the well is known, the exhaustive method can be used to solve the optimization problem; however, because of the large amount of calculations involved, the genetic algorithm is used to solve the location problem. The genetic algorithm is a simulated evolutionary process of the random search algorithm. First, randomly generated values that satisfy the restrictions of the actual problem are used as the initial solution set (parent), and then through genetic coding, genetic operations such as selection, crossover, and mutation determine the offspring solution from the parent solution to better adapt to the given fitness function and determine the optimal solution to the problem, or to approximate the optimal solution [35]. The basic process of solving the genetic algorithm is shown in Figure 2.

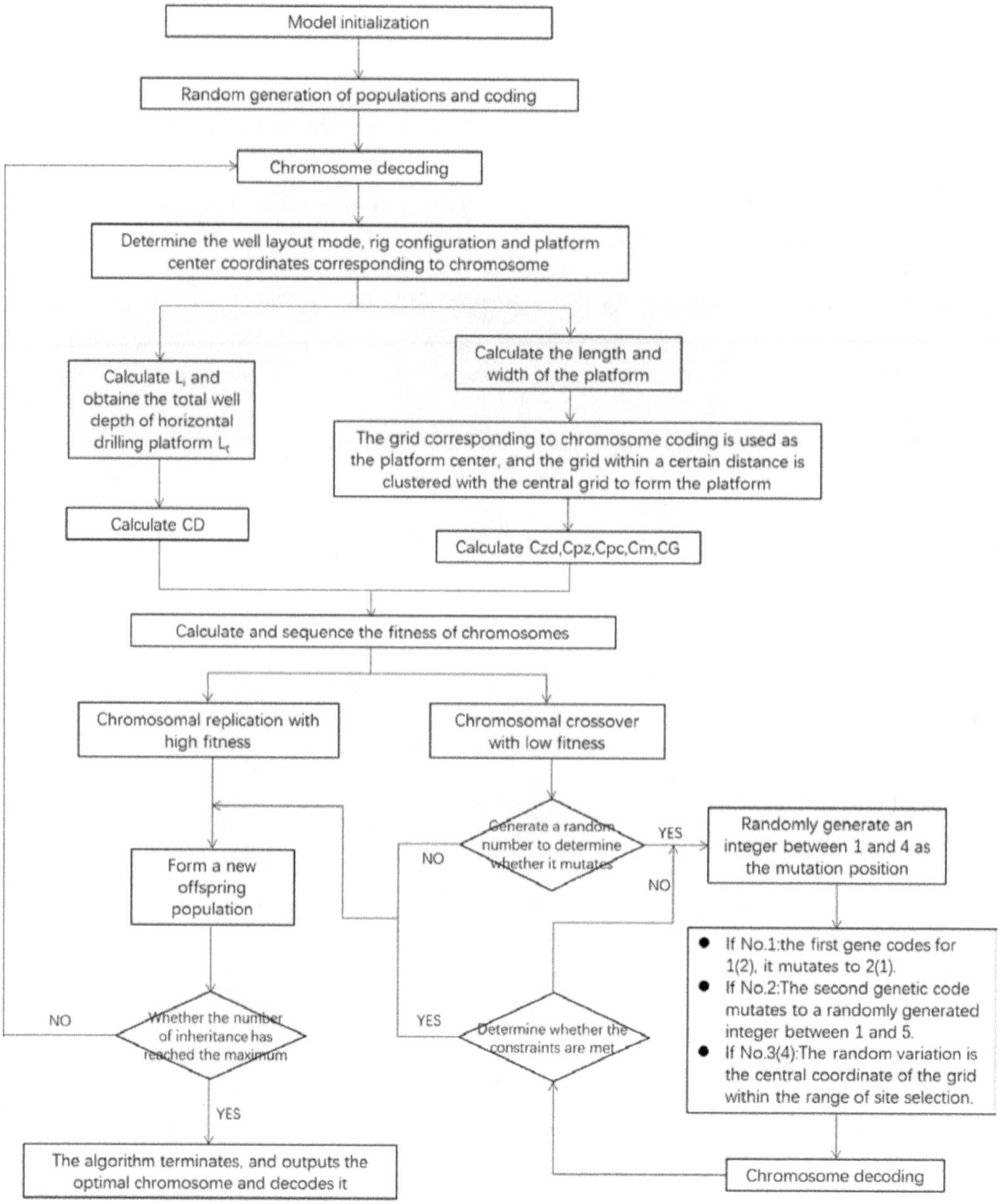

Figure 2. The process of determining the genetic algorithm solution.

As shown in Figure 2, firstly, the model is initialized, and the chromosome encoding and decoding rules are determined. Secondly, a population is randomly generated, and the chromosomes in the population are taken as the primary solution. Thirdly, the required length and width of the well site are determined according to the well layout method corresponding to the initial solution, and the location of the well site is obtained by grid clustering based on the center coordinates of the initial solution according to certain rules. Fourth, the platform construction cost model and drilling cost model are used to calculate the platform construction and drilling costs, respectively, which are put into the site selection model to obtain the platform location and the total cost of well construction corresponding to the initial solution. Fifth, the fitness of individuals in each initial population is calculated according to the fitness function, and the fittest individuals are selected to pass on to the next generation. All of the other chromosomes randomly generate mutation and crossover positions for the offspring solution set and retrieve the level of fitness. Sixth, the fitness of the new generation of individuals is recalculated. The best chromosome is selected to move to the next generation, whilst the remaining the chromosomes cross and mutate. The third step is performed until the condition indicating the end of the algorithm is reached. Finally, the optimal solution decoded is used as the optimal site selection scheme.

3.1. Initialization and Encoding

The initial population size is generated by random selection. The encoding method is mixed, and the chromosome form is "well layout + rig configuration + well site location (the grid position of the center of the well site)". The specific encoding method is as follows:

In terms of the well layout code, a single layout well is indicated by "1"; double-arranged wells are indicated by "2".

The drilling rig configuration codes are as follows: "1" for "*ZJ50*", "2" for "*ZJ70*", "3" for "*ZJ50* + *ZJ50*", "4" for "*ZJ50* + *ZJ70*", and "5" for "*ZJ70* + *ZJ70*".

In terms of platform center position coding, the grid number is coded from any vertex in the feasible site area. For example, in the double-layout well mode, the "*ZJ50* + *ZJ70*" drilling rig is used, and the grid number of the platform center is (113,230), so the chromosome code is "1301130230".

3.2. Cluster

The length and width of the platform were calculated according to the well distribution mode corresponding to the chromosome and rig configuration. The grid corresponding to the chromosome code was taken as the platform center, and the grid within a certain distance was clustered with the central grid to form the platform. Then, the grid covered by the platform was

$$\begin{cases} O_R \pm (X/2\varepsilon + 1) \\ O_C \pm (Y/2\varepsilon + 1) \end{cases} \text{or} \begin{cases} O_R \pm (Y/2\varepsilon + 1) \\ O_C \pm (X/2\varepsilon + 1) \end{cases} \tag{21}$$

where O_R and O_C are the row and column numbers of the platform's center grid, respectively.

3.3. Selection

Taking the grid corresponding to the platform center position as the platform center, the length and width of the platform, the platform construction cost for all grids within the platform range, and the drilling and completion costs from the target point to the platform center position were determined to form the fitness function. Optimal preservation strategies were adopted to ensure that the chromosome with the highest fitness function value in each generation could be further inherited:

$$f_i = M - CG_i - CD_i \tag{22}$$

where f is the fitness of the chromosome and M is an infinite number.

3.4. Crossover

The crossover operation can randomly exchange some genes from two chromosomes and generate new chromosomes. After crossover of the genes, it may appear that individuals do not meet the constraints of site selection. Therefore, the strategy of partial matching crossover is adopted. The specific method used to determine the intersection position is as follows:

$$P_c = \text{random}(1,\ 4) \tag{23}$$

where P_c is the cross location and random(1, 4) represents the random generation of an integer between 1 and 4.

3.5. Mutation

The mutation operation can randomly change the selected chromosome genes, which can avoid the solution process falling into local convergence. For every part of each offspring in the cross offspring set, a random number is generated. If the random number is less than the mutation probability, the location of the mutant gene is determined according to Equation (23), and the corresponding mutant gene is randomly determined within the corresponding gene boundary range to form a new chromosome.

3.6. Calculated Fitness

The fitness levels of all chromosomes were recalculated. The chromosomes with the greatest fitness levels in the father generation were copied to the offspring, and the chromosomes with the least fitness in the offspring generation were eliminated, forming a new offspring population. We then returned to step (3.2).

3.7. Termination of the Algorithm

The value of the chromosome fitness function in the population tends to be stable, which represents the convergence of the algorithm, and the algorithm terminates.

4. Results and Discussion

According to the established site selection model, the location of the multi-well drilling pad, the arrangement of the platform wellhead, and the configuration of the drilling rig can be obtained with a minimal well construction cost by solving the site selection model under the conditions of the known target coordinates of the horizontal section, the coordinates of the building deviation point, and the relevant parameters of the directional well. The parameters involved in the model solution are shown in Table 1.

Table 1. Basic calculation parameters obtained from experience.

Parameter	Value	Parameter	Value
C_{zds} (\$/m^2)	0.01	P_{vs} (\$/m)	500
C_{pzs} (\$/mp)	150	P_{ks} (\$/m)	1500
C_{pcs} (\$/m^2)	160	P_{hs} (\$/m)	2500
C_{m1} (\$)—ZJ50	120,000	C_{m1} (\$)—ZJ70	160,000
C_{m2} (\$)—ZJ50	19,000	C_{m2} (\$)—ZJ70	28,000
C_{m3} (\$)—ZJ50	33,000	C_{m3} (\$)—ZJ70	45,000
K (barren land)	1	K (forest)	1.2

4.1. The First Case

The Y well group is located in Xuyong County, Luzhou City, Sichuan Province, which is located in the southern margin of the Sichuan Basin. The highest altitude in this area is 1304 m; the lowest altitude is 480 m. These altitudes represent a mountain landform with a deeply local topography cut. The area has convenient transportation and lush trees throughout. It is an environmentally sensitive area. There are five horizontal wells in the well group. The target layer is the Lower Silurian Longmaxi Formation, which is generally distributed in a comb shape on the plane and has a horizontal section length of 1000–1200 m, a well spacing of about 340 m in the horizontal section, and a depth of 500 m for the inclination point. Table 2 shows the specific drilling design of the Y well group.

Table 2. Y well group drilling design.

Npw	Target Coordinates			L_h
	x_i	y_i	z_i	
Y1	1664	1456	909	1200
Y2	1393	1247	934	1030
Y3	1084	1145	961	1000
Y4	763	1089	986	1100
Y5	396	1141	1016	1100

According to the geological design data from the well group, a single well arrangement mode was adopted, and the length and width of the well site were determined to be 85 m and 50 m by the platform area model. The center coordinates of the platform are (1207,1483) and the corresponding rig model is "ZJ50" under the condition of the lowest cost obtained by the genetic algorithm. As shown in Figure 3., the selected location is easily accessible for engineering vehicles and is nonagricultural, which meets the HSE requirements for shale gas drilling. The drilling design data show that the actual drilling platform centers are located at (1496,1223), and the center of the selected platform is 388 m away from the actual drilling platform center; however, there are many farmers around the actual drilling platform, and the nearest farmer anomaly platform is only 80 m, which proves that the site selection method in this paper is feasible and effective.

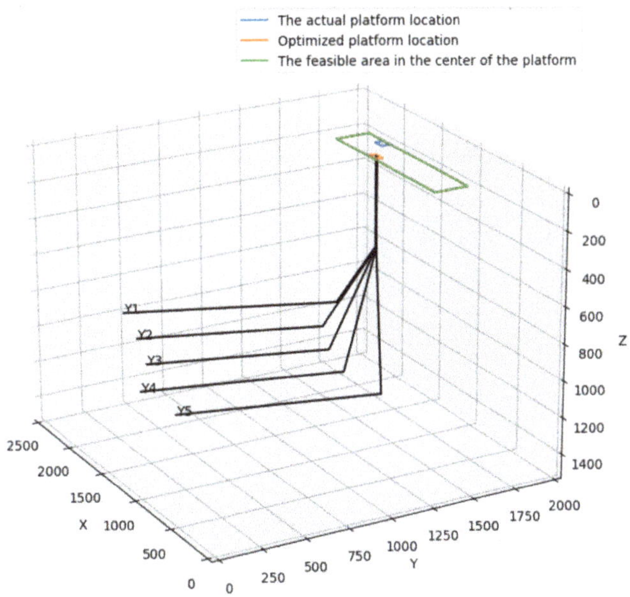

Figure 3. Y well group drilling platform optimization.

4.2. The Second Case

The J59 well group is located in Fuling District, Chongqing City in China, which is located in the transition zone between basins. The terrain is mainly low mountains and hills and is generally high in the southeast and low in the northwest, with the highest elevation being 1977 m and the lowest being 138 m, mostly ranging from 200 to 800 m. The surface conditions are extremely complex. A total of six horizontal wells were deployed and designed in the well group. The target layer was shale gas intervals of Upper Ordovician Wufeng Formation–Lower Silurian Longmaxi Formation, and the drilling rig was configured as a dual drilling rig. The specific design is shown in Table 3.

Table 3. The drilling design of the J59 well group.

Npw	z_k	Target Coordinates			L_h
		x_i	y_i	z_i	
J59-1	1000	3427	79	2533	1300
J59-2	1500	3427	679	2860	1300
J59-3	1450	3427	1279	2478	1300
J59-4	1700	1730	1279	2548	1700
J59-5	1600	1550	679	2566	1500
J59-6	900	1230	79	2618	1200

In order to avoid complex wellbore orbits and reduce the difficulty of drilling engineering, different well layout methods in the feasible platform location regions can be preliminarily determined according to the drilling design data. The area of the following platform is then determined according to the actual demands of the well site, as shown in Table 4.

Table 4. Possible wellhead alignment, combined with the rig configuration.

No.	N_{pw}	A	*DM*	S (m^2)	X_R (m)	Y_A (m)	Feasibility
1	6	1	*ZJ50 + ZJ50*	6500	130	50	Infeasible
2	6	1	*ZJ50 + ZJ70*	7000	140	50	Feasible
3	6	1	*ZJ70 + ZJ70*	7000	140	50	Feasible
4	6	2	*ZJ50 + ZJ50*	9200	115	80	Infeasible
5	6	2	*ZJ50 + ZJ70*	10,000	125	80	Feasible
6	6	2	*ZJ70 + ZJ70*	10,000	125	80	Feasible

As shown in Figure 4., the maximum designed drilling depth of the J59 well group is 4710 m. According to the principles of drilling rig selection, "*ZJ50*" and larger drilling rigs should be selected. The load capacity and configuration of drilling rig equipment should meet the required well drilling depth of 5000 m, so the "*ZJ50 + ZJ50*" drill rig configuration scheme was deleted. According to the above model, the length of the well site was determined to be 125 m and the width was determined to be 80 m. Combined with the *DEM* of the well area, the feasible platform site selection area was determined. Based on the model of minimum well construction costs for the platform, the central coordinate of the platform (the grid position of the center of the well site) with the double-layout well and "*ZJ50 + ZJ70*" rig configuration was determined to be located at (2475,949), with civil houses located 580 m southeast of the selected location. Transportation in the area is convenient, the nonagricultural land is nearly 350 m away from the actual drilling platform center, and the area meets the HSE requirements for shale gas drilling.

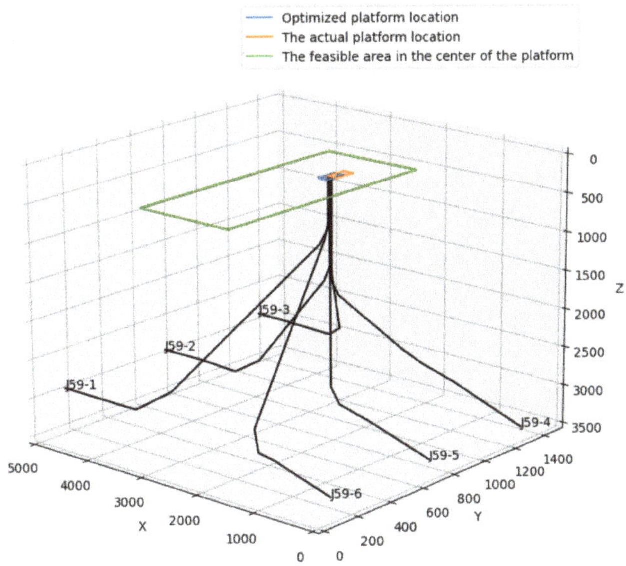

Figure 4. J59 well group drilling platform optimization.

5. Conclusions

Shale gas development is a capital-intensive project, and well construction investment accounts for the largest proportion of total oilfield construction investment. Rapid site selection under complex terrain conditions can effectively shorten the time required to formulate development plans, and making shale gas to be rapidly developed and put into production. From the perspective of the time value of capital, rapid well construction and initiation of production can significantly improve development benefits.

By analyzing the factors influencing well construction cost, a construction cost model of the shale gas factory multi-well pad and a drilling cost model were established and a site selection scheme with minimum construction costs for the pad was proposed. The drilling data from the first case were used to determine the location of the drilling platform, which was 388 m away from the platform's location; the selected location was found to meet the requirements of drilling engineering and HSE, proving the feasibility of the location selection model. In the second case, a quick site selection method was used to determine the characteristics of the single pad and double layout well, where the length and width of the well field were 125 m and 80 m, respectively. The selected rig configuration was "ZJ50 + ZJ70", and the selected wellsite location was 350 m away from the actual location. The analysis shows that the method is not only applicable to platforms with known well group relations, it is also applicable to symmetry relations of unknown situations, as it is able to quickly choose a platform under complex initial position surface conditions on the basis of further evaluation of the multi-well drilling pad position, and it can greatly reduce the exploration field work required, thereby improving productivity.

This paper discussed the feasibility of rapidly identifying the location of a shale gas multi-well drilling pad using the DEM, but there are still some factors that have not been fully considered, such as the different angle penetration lengths in the strata, complex geological structures, the sensitivity of the factors that affect costs, and the speed analysis and comparison of algorithms. Moreover, the location methods for the resolution of terrain data used in the article have a higher level of demand, and if the resolution increases, the corresponding lattice grid number increases four-fold. Thus, the algorithm should be improved to improve its speed and accuracy.

Author Contributions: Conceptualization, J.Z.; methodology, software, validation, N.H.; formal analysis, investigation, W.L. All authors have read and agreed to the published version of the manuscript.

Funding: This work was supported by The Humanities and Social Science Special Fund—Scientific Research Starting Project of SWPU (No. 2018QHR006) and The Special Project of Science and Technology Strategic Cooperation between Nanchong City and Southwest Petroleum University (No. SXHZ012).

Institutional Review Board Statement: Not applicable.

Informed Consent Statement: Not applicable.

Data Availability Statement: Not applicable.

Conflicts of Interest: The authors declare no conflict of interest.

References

1. Zou, C.; Zhao, Q.; Dong, D.; Yang, Z.; Qiu, Z.; Liang, F.; Wang, N.; Huang, Y.; Duan, A.; Zhang, Q.; et al. Geological characteristics, main challenges and future prospect of shale gas. *Nat. Gas Geosci.* **2017**, *28*, 1781–1796. [CrossRef]
2. Long, S.; Cheng, Z.; Xu, H.; Chen, Q. Exploration domains and technological breakthrough directions of natural gas in SINOPEC exploratory areas, Sichuan Basin, China. *J. Nat. Gas Geosci.* **2020**, *5*, 307–316. [CrossRef]
3. Feng, X.; Liao, X. Study on Well Spacing Optimization in a Tight Sandstone Gas Reservoir Based on Dynamic Analysis. *ACS Omega* **2020**, *5*, 3755–3762. [CrossRef]
4. Qi, W.; Xing, L.; Chenggang, X.; Xun, L. Geoscience-to-Production Integration Ensures Effective and Efficient South China Marine Shale Gas Development. *China Pet. Explor.* **2015**, *20*, 1–23.
5. Caineng, Z.O.U.; Guangming, Z.H.A.I.; Zhang, G.; Hongjun, W.; Zhang, G.; Jianzhong, L.; Wang, Z.; Wen, Z.; Ma, F.; Liang, Y.; et al. Formation, distribution, potential and prediction of global conventional and unconventional hydrocarbon resources. *Pet. Explor. Dev.* **2015**, *42*, 13–25.
6. Yue, Y.; Wu, X.; Zhang, Q.; Zhao, H.; Jiang, W. Integration and scale application of shale gas exploration and development engineering technologies in Sichuan and Chongqing areas. *Nat. Gas Ind. B* **2018**, *5*, 523–532. [CrossRef]
7. Devine, M.D.; Lesso, W.G. Models for the Minimum Cost Development of Offshore Oil Fields. *Manag. Sci.* **1972**, *18*, 378–387. [CrossRef]
8. Frair, L.; Devine, M. Economic Optimization of Offshore Petroleum Development. *Manag. Sci.* **1975**, *21*, 1370–1379. [CrossRef]
9. Costa, L.R. Optimization Models for Offshore Oil Field Development. Ph.D. Thesis, Case Western Reserve University, Cleveland, OH, USA, 1975.
10. Dogru, S. Selection of Optimal Platform Locations. *SPE Drill. Eng.* **1987**, *2*, 382–386. [CrossRef]
11. Grimmett, T.; Startzman, R. Optimization of Offshore Field Development to Minimize Investment. *SPE Drill. Eng.* **1988**, *3*, 403–410. [CrossRef]
12. Garcia-Diaz, J.C.; Startzman, R.; Hogg, G.L. A New Methodology for Minimizing Investment in the Development of Offshore Fields. *SPE Prod. Facil.* **1996**, *11*, 22–29. [CrossRef]
13. Rodrigues, H.W.L.; Prata, B.A.; Bonates, T.O. Integrated optimization model for location and Sizing of offshore platforms and location of oil wells. *J. Pet. Sci. Eng.* **2016**, *145*, 734–741. [CrossRef]
14. Walesca, H.; Rodrigues, L.; De, B.; Prata, A.; De Oliveira, T.; Assistant, B. New integrated approach for location and sizing of offshore multicapacitated platforms and location of oil wells. *Petrol* **2016**, *145*, 734–741.
15. Mohammed, K.; Almedallah, S.D.C. Walsh Integrated well-path and surface-facility optimization for shallow-water oil and gas field developments. *J. Pet. Sci. Eng.* **2019**, *174*, 859–871.
16. Tariq, Z.; Mahmoud, M.; Abdulraheem, A.; Al-Shehri, D.; Khan, M.R.; Janjua, A.N. An Intelligent Solution To Forecast Pressure Drop in a Vertical Well Having Multiphase Flow Using Functional NetworkTechnique. In Proceedings of the PAPG/SPE Pakistan Section Annual Technical Conference and Exhibition Islamabad, Islamabad, Pakistan, 10–12 December 2018.
17. Tariq, Z.; Mahmoud, M.; Al-Shehri, D.; Sibaweihi, N.; Sadeed, A.; Hossain, M.E. A Stochastic Optimization Approach for Profit Maximization Using Alkaline-Surfactant-Polymer Flooding in Complex Reservoirs. In Proceedings of the SPE Kingdom of Saudi Arabia Annual Technical Symposium and Exhibition, Dammam, Saudi Arabia, 23–26 April 2018.
18. Tariq, Z.; Mahmoud, M.; Abdulraheem, A. Real-time prognosis of flowing bottom-hole pressure in a vertical well for a multiphase flow using computational intelligence techniques. *J. Pet. Explor. Prod. Technol.* **2019**, *10*, 1411–1428. [CrossRef]
19. Tariq, Z.; Aljawad, M.S.; Hasan, A.; Murtaza, M.; Mohammed, E.; El-Husseiny, A.; Sulaiman, A.A.; Mohamed, M.; Abdulazeez, A. A systematic review of data science and machine learning applications to the oil and gas industry. *J. Pet. Explor. Prod. Technol.* **2021**, *11*, 4339–4374. [CrossRef]
20. Tariq, Z.; Mahmoud, M.; Abouelresh, M.; Abdulraheem, A. Data-Driven Approaches to Predict Thermal Maturity Indices of Organic Matter Using Artificial Neural Networks. *ACS Omega* **2020**, *5*, 26169–26181. [CrossRef] [PubMed]
21. Mahmoud, M.; Tariq, Z.; Kamal, M.S.; Al-Naser, M. Intelligent prediction of optimum separation parameters in the multistage crude oil production facilities. *J. Pet. Explor. Prod. Technol.* **2019**, *9*, 2979–2995. [CrossRef]

22. Khan, M.R.; Sadeed, A.; Asad, A.; Tariq, Z.; Tauqeer, M. Maximizing Oil Recovery in a Naturally Fractured Carbonate Reservoir Using Computational Intelligence based on Particle Swarm Optimization. In Proceedings of the PAPG/SPE Pakistan Section Annual Technical Conference and Exhibition, Islamabad, Pakistan, 10–12 December 2018.
23. Sadeed, A.; Tariq, Z.; Janjua, A.N.; Asad, A.; Enamul Hossain, M. Smart water flooding: An economic evaluation and optimization. In Proceedings of the SPE Kingdom of Saudi Arabia Annual Technical Symposium and Exhibition, Dammam, Saudi Arabia, 23–26 April 2018.
24. Yunhua, G.; Aimin, Y.; Yongrong, G.; Xueping, T. Drilling pad optimization for oilfield development by cluster horizontal wells. *Pet. Explor. Dev.* **2005**, *32*, 94–100.
25. Tie, Y.; Wenfu, J.; Xueliang, B.; Ting, X.; Yang, Y.; Guangdong, L. An allocation method for platform wellheads of cluster well based on ACA. *Fault-Block Oil Gas Field* **2014**, *21*, 224–227.
26. Zhen, L.; Bin, P.; Lianbao, W. Position optimization for central platform of offshore oilfield. *China Offshore Platf.* **2002**, *17*, 28–30.
27. Yucai, S.; Zhichuan, G.; Qiuyan, C.; Yanjun, Z. Location optimization method for drilling platform. *J. China Univ. Pet.* **2007**, *31*, 44–47.
28. Wenfei, L.; Kuanliang, Z.; Zhichuan, G.; Ming, C.; Yunpeng, L.; Yucai, S. Location optimization for the drilling platform of large-scale cluster wells. *Acta Pet. Sin.* **2011**, *32*, 162–166.
29. Yuchen, Z.; Hai, L.; Likong, C.; Hailong, L. Research and practice of position selection methods of wellhead platforms for offshore cluster wells. *J. Yangtze Univ. (Nat. Sci. Ed.)* **2016**, *13*, 36–39.
30. Zhiyue, W.; Deli, G.; Binbin, D.; Degao, H. Optimization of platform positioning considering the learning effect in the "well factory" mode. *Nat. Gas Ind.* **2018**, *38*, 102–108.
31. Hongbo, H.; Zhiheng, L.; Pinghua, D. The platform location optimization technology for the exploration-development integration of middle and deep reservoirs in the Bohai Sea. *Oil Drill. Prod. Technol.* **2018**, *40*, 84–86.
32. Ping, C. *Drilling & Completion Engineering*; Petroleum Industry Press: Beijing, China, 2005; pp. 246–276.
33. Xiushan, L. *Well Trajectory Geometry*; Petroleum Industry Press: Beijing, China, 2006.
34. Xiushan, L.; Jun, G. The description and calculation of space circular orbit. *Nat. Gas Ind.* **2000**, *20*, 44–47.
35. Gen, M. *Genetic Algorithm and Engineering Optimization*; Qinjie, Y., Gengui, Z., Eds.; Tinghua University Press: Beijing, China, 2016.

processes

Article

Optimization of the Lateral Length of Shale-Gas Horizontal Wells Based on Geology–Engineering–Economy Integration

Jialin Zhu [1,*], Sha He [1] and Lin Lin [2]

1 School of Economics and Management, Southwest Petroleum University, Chengdu 610500, China
2 Southwest Oil & Gas Co. of PetroChina, Chengdu 610051, China
* Correspondence: zhujl@swpu.edu.cn; Tel.: +86-159-8212-9597

Abstract: Horizontal wells with extended lateral lengths and large-scale hydraulic fracturing are a key technology for shale gas development. Lateral length is the key factor in determining the production and economic benefits of horizontal wells. Therefore, based on geology–engineering–economy integration, a method for optimizing the lateral length of shale-gas horizontal wells is established. Through fracture-shape prediction, productivity simulation and input–output analysis, the net present-value model of the technical–economic evaluation of the economic lateral length is established. A comprehensive evaluation of lateral lengths in Changning Block is then conducted. The results show that, under the current geological, engineering, and economic conditions in Changning Block, a horizontal well with a lateral length between 175 m and 3508 m is economically viable, and the optimal economic lateral length is 2000 m. The porosity and thickness of the reservoir matrix, the production time, the drilling investment, and the price of the natural gas wellhead in the first year have a great impact on the economic lateral length. On one hand, we can increase the drilling rate by increasing the technical research and development efforts. On the other hand, we can improve the construction management level to reduce investment and reasonably increase the price subsidy to optimize the lateral length of shale-gas horizontal wells.

Keywords: shale gas; horizontal well; economic lateral length; geology–engineering–economy integration

Citation: Zhu, J.; He, S.; Lin, L. Optimization of the Lateral Length of Shale-Gas Horizontal Wells Based on Geology–Engineering–Economy Integration. *Processes* **2023**, *11*, 249. https://doi.org/10.3390/pr11010249

Academic Editors: Tianshou Ma and Yuqiang Xu

Received: 19 November 2022
Revised: 7 January 2023
Accepted: 10 January 2023
Published: 12 January 2023

1. Introduction

Energy is the blood of modern society and industrial civilization. China is gradually transforming into a green, low-carbon-energy society, taking steps towards human development as its economy is in a period of steady development. However, at this stage, fossil energy remains the pillar of China's energy consumption, with coal, oil, and natural gas accounting for 85.7% of its primary energy consumption [1]. Natural gas is a relatively clean, low-carbon fossil fuel. It is a practical alternative to low-carbon sources. According to the resource evaluation results, China's shale-gas resource endowment is outstanding, with 21.8×10^{12} m^3 of technically recoverable resources. However, the current proven rate is only 4.79%. The industrial exploitation of shale gas cannot be realized because of its ultra-low porosity, low permeability, and a lack of production capacity depending on the natural energy of the formation. The horizontal well, combined with large-scale hydraulic fracturing technology, has enabled the industrial exploitation of shale gas, which resulted in the shale-gas revolution in the U.S.A. and demonstrated the great value of shale-gas development in China.

After more than ten years of technology tracking, technology transplantation, and independent research and development, China has mastered a series of technologies for shale-gas development using horizontal wells and large-scale hydraulic fracturing. Horizontal wells are a core technology utilized to realize the commercial development of shale gas. The lateral length, which determines the contact area with shale-gas reservoir and the effective length of hydraulic fracturing, is an important technical parameter in determining the productivity of a horizontal well [2,3]. Therefore, it is necessary to determine the economic lateral length of horizontal wells in shale-gas development. Most existing studies focus on the limited technical length of a shale-gas horizontal well from the perspective of drilling or oil production, considering wellbore stability, pump pressure and wellbore friction. Few studies focus on the economic lateral length of shale-gas horizontal wells.

Based on a genetic algorithm, Liu Yusong et al. [4] optimized the lateral length of a horizontal well using the coupling model of the horizontal wellbore and reservoir, taking into account the flow state of fluid in the wellbore, the friction pressure drop, and other factors. Guo Xiaole et al. [5] established the hydraulic extension limit model of extended-reach wells, which showed that the formation safety density window and the maximum allowable total pressure loss were objective constraints on the hydraulic extension limit of extended-reach wells in the Liuhua Oilfield in the South China Sea. Under the existing equipment conditions, the hydraulic extension limit was mainly limited by the pump pressure in the slope-stabilizing section, and the hydraulic extension limit length of the slope-stabilizing section was 6500~7200 m. The horizontal extension length of horizontal wells were determined by different factors and could be divided into three parts, including the acceptable cutting bed height, cutting lifting efficiency, pump pressure, total circulation pressure loss, well bottom pressure, and formation fracture pressure [6]. Based on the established calculation model of the circulating pressure loss of horizontal wells, Xu Kunji et al. [7] studied the lateral extension ability of horizontal wells and found that the hydraulic extension ability of horizontal sections was affected by factors such as drilling fluid density, the rated pressure of drilling pumps, the height of cutting beds, and the drilling-fluid displacement. Jin Xiuju et al. [8] suggested that the lateral length of horizontal wells in the Puguang Gas Field should be maintained within 400–600 m according to the productivity data obtained from a single-well geological model simulation and the analysis of factors such as reservoir thickness and productivity requirements. Yuan Junliang et al. [9] conducted an integrated-rock mechanical study to evaluate the wellbore stability of a shale horizontal well. The results showed that wellbore instability was related to the bedding plane and azimuth angle.

Based on the horizontal-well productivity model, Chen Yaohui et al. [10] built an equation demonstrating the relationship between the cost of the lateral section and the production of a horizontal well to determine the optimal lateral length of a horizontal well in a block in Daqing, combined with economic indicators such as the investment payback period and net present value. Zhou Yingjie [11,12] proposed that the optimal lateral length of oil wells in the area should be 200~250 m, determined via statistical analysis, numerical simulation, analyzing actual data from the Shengli Oil Region, as well as a comprehensive economic evaluation. T. Ariadji et al. [13] established the relationship function between the cumulative gas production and the lateral length to quickly determine the economic lateral length of a horizontal well, combined with the economic evaluation method. Based on the model proposed by Fan Zifei, Zeng Xiaojing et al. [14] reasonably assumed and simplified the flow of fluid in a reservoir to establish a new model to determine the optimal lateral length of a horizontal well, with the impact of various drilling costs and oil prices fully considered. Hu Junkun et al. [15] established the relationship curve between the net present value and the lateral length of a horizontal well by determining the reasonable production of the gas well and using numerical simulation software to dynamically predict the production capacity of the gas well. They then determined the optimal economic value of the lateral length of the horizontal gas well and conducted a sensitivity analysis of factors such as the formation coefficient, natural gas price, and operating cost. Rammay et al. [16] pointed out that the NPV of a shale-gas horizontal well can be maximized by optimizing the lateral length and hydraulic fracturing operation parameters. Dosummu et al. [17] determined the optimal lateral length and well parameters of a specific reservoir based on the net present value method. They also discussed the impact of crude-oil viscosity, horizontal permeability, and well diameter on the productivity of horizontal wells. Kalantari et al. [18] studied the effects of porosity, permeability, reservoir pressure, production layer thickness, horizontal well length, natural fracture length, and density on shale-gas production by using a numerical simulation method. The above parameters were prioritized with the net-present-value theory and method. Lu Honglin et al. [19] built a model for evaluating the technical and economic limit of the lateral length of a horizontal well in a low-permeability gas reservoir based on the input–output analysis method and discussed the influence of production parameters, gas reservoir parameters, and economic parameters. Based on the productivity prediction of horizontal wells, Wang Dawei et al. [20] and Sun Huachao et al. [21] built a model for evaluating the economic lateral length of a horizontal well under specific parameters, taking into account fixed investments such as drilling and completion, operating costs, and oil and gas sales revenue. The simulation results of an integrated geological and engineering model and the calculation results of economic limit production showed that only gas wells with a horizontal section length of 2000 m could realize economic development, and a single-well EUR had to be more than 154 million cubic meters [22]. He Chang et al. [23,24] used historical production data from the Weiyuan Shale Gas Field to predict future production by analyzing empirical production decline, and then carried out an input–output analysis to obtain the economic lateral length of the horizontal well in the Weiyuan Gas Field. Zhang Jijun et al. [25] optimized the drilling-platform location of a shale-gas multi-well pad based on a digital elevation model and construction cost analysis.

Previous studies have laid an excellent foundation for this study, but there are several problems. (1) The research on the lateral length of shale-gas horizontal wells mainly focuses on two aspects of the technical length: drilling engineering and gas production. (2) Only a few studies focus on the economic lateral length of shale-gas horizontal wells, and they aim to obtain the economic lateral length of horizontal wells in specific gas fields through the statistical analysis of historical production data and economic evaluation, which usually ignore the influence of different hydraulic-fracturing designs [15] and require historical production data from the block [23,24]. This study establishes a universal method for determining the economic lateral length of shale-gas horizontal wells, which considers the impact of hydraulic-fracturing design based on the geological parameters of the blocks. This method does not need to use historical production data and can be used in new shale-gas production areas.

2. Problem Statement and Formulation

In the context of the dual-carbon strategy, oilfield enterprises should accelerate the establishment of a method for optimizing shale-gas development based on geology–engineering–economy integration [26], which will not only reduce the cost but can also keep the energy supply safe and stable [27]. However, scholars have not yet established a clear method for evaluating the economic lateral length of horizontal wells in different shale-gas blocks from the perspective of geology–engineering–economy integration [28], so it is necessary to carry out targeted research to realize the economic development of shale-gas fields, especially new production blocks.

Integrated development is focused on geological research, optimizing the quality of drilling and completion and pursuing optimal economic benefits. Based on previous studies [15,18,23], this paper proposes a method of optimizing the lateral length of shale-gas horizontal wells based on geology–engineering–economy integration. The process (Figure 1) is as follows. Firstly, the geological stratification and formation mechanics parameters are obtained by interpreting the logging curves of the target well with logging software. Secondly, based on the geological and mechanical parameters of the reservoir in the target work area, the Stimplan software developed by NSI is used to simulate the hydraulic-fracture distribution with the actual fracturing process parameters, so the hydraulic fracture half-length and conductivity values under certain fracturing operation parameters can be obtained. Thirdly, the dynamic productivity of shale-gas horizontal wells is predicted using the CMG software developed by the Canada Computer Simulation

Software Group. Fourthly, an input–output analysis of shale-gas horizontal wells is conducted to build a technical–economic evaluation model for the economic lateral length of shale-gas horizontal wells. The economic lateral length is then calculated under specific reservoir, technical, and economic parameters. Lastly, the influence of reservoir geological parameters, technical parameters, and economic parameters on the economic lateral length of horizontal wells is further analyzed and corresponding optimization strategies are given.

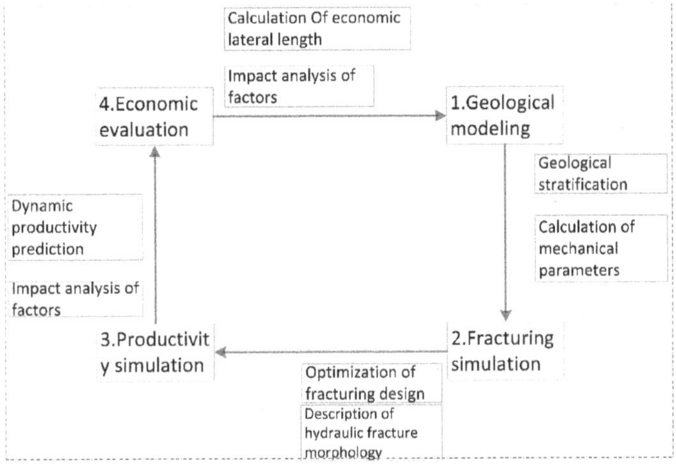

Figure 1. Flow chart detailing the process for the optimization of the lateral length of shale-gas horizontal well based on geology–engineering–economy integration.

3. Model for Evaluating the Economic Lateral Length of Shale-Gas Horizontal Well

In order to establish a model for evaluating the economic lateral length of shale-gas horizontal wells, an input–output analysis is required. The investment in shale-gas horizontal wells mainly refers to the fixed-asset investment in well construction before production, as well as the operation cost and taxes paid after production. The output refers to the revenue from the sales of natural gas, not including the revenue from natural gas by-products.

3.1. Input Analysis

Shale-gas horizontal wells mainly have fixed-asset investments such as drilling, hydraulic-fracturing construction, and surface-engineering construction, as well as operating costs and taxes after production. Drilling and hydraulic fracturing have a great impact on the economic viability of shale-gas horizontal wells [29].

3.1.1. Drilling Investment

For convenience of calculation, referring to previous studies [19,30], the drilling investment is expressed as the drilling length, which is divided into the vertical section, inclined section, and lateral section. The drilling investment of horizontal well is

$$C_d = C_{vd} + C_{sd} + C_{hd} \tag{1}$$

where C_d is the horizontal well-drilling investment, C_{vd} is the drilling investment in the vertical section, C_{sd} is the drilling investment in the inclined section, and C_{hd} is the drilling investment in the lateral section.

Each investment in Formula (1) can be given as an expression related to length.

The drilling investment in the vertical well section can be expressed as Formula (2) and the drilling investment in the inclined section can be expressed as Formula (3).

$$C_{vd} = C_v L_v \tag{2}$$

$$C_{sd} = C_s L_s \tag{3}$$

where C_v is the drilling investment per unit length of the vertical well section, L_v is the length of the vertical section, C_s is the drilling investment per unit length of the inclined section, and L_s is the length of the inclined section.

The drilling-rig rental fee, wage cost, and rotary steering cost increase with the growth of the lateral length of a shale-gas well, as larger lengths require bigger drilling rigs and longer construction periods, leading to an increase in the drilling investment per unit length in the lateral section. To facilitate this calculation, the investment in the horizontal section is expressed as the linear quadratic equation of the length of the horizontal section, namely:

$$C_{hd} = a_0 + a_1 L_h + a_2 {L_h}^2 \tag{4}$$

where a_0, a_1 and a_2 are coefficients and L_h is the lateral length.

Combined with the drilling investment data from 16 typical horizontal wells in the Changning Block with a well depth below 5000 m and expert suggestions, the vertical length is determined to be 3000 m, the unit investment is determined to be 3600 CNY/m, the inclined length is determined to be 1000 m, and the unit investment is determined to be 6000 CNY/m. The values of three constants are fitted using the least-square method, so the relationship between the horizontal section investment and lateral length is as follows:

$$C_{hd} = 1039.78 + 0.21 L_h + 0.0005 {L_h}^2 \tag{5}$$

Since a 50-type drilling rig is generally used for drilling horizontal wells less than 5000 m in depth and a 70-type or even 90-type drilling rig is required for drilling the other ten horizontal wells greater than 5000 m in depth, the investment is increased when compared with a 50-type drilling rig. In this study, the vertical length and inclined length are fixed, so different lateral lengths determine different types of drilling rigs. When $L_h < 900$ m, a 50-type drilling rig is selected; when 900 m $\leq L_h <$ 2500 m, a 70-type drilling rig is selected; and when $L_h \geq 2500$ m, a 90-type drilling rig is selected. Through investment data fitting, it is found that the investment of a 70-type drilling rig is 1.1 times that of a 50-type drilling rig, and the investment of a 90-type drilling rig is 1.3 times that of a 50-type drilling rig. Therefore, the final relationship between horizontal well investment and lateral length is as follows:

$$\begin{cases} C_d = (C_{vd} + C_{sd} + C_{hd}) \times 1.0, L_h < 900 \text{ m} \\ C_d = (C_{vd} + C_{sd} + C_{hd}) \times 1.1, 900 \text{ m} \leq L_h < 2500 \text{ m} \\ C_d = (C_{vd} + C_{sd} + C_{hd}) \times 1.3, L_h \geq 2500 \text{ m} \end{cases} \tag{6}$$

3.1.2. Investment in Hydraulic Fracturing

The number of hydraulic-fracturing sections, sand consumption and liquid consumption will vary with the lateral length. Additionally, the economic lateral length will vary with the half-length and conductivity of the hydraulic fracture. Therefore, referring to the research results of Guo Jianchun et al. [31], the hydraulic-fracturing investment is expressed as the relationship between sand consumption, liquid consumption, and other parameters, as follows:

$$C_f = C_w V_L + C_P Q_{sd} + C_{cz} \tag{7}$$

where C_f is the fracturing investment, C_w is the unit price of the fracturing fluid, V_L is liquid consumption, C_p is the unit price of the proppant, Q_{sd} is sand consumption, and C_{cz} is the cost of the fracturing truck set, packer, perforating gun, and personnel.

In order to facilitate the calculation, using the hydraulic-fracturing investment data of 26 typical wells in Changning Block, the relationship between C_{cz} and fluid consumption is derived so that the hydraulic-fracturing investment can be expressed as

$$C_f = 0.0223V_L + 0.0713Q_{sd} + 0.85 \times 10^{-6}{V_L}^2 + 61.82 \tag{8}$$

3.1.3. Other Investments

The operation cost is directly related to the output, so the operation cost is expressed considering its the relationship with the output. Considering the change in the operation cost per unit output of horizontal wells over time and the benchmark discount rate, the present value of the operation cost of fracturing horizontal wells [19] can be expressed as

$$C_p(t) = \sum_{j=1}^{t} Q_j C_{mg}(1+r_m)^{j-1}\frac{1}{(1+i)^{j-1}}, \tag{9}$$

where $C_p(t)$ is the net present value of horizontal well operating cost from year one to year t, t is the production life of the horizontal well, C_{mg} is the unit operating cost in the first year, Q_j is the annual output of the horizontal well, r_m is the annual growth rate of the operating cost, and i is the benchmark discount rate.

Comprehensive tax on a shale-gas horizontal well mainly includes the value-added tax, urban maintenance and construction tax, education surcharge, resource tax, and income tax. For the sake of convenience, the comprehensive tax in this paper is expressed as the natural-gas sales revenue multiplied by the comprehensive tax rate, so the net present value of the comprehensive tax for horizontal wells is

$$C_t(t) = CI(t)r_{tax} \tag{10}$$

where $C_t(t)$ is the net present value of comprehensive taxes required for horizontal-well production from year one to year t, $CI(t)$ is the net present value of natural-gas sales revenue from year one to year t, and r_{tax} is the comprehensive tax rate.

According to the estimation of surface-engineering construction investment in Changning Block, the surface-engineering construction investment shared by a single shale-gas well is CNY 1~3 million.

3.2. *Output Analysis*

In the process of natural-gas production, considering the change in the horizontal well production, natural-gas sales over time, and the benchmark discount rate, the net present value of horizontal-well sales income is

$$CI(t) = \sum_{j=1}^{t} Q_j P(1+C_p)^{j-1}\frac{1}{(1+i)^{j-1}} \tag{11}$$

where $CI(t)$ is the net present value of natural-gas sales revenue from year one to year t, Q_j is the annual output of the horizontal well, P is the natural-gas wellhead price in the first year, and C_p is the annual growth rate of the natural-gas sales.

3.3. *Evaluation Model of Economic Lateral Length*

The cash inflow and outflow of the shale-gas horizontal well are clarified through input–output analysis. The cash inflow is the sales revenue from natural gas produced by the horizontal well. The cash outflow, which is the sum of the fixed-asset investments, operating costs, and the comprehensive taxes of the horizontal well, is as follows:

$$CO(t) = C_d + C_f + C_{dm} + C_p(t) + C_t(t) \tag{12}$$

The NPV is the difference between the cash inflow and the cash outflow of the horizontal well, which is

$$NPV = CI(t) - CO(t) \quad (13)$$

The productivity numerical simulation model, established using the CMG software, can accurately simulate the dynamic productivity of horizontal wells with different lateral lengths under specific hydraulic-fracture half-length and conductivity conditions. Combined with the established model for evaluating the economic lateral length of a shale-gas horizontal well, it can calculate and analyze the change law of the net present value of shale-gas horizontal wells with different lateral lengths. On this basis, the sensitivity analysis method is used to analyze the influence of reservoir geological parameters, technical parameters (production time), and economic parameters on the extreme economic lateral length and the optimal economic lateral length.

4. Calculation Results and Discussion

Due to space limitations, this paper only calculates the economic lateral length of one horizontal well: Well A. It is particularly important to point out that changes in various input parameters will cause changes in the economic lateral length, which can lead to other changes. Well A in Changning Block is a horizontal well in a county in southern Sichuan, located in the south wing of the Ordovician top structure in the Changning anticline structure. The completed well depth is 5000 m, and the artificial bottom is 4950 m, completed with casing and the completed horizon located in the Longmaxi Formation. Table 1 presents a list of geological parameters such as porosity and permeability. Table 2 provides a list of technical parameters such as horizontal-well parameters and production parameters. Table 3 presents a list of economic parameters such as the drilling investment, hydraulic-fracturing investment, and operation cost.

Table 1. Geological stratification results.

No.	VD (m)		MD (m)		Formation Property
	Top	Bottom	Top	Bottom	
1	3333.7	3336.6	3590.2	3600.7	Gas
2	3336.6	3338.7	3600.7	3608.3	Gas
3	3338.7	3339.7	3608.3	3611.8	Gas
4	3339.7	3341.2	3611.8	3615.3	Gas
5	3341.2	3342.8	3615.3	3622.2	Gas
6	3342.8	3345.1	3622.2	3631.5	Gas
7	3345.1	3347.0	3631.5	3645.4	Gas
8	3347.0	3349.9	3645.4	3657.5	Gas
9	3349.9	3351.5	3657.5	3670.0	Gas
10	3351.5	3353.6	3670.0	3789.3	Gas
11	3353.6	3355.9	3789.3	3818.2	Gas
12	3355.9	3358.0	3818.2	3832.5	Gas
13	3358.0	3360.3	3832.5	3975.0	Gas
14	3360.3	3363.8	3975.0	4975.0	Gas

Table 2. Reservoir geological parameters of the shale gas horizontal well.

No.	Parameter Name	Quantity Sign	Unit	Value		
1	Matrix permeability	K	mD	0.0005	0.0010	0.0015
2	Matrix porosity	ϕ	Dimensionless	5.0%	5.6%	6.2%
3	Reservoir thickness	h	m	30	40	50
4	Reservoir temperature	T	K		365.15	
5	Deviation factor of natural gas	Z	Dimensionless		0.87	
6	Relative density of natural gas	γ_g	Dimensionless		0.6	
7	Viscosity of natural gas	μ_g	mPa.s		0.025	

Table 3. Technical parameters of the shale-gas horizontal well.

No.	Parameter Name	Quantity Sign	Unit	Value		
1	Production time	t	year	6	8	10
3	Production differential pressure	ΔP	MPa	3	5	7
4	Vertical length	L_v	m		3000	
5	Inclined length	L_s	m		1000	
6	Lateral length	L_h	m		100~3600	
7	Cluster spacing	L_c	m		20	
8	Number of perforating clusters	S_{pf}	cluster		3	
9	Well control radius	R_e	m		400	

4.1. Calculation Example

4.1.1. Geological Modeling

Logging software is used to interpret the density (DEN), compression wave (DTC), shear wave (DTS), and natural gamma ray (GR) logging curves of Well A so that geological stratification can be conducted, which is shown in Table 1. The static Young's modulus (E), Poisson's ratio (PR), and in situ stress (STRESS) are calculated using a built-in program, which will be used in subsequent fracturing simulations.

4.1.2. Fracturing Simulation

The lateral length of the horizontal well is 1250 m, 1200 m of which will be fractured, and the thickness of the reservoir is approximately 20 m. The Stimplan software is used for fracturing simulation. The length of a single fracturing section is 60 m, the number of fracturing sections is twenty, the number of perforation clusters in each section is three, and the cluster spacing is approximately 20 m. The average half-length of the fracture is 180 m and the conductivity is 300 mD · m under the pumping procedure conditions of 1364.7 m^3 liquid consumption and 467.1 t sand consumption in a single section. The hydraulic-fracture morphology in a single fracturing section is shown in Figure 2.

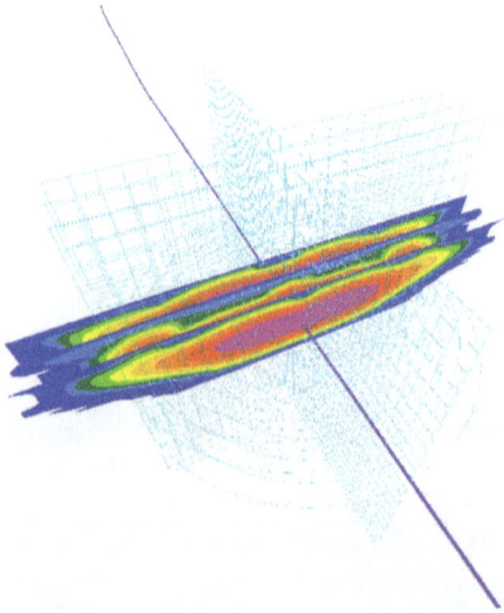

Figure 2. The hydraulic-fracture morphology in a single fracturing section.

4.1.3. Capacity Simulation

The CMG software is used to simulate the productivity of shale-gas horizontal wells with a horizontal lateral length ranging from 100 m to 3600 m.

4.1.4. Calculation of the Economic Lateral Length

The net present value of a shale-gas horizontal well with a specific lateral length is calculated using the methodology shown in Figure 3, with the second parameter in the "parameter value" field in Tables 2–4 [16,19]. The above process is repeated to obtain the relationship curve between the NPV of the shale-gas horizontal well and its lateral length, as is shown in Figure 4. The average half-length of the fracture is 180 m and the conductivity is 300 mD · m.

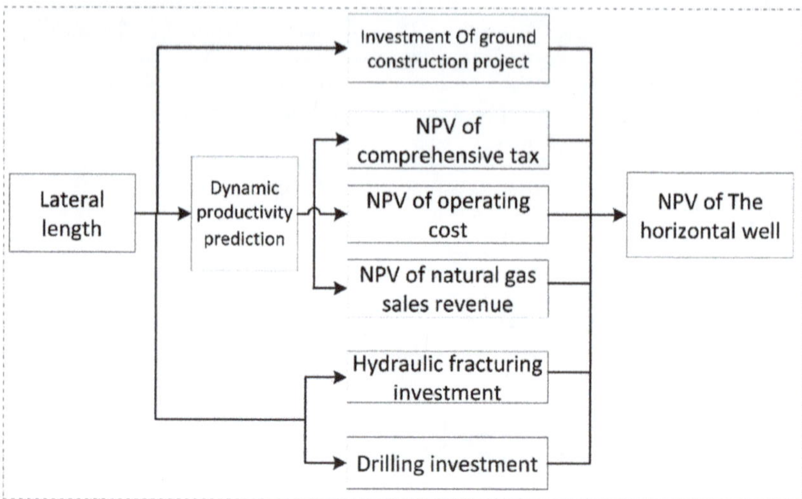

Figure 3. Calculation methodology used to find the net present value of a horizontal well with a specific lateral length.

Table 4. The economic parameters of the shale-gas horizontal well.

No.	Parameter Name	Quantity Sign	Unit	Value		
1	Natural-gas wellhead price	P	CNY/m^3	1.159	1.275	1.402
2	Annual growth rate of natural-gas wellhead price	C_p	Dimensionless	5%	10%	15%
3	Benchmark discount rate	i	Dimensionless	6%	8%	10%
4	Investment in surface engineering construction	C_{dm}	10^4 CNY	100	200	350
5	Unit drilling investment in vertical section	C_v	10^4 CNY/m		0.36	
6	Unit drilling investment in inclined section	C_s	10^4 CNY/m		0.60	
7	Unit drilling investment in lateral section	C_h	10^4 CNY		Equation (6)	
8	Unit operating cost in the first year	C_{mg}	CNY/m^3	0.125	0.138	0.152
9	Annual growth rate of operating cost	r_m	Dimensionless	5%	10%	15%
10	Comprehensive tax rate	r_{tax}	Dimensionless	8%	10%	12%
11	Hydraulic-fracturing investment	C_f	10^4 CNY		Equation (8)	

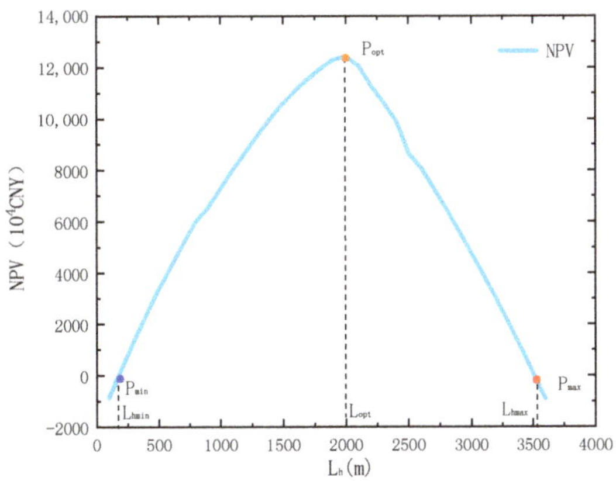

Figure 4. Relationship curve between the NPV and lateral length when the half-length of the fracture is 180 m.

Figure 4 has three points: P_{min}, P_{opt}, and P_{max}. The NPV corresponding to the P_{min} point is 0, and the corresponding lateral length is the minimum extreme economic length L_{min}, which is 175 m. The NPV corresponding to the P_{max} point is 0, and the corresponding lateral length is the maximum extreme economic length L_{max}, which is 3508 m. The NPV corresponding to the P_{opt} point is the largest, and the corresponding lateral length is the optimal economic length L_{opt}, which is 2000 m. The selection range of the actual lateral length is 175 m $\leq L \leq$ 3518 m under the parameters used in this example. The lateral length of the 336 production wells in Changning Block range from 800 m to 3500 m [32], which confirms the accuracy of this study to some extent.

The optimal lateral length is largely dependent on the hydraulic-fracturing design. Therefore, the effects of a change in the economic length of the horizontal section are investigated, when the half-length of the hydraulic fracture is 300 m and the conductivity is 300 mD · m, as is shown in Figure 5. The optimal lateral length is still 2000 m long, but the minimum extreme economic length decreases and the maximum extreme economic length increases. The increase of the half-length of the hydraulic fracture increases the optional range of the economic lateral length.

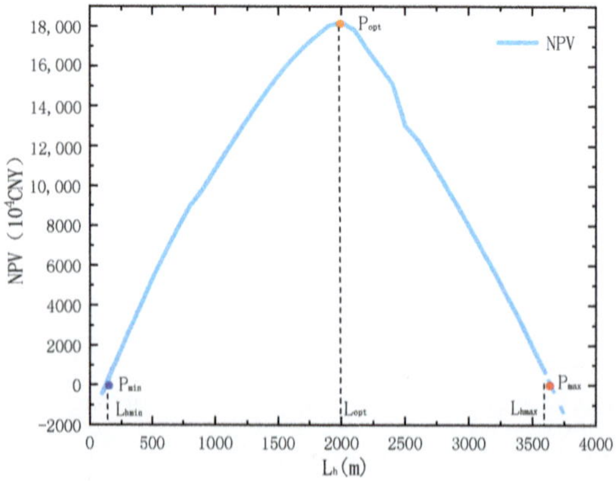

Figure 5. Relationship curve between the NPV and the lateral length when the half-length of the fracture is 300 m.

4.2. Sensitivity Analysis and Discussion

The control variable method is adopted to study the influence of six factors on the economic lateral length, including reservoir matrix porosity, thickness, production time, drilling investment, natural-gas wellhead price, and benchmark discount rate. The results are shown in Figure 6.

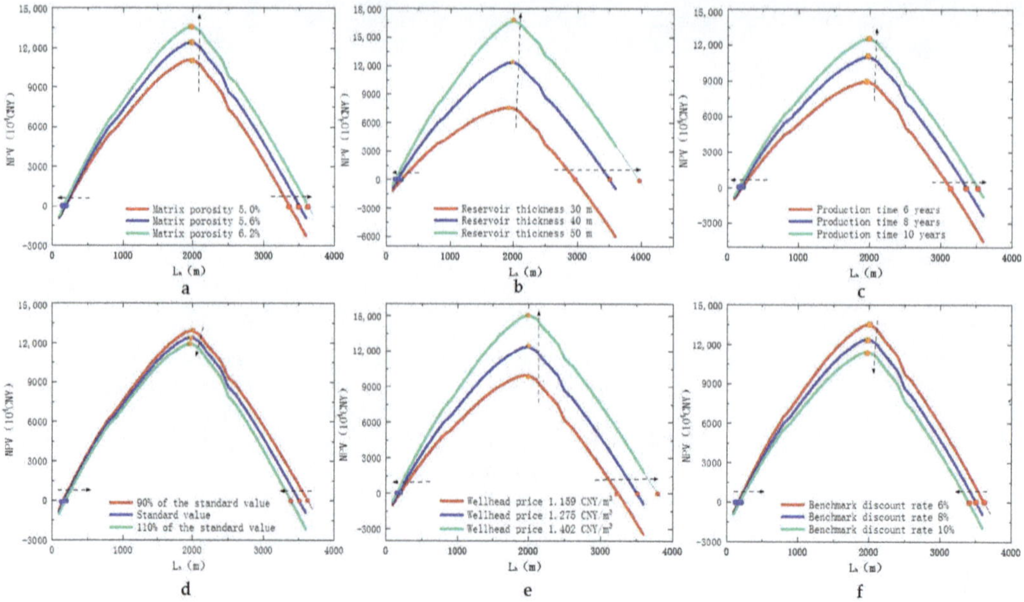

Figure 6. Relationship curves between the NPV and lateral length of a shale-gas horizontal well with six changing factors. (**a**) the relationship curve between the NPV and lateral length with different matrix porosities; (**b**) relationship curve between the NPV and lateral length with different reservoir thicknesses; (**c**) relationship curve between the NPV and lateral length with different production time; (**d**) relationship curve between the NPV and lateral length with different drilling investment values; (**e**) relationship curve between the NPV and lateral length with different wellhead prices; and (**f**) relationship curve between the NPV and lateral length with different benchmark discount rates.

4.2.1. Sensitivity Analysis of Reservoir Matrix Porosity

Generally speaking, the production of a shale-gas horizontal well increases with an increased reservoir matrix porosity, which means that a shorter lateral length is more economical. At the same time, the maximum extreme economic lateral length will also increase because the increase in production is enough to offset the increase in drilling and completion investment. Figure 6a shows that, with an increase in reservoir matrix porosity, the curve moves to the upper right, the minimum extreme economic lateral length decreases, the maximum extreme economic lateral length increases, and the economic benefits of a horizontal well with the same lateral length improve.

4.2.2. Sensitivity Analysis of Reservoir Thickness

There is a strong positive correlation between the productivity of a shale-gas horizontal well and its reservoir thickness. Therefore, an increase in reservoir thickness improves the economic benefits of horizontal wells with a shorter lateral length. Figure 6b shows that, with an increase in reservoir thickness, the curve moves to the upper right, the minimum extreme economic lateral length decreases, the maximum extreme economic lateral length increases, and the economic benefits of a horizontal well with the same lateral length improve.

4.2.3. Sensitivity Analysis of Production Time

Both production practice and academic research show that 20 years is a reasonable evaluation period for a shale-gas horizontal well [33]. The economic viability of a horizontal well with the same lateral length increases with production time, which leads to a larger range of lateral length. Figure 6c shows that, with an increase in production time, the curve moves to the upper right, the minimum extreme economic lateral length decreases, the maximum extreme economic lateral length increases, and the economic benefits of horizontal well with the same lateral length improve.

4.2.4. Sensitivity Analysis of Drilling Investment

Production is certain for a horizontal well with a specific lateral length. An increase in drilling investment indicates worsening economic benefits, leading to a smaller range of lateral length. Figure 6d shows that, with increasing drilling investment, the curve moves to the lower left, the minimum extreme economic lateral length increases, the maximum extreme economic lateral length decreases, and the economic benefits of a horizontal well with the same lateral length worsen.

4.2.5. Sensitivity Analysis of Natural-Gas Wellhead Price

For a horizontal well with a specific lateral length in a specific reservoir, the production is certain. An increase in the natural-gas wellhead price means an increase in the gas sales revenue, which makes the horizontal well more economical and leads to a larger range of lateral length. Figure 6e shows that with an increase in production time the curve moves to the upper right, the minimum extreme economic lateral length decreases, the maximum extreme economic lateral length increases, and the economic benefits of horizontal well with the same lateral length improve.

4.2.6. Sensitivity Analysis of Benchmark Discount Rate

The benchmark discount rate represents the expected return on investment of the project. In terms of the investment in a shale-gas horizontal well, a high-benchmark discount rate will lead to a smaller range of lateral length, while a low-benchmark discount rate will lead to a larger range of lateral length. Figure 6f shows that, with an increase in the benchmark discount rate, the curve moves to the lower left, the minimum extreme economic lateral length increases, the maximum extreme economic lateral length decreases, and the economic benefits of the horizontal well with the same lateral length worsen.

Table 5 shows the minimum extreme economic lateral length, the maximum extreme economic lateral length, and the optimal economic length under different factors. The influence of the natural-gas wellhead price in the first year, the reservoir thickness, matrix porosity, production time, drilling investment, the benchmark discount rate on the economic lateral length, and the economic benefits of the horizontal well are weakened in turn.

Table 5. Economic lateral length of the horizontal well with six changing factors.

No.	Parameter	Value	Minimum Extreme Economic Length/m	Maximum Extreme Economic Length/m	Optimal Economic Length/m
1	Matrix porosity	5%	189	3373	1993
		5.6%	175	3508	2000
		6.2%	165	3633	2004
2	Reservoir thickness	30 m	238	2972	1875
		40 m	175	3508	2000
		50 m	142	3962	2009
3	Production time	6 years	203	3143	1957
		8 years	178	3350	1984
		10 years	175	3508	2000
4	Drilling investment	90% of the standard value	159	3635	2015
		Standard value	175	3508	2000
		110% of the standard value	191	3398	1980
5	Natural-gas wellhead price	1.159 CNY/m^3	201	3229	1986
		1.275 CNY/m^3	175	3508	2000
		1.402 CNY/m^3	154	3787	2012
6	Benchmark discount rate	6%	165	3626	2014
		8%	175	3508	2000
		10%	186	3408	1987

5. Conclusions

The lateral length of horizontal wells is an important parameter in shale-gas development. Advances in engineering technology, the optimization of management measures, and changes in the natural-gas wellhead price will alter the economic lateral length and the economic benefits of horizontal wells. Therefore, it is necessary to adhere to the idea of geology–engineering–economy integration and take the following measures to optimize lateral length.

Applying the theory of technical economics to conduct interdisciplinary research with the aim of achieving optimized engineering technical parameters and management measures is a practical way to reduce costs and increase the efficiency of oil and gas enterprises, especially as drilling technology and reservoir-stimulation technology in petroleum engineering have peaked and revolutionary breakthroughs are rare. The optimal lateral length of horizontal wells in Changning Block is 175~3508 m and the optimal economic length is 2000 m under the current geological understanding, engineering technology and production-management level. The R&D of drilling technology and management optimization should be a point of focus to further improve drilling engineering and to reduce drilling investment. On one hand, we should actively pursue the continuous improvement of technology to achieve technical breakthroughs and continue to tackle the key problems that long horizontal sections present to drilling technologies, especially issues with geological-steering and rotary-steering technologies, environmentally friendly anti-collapse drilling fluids, the reduction of drilling accidents, and an improvement in penetration rates. On the other hand, the batch drilling mode of well plants should be further promoted, and the construction links should be optimized according to the drilling investment list. We should promote the marketization reform of natural-gas prices and reasonably increase the price of natural gas in non-livelihood areas through the reasonable evaluation of the contribution of shale gas to the "dual-carbon" strategy. Subsidies should be established considering the price of shale gas in high-risk blocks or upstream production blocks to maintain competitive shale-gas wellhead prices and a certain annual growth rate.

This article discusses a method for optimizing the economic lateral length of shale-gas horizontal wells. However, there are still several factors that have not been considered, such as reservoir heterogeneity, early exploration investment, and workover investment. Moreover, only a single-factor sensitivity analysis has been carried out, which cannot fully meet the actual requirements on-site. The estimation of drilling investment and hydraulic-fracturing investment is relatively rough, affecting the accuracy of the economic length of the horizontal section. Therefore, a multifactor sensitivity analysis should be conducted and the drilling investment and hydraulic-fracturing investment should be refined in the next step to further improve engineering practices.

Author Contributions: Conceptualization, S.H. and J.Z.; methodology, software, validation, J.Z.; formal analysis, investigation, L.L. All authors have read and agreed to the published version of the manuscript.

Funding: This research received no external funding.

Institutional Review Board Statement: Not applicable.

Informed Consent Statement: Not applicable.

Data Availability Statement: Not applicable.

Conflicts of Interest: The authors declare no conflict of interest.

References

1. Jiangtao, L.; Yuyan, W.; Chuncheng, Z. Talking about energy: China and the world. *China Pow. Enter. Manag.* **2021**, *7*, 47–50.
2. Zhou, Q.; Dilmore, R.; Kleit, A.; Wang, J.Y. Evaluating gas production performances in marcellus using data mining technologies. *J. Nat. Gas Sci. Eng.* **2014**, *20*, 109–120. [CrossRef]
3. Wu, Y.; Pan, Z.; Zhang, D.; Lu, Z.; Connell, L.D. Evaluation of gas production from multiple coal seams: A simulation study and economics. *Int. J. Min. Sci. Technol.* **2018**, *28*, 359–371. [CrossRef]
4. Yusong, L.; Peiqing, L.; Dengke, T.; Ji, T. Optimization design of horizontal section length of horizontal well using genetic algorithm. *J. Petrol.* **2008**, *2*, 296–299.
5. Xiaole, G.; Zhiming, W. The hydraulic extended limit of mega-extended-reach well at Liuhua Field in South China Sea. *Oil Drill. Prod. Technol.* **2009**, *31*, 10–13.
6. Lu, N.; Zhang, B.; Wang, T.; Fu, Q. Modeling research on the extreme hydraulic extension length of horizontal well: Impact of formation properties, drilling bit and cutting parameters. *J. Pet. Explor. Prod. Technol.* **2021**, *11*, 1211–1222. [CrossRef]
7. Kunji, X.; Jiyou, X.; Jun, C.; Dawei, Q.; Honglin, X. The evaluation and analysis of hydraulic extensions ability of horizontal section in deep horizontal wells. *J. Southwest Pet. Univ. (Sci. Technol. Ed.)* **2012**, *34*, 101–106.
8. Xiuju, J.; Jianxia, B.; Honglei, L.; Xiuzhi, W.; Jiqiang, L. An optimal design of horizontal wells applied in the productivity construction of the Puguang gas field. *Nat. Gas Ind.* **2011**, *31*, 58–60.
9. Yuan, J.; Zhao, K.; Feng, Y. Research on wellbore instability of shale formation in extremely complex geo-mechanical environment. *Processes* **2022**, *10*, 1060. [CrossRef]
10. Yaohui, C.; Tie, Y.; Yin, L.; Xueliang, B. Research on the reasonable horizontal interval length of horizontal wells. *Nat. Gas Geosci.* **2006**, *26*, 151–153, 177–178.
11. Yingjie, Z. Economic and technical policy limits of horizontal wells in thick bottom water fault block reservoirs in shengli oilfield. *Petrol. Geol. Rec. Effic.* **2007**, *14*, 59–61.
12. Yingjie, Z. Technical and economic limits for horizontal well development of heavy oil reservoir in shengli oil field. *Xinjiang Pet. Geol.* **2008**, *29*, 76–78.
13. Ariadji, T.; Aziz, P.A.; Soewono, E.; Syifa, A.A.; Riza, L.S.; Sidarto, K.A.; Sukarno, P. A robust method for determining the optimum horizontal well direction and length for a petroleum field development using genetic algorithm. In Proceedings of the 5th International Conference on Research and Education in Mathematics (Icrem5), Portland, OR, USA, 23–25 February 2012; Volume 1450, pp. 319–325.
14. Xiaojing, Z.; Dengke, T. An improvement of the design method for optimal horizontal wellbore length. *Pet. Explor. Dev.* **2011**, *38*, 216–220.
15. Junkun, H.; Zhibin, Z.; Xiaoping, L. A method of determining the horizontal-well lateral length with optimal economic value. *Nat. Gas Ind.* **2014**, *34*, 142–146.

16. Rammay, M.H.; Awotunde, A.A. Stochastic optimization of hydraulic fracture and horizontal well parameters in shale gas reservoirs. *J. Nat. Gas Sci. Eng.* **2016**, *36*, 71–78. [CrossRef]
17. Lin, H. Road Operation Safety Risk Analysis based on Data Mining. Ph.D. Thesis, Dalian Maritime University, Dalian, China, June 2012.
18. Kalantari Dahaghi, A.; Mohaghegh, S.D. Economic impact of reservoir properties, horizontal well length and orientation on production from shale formations: Application to New Albany shale. In Proceedings of the SPE Eastern Regional Meeting 2009: Limitless Potential/Formidable Challenges, Charleston, WV, USA, 23–25 September 2009; Society of Petroleum Engineers (SPE): Charleston, WV, USA, 2009; pp. 312–324.
19. Honglin, L. Research on the Technology and Economic Boundary of the Horizontal Well in Low Permeability Gas Reservoir. Ph.D. Thesis, Southwest Petroleum University, Chengdu, China, October 2017.
20. Dawei, W.; Xiaohong, L.; Chunxiao, D.; Zhigang, G.; Zhennan, G. Research on horizontal well horizontal section length optimization and economic limit. *Xinjiang Oil Gas* **2017**, *13*, 19–24.
21. Huachao, S.; Chunya, L.; Zeying, T.; Ruiyang, W. Determination of economic optimal lateral length in horizontal well development of gas reservoirs—A case study of cn gas reservoir in country a. *Nat. Gas Technol. Econ.* **2017**, *11*, 24–25, 39, 82.
22. Wuguang, L.; Hong, Y.; Yongpeng, S.; Yu, G.; Tianpeng, W.; Nanqiao, Z.; Yue, C. Development evaluation and optimization of deep shale gas reservoir with horizontal wells based on production data. *Geofluids* **2021**, *2021*, 4815559.
23. Chang, H.; Xiaoyan, G.; Yujin, W.; Yunhe, S.; Xiaowei, Z. Analysis on economic length of shale gas horizontal well. In Proceedings of the 32nd National Natural Gas Academic Annual Conference (2020), Chongqing, China, 13–14 November 2020; pp. 1280–1287.
24. Chang, H.; Jinyu, W.; Nan, W.; Hang, Z.; Yunsheng, W.; Xiaowei, Z.; Yunhe, S. Optimization of horizontal well length in weiyuan shale gas field considering technical and economic conditions. *Pet. Geol. Oilfield Dev. Daqing* **2021**, *40*, 158–166.
25. Zhang, J.; Hu, N.; Li, W. Rapid site selection of shale gas multi-well pad drilling based on digital elevation model. *Processes* **2022**, *10*, 854. [CrossRef]
26. Rui, Y.; Cheng, C.; Jianfa, W.; Haoyong, H.; Daijiao, J.; Jian, Z. Optimization of shale-gas horizontal well spacing based on geology–engineering–Economy integration: A case study of Well block Ning 209 in the National Shale Gas Development Demonstration Area. *Nat. Gas Ind.* **2020**, *40*, 42–48.
27. Opinions of the CPC Central Committee and the State Council on Completely, Accurately and Comprehensively Implementing the New Development Concept and Doing a Good Job in Carbon Peak and Carbon Neutralization. Available online: http://www.gov.cn/zhengce/2021-10/24/content_5644613.htm (accessed on 25 October 2022).
28. Liehui, Z.; Xiao, H.; Xiaogang, L.; Kuncheng, L.; Jiang, H.; Zhi, Z.; Jingjing, G.; Yinan, C.; Wenshi, L. Shale gas exploration and development in the Sichuan Basin: Progress, challenge and countermeasures. *Nat. Gas Ind.* **2021**, *41*, 143–152.
29. Mahdi, S.; Wang, X.; Shah, N. Interactions between the Design and Operation of Shale Gas Networks, Including CO_2 Sequestration. *Engineering* **2017**, *3*, 244–256. [CrossRef]
30. Dong, Z. Optimum Design of Horizontal Section Length of the Horizontal Wells in Sebei-2 Gas Field. Master's Thesis, China University Pet, Beijing, China, May 2010.
31. Jianchun, G.; Bo, L.; Cong, L.; Weigang, D.; Li, C.; Baoyun, J.; Yulong, Z.; Xin, X.; Senwen, X.; Mingming, Z. A Method for Optimizing the Stages of Stratified Fracturing in Low Permeability and Tight Reservoirs. CN 105735961 A, 6 June 2017.
32. Cheng, S.; Jianfa, W.; Yongqiang, F.; Bo, Z. Integrated dynamic evaluation of long lateral fracturing in shale gas wells: A case study on the Changning National Shale Gas Demonstration Area. *Nat. Gas Ind.* **2022**, *42*, 123–132.
33. Xiaoli, C.; Jie, Y.; Jiaojiao, X.; Long, D. Exploration and application of investment benefit evaluation method for shale gas exploration and development-taking A oil and gas engineering company as an example. *China Chief Financ. Officer.* **2019**, *10*, 59–61.

processes

MDPI

Article

Research on Pore Pressure Detection While Drilling Based on Mechanical Specific Energy

Hu Yin *, Hongwei Cui and Jiajia Gao

School of Petroleum Engineering, Southwest Petroleum University, Chengdu 610500, China; 202121000854@stu.swpu.edu.cn (H.C.); yiyingtuoli@163.com (J.G.)
* Correspondence: huyinswpu@outlook.com

Abstract: The detection of the formation of pore pressure while drilling is of great importance to ensure safe drilling operations. At present, the dc-exponent concept is mainly used to detect pore pressure while drilling. The dc-exponent concept is based on the theory of shale compaction, which is limited when used in carbonate rocks. A mechanical specific energy (MSE)-based method is proposed to detect pore pressure in deep, complex intervals. The method is based on the theory that the energy consumed by the bit to break and remove a unit volume of rock can reflect the effective stress and pore pressure of the rock in situ. In this paper, a torque and weight on bit (WOB) transfer model is proposed for estimating the downhole torque and WOB using drill string mechanics. Meanwhile, the rotary speed and torque of the positive displacement motors under compound drilling are considered, and the model of total MSE under compound drilling is modified. The MSE-based method was used to estimate the pore pressure in a region in western Sichuan, and there is a good agreement between the detected and measured pore pressure. The results demonstrate that the accurate computed MSE-based method is useful in detecting pore pressure in deep complex intervals.

Keywords: formation pore pressure; mechanical specific energy; drill string mechanics; effective stress; compound drilling

Citation: Yin, H.; Cui, H.; Gao, J. Research on Pore Pressure Detection While Drilling Based on Mechanical Specific Energy. *Processes* **2022**, *10*, 1481. https://doi.org/10.3390/pr10081481

Academic Editors: Alfredo Iranzo and Yidong Cai

Received: 15 June 2022
Accepted: 26 July 2022
Published: 28 July 2022

1. Introduction

Pore pressure is the pressure of the formation fluid within the pores of the soil or rock. In drilling engineering, accurate knowledge of pore pressure changes can provide a basis for the optimization of drilling fluid density and well structure design to avoid drilling incidents (e.g., well kicks/blowout and borehole collapse). There are three aspects of pore pressure analysis: pre-drill prediction, detection while drilling, and post-drill analysis [1]. The pre-drill prediction of pore pressure is usually based on seismic data combined with logging data from the offset wells. Detection while drilling mainly relies on logging while drilling (LWD), measurement while drilling (MWD) and drilling parameters. Post-drill analysis summarizes pore pressure in a developed region using logging data from drilled wells, which is an after-the-fact technique that can be used for pre-drill prediction in future wells.

Hottmann and Johnson [2] were probably the first to use logging data (sonic and resistivity) to make predictions of overpressure in shales. They found a linear relationship between the common logarithm of sonic transit time or resistivity and depth in hydrostatic pressure intervals, which they called the normal compaction trend (NCT). When entering anomalous pressure intervals, the sonic transit time or resistivity trend will be deflected. Pennebaker [3] was the first to apply seismic interval velocity data to the prediction of pore pressure. Under normal conditions, interval velocity increases with depth, and when an overpressure interval appears, it is often accompanied by a decrease in interval velocity.

Based on data provided by Hottmann and Johnson, Gardner et al. [4] proposed an empirical equation that can be expressed as a direct form of pore pressure (Equation (1))

$$P_{\mathrm{p}} = \sigma_{\mathrm{v}} - \frac{(\mathrm{A} - \mathrm{B}\ln \Delta t)^3 (OBG - P_{\mathrm{ng}})}{Z^2} \tag{1}$$

where P_{p} is the formation pore pressure (psi); σ_{v} is the normal overburden pressure (psi); Δt is the sonic transit time (μs/ft); OBG is the overburden pressure gradient (psi/ft); P_{ng} is the normal fluid pressure gradient (psi/ft); Z is the depth (ft-TVD).

Eaton [5] proposed three sets of prediction equations for pore pressure based on resistivity, sonic transit time and the dc-exponent, respectively (Equations (2)–(4)). Estimation of pore pressures using Eaton's model requires an accurate NCT and reliable input data, so that accurate pore pressure can be obtained.

$$P_{\mathrm{pg}} = OBG - (OBG - P_{\mathrm{ng}}) \times \left(\frac{R_{\mathrm{o}}}{R_{\mathrm{n}}}\right)^{1.2} \tag{2}$$

$$P_{\mathrm{pg}} = OBG - (OBG - P_{\mathrm{ng}}) \times \left(\frac{\Delta t_{\mathrm{n}}}{\Delta t_{\mathrm{o}}}\right)^{3} \tag{3}$$

$$P_{\mathrm{pg}} = OBG - (OBG - P_{\mathrm{ng}}) \times \left(\frac{d_{\mathrm{co}}}{d_{\mathrm{cn}}}\right)^{1.2} \tag{4}$$

where P_{pg} is the formation pore pressure gradient (psi/ft); R_{o} is the observed shale resistivity at a given depth (ohm-m); R_{n} is the normal compaction shale resistivity at a given depth (ohm-m); Δt_{o} is the observed shale transit time at a given depth (μs/ft); Δt_{n} is the normal compaction shale transit time at a given depth (μs/ft); d_{co} is the computed dc-exponent from the measured data at a given depth; and d_{cn} is the dc-exponent from the normal compaction trend at a given depth.

Fillippone [6] proposed an empirical equation for predicting pore pressures through a comprehensive study of seismic, logging and drilling data in the Gulf of Mexico region (Equation (5)). The new empirical equation does not rely on the NCT and can be used as an effective pre-drill pore pressure prediction method [7].

$$P_{\mathrm{pg}} = \frac{v_{\mathrm{max}} - v}{v_{\mathrm{max}} - v_{\mathrm{min}}} OBG \tag{5}$$

where v is the seismic interval velocity (ft/s); v_{min} is the seismic interval velocity when rock rigidity is 0 (ft/s); and v_{max} is the seismic interval velocity when formation porosity is 0 (ft/s).

Bowers [8] derived the empirical equations between effective stress and sonic velocity based on the effective stress principle. In this way, the pore pressure can be estimated from the sonic velocity. He considered the relationship between effective stress and sonic velocity from soil mechanics for different overpressure genesis. When the overpressure is caused by uncompaction, the relationship follows the original curve (Equation (6)), while when it is caused by the expansion of the fluid, the relationship follows the unloading curve (Equation (7)). The values of parameters A and B can be calibrated with offset sonic velocity versus effective stress data.

$$V = 5000 + \mathrm{A}\sigma_{\mathrm{e}}^{\mathrm{B}} \tag{6}$$

$$V = 5000 + \mathrm{A}\left[\sigma_{\mathrm{max}}\left(\frac{\sigma_{\mathrm{e}}}{\sigma_{\mathrm{max}}}\right)^{\frac{1}{U}}\right]^{\mathrm{B}} \tag{7}$$

$$\sigma_{\mathrm{max}} = \left(\frac{V_{\mathrm{max}} - 5000}{\mathrm{A}}\right)^{\frac{1}{\mathrm{B}}} \tag{8}$$

where V is the sonic velocity (ft/s); V_{max} is the velocity at the onset of unloading (ft/s); σ_e is the vertical effective stress (psi); and σ_{max} is the vertical effective stress at the onset of unloading (psi).

Most current pore pressure prediction methods are applicable to shale intervals, where there is a strong relationship between porosity and pore pressure. In contrast, carbonate rocks are harder, and overpressure has a weak effect on porosity. Atashbari and Tingay [9] and Azadpour et al. [10] derived the relationship between the pore pressure and compressibility to estimate pore pressure in carbonate rocks (Equation (9)). By further studying the numerical empirical relationship between compressibility correlations and porosity, a simplified approximation of Equation (10) with only porosity and effective stress was obtained [11]. This new method requires sufficient logging and core data to obtain regional formation porosity and pore volume compressibility.

$$P_p = \left(\frac{(1-\phi)C_b\sigma_{eff}}{(1-\phi)C_b - \phi C_p} \right)^{\gamma} \tag{9}$$

$$P_p = \left(\frac{(1-\phi)\sigma_{eff}}{(1-2\phi)} \right)^{\gamma} \tag{10}$$

where ϕ is the formation porosity (fraction); C_b is the bulk compressibility (psi^{-1}); C_p is the pore compressibility (psi^{-1}); and σ_{eff} is the effective overburden pressure (overburden pressure-hydrostatic pressure) (psi).

Due to complicated geological factors and variable causes of overpressure, the development of deep and ultra-deep wells makes it difficult to accurately predict the pore pressure simply by pre-drill prediction. Therefore, during the drilling process of deep and complicated formations, it is inevitable to carry out the detection of pore pressure while drilling. Real-time detection of pore pressure usually uses LWD and drilling parameters. LWD generally offers more accurate results than methods based on drilling parameters. However, the real-time performance of the LWD is not ideal because the LWD tool is positioned before the drill collar, slightly far from the bit. Moreover, there is a time delay in bottom hole data transmission [12]. Conversely, drilling parameters are real-time field data that are easily obtained in the drilling process. This paper is concerned with drilling parameter-based pore pressure detection.

Jorden and Shirley [13] proposed the concept of the d-exponent for pore pressure detection by normalizing the factors (WOB, rotary speed, bit diameter) affecting the rate of penetration (ROP). Making a normal compaction trend (NCT) in the shales, the d-exponent will increase linearly with depth in normal pressure intervals. When entering the overpressure intervals, the d-exponent will tend to deviate from the NCT in a decreasing trend. The degree of deviation of the d-exponent could reflect the degree of overpressure. During drilling, the increase in drilling fluid density will cover up the change in the d-exponent, and a corrected d-exponent was proposed (Equation (11)) [14]. However, based on shale compaction, it is not applicable for the d-exponent concept to detect pore pressure in carbonate formations.

$$d_c = \frac{\log\left(\frac{0.0547ROP}{N}\right)}{\log\left(\frac{0.0684WOB}{D_b}\right)} \times \frac{NPP}{ECD} \tag{11}$$

where ROP is the rate of penetration (m/hr); N is the rotary speed (rpm); WOB is the weight on bit (kN); D_b is the bit diameter (mm); NPP is the normal pore pressure (sg); and ECD is the equivalent circulating density (sg).

Belotti and Gerard [15] used the concept of sigmalog (rock strength parameter) to detect the pore pressure (Equation (12)), which was based on the relationship between rock strength, pore pressure and formation lithology. Likewise, the variation of sigmalog with depth can be plotted, and a trend line can be created. There are a number of factors that affect sigmalog, which can be divided into two aspects: geological factors and drilling parameters. The effect of factors other than pore pressure and porosity on the sigmalog can be excluded by correcting the equation and adjusting the trend line intercept. In this way, the pore pressure abnormality can be reflected by the inversion of the sigmalog curve. Based on the drilling strength of the rock, it is usually not obstructed by lithology in the detection of pore pressure. However, the application of the sigmalog method in carbonate rocks is practically not ideal [16].

$$\sigma = \left(1 + \frac{1 - \sqrt{1 + n^2 \times \Delta p^2}}{n \times \Delta p}\right) \times \left[\frac{WOB^{0.5} \times ROP^{0.25}}{D_b \times N^{0.25}} + 0.028 \times \left(7 - \frac{h}{1000}\right)\right] \quad (12)$$

where σ presents the rock strength parameter; n is the hole cleaning coefficient; Δp is the pressure difference between the drilling fluid and the hydrostatic pressure, kg/L; and h is the drilling depth, m.

Recently, research on pore pressure detection about drilling parameters has begun to use the mechanical specific energy (MSE) model. The original model of MSE was proposed by Teale [17], which reflects the energy required to break and remove a unit volume of rock. It is widely used for real-time detecting of drilling efficiency, optimization of drilling parameters, downhole identification of inefficient working conditions, and lithology identification [18–27].

Cardona [28] was probably the first to use the MSE concept to detect pore pressure. Subsequently, Majidi et al. [29] proposed a method to detect pore pressure through downhole drilling parameters and in situ rock data using the concept of drilling efficiency and mechanical specific energy (DE-MSE). The pore pressure was expressed as a function of equivalent circulating density (ECD), DE, MSE, angle of internal friction, and uniaxial compressive strength (UCS) of the rock (Equation (13)). Among them, the angle of internal friction and UCS of the rock are computed from the compressional velocity and based on empirical equations (Equations (14) and (15)).

$$P_p = ECD - (DE_{trend} \times MSE - UCS) \times \left(\frac{1 - \sin\theta}{1 + \sin\theta}\right) \quad (13)$$

$$DE_{trend} = a\phi_n^b \quad (14)$$

$$\theta = 18.53V_p^{0.5148} \quad (15)$$

$$UCS = 145 \times 0.43V_p^{3.2} \quad (16)$$

where P_p is the formation pore pressure (psi); DE_{trend} is the normal drilling-efficiency trendline; MSE is the mechanical specific energy (psi); UCS is the uniaxial compressive strength of the rock (psi); θ is the angle of internal friction (degrees); ϕ_n is the normal compaction porosity trendline(fraction); and V_p is the compressional velocity (km/s).

Not taking into account the energy of hydraulic shock-assisted rock breaking, the MSE model may have some limitations in estimating pore pressure in a soft rock environment. An actual calculation case of a shallow interval shows that the contribution of hydraulic energy to the total energy is up to 20% [30]. Therefore, Oloruntobi et al. [31] and Oloruntobi and Butt [32] applied the hydro-rotary specific energy (HRSE) model (Equation (17)) and the hydro-mechanical specific energy (HMSE) model (Equation (18)) to overcome this limitation. The HMSE model was originally proposed by Mohan et al. [33], which includes axial, torsional, and hydraulic energy. However, in conventional drilling, the assisted breaking effect of hydraulic energy is relatively weak, so the HMSE model is more suitable for high-pressure jet drilling. Moreover, for some deep wells and ultra-deep wells, the deep rocks are harder. The role of hydraulic rock breaking is rather limited compared to torsional rock breaking, and the MSE should be the main focus.

$$HRSE = \frac{120\pi \times N \times T + 6 \times 10^4 \eta \times \Delta P_b \times Q}{A_b \times ROP} \quad (17)$$

$$HMSE = \frac{WOB_e \times ROP + 120\pi \times N \times T + 6 \times 10^4 \eta \times \Delta P_b \times Q}{A_b \times ROP} \quad (18)$$

where $HMSE$ is the hydro-mechanical specific energy (MPa); WOB_e is the effective weight on bit (N); T is the torque (N-m); A_b is the bit area (mm^2); η is the hydraulic energy reduction factor; ΔP_b is the bit pressure drop (MPa); and Q is the flow rate (L/min).

Cui et al. [34] considered the MSE model under compound drilling (Equation (19)), in which the bit is driven by both surface drive and a positive displacement motor (PDM). The rotary speed and torque on the bit are both a combination of surface drive and PDM. In addition, PDM is powered by drilling fluid and transmits power to the bit. The contribution of PDM to total energy may indirectly account for the hydraulic energy term.

$$MSE = E_f \times \left[\frac{WOB}{A_b} + \frac{120\pi \times \left(N + \frac{Q}{q}\right) \times (T + 159.24q \times \Delta P)}{A_b \times ROP} \right] \quad (19)$$

where E_f is the Energy efficiency (fraction); q is the flow rate of the hollow rotor (L/r); and ΔP is the PDM pressure drop (MPa).

Torque and WOB are the main variables of the MSE, and in the absence of reliable downhole measurements, torque is usually calculated using the model proposed by Pessier and Fear [35] for estimating torque from WOB, the bit sliding friction coefficient and the bit diameter (Equation (20)). However, the torque estimated in this way is essentially a function of WOB, which may not take into account the role of torque. Majidi et al. [29] emphasized the importance of downhole drilling parameters, especially the torque, by comparing MSE calculated from the surface and downhole parameters. In the absence of reliable downhole measurements, some studies have attempted to estimate friction losses along the drill string by considering wellbore geometry and drilling equipment parameters [36–38]. These new technologies use surface drilling parameters and drill string mechanics to estimate more accurate drilling parameters at the bit in real-time.

$$T = \frac{\mu \times WOB \times D_b}{3000} \quad (20)$$

where μ is the bit coefficient of sliding friction.

In this paper, a pore pressure detection method based on MSE is proposed. The new method estimates the downhole parameters from surface measurements when reliable downhole measurements are not available. Meanwhile, the model of total MSE under compound drilling was modified by considering the effects of the PDM.

2. Theory

Teale defined MSE as the energy required to break and remove a unit volume of rock by the bit, which is related to the strength of the rock (Equation (21)). It was confirmed by experiments that the minimum MSE is roughly equal to the compressive strength of the material drilled [17]. The effective stress is also related to the compressive strength of the rock [39]. Moreover, the dependence of MSE on pore pressure was experimentally verified in rock samples [40,41]. Therefore, the MSE obtained while drilling can reflect the effective stress of the rock to some extent. The higher the effective stress, the greater the strength of the rock, and the more MSE is required. At the same depth, overpressure intervals with lower effective stress require less MSE than normal pore pressure intervals, and then the pore pressure can be reflected indirectly.

$$MSE = \frac{WOB}{A_b} + \frac{120\pi \times N \times T}{A_b \times ROP} \quad (21)$$

2.1. Downhole Parmeters

In the model of MSE, WOB and torque are the main variables. For conventional drilling, there is generally a lack of the downhole WOB and torque measurements, so they are considered to be computed from surface measurements (hook load, wellhead torque). In some high-angle wells, especially directional and horizontal wells, the friction between the tubular and wellbore wall is high, and there might be beds of cuttings. Therefore, the influence of friction during the transmission of WOB and torque should be considered.

The following assumptions can be made for obtaining a simplified soft rod model for estimating the friction: (1) the force and deformation of the drill string are within the elastic range; (2) the borehole curvature of the calculation unit is a constant; (3) the drill string touches the upper or lower side of the well wall with the same curvature; (4) the shear forces on the cross-section of the drill string are neglected, and the effect of the drill string stiffness is disregarded, but the drill string can withstand axial pressure; (5) the influence of the dynamic effect of the drill string is neglected. The downhole WOB and torque are calculated by Equation (22) and Equation (23), respectively [42]. Here, in the Figure 1, the models for both WOB and torque are transcendental equations that need to be solved using numerical calculations. A trial value is first taken at the bit, and the WOB and torque are calculated from the bit to the wellhead by unit. Then, the trial results are compared with the actual torque and hook load at the wellhead, respectively. Based on the error, the trial value at the bit is adjusted for iteration.

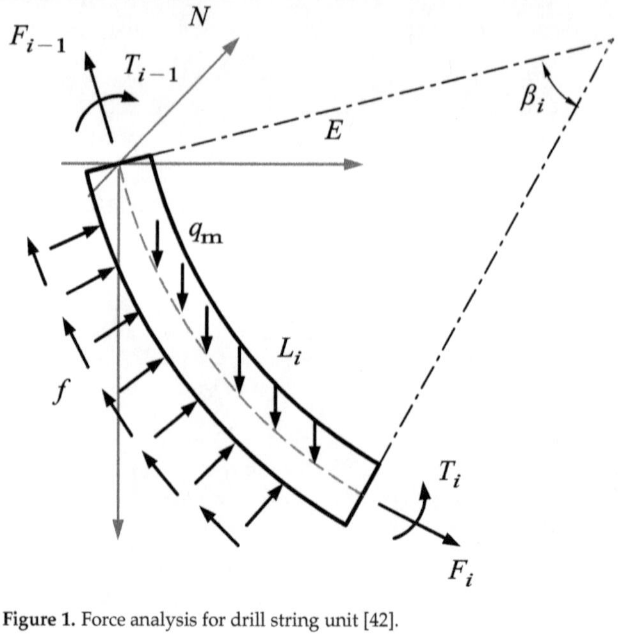

Figure 1. Force analysis for drill string unit [42].

Where F_i is the axial force at the lower end of the unit (N); F_{i-1} is the axial force at the upper end of the unit (N); q_{m} is the floating weight of the unit in the drilling fluid (N/m); L_i is the length of the i-th unit (m); β_i is the overall angle change rate of the i-th unit (rad); T_i is the torque at the lower end of the unit (N-m); and T_{i-1} is the torque at the upper end of the unit (N-m).

$$F_{i-1} = F_i + \frac{q_{\mathrm{m}} L_i \cos \overline{\alpha}_i - \mu_i |N_i|}{\cos \frac{\beta_i}{2} \cos \frac{\Delta \alpha_i}{2}} \tag{22}$$

$$T_{i-1} = T_i + \mu_t r_i |N_i| \tag{23}$$

$$\beta_i = \cos^{-1}(\cos\alpha_{i-1}\cos\alpha_i + \sin\alpha_{i-1}\sin\alpha_i\cos\Delta\phi) \tag{24}$$

$$N_i = \sqrt{(F_i\Delta\phi\sin\overline{\alpha}_i)^2 + (F_i\Delta\alpha_i + q_m L_i\sin\overline{\alpha}_i)^2} \tag{25}$$

where $\overline{\alpha}_i$ is the average of the well slope angles at the ends of the i-th unit (rad); μ_i is the friction coefficient of the i-th unit; N_i is the radial support force on the i-th unit (N); $\Delta\alpha_i$ is the incremental well slope angle of the i-th unit (rad); μ_t is the circumferential friction coefficient; r_i is the radius of curvature of the i-th unit (m); α_i is the well slope angle at the lower end of the unit (rad); α_{i-1} is the well slope angle at the upper end of the unit (rad); $\Delta\phi$ is the azimuth increment (rad); and $\Delta\alpha_i$ is the incremental well slope angle of the i-th unit (rad).

2.2. Drilling Parameters under Compound Drilling

There are two sources of power for the bit: one is the surface drive, usually the rotary table (RT) or the top drive system (TDS), and the other is the positive displacement motor (PDM). Under compound drilling, the bit is driven by both the surface drive and the PDM (RT/TDS + PDM). Therefore, the torque of the bit consists of the torque of the RT and the PDM, and the rotary speed of the bit is calculated in the same way [34]. The PDM is a volumetric power drilling tool driven by high-pressure drilling fluid. It is characterized by an output shaft torque proportional to the pressure drop of the drilling fluid in it and a rotary speed proportional to the flow rate of the drilling fluid. The total rotary speed and total torque of the bit are calculated by Equation (26) and Equation (27), respectively.

$$N_t = N + \frac{Q}{q} \tag{26}$$

$$T_t = 159.24q\Delta P + T_0 - \sum_{i=1}^{n}\mu_t r_i|N_i| \tag{27}$$

where N_t is the total rotary speed under compound drilling (rpm); T_t is the total torque under compound drilling (N-m); and T_0 is the wellhead torque (N-m).

The pressure drop of the PDM can be calculated by Equations (28)–(32) [43].

$$\Delta P = \left(\frac{Q}{Q_0}\right)^{1.8} P_0 + k\left(aw + bw^2\right)^f \tag{28}$$

$$a = \frac{T_1 - T_{max}\left(\frac{W_1}{W_{max}}\right)^2}{W_1 - \frac{W_1^2}{W_{max}}} \tag{29}$$

$$b = \frac{T_1 - aW_1}{W_1^2} \tag{30}$$

$$f = \frac{\ln\frac{P_1 - P_0}{P_{max} - P_0}}{\ln\frac{T_1}{T_{max}}} \tag{31}$$

$$k = \frac{P_1 - P_0}{T_1^f} \tag{32}$$

where Q_0 is the maximum permissible flow rate of PDM (L/min); P_0 is the pressure drop at idle at a maximum flow rate of PDM (MPa); w is the weight on bit (kN); T_1 is the output torque of PDM (kN-m); T_{max} is the maximum torque of PDM (kN-m); W_1 is the working weight on bit (kN); W_{max} is the maximum permissible weight on bit for PDM (kN); P_1 is the recommended working pressure drop of the PDM (MPa); P_{max} is the maximum allowable pressure drop of the PDM (MPa).

2.3. Normal Trendline

During drilling, excessive pressure overbalance conditions will increase the resistance of rocks and cuttings to the bit, which may result in a reduction in ROP and increase the MSE [32]. Therefore, the MSE was corrected with the regional normal pore pressure ratio on equivalent circulation density (Equation (33)).

$$MSE = \left(\frac{WOB}{A_b} + \frac{120\pi \times N_t \times T_t}{A_b \times ROP} \right) \times \frac{NPP}{ECD} \quad (33)$$

We plot the MSE with depth on a semi-log. The interval with clean shale and reliable drilling parameters data is selected as the target interval for establishing the normal trendline. For the abnormal points on the curve of the target interval, the values of MSE are smoothed out by moving the average filtering of the five-point bell-shaped function (Equation (34)) [44]. The coefficients β, γ and ε take the values 0.11, 0.24 and 0.3, respectively.

$$MSE_i = \beta(MSE_{i-2} + MSE_{i+2}) + \gamma(MSE_{i-1} + MSE_{i+1}) + \varepsilon MSE_i \quad (34)$$

2.4. Estimation of Pore Pressure

For the assessment of the formation of pore pressure, Eaton's model based on MSE is used (Equation (35)). The Eaton exponent (m) is a regional factor, and it can be obtained from any known overpressure intervals in the offset or current wells [32]. The Eaton exponent is estimated by substituting the overburden pressure, pore pressure, regional normal pore pressure, MSE, and the MSE from the corresponding normal trendline into Equation (36) for the target calibration interval, where the overburden pressure is obtained by integrating the formation bulk density logs.

$$G_{pp} = G_{ob} - (G_{ob} - G_{np}) \times \left(\frac{MSE}{MSE_n} \right)^m \quad (35)$$

$$m = \frac{\log\left(\frac{G_{ob} - G_{pp}}{G_{ob} - G_{np}} \right)}{\log\left(\frac{MSE}{MSE_n} \right)} \quad (36)$$

where G_{pp} is the formation pore pressure gradient (sg); G_{ob} is the overburden formation pressure gradient (sg); G_{np} is the normal pore pressure gradient (sg); MSE_n is the MSE from the normal trendline at a given depth (MPa); and m is the Eaton exponent.

2.5. Methodology

(1) The downhole parameters were numerically solved using Equations (22)–(25) based on the well geometry, bottom hole assembly (BHA), and surface measurements. When using PDM, we estimated its contribution to bit torque and rotary speed using Equations (26)–(32) based on the permissible and recommended parameters of the PDM.

(2) We calculated the MSE at a given depth using Equation (33) and plotted the MSE against depth on a semi-log.

(3) We selected a clean shale interval with reliable drilling parameters to establish the normal trend line, which can be appropriately filtered for abnormal points.

(4) The Eaton exponent can be corrected by Equation (36) in overpressure intervals with clear pore pressure from the offset or current well. Then, the pore pressure at a given depth can be estimated using Eaton's model.

3. Results

To demonstrate the applicability of the new technique in deep and complicated lithological formations, we took well L and well M in a region in western Sichuan as case studies. In this paper, all depths were referenced to the true vertical depth (TVD) below the rotary table. The marine strata start from the Leikoupo Formation at about 3850 m in this region, and the strata above the Leikoupo Formation are dominated by clastic rocks such as mud, shale, and sandstone, while limestone and dolomite gradually become the main lithology in the following strata. According to the actual pore pressure measurements in the overpressure intervals of the offset wells, the pore pressure gradient is about 1.70 sg in this region near 4925 m. Therefore, the interval in well L near this depth was selected as the calibration interval for the Eaton exponent. As shown in Figure 2, the average Eaton exponent was fitted to 0.226.

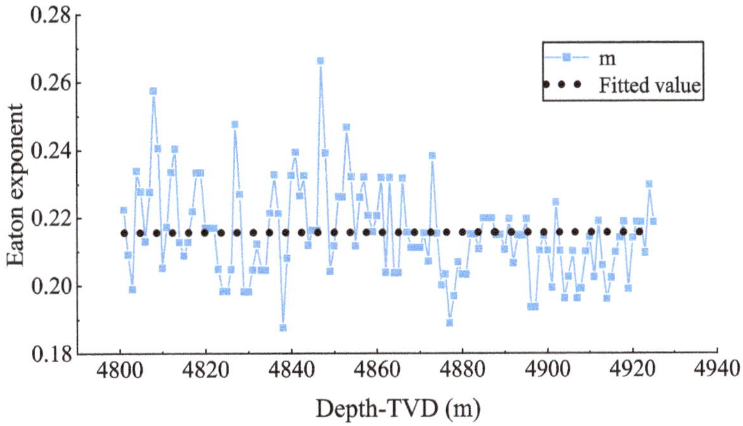

Figure 2. Estimation of the Eaton exponent from well L.

The well M is a sidetracking well, 6510 m deep, and was sidetracked from 5195 m with a maximum inclination of 20.12 degrees. In this well, PDM was used extensively for compound drilling, so both the influence of wellbore friction and the contribution of PDM to the total MSE should be considered. The most realistic downhole parameters can be obtained using MWD tools, but they are costly. Here, the downhole torque was computed from surface measurements (hook load and wellhead torque), and the results are plotted in the Figure 3b. The overall compliant trend of downhole torque is lower than the wellhead torque due to the friction of the drill string with the wellbore wall. Moreover, as the depth increases, the friction gradually increases. In the absence of reliable torque measurements, some studies have estimated torque from WOB, bit coefficient of sliding friction and bit size. The same operation was made here. The sliding friction coefficient of the bit is related to the type of bit, lithology, rock strength, mud weight and bit wear. To minimize the error in calculating MSE, it is reasonable to take 0.25 for roller-cone bits and 0.5 for poly-crystalline-diamond-compact (PDC) bits [45].

Since the friction coefficient between the drill string and the well wall was not sufficiently known in well M, the WOB was calculated using the data from the field. The WOB recorded in the field used the total weight of the drill string minus the hook load measured while drilling, which may have some error with the real WOB; we simply take a trend of variation. In fact, the contribution of WOB to MSE is rather limited compared to torque, especially in deep intervals where PDC bits are mainly used, and rocks are largely broken by the shearing of the bit.

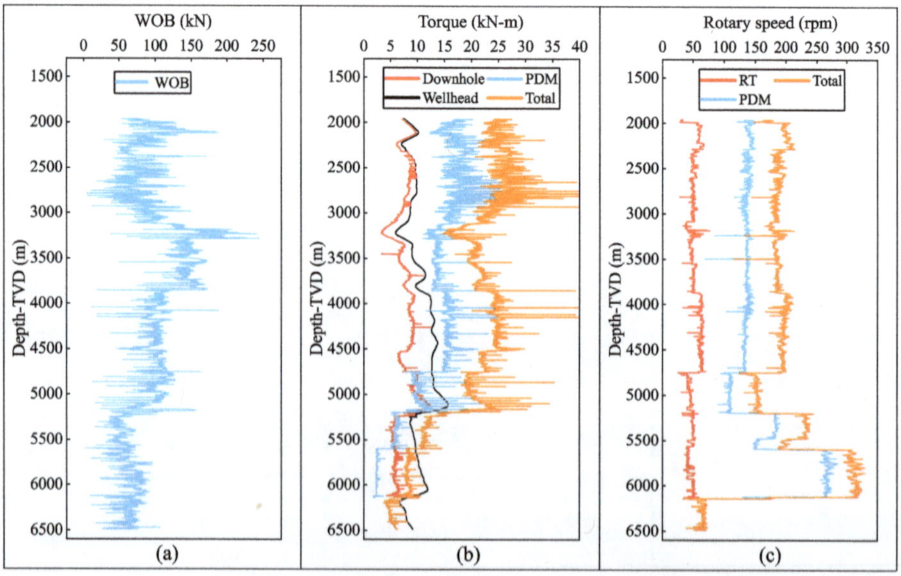

Figure 3. Drilling parameters under compound drilling for well M. (**a**) shows the weight on bit data; (**b**) plots the theoretical total torque on the bit, considering downhole friction and PDM; (**c**) presents the theoretical rotary speed of the bit under compound drilling.

The section from 1950 m to 6125 m was under compound drilling conditions. Based on the type of PDM and flow rate used in each section, the torque and rotary speed of the PDM were calculated. In contrast to the rotary table, the PDM drive is the main power for the bit during compound drilling. Combining the torque and rotary speed of the rotary table, the total torque and total rotary speed were calculated separately, as plotted in Figure 3.

Taking the sidetrack section in the well M as an example, the MSE was computed by using the wellhead torque, the downhole torque, and the torque estimated from WOB, respectively. Figure 4 illustrates the comparison of the pore pressure detection results for these three cases.

The differences in the comparison of the MSE plotted in Figure 4b are striking, especially since the MSE computed from the torque estimated by the WOB is significantly lower than in the other two cases, indicating the high sensitivity of the MSE to torque. Typically, the mud weight (MW) has a safety adder of 0.07 to 0.15 sg compared to pore pressure. In Figure 4c, the MW was reduced by 0.15 sg to serve as a rough reference for the comparison of pore pressure detection values. It can be observed that the overall low pore pressure reflected in the case of using wellhead torque, the detected values range from 1.27 to 2.35 sg with an average of 1.85 sg. The MSE computation based on the wellhead torque does not take into account the effect of downhole friction, resulting in a high reflected effective stress and low pore pressure. In contrast to the former, when calculating MSE based on downhole torque, downhole friction is taken into account, and pore pressure detection is fairly close to the reference line of pore pressure estimated from MW, ranging from 1.34 to 2.38 sg with an average of 1.92 sg. Interestingly, in the case of torque estimated from WOB, since the MSE values are overall significantly lower than the previous two cases, the final indirectly reflected pore pressure detection trend is instead closer to the reference line, especially in the interval from 5395 m to 5944 m. Similarly, acceptable downhole torque appears to be achieved with a more reasonable bit sliding friction coefficient.

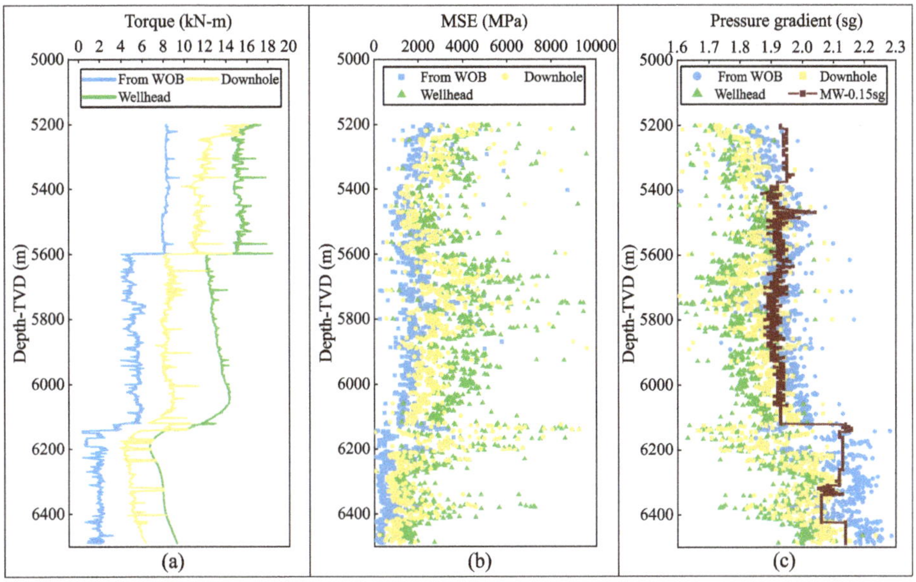

Figure 4. Comparison of detection results using torque from the wellhead, downhole and WOB for well M. (**a**) shows the torque from the wellhead, downhole and WOB; (**b**) plots the MSE computed from the different torque source; (**c**) presents the pore pressure detection values for the three cases.

4. Discussion

Here, a comparison between the dc-exponent method and the MSE-based method is performed for the entire section from 1960 m to the bottom of the well. Gas drilling was performed in the 361 m to 1950 m section of the well M, and the overall low dc-exponent values were tried to correct. Of these, a normally compacted, clean mudstone section from 430 m to 960 m was used for the establishment of NCT for the dc-exponent. Simultaneously, large changes in bit diameter and drilling fluid density in subsequent sections usually cause significant changes in the dc-exponent, which are easily misinterpreted as significant changes in pore pressure. Therefore, an attempt was made to correct the bit diameter and drilling fluid density for the NCT-established section to re-establish the NCT for the new section.

In well M, actual pore pressure measurements from two offset wells were used as a rough reference because the measured pore pressure data were not sufficient. In the section above 3850 m, the main lithologies are mudstone, sandstone and shale. As can be observed in Figure 5, both the dc-exponent and MSE-based methods provide reasonable estimates of pore pressure, and the MSE-based method has less scattering in its estimates. However, in the section above 2891 m, the MSE-based method appears to over-detect the pore pressure, and the dc-exponent provides more practical estimates. In the section below 3850 m, where the predominant lithologies are limestone, dolomite and gypsum, the pore pressure estimated by the MSE-based method matches better with the measured values in the current and offset wells. Meanwhile, it can be found that the dc-exponent approximately presents a tendency to a stable value at each hole section in the marine carbonate formation. The carbonate skeleton is stiffer, and its porosity does not necessarily tend to decrease with increasing burial depth, which may lead to the dc-exponent not being sensitive enough to changes in pore pressure.

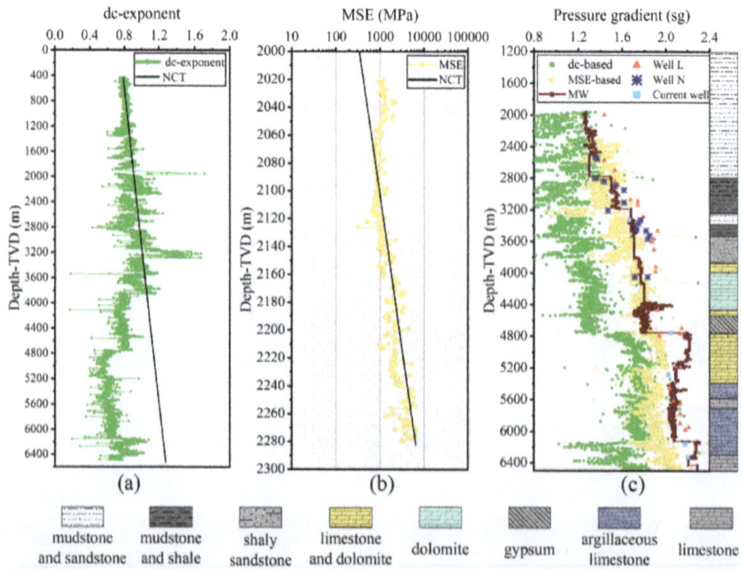

Figure 5. Comparison of the MSE-based method and dc-exponent method for the pore pressure detection in well M. (**a**) The NCT of dc-exponent method; (**b**) the NCT of MSE-based method; (**c**) comparison of pore pressure detection between the dc-exponent method and MSE-based method.

Notably, the entire section from 1960 m to the bottom of the well was drilled extensively with PDM in conjunction with PDC bits. Both sections from 3174 m to 3850 m and 6141 m to 6189 m were replaced with roller-cone bits. Additionally, for both dc-exponent and MSE, the pore pressure estimates near the freshly replaced bits show varying degrees of scattering. The type of bits and the degree of wear on the bits both have a significant impact on the drilling efficiency and MSE. Therefore, when the bit is changed frequently and the bit wear is high, the MSE may not accurately reflect the effective stress, and the MSE-based method should be used with caution. Especially when drilling through pressure transition zones, the above factors can mask overpressure conditions.

5. Conclusions

(1) A mechanical-specific energy-based drilling parameter method is proposed and validated to provide reasonable pore pressure estimates in deep complex lithologic intervals. The computation of MSE uses downhole parameters as far as possible. When reliable downhole measurements are lacking, a method based on drill string mechanics is proposed to estimate friction losses along the drill string by considering wellbore geometry and surface drilling parameters. In this way, more accurate drilling parameters at the bit can be obtained in real-time.

(2) Torsional energy is the primary source of contribution to MSE, and MSE-based pore pressure detection is highly sensitive to downhole torque. More attention and effort should be paid to the measurement and estimation of downhole torque. It should also be noted that when using PDM, the contribution of the PDM to the total MSE needs to be fully considered.

(3) The new method relies on trend lines and requires correction of the Eaton exponent, which has the influence of subjectivity. Moreover, all factors that are not related to pore pressure but can lead to significant changes in MSE may cause the overpressure conditions to be masked. At this point, the MSE-based method needs to refer to log data and actual drilling conditions from offset wells.

Author Contributions: Methodology, H.Y., H.C. and J.G.; investigation, H.Y., H.C. and J.G.; resources, H.Y.; writing—original draft preparation, H.C.; writing—review and editing, H.Y. and H.C. All authors have read and agreed to the published version of the manuscript.

Funding: This research was funded by the National Key Research and Development Program of China (grant number 2019YFA0708302).

Institutional Review Board Statement: Not applicable.

Informed Consent Statement: Not applicable.

Data Availability Statement: Not applicable.

Conflicts of Interest: The authors declare no conflict of interest.

References

1. Zhang, J. Pore pressure prediction from well logs: Methods, modifications, and new approaches. *Earth-Sci. Rev.* **2011**, *108*, 50–63. [CrossRef]
2. Hottmann, C.E.; Johnson, R.K. Estimation of formation pressures from log–derived shale properties. *J. Pet. Technol.* **1965**, *17*, 717–722. [CrossRef]
3. Pennebaker, E.S. Seismic data indicate depth, magnitude of abnormal pressure. *World Oil* **1968**, *166*, 73–78.
4. Gardner, G.H.; Gardner, L.W.; Gregory, A.R. Formation velocity and density—The diagnostic basics for stratigraphic traps. *Geophysics* **1974**, *39*, 770–780. [CrossRef]
5. Eaton, B.A. The equation for geopressure prediction from well logs. In Proceedings of the Fall Meeting of the Society of Petroleum Engineers of AIME, Dallas, TX, USA, 1 January 1975.
6. Fillippone, W.R. On the prediction of abnormally pressured sedimentary rocks from seismic data. In Proceedings of the Offshore Technology Conference, Houston, TX, USA, 30 April 1979.
7. Wang, L.; Yang, R.; Sun, Z.; Wang, L.; Guo, J.; Chen, M. Overpressure: Origin, Prediction, and Its Impact in the Xihu Sag, Eastern China Sea. *Energies* **2022**, *15*, 2519. [CrossRef]
8. Bowers, G.L. Pore pressure estimation from velocity data: Accounting for overpressure mechanisms besides undercompaction. *SPE Drill. Complet.* **1995**, *10*, 89–95. [CrossRef]
9. Atashbari, V.; Tingay, M.R. Pore pressure prediction in carbonate reservoirs. In Proceedings of the SPE Oil and Gas India Conference and Exhibition, Mumbai, India, 4–6 April 2012.
10. Azadpour, M.; Manaman, N.; Kadkhodaie-Ilkhchi, A.; Sedghipour, M. Pore pressure prediction and modeling using well-logging data in one of the gas fields in south of Iran. *J. Pet. Sci. Eng.* **2015**, *128*, 15–23. [CrossRef]
11. Atashbari, V. Origin of Overpressure and Pore Pressure Prediction in Carbonate Reservoirs of the Abadan Plain Basin. Ph.D. Thesis, The University of Adelaide, Adelaide, Australia, November 2016.
12. Chen, X.; Cao, W.; Gan, C.; Wu, M. A hybrid partial least squares regression-based real time pore pressure estimation method for complex geological drilling process. *J. Pet. Sci. Eng.* **2022**, *210*, 109771. [CrossRef]
13. Jorden, J.R.; Shirley, O.J. Application of drilling performance data to overpressure detection. *J. Pet. Technol.* **1966**, *18*, 1387–1394. [CrossRef]
14. Rehm, B.; McClendon, R. Measurement of formation pressure from drilling data. In Proceedings of the Fall Meeting of the Society of Petroleum Engineers of AIME, New Orleans, LA, USA, 3 October 1971.
15. Belotti, P.; Gerard, R.E. Istantaneous log indicates porosity and pore pressure. *World Oil* **1976**, *183*, 90–94.
16. Wang, Z. Detection of abnormal pressure while drilling in carbonate formations of northeastern Sichuan Basin. *Acta Pet. Sin.* **2012**, *33*, 1068–1075.
17. Teale, R. The concept of specific energy in rock drilling. International Journal of Rock Mechanics and Mining Sciences and Geomechanics Abstracts. *Pergamon* **1965**, *2*, 57–73. [CrossRef]
18. Pinto, C.N.; Lima, A.L.P. Mechanical specific energy for drilling optimization in deepwater Brazilian salt environments. In Proceedings of the IADC/SPE Asia Pacific Drilling Technology Conference, Singapore, 24 August 2016.
19. Wei, M.; Li, G.; Shi, H.; Shi, S.; Li, Z.; Zhang, Y. Theories and applications of pulsed-jet drilling with mechanical specific energy. *SPE J.* **2016**, *21*, 303–310.
20. Zhou, Y.; Zhang, W.; Gamwo, I.; Lin, J.S. Mechanical specific energy versus depth of cut in rock cutting and drilling. *Int. J. Rock Mech. Min. Sci.* **2017**, *100*, 287–297. [CrossRef]
21. Hassan, A.; Elkatatny, S.; Al-Majed, A. Coupling rate of penetration and mechanical specific energy to Improve the efficiency of drilling gas wells. *J. Nat. Gas Sci. Eng.* **2020**, *83*, 103558. [CrossRef]
22. Mazen, A.Z.; Rahmanian, N.; Mujtaba, I.M.; Hassanpour, A. Effective mechanical specific energy: A new approach for evaluating PDC bit performance and cutters wear. *J. Pet. Sci. Eng.* **2021**, *196*, 108030. [CrossRef]
23. Nystad, M.; Aadnøy, B.S.; Pavlov, A. Real-time minimization of mechanical specific energy with multivariable extremum seeking. *Energies* **2021**, *14*, 1298. [CrossRef]

24. Yu, B.; Zhang, K.; Niu, G. Rock strength determination based on rock drillability index and drilling specific energy: Numerical simulation using discrete element method. *IEEE Access* **2021**, *9*, 43923–43937. [CrossRef]
25. Oloruntobi, O.; Butt, S. Application of specific energy for lithology identification. *J. Pet. Sci. Eng.* **2020**, *184*, 106402. [CrossRef]
26. Khalilidermani, M.; Knez, D. A Survey of Application of Mechanical Specific Energy in Petroleum and Space Drilling. *Energies* **2022**, *15*, 3162. [CrossRef]
27. Yassien, M.A.; Sayed, M.A.; Boghdady, G.Y.; Ali, M.A.M.; Mohamed, A.S. Experimental research into the effect of some operation factors and rock properties on the rate of penetration. *Min. Miner. Depos.* **2020**, *14*, 38–43. [CrossRef]
28. Cardona, J. Fundamental Investigation of Pore Pressure Prediction during Drilling from the Mechanical Behavior of Rocks. Ph.D. Thesis, Texas A&M University, College Station, TX, USA, 2011.
29. Majidi, R.; Albertin, M.; Last, N. Pore-pressure estimation by use of mechanical specific energy and drilling efficiency. *SPE Drill. Complet.* **2017**, *32*, 97–104. [CrossRef]
30. Mohan, K.; Adil, F.; Samuel, R. Comprehensive hydromechanical specific energy calculation for drilling efficiency. *J. Energy Resour. Technol.* **2015**, *137*, 1–8. [CrossRef]
31. Oloruntobi, O.; Adedigba, S.; Khan, F.; Chunduru, R.; Butt, S. Overpressure prediction using the hydro-rotary specific energy concept. *J. Nat. Gas Sci. Eng.* **2018**, *55*, 243–253. [CrossRef]
32. Oloruntobi, O.; Butt, S. Energy-based formation pressure prediction. *J. Pet. Sci. Eng.* **2019**, *173*, 955–964. [CrossRef]
33. Mohan, K.; Adil, F. Tracking Drilling Efficiency Using Hydro-Mechanical Specific Energy. In Proceedings of the SPE/IADC Drilling Conference and Exhibition, Amsterdam, The Netherlands, 17–19 March 2009.
34. Cui, M.; Li, J.; Ji, G.; Chen, Y. Optimize method of drilling parameter of compound drilling based on mechanical specific energy theory. *Pet. Drill. Tech.* **2014**, *42*, 66–70.
35. Pessier, R.C.; Fear, M.J. Quantifying common drilling problems with mechanical specific energy and a bit-specific coefficient of sliding friction. In Proceedings of the SPE Annual Technical Conference and Exhibition, Washington, DC, USA, 4–7 October 1992.
36. Jacques, A.; Ouenes, A.; Dirksen, R.; Paryani, M.; Rehman, S.; Bari, M. Completion Optimization While Drilling–Geomechanical Steering Towards Fracable Rock Using Corrected Mechanical Specific Energy. In Proceedings of the SPE/AAPG/SEG Unconventional Resources Technology Conference, Austin, TX, USA, 24–26 July 2017.
37. Ouenes, A.; Dirksen, R.; Paryani, M.; Rehman, S.; Bari, M. Completion Optimization While Drilling–Geomechanical Steering towards Fracable Rock for Optimal Selection of Stage Spacing and Cluster Density in Unconventional Wells. In Proceedings of the SPE Kingdom of Saudi Arabia Annual Technical Symposium and Exhibition, Dammam, Saudi Arabia, 24–27 April 2017.
38. Kalinec, J.; Paryani, M.; Ouenes, A. Estimation of 3D Distribution of Pore Pressure from Surface Drilling Data–Application to Optimal Drilling and Frac Hit Prevention in the Eagle Ford. In Proceedings of the SPE/AAPG/SEG Unconventional Resources Technology Conference, Denver, CO, USA, 22–24 July 2019.
39. Chen, T.; Guan, Z. *Drilling Engineering Theory and Technology*, 1st ed.; Petroleum University Publishing House: Dongying, China, 2000; pp. 35–36.
40. Akbari, B.; Miska, S.; Mengjiao, Y.; Ozbayoglu, E. Effect of rock pore pressure on mechanical specific energy of rock cutting using single PDC cutter. In Proceedings of the 47th US Rock Mechanics/Geomechanics Symposium, San Francisco, CA, USA, 23–26 June 2013.
41. Akbari, B.; Miska, S.; Yu, M.; Ozbayoglu, M. Experimental investigations of the effect of the pore pressure on the MSE and drilling strength of a PDC bit. In Proceedings of the SPE Western North American and Rocky Mountain Joint Meeting, Denver, CO, USA, 16–18 April 2014.
42. Li, Q.; Chen, Z. Frictional resistance computation of setting casing in highly deviated well. *Nat. Gas Ind.* **1993**, *13*, 50–54.
43. Zhang, X.; Liu, X.; Xia, H.; Tao, Q.; Peng, M. Drawdown analysis and calculation of helicoid hydraulic motor. *Fault-Block Oil Gas Field* **2006**, *13*, 60–61.
44. Hu, Z.; Yuan, B.; Zhang, X.; Han, B.; Li, Y.; Lai, F.; Li, Y. Application of evaluation technique of bit efficiency while drilling. *Mud Logging Eng.* **2019**, *30*, 1–7.
45. Guan, Z.; Hu, H.; Wang, B.; Sun, M.; Liu, Y.; Xu, Y. Experimental study on rock-breaking efficiency of PDC bit based on mechanical specific energy and sliding frictional coefficient. *J. China Univ. Pet. (Ed. Nat. Sci.)* **2019**, *43*, 92–100.

processes

MDPI

Article

Warning Signs of High-Pressure Formations of Abnormal Contour Pressures When Drilling for Oil and Natural Gas

Tomas Huszar [1], Gabriel Wittenberger [2] and Erika Skvarekova [2,*]

1 District Mining Office in Kosice, Timonova 23, 041 57 Kosice, Slovakia; huszartomas@gmail.com
2 Department of Montaneous Sciences, Faculty of Mining, Ecology, Process Control and Geotechnologies, Institute of Earth's Resources, Technical University of Kosice, Park Komenskeho 19, 040 01 Kosice, Slovakia; gabriel.wittenberger@tuke.sk
* Correspondence: erika.skvarekova@tuke.sk; Tel.: +421-55-6023148

Abstract: When drilling to obtain hydrocarbons (oil and natural gas), we cannot underestimate the anomalously high pressures in the deposit layers, as these pressures can cause an uncontrollable eruption. Therefore, it is important to look for signs of anomalous high contour pressures over time, which, according to a detailed analysis, could be used to predict and quantify high formation pressures. These arise under conditions of intense vertical migration of formation fluids, where the liquids in the well have to carry part of the weight of overlying rocks and are often also related to tectonic activity. The main aim of the present study was to detect the emergence of a gas kick, which, as a result of an improper technological procedure, can cause an uncontrollable eruption, which can lead to a total accident of the well. In this article, we describe the use of modern drilling technology and sophisticated software that displays the current status inside the well. These can reveal impending pressure anomalies that can cause complications in managing the gas kick in oil and natural gas drilling. We analysed the most appropriate procedure for well control in a hydrocarbon well using the "driller's method" and the "wait and weight method". On the basis of theoretical background, we verified the correctness of the procedure for well control and compared it with the reaction to gas kick from a well drilled in Hungary. In the article, we highlight mistakes, as well as the particular importance of properly managing gas kick and its early prediction. Proper management of gas kick and its early prediction highlight the particular importance of implementing safe and effective procedures in well drilling.

Keywords: modern drilling technology; procedure for well control; proper management of gas kick; early prediction

Citation: Huszar, T.; Wittenberger, G.; Skvarekova, E. Warning Signs of High-Pressure Formations of Abnormal Contour Pressures When Drilling for Oil and Natural Gas. *Processes* **2022**, *10*, 1106. https://doi.org/10.3390/pr10061106

Academic Editors: Tianshou Ma and Yuqiang Xu

Received: 26 April 2022
Accepted: 30 May 2022
Published: 1 June 2022

1. Introduction

Very often, gas kick and related complications are manifested in oil and natural gas wells and during drilling on geothermal water under pressure. According to theoretical knowledge, the behaviour of these gas kicks can be predicted in advance and safely eliminated. During well control by the "driller's method" or by the "wait and weight method" essentially assumes the pumping of a new drilling mud with a new adjusted density. The gas kick is most often due to the fact that the current pressure ratios on the bottom hole are disrupted. There is an uncontrolled release of oil or natural gas from an oil or gas well after the failure of the drilling pressure control systems. Hydrocarbon wells have systems (BOP systems: blowout preventer) to prevent gas eruptions. Random sparks during an eruption can lead to a catastrophic oil or natural gas fire. Unmanaged gas kick is one of the most tragic and expensive technical accidents that can occur in the oil and gas industry. When an accident occurs, it is an immediate emergency situation that endangers life, the environment and all related equipment. Gas kicks in deep sea drilling for oil and natural gas are becoming serious with increasing drilling depth and a more complex geological environment. Their prediction and prevention represent one of the main challenges for drilling companies and rock engineers. Gas kick manifestations of hydrocarbon deposits are less documented, such as gas kicks and explosions of coal gas. Therefore, the exchange of information is important, especially in the area of prediction and prevention of gas kicks in oil and gas drilling. This contribution clarifies possible differences between theoretical prediction and practical solutions to actual gas kick events. In this article, we deal with the issue of theoretical and technical research concerning the prediction and prevention of hydrocarbon gas kick, taking the example of a D-1 well in Hungary. A comprehensive prediction method is proposed for a particular hydrocarbon well, and we highlight the incorrect actions and decisions taken by the driller in an attempt to resolve the situation. Such erroneous decisions in dealing with well control can cause uncontrolled eruption, with the possibility of an accident at the well. Therefore, it is necessary to establish a clear concept and procedure for dealing with such complications. InfoDrill and Drill Lab MasterLog software were used in this case to process the detected data. Data were collected and compared with the results of theoretical and actual well control situations, with consideration of terrain monitoring, laboratory tests, and measurements, in order to guarantee the accuracy of the well control forecast. The principle and scheme of the hydrocarbon well were studied and modelled, on the basis of which it was possible to deduce the correct technological procedure for well control in the hydrocarbon well [1]. The global oil and gas market has been changing dynamically in recent decades. The exploration and production of newly discovered high-quality oil and gas deposits, as well as government involvement in the development of unconventional low-carbon energy resources, have led to the global market presently being shaped not only by supply but also by demand [2]. Not all hydrocarbon deposits that are discovered are cost-effective. Mined oil or gas deposits can also be used for other mining activities. The use of already extracted oil and gas deposits is mainly aimed at the construction of underground natural gas storage facilities, as well as the pumping of mining waters, sulphides, CO_2, etc. [3]. These mined wells must be deepened, cleaned, equipped and installed using underground technological equipment [4]. The disposal of economically inefficient wells is an economically and technically challenging task. Such situations require technical knowledge and necessitate the use of quality materials, with emphasis on environmental protection [5,6]. For oil and natural gas extraction itself, the casing has to be accordingly designed, tested, operated, installed and monitored; failure to do so can cause notable problems and reduce the durability of the well. In the lifespan of a hydrocarbon well, various processes are carried out, including liner installation, drilling, pressure and temperature exploration, production and pumping tests. In order to achieve technical success, safe management and well control, a systematic understanding of the parameters during drilling and equipping of the well is necessary [7]. Horizontal or direct natural gas and oil wells are difficult to evaluate owing to the challenges associated with collecting data on formation pressure using a wire line [8,9].

Well deviation control has become a bottleneck preventing the development of gas drilling. Without greasing of the drilling fluid and the actuality of the negative pressure differential, the cause of well deviation in gas drilling differs from that in mud drilling. In this article, we analyse and consider the consequences of the stress distribution difference, well bore enlargement, the rock-breaking mechanism and water export in well deviation during both mud drilling and gas drilling, with reference to previous studies. Our conclusion is that the uneven crater formed by rock breaking, the new stress state in the bottom rock and borehole enlargement are the main reasons for well deviation during gas drilling [10]. According to the characteristics of horizontal wells, a multiphase flow theoretical model of a horizontal well was established. The finite difference method was adopted for solving mathematical equations. The bottom hole pressure of a horizontal well during gas kick was analysed. The results indicated that the bottom hole pressure of a vertical well decreases quickly, whereas the pressure of a horizontal well decreases after a given time. In a horizontal well with a large curvature radius, the bottom hole pressure decreases less than that in a well with a small curvature radius with a horizontal section of the same length. Comparing two wells with equivalent curvature radii, the bottom hole pressure of that with a longer horizontal section reduces slower than that with the shorter horizontal section. Increased build-up rate has no effect on the bottom hole pressure of horizontal wells. In wells with a small hole, the bottom hole pressure first increases and then reduces quickly after a given time. The pump output has little effect on the bottom hole pressure of horizontal wells [11]. The differential pressure on the bottom hole is one of the most important factors influencing mining speed [12]. Rock stress analyses for induced seismicity monitoring are necessary operating procedures for the safe and effective production of oil and gas. Analysis of the rock stresses on the bottom hole, as well as mechanisms of clearing drill cuttings, suggests that hydraulic pulsation in the well can reduce pressure, reduce the quarry strength of the rock and improve purification at the bottom hole, thus improving drilling force and efficiency. The higher the pulsation value, the more effective the acceleration of drilling speed; in contrast, as the depth of the well increases, the effect of acceleration gradually decreases [13]. A series of steps can helps to ensure safe and efficient production of natural gas and oil [14]. A stochastic drilling approach employs well performance algorithms to determine the optimal drilling profundity measured in vertical wells in three-dimensional space. The generated algorithms are then used to display the measured depth of vertical wells in which the pressure expression is assumed. A total of 15 hydrocarbon wells were designed and carried out in applicable rock environments, most often in sandstones [15]. Due to its rock form, sandstone allows for the aggregation and migration of hydrocarbons, such as natural gas and oil. To evaluate drilling cases and applications, including drilling unpredictability and many other drilling issues, it is necessary analyse the effects of hydrostatic and bearing pressure in the extracted rocks [16,17]. In order to increase the pressure, it is necessary to perform constant optimization and analyse the measurement history. It is appropriate to decrease the number of measurements in wells where issues have not been reported for a long period of time, focusing instead on potentially troublesome wells [18]. Natural gas is also possible solution to the usage of low-carbon fuel resources. One of the options for effective use of this energy source is underground storage [19,20]. One possibility for acquiring natural gas from unusual sources, especially from shale, is hydraulic fracturing.

Hydraulic fracturing techniques are especially effective in triggering the production of hydrocarbons from oil formations or shale gas [21,22]. In this article, we call attention to the significance of the dynamic increase in financial expenditure acquired by many countries in recognizing and investigating gas deposits contained in rocks, hydrates or aquifers. In terms of geological exploration, the volume of overall proven geological sources of unconventional gas has increased by 66% in recent years [1]. In this article, we characterize the consequences of utilizing innovative methods of natural gas exploitation in order to achieve increased sustainability and balanced global economic development, as well as a radical reduction in the cost of gas transport on a global scale. We are currently witnessing a dynamic increase in financial expenditures of countries on all continents with respect to exploration and analysis of gas deposits contained in rock or aquifer layers.

The aim of this article is to demonstrate how to proceed correctly in response to the first warning signs of pressure expression in hydrocarbon wells. By using modern software support, can detect deviations from the correct drilling mode. In this article, we present a case study to demonstrate how a software system can alert the driller and monitor any anomalies that may arise during the drilling process. This realistic example of the hydrocarbon drilling process also draws attention to important functional parameters, e.g., drilling depth, temperature and drilling pressure, weight on the hook, rotation per minute on the drill string, weight on the bit, etc., which exert a fundamental influence on the drilling process. Practical information and advice on procedures can help other drillers in terms of safety and future prospects.

2. Materials and Methods

A hydrocarbon well in Hungary was selected for our research (the Company Oil and Gas Development Kft, Figure 1). The well was drilled by a Bentec AC 250 heavy electric-hydraulic drilling rig with a load capacity of 250 tone (1600 horsepower), BOP ($13^{5/8}$ ", 10,000 psi), double, single and Cameron annular.

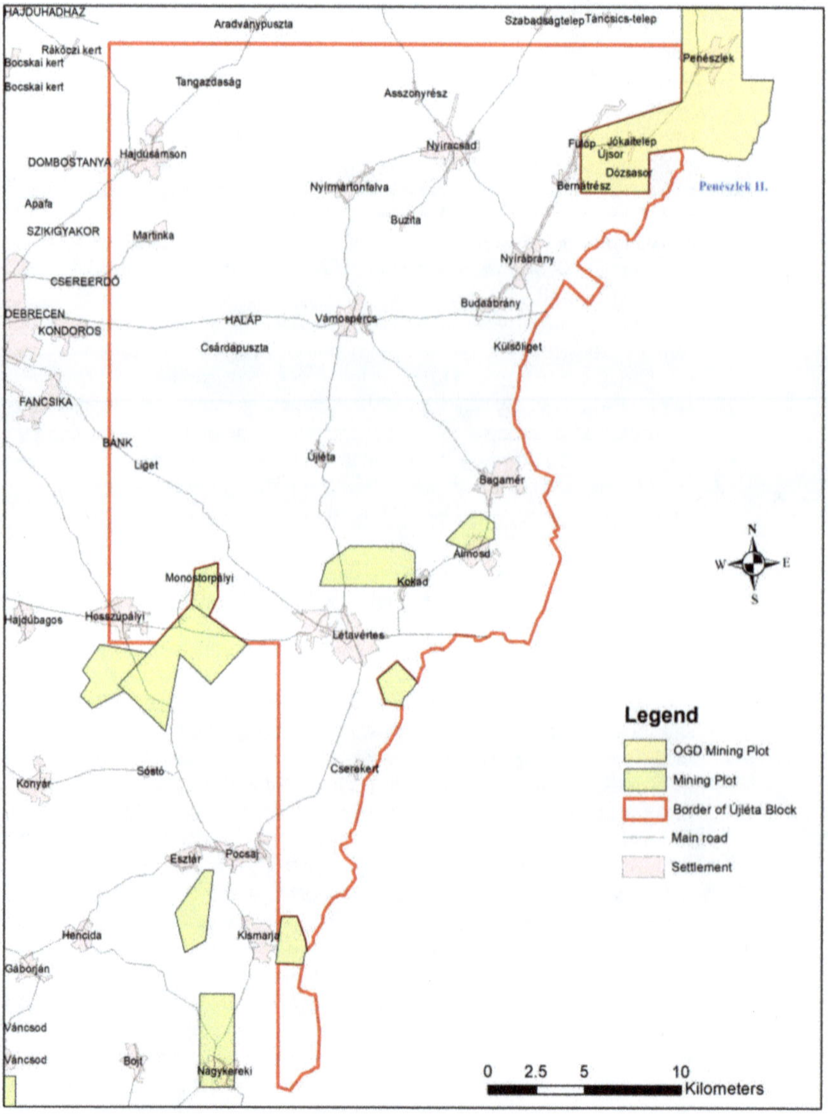

Figure 1. Area of interest for oil and gas development Kft [19].

The following steps were carried out:

- Assessment of the correctness and suitability of standard theoretical procedures according to the real in situ technological process under specific conditions;

- Analysis, comparison and reference to frequent technological errors and decisions with respect to recovery of well control in the drilling of hydrocarbon deposits; and
- Software modelling of a particular well control strategy under specific conditions.

The main cause of gas kick is the existence of a pressure difference between the formation pressure and the pressure in the well (hydrostatic), which is transferred from the formation layer to the well.

The accompanying phenomena of possible abnormal gas kick are:

1. Rate of penetration (ROP);
2. Change in drill cutting size;
3. Higher values of gas types:
    (a) background gas;
    (b) trip gas;
    (c) connection gas;
4. Higher salinity or chloride values in drilling mud;
5. Higher temperatures of drilling mud;
6. Gas-saturated drilling mud;
7. Change of "D exponent".

- Rate of penetration (ROP)

The speed of drilling is directly related to the type of drilled formations and the drilling bit. Hydrocarbons are predominantly found in porous formations, so the drilling of such formations is much easier, with intense speed increases compared to compact clay overlying formations.

- Change of the drill cuttings

An increase in the size of drill cuttings can occur in hard formations (Figure 2) due to increased pressure. Drill cuttings, on the other hand, may disappear altogether in soft coastal and marine sediments. Occasionally, shale shakers can be completely blinded by fragments of drilling cuttings.

Figure 2. Cuttings on shale shakers.

- Higher values of gas types: (a) background gas

The basic or normal curve of hydrocarbon gases that are discharged from the drilling mud can be determined by means of a gas detector. In general, gas may appear as a background gas. During drilling operations, the gas, together with the cuttings, reaches the surface and separates from the drilling mud. According to this principle, it is possible to obtain a curve of this gas, which occurs in small amounts (Figure 3).

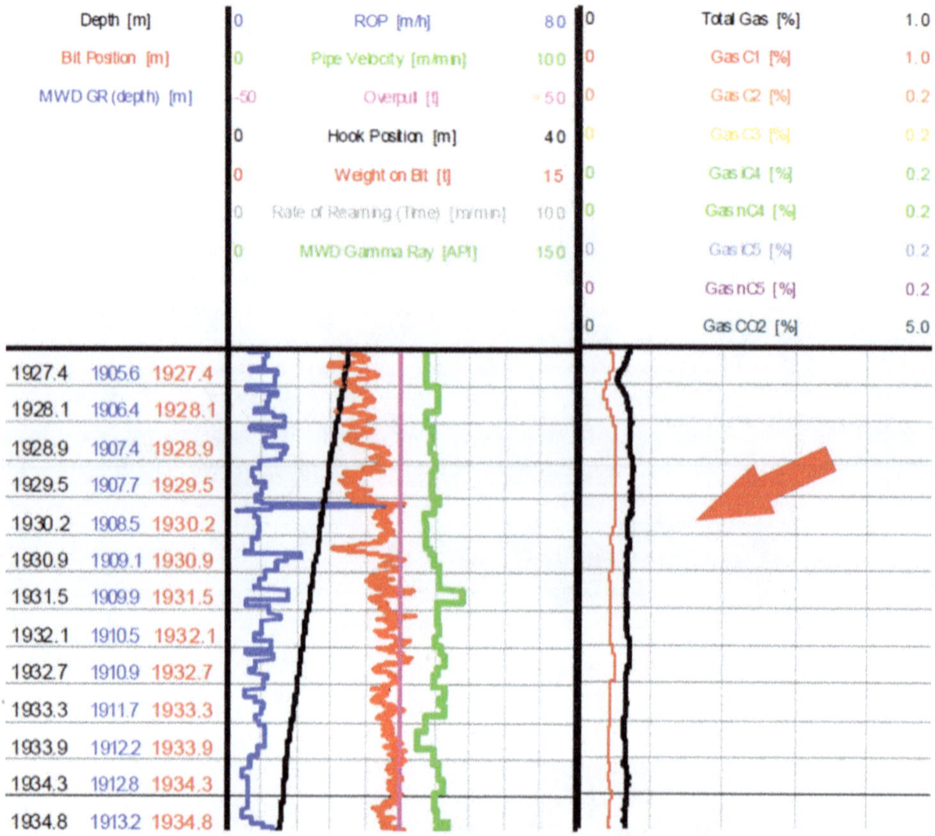

Figure 3. Arrow indicates background gas.

- Higher values of gas types: (b) trip gas

During swabbing or snubbing into or out of the well, pressure in the annulus may be reduced, and the pressures in the well may become unbalanced, with gas appearing in the drilling mud, which is easily identifiable on the curve (Figure 4).

- Higher values of gas types: (c) connection gas

When connecting another stand, the pumps are switched off, resulting in a sudden reduction in the pressure in the well due to a loss of pressure in the annulus. With reduced pressure in the well, the pressure of the formation releases gas into the column, resulting in a noticeable increase in gas on the surface separated from the drilling mud (Figure 5).

In the case of the occurrence of the mentioned manifestations of gases, it is recommended to implemented the following technological procedures:

- Use of a vacuum degasser;
- Change of job operations;
- Flow check.

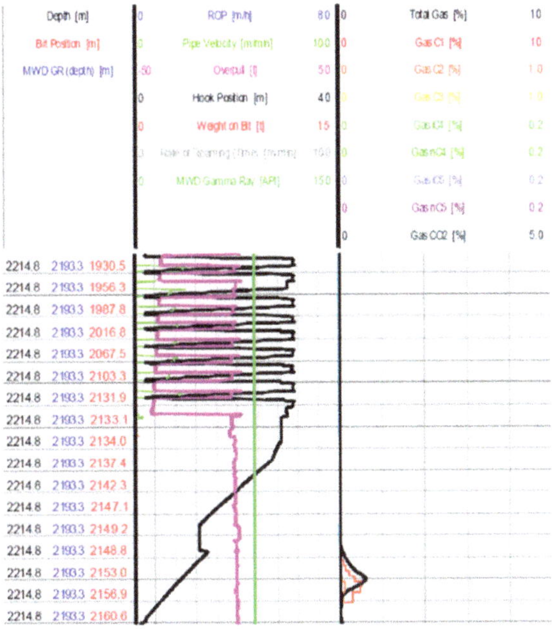

Figure 4. Arrow indicates trip gas.

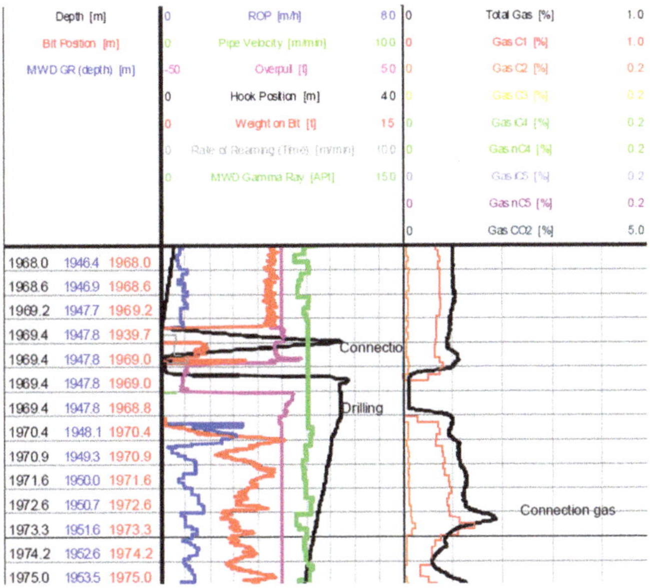

Figure 5. Arrow indicates connection gas.

- Higher salinity or chloride values in drilling mud

When the pressure in formations increases, water is squeezed from the formation, resulting in a higher concentration of salts in the remaining formations. Therefore, when drilling an overpressurized zone, the salinity value increases.

- Higher values of drilling mud temperatures

Because temperature and pressure are related to each other, the thermal gradient can drop intensely just above the transition zone, increasing sharply in formations with abnormally high pressure. In such cases, the normal curve of the thermal gradient can be determined and, in cases of anomalously high pressure, be used to determine the formations.

The most common parameters that affect temperature are:

- The density of the drilling mud;
- The amount of solid particles in the drilling mud;
- Flow properties and yield point;
- Circulation rates; and
- The geometry of the well.

- Gas-saturated drilling mud

Gas-cut mud during drilling may not be a signal of pressure expression, as the pressure on the bottom hole is not significantly reduced. Gas-saturated drilling mud can occur for a variety of reasons. An example is the collector rock, which releases gas into the drilling mud as it is crushed by the drill bit. This is a sign that a formation has been indicated. However, this will not cause a decrease in the density of the drilling mud and will not cause an imbalance of pressures in the well. In the case of any doubt, it is necessary to stop the pumps and perform a flow check. Gas-cut mud also occurs when drilling low-transmittance formations, which contain gas at a higher pressure than the hydrostatic pressure. Because such formations are not very permeable, the gas inflow is slow (Figure 6).

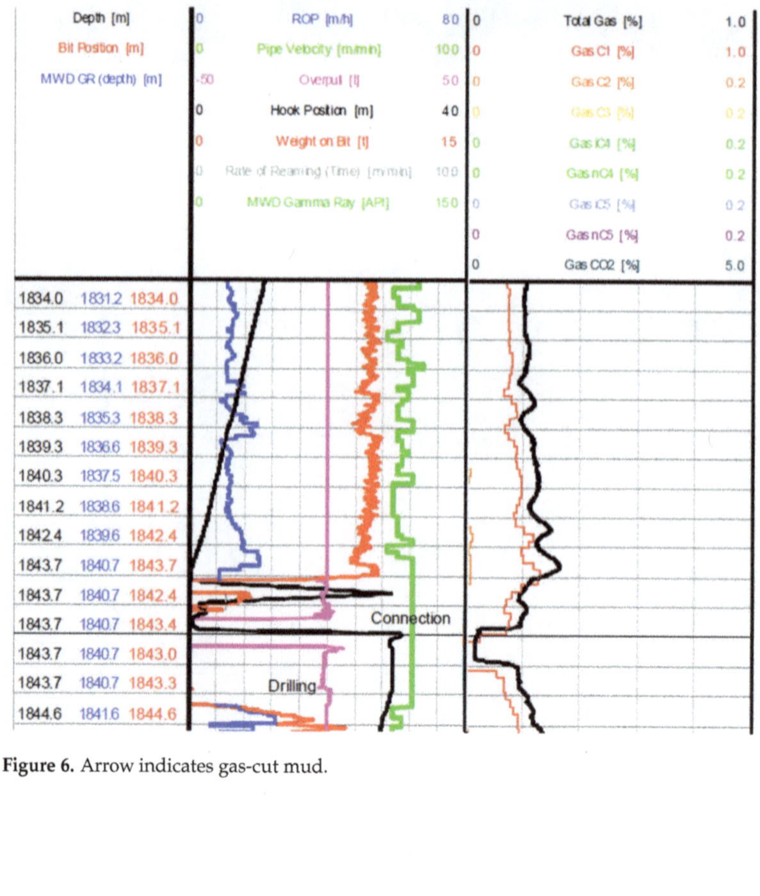

Figure 6. Arrow indicates gas-cut mud.

- Change of "D exponent"

"D exponent" is the value to traverse formations with abnormal pressure on the Gulf coast and was designed in 1966 by Jorden and Shirley [23] (Figure 7).

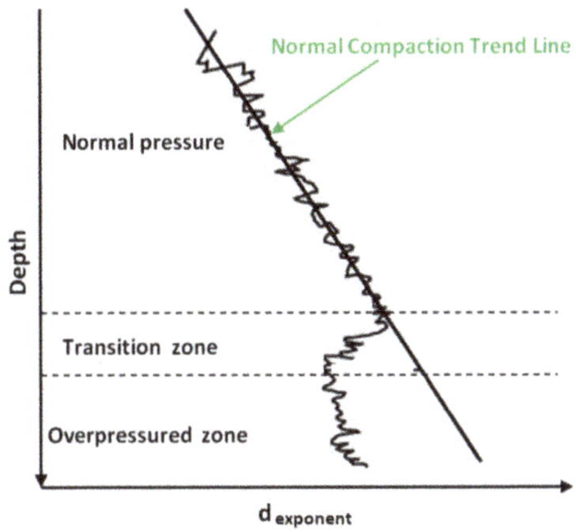

Figure 7. "D exponent".

"D exponent" is an approximate calculation of drilling parameters to obtain a trend while drilling into overpressurized zones. Usually, mud logger will correct all data, calculate the D exponent and plot the D-exponent valve on the curve. The D exponent can be utilized to survey the transition from a normal pressure regime to an abnormal formation pressure. A change in the drilling trend warns rig supervisors to exercise caution, as this is one of the possible indicators for drilling control [24].

$$d = \frac{\log \frac{R}{N}}{\log \frac{W}{Db}} \quad (1)$$

$$d_c = d\frac{MW1}{MW2} \quad (2)$$

where:

R = ROP (m/h)
K = const. 1
N = rate per minute (RPM) 1/min
E = RPM exponent = 1
W = weight on bit (WOB), (kg)
D_b = diameter of bit (mm)
MW_1 = original mud density (SG)
MW_2 = new mud density (SG).

2.1. Warning Signs of Gas Kick during Drilling

The most common symptoms of emerging kick gas include the following:

- Change in drilling rate;
- Increase in flow of drilling mud from the well;
- Decreased mud pump pressure and increased strokes;
- Increase in rotary torque;
- Increased drilling string weight.

- Change in drilling rate

Change in drilling rate is the first and fastest indicator that sand and shale formations were drilled. When a drilling bit penetrates layers with abnormally high pressures, the drilling speed increases, but when using OBM (oil-based mud) rinses, the speed can also slow down. With experience, it is easy to identify a sudden increase in speed.

Procedure for a sudden increase in drilling rate:

1. Stop the rotation on the drilling string;
2. Pick up the drilling string to correct the position of the tool joint above the drilling table;
3. Stop the mud pumps;
4. Perform a flow check;
5. If there is no leak, continue to drill. When the well flows, the pressure manifestation is declared, and the well is closed. Subsequently, stand-pipe pressure (SIDPP) and SICP (casing pressure) values are subtracted.

The flow check takes approximately 5 min for water-based mud and approximately 30 min with OBM.

- Increase in flow of drilling mud from the well

Another indication of gas kick is increasing discharge of gain from the well. With a device to measure the discharge from the float valve, it is possible to react in a timely manner to the amount of drilling mud returning to the surface. In the event of low values after subtraction, it is generally sufficient to increase the density of the drilling mud so that the hydrostatic pressure is higher than the pressure of the formation. However, if the subtracted parameters continue increase continuously, it is necessary to shut down the well. If there is a gas kick, the amount and pressure of the influx from the deposits depends on how quickly the well is closed.

- Decreased mud pump pressure and increased strokes

Circulation pressure is related to losses in the circulation circuit in the form of liquid friction in the DP (drill pipe), DC (drill collar), jets in the drill bit, MM (mud motor) and in the annulus. In addition, the circulating pressure is affected by the imbalance of hydrostatic pressure between the inside and the outside of the DP. When drilling gas, the gas rises and expands in the annulus. Depending on the weight of the liquid, the circulating pressure gradually decreases, and the number of pump strokes increases.

- Increase in rotary torque

Torque increases depending on the depth. In the case of drilling a formation with abnormally high pressure, the fragments are displaced more quickly from the formation, exerting an increased influence on the drill bit and the string. Consequently, torque suddenly increases sharply.

- Increase in drilling string weight

With a gas kick, layered liquids disrupt the homogeneity of the drilling mud and become lighter; therefore, the weight of the string is increased. Not every indicator signal provides an immediate warning about "drilling break", but if multiple signals appear simultaneously, it is necessary to react immediately [25].

2.2. Shut-In Procedures

Before a well is drilled, the drilling company must establish a procedure for shutting the well. If any sign of pressure manifestation is observed, such as an increase in the flow of mud from a well or the amount of drilling mud in the tanks, then the well has to be closed immediately.

It is important to distinguish:

- A soft shut-in of a well during drilling;
- A hard shut-in of a well during drilling;

- A soft shut-in of a well during tripping; and
- A hard shut-in of a well during tripping.

2.2.1. Reading and Recording SIDPP and SICP

Once the well is closed, it is important to record values from manometers on the drill string and in the annulus for well control situations. Because SIDPP pressure values lead to the drill bit, they indicate the pressure on the bottom hole. In the event that there is a float valve in the string, the pressure in DP is 0. Given that the mud in the annulus contains drill cuttings, which are suspended particles that enter the well due to increased pressure, this pressure is not suitable for calculating the exact value to increase the density of drilled mud.

2.2.2. Shutting Down the Well during Gas Kick

Before drilling for oil and natural gas or water under pressure, it is essential that drilling workers have a clear understanding of which type of well shutting will be implemented in the event of increased pressure.

There are two types of well shutting during drilling: soft shut-in and hard shut-in:

(a) Hard shut-in

If hard shut-in is selected, the choke in the choke manifold and the HCR (hydraulic choke valve) are set to the closed position.

Steps for hard well shutting:

- Close the BOP;
- Open the HCR (hydraulic choke valve);
- Read and record the SIDPP and SICP values from the manometers after allowing them to stabilize;
- Read and record gain parameters before closing the BOP.

Advantages:

- Fast shut-in influx volume;
- The pressure in the annulus is lower;
- No need for additional procedures.

Disadvantages:

- A pressure pulse or "water hammer" effect is produced in the well bore when the BOP is closed;
- Possible damage to the formation.

(b) Soft shut-in

During soft shutting the choke in the choke manifold is fully opened, and the HCR (hydraulic choke valve) is closed.

Steps for soft well shutting:

- Open the HCR;
- Close the BOP;
- Close the choke;
- Read and record the SIDPP and SICP values from the manometers after allowing them to stabilize;
- Read and record gain parameters before closing the BOP and before closing choke.

Advantages:

- A pressure pulse or "water hammer" effect is not significant when the BOP is closed.

Disadvantages:

- It takes longer to stop the penetration of the influx into the well;
- Higher pressure in the annulus;
- More steps need to be taken to shut down the well.

2.3. Well Control

A number of procedures and methods are available for well control, with the aim of killing the gas kick and bringing the well back under control.

Basic methods of well control:

- Driller's method;
- Wait and weight method;
- Concurrent method;
- Volumetric method;
- Bullheading.

In our case, we chose to deal with pressure expression with the use of the following two methods:

The driller's method is the most basic of all methods and can be employed in a number of well-controlled situations. Because it involves the use of many techniques common to other well control methods, the driller's method can be studied to learn basic well-control procedures. To initiate the procedure, start circulating, open the choke, slowly bring the pump up to the kill rate and hold the SICP at a constant value by adjusting the choke. Keeping SICP constant for this short period of time maintains constant bottom-hole pressure. When the pump is at kill-rate speed, observe the drill pipe gauge, which shows the ICP. Circulate the influx out, holding SIDPP constant at ICP. When the pits are full of kill-weight mud, open the choke and slowly bring the pump up to the kill rate, holding the casing pressure and pump rate constant. If KRP is known, the final circulating pressure can be calculated (Figure 8).

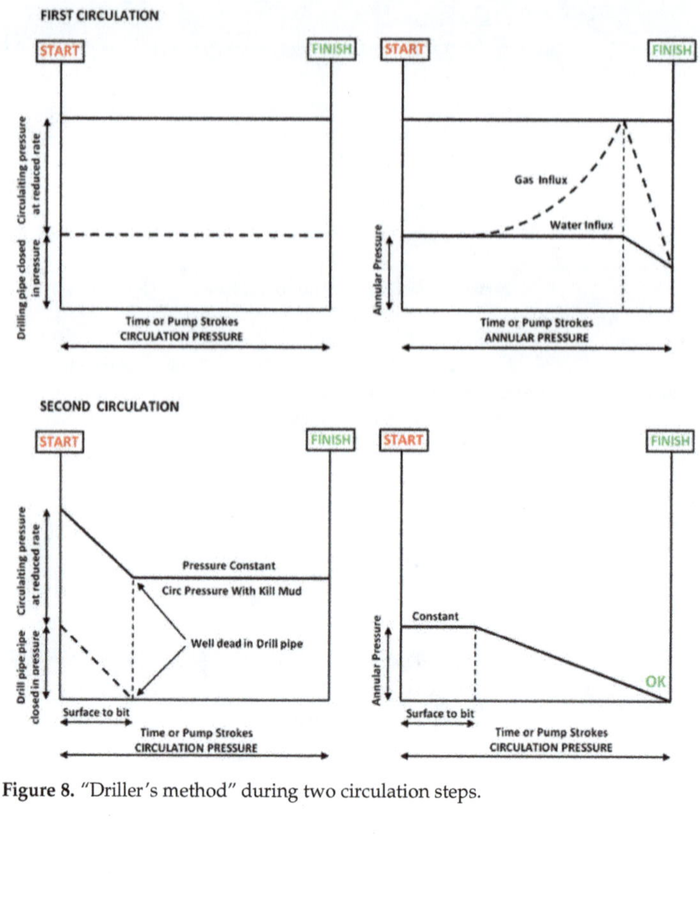

Figure 8. "Driller's method" during two circulation steps.

The "wait and weight" method is so named because the crew first shuts the well down, waits for kill-weight mud to be prepared and then circulates the new, weighted-up mud into the hole. At the same time, new mud is pumped in and old-weight mud and kick fluids are removed through the choke. Pumping in new mud while removing old mud and kick fluids may result in lower surface or casing pressure than when first circulating the kick out with old mud and then circulating in new mud.

Both methods lead to successful management of pressure manifestation. The main differences between these methods are that the "driller method" requires two cycles, whereas the "wait and weight" method requires only one cycle. The advantages of the "wait and weight" method are lower pressures at the bottom of the casing column during circulation and lower pressure at the mouth of the well surface. The disadvantages include the time required, the associated longer gas migration from the ground to the surface and the required recalculation of the kill sheet. In practice, these two methods are most often used to safely and successfully eliminate pressure sores. Therefore, in Section 3, we mainly describe and analyse the "wait and weight" method [26].

3. Results and Discussion

3.1. Well Control In Situ

Figure 9 shows the two hydrocarbon formation areas of individual wells in the terrain according to the well project, Company Oil and Gas Development Kft, in Hungary on the D-1 well [19].

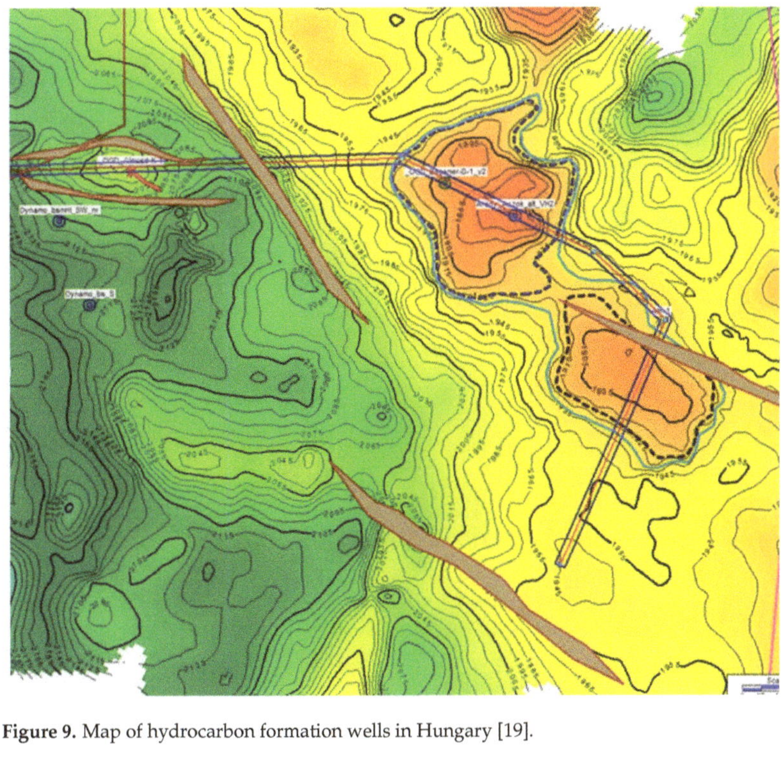

Figure 9. Map of hydrocarbon formation wells in Hungary [19].

3.2. *Origin and Analyses of the Problem*

Figure 10 shows a real example of gas kick. In the first phase, from a depth of 1290–1295 m between 12:00 to 12:15 h, drilling takes place without an anomaly. At a depth of 1298 m at approx. 12:23 p.m., a sudden increase ROP and an a decrease WOB are observed. Consequently, SPP decreases and WOH increases.

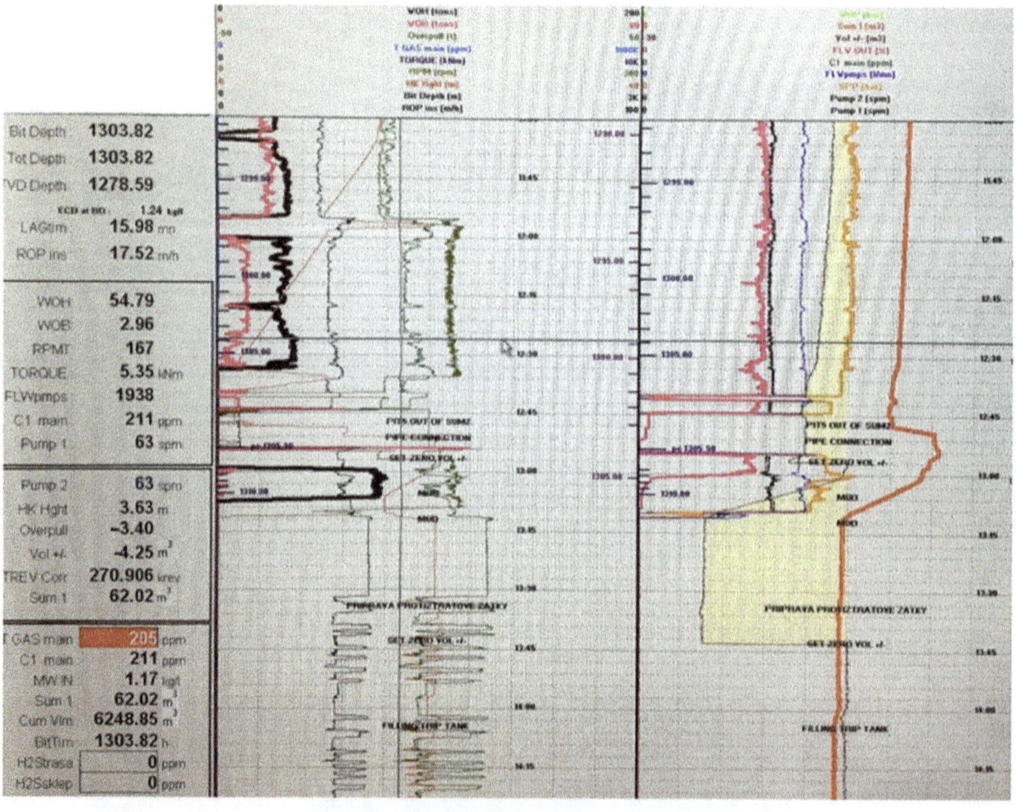

Figure 10. Software data "InfoDrill" kick.

In this case the driller has to react immediately:

- Pick up drilling tools;
- Switch off the mud pumps;
- Perform a flow check.

The error occurred at the moment when the driller decided to drill the stand of DP and decided to connect another stand of DP. At 12:45 p.m., a continuous inflow of mud gain into the tanks for 10 min is observed. After the conception, the driller switched on the pumps and continued drilling. Figure 10 shows that ROP is two times higher than when drilling the previous stand of DP where the drilling break occurred. At 1:10 p.m., the driller switched off the pumps and closed the well. Values were been read and recorded from the manometers: SIDPP = 28 bar and SICP = 38 bar, with a pit gain of approximately 1000 L, which was subsequently used to calculate the kill mud. These calculations are the most important step in the process of well control during oil and gas drilling.

3.3. Shutting Down the Well

Based on the situation at the well, we proposed the use of the "hard shut" method (Figure 11). Figure 11 shows the manual valves and their positions when the well is closed by BOP. Annular BOP was used to close the well. Subsequently, the HCR and the manual valve were opened to divert drilling mud into the choke manifold with the closed choke. Behind the choke, the valves were opened and diverted to the trip tank, shale shakers or via "poorboy" [27,28].

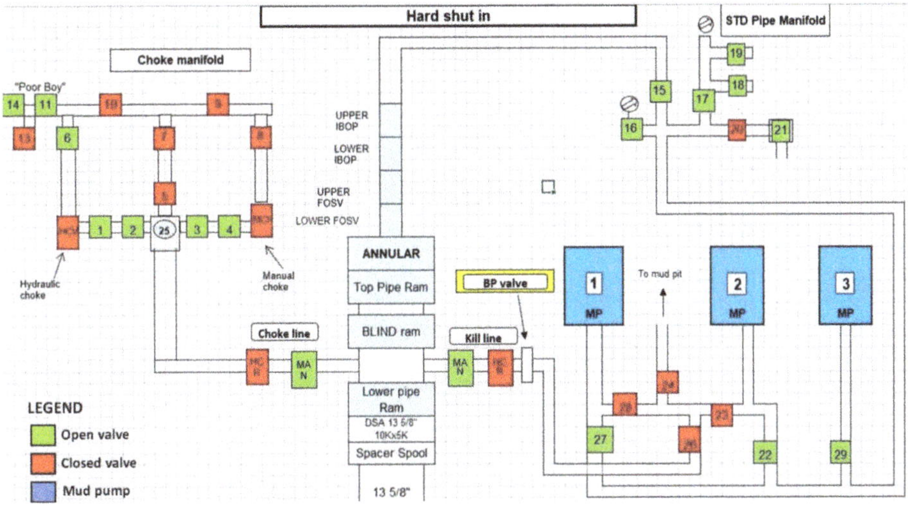

Figure 11. Schematic of hard shut-in.

3.4. Measured Values of the Gas Kick and Their Calculation

Software was used to calculate the well control values. With this program, the individual values for safe well control were calculated (Figures 12–14). The leak-off pressure of the previous casing shoe and the reduced slow pump rate are entered (formation strength data: A, B, C) in the upper-left side of the worksheet. The well verticality values are entered (current well data) in the upper-right section. The capacity of strings, the recalculated strokes of mud pumps and their lag times are entered in lower section, (calculated Data: D, F, G, H, I) (Figure 12).

The kick data are entered on the second worksheet (Figure 13) after well shut-in, namely:

- Shut-in drip pipe pressure (SIDPP),
- Shut-in casing pressure (SICP),
- Pit gain value (PG).

After entering these values kill fluid density for well control is calculated. This calculate obtains the values of the initial circulating pressure (ICP) and the final circulation pressure (FCP) for the successful killing of the well.

The third page of the worksheet (Figure 14) shows the dependence of pressure on the annulus and the number of pump strokes, as well as how to control the choke (open and close) for well control.

The wait and weight method was suggested because of the long open hole section and the low formation strength pressures in the casing shoe ((A)—23 bar, Figure 12).

After reaching the FCP pressure and completing the well inspection, the total volume of the well system is circulated for one cycle (Figure 12, point I). Then, the sludge pumps are turned off, the choke is left completely closed and the pressure increase in the annulus is monitored. If the value is zero, the BOP opens, and a flow check is performed [29].

International Well Control Forum	DATE :	4.4.22	6:54 PM
SURFACE BOP (Deviated Well) KILL SHEET - 1	NAME :	Tomas Huszar	
	UNITS :	S.I. (bar & liter)	PRINT

FORMATION STRENGTH DATA:			
SURFACE LEAK-OFF PRESSURE FROM FORMATION STRENGTH TEST	**(A)**	23	**bar**
DRLG FLUID DENSITY AT TEST	**(B)**	1,08	**kg/l**
		0,1059	bar/m
MAX. ALLOWABLE DRILLING FLUID DENSITY: **(B) +** $\frac{\textbf{(A) x 10.2}}{\text{SHOE T.V.DEPTH}}$	**(C)**	1,55	**kg/l**
		0,1519	bar/m
INITIAL MAASP: $\frac{\text{[(C) - CURR. DENSITY] x SHOE T.V.D.}}{10.2}$ =		18,59	**bar**

CURRENT WELL DATA:		
DRILLING FLUID DATA:		
DENSITY	1,17	**kg/l**
GRADIENT	0,1147	**bar/m**
DEVIATION DATA:		
KOP M.D.	1103,00	m
KOP T.V.D.	1103,00	m
EOB M.D.	1250,00	m
EOB T.V.D.	1230,00	m
CASING & SHOE DATA:		
SIZE	9,625	inch
ID	8,921	inch
M. DEPTH	500,00	m
T.V. DEPTH	500,00	m
HOLE DATA:		
SIZE	8 1/2	inch
M. DEPTH	1303,00	m
T.V.DEPTH	1278,00	m

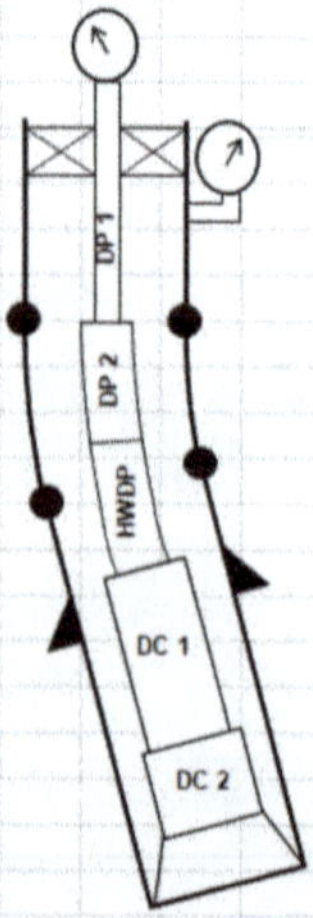

PUMP 1 DISPLACEMENT	PUMP 2 DISPLACEMENT
13,80 l/stk	13,80 l/stk

SLOW PUMP RATE DATA SPM	DYNAMIC PRESSURE LOSS PUMP NO. 1		PUMP NO. 2	
50 SPM	23	bar	22	bar
60 SPM	27	bar	28	bar

Take values from PUMP data

PUMP 1 or 2: 13,80 l/stk — 50 SPM

PRE-RECORDED VOLUME DATA			LENGTH m		CAPACITY l/m		VOLUME liter			PUMP STROKES strokes		TIME minutes	
DP 1	5	inch	1077,14	x	8,69								
- *SURFACE to KOP*			*1077,14*	*x*	*8,69*	=	9 360		(L1)	678	stks		
- *KOP to EOB*			*0*	*x*	*8,69*	=	0	+	(M1)	0	stks		
- *EOB to DP2/BHA*			*0*	*x*	*8,69*	=	0	+	(N1)	0	stks		
DP 2	5	inch		x									
- *SURFACE to KOP*			*0*	*x*	*0*	=	0	+	(L2)	0	stks		
- *KOP to EOB*			*0*	*x*	*0*	=	0	+	(M2)	0	stks		
- *EOB to BHA*			*0*	*x*	*0*	=	0	+	(N2)	0	stks		
HWDP	5	inch	145,38	x	4,61	=	670	+	(N3)	49	stks		
DC 1	0	inch		x		=	0	+	(N4)	0	stks		
DC 2	6 1/2	inch	80,48	x	4,01	=	323	+	(N5)	23	stks		
DRILL STRING VOLUME						(D)	10 353	ltr		750	stks	15,0	min
DC 2 x OPEN HOLE			80,48	x	15,20	=	1 223						
DC 1 x OPEN HOLE			0	x	36,61	=	0	+					
DP2/HW x OPEN HOLE			145,38	x	23,94	=	3 481	+					
DP 1 x OPEN HOLE			577,14	x	23,94	=	13 818	+					
OPEN HOLE VOLUME						(F)	18 522	ltr		1342	stks	26,8	min
DC 2 x CASING			0	x	18,92	=	0	+					
DC 1 x CASING			0	x	40,33	=	0	+					
DP2/HW x CASING			0	x	27,66	=	0	+					
DP1 x CASING			500	x	27,66	=	13 829	+					
TOTAL CASING VOLUME						(G)	13 829	ltr		1002	stks	20,0	min
TOTAL ANNULUS VOLUME					(F + G) = (H)		32 351	ltr		2344	stks	46,9	min
TOTAL WELL SYSTEM VOLUME					(D + H) = (I)		42 704	ltr		3095	stks	61,9	min
ACTIVE SURFACE VOLUME					(J)		49 000	ltr		3551	stks		
TOTAL ACTIVE FLUID SYSTEM					(I+J)		91 704	ltr		6645	stks		

Figure 12. First page of the "kill sheet" worksheet.

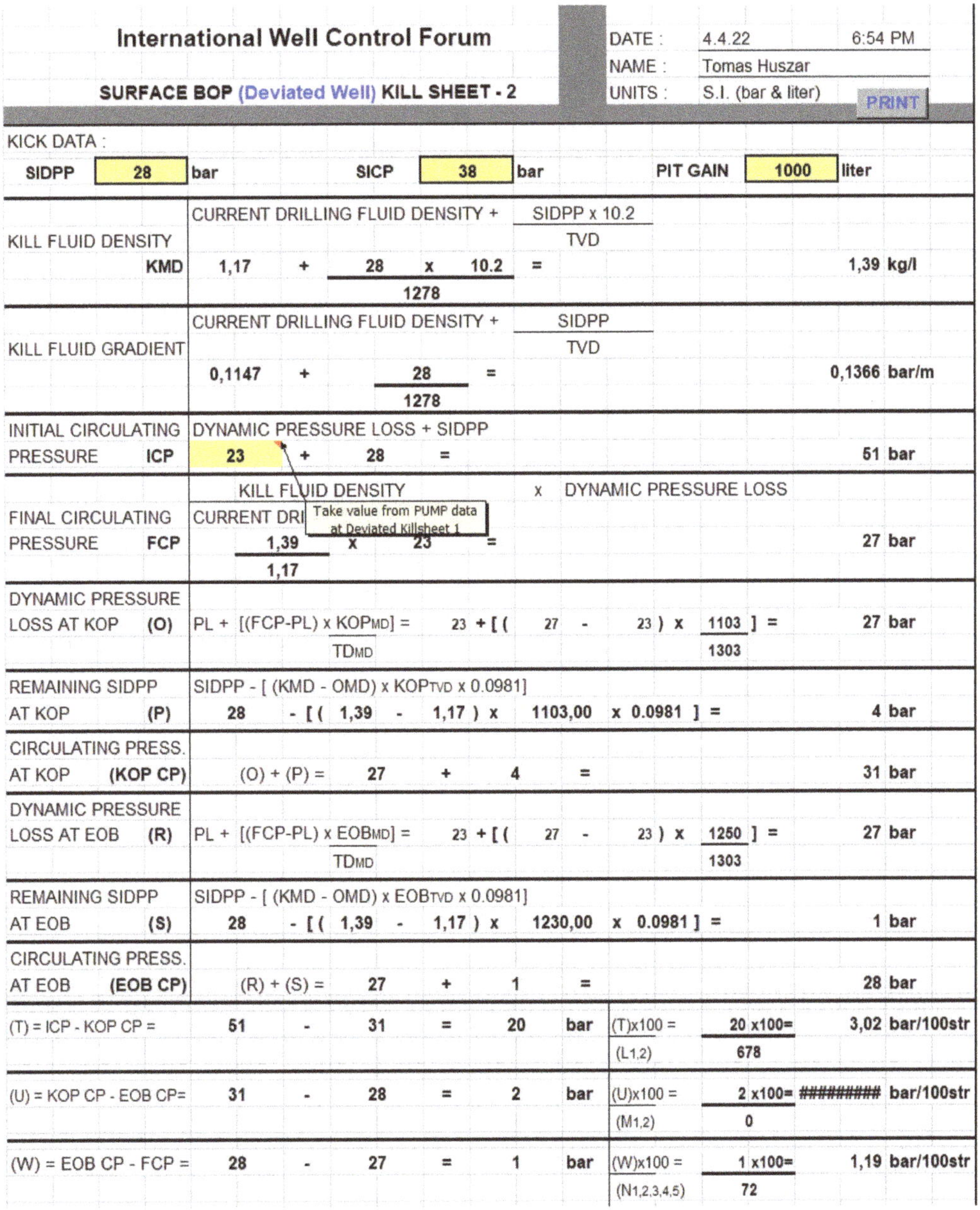

International Well Control Forum

SURFACE BOP (Deviated Well) KILL SHEET - 2

DATE : 4.4.22 6:54 PM
NAME : Tomas Huszar
UNITS : S.I. (bar & liter)
PRINT

KICK DATA :

SIDPP	28	bar	SICP	38	bar	PIT GAIN	1000	liter

KILL FLUID DENSITY **KMD**	CURRENT DRILLING FLUID DENSITY + SIDPP x 10.2 / TVD 1,17 + 28 x 10.2 / 1278 =	1,39 kg/l
KILL FLUID GRADIENT	CURRENT DRILLING FLUID DENSITY + SIDPP / TVD 0,1147 + 28 / 1278 =	0,1366 bar/m
INITIAL CIRCULATING PRESSURE **ICP**	DYNAMIC PRESSURE LOSS + SIDPP 23 + 28 =	51 bar
FINAL CIRCULATING PRESSURE **FCP**	KILL FLUID DENSITY / CURRENT DRI[LLING FLUID DENSITY] x DYNAMIC PRESSURE LOSS 1,39 / 1,17 x 23 = (Take value from PUMP data at Deviated Killsheet 1)	27 bar
DYNAMIC PRESSURE LOSS AT KOP **(O)**	PL + [(FCP-PL) x KOP_{MD}] / TD_{MD} = 23 + [(27 - 23) x 1103 / 1303] =	27 bar
REMAINING SIDPP AT KOP **(P)**	SIDPP - [(KMD - OMD) x KOP_{TVD} x 0.0981] 28 - [(1,39 - 1,17) x 1103,00 x 0.0981] =	4 bar
CIRCULATING PRESS. AT KOP **(KOP CP)**	(O) + (P) = 27 + 4 =	31 bar
DYNAMIC PRESSURE LOSS AT EOB **(R)**	PL + [(FCP-PL) x EOB_{MD}] / TD_{MD} = 23 + [(27 - 23) x 1250 / 1303] =	27 bar
REMAINING SIDPP AT EOB **(S)**	SIDPP - [(KMD - OMD) x EOB_{TVD} x 0.0981] 28 - [(1,39 - 1,17) x 1230,00 x 0.0981] =	1 bar
CIRCULATING PRESS. AT EOB **(EOB CP)**	(R) + (S) = 27 + 1 =	28 bar

(T) = ICP - KOP CP =	51 - 31 =	20	bar	(T)x100 / (L1,2) = 20 x100= / 678	3,02 bar/100str
(U) = KOP CP - EOB CP=	31 - 28 =	2	bar	(U)x100 / (M1,2) = 2 x100= / 0	######### bar/100str
(W) = EOB CP - FCP =	28 - 27 =	1	bar	(W)x100 / (N1,2,3,4,5) = 1 x100= / 72	1,19 bar/100str

Figure 13. Second page of the "kill sheet" worksheet.

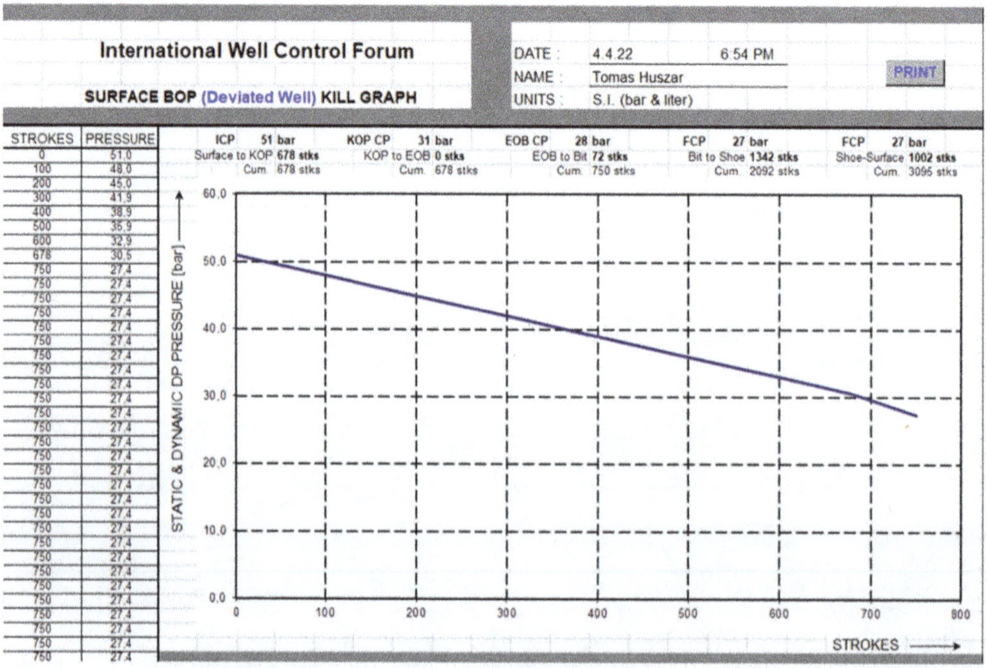

Figure 14. Third page of the "kill sheet worksheet.

3.5. *Algorithm of Kill-Well Decision Tree during Drilling*

Figure 15 shows a decision-making algorithm to simplify the following/pressure/kick operations with the three most common well control methods (wait and weight method, driller's method and volumetric method).

The algorithm was created on the basis of experience of reactions and internal regulations all over the world. The speed of the response to pressure is paramount. This algorithm provides helpful information for supervisors in the form of a checklist.

Based on the advantages of the InfoDrill and Drill Lab MasterLog software technology [30] and the solution of key problems, this software represents a flexible technique in the search for and extraction of hydrocarbon and geothermal deposits [31,32]. The data from InfoDrill and Drill Lab MasterLog software are not compatible with other similar software due to the paid license.

Other authors have reported new techniques that can be used for the drilling process, such as slot impingement techniques, as well as techniques to cool materials [33,34].

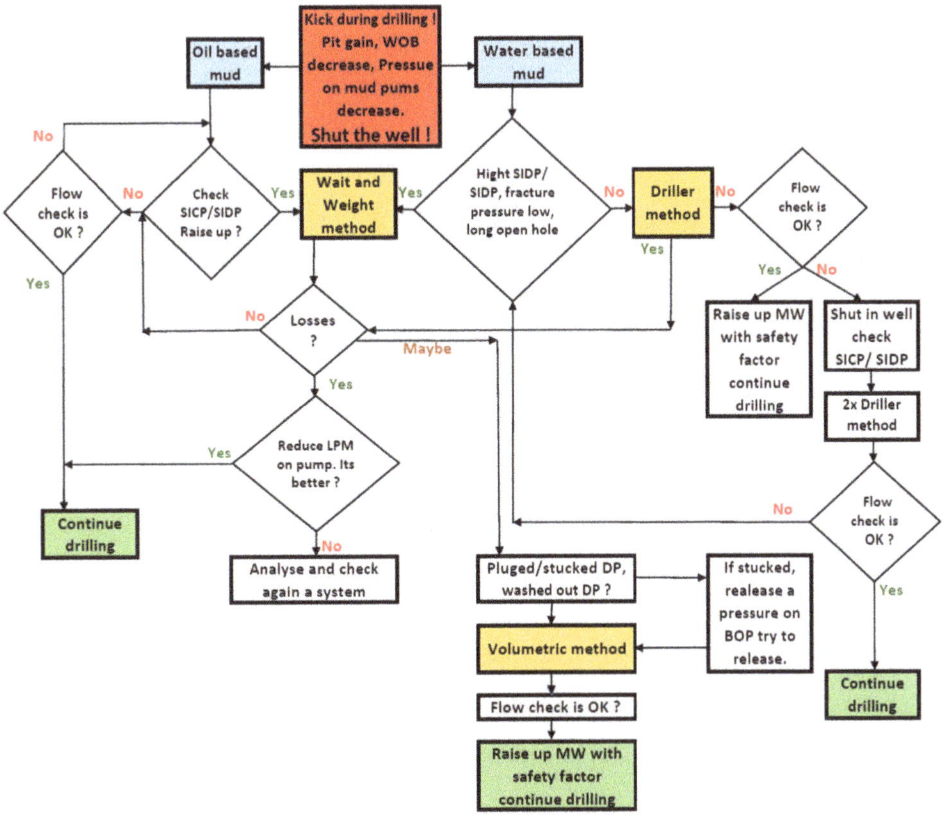

Figure 15. Kill-well decision tree during drilling.

4. Conclusions

Gas kick during hydrocarbon drilling occurs in formations with highly overpressurized zones. It is essential to establish which technological processes (fastest and safest) will be used to eliminate such pressure events. By predicting steps in a timely manner and implementing the appropriate technological procedures for well control, undesirable consequences can be safely predicted and eliminated, preventing a possible accident.

The algorithm presented in Figure 15 was developed to assist in decision making with regards to determining the appropriate choice of procedure and steps (yes, no, maybe) for well control.

By correctly predicting the development of the drilling process in high-pressure formations, undesirable consequences it can be safely eliminated and prevented, avoiding possible accidents. By linking appropriate methods using sophisticated software support, it is possible to create an effective technological process to be implemented for emergency well control. In terms of the oil and gas industry, we emphasise the need to use hydrocarbon fuels (natural gas, oil and their products oils, i.e., gasoline, lubricants, etc.) in individual countries. We therefore addressed this issue, considering that accurate prediction can be carried out safely, efficiently and economically, with minimal impact on the environment. The choice of an appropriate drilling method and the use of a kill sheet require considerable expertise and practical experience in the field.

Author Contributions: Conceptualization, T.H. and G.W.; methodology T.H. and G.W.; software, T.H.; validation, T.H., G.W. and E.S.; formal analysis, G.W. and E.S.; investigation, T.H.; resources, T.H.; data curation, T.H.; writing—original draft preparation, T.H., G.W. and E.S.; writing—review and editing, T.H., G.W. and E.S.; visualization, G.W.; supervision, G.W.; project administration, E.S.; funding acquisition, G.W. All authors have read and agreed to the published version of the manuscript.

Funding: This research was funded by the Scientific Grant Agency of the Ministry of Education, Science, Research and Sport of the Slovak Republic, VEGA: 1/0588/21, The research and development of new methods based on the principles of modelling, logistics and simulation in managing the interaction of mining and backfilling processes with regard to economic efficiency and the safety of raw materials extraction.

Institutional Review Board Statement: Not applicable.

Informed Consent Statement: Not applicable.

Data Availability Statement: The drilling company consented to the publication of data obtained during the survey project (attached).

Conflicts of Interest: The authors declare no conflict of interest.

Abbreviations

BOP	blowout preventer
ROP	rate of penetration
WOB	weight on bit
SIDP	shut-in drill pipe pressure
SICP	shut-in casing pressure
HCR	hydraulic choke valve
OBM	oil-based mud
SIDPP	shut-in drip pipe pressure
SICP	shut-in casing pressure
DP	drill pipe
DC	drill collar
MM	mud motor
LOT	leak-off test

References

1. Li, X.Z.; Hua, A.Z. Prediction and prevention of sandstone-gas outbursts in coal mines. *Int. J. Rock Mech. Min. Sci.* **2006**, *43*, 2–18. [CrossRef]
2. Failler, P. Special Issue on Global Market for Crude Oil. *Energies* **2021**, *14*, 1199. [CrossRef]
3. Pinka, J.; Wittenberger, G.; Engel, J. *Borehole Mining*, 1st ed.; FBERG TU: Kosice, Slovakia, 2007; p. 233.
4. Jung, J.; Han, D.; Kwon, S. Well placement optimisation using a productivity potential area map. *Int. J. Oil Gas Coal Technol.* **2021**, *27*, 41–53. [CrossRef]
5. Atashnezhad, A.; Wood, A.D.; Fereidounpour, A.; Khosravanian, R. Designing and optimizing deviated wellbore trajectories using novel particle swarm algorithms. *J. Nat. Gas Sci. Eng.* **2014**, *21*, 1184–1204. [CrossRef]
6. Klempa, M.; Malis, J.; Sancer, J.; Slivka, V. Research of Stowing Material for Filling of Free Underground Spaces of Old Mine Works. *Inz. Miner. J. Pol. Miner. Eng. Soc.* **2018**, *2*, 321–325. [CrossRef]
7. Marbun, B.; Sinaga, S.; Purbantanu, B.; Ridwan, R. Improvement of Loads Calculation of the Perforated Liner in a Geothermal Production Well. Renew. *Energy* **2021**, *174*, 468–486. [CrossRef]
8. Bildstein, J.; Strobel, J. Formation Pressure while Drilling, a valuable alternative in difficult drilling environment in Northern Germany. In Proceedings of the 67th EAGE Conference & Exhibition, Madrid, Spain, 13–16 June 2005. [CrossRef]
9. Sun, S.H.; Yan, T.; Bi, X.L.; Wang, P. Analysis on the Influence Factors of Well Deviation in Gas Drilling. *Adv. Mater. Res.* **2012**, *577*, 132–136. [CrossRef]
10. Su, K.H. Research on Bottom Hole Pressure of Horizontal Well during Gas Kick. In Proceedings of the International Conference on Environmental Systems Science and Engineering, Dalian, China, 6 August 2010; Volume 1, pp. 318–325.
11. Chang, D.; Li, G.; Shen, Z.; Huang, Z.; Tian, S.; Shi, H.; Song, X. A Study on the Effect of Bottom-hole Differential Pressure on the Rock Stress Field. *Energy Sources Part A Recovery Util. Environ. Eff.* **2014**, *36*, 275–283. [CrossRef]
12. Ni, H.; Wang, R.; Song, W.; Song, H. Study on the Mechanism of Bottom Hole Hydraulic Pulse Increasing Drilling Rate. *Adv. Mater. Res.* **2012**, *354–355*, 621–626. [CrossRef]

13. Pindor, T. Innovative methods of natural gas exploitation as a factor of sustainable development of world economy. *Ekon. I Srodowisko-Econ. Environ.* **2017**, *4*, 205–217.
14. Li, L.; Tan, J.; Wood, A.D.; Zhao, Z.; Becker, D.; Lyu, Q.; Shu, B.; Chen, H. A review of the current status of induced seismicity monitoring for hydraulicfracturing in unconventional tight oil and gas reservoirs. *Fuel* **2019**, *242*, 195–210. [CrossRef]
15. Geertsma, J. Land subsidence above compacting oil and gas reservoirs. *Pet Technol.* **1973**, *25*, 734–744. [CrossRef]
16. Knez, D.; Rajaoalison, H. Discrepancy between measured dynamic poroelastic parameters and predicted values from Wyllie's equation for water-saturated Istebna sandstone. *Acta Geophys.* **2021**, *69*, 673–680. [CrossRef]
17. Malvić, T.; Barudžija, U.; Pašić, B.; Ivšinović, J. Small Unconventional Hydrocarbon Gas Reservoirs as Challenging Energy Sources, Case Study from Northern, Croatia. *Energies* **2021**, *14*, 3503. [CrossRef]
18. Rubesova, M.; Bujok, P.; Klempa, M. Application of production of production well logging for the monitoring of tightness underground gas storage. In Proceedings of the 17th International Multidisciplinary Scientific GeoConference SGEM 2017, Vienna, Austria, 27–29 November 2017; Volume 17, pp. 27–32. [CrossRef]
19. *Project: Drilling Programm of Company O&GD Central Kft, Exploration Well D-1*; Company O&GD Central Ltd.: Budapest, Hungary, 2017.
20. Zadsarb, M.; Zareipourc, H.; Kazemid, M. Resilient operation planning of integrated electrical and natural gas systems in the presence of natural gas storages. *Int. J. Electr. Power Energy Syst.* **2021**, *130*, 106936. [CrossRef]
21. Quosay, A.A.; Knez, D.; Ziaja, J. Hydraulicfracturing: New uncertainty based modeling approach for process design using Monte Carlo simulation technique. *PLoS ONE* **2020**, *15*, e0236726. [CrossRef] [PubMed]
22. Mavko, G.; Mukerji, T.; Dvorkin, J. Tools for seismic analysis of porous media. In *The Rock Physics Handbook*; Cambridge University Press: Cambridge, UK, 2010.
23. DrillingFormulas.Com. D Exponent Calculation. Available online: https://www.drillingformulas.com/d-exponent-calculation (accessed on 11 March 2022).
24. Lyons, W.C.; Carter, T.; Lapeyrouse, N.J. *Formulas and Calculations for Drilling, Production and Workover*, 4th ed.; Elsevier: Amsterdam, The Netherlands, 2016.
25. Baker, R. *Practical Well Control*; University of Texas at Austin: Austin, TX, USA, 1998; Volume 2, p. 8040.
26. Strnište, K.; Šmolík, S. *Hĺbinné Vŕtanie, Vysokoškolská Učebnica*; Vydavateľstvo Alfa: Bratislava, Slovakia, 1992.
27. Carlos, M. Driller's Method vs Wait and Weight Method on Deepwater, Linkedin. Available online: https://www.linkedin.com/pulse/drillers-method-vs-wait-weight-deepwater-carlos-moura (accessed on 11 March 2022).
28. Drilling Manual. Hard Shut In Procedure & Lineup in Well Control. Available online: https://www.drillingmanual.com/hard-shut-in-procedure-while-drilling/ (accessed on 11 March 2022).
29. Cormack, D. *An Introduction to Well Controll Calculation for Drilling Operations*; Springer: New York, NY, USA, 2017. [CrossRef]
30. Drill-Lab. Software. Available online: https://www.drill-lab.com.pl/en/our-services/software (accessed on 17 May 2022).
31. Huang, Z.; Huang, Z. Review of Radial Jet Drilling and the key issues to be applied in new geo-energy exploitation. *Energy Procedia* **2019**, *158*, 5969–5974. Available online: http://creativecommons.org/licenses/by-nc-nd/4.0/ (accessed on 11 March 2022). [CrossRef]
32. Wittenberger, G.; Sofranko, M. Formation and protection against incrustation on the geothermal pipe by utilizing of geothermal water in the area of Durkov (Eastern Slovakia). *Acta Montan. Slovaca* **2015**, *20*, 10–15. Available online: https://actamont.tuke.sk/pdf/2015/n1/2wittenberger.pdf (accessed on 11 March 2022).
33. Huang, H.; Sun, T.; Zhang, G.; Li, D.; Wei, H. Evaluation of a developed SST k-ω turbulence model for the prediction of turbulent slot jet impingement heat transfer. *Int. J. Heat Mass Transf.* **2019**, *139*, 700–712. [CrossRef]
34. Cademartori, C.; Cravero, C.; Marini, M.; Marsano, D. CFD Simulation of the Slot Jet Impingement Heat Transfer Process and Application to a Temperature Control System for Galvanizing Line of Metal Band. *Appl. Sci.* **2021**, *11*, 1149. [CrossRef]

processes

Article

Calculation Model of the Effect of Periodic Change of Normal Force on Sliding Friction Characteristics between the Planes

Yongwang Liu [1,*], Yanan Liu [2], Zhichuan Guan [1] and Yixiang Niu [1]

1 School of Petroleum Engineering, China University of Petroleum, Qingdao 266580, China; guanzhch@upc.edu.cn (Z.G.); z21020055@s.upc.edu.cn (Y.N.)
2 China Eastern Airlines Co., Ltd., Shanghai 201100, China; cookie125@163.com
* Correspondence: liuyongwang2003@163.com

Abstract: How to reduce the friction resistance between two planes is a difficult problem that has been studied in the field of tribology. Aiming at this problem, the concept of reducing the friction resistance by exciting the periodic change of normal force between the planes is proposed. A calculation model of the displacement amplitude distribution of the plate is established based on the principle of reciprocity of work, and the influence of the periodic change of normal force on sliding friction between the planes is studied. Additionally, an experimental device for analysis of friction between the planes under the periodic change of normal force is established. The calculation model of the plate's displacement amplitude distribution considering the change frequency of normal force is verified and modified by experiments. The research results mainly show two aspects. On the one hand, the calculation model of the displacement amplitude distribution of the plate is in good agreement with the experimental results, which can effectively help to study the effect of periodic change of normal force on sliding friction between the planes. On the other hand, the change of amplitude and frequency of the normal force have an influence on the sliding friction between the planes. That is, the latter decreases with the increase of the former. The above conclusions have great reference significance for the study of vibration drag reduction in engineering production.

Keywords: normal force; vibration drag reduction; sliding friction; displacement amplitude; friction change

Citation: Liu, Y.; Liu, Y.; Guan, Z.; Niu, Y. Calculation Model of the Effect of Periodic Change of Normal Force on Sliding Friction Characteristics between the Planes. *Processes* **2022**, *10*, 1138. https://doi.org/10.3390/pr10061138

Academic Editors: Tianshou Ma and Yuqiang Xu

Received: 30 April 2022
Accepted: 2 June 2022
Published: 6 June 2022

1. Model of Influence of Normal Force's Periodic Change on Sliding Friction Characteristics between the Planes

Reducing the friction resistance of moving surfaces is an important problem in the field of tribology [1–3]. In order to solve this problem, a series of engineering methods have been formed, which are as follows: bionic drag reduction [4], chemical drag reduction [5], vibration drag reduction [6], paint drag reduction [7], etc. Research and practice show that vibration drag reduction is one of the most effective methods. To this end, various engineering processes for vibration drag reduction have been formed, such as vibration pile pulling [8], vibration conveying [9], and vibration screening [10]. However, as there are many forms of vibration, the matching relationship between different forms of vibration drag reduction and drag reduction's working conditions has not been clear, and the research on the influence of different vibration forms on characteristics of friction and drag reduction has not been in-depth. Jae Hyeok Choi et al. studied the influence of contact vibration on nano friction among graphene, Pt, Au and SiOx from the micro mechanism [11], but for the actual engineering requirements, it is also necessary to study the influence of friction between the two Q235A plates on vibration drag reduction under macro conditions. For example, the vibration drag reduction between the drill string and the inner wall of the casing during sliding guidance of oil and gas wells is not enough from the micro perspective, but needs to be studied from the macro perspective. Therefore, in view of the problem of how to reduce friction and drag between two planes, the author proposes the concept of stimulating the periodic change of normal force between the planes to change the friction resistance, establishes models and carries out related experiments so as to provide theoretical support for using normal vibration to reduce drag in engineering.

1.1. Establishment of Calculation Model for the Influence of Normal Force's Periodic Change on Sliding Friction between the Planes

The study shows [12–14] that when the normal force of two plates in contact with each other changes periodically, the high frequency mechanical collision will occur, resulting in the constant change of real contact area. As shown in Figure 1a, when the plate is not subjected to periodic change of normal force, no deformation occurs on the plate. At this time, if there is relative movement between contact surfaces, the friction resistance on the plate is the product of the positive pressure on the upper plate and the friction coefficient.

$$f_1 = \mu N_1 \quad (1)$$

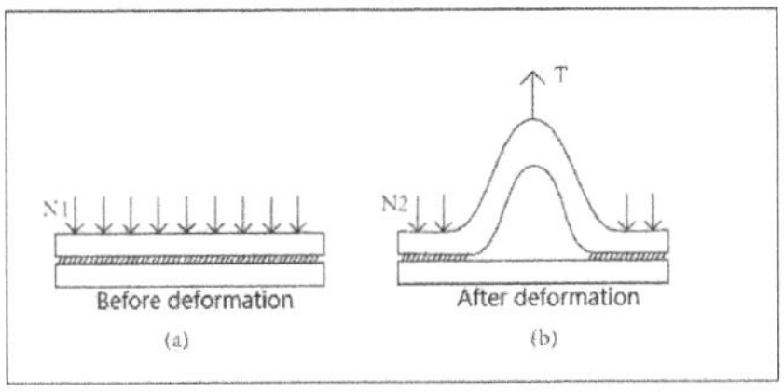

Figure 1. The deformation diagram of displacement amplitude of the plate under the force.

Figure 1b shows the deformation of the plate under the action of the changing normal force. When the plate is subjected to the normal force whose changing amplitude is less than the plate's gravity, the displacement amplitude of the plate in a certain area, near where the normal force is applied, will be generated. In this area the two plates do not contact and are still contact with each other in the rest of areas, which means that the contact area of the two plates decreases compared with that before the deformation. When the plate is subjected to the periodically changing normal force, the displacement amplitude is in a dynamic change motion. Besides, the upper plate with displacement amplitude is in a "weightless state", that is, when the plate with displacement amplitude is deformed, the positive pressure on the plate and the average friction resistance in relative motion all decrease. At this time, the friction resistance between two plates is:

$$f_2 = \mu N_2 \quad (2)$$

1.2. Model Building and Solving

According to the above analysis, as long as the distribution of the displacement amplitude on the contact surface of the plate is obtained, the change of the friction resistance under the excitation state can be got so as to analyze the influence of vibration on the friction resistance.

In many engineering problems, one of the two objects in friction is elastic, such as the stratum in the vibration pile pulling process mentioned above. Therefore, this kind of research problem can be transformed into the force bending model of four-sided free plate on elastic foundation. As shown in Figure 2, a flat plate with free four sides on an elastic foundation is subjected to an arbitrary load (P) on the position (ζ, η), and the displacement of the plate edge is assumed to be:

$$W(x,0) = D_1 + (D_2 + D_1)x/a + \sum_{m=1}^{\infty} B_{1m} \sin k_m x \quad (3)$$

$$W(x,b) = D_3 + (D_4 + D_3)x/a + \sum_{m=1}^{\infty} B_{2m} \sin k_m x \quad (4)$$

$$W(0,y) = D_1 + (D_1 + D_3)y/b + \sum_{m=1}^{\infty} B_{1n} \sin k_m y \quad (5)$$

$$W(a,y) = D_2 + (D_4 + D_2)y/b + \sum_{m=1}^{\infty} B_{2n} \sin k_m y \quad (6)$$

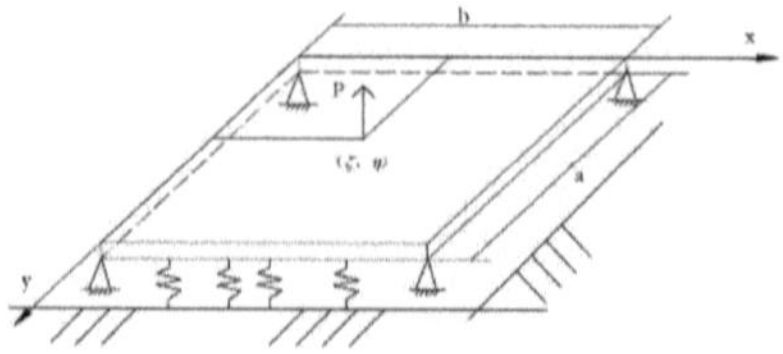

Figure 2. Vibration plate with four edges free.

In the formula, D_1, D_2, D_3, D_4 represents the displacement at the plate angle ($(0,0)$, $(a,0)$, $(0,b)$, (a,b)) respectively, and their units are mm.

At present, numerical method and analytical method are the main methods to deal with plate bending. Numerical calculation methods mainly include finite element method and boundary method [15] which lack universality. While in the analytical method, the displacement function needs to be determined first, and then the unknown parameters need to be calculated according to the boundary conditions, whose process is complicated and is not conducive to engineering application. The use of the reciprocity theorem of work can simplify the solution process [16], and the results have high reliability and strong applicability. The theorem of reciprocity of work means that when two sets of generalized forces act on a linear elastic body, the work done by the first set of force on the displacement caused by the second set of force is equal to the work done by the second set of force on the displacement caused by the first set of force.

Figure 3 shows a four-sided simply supported plate on an elastic foundation. According to the theorem of reciprocity of work, the work done on the displacement caused by the periodically changing normal force in the system of Figure 2 is equal to the work done by the displacement caused by the periodically changing normal force in the system of Figure 3. According to the theory of elastic thin plate, the equilibrium equation of four-sided simply supported plate can be obtained as:

$$\frac{\partial^4 w_1}{\partial x^4} + 2\frac{\partial^4 w_1}{\partial x^2 \partial y^2} + \frac{\partial^4 w_1}{\partial y^4} + K w_1 = \frac{\delta(x-\zeta, y-\eta)}{D} \quad (7)$$

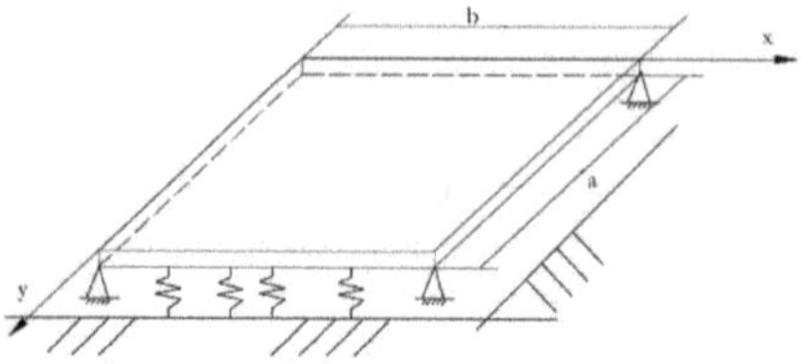

Figure 3. Four-sided simply supported plate.

In the formula, w_1 is the vertical displacement of the plate and its unit is mm; K is the stiffness coefficient of the foundation reaction and its unit is N/mm; D is the bending stiffness of the plate and its unit is N/mm; h is the thickness of the plate and its unit is mm; $\delta(x-\zeta, y-\eta)$ is the Dira Delta function.

By applying the reciprocity principle of work to the systems in Figures 2 and 3, it can be obtained that:

$$W(\zeta,\eta) = \int_0^a [V_{1y}(x,0,\zeta,\eta)W(x,0) - V_{1y}(x,b,\zeta,\eta)W(x,b)]dx + \int_0^b [V_{1x}(0,y,\zeta,\eta)W(0,y) - V_{1x}(a,y,\zeta,\eta)W(a,y)]dy + [R_1(a,b,\zeta,\eta)D_4 - R_1(0,b,\zeta,\eta)D_3] + [R_1(0,0,\zeta,\eta)D_1 - R_1(a,0,\zeta,\eta)D_2] \tag{8}$$

In the formula, $V_{1y}(x,0,\zeta,\eta)$, $V_{1x}(0,y,\zeta,\eta)$, $R_1(x,y,\zeta,\eta)$ are the distributed shear force at the plate edge and plate angular reaction force along the x-axis and y-axis respectively. The boundary condition of the four-sided free vibrating plate is:

$$\frac{\partial^3 W(\zeta,0)}{\partial\eta^3} + (2-v)\frac{\partial^3 W(\zeta,0)}{\partial\eta\partial\varsigma^2} = 0 \tag{9}$$

$$\frac{\partial^3 W(\zeta,b)}{\partial\eta^3} + (2-v)\frac{\partial^3 W(\zeta,b)}{\partial\eta\partial\varsigma^2} = 0 \tag{10}$$

$$\frac{\partial^3 W(0,\eta)}{\partial\zeta^3} + (2-v)\frac{\partial^3 W(0,\eta)}{\partial\eta^2\partial\varsigma} = 0 \tag{11}$$

$$\frac{\partial^3 W(a,\eta)}{\partial\zeta^3} + (2-v)\frac{\partial^3 W(a,\eta)}{\partial\eta^2\partial\varsigma} = 0 \tag{12}$$

$$\frac{\partial^2 W(0,0)}{\partial\zeta\partial\eta} = 0 \tag{13}$$

$$\frac{\partial^2 W(a,0)}{\partial\zeta\partial\eta} = 0 \tag{14}$$

$$\frac{\partial^2 W(0,b)}{\partial\zeta\partial\eta} = 0 \tag{15}$$

$$\frac{\partial^2 W(a,b)}{\partial\zeta\partial\eta} = 0 \tag{16}$$

Substitute (3)–(6) into (8) to get:

$$\begin{aligned}
&\sum_{m=1}^{\infty} \frac{B_{1n}}{2\lambda^2}\left[\frac{\phi(a_n,n)\sinh a_n(a-\zeta)}{\sinh a_n a} - \frac{\phi(\beta_n,n)\sinh\beta_n(a-\zeta)}{\sinh\beta_n a}\right]\sin k_n\eta + \\
&\sum_{m=1}^{\infty} \frac{B_{2n}}{2\lambda^2}\left[\frac{\phi(a_n,n)\sinh a_n\zeta}{\sinh a_n a} - \frac{\phi(\beta_n,n)\sinh\beta_n\zeta}{\sinh\beta_n b}\right]\sin k_n\eta + \\
&D_1\left[\frac{(a-\zeta)(b-\eta)}{ab} - \sum_{m=1}^{\infty}\frac{\lambda^2}{m\pi}\left[\frac{\sinh a_m(b-\eta)}{a^2{}_m\sinh a_m b} - \frac{\sinh\beta_m(b-\eta)}{\beta^2{}_m\sinh\beta_m b} + \frac{2\lambda^2(b-\eta)}{a^2{}_m\beta^2{}_m b}\right]\sin k_m\zeta\right] + \\
&D_2\left[\frac{\zeta(b-\eta)}{ab} - \sum_{m=1}^{\infty}\frac{\lambda^2(-1)^m}{m\pi}\left[\frac{\sinh a_m(b-\eta)}{a^2{}_m\sinh a_m b} - \frac{\sinh\beta_m(b-\eta)}{\beta^2{}_m\sinh\beta_m b} + \frac{2\lambda^2(b-\eta)}{a^2{}_m\beta^2{}_m b}\right]\sin k_m\zeta\right] + \\
&D_3\left[\frac{(a-\zeta)\eta}{ab} - \sum_{m=1}^{\infty}\frac{\lambda^2}{m\pi}\left[\frac{\sinh a_m\eta}{a^2{}_m\sinh a_m b} - \frac{\sinh\beta_m\eta}{\beta^2{}_m\sinh\beta_m b} + \frac{2\lambda^2\eta}{a^2{}_m\beta^2{}_m b}\right]\sin k_m\zeta\right] + \\
&D_4\left[\frac{\zeta\eta}{ab} - \sum_{m=1}^{\infty}\frac{\lambda^2(-1)^m}{m\pi}\left[\frac{\sinh a_m\eta}{a^2{}_m\sinh a_m b} - \frac{\sinh\beta_m\eta}{\beta^2{}_m\sinh\beta_m b} + \frac{2\lambda^2\eta}{a^2{}_m\beta^2{}_m b}\right]\sin k_m\zeta\right]
\end{aligned} \tag{17}$$

Eight algebraic equations can be obtained by substituting (17) into (9)–(16) respectively so as to get the unknown parameter (D_1, D_2, D_3, D_4, B_{1m}, B_{2m}, B_{1n}, B_{2n}) and then obtain the theoretical solution of the equation. The value of m,n depends on the requirements of calculation accuracy.

1.3. Instance Calculation

In practical engineering, steel and rock are the main materials that generate friction and friction pair. The choice of steel for the plate in the model can make the results more practical. Q235A steel is taken as an example to calculate. When elastic modulus $E = 2.0 \times 10^6$ t/m^2, Poisson's ratio $v = 0.312$, plate size $a = b = 0.18$ m, plate thickness $h = 0.03$ m, the displacement amplitude distribution under load (P) of 5.7N is calculated using the model.

Figure 4 shows the amplitude distribution of plate displacement under the action of periodic normal force.

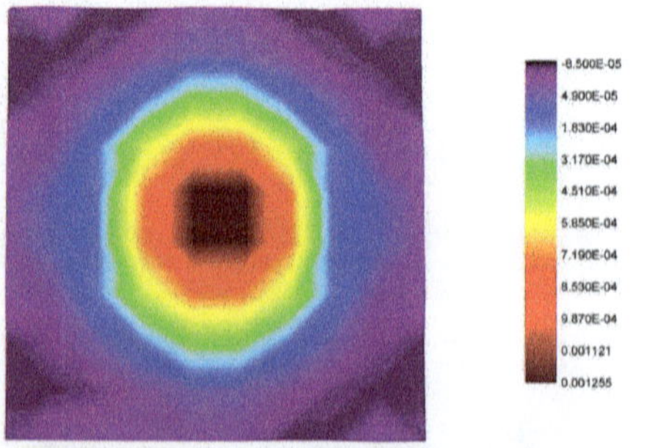

Figure 4. Displacement amplitude distribution of the plate on loading (P) of 5.7N.

It can be seen that under the action of periodically changing normal force, different degrees of displacement amplitudes with the action point of the normal force as the center occur on the plate, which decrease along the action point to the four edges. Considering the area of the plate as S, the friction resistance before excitation as f, the area of the displacement amplitude of the plate in the figure is 0.20 S, the friction resistance under the corresponding conditions can be calculated to be 0.80 f according to the relationship between the displacement amplitude of the plate and the friction resistance established in Section 1.1.

2. Establishment and Analysis of an Experimental Device for the Analysis of Friction between Planes under the Action of Periodically Changing Normal Force

In order to verify the reliability of the model, an experimental device for the analysis of friction resistance between planes under the action of periodically changing normal force is established, whose schematic diagram is shown in Figures 5 and 6.

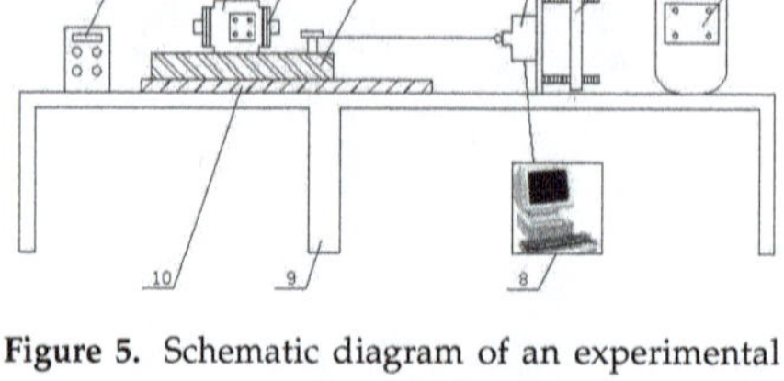

Figure 5. Schematic diagram of an experimental device for the analysis of frictional resistance between planes under the action of periodically changing normal force. 1. electric motor; 2. threaded screw device; 3. tension sensor; 4. upper sample; 5. eccentric block of shaker; 6. single axis inertial shaker; 7. inverter; 8. data collection system; 9. Shaker; 10. lower sample.

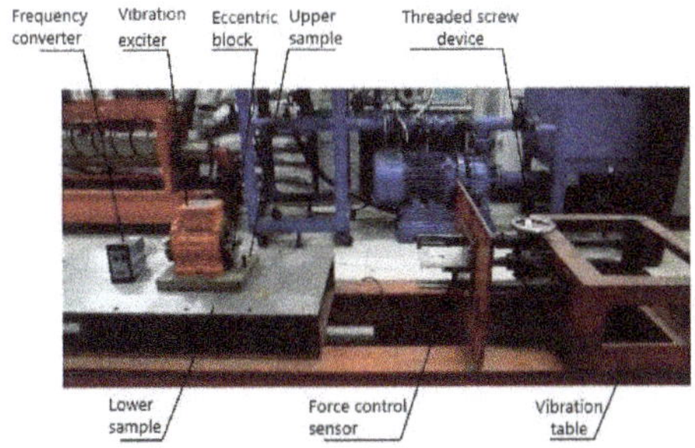

Figure 6. Experimental device picture.

Figure 7 shows the variation curve of Q235A steel and −Q235A steel friction pair's friction reduction percentage with the amplitude, under different action frequency of normal force. It can be seen that under the same action frequency, the percentage of friction reduction decreases with the increase of the normal force amplitude. Under the action frequency of 40 Hz, when the normal force amplitude is 5.7 N, the percentage of friction reduction is 86.03%; when the normal force amplitude is 22.7 N, the percentage of friction reduction is 26.02% and the friction resistance is reduced by 69.75%. The trend of the curve shows that a higher normal force amplitude is beneficial to improve the drag reduction effect, because a higher normal force amplitude can cause a larger range of displacement amplitudes on the contact surface.

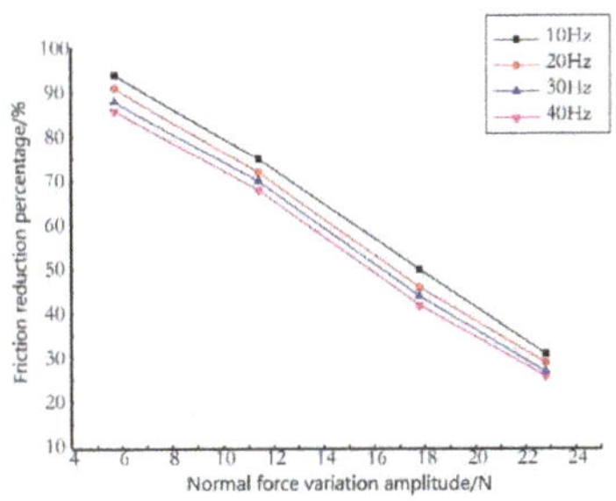

Figure 7. Graph showing Q235A steel and -steel friction pair's percentage of friction reduction varying with the normal force amplitude's change.

Under the same impact force, the percentage of friction reduction decreases with the increase of the action frequency. When the normal force change amplitude is 5.7 N, the percentage of friction reduction is 94.01%. When the action frequency is 10 Hz, the percentage of friction reduction is 86.07%. When the action frequency is 40 Hz and the friction resistance is reduced by 8.44%. The trend of the curve shows that the drag reduction effect increases with the increase of the normal force's change frequency, which is because the change of the normal force, which belongs to the simple harmonic force, is generated by the inertial vibration exciter. The measured friction is the average value in a measurement period. When the action frequency increases, the number of actions per unit time also increases, which means that the number of displacement amplitudes of the plate increases so as to be beneficial to the friction reduction.

3. Model Validation and Modification

According to the friction reduction experiment of the periodically changing normal force, it can be seen that the action frequency of the normal force has a certain effect on the friction reduction. However, the displacement amplitude distribution model of the plate does not consider the situation under the action frequency, so the model needs to be modified.

The plate has a displacement amplitude due to the changing normal force, which reduces the friction resistance, so there are:

$$f_0 - f_d \times C_f = f_{1(N,f)} \cdot (C_f > 0) \tag{18}$$

In the formula, f_0 represents the friction resistance of the plate without the action of the periodically changing normal force and its unit is N; f_d represents the reduced friction resistance of the plate subjected to periodically changing normal force and its unit is N. C_f represents correction factor of action frequency; $f_{1\ (N,f)}$ represents frictional resistance of the plate subjected to periodically changing normal force and its unit is N.

For the convenience of calculation, each item in the formula is expressed as the percentage of friction reduction. That is, all the terms in formula (18) are divided by the friction resistance not affected by the periodically changing normal force.

The frequency correction coefficient is calculated by taking the Q235A steel and - Q235A steel friction pair, impact force of 5.7 N, impact frequency of 40 Hz as example. Under this condition, when f_d/f_0 is 0.80 and $f_{1(5,40)}$ is 0.86, C_f can be got as 0.175. In the same way, the correction coefficient curves with different frequencies under different amplitude changes of normal force can be obtained, as shown in Figure 8.

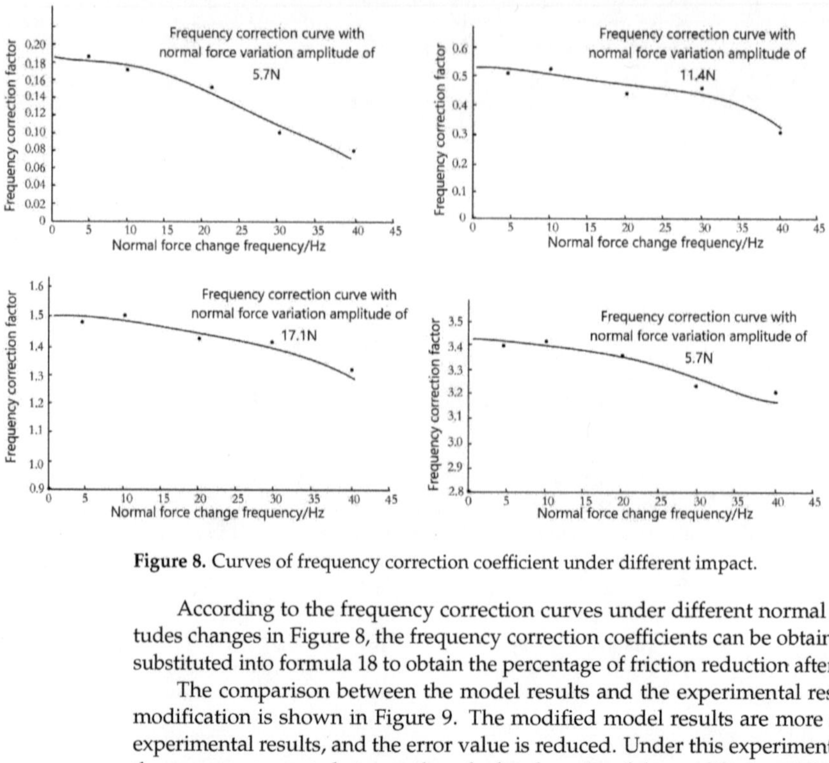

Figure 8. Curves of frequency correction coefficient under different impact.

According to the frequency correction curves under different normal force's amplitudes changes in Figure 8, the frequency correction coefficients can be obtained, which are substituted into formula 18 to obtain the percentage of friction reduction after modification.

The comparison between the model results and the experimental results after the modification is shown in Figure 9. The modified model results are more similar to the experimental results, and the error value is reduced. Under this experimental conditions, the average error rate between the calculated results of the modified model and the experimental results is 3.28%, so the model used to explain the friction reduction mechanism of periodically changing normal force is reliable.

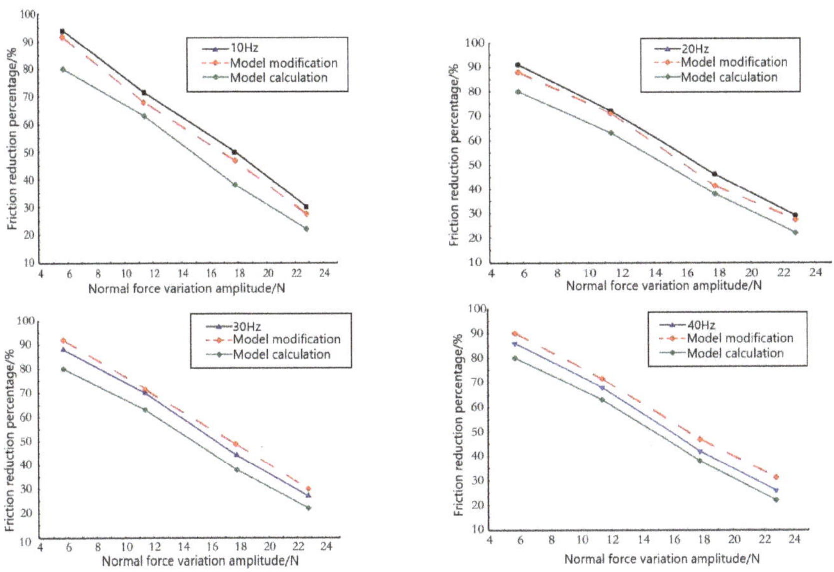

Figure 9. Comparison of modified model results and experimental results.

4. Conclusions

1. The amplitude distribution model of plate displacement is established and the influence of periodically changing normal force on plane sliding friction is studied. According to the model, the change of displacement amplitude of contact surface under the action of periodically changing normal force is the main reason for the change of friction resistance of contact surface. Under the action of the periodically changing normal force, the displacement amplitude of the upper plate deviates from the contact surface and is in the state of "weightlessness in the air". Then the positive pressure on the contact surface decreases, which is finally manifested as the reduction of friction resistance.
2. An experimental device for friction analysis between plates under the action of periodically changing normal force is established to study the influence of normal force's amplitude change and action frequency on friction. The experiments show that under the action of the periodically changing normal force, the friction decreases with the increase of the action frequency, and decreases with the increase of the normal force's amplitude change.
3. The frequency modification of the model is carried out through the experimental results. The calculation results of modified model are in good agreement with the experimental results, and the error between them is low. Under the experimental conditions, the average error value is 3.28%.
4. The model is reliable for calculating the friction resistance between planes under the action of the periodically changing normal force, which can provide a theoretical basis for the technology of using the excited normal force to reduce friction in engineering.

Author Contributions: Writing—original draft, Y.L. (Yongwang Liu), Y.L. (Yanan Liu), Z.G. and Y.N.; Writing—review & editing, Y.L. (Yongwang Liu). All authors have read and agreed to the published version of the manuscript.

Funding: China's national key R&D plan project (2021YEE0111400); National Natural Science Foundation of China (52074324, 51674284); Maior Science and Technology Project of PetroChina (ZD2019-183-005).

Institutional Review Board Statement: Not applicable.

Informed Consent Statement: Not applicable.

Conflicts of Interest: The authors declare no conflict of interest.

References

1. Prater, T.; Gibson, B.; Cox, C.; Cook, G.E.; Strauss, A.; Longhurst, W. Evaluation of torque as a means of in-process sensing of tool wear in friction stir welding of metal matrix composites. *Ind. Robot. Int. J.* **2015**, *42*, 120–131. [CrossRef]
2. Musalek, R.; Kovarik, O.; Medricky, J.; Curry, N.; Bjorklund, S.; Nylen, P. Fatigue Testing of TBC on Structural Steel by Cyclic Bending. *J. Therm. Spray Technol.* **2015**, *24*, 215–223. [CrossRef]
3. Taghavifar, H.; Mardani, A. A comparative trend in forecasting ability of artificial neural networks and regressive support vector machine methodologies for energy dissipation modeling of off-road vehicles. *Energy* **2014**, *44*, 135–141. [CrossRef]
4. Zhang, J.; Tong, J.; Ma, Y. The bionic drag reduction deep loosening shovel design and test. *Trans. Chin. Soc. Agric. Mach.* **2014**, *45*, 141–145.
5. Wang, X.; Di, Q.; Zhang, R. Nanoparticles aqueous dispersion in the rock core channels of the dual mechanism of drag reduction and its experimental verification. *Acta Phys. Sin.* **2012**, *61*, 336–342.
6. Wang, P.; Ni, H.; Wang, R. Based on micro convex body contact longitudinal vibration of drill string friction resistance. *J. China Univ. Pet. Nat. Sci. Ed.* **2015**, *39*, 88–94.
7. Li, Y.; Gu, Y.; Mou, J. Low surface energy antifouling anti-friction research status of biomimetic coating. *Mater. Prot.* **2014**, *47*, 48–51.
8. Li, X.; Zhao, G.; Liang, Y. Rotary valve type hydraulic vibration pile drawing machine experiment system characteristics analysis. *J. Vib. Shock* **2014**, *33*, 90–95.
9. Guo, Y.; Hou, B. Flexible transmission and load change ammunition transmission mechanical arm position control and flexible vibration suppression. *J. Vib. Shock* **2015**, *34*, 1–8.
10. Huang, T.; Shen, H.; Li, M. Continuous feeding parallel sieve sieve performance experimental research. *Mech. Des.* **2015**, *32*, 91–95.
11. Du, J.; He, H.; Yang, T. Dense oil exploration and development in China and the challenge. *China Pet. Explor.* **2014**, *19*, 1–9.
12. Zhang, H.; Guan, Z.; Liu, Y. Based on rotating excitation vibration of drill string drag reduction tools. *Pet. Mach.* **2015**, *43*, 9–12.
13. Teng, Y.; Zeng, Q.; Yao, H. Vibration friction mechanism and nonlinear dynamic characteristics. *J. Tribol.* **2010**, *30*, 197–202.
14. Xia, M.; Liu, C. Application of large displacement amplitude of thin plate work reciprocal theorem to solve a pair of two opposite edges simply supported edges free displacement amplitude equation of the rectangular plate. *J. Hei Longjiang Univ.* **2006**, *23*, 359–362.
15. Li, W.; Fu, B.; Yang, Z. Bending of rectangular plates with free edges on elastic foundation thick solution. *J. Tianjin Univ.* **2000**, *33*, 73–77.
16. Choi, J.H.; Kwan, S.K.; Ko, H.E.; Park, J.H.; Kim, D.K.; Park, H.W.; Caron, A. Effect of Normal Contact Vibration on Nano-Scale Friction. *Lubricants* **2019**, *7*, 99. [CrossRef]

processes

MDPI

Article

Design of Downhole Robot Actuator System and Mechanical Behavior Analyses of the PRSM by Considering Elastic Errors and Radial Loads

Xuelian Dong, Haiyan Zhu, Qingyou Liu, Qiaozhu Wang * and Xingming Wang

State Key Laboratory of Oil and Gas Reservoir Geology and Exploitation, Chengdu University of Technology, Chengdu 610059, China; dongxuelian11@163.com (X.D.); zhuhaiyan040129@163.com (H.Z.); liuqy66@aliyun.com (Q.L.); upjzk@sina.com (X.W.)
* Correspondence: wangqiaozhu317103@163.com

Abstract: This paper designs a new class of actuators for the downhole traction robot system to achieve high-accuracy transmission, which is realized by the planetary roller screw mechanism (PRSM). As the downhole environment is a non-structure one, which increases the difficulty of the load analyses and distributions of the downhole robot system to complete a predesigned mission. Traditional achievements about the mechanical behavior analyses of PRSM ignore the effects of radial loads and torque elastic deformation errors, which are inevitable for the downhole robot actuator, and the results of which would affect the load distribution and fatigue life of the PRSM-aided actuator. To assist the complex task, in this study the mechanical behavior analyses of PRSM for the downhole robot system are investigated by considering axial loads, torque elastic deformation errors, and radial loads. Moreover, the calculation models for contact load distribution and fatigue life are established by utilizing the equivalent contact load and Hertz contact theory. Two cases for the robot actuator in the downhole environment are addressed, the results of which indicate that the contact load change and decrease with the thread growth direction of the PRSM, the first several threads bore most of the loads, and the last several threads only took a few loads. Additionally, the fatigue life reduces sharply under the condition that the axial loads, radial loads, and rotation speeds increase. Compared with the other two effectors, the fatigue life is more sensitive to the radial loads. The results show the sustainability of the presented screw–roller–nut and provide a potential reference for the downhole robot actuator motion analyses.

Keywords: downhole robot actuator; PRSM; elastic deformation errors; axial loads and radial loads; load distribution; fatigue life

Citation: Dong, X.; Zhu, H.; Liu, Q.; Wang, Q.; Wang, X. Design of Downhole Robot Actuator System and Mechanical Behavior Analyses of the PRSM by Considering Elastic Errors and Radial Loads. *Processes* **2022**, *10*, 1520. https://doi.org/10.3390/pr10081520

Academic Editor: Raul D.S.G. Campilho

Received: 25 June 2022
Accepted: 27 July 2022
Published: 2 August 2022

1. Introduction

Horizontal wells are widely used in the exploration and development of deep sea complex oil and gas resources. As the depth and the length of horizontal wells continue to increase, it is urgent to create a new model to deal with the increasingly complex mission of resource explorations, which can be realized based on horizontal wells and downhole robot technologies.

The concept of the downhole robot was first proposed by H. Jørgen in 1987 [1], which can be divided into two categories according to its structural characteristics and application conditions: downhole traction robots and downhole drilling robots [2]. Downhole robots play an indispensable role in drilling and completion operations to transport downhole tools to designated locations, and to achieve explosive fracturing. Additionally, the downhole robot can solve the buckling deformation of coiled tubing, the difficulty of running downhole tubing strings in horizontal well logging, fracturing, and other operations. Nowadays, the downhole robot has been widely used in downhole operations in long horizontal complex wells.

The structure of the downhole robot mainly includes a right and left body-support mechanism, traction mechanism, power mechanism, and so on. As shown in Figure 1, according to their structures and different actuators, downhole traction robots can be divided into three categories, which are the wheeled traction robot, pedrail traction robot, and telescopic traction robot. Additionally, more concretely, the driving forces of the downhole robot can be founded as two kinds, i.e., hydraulic cylinder and motor. Compared with the other two traction robots, the telescopic type has a good performance in terms of passing over the obstacle. From the perspective of driving power in the hydraulic cylinder scheme, some complex oil circuits should be designed and arranged in a small robot system, which decreases the flexibility of the downhole robot system. On the other hand, the unnecessary oil circuit can be omitted in the motor driving framework, and the power supply and communication can be integrated. This motivates the motor-driven actuator design of this study.

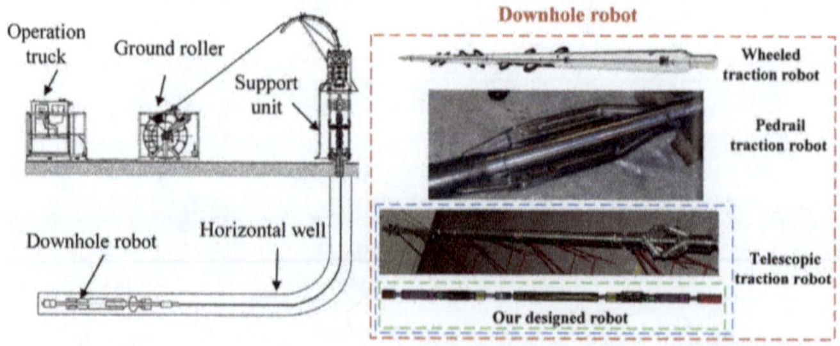

Figure 1. Brief introduction of the downhole robot system.

It should be noted that the traction devices used by the previous downhole robots are hydraulic telescopic and wheeled [3,4]. However, the wheeled downhole robots have a large diameter and cannot be used in tiny boreholes [5]. On the other hand, signal feedback is troublesome during the operation of the robots for the hydraulic power traction condition. Additionally, the solenoid valve is limited to withstand high temperatures; indeed, the solenoid valve and hydraulic oil passage are easy to enter and block impurities, which limits the application of the hydraulic telescopic power mechanism [6–8]. From the transmission accuracy and control response speed perspective, the planetary roller screw mechanism (PRSM) has been widely used in aerospace, ships, medical equipment, special machinery, weaponry, and other precision fields [9,10] owing to its high transmission accuracy in recent years. However, only a few studies have investigated the application of PRSM to downhole robot systems. This work presents a class of downhole robot systems, the traction actuator mechanism of which is a planetary roller screw traction mechanism one. The PRSM is a power unit that converts the rotational motion of the lead screw into the linear motion of the nut, which can achieve a precision transmission task. With the PRSM actuator mechanism, the actuator is driven by a motor rather than a hydraulic ram, which implies that the structure is simplified due to the omission of some hydraulic oil circuits. With the continuous development and application of long horizontal wells, the well structure will have local bending an inclination due to the change of deflection parameters in long-horizontal complex-structure wells. When the robot works in the downhole, the non-structural shaft wall will cause different degrees of extrusion to affect the robot. In addition, the high temperature and pressure in the downhole have also posed challenges to its mechanical behaviors and fatigue life.

There are many factors affecting the safety and stability of PRSM. The existing achievements could provide a theoretical basis for the load distribution calculation and fatigue life analysis. For example, in [11], the authors studied the effects of uniformly distributed loads on the carrying capacity of PRSM. Furthermore, an improved iterative algorithm was proposed for multi-objective optimization, such as for contact deformation and finding the best pitch. In [12], Fu et al. studied the influences of unevenly distributed load on the transmission accuracy and life of PRSM; the results show that it will accelerate the destruction of the thread contact surface and reduce the transmission stability when the screw rotates at a high speed. The results in [13,14] studied the effects of assembly errors and external loads on PSRM, and safe transmission mechanisms were discussed. Aur′egan and Yao, et al., respectively, investigated the effects of wear behavior, uncertain factors of the elastic–plastic contact characteristics, and machining errors on the PRSM in [15,16]. In [17], the nature of the contact with friction between the threaded surfaces in a PRSM was investigated, and the calculation model was established and simulated. The achievements in [18,19] discussed the load sharing and pitch deviation of the PSRM. The results show that the deformation also has a great influence on the fatigue life. More recently, the transmission efficiency of PRSM was studied in [20–22]; for instance, Mamaev, et al. reached the conclusion that the stiffness and bearing capacity will increase the transmission efficiency of PSRM in [20]. Auregan, et al. in [21], verified that the reason for reducing the transmission efficiency of PSRM was the complete sliding of the PRSM pair, which was caused by the plastic–flow ratio between the thread engagements. In [22], the author discussed Hertz contact model and a multi-axial fatigue criterion of PSRM, and the results were applied to two cases to verify the reliability. However, the aforementioned results have rarely discussed the PRSM utilized in the downhole environment, which is more complex and has some new challenges. Furthermore, the coupled torque elastic deformations and radial loads for the PRSM of the downhole robot system not have been explored.

As aforementioned, the performance of PRSM were affected by multiple factors, including the external loads, elastic errors, and even the complex working conditions. In this note, a class of traction robot systems was designed. The PRSM is introduced as the actuator. Additionally, the mechanical behavior of the PRSM used in the downhole robot system was investigated. The load distribution and fatigue life calculation models are established. The contributions of this study can be highlighted as follows:

(1) A new actuator is designed for the downhole traction robot system. Compared with the aforementioned downhole robot in [6–8], in this note, the PRSM is introduced. Moreover, the driver power can be supported by a motor instead of the hydraulic cylinder, which indicates that some hydraulic oil circuits can be omitted and then the structure of the robot system can be simplified.
(2) The mechanical analyses of the PRSM in the downhole environment is investigated. This is different from the existing achievements in [18–20], where the load distributions are analyzed by considering axial loads and external deformations. For the downhole traction robot system, the working environment is a non-structure one, which means that the radial loads and torque deformations cannot be ignored, and the load distributions and fatigue life will be affected. In this note, we establish the calculation and fatigue life calculation models for the presented actuator by considering the axial loads, radial loads and torque elastic deformations, simultaneously, with the help of the equivalent contact load and Hertz contact theory.

The remainder of this study is organized as follows. In Section 2, the mechanical model is discussed and the problem is formulated. In Section 3, the elastic deformation error is analyzed. In Section 4, the mechanical behavior analysis is verified. In Section 5, some simulation results are given. Finally, we conclude this study in Section 6.

2. Mechanical Model and Problem Formulation

2.1. Downhole Robot System Actuator Model Analysis and Design

As mentioned before, two driving power mechanisms are contained in the telescope traction robots. In this subsection, the comparison results of the two driving power mechanisms in the telescopic traction robot is expressed in Table 1. Some more detail results of the downhole robots can be found in the work [1].

Table 1. Comparisons of two driving powers in downhole telescopic traction robot mode.

Power	Actuator	Model	Advantages and Analyses	Disadvantages
Hydraulic	Hydraulic cylinder	Telescopic traction robot	Accessories are necessary; less structure design freedom and the control of the cylinder is difficult	The sensor requirement is high and lacks of feedback data; control precision is not so high; signal transmission is difficult; control logic is complicated
Electro-hydraulic	Motor and hydraulic cylinder	Wheeled and Telescope	More structure flexibility, higher traction force; good performance of trade off velocity range	Difficult to control; larger size; requirement of sealing performance; more parameters to be adjusted
Motor (Our presented)	PRSM (Our presented)	Telescopic traction robot	Easy to control; more design freedom; integration of power supply and communication; sensor data can be transmitted	The mechanical behavior of the PRSM should be analyzed

In this study, we designed a new actuator for the telescopic traction robot to complete complex tasks in the downhole, as shown in Figure 2. The PRSM is introduced in the downhole robot system design scheme. In our presented robot, the power is supported by the motor, and the PRSM is driven by the motor. It should be mentioned that the PRSM in the robot system can help the robot walk forward in the downhole. However, as the downhole environment is complex, the PRSM will suffer from radial force and elastic deformation, which will affect the load distribution on the threads of the PRSM. To guarantee high transmission accuracy, the mechanical behavior analysis of the actuator in the downhole should be made.

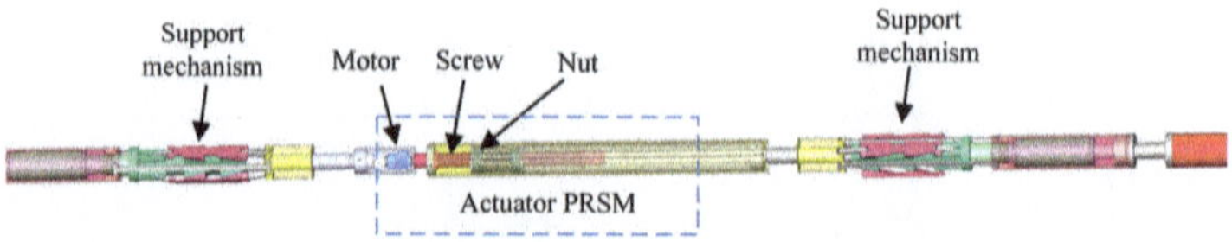

Figure 2. The presented downhole robot system

2.2. Analysis and Problem Formulation

The three dimensional model of the downhole robot system is shown in Figure 3a. The downhole robot system works in a retractable mode, as shown in Figure 3b, to complete the downhole task, eight steps are consisted in one cycle motion. The motion procedure of the robot system can be expressed as follows.

(a) Initial position. The left support cylinder, left support arm, PSRM, right support arm, and right support cylinder are all in the initial states, which are shown as in Figure 3;
(b) The left support mechanism of the robot system works with the cylinder;
(c) The motor rotates forward; the nut moves forward to drive the right support arm and the right main body to move forward;
(d) The left support arm contracts and moves forward with the left retractable cylinder;
(e) The right support arm completes the operation of the contract with the right support cylinder;
(f) The motor rotates in reverse and nut fixed; the screw draws the right support cylinder and the right main body to move forward;
(g) The right support arm completes the support operation with the right support cylinder;

(h) The left support arm completes the operation of the contract with the left support cylinder, and then goes back to state b;
Repeat steps b–h.

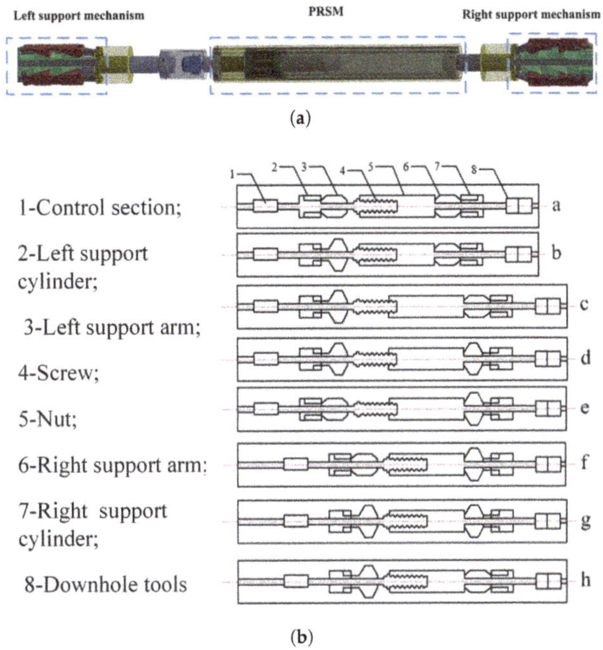

Figure 3. (**a**) The three-dimensional model of the downhole robot system; (**b**) the working mechanism of the downhole robot system.

It should be noted that during the working procedure, the mechanical behavior of PRSM is one of the most important parts of the robot system. In what follows, the contact load and the deformation of the PRSM are analyzed when the downhole robot works. The scheme of the PRSM of the robot system is depicted in Figure 4. We number the threads of each roller individually along with the axial direction. For the downhole robot system, two cases are contained to complete the predesigned task.

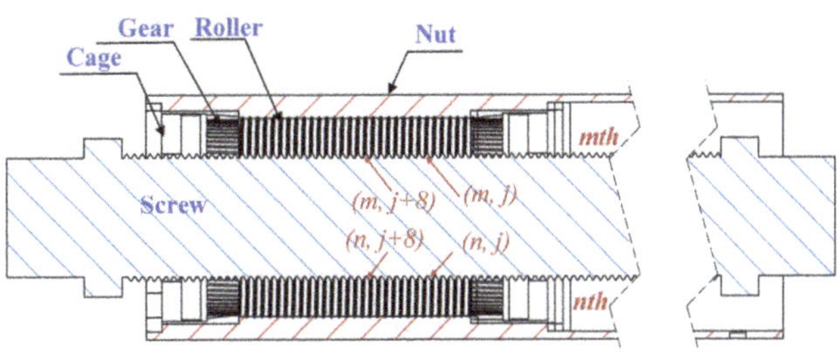

Figure 4. The threads number of each roller for the downhole robot system.

Case 1: The robot moves forward through the screw and roller, which are driven by the motor. In this condition, the nut is fixed and the screw is rotating and moving.

Case 2: The forward motion of the accessory support frame happens through the movement of the nut. In this condition, the screw is driven by the motor and only rotates without moving.

Before we give our main results, the following assumption is made.

Assumption 1 ([23]). *The deformations considered of the screw to roller and roller to the nut are elastic and satisfy the Hertz contact theory.*

Assumption 2. *The centrifugal force and gyroscopic moment are neglected.*

3. Elastic Deformation Error Analysis of the PRSM

3.1. Robot Elastic Deformation Error in Case 1

When the screw works in condition 1 in the downhole robot, it means that the motor drives the screw to rotate and move, and the axial load F_a is shown in Figure 5. In this condition, an additional radial load F_ζ exists as the screw drags some accessories and the downhole environment is non-structural. Additionally, a torsion deformation occurs when the screw works.

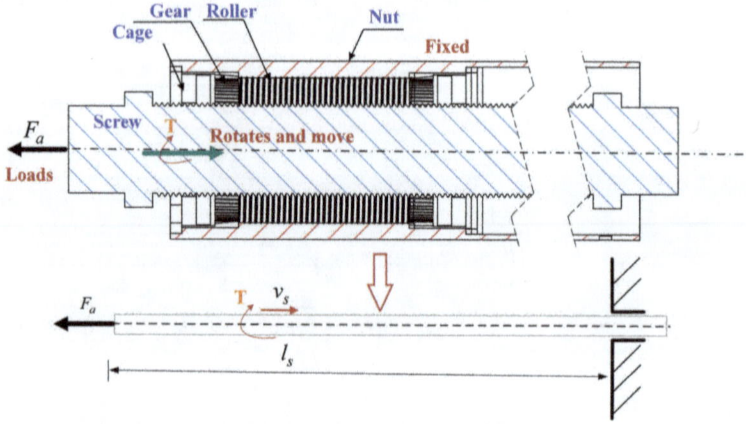

Figure 5. The diagram of the deformation in case 1.

For the downhole robot system in case 1, when the screw rotates and moves, the thread plays the main role in power and motion transmission. From the results in [24,25], the deformations caused by axial loads F_a and torsional T in the axis direction can be obtained as follows,

$$\lambda_{a1} = \frac{4F_a l_x}{E\pi d_z} \tag{1}$$

where λ_{a1} is the axial deformation caused by the tension load. l_x denotes the nut working position. E represents the elastic modulus of materials, and d_z is the diameter of screw thread.

$$\lambda_b = \frac{16Tl_s l_z}{\pi^2 d_z^4 G} \tag{2}$$

where λ_b represents the axial deformation caused by the driving torque. T is the driving torque. l_z denotes the lead of the screw. l_s is the length of screw against the torque, and G is the shear modulus.

In order to drag the accessories, the driving torque can be calculated as

$$T = \frac{F_a l_z}{2\pi\mu} \tag{3}$$

where μ denotes the transmission efficiency. From the Equation (3), Equation (2) can be reformulated as

$$\lambda_b = \frac{8F_a {l_z}^2 l_d}{\pi^3 d_z^4 G\mu} \tag{4}$$

Then, the total deformation error of the screw in the axial direction can be obtained as

$$\lambda_s = \lambda_{a1} + \lambda_b = \frac{4F_a l_x}{E\pi d_z} + \frac{8F_a {l_z}^2 l_d}{\pi^3 d_z^4 G\mu} \quad (5)$$

3.2. Robot Elastic Deformation Error in Case 2

As expressed in Figure 6, in this condition, only pressure deformation occurs in the nut, which means that the total deformation $\lambda_s = \lambda_{a2}$, and the deformation can be obtained as

$$\lambda_{a2} = \frac{4F_a l_x}{E\pi d_z} \quad (6)$$

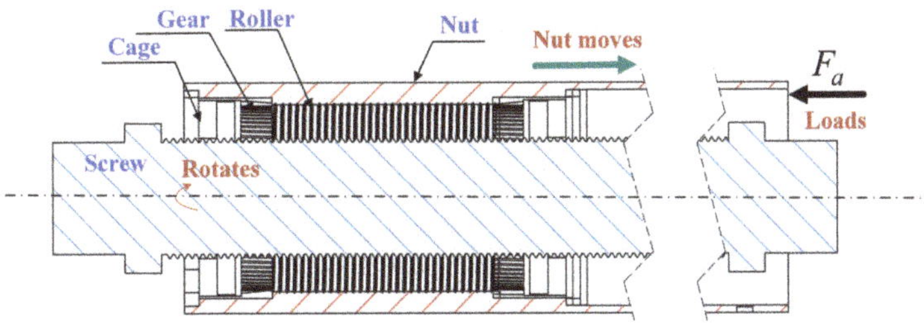

Figure 6. The diagram of the deformation in case 2.

Now we are in the position to analyze the contact deformation of the threads of screw-to-roller and roller-to-nut of the downhole robot system.

4. Mechanical Behavior Analysis

4.1. Robot Contact Deformation Analysis

As described in Figure 7, F_ζ denotes the radial loads. When the robot moves in the downhole, the PRSM is subjected to a non-structure radial force. As a result, a non-structure radial deformation of the roller at different position is generated correspondingly when the roller rotates. As expressed in Figure 7, the solid line denotes the initial position without the non-structure radial force, and the dashed line represents the new position when the nut is subjected to the radial loads. For the downhole robot system, the axial loads of screw and nut are the prior known knowledge and then the axial deformation can be obtained. It should be noted that the axial loads of each roller are not equal and difficult to calculate when the robot moves in the hole through the PRSM. In this study, the axial deformations of each roller of all threads are identical, and the radial deformations can be calculated by [26]

$$\zeta_{r(n,j)} = \zeta_r \cos \varphi_i \quad (7)$$

where ζ_r is the radial deformation, φ_i is the so-called position angle of the *ith* roller. By defining the angle between two adjacent rollers $\phi = \frac{2\pi n_a}{z}$, where z and n_a represent the number of rollers and work cycles. Then, φ_i can be calculated as

$$\varphi_i = \theta + (i-1)\phi \quad (8)$$

By changing the value of θ, the radial deformations of all the threads can be calculated.

In this study, the equivalent deformation concept is adopted, i.e., the normal deformations of the screw-to-roller and roller-to-nut are calculated by

$$\zeta_{e(n,j)} = \zeta_{a(n,j)} \sin\beta / \cos\varpi + \zeta_{r(n,j)} \cos\beta \cos\phi \quad (9)$$

where $\zeta_{e(n,j)}$ denotes the equivalent normal deformation of the thread, $\zeta_{a(n,j)}$ is the axial deformation, β and ϖ are the contact angle and helix angle, respectively.

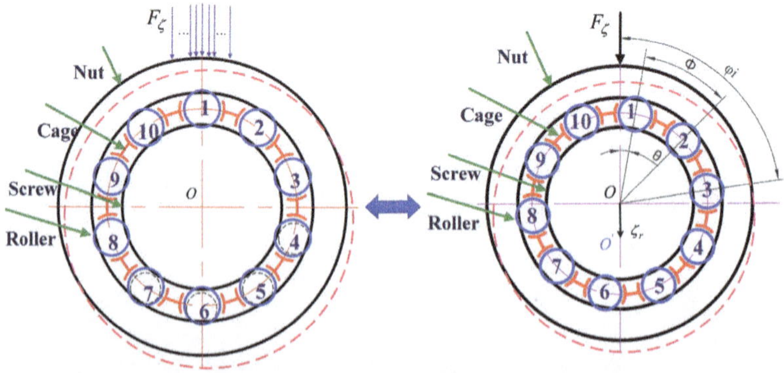

Figure 7. The diagram of deformation when subjected to radial loads.

From the Hertzian contact theory, the equivalent contact load of the contact point can be calculated via:

$$\Re_{e(n,j)} = \kappa \left(\frac{1}{\mathcal{G}_{ra} + \mathcal{G}_{rb}} \zeta_{e(n,j)} \right)^{\frac{3}{2}} \tag{10}$$

where κ is an indicator with $\kappa \in \{0, 1\}$, which means that the threads are contacting when $\kappa = 1$, and the threads contact are abnormal when $\kappa = 0$. $\mathcal{G}_{ra}$ and $\mathcal{G}_{rb}$ represent the stiffness coefficients in the contact threads of the screw-to-roller and roller-to-nut, which can be obtained as [27]

$$\begin{cases} \mathcal{G}_{ra} = \frac{Q_{es}}{\pi m_{vs}} \sqrt[3]{9E^2 \sum \rho_{as}/4} \\ \mathcal{G}_{rb} = \frac{Q_{en}}{\pi m_{vn}} \sqrt[3]{9E^2 \sum \rho_{an}/4} \end{cases} \tag{11}$$

where Q_{es} and Q_{en} are two elliptic integral parameters of the screw-to-roller and roller-to-nut contact threads. m_{vs} and m_{vn} denote two semi-axis coefficients of the two contact threads. $\sum \rho_{as}$ and $\sum \rho_{an}$ represent the curvature sum of the two contact interfaces.

Remark 1. *In this study, the PRSM will be subjected to radial loads and axial loads during the working process in the downhole environment. However, in the existed achievements, most of the studies have only considered axial loads and the corresponding load distributions. There are rare results that have discussed the axial loads and radial loads simultaneously, and it should be noted that the loads of each roller of the PRSM are not equal when considering the radial loads. From this point of view, the equivalent contact load and Hertz contact theory are used to calculate the loads of all threads in this work.*

By combing Equations (7)–(11), the equilibrium equations in the radial and axial direction can be expressed by

$$\begin{cases} F_{\zeta} - \sum_{n=1}^{N} \sum_{j=1}^{P_z} \zeta_{e(n,j)} \cos\beta \cos\phi = 0 \\ F_a - \sum_{n=1}^{N} \sum_{j=1}^{P_z} \zeta_{e(n,j)} \sin\beta \cos\varpi = 0 \end{cases} \tag{12}$$

where N and P_z stand for the numbers of rollers and threads. Furthermore, the axial load of the nth roller can be calculated by $F_{an} = \sum_{j=1}^{P_z} \zeta_{e(n,j)} \sin\beta \cos\varpi$.

4.2. Robot Contact Load and Deformation Analysis

As described above, two cases are considered for the downhole robot system. In case 1, the screw is moving and the nut is fixed, which implies that the screw is in tension and the nut is in compression $(S_c - N_t)$. In case 2, the screw rotates and the nut moves, which means that both the screw and nut are in compression $(S_c - N_c)$.

4.2.1. Deformation in $S_c - N_t$

As shown in Figure 8, denoting $\Im_{S(n,j)}$ and $\Im_{N(n,j)}$ as the thread contact loads of the screw and nut, when the PRSM works on the equilibrium point, the axial load for the case can be obtained as [23]

$$\Im_{S(n,j)} = \Im_{N(n,j)} = F_{an} - \sum_{h=1}^{j-1} \zeta_{e(n,h)} \sin\beta \cos\varpi \tag{13}$$

where $\zeta_{e(n,h)}$ can be solved by (12).

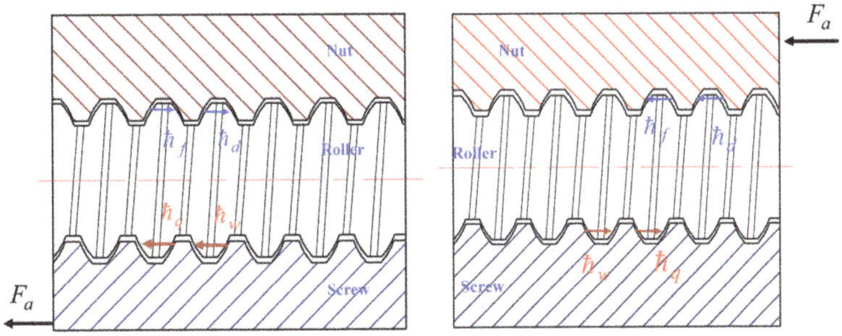

(a) (b)

Figure 8. The sketches of the axial deformations in two cases.

Furthermore, when the screw is driven by a motor and subjected to external loads, some errors are evitable for the PRSM. Additionally, correspondingly, the following equations are derived

$$\hbar_{SZ(n,j)} = \hbar_w - \hbar_q = \frac{1}{\sin\beta\cos\varpi}\left(\upsilon_{SZ(n,j-1)} - \upsilon_{SZ(n,j)}\right) - \left(\Re_{SZ(n,j-1)} - \Re_{SZ(n,j)}\right) \tag{14}$$

where $\hbar_{SZ(n,j)}$ denotes the contact deformation in the screw to roller interface, which satisfies the Hertz condition. $\upsilon_{SZ(n,j-1)}$ and $\upsilon_{SZ(n,j)}$ represent two consecutive contact surfaces of the screw-to-roller. $\Re_{SZ(n,j-1)}$ and $\Re_{SZ(n,j)}$ are the contact interface axial elastic deformation errors of jth and $(j-1)th$ threads caused by driven torque and loads.

Similarly, the deformation occurs in the roller-to-nut with elastic errors can be calculated by

$$\hbar_{ZN(n,j)} = \hbar_d - \hbar_f = \frac{1}{\sin\beta\cos\varpi}\left(\upsilon_{ZN(n,j-1)} - \upsilon_{ZN(n,j)}\right) - \left(\Re_{ZN(n,j-1)} - \Re_{ZN(n,j)}\right) \tag{15}$$

where $\upsilon_{ZN(n,j-1)}$ and $\upsilon_{ZN(n,j)}$ express two consecutive contact surfaces deformations of roller to nut. $\Re_{ZN(n,j-1)}$ and $\Re_{ZN(n,j)}$ depict the elastic deformation errors of jth and $(j-1)th$ threads in the roller to nut interface caused by driven torque and loads.

From the Equations (14) and (15), the combined axial deformation errors by considering the elastic deformation errors can be obtained as

$$\begin{aligned}\hbar_{(n,j)} &= \hbar_w - \hbar_q + \hbar_d - \hbar_f \\ &= \tfrac{1}{\sin\beta\cos\varpi}\left(\upsilon_{SZ(n,j-1)} - \upsilon_{SZ(n,j)}\right) - \left(\Re_{SZ(n,j-1)} - \Re_{SZ(n,j)}\right) \\ &\quad + \tfrac{1}{\sin\beta\cos\varpi}\left(\upsilon_{ZN(n,j-1)} - \upsilon_{ZN(n,j)}\right) - \left(\Re_{ZN(n,j-1)} - \Re_{ZN(n,j)}\right) \\ &= \tfrac{1}{\sin\beta\cos\varpi}\left[\upsilon_{r(n,j-1)} - \upsilon_{r(n,j)}\right] - \left[\Re_{r(n,j-1)} - \Re_{r(n,j)}\right]\end{aligned} \tag{16}$$

where $v_{r(n,j-1)} = v_{SZ(n,j-1)} + v_{ZN(n,j-1)}$, $v_{r(n,j)} = v_{SZ(n,j)} + v_{ZN(n,j)}$, $\Re_{r(n,j-1)} = \Re_{SZ(n,j-1)} + \Re_{ZN(n,j-1)}$, $\Re_{r(n,j)} = \Re_{SZ(n,j)} + \Re_{ZN(n,j)}$.

For the downhole robot system in case 1, the axial deformation can also be formulated as

$$\begin{cases} \hbar_{S(n,j)} = \frac{\Im_{S(n,j)}\rho_a}{2E_sA_s} \\ \hbar_{N(n,j)} = \frac{\Im_{N(n,j)}\rho_a}{2E_NA_N} \end{cases} \tag{17}$$

where $\hbar_{S(n,j)}$ and $\hbar_{N(n,j)}$ represent the deformations in the screw-to-roller and roller-to-nut contact threads. ρ_a stands for the pitch, E_s. E_N , A_s, and A_N are the elastic modulus and cross sectional areas of the screw and nut, respectively. A_s and A_N can be calculated as $A_s = \pi {R_s}^2$, $A_N = \pi(R_{N1}^2 - R_{N2}^2)$, where R_s is the radius of the screw. R_{N1} and R_{N2} are the external and normal radius of nut.

In accordance with (17), similar to (16), the combined axial deformation can be obtained as

$$\hbar_{(n,j)} = \hbar_{S(n,j)} + \hbar_{N(n,j)} = \frac{\Im_{S(n,j)}\rho_a}{2E_sA_s} + \frac{\Im_{N(n,j)}\rho_a}{2E_NA_N} \tag{18}$$

By combing (16) and (18) yields,

$$\frac{\Im_{N(n,j)}\rho_a(A_N + A_s)\sin\beta\cos\varpi}{2E_sA_sA_N} = \left[v_{r(n,j-1)} - v_{r(n,j)}\right] - \sin\beta\cos\varpi\left[\Re_{r(n,j-1)} - \Re_{r(n,j)}\right] \tag{19}$$

According to the Hertzian contact theory, the following recursive condition can be derived

$$\begin{aligned} \Re_{e(n,j-1)}{}^{\frac{2}{3}} - \Re_{e(n,j)}{}^{\frac{2}{3}} = & \frac{\sin\beta\cos\varpi\left[\Re_{r(n,j-1)} - \Re_{r(n,j)}\right]}{(\mathcal{G}_{ra}+\mathcal{G}_{rb})} \\ & + \frac{\left(\sum\limits_{j=1}^{P_z}\zeta_{e(n,j)}\sin\beta\cos\varpi - \sum\limits_{h=1}^{j-1}\zeta_{e(n,h)}\sin\beta\cos\varpi\right)\rho_a(A_N+A_s)\sin\beta\cos\varpi}{2E_sA_sA_N(\mathcal{G}_{ra}+\mathcal{G}_{rb})} \end{aligned} \tag{20}$$

Now, we are in the position to analyze the deformation of case 2.

4.2.2. Deformation in $S_c - N_c$

For the $S_c - N_c$ condition, the axial load $\Im_{S(n,j)}$ and $\Im_{N(n,j)}$ of the screw and nut can be reorganized as [23]

$$\begin{cases} \Im_{N(n,j)} = F_{an} - \sum\limits_{h=1}^{j-1}\zeta_{e(n,h)}\sin\beta\cos\varpi \\ \Im_{S(n,j)} = \sum\limits_{h=1}^{j}\zeta_{e(n,h)}\sin\beta\cos\varpi \end{cases} \tag{21}$$

Moreover, accordingly, the deformations in the threads of the screw-to-roller and roller-to-nut in the axial direction can be calculated similarly as in Equations (14) and (15), while the roller-to-nut deformation is revised as

$$\hbar_{ZN(n,j)} = \hbar_f - \hbar_d = \frac{1}{\sin\beta\cos\varpi}\left(v_{ZN(n,j)} - v_{ZN(n,j-1)}\right) - \left(\Re_{ZN(n,j)} - \Re_{ZN(n,j-1)}\right) \tag{22}$$

From (14) and (22), the total axial deformation by considering the elastic deformation error can be obtained as

$$\begin{aligned} \hbar_{(n,j)} &= \hbar_w - \hbar_q + \hbar_f - \hbar_d \\ &= \frac{1}{\sin\beta\cos\varpi}\left(v_{SZ(n,j-1)} - v_{SZ(n,j)}\right) - \left(\Re_{SZ(n,j-1)} - \Re_{SZ(n,j)}\right) \\ &\quad + \frac{1}{\sin\beta\cos\varpi}\left(v_{ZN(n,j)} - v_{ZN(n,j-1)}\right) - \left(\Re_{ZN(n,j)} - \Re_{ZN(n,j-1)}\right) \\ &= \frac{1}{\sin\beta\cos\varpi}\left[\bar{v}_{r(n,j-1)} - \bar{v}_{r(n,j)}\right] - \left[\bar{\Re}_{r(n,j-1)} - \bar{\Re}_{r(n,j)}\right] \end{aligned} \tag{23}$$

where $\bar{v}_{r(n,j-1)} = v_{SZ(n,j-1)} + v_{ZN(n,j)}$, $\bar{v}_{r(n,j)} = v_{SZ(n,j)} + v_{ZN(n,j-1)}$, $\bar{\Re}_{r(n,j-1)} = \Re_{SZ(n,j-1)} + \Re_{ZN(n,j)}$, $\bar{\Re}_{r(n,j)} = \Re_{SZ(n,j)} + \Re_{ZN(n,j-1)}$.

By combing (17) and (21), it is straightforward that

$$\begin{aligned}\hbar_{(n,j)} &= \hbar_{S(n,j)} + \hbar_{N(n,j)} = \frac{\Im_{S(n,j)} A_N \rho_a}{2E_s A_s A_N} + \frac{\Im_{N(n,j)} A_s \rho_a}{2E_N A_s A_N} \\ &= \frac{\left(F_{an} - \sum_{h=1}^{j-1} \zeta_{e(n,h)} \sin\beta \cos\varpi\right) A_N \rho_a}{2E_s A_s A_N} + \frac{\sum_{h=1}^{j} \zeta_{e(n,h)} \sin\beta \cos\varpi \rho_a A_s \rho_a}{2E_N A_s A_N}\end{aligned} \tag{24}$$

Furthermore, according to the Hertzian contact theory, the corresponding recursive equation in case 2 can be obtained as

$$\begin{aligned}&\Re_{e(n,j-1)}^{\frac{2}{3}} - \Re_{e(n,j)}^{\frac{2}{3}} \\ &= \frac{\sin\beta\cos\varpi\left[\bar{\Re}_{r(n,j-1)} - \bar{\Re}_{r(n,j)}\right]}{\mathcal{G}_{ra} + \mathcal{G}_{rb}} + \frac{\sum_{h=1}^{j} \zeta_{e(n,h)} \sin\beta\cos\varpi\rho_a A_s \rho_a \sin\beta\cos\varpi}{2E_N A_s A_N (\mathcal{G}_{ra} + \mathcal{G}_{rb})} \\ &+ \frac{\left(F_{an}\sin\beta\cos\varpi - \sum_{h=1}^{j-1} \zeta_{e(n,h)} \sin^2\beta\cos^2\varpi\right) A_N \rho_a}{2E_s A_s A_N (\mathcal{G}_{ra} + \mathcal{G}_{rb})}\end{aligned} \tag{25}$$

In this subsection, the mechanical analyses of the presented PRSM have been made by referencing the work in [23] and some ball rollers in [26]. The calculation process seems similar to the results in [23]. For example, in Section 3.2, the recursive equations of the contact load on the threads of two cases have been calculated; for details see (20) and (25). In Sections 2.2.2 and 2.2.3 of the work in [23], the contact deformations are also calculated. The calculation ideas of the work in our study and [23] may be similar, and this can be explained as follows. On the one hand, the force analysis model after simplifying for the downhole traction robot actuator is similar to the work in [23], i.e., both the external loads are axial load and radial load, and the difference lies in that the driven torque elastic deformation is considered. Additionally, the parameters of the PRSM are designed according to the downhole traction robot working condition, which are different from the work in [23]. On the other hand, the contact theory used in the two achievements are similar. In our study and the work in [23], the equivalent Hertz contact theory is utilized to analyze the contact load distribution, which in return implies that some equations and calculation idea may similar.

4.3. Downhole Robot Contact Load and Deformation Coefficient Analysis

As depicted in the aforementioned part, for the downhole robot system, the driven torque and external loads will cause elastic deformations and even errors, which will degrade the transmission accuracy in return. It should be also noted that elastic deformations will cause the change of load distribution. By considering the elastic deformation, the deformation coefficient is investigated in this note. Similar to [23], to express the deformation rate, we define this coefficient as the ratio of the normal contact deformation in the condition without and with the elastic deformation errors.

$$\lambda_{(n,j)} = \frac{\zeta_{sw(n,j)}}{\zeta_{st(n,j)}} \tag{26}$$

where $\zeta_{sw(n,j)}$ and $\zeta_{st(n,j)}$ denote the normal contact deformation in the condition without and with the elastic deformation errors.

Here, by considering the contact load and the Hertzian contact theory, we have

$$\lambda_{(n,j)} = \left(\frac{\Re_{sw(n,j)}}{\Re_{st(n,j)}}\right)^{\frac{2}{3}} \tag{27}$$

where $\Re_{st(n,j)}$ and $\Re_{sw(n,j)}$, respectively, represent the contact loads by considering the elastic deformation errors or not.

When the PRSM works, the load distribution is very important. We introduce the load coefficient to better show the mechanical behavior of the PRSM, which is defined as follows

$$\varepsilon_{(n,j)} = \frac{\Re_{sw(n,j)}}{\min\left(\Re_{sw(n,j)}\right)} \tag{28}$$

where $\varepsilon_{(n,j)}$ is the load coefficient, $\Re_{sw(n,j)}$ denotes the minimum value of contact load $\min\left(\Re_{sw(n,j)}\right)$.

4.4. Fatigue Life Analysis

For the downhole robot system, the non-structured environment increases the difficulty of analysis of the PRSM. As the PRSM would work for a long time in the downhole, the PRSM switches back and forth between the two cases to guarantee the complex task. Moreover, the fatigue life is a crucial performance that should be emphasized. From the stress life method, the estimation of fatigue lifetime is given as follows [23,26]

$$L_S = \frac{N_s(\kappa_s + 1)}{60 w_s \kappa_s(\kappa_s + 2)} \tag{29}$$

where $N_s = k_\alpha{}^{m_s} N_0$, $\kappa_s = \frac{d_s}{d_r}$, w_s is the rotating speed. k_α and N_0 are the stress ratio and circulation of the material, m_s is an index corresponding to the material, and d_s and d_r are the diameters of the screw and roller, respectively.

Remark 2. *It should be noted that the effects of the fatigue of roller screws can be listed as materials, stress, working conditions, and other effectors, such as lubrication. Here, we use the fatigue life equation of ball screws to estimate the fatigue life of roller screws, which means that (29) is not an accurate result, it is an estimation of fatigue life.*

Remark 3. *In Section 2.4 of the work in [23], the authors have considered the lifetime and mechanical analyses for the PRSM when subjected to axial load, radial load, and machine errors. However, in this study, the considered PRSM is utilized in the downhole environment, the parameters and working conditions are different from the case in [23] but somewhat similar. We established our estimation lifetime model by referencing [23,26], and the ball rollers. The idea of the lifetime estimation model is similar to [23], and we hope we can consider some more complex conditions for the downhole robot actuator, such as the lubrication analysis and thermal analysis in our future work, and that we can establish some more accurate calculation model for the lifetime by carrying out some experimentation.*

5. Analytical Calculations and Discussions

In this section, some numerical analyses are given to show the validity of the presented model for the downhole robot system. The calculation flow can be seen in Figure 9, and the detail analysis parameters are given in Table 2.

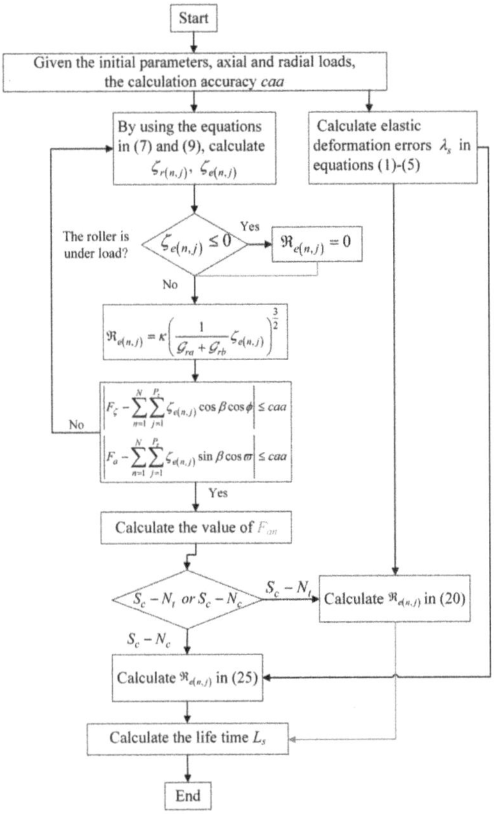

Figure 9. The mechanical behavior analysis flow of the downhole robot system.

Table 2. The designed parameters of the PRSM for the downhole robot system.

Parameters (Unit)	Screw	Roller	Nut
Radius (mm)	15	5	25
Contact angle (°)	45	45	45
Starts	5	1	5
Pitch (mm)	0.8	0.8	0.8
Number	1	10	1
External radius (mm)	/	/	30

Additionally, the left parameters are given as $E = 2.12 \times 10^{11}$ (Pa), $N_0 = 2.5 \times 10^8$, $m_s = 6$.

5.1. The Effects of Axial and Radial Loads

For the downhole robot system, two cases are contained in a complete work procedure. In case 1, the axial loads are chosen to be $F_a = 20\,\text{KN}$, $F_r = 4\,\text{KN}$. Additionally, $F_a = 30\,\text{KN}$, $F_r = 4\,\text{KN}$ in case 2.

From the calculation flow in Figure 9, the corresponding relationship of the contact load and thread number of the two cases can be obtained. The results can be found in Figures 10–15. As depicted in Figures 10–12, the contact load distribution of case 1 for the designed PRSM actuator of the downhole robot system is presented when the PRSM is subjected to axial loads, radial loads, and elastic deformation errors. From Figure 10, the contact loads of all the rollers decreases along the roller axial direction. The results are consistent with the real application and can be explained as follows. The first several threads bear most of the loads, and the last several threads take only a few loads. Additionally, the #1 and the #10 rollers bear the two largest loads, and the loads that are distributed on other rollers reduce gradually from the radial direction. One can also find that the roller that without radial loads bears the lowest contact load. The result indicates that the radial loads should act on the PRSM as uniformly as possible, which in return can help to design and explore good trajectory planning. It should also be noted that the last two threads of the screw, roller, and nut decrease dramatically for the downhole robot system.

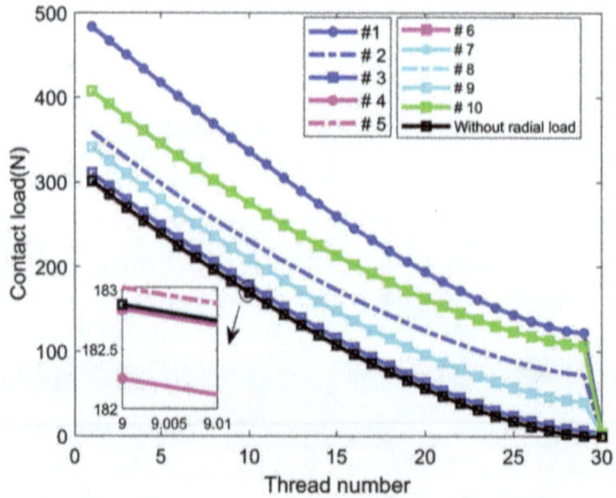

Figure 10. Contact load distribution in different threads of the robot system of case 1.

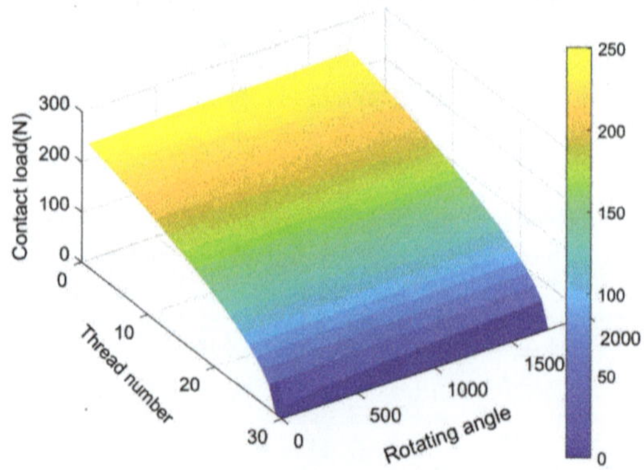

Figure 11. Contact load distribution cycling process without radial load of case 1.

To show the effects of the radical loads on the PRSM, Figures 11 and 12 express the load behaviors in several cycling processes. From Figure 11, we take the first roller as an example. When there are no radial loads, the contact load stays the same in the cycling process, and different threads have their loads. As each roller has the similar load distribution, we only provide the #1 roller of the PRSM. Shown in Figure 12, one can see that the contact load varies when the radial loads are applied to the PRSM. Different from the results in Figure 11, when the radial loads are considered, the bearing force of the roller changes periodically, and different values appear alternately at different positions on the circumference.

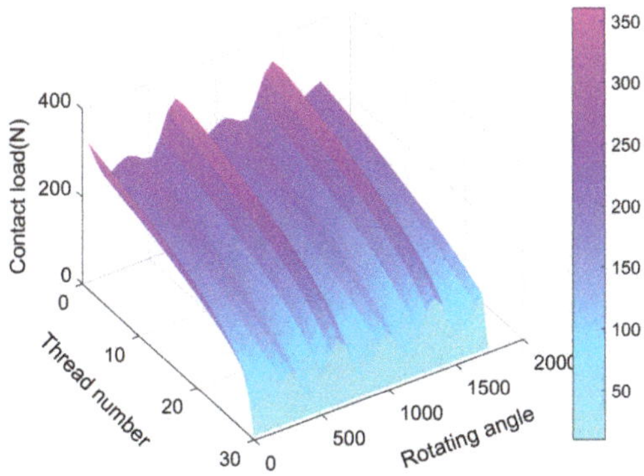

Figure 12. Contact load distribution cycling process with radial load of case 1.

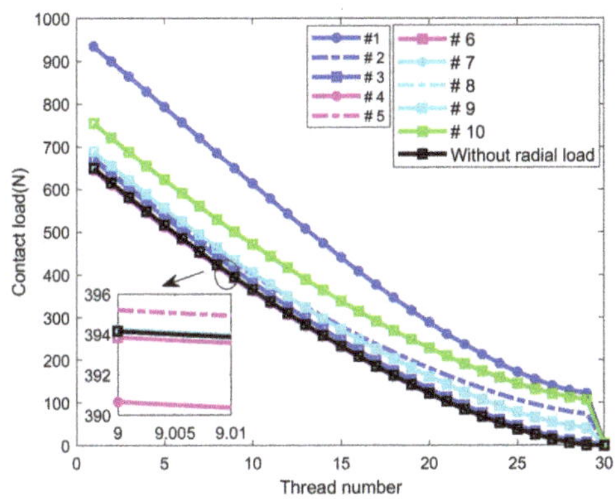

Figure 13. Contact load distribution in different threads of the robot system of case 2.

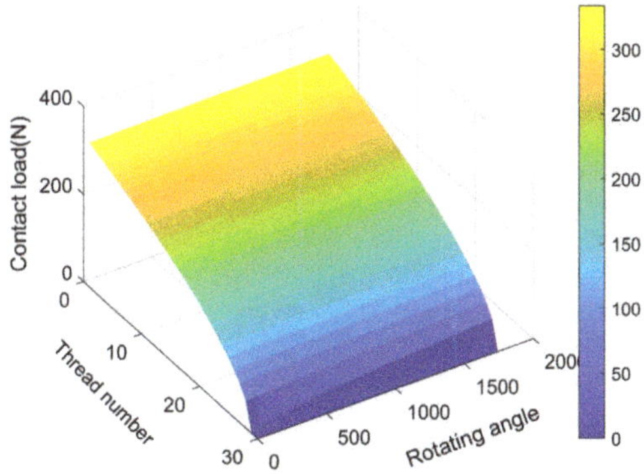

Figure 14. Contact load distribution cycling process without radial load of case 2.

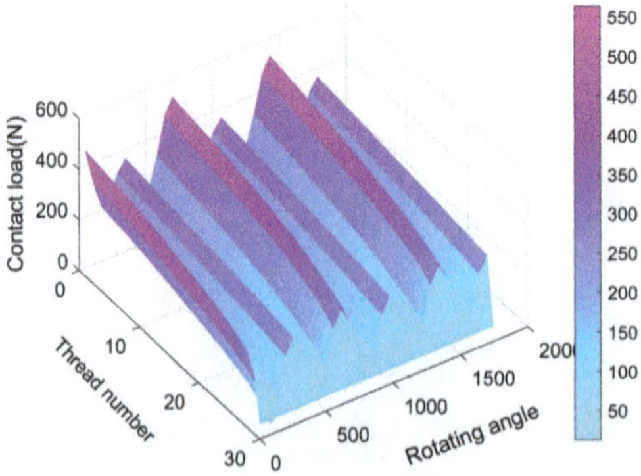

Figure 15. Contact load distribution cycling process with radial load of case 2.

Figures 13–15 show the load distribution behaviors of case 2 for the robot system, where the contact load behaviors are similar to the case 1. For example, the contact load decreases in the axial direction of each roller, and the mechanical behavior of the first and last rollers are the closest but there are still differences. This phenomenon can be understood as follows: on the one hand, the effect of axial loads on the rollers is relatively symmetrical; however, for the robot system, the downhole is not a structural environment, which means that the radial load is not an ideal symmetrical one for the rollers in the circumferential direction. Meanwhile, an angle exists in the first and last roller. As observed from Figures 14 and 15, one can see that the contact load of each thread maintains in the same without radial loads and varies when the external radial loads are applied to the PRSM. As the load distribution behavior of each roller is similar, in this note, we take the first roller as an example. The results show that the radial loads have significant effects on the load distribution, which means that the contact load may be different once the radial loads become inconstant.

5.2. *The Effects of Axial Load*

Figures 16 and 17 show the load distributions on the threads with different axial loads and the same radial loads. As each roller has a similar load distribution, we take one of them as an example in this note. From Figures 16 and 17, we can see that each load has a corresponding curve. When the PRSM works, the load distribution on the threads will increase on the condition that the axial load increases. On the other hand, the total load reduces along the thread number increasing direction, which has the same tendency in Figures 10 and 13. We can also see from the two figures that the curve has some partial overlap, which means that each thread could be subjected to the same load with the increasing axial load when the PRSM works in both cases of the downhole robot system. The results show us the contact load law of the roller, which gives us the idea that the load optimization could be made based on the results in the two figures, and this part can also extend to our future work.

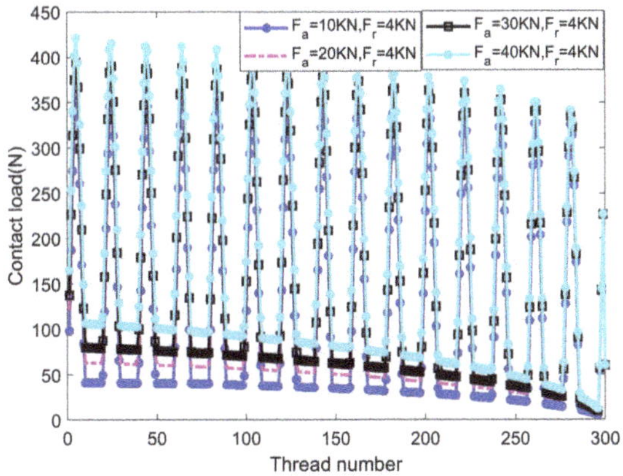

Figure 16. The effects of different axial loads in case 1.

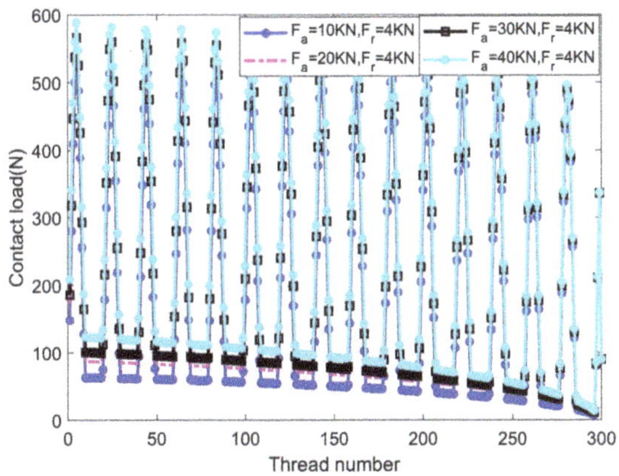

Figure 17. The effects of different axial loads in case 2.

5.3. The Effects of Radial Loads

Figures 18 and 19 express the effects of the radial loads. As depicted before, the downhole is a non-structural environment, and the radial loads are inevitable which have important influences on the PRSM. As expressed in the two figures, when the screw drags the tools to walk forward in the hole or the nut pushes the tools forward, with the growth of the thread number, the load distribution decreases when the PRSM is subjected to a fixed radial load. In Figures 18 and 19, one can see that when there are no radial loads, the total loads are distributed only on the first dozen threads, and the subsequent threads hardly share the loads, which has the same tendency in Figures 10 and 13. From the two figures, we can also see that the total load distribution decreases in the threads' growth direction, and only the radial loads are distributed on the last dozen threads. This means that the radial loads influence each thread, while the axial loads mainly affect the front threads of the PRSM.

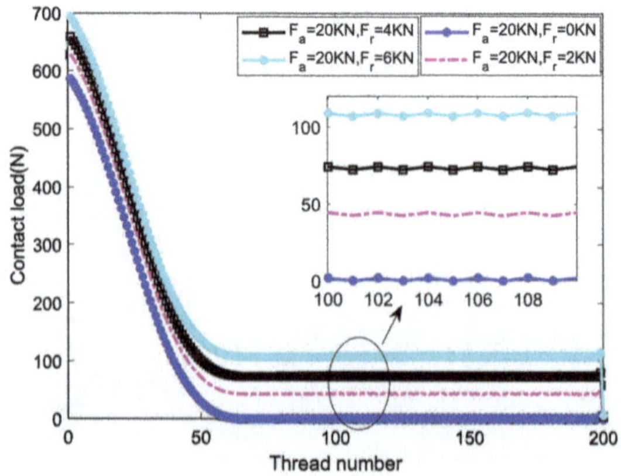

Figure 18. The effects of different radial loads in case 1.

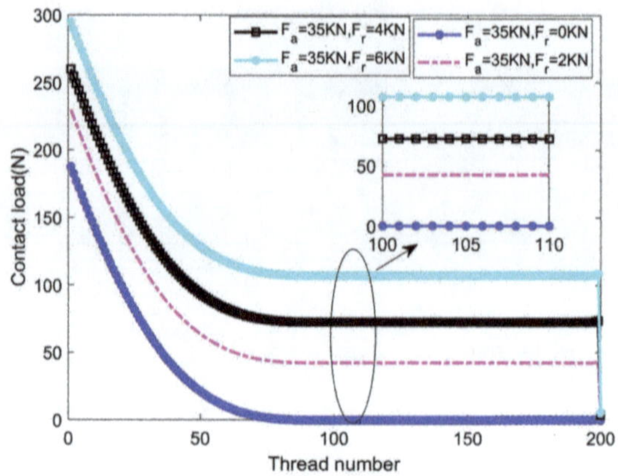

Figure 19. The effects of different radial loads in case 2.

5.4. The Results of the Fatigue Life

Figures 20 and 21 show the results of the fatigue life of the PRSM in case 1, to investigate the effects of external loads on the threads and explore the influences of rotation speeds on the fatigue life. In these cases, we set $w_s = 100$ rpm, $w_s = 300$ rpm, $w_s = 350$ rpm, $w_s = 500$ rpm, and $w_s = 700$ rpm. As shown in Figure 20, we assume that $F_r = 4$ KN, and $F_a = 10$ KN, $F_a = 20$ KN, $F_a = 30$ KN, $F_a = 42$ KN, and $F_a = 55$ KN. We can see from Figure 20 that the fatigue life decreases rapidly first and then decreases slowly with the increase of axial load. More specifically, when the axial loads increase from 10 KN to 20 KN, the fatigue life reduces from nearly 12,000 h to nearly 7000 h. Moreover, when the axial loads increase from 20 KN to 55 KN, the fatigue life reduces from nearly 7000 h to nearly 4000 h, which means that the fatigue life is more sensitive when the axial loads are relatively small and not so sensitive when the axial loads become large. This tendency can also be found in (30). Additionally, from Figure 20, one can also see that the fatigue life reduces sharply with the growth of rotation speed and the same external load. One can also see that the fatigue reduces faster when the axial loads are small and slower when the axial loads become bigger, which implies that the fatigue life is more sensitive under relatively small axial loads and low rotation speed changes.

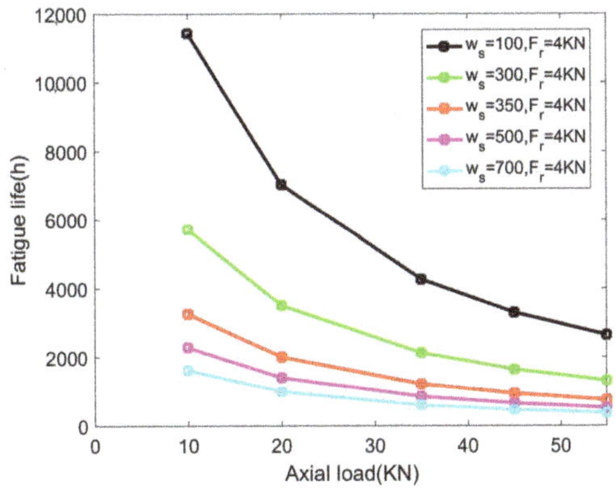

Figure 20. The fatigue life with different axial loads of case 1.

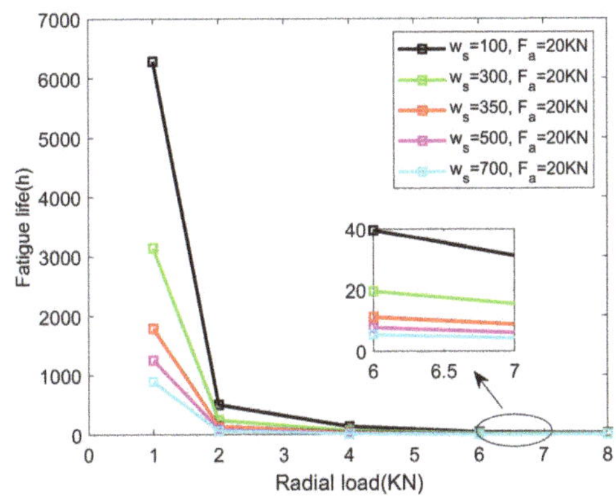

Figure 21. The fatigue life with different radial loads of case 1.

From Figure 21, one can observe that the fatigue life reduces sharply under the condition that the axial loads remain the same while the radial loads increase gradually. More precisely, the fatigue life reduces fast when the radial loads increase from 1 KN to 2 KN, and the fatigue life reduces slower when the radial loads turn 4 KN to 8 KN. Compared with the result in Figure 20, we can see that the fatigue life is more sensitive to the radial loads than the axial loads. Figures 22 and 23 show the results of the fatigue life of the PRSM in case 2. As each roller has a similar tendency, in this subsection, we take one of the rollers as an example. Similar to the results in case 1, the influences of different axial loads, radial loads, and rotation speeds are considered. The results show that the fatigue life reduces sharply at first and then slower, and the variation tendency is similar to case 1. The results in turn indicate that excessive radial force should be avoided as far as possible in the operations of the downhole robot system.

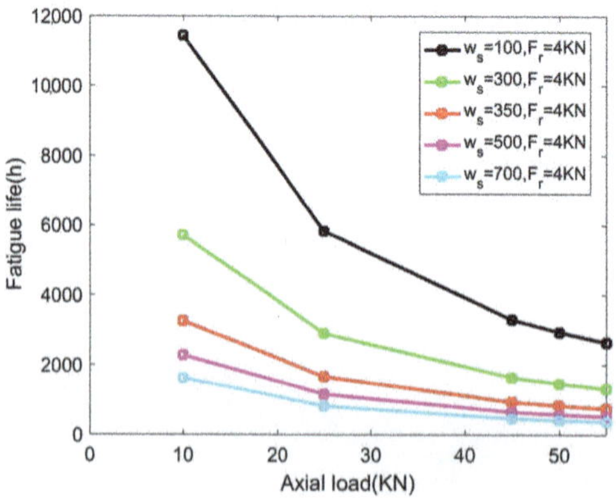

Figure 22. The fatigue life with different axial loads of case 2.

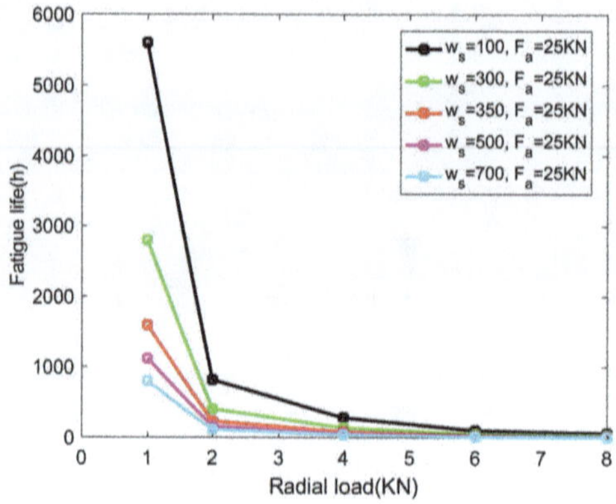

Figure 23. The fatigue life with different radial loads of case 2.

6. Conclusions

In this work, a new downhole robot actuator is presented by introducing the PRSM, and the mechanical behavior analysis is developed for the PRSM during the operations in the downhole environment. The torque and axial elastic deformation errors and the influence of the radial loads are considered when establishing the calculation load distribution model. The fatigue life analysis is also executed. The effects of the axial loads and radial loads on the contact load distribution and the fatigue life are simulated. The following conclusions can be obtained:

(1) The contact load decreases with the thread growth in the downhole robot system when it subjected to radial and axial loads. The first several threads bear most of the loads, and the last several threads take only a few loads. The axial loads are almost distributed on the first several threads while the effects of the radial loads are distributed on each thread.

(2) In the condition that the PRSM works in a different cycling process, when there is no radial load, the contact load stays the same in the cycling process, and different threads have their loads. When the PRSM is subjected to a radial load, the contact load varies periodically, and the mechanical behaviors have similar properties.

(3) For the condition in which the axial loads are different and that in which the axial loads are consistent, the contact load distribution varies and decreases along the axial direction. The tendency in two cases of the downhole robot system stay similar to some degree. Additionally, when the axial loads stay constant and the radial loads

increase, the contact load is distributed only on the first dozen threads while the subsequent threads hardly share the loads.

(4) The rotation speed and external load would affect the fatigue life of the PRSM of the robot system. The fatigue life reduces sharply under the condition that the axial loads, radial loads, and rotation speeds increase gradually. Compared with the axial load, the fatigue life is more sensitive to the radial load, which indicates that the radial load should be as small as possible, and the load distribution optimization could be investigated.

However, the numerical analyses of the stress and strain contours are not simulated in this study, which can help to verify the results of the designed actuator. Indeed, this remains one of the working directions of our study. Additionally, the dynamic analyses, lubrication, and load distribution optimization can also extend to our future work.

Author Contributions: X.D., Conceptualization, methodology, writing—original draft, writing—review and editing; H.Z., Formal analysis, data curation; Q.L., writing—review and editing, investigation, supervision, funding acquisition; Q.W., methodology, review and editing; X.W., formal analysis, resources. All authors have read and agreed to the published version of the manuscript.

Funding: This study was supported by the National Natural Science Foundation of China of FUNDER grant number No. U19A2097.

Institutional Review Board Statement: Not applicable.

Informed Consent Statement: Informed consent was obtained from all subjects involved in the study.

Data Availability Statement: There is no associate data in this study.

Conflicts of Interest: The authors declare no conflict of interest.

Nomenclatures

F_ζ	the radial load
F_a	the axial load
T	driving torque
$\lambda_{a1}, \lambda_{a2}$	the axial deformation caused by the tension load in case 1 and in case 2
λ_b	the axial deformation caused by the driving torque
λ_s	the total deformation error of the screw in the axial direction
ζ_r	the radial deformation
β, ϖ	the contact angle and helix angle
$\mathcal{G}_{ra}, \mathcal{G}_{rb}$	the stiffness coefficients in the contact threads of the screw to roller and roller to nut
$\Im_{S(n,j)}, \Im_{N(n,j)}$	the thread contact loads of the screw and nut
$\hbar_{SZ(n,j)}, \hbar_{ZN(n,j)}$	the contact deformation in the screw to roller interface and the roller to nut with elastic errors
ρ_a	the pitch
E_s, E_N	the elastic modulus of screw and nut
A_s, A_N	the cross sectional areas of screw and nut
$\upsilon_{SZ(n,j-1)}, \upsilon_{SZ(n,j)}$	two consecutive contact surfaces of the screw to roller
$\Re_{SZ(n,j-1)}, \Re_{SZ(n,j)}$	the elastic deformation errors of jth and $(j-1)th$ threads in the screw to roller t interface
$\upsilon_{ZN(n,j-1)}, \upsilon_{ZN(n,j)}$	two consecutive contact surfaces deformations of roller to nut
$\Re_{ZN(n,j-1)}, \Re_{ZN(n,j)}$	the elastic deformation errors of jth and $(j-1)th$ threads in the roller to nut interface
F_{an}	the axial load of the nth roller
N	the numbers of rollers
P_z	the numbers of threads

$\zeta_{sw(n,j)}, \zeta_{st(n,j)}$	the normal contact deformation in the condition without and with the elastic deformation errors
$\lambda_{(n,j)}$	the deformation coefficient
$\varepsilon_{(n,j)}$	the load coefficient
L_S	the fatigue lifetime
w_s	the rotating speed
d_s	the diameter of screw
d_r	the diameter of roller

References

1. Liu, Q.Y.; Zhao, J.G.; Zhu, H.Y.; Wang, G.R.; D.Mclennan, J. Review, classification and structural analysis of downhole robots. *Rob. Auton. Syst.* **2019**, *115*, 104–120. [CrossRef]
2. Zhao, J.G.; Liu, Q.Y.; Zhu, H.Y.; Wang, Z.D.; Liu, W.Q. Nonlinear dynamic model and characterization of coiled tubing drilling system based on drilling robot. *J. Vib. Eng. Technol.* **2019**, *9*, 541–561. [CrossRef]
3. Yue, Q.B.; Liu, J.B.; Zhang, L.G.; Zhang, Q. The posting-buckling analysis and evaluations of limit drilling length for coiled tubing in the sidetrack horizontal well. *J. Pet. Sci. Eng.* **2018**, *164*, 559–570.
4. Abdo, J.; Al-Shabibi, A.; Al-Sharji, H. Effects of tribological properties of water-based drilling fluids on buckling and lock-up length of coiled tubing in drilling operations. *Tribol. Int.* **2015**, *82*, 493–503. [CrossRef]
5. Shang, J.; Fang, d.; Luo,Z.; Wang, R.; Li,X.; Yang,J. Design and analysis of a hydraulic drive downhole traction in-pipe robot based on flexible support structure. *J. Mech. Eng. Sci.* **2021**, *235*, 18–27. [CrossRef]
6. Liu, Q.Y.; Zhao, J.G.; Zhu, H.Y.; Zhang, W. Mechanical model of drilling robot driven by the differential pressure of drilling fluid. *Arab. J. Sci. Eng.* **2019**, *44*, 1447–1458. [CrossRef]
7. Xu, Y.; Wan, Q.F.; Lu, G.; Zhang, B.S. Dynamic characteristics of the end-effector of a drilling robot for aviation. *Int. J. Mater. Sci. Appl.* **2018**, *7*, 192–198. [CrossRef]
8. Duthie, L.; Saeed, A.; Shaheen, S.; Saiood, H.; Moore, B.; Krueger, E. Design transformation of hydraulically powered coiled tubing tractors for matrix acidizing stimulations in extended reach carbonate reservoirs. In Proceedings of the Abu Dhabi International Petroleum Exhibition & Conference, Abu Dhabi, UAE, 13–16 November **2017**.
9. Yao, Q.; Zhang, M.C.; Ma, S.J. Structural design for planetary roller screw mechanism based on the developed contact modelling. *Tribol. Int.* **2022**, *171*, 107570. [CrossRef]
10. Zhou, G.W.; Zhang, Y.H.; Wang, Z.Z.; Pu, W. Analysis of transient mixed elastohydrodynamic lubrication in planetary roller screw mechanism. *Tribol. Int.* **2021**, *163*, 107158. [CrossRef]
11. Yao, Q.; Zhang, M.C.; Liu, Y.S.; Ma, S.J. Multi-objective optimization of planetary roller screw mechanism based on improved mathematical modelling. *Tribol. Int.* **2021**, *161*, 107095. [CrossRef]
12. Fu, X.J.; Liu, G.; Ma, S.J.; Tong, R.T.; C.Lim, T. A comprehensive contact analysis of planetary roller screw mechanism. *J. Mech. Des.* **2017**, *139*, 012302. [CrossRef]
13. Abevi, F.; Daidie, A.; Chaussumier, M. Sartor, M. Static load distribution and axial stiffness in a planetary roller screw mechanism. *J. Mech. Des.* **2016**, *138*, 012301. [CrossRef]
14. Ma, S.J.; Cai, W.; Wu, L.P.; Liu, G.; Peng, C. Modelling of transmission accuracy of a planetary roller screw mechanism considering errors and elastic deformations. *Mech. Mach. Theory* **2019**, *134*, 151–168. [CrossRef]
15. Auregan, G.; Fridrici, V.; Kapsa, P.; Rodrigues, F. Wear behavior of martensitic stainless steel in rolling-sliding contact for planetary roller screw mechanism: Study of the wc/c solution. *Tribol. Online* **2016**, *11*, 209–217.
16. Yao, Q.; Liu, Y.S.; Zhang, M.C.; Liu, G.; Ma, S.J. Investigation on the uncertain factors of the elastic-plastic contact characteristics of the planetary roller screw mechanism. *Proc. Inst. Mech. Eng. Part C J. Mech. Eng. Sci.* **2018**, *233*, 1795–1806. [CrossRef]
17. Ma, S.J.; Wu, L.P.; Fu, X.J.; Li, Y.J.; Liu, G. Modelling of static contact with friction of threaded surfaces in a planetary roller screw mechanism. *Mech. Mach. Theory* **2019**, *139*, 212—236. [CrossRef]
18. Fu,X.; Li,X.; Ma,S.; Gerada, D.; Liu, G.; Gerada, C. A multi-roller static model of the planetary roller screw mechanism considering load sharing. *Tribol. Int.* **2022**, *173*, 107648. [CrossRef]
19. Zhang, W.; Liu, G.; Ma, S.; Tong, R. Load distribution over threads of planetary roller screw mechanism with pitch deviation. *Proc. IME C J. Mech. Eng. Sci.* **2019**, *233*, 4653–4666. [CrossRef]
20. Mamaev, I.M.; Morozov, V.V.; Fedotov, O.V.; Filimonov, V.N. Harmonic analysis of the kinematic error in a planetary roller screw. *Russ. Eng. Res.* **2016**, *36*, 515–519. [CrossRef]
21. Auregan, G.; Fridrici, V.; Kapsa, P.; Rodrigues, F. Experimental simulation of rolling sliding contact for application to planetary roller screw mechanism. *Wear* **2015**, *332–333*, 1176–1184. [CrossRef]
22. Lepagneul, J.; Tadrist, L.; Sprauel, J.-M.; Linares, J.-M. Fatigue lifespan of a planetary roller-screw mechanism. *Mech. Mach. Theory* **2022**, *172*, 104769. [CrossRef]
23. Du, X.; Chen, B.K.; Zheng, Z.D. Investigation on mechanical behavior of planetary roller screw mechanism with the effects of external loads and machining errors. *Tribol. Int.* **2021**, *154*, 106689. [CrossRef]

24. Jones, M.H.; Velinsky, S.A. Stiffness of the roller screw mechanism by the direct method. *Mech. Based Des. Struct. Mach* **2013**, *42*,17–34. [CrossRef]
25. Zhang, W.; Geng, L.; Tong, R.; Ma, S. Load distribution of planetary roller screw mechanism and its improvement approach. *Proc. Inst. Mech. Eng. Part C J. Mech. Eng. Sci.* **2016**, *230*,3304–3318. [CrossRef]
26. Zhen, N.; An, Q. Analysis of stress and fatigue life of ball screw with considering the dimension errors of balls. *Int. J. Mech. Sci.* **2018**, *137*, 68–76. [CrossRef]
27. Xie, Z.; Xue, Q.; Wu, J.; Gu, L.; Wang, L.; Song, B. Mixed-lubrication analysis of planetary roller screw. *Tribol. Int.* **2019**, *140*, 105883. [CrossRef]

processes
MDPI

Article

Numerical Simulation and Field Test Research on Vibration Reduction of PDC Cutting of Pebbled Sandstone under Composite Impact Load

Heng Zhang [1,2,*], Hongjian Ni [2,3,*], Henglin Yang [1,2], Li Fu [1,2], Yuan Wang [1,2], Shubin Liu [2,3], Bin Huang [2,3], Zixin Wang [1,2] and Gang Chen [1,2]

1 CNPC Engineering Technology R & D Company Limited, Beijing 102206, China
2 National Engineering Research Center of Oil & Gas Drilling and Completion Technology, Beijing 102206, China
3 School of Petroleum Engineering, China University of Petroleum (East China), Qingdao 266580, China
* Correspondence: zhhengdr@cnpc.com.cn (H.Z.); nihj@upc.edu.cn (H.N.)

Abstract: Downhole vibrations caused by rock breaking when drilling through pebbled sandstone formations negatively affect the rate of penetration (ROP) and the safety of downhole tools. Therefore, it is of great significance to study the cutting characteristics of pebbled sandstone and find a method of reducing the drilling vibrations of pebbled sandstone formations. Based on the DEM (discrete element method), a simulation model of pebbled sandstone considering the random filling of high-strength gravels was established by using the random polygon distribution method. The influence of gravel content on the strength parameters and the breaking state of the pebbled sandstone samples was analyzed. Additionally, a DEM model of PDC cutting rocks loaded by a spring–mass system was established, and the Stribeck effect of contact friction between the PDC cutter and the rock was analyzed. The periodic vibration and the stick–slip phenomenon of the cutting system during the drilling process were presented by this model. The model was employed to simulate and explore the influence of composite impact load on stick–slip vibration during PDC cutting of pebbled sandstone. The simulation results showed that the composite impact load had a more obvious effect on mitigating the vibration of PDC cutting of pebbled sandstone under the condition of a higher horizontal impact amplitude coefficient (q_h = 40%). Based on the simulation results, a composite impactor with a large impact angle α = 70° was selected to conduct the field tests in the pebbled sandstone formation of Well T1. The results showed that, compared to conventional drilling, the average WOB (weight on bit) of the section drilled with the composite impactor decreased by 57.13%, the standard deviation of the WOB decreased by 57.29%, and the average ROP increased by 98.31%. The employing of composite impactors in pebbled sandstone formations can significantly reduce drilling vibration, improve ROP, and protect bits and downhole instruments.

Keywords: discrete element method; pebbled sandstone; impact drilling; composite impactor; PDC cutter

Citation: Zhang, H.; Ni, H.; Yang, H.; Fu, L.; Wang, Y.; Liu, S.; Huang, B.; Wang, Z.; Chen, G. Numerical Simulation and Field Test Research on Vibration Reduction of PDC Cutting of Pebbled Sandstone under Composite Impact Load. *Processes* **2023**, *11*, 671. https://doi.org/10.3390/pr11030671

Academic Editors: Tianshou Ma and Yuqiang Xu

Received: 16 January 2023
Revised: 16 February 2023
Accepted: 21 February 2023
Published: 22 February 2023

1. Introduction

Increasing drilling rate of penetration (ROP) is the primary way to save drilling investment and increase the benefit of oil and gas resource development. However, with long-term, large-scale exploration and development of oil and gas resources, the engineering environment of newly discovered oil fields is very unfriendly to drilling, which makes it increasingly difficult to increase the ROP [1]. For example, more and more gravel-bearing formations are being drilled, but the economic and time costs of drilling operations in these formations are generally higher due to highly heterogeneous rocks and strong and abrasive gravel particles. Drilling into formations bearing high-strength gravel particles is easy to cause severe vibrations of the bit, including stick–slip and bit bouncing, which then lead to serious damage, fracture, or even collapse of the bit cutters. The working time of a bit is very short, and the trip time will significantly increase. Besides, when drilling into these formations, it is common to weaken the drilling parameters to reduce downhole vibrations, which further restricts ROP improvement. Therefore, it is crucial to study the characteristics and performance of cutting highly heterogeneous gravel-bearing rocks to improve ROP and reduce drilling risks related to vibrations.

Polycrystalline Diamond Compact (PDC) bits have been used in rotary drilling since about the 1900s and have become the most commonly used type of bits in oil and gas drilling [2]. In China and the United States, more than 85% of the drilled footage in recent years used PDC bits. PDC cutters are in direct contact with formation rocks during drilling, and the study of PDC rock-breaking characteristics has the most direct guiding role in realizing efficient rock breaking. Aiming at the problem of PDC bit rock breaking, researchers have carried out a lot of theoretical and numerical simulation research. In the early research, the relation between the load state of a PDC bit and the rock-breaking volume was examined, aiming to predict the ROP and the wear of the bit [3–7]. Warren and Sinor [8] (1989) proposed a drag-bit performance model to predict the removed volume of rock penetrated by the designed layout of bit cutters. The model predictions compared well to laboratory drilling tests for four radically different bit designs when used on four different rocks. The relationship model between bit force and ROP is widely adopted as the bit boundary conditions in the process of drill-string mechanical modeling [9,10]. With the development of computer technology and the demand for rock-breaking enhancement of PDC bits, numerical simulation has become one of the important ways to analyze the rock-breaking performance of the bit. In the beginning, rock-breaking simulations of tools were not included in the cutting behavior, and the focuses were on the process of the tools when invading a rock under a static load. Chiaia [11] (2001) built a lattice model to discuss the process of indentation of brittle and quasi-brittle materials and pointed out that increasing the size of the indenters improves the local fracture mechanisms. The number of indenters also has a significant effect on the rock-breaking process and the generation of cracks in the rock. Simultaneous loading of multiple indenters with an appropriate line spacing seems to provide a possibility of forming larger rock chips, controlling the direction of subsurface cracks, and consuming a minimum total specific energy [12]. Later, the rock-breaking behavior of PDC cutters was studied by using the finite element method (FEM) and discrete element method (DEM). The effects of rock type and material model on PDC cutting force and cutting depth were studied to verify the simulation method and obtain reliable modeling parameters [13–16]. Jaime [17,18] (2011) employed LS-DYNA to simulate the rock-cutting process of PDC cutters; introduced the modeling method, material model selection, and parameter modification process in detail; and analyzed the sensitivity of rock-breaking behavior to loading parameters and material parameters. In addition, the effects of cutter velocity, friction coefficient, and rake angle on rock cutting were investigated to explore the rock-breaking mechanism of the bit and to optimize the bit design [19]. These simulations demonstrated the importance of the explicit finite element model for simulating the rock cutting and fragmentation process.

With the development of bit machining and design technology, the techniques of percussion drilling and non-plane cutter bit are widely used at present, and corresponding numerical simulation methods have been explored. In addition, the temperature and pressure of a wellbore have a significant influence on rock failure [20–22]. Therefore, various factors, such as bottom hole pressure and temperature influencing rock-breaking performance, are considered [23]. Guarin et al. [24] (1949) introduced the first case, in the oil field history, of rotary drilling with an impact tool that accomplished extended intervals of formation for sustained periods of time. Melamed et al. [25] summarized several early percussion drilling tools. Subsequently, numerous studies and experiments were conducted to explore impact drilling techniques. The effects of cutter geometry parameters, cutting angle, impact load, and cutting speed on crack propagation, debris formation, and damage evolution of rock were extensively studied [26,27]. Xiong et al. [28]

(2021) investigated mixed tool cutting of granite with stinger PDC cutters and conventional PDC cutters, and they pointed out that the hybrid PDC bits that combine conventional and stinger PDC cutters have significant advantages in drilling hard, interbedded, and highly abrasive rocks. Dong and Chen [29] (2018) conducted a 3D full-scale PDC bit model to study the dynamic damage characteristics of anisotropic shale during impact rotary drilling. The aim of these studies is to explore the mechanism of PDC cutting rock under an axial impact load or a torque impact load and find realistic ways to improve the efficiency of rock drilling [30]. In recent years, improving the rock-breaking efficiency of PDC bits and reducing bit vibrations have been considered equally important in dealing with complex downhole drilling conditions [31,32]. Additionally, a compound impact load is considered to be an effective method to achieve this goal [33].

The heterogeneity of rock is one of the main causes of bit vibration and significantly impacts the strength and fracture characteristics of rock [34]. Therefore, it is necessary to consider the heterogeneity of rock when simulating rock cutting under a compound impact load [34]. The possibility of simulating heterogeneous rock failure using the finite element method (FEM) or the discrete element method (DEM) has been confirmed [35]. It is more convenient to establish a DEM heterogeneous model by changing the bond strength between the contact particles [36]. However, in the aforementioned studies of rock-breaking mechanisms, the investigation of gravel-bearing rock, which has stronger heterogeneity caused by the huge strength difference between the matrix and the gravel broken by a PDC cutter, under a compound impact load is still in the exploratory stage. Furthermore, the simulation method using a constant boundary of bit loading or cutting depth to characterize such cutters has obvious defects. The stick–slip phenomenon and continuous cutting in the process of rock breaking cannot be simulated at the same time.

In this paper, a heterogeneous rock modeling method based on the discrete element theory is proposed. In this method, a random polygonal aggregate is used to characterize the gravel particles inside the rock. The fracture characteristics of a PDC cutting heterogeneous gravel-bearing rock under the action of a compound impact load are simulated and studied. On this basis, field tests to reduce the rock-breaking vibration of gravel-bearing formation with a composite impactor are carried out. The method and conclusions of this paper provide theoretical basis and technical reference for ROP improvement and safe drilling in gravel-bearing formations.

2. Simulation Method

The DEM was first introduced into rock mechanics by Cundall, and a particle assembly program based on the DEM was proposed in 1979 [37]. It is based on the idea that the object researched is composed of a certain number of arbitrary particles or particle clusters with mass and volume in the particle assembly procedure. The particles are specified as disks with unit thickness in the two-dimensional (2D) model, whereas the particles are assumed to be spherical in the three-dimensional (3D) model. The particles interact with each other through the prescribed contact model and follow Newton's law and Hooke's law. The macro performance of the object is described by calculating the motion and stress state of each particle. It has high operability in rock microstructure modeling. It is simpler and more intuitive than the finite element method during the mechanical behavior simulation of discontinuous rock mass. Additionally, it has been widely used to solve rock mechanics problems [38], as shown in Figure 1.

Using a particle flow program is an essential method of solving practical problems using the discrete element theory. It is assumed that particles (clusters) are rigid bodies with a certain mass and interface, and each particle (cluster) can translate or rotate independently. Additionally, the particles (clusters) are allowed to overlap with one another in a relatively small area of the interface contact. There is a finite stiffness bonding at the contact, which can bear a specific load. The bonding will be removed when the load reaches the bonding strength. The interaction law between the particles (clusters) and between the particles (clusters) and the boundary is specified in the contact model, which is used to calculate the relative motion and the load transfer between the particles (clusters) [39].

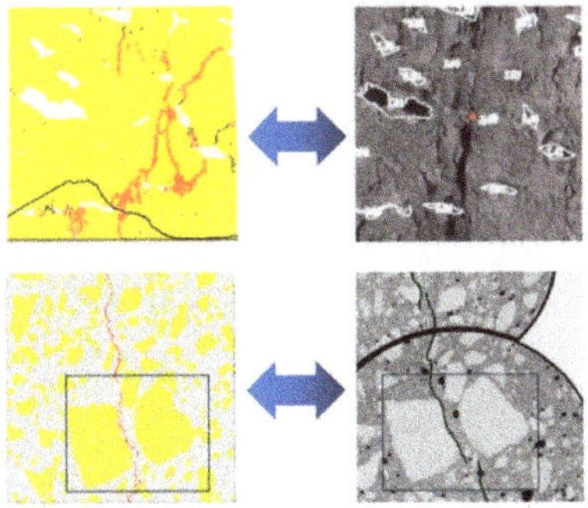

Figure 1. The assembly modeling of rock pore, fracture, and rubble using the DEM.

The contact between particles in a 2D DEM model is represented in Figure 2. The generated contact particles have a definite radius R, a centroid vector $\mathbf{x}$, a translational velocity $\dot{\mathbf{x}}$, and an angular velocity relative to the centroid $\boldsymbol{\omega}$. The position vector $\mathbf{x}_c$, the normal vector $\hat{\mathbf{n}}_c$, the relative translational velocity $\dot{\boldsymbol{\delta}}$, and the relative angular velocity $\dot{\boldsymbol{\theta}}$ of the contact surface can be calculated using Equations (1)–(4), respectively:

$$\mathbf{x}_c = \mathbf{x}^{(A)} + \left(R^{(A)} + \frac{g_c}{2}\right)\hat{\mathbf{n}}_c \tag{1}$$

$$\hat{\mathbf{n}}_c = \frac{\mathbf{x}^{(B)} - \mathbf{x}^{(A)}}{d} \tag{2}$$

$$\dot{\boldsymbol{\delta}} = \dot{\mathbf{x}}_c{}^{(B)} - \dot{\mathbf{x}}_c{}^{(A)} \tag{3}$$

$$\dot{\boldsymbol{\theta}} = \boldsymbol{\omega}^{(B)} - \boldsymbol{\omega}^{(A)} \tag{4}$$

where g_c is the minimum gap length between the contact disks, with $g_c = d - \left(R^{(A)} + R^{(B)}\right)$; d is the center distance between two contact parts, with $d = \|\mathbf{x}^{(B)} - \mathbf{x}^{(A)}\|$; and $\dot{\mathbf{x}}_c$ is the translational velocity at the contact, with $\dot{\mathbf{x}}_c = \dot{\mathbf{x}} + \boldsymbol{\omega}(\mathbf{x}_c - \mathbf{x})$.

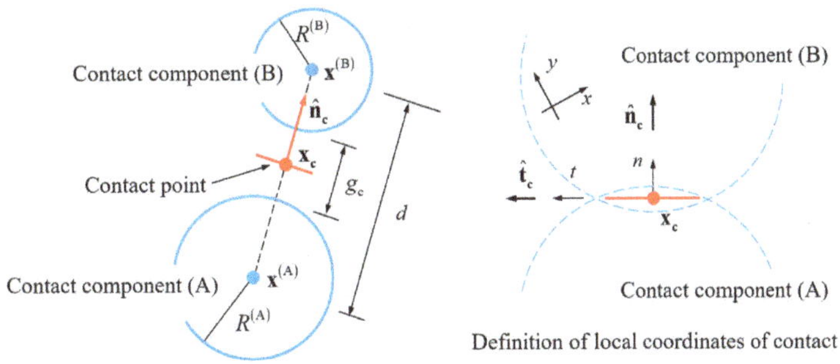

Figure 2. Diagram for the contact detected in a 2D particle flow model.

The normal unit vector of $n \times t$ surface is defined as $\hat{\mathbf{k}} = \hat{\mathbf{n}}_c \times \hat{\mathbf{t}}_c$ for convenient calculation. Equations (5) and (6) can be obtained by decomposing the relative translational velocity $\dot{\boldsymbol{\delta}}$ and the relative angular velocity $\dot{\boldsymbol{\theta}}$ along $\hat{\mathbf{n}}_c$ and $\hat{\mathbf{t}}_c$, respectively, as shown in Figure 3.

$$\dot{\boldsymbol{\delta}} = \dot{\boldsymbol{\delta}}_n + \dot{\boldsymbol{\delta}}_t \quad (5)$$

$$\dot{\boldsymbol{\theta}} = \dot{\boldsymbol{\theta}}_{bk} \quad (6)$$

where $\dot{\boldsymbol{\delta}}_n$ is the relative translational velocity of the contact along $\hat{\mathbf{n}}_c$, with $\dot{\boldsymbol{\delta}}_n = \dot{\delta}_n \cdot \hat{\mathbf{n}}_c$; $\dot{\boldsymbol{\delta}}_t$ is the relative translational velocity of the contact along $\hat{\mathbf{t}}_c$, with $\dot{\boldsymbol{\delta}}_t = \dot{\boldsymbol{\delta}} - \dot{\boldsymbol{\delta}}_n$; and $\dot{\boldsymbol{\theta}}_{bk}$ is the angular velocity around $\hat{\mathbf{k}}$ at the contact.

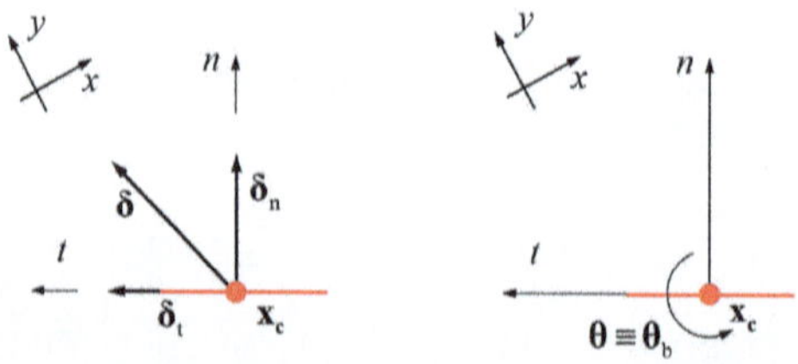

Figure 3. Kinematics of contact with relative rotation and motion of piece surfaces.

Assuming that the time step is Δt, the relative displacement increment $\Delta\boldsymbol{\delta}$ and the relative rotation angle increment $\Delta\boldsymbol{\theta}$ of the contact can be expressed using Equations (7) and (8), respectively:

$$\Delta\boldsymbol{\delta} = \Delta\delta_n \hat{\mathbf{n}}_c + \Delta\delta_t \hat{\mathbf{t}}_c \quad (7)$$

$$\Delta\boldsymbol{\theta} = \Delta\boldsymbol{\theta}_b = \dot{\boldsymbol{\theta}}_{bk}\Delta t \quad (8)$$

where $\Delta\delta_n$ is the displacement increment along the coordinate axis $\hat{\mathbf{n}}_c$, with $\Delta\delta_n = \dot{\delta}_n\Delta t$, and $\Delta\delta_t$ is the displacement increment along the coordinate axis $\hat{\mathbf{t}}_c$, with $\Delta\delta_t = \dot{\delta}_t\Delta t$.

The contact kinematics law obtained from Equations (1)–(8) is the basis for judging the contact state and calculating the load transfer. It is assumed that the contact components contact each other and activate the contact model, when the traverse contact gap, $g_s \leq 0$, is based on the reference contact distance of g_r, as shown in Figure 4. The contact in the model can be regarded as a combination of linear contact (the contacts only transfer compression, but not tension and torque) and adhesive contact (the contacts are rigid connection) when the failure of cemented materials (such as rock) are calculated.

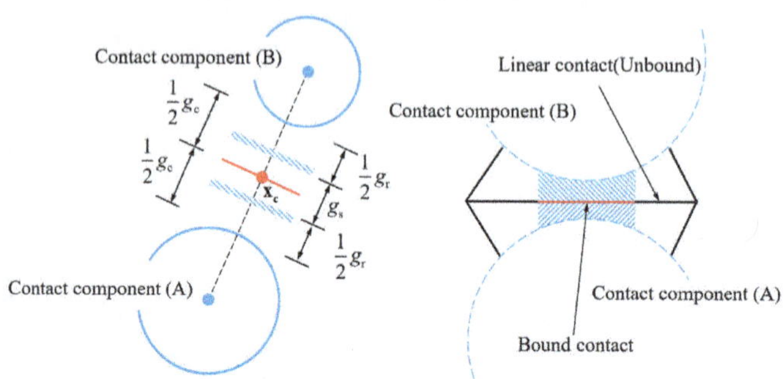

Figure 4. Sketch of the bound contact model.

The linear contact model and the parallel-bond model are shown in Figure 5 [40]. The parallel-bond force and torque calculation formula is expressed in Equation (9):

$$\begin{cases} \mathbf{F}_c = \mathbf{F}^l + \mathbf{F}^d + \overline{\mathbf{F}} \\ \mathbf{M}_c = \overline{\mathbf{M}} \end{cases} \quad (9)$$

where $\mathbf{F}_c$ and $\mathbf{M}_c$ are the contact force and contact torque, respectively; $\mathbf{F}^l$ is the linear contact force; $\mathbf{F}^d$ is the dashpot force; $\overline{\mathbf{F}}$ is the parallel-bond force; and $\overline{\mathbf{M}}$ is the parallel-bond moment.

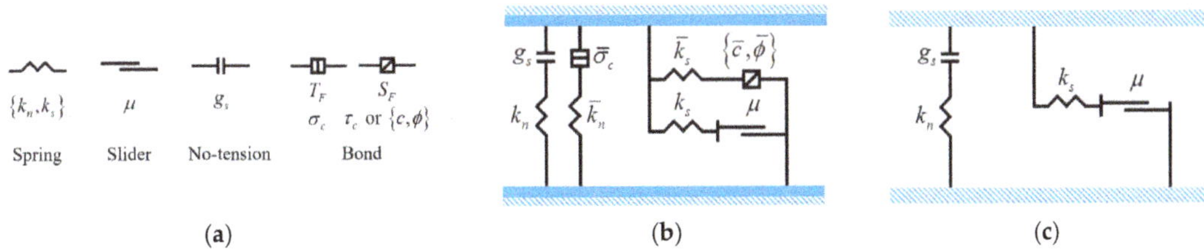

Figure 5. Contact rheological components (**a**) of the linear parallel-bond model (**b**) and the linear contact model (**c**).

According to the definition of the parallel-bond model, the parallel-bond stress $\overline{\sigma}$ (tensile stress if $\overline{\sigma} > 0$) is calculated based on Equation (10). The bond will be broken when $\overline{\sigma} > \overline{\sigma}_c$, and $\overline{\sigma}_c$ is defined as the bond strength. In this case, $\overline{\mathbf{F}} = 0$, $\overline{\mathbf{M}} = 0$, and the parallel-bond contact model changes to the linear contact model.

$$\overline{\sigma} = \frac{\overline{F}_n}{\overline{A}} + \overline{\beta} \frac{\|\overline{\mathbf{M}}_b\| \overline{R}}{\overline{I}} \quad (10)$$

Here, $\overline{F}_n$ is the normal parallel-bond force; $\overline{A}$ is bonding cross-sectional area, with $\overline{A} = 2\overline{R}t$, $t = 1$, and $\overline{R} = \min(R^{(A)}, R^{(B)})$ in the 2D model; $\overline{\beta}$ is the contribution coefficient of torque to contact stress, with $\overline{\beta} \in [0, 1]$; $\overline{\mathbf{M}}_b$ is the parallel-bond moment; and $\overline{I}$ is the moment of inertia, with $\overline{I} = \frac{2}{3}t\overline{R}^3$.

3. Modeling of Pebbled Sandstone

3.1. Modeling Steps

The inhomogeneity of pebbled sandstone is mainly manifested in the material inhomogeneity between the sandstone matrix and the filled gravel. The matrix area and the gravel area are separated by generating a stochastic regular polygon closed boundary. These polygons are set as the gravel material and are surrounded by sandstone material. The modeling process is shown in Figure 6, which includes the following steps:

(1) Establish the polygon-generating area according to the size of the rock and generate points $p_0^{(i)}\left(x_0^{(i)}, y_0^{(i)}\right)$ randomly in this area.

(2) Take a random radius $r^{(i)}$ within the set granularity range $(r_{\min}, r_{\max})$ to generate a circle $\Theta^{(i)}$ with $p_0^{(i)}$ as the center and $r^{(i)}$ as the radius.

(3) Judge whether $\Theta^{(i)}$ intersects with the circumscribed circle $\Theta^{(j)}(j = 1, 2, \cdots, i-1)$ of all the polygons generated, and return to Step 1 in the case of intersection.

(4) Take point $p_1^{(1)}\left(x_1^{(1)}, y_1^{(1)}\right)$ on the circle to generate an arbitrary regular $n^{(i)}$ polygon with circle $\Theta^{(i)}$ as the circumscribed circle and $p_1^{(i)}$ as the vertex. The included angle

between the line between point $p_0^{(i)}$ and point $p_1^{(i)}$ and the x-axis square is a random angle $\alpha^{(i)}$.

(5) Calculate the sum of the areas of all polygons currently generated and the volume ratio of the rock sample, $P_g^{(i)} = \frac{\sum_{j=1}^{i} A_p{}^{(j)}}{A} \times 100\%$, to characterize the gravel content. If $P_g^{(i)} \geq P_{set}$, the cycle ends, where A_p is the polygon area, A is the rock model area, and P_{set} is the gravel content.

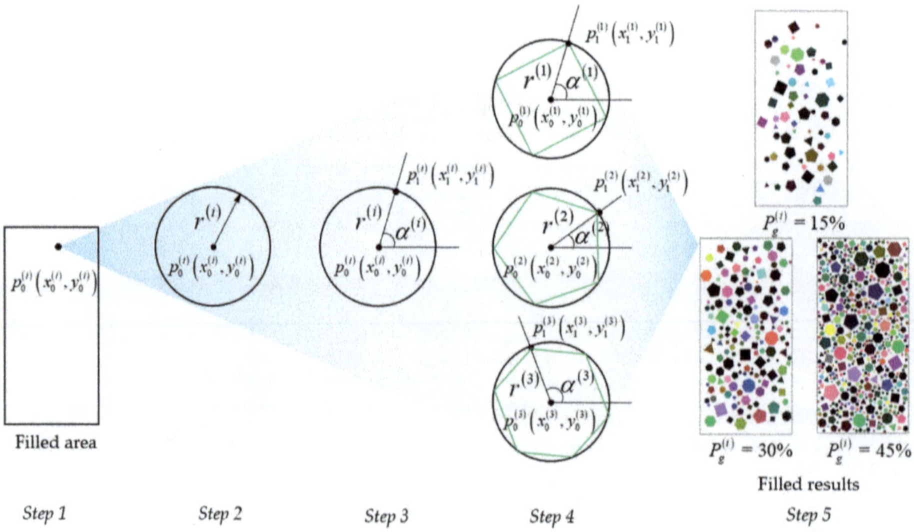

Figure 6. Generation process of pebbled sandstone samples.

In the modeling program, the coordinates of the polygon center point $\left(x_0^{(i)}, y_0^{(i)}\right)$, the radius $r^{(i)}$ of the circumscribed circle, the polygon rotation angle $\alpha^{(i)}$, and the number $n^{(i)}$ of polygon sides can be adjusted by establishing different random distribution models to choose the distribution law of the generated random numbers.

3.2. Comparison and Validation

In the process of rock-breaking simulation using the particle flow code, the mechanical properties of particles and bonds are characterized by micro-mechanical parameters. The micro-mechanical parameters adopted in the simulation are not directly related to the macro-mechanical parameters of the natural rock, but they can be determined by the simulation tests. The simulation tests adopt the trial-and-error method to compare the results of the simulated macro-mechanical parameters of the rock obtained from each test with the actual rock sample until the error is acceptable. At this time, the adopted micro-mechanical parameters are the DEM modeling parameters of the actual rock sample. In the calibration process of the micro-mechanical parameters of the discrete element rock model, the filling gravel of the rock samples refers to the test data of the actual rock samples in the Tabei area, and the material parameters of the pebbled sandstone matrix refer to the pure sandstone rock samples [41]. The macro-mechanical parameters of the rocks are shown in Table 1. The micro-mechanical parameters of the DEM rock models are calibrated using the UCS (uniaxial compression strength) simulation tests, and the results are shown in Table 2.

Table 1. Comparison of mechanical parameters between the rock samples and the DEM rock model.

Rock Samples	Compressive Strength/MPa	Elastic Modulus/GPa	Poisson's Ratio
Sandstone	33.10	8.25	0.33
Sandstone model	32.97	8.24	0.32
Gravel	203.45	70.47	0.177
Gravel model	206.95	70.41	0.177

Table 2. Material parameters of the DEM pebbled sandstone models.

Rock Composition	Rock Matrix	Gravel
Particle density/(kg/cm^3)	2000	2650
Effective modulus/GPa	4.33	33.82
Normal-to-shear stiffness ratio	1.50	1.58
Bonding effective modulus/GPa	4.33	33.82
Bond normal-to-shear stiffness ratio	1.50	1.58
Tensile strength/MPa	9.98	67.20
Cohesion/MPa	11.96	78.33
Friction angle/°	32	25
Friction coefficient	0.65	0.65

In order to obtain the mechanical properties of gravelly sandstone with different gravel content, the uniaxial compression and crushing process of pebbled sandstone with 5–45% P_g is simulated. The physical model is established by using the stochastic regular polygon filling method. Some rock samples after particle filling are shown in Figure 7. The light parts in the figure are pure sandstone material, and the dark parts are filled gravel material. The rock sample used in the simulation is a standard core column of 50 mm × 100 mm in size. The minimum radius of the modeled particles is 0.3 mm, and the maximum particle size is 0.5 mm. A total of 8753 particles are generated.

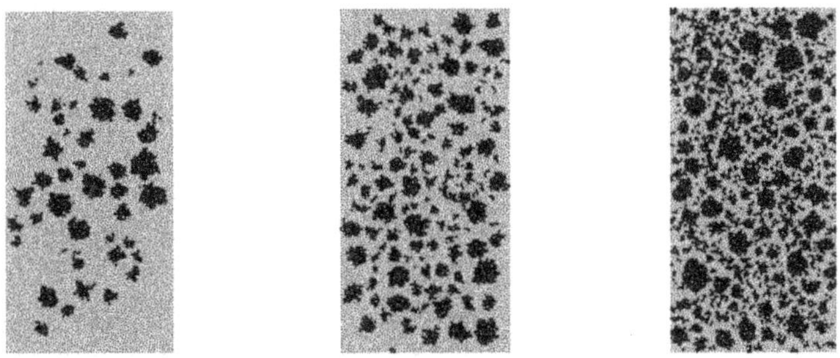

$P_g = 15\%$ $P_g = 30\%$ $P_g = 45\%$

Figure 7. Pebbly sandstone samples with different gravel contents.

The uniaxial compression stress–strain curves of the rock samples with different gravel contents are shown in Figure 8a, and the change curves of the macro-mechanical parameters are shown in Figure 8b. The simulation results show that the elastic modulus of pure sandstone is the lowest. With an increase in gravel content in the rock models, the elastic modulus of the rock samples increases linearly, while the peak strength of the rock samples decreases slowly at first and then increases gradually. The UCS test results conducted on the pebbled sandstone model established in this paper show that the gravel content in the rock increases from 0 to 40%, the elastic modulus of the rock increases by 87.86%, and the peak strength of the rock increases by 12.41%, indicating that a rise in the gravel content has a greater impact on the elastic modulus of the rock samples compared to the peak strength.

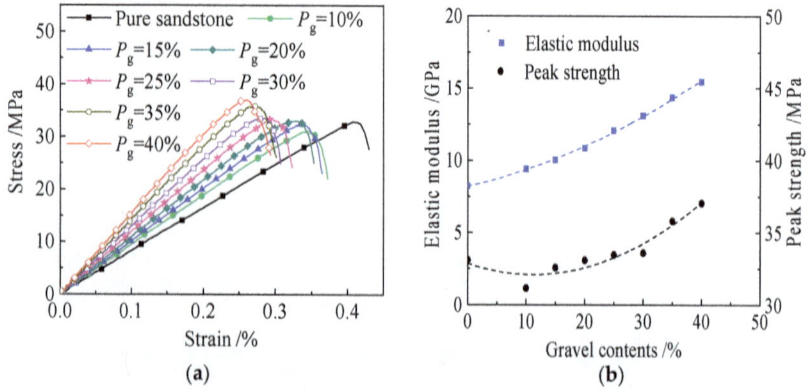

Figure 8. Simulated UCS tests results of DEM pebbled sandstone samples, (**a**) strain–stress plot and (**b**) trend analysis of rock Elastic Modulus and peak strength.

The fracture characteristics of the simulated rock samples are shown in Figure 9. The gray part and the black part in the figure represent the split rock fragments; the blue part represents the gravels; and the broken red lines indicate the crack distribution. These characteristics are in good agreement with the experimental results and the numerical simulation results of previous researchers [34,42]. The figure shows that the distribution of gravel has a significant impact on the formation and propagation of cracks. Cracks pass through gravel, bypass gravel, and cut off after encountering gravel. In addition, the elastic modulus of the rock is greatly increased due to gravel packing, which leads to an increase in rock brittleness. The number of cracks in the final wreck tends to increase.

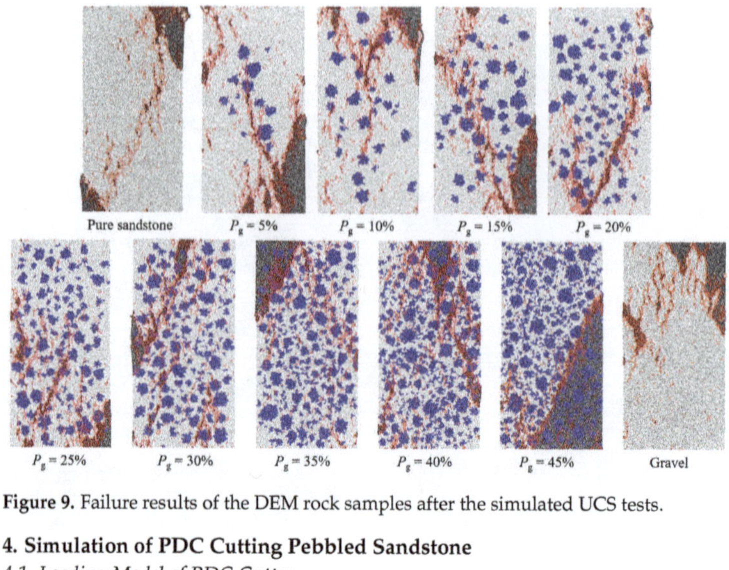

Figure 9. Failure results of the DEM rock samples after the simulated UCS tests.

4. Simulation of PDC Cutting Pebbled Sandstone

4.1. Loading Model of PDC Cutter

The rotary table or top drive transmits torque to the bit through the drill pipe and drill collar while drilling. At the same time, part of the gravity of the drill collar also acts on the bit so that the bit can cut and break rocks. In the previous modeling process of rock breaking using PDC single-tooth cutting, the cutting speed of the PDC cutter is set to be uniform, and the cutting depth of the cutter in the cutting process is fixed. However, for PDC bits, the penetration of the cutter in the process of rock breaking is mainly controlled by the vertical load acting on them. In the process of movement, due to the great slenderness ratio of the drill string, the torsional stiffness of the cutter on the cutting plane becomes smaller, and the cutting speed is not fixed, which will cause fluctuations and stick–slip.

The force state of the PDC bit and the cutter during drilling is shown in Figure 10. When the drill bit rotation angle $\Delta\Phi$ is small enough, the movement of the cutter can be approximately regarded as a linear movement. In the rock-breaking process, it receives the support force F_n from the rock, the force F_{wob} of weight on the bit, the friction force F_f, and the cutting-resistance forces F_{cx} and F_{cy} of the rock to the cutter. F_t refers to the thrust formed by the top drive torque transmitted to the bit along the drill string when the top drive rotates. Under the action of F_t, the cutter breaks the rock at the speed of v_{cx} and v_{cy}. In the process of rock cutting, because torsional vibration always exists, the speed of the cutter is not uniform. Therefore, the spring–mass unit is considered in the model to simulate the process of the loading torque on the bit by the top drive through the drill string.

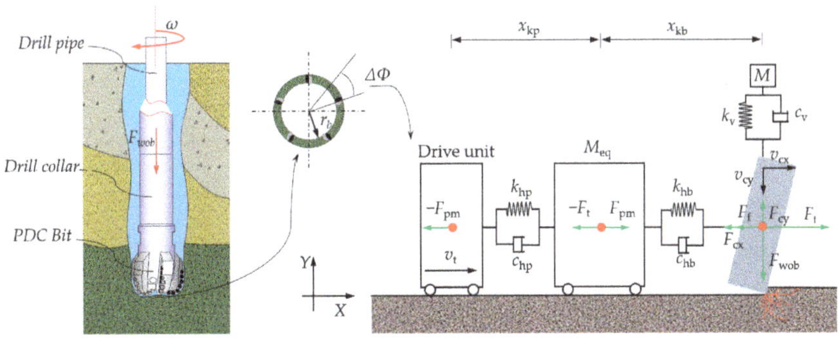

Figure 10. Schematic of the load state of the PDC bit and the cutter during drilling.

Based on the above analysis, a rock cutting model is established in this paper. In Figure 10, x_{kp} and $x_{kb,}$ respectively, represent the initial length of the simulated spring; v_t represents the system driving speed; v_m represents the movement speed of the mass unit; v_{cy} and v_{cx} represent the vertical speed and the horizontal speed of the cutter, respectively; and M_{eq} represents the mass unit. This model can simulate the vibration of the cutter, especially the formation and development of stick–slip vibration, and is used to study the vibration suppression method and its impact on the cutter. Since stick–slip vibration is the main reason for the damage to drill components, the control model is simplified, ignoring the elasticity and damping in the vertical direction. In order to facilitate the comparison test, the proportional coefficient is defined. The expression form is shown in Equation (11), and the model parameters are shown in Table 3.

Table 3. Parameter values used for spring-mass-damping system.

Parameters	Value	Parameters	Value
v_t (m·s^{-1})	1	k_{hb} (N·m^{-1})	1×10^9
x_{max} (m)	0.16	c_{hp} (N·m·s^{-1})	5×10^4
x_{kp} (m)	10	α	0.005
x_{kb} (m)	10	β	0.01
F_{wob} (N)	200	γ	0.001

$$\begin{cases} \alpha = \frac{m_{\text{pdc}}}{M_{\text{eq}}} \\ \beta = \frac{k_{\text{hp}}}{k_{\text{hb}}} \\ \gamma = \frac{c_{\text{hp}}}{c_{\text{hb}}} \end{cases} \tag{11}$$

Here, α, β, and γ are the dimensionless proportional coefficients; m_{pdc} is the mass of the PDC cutter, in kg; M_{eq} is the mass of the mass element, in kg; k_{hp} and k_{hb} represent the elastic coefficients of the first and second spring damping systems, respectively, in $\text{N}\cdot\text{m}^{-1}$; and c_{hp} and c_{hb} represent the damping coefficients of the first- and second-stage spring damping systems, respectively, in $\text{N}\cdot\text{s}\cdot\text{m}^{-1}$.

In addition, the effect of Stribeck is considered in the friction calculation, that is, in the process of low-speed movement, the friction coefficient decreases with an increase in the relative speed between the contact surfaces [43,44]. The friction model expression is shown in Equation (12):

$$\mu = \mu_{\text{d}} + (\mu_{\text{s}} - \mu_{\text{d}})e^{-\left(\frac{|\Delta v|}{V_{\text{r}}}\right)^2} \tag{12}$$

where the friction coefficient between the research objects is μ; the static friction coefficient is μ_{s}, and for the contact between the rock and the cutter, μ_{s} = 0.82 [9]; the dynamic friction coefficient is μ_{d}, where $\mu_{\text{d}} = \mu_{\text{s}}\cdot V_{\text{s}}$, V_{s} is the proportional coefficient, and in this paper, V_{s} = 0.7; Δv is the relative velocity between the contact objects, in $\text{m}\cdot\text{s}^{-1}$; and V_{r} represents the characteristic parameter, with $V_{\text{r}} = 0.2\ \text{m}\cdot\text{s}^{-1}$.

4.2. Physical Model

As shown in Figure 11, a 2D DEM rock cutting model is established, and the PDC cutter is modeled by particle clusters. The density of the cutter is 3200 kg/m^3, the diameter is 16 mm, the thickness is 8 mm, and the cutting angle is 15°. The length of the rock sample is 200 mm, and the height is 40 mm. For the convenience of calculation, the initial cutting depth is set as 1 mm, and the maximum cutting displacement of the cutter is 160 mm. The verification process of the model has been elaborated in previous studies [45].

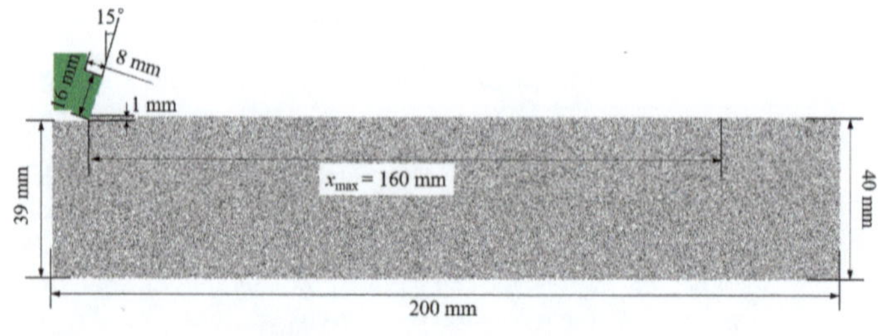

Figure 11. The PDC cutter and rock sample geometry.

In order to study the rock-breaking performance and vibration characteristics of the cutter when the vertical impact and torsional impact act on the cutter at the same time, the impact loads are set in both the horizontal and vertical directions of the cutter, as shown in Figure 12. Assuming that the impact load changes in a sinusoidal form, the vertical load of the cutter is $F_{\text{y}} = f_{\text{y}}(t)$, and the horizontal load is $F_{\text{x}} = f_{\text{y}}(t)$. It is also assumed that the impact frequency and phase of the horizontal impact load and the vertical impact load are the same. Since different loading angles have obvious influences on rock failure [46], the proportional coefficient q_{v} of the axial impact amplitude and F_{wob} and the proportional coefficient q_{h} of the torsional impact amplitude and F_{wob} are defined, respectively. See Equation (13) for the expression.

$$\begin{cases} F_{\text{x}} = F_{\text{t}} + \frac{q_{\text{h}}\cdot F_{\text{wob}}}{2}[1 + \sin(2\pi \cdot f \cdot t)] \\ F_{\text{y}} = F_{\text{wob}} + \frac{q_{\text{v}}\cdot F_{\text{wob}}}{2}[1 + \sin(2\pi \cdot f \cdot t)] \end{cases} \tag{13}$$

where F_x and F_y are, respectively, the horizontal thrust and vertical compression on the cutter, in N; F_t is the thrust of the mass unit to the cutter, in N; q_h and q_v are proportional coefficients, in %; f is the impact load frequency, in Hz; and t is the action time of the impact load, in s.

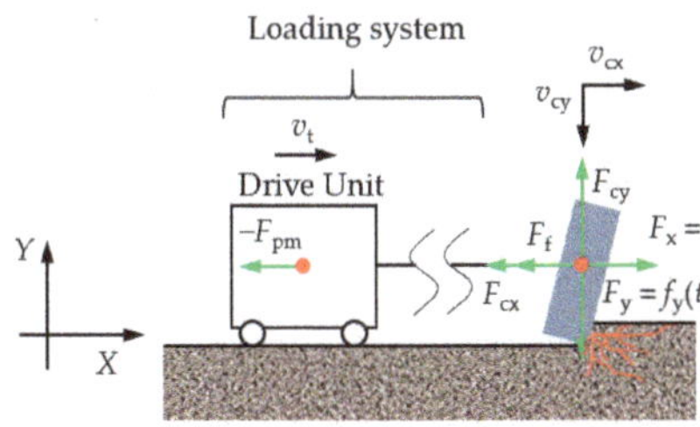

Figure 12. Schematic of the load state of the PDC cutter under vertical and horizontal impact loading.

4.3. Numerical Simulation

The rock-cutting simulations are carried out on a pebbled sandstone with 40% gravel content, the F_{wob} is set to 2 kN, the driving speed is set to 1 m/s, and the impact frequency is assumed to be 200 Hz. The pebbled sandstone cutting is calculated under different impact amplitude proportional coefficients q_h and q_v. All parameters of the test set are summarized in Table 4. The impact load parameters are specified with regard to the performance parameters of the multi-dimensional impactor.

Table 4. Parameters of the simulated test set.

Case	q_h/%	q_v/%
T-1	0	0
T-2	20	40
T-3	40	20
T-4	40	40

The comparison of the cutting and crushing states of the rock under static and impact loads is shown in Figure 13, where the broken red lines indicate the cracks and the colored blocks indicate the rock-crushing blocks. The broken state figure shows that multiple cracks extending to the interior of the rock are generated when the rock is cut under the impact load. The formation of such cracks is helpful for subsequent cutters to break the rock continually. Different from homogeneous rocks, more gravel is stripped as a whole during the cutting process of pebbled sandstone, as outlined in Figure 13. In particular, the case of T-2 has the largest number of stripped gravels. After the vertical impact amplitude and the horizontal impact amplitude are increased simultaneously, the largest number of cracks extending to the interior of the rock are generated in the case of T-4.

The velocity and force signal simulation results of each component of the cutter loading model are shown in Figure 14. When the cutter breaks the pebbled sandstone, the velocity fluctuation range is significant, and there are data points of $v_m = 0$ and $v_{cx} = 0$ in each simulation case. That is, the cutter has stick–slip vibration in varying degrees during rock breaking. The stick–slip vibration of drag bits has been widely investigated in theoretical calculations and field tests [47]. The results from the simulated tests in this study are in agreement with Richard et al. [48] (2007) and Germay et al. [49] (2009). In the field, this phenomenon is reflected in the torsional vibration of the drill string caused by top drive's startup [50]. However, in the range of the simulation parameters, the case of T-4, with the highest amplitude of composite impact load, has the lowest stick–slip degree, and the fluctuation of the driving thrust on the mass unit is relatively minimal.

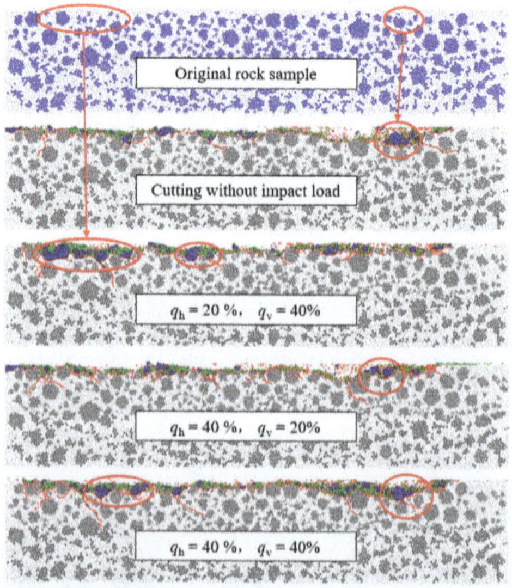

Figure 13. Broken state of the rock samples under different impact loading.

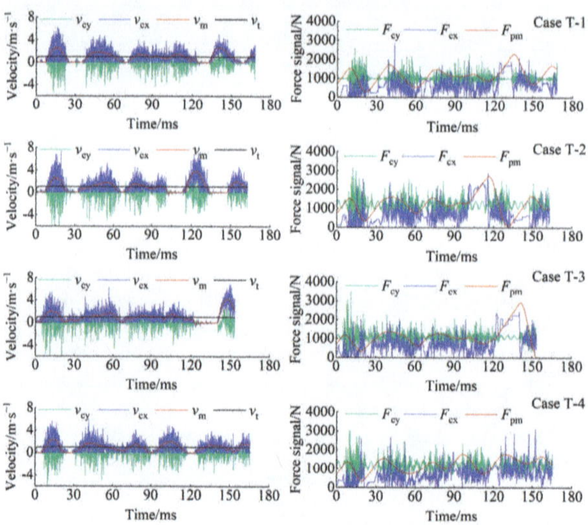

Figure 14. Simulation results of velocity and force signals.

The boxplot and standard deviation analysis of the horizontal and vertical velocities of the PDC cutter are shown in Figure 15. The fluctuation in the cutting velocity and force on the cutter tends to increase at first and then decrease. Increasing the amplitude of the vertical impact load alone may lead to an intensification of the cutting vibration. The analytical results of the case of T-4 show that, within the range of the simulation parameters, increasing the amplitude of impact load in the horizontal and vertical directions to $q_h = q_v = 40\%$ at the same time can effectively alleviate the fluctuation in the cutter velocities. At this point, the peak value of each parameter is relatively small, so it can be presumed that the cutting vibration and stick–slip are effectively suppressed in this case. According to the rock-breaking specific energy and average cutting depth illustrated in Figure 16, compared to increasing the vertical impact amplitude, a reduction in cutting stick–slip and rock-breaking *MSE* (mechanical specific energy) can be achieved by increasing the horizontal impact amplitude, and the cutting depth is also significantly increased.

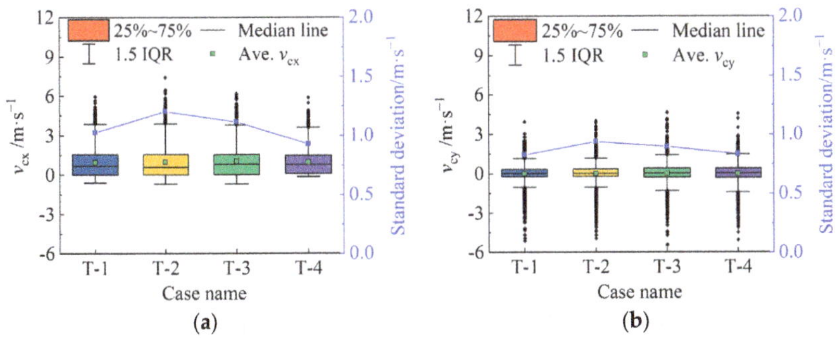

Figure 15. Analytical results of (**a**) cutter horizontal velocity signals and (**b**) cutter vertical velocity signals.

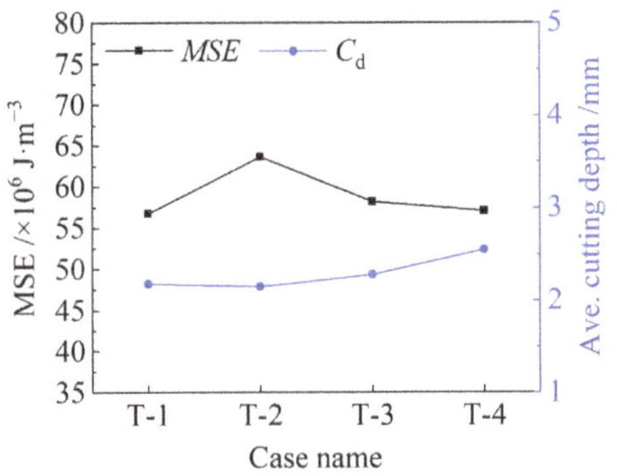

Figure 16. Statistical results of MSE and average cutting depth.

5. Field Test of Composite Impactor

5.1. Structure Introduction

The structure diagram of the composite impactor is shown in Figure 17. The main components of this impactor are the upper sub, pulse nozzle, shell, load transmission rod, and lower sub. A periodic fluid pulsation is generated when the drilling fluid flow in the Helmholtz cavity I and II of the impactor [51,52]. The fluid pulsation acts on the upper end-face of the load transmission rod to form the composite impact. The composite impactor is connected to the bit while drilling, which can simultaneously exert axial impact and torsional impact on the bit.

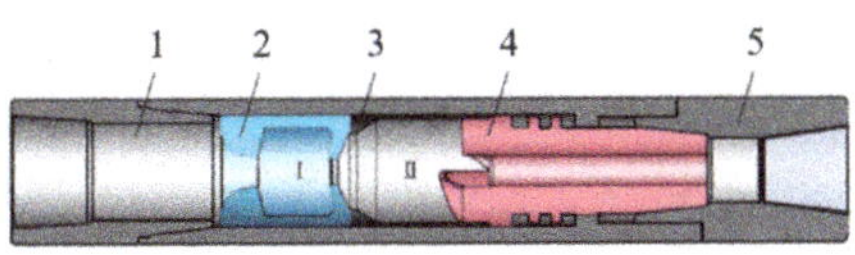

Figure 17. Schematic of the composite impactor. 1—upper sub; 2—nozzle; 3—shell; 4—load transmission rod; 5—lower sub; I, and II—Helmholtz cavity.

The impact amplitude of the composite impactor can be tuned by selecting Helmholtz cavities and nozzles of different sizes. The distribution ratio of the axial impact component and the torsional impact component can be modified by changing the spiral angle of the upper end-face of the load transmission rod, as shown in Figure 18. At present, there are three types of load transmission rod with helix angles of 10°, 50°, and 70°. According to the conclusions obtained from the simulation study in Section 4, a composite impactor with a large torsional impact component and an impact angle α = 70° was selected for the field test in order to reduce the axial vibration and stick–slip of the bit when drilling encounters the gravel layer, and the detailed performance parameters of the impactor are shown in Table 5.

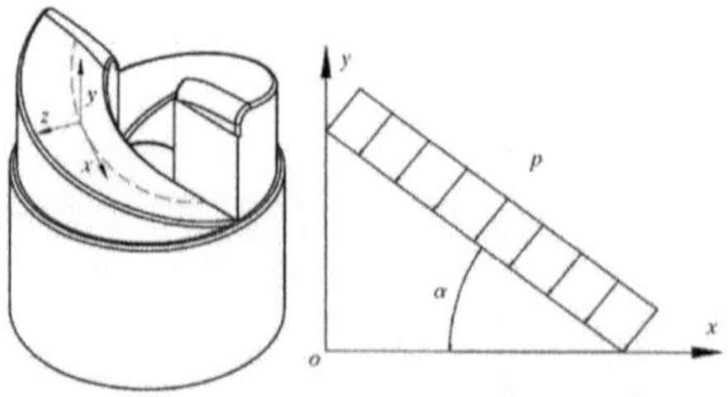

Figure 18. Schematic diagram of impact spiral surface.

Table 5. Key parameters of the composite impactor.

Parameters	Value
Diameter/mm	245
Spiral angle α/°	70
Number of spiral faces	5
Pressure/MPa	2
Axial load/kN	24
Torsional load/N·m	3479

5.2. Field Drilling Test

A field test was carried out in shallow formations in Well T1 to verify the vibration control effect of the impactor. The Well T1 is a straight well, and the test drilling layer is a shallow pebbled sandstone formation. The formation restricts the enhancement of drilling parameters and has a negative effect on the use of downhole tools. Therefore, this application aims to reduce the severe vibration of the bit, restrain the severe fluctuation of bit weight and torque, and maintain the stability of the bit and the drill string.

Before the test section, the well section ① was drilled to 2850 m using a conventional double-stabilizer BHA. Due to the severe drill string vibrations caused by the pebbled sandstone, a turbodrill was selected to continue drilling to 3207 m of section ②. The ROP (rate of penetration) remained at a low level while drilling the well section ① and section ②. Therefore, in order to verify the damping capacity of the composite impactor and protect the downhole instruments, the drilling test of the composite impactor was carried out in this well section. The test well section ③ was 3207–3569 m, the total footage of the impactor was 362 m, and the pure drilling time of logging was 103.12 h. The recorded WOB signal and ROP are shown in Figure 19.

The test results are shown in Table 6, and the standard deviation is used to evaluate the fluctuation amplitude of the bit vibration. The data evaluation demonstrates that the volatility of the WOB in the test section of the composite impactor is similar to that of the section drilled with the turbine, both with lower vibration than the section drilled with a conventional drill string assembly. Compared to the conventional drilling section, the average WOB of the well section drilled by composite impactor is reduced by 57.13%, the standard deviation of the WOB is reduced by 57.29%, and the average ROP is increased by 98.31% and 56% compared to the conventional and turbine drilled sections, respectively.

These results show that the composite impactor that can simultaneously generate axial and torsional impacts has the ability to suppress drilling vibration and improve ROP.

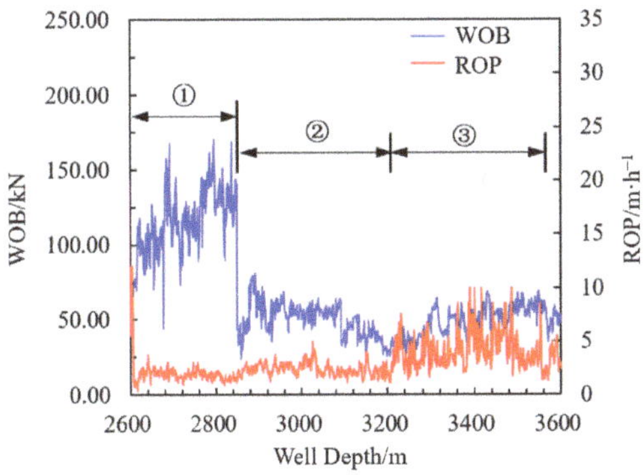

Figure 19. Comparison of WOB and ROP in the applied well interval.

Table 6. Comparison of field test results of different drilling methods.

Drilling Method	Conventional	Turbine Drill	Composite Impactor
Test section/m	2612–2850	2850–3207	3207–3569
WOB/kN	115.27	49.95	49.42
Standard deviation of WOB/kN	22.97	11.21	9.81
ROP/(m/h)	1.77	2.25	3.51
ROP enhancement rate/%	/	27.12	98.31

6. Conclusions

This paper proposes a heterogeneous rock modeling method based on the discrete element theory and presents a rock-breaking model for a single PDC cutter. Using the proposed methods, the mechanisms and characteristics of cutting a heterogeneous gravel-bearing rock under a compound impact load are simulated and studied. Finally, field tests of the composite impactor in a gravel-bearing formation are carried out. The main conclusions for this paper are as follows:

The random regular polygon filling method is a simple and feasible modeling method for heterogeneous rocks, which can simulate and analyze the crack growth, crushing state, and strength characteristics of heterogeneous rocks, such as gravel-bearing and hole-bearing rocks, by changing the distribution law of polygon, the geometric size, and the filling material parameters. The results show that the peak strength of the rock samples decreases slowly at first and then increases with an increase in the gravel content in the rock model. Compared to the peak strength, the increase in gravel content has a greater impact on the elastic modulus of the rock samples.

In the cutting process with a compound impact load, a higher horizontal impact amplitude coefficient (q_h = 40%) has a more obvious effect on alleviating the vibration induced by the PDC cutting gravel-bearing rock. The field tests show that the composite impactor can remarkably reduce the vibration level in the gravel-bearing formation. The average WOB is reduced by 57.13%, the standard deviation of the WOB is reduced by 57.29%, and the average ROP is increased by 98.31%, compared to a conventional double-stabilizer BHA.

Improving the rock-breaking efficiency and the working life of PDC bits has always been the goal pursued by drilling engineering researchers. Optimizing the amplitude–frequency characteristics and the component ratio of the combined impact load plays a critical role in improving the ROP and suppressing bit vibration. In the future, the development of higher-performance composite impact tools and bits will be an important research direction to further improve drilling efficiency. For example, tools that automatically adjust the impact parameters will allow rapid adaptive drilling in different formations. Furthermore, with the mutual exchange and integration of multi-disciplines, it will become an important direction to carry out research on transformative rock-breaking methods when mechanical rock-breaking methods gradually reach their limit.

Author Contributions: Conceptualization, H.N.; methodology, validation, investigation, writing—review and editing, H.Z.; resources, software, B.H. and S.L.; data curation, Z.W.; writing—original draft preparation, S.L. and H.Z.; visualization, Y.W. and G.C.; supervision, H.Y.; project administration, H.N.; funding acquisition, H.N. and L.F. All authors have read and agreed to the published version of the manuscript.

Funding: This research was funded by the CNPC Scientific Research and Technology Development Project (Grant No. 2021DJ4202, 2021DJ2003) and the Science and Technology Major Project of the CNPC (No. ZD2019-183-005).

Institutional Review Board Statement: Not applicable.

Informed Consent Statement: Not applicable.

Data Availability Statement: The data used to support the findings of this study are available from the corresponding author upon request.

Acknowledgments: This research received financial support from the CNPC Scientific Research and Technology Development Project (Grant No. 2021DJ4202, 2021DJ2003) and the Science and Technology Major Project of the CNPC (No. ZD2019-183-005).

Conflicts of Interest: The authors declare no conflict of interest.

References

1. Wang, P.; Ni, H.; Wang, R.; Li, Z. Modulating downhole cuttings via a pulsed jet for efficient drilling-tool development and field testing. *J. Nat. Gas Sci. Eng.* **2015**, *27*, 1287–1295. [CrossRef]
2. Detournay, E.; Richard, T.; Shepherd, M. Drilling response of drag bits: Theory and experiment. *Int. J. Rock Mech. Min. Sci.* **2008**, *45*, 1347–1360. [CrossRef]
3. Glowka, D.A. Use of single-cutter data in the analysis of PDC bit designs: Part 1-development of a PDC cutting force model. *J. Pet. Technol.* **1989**, *41*, 797–849. [CrossRef]
4. Wojtanowicz, A.; Kuru, E. Mathematical modeling of PDC bit drilling process based on a single-cutter mechanics. *J. Energy Resour. Technol.* **1993**, *115*, 247–256. [CrossRef]
5. Mishnaevsky, L.L., Jr. Physical mechanisms of hard rock fragmentation under mechanical loading: A review. *Int. J. Rock Mech. Min. Sci. Geomech. Abstr.* **1995**, *32*, 763–766. [CrossRef]
6. Black, A.D.; Walker, B.H.; Tibbitts, G.A.; Sandstrom, J.L. PDC Bit Performance for Rotary, Mud Motor, and Turbine Drilling Applications. *SPE Drill. Eng.* **1986**, *1*, 409–416. [CrossRef]
7. Slnor, A.; Warren, T. SPE Annual Technical Conference and Exhibition. In *Drag Bit Wear Model*; Society of Petroleum Engineers: Dallas, TX, USA, 1987.
8. Warren, T.; Sinor, A. Drag-bit performance modeling. *SPE Drill. Eng.* **1989**, *4*, 119–127. [CrossRef]
9. Detournay, E.; Defourny, P. International journal of rock mechanics and mining sciences & geomechanics abstracts. In *A Phenomenological Model for the Drilling Action of Drag Bits*; Elsevier: Amsterdam, The Netherlands, 1992; pp. 13–23.
10. Nandakumar, K.; Wiercigroch, M. Stability analysis of a state dependent delayed, coupled two DOF model of drill-string vibration. *J. Sound Vib.* **2013**, *332*, 2575–2592. [CrossRef]
11. Chiaia, B. Fracture mechanisms induced in a brittle material by a hard cutting indenter. *Int. J. Solids Struct.* **2001**, *38*, 7747–7768. [CrossRef]
12. Liu, H.Y.; Kou, S.Q.; Lindqvist, P.A.; Tang, C.A. Numerical simulation of the rock fragmentation process induced by indenters. *Int. J. Rock Mech. Min. Sci.* **2002**, *39*, 491–505. [CrossRef]
13. Bilgesu, I.; Sunal, O.; Tulu, I.; Heasley, K. The 42nd US Rock Mechanics Symposium (USRMS). In *Modeling Rock and Drill Cutter Behavior*; American Rock Mechanics Association: San Francisco, CA, USA, 2008.
14. Jalali, S.M.E.; Naghadehi, M.Z. An analytical pre-feasibility study to generate rotary-percussive concept in hard rock TBMs. *J. Geol. Min. Res.* **2009**, *1*, 140–148.
15. Tulu, I.; Heasley, K. 43rd US Rock Mechanics Symposium & 4th US-Canada Rock Mechanics Symposium. In *Calibration of 3D Cutter-Rock Model with Single Cutter Tests*; American Rock Mechanics Association: Asheville, NC, USA, 2009.

16. Mendoza, J.; Gamwo, I.; Zhang, W.; Lin, J. 44th US Rock Mechanics Symposium and 5th US-Canada Rock Mechanics Symposium. In *Discrete Element Modeling of Rock Cutting Using Crushable Particles*; American Rock Mechanics Association: Salt Lake City, UT, USA, 2010.
17. Jaime, M.C. Numerical Modeling of Rock Cutting and Its Associated Fragmentation Process Using the Finite Element Method. Ph.D. Thesis, University of Pittsburgh, Pittsburgh, PA, USA, 2011.
18. Jaime, M.C.; Zhou, Y.; Lin, J.-S.; Gamwo, I.K. Finite element modeling of rock cutting and its fragmentation process. *Int. J. Rock Mech. Min. Sci.* **2015**, *80*, 137–146. [CrossRef]
19. Menezes, P.L. Influence of rock mechanical properties and rake angle on the formation of rock fragments during cutting operation. *Int. J. Adv. Manuf. Technol.* **2017**, *90*, 127–139. [CrossRef]
20. Cao, L.; Sun, J.; Zhang, B.; Lu, N.; Xu, Y. Sensitivity analysis of the temperature profile changing law in the production string of a high-pressure high-temperature gas well considering the coupling relation among the gas flow friction, gas properties, temperature, and pressure. *Front. Phys.* **2022**, *10*, 1112. [CrossRef]
21. Zhong, R.; Miska, S.; Yu, M.; Meng, M.; Ozbayoglu, E.; Takach, N. Experimental investigation of fracture-based wellbore strengthening using a large-scale true triaxial cell. *J. Pet. Sci. Eng.* **2019**, *178*, 691–699. [CrossRef]
22. Galkin, V.I.; Martyushev, D.A.; Ponomareva, I.N.; Chernykh, I.A. Developing features of the near-bottomhole zones in productive formations at fields with high gas saturation of formation oil. *J. Min. Inst.* **2021**, *249*, 386–392. [CrossRef]
23. Saksala, T. Numerical study of the influence of hydrostatic and confining pressure on percussive drilling of hard rock. *Comput. Geotech.* **2016**, *76*, 120–128. [CrossRef]
24. Guarin, P.L.; Arnold, H.E.; Harpst, W.E.; Davis, E.E. Rotary Percussion Drilling. In *Drilling and Production Practice*; American Petroleum Institute: New York, NY, USA, 1949; p. 11.
25. Melamed, Y.; Kiselev, A.; Gelfgat, M.; Dreesen, D.; Blacic, J. Hydraulic hammer drilling technology: Developments and capabilities. *J. Energy Resour. Technol.* **1999**, *122*, 1–7. [CrossRef]
26. Liu, S.; Ni, H.; Zhang, H.; Wang, Y.; Xie, H. Numerical study on optimal impact angle of a single PDC cutter in impact rock cutting. *Energy Rep.* **2021**, *7*, 4172–4183. [CrossRef]
27. Akbari, B.; Butt, S.; Munaswamy, K.; Arvani, F. 45th US Rock Mechanics/Geomechanics Symposium. In *Dynamic Single PDC Cutter Rock Drilling Modeling and Simulations Focusing on Rate of Penetration Using Distinct Element Method*; American Rock Mechanics Association: San Francisco, CA, USA, 2011.
28. Xiong, C.; Huang, Z.; Shi, H.; Shi, X.; Zhang, B.; Chalaturnyk, R. Experimental Investigation into Mixed Tool Cutting of Granite with Stinger PDC Cutters and Conventional PDC Cutters. *Rock Mech. Rock Eng.* **2021**, *55*, 813–835. [CrossRef]
29. Dong, G.; Chen, P. 3D Numerical Simulation and Experiment Validation of Dynamic Damage Characteristics of Anisotropic Shale for Percussive-Rotary Drilling with a Full-Scale PDC Bit. *Energies* **2018**, *11*, 1326. [CrossRef]
30. Xi, Y.; Wang, W.; Fan, L.; Zha, C.; Li, J.; Liu, G. Experimental and numerical investigations on rock-breaking mechanism of rotary percussion drilling with a single PDC cutter. *J. Pet. Sci. Eng.* **2021**, *208*, 109227. [CrossRef]
31. Tian, K.; Detournay, E. Influence of PDC bit cutter layout on stick–slip vibrations of deep drilling systems. *J. Pet. Sci. Eng.* **2021**, *206*, 109005. [CrossRef]
32. Song, H.; Shi, H.; Ji, Z.; Wu, X.; Li, G.; Zhao, H.; Wang, G.; Liu, Y.; Hou, X. The percussive process and energy transfer efficiency of percussive drilling with consideration of rock damage. *Int. J. Rock Mech. Min. Sci.* **2019**, *119*, 1–12. [CrossRef]
33. Liu, S.; Ni, H.; Jin, Y.; Zhang, H.; Wang, Y.; Huang, B.; Hou, W. Experimental study on drilling efficiency with compound axial and torsional impact load. *J. Pet. Sci. Eng.* **2022**, *219*, 111060. [CrossRef]
34. Wang, W.; Liu, G.; Li, J.; Zha, C.; Lian, W. Numerical simulation study on rock-breaking process and mechanism of compound impact drilling. *Energy Rep.* **2021**, *7*, 3137–3148. [CrossRef]
35. He, X.; Xu, C. Discrete element modelling of rock cutting: From ductile to brittle transition. *Int. J. Numer. Anal. Methods Geomech.* **2015**, *39*, 1331–1351. [CrossRef]
36. He, Q.; Zhu, L.; Li, Y.; Li, D.; Zhang, B. Simulating hydraulic fracture re-orientation in heterogeneous rocks with an improved discrete element method. *Rock Mech. Rock Eng.* **2021**, *54*, 2859–2879. [CrossRef]
37. Cundall, P.A.; Strack, O.D.L. A discrete numerical model for granular assemblies. *Geotechnique* **1979**, *29*, 47–65. [CrossRef]
38. Potyondy, D.O.; Cundall, P.A. A bonded-particle model for rock. *Int. J. Rock Mech. Min. Sci.* **2004**, *41*, 1329–1364. [CrossRef]
39. Potyondy, D.O. The bonded-particle model as a tool for rock mechanics research and application: Current trends and future directions. *Geosyst. Eng.* **2015**, *18*, 1–28. [CrossRef]
40. Itasca Consulting Group. *PFC (Particle Flowcodein 2 and 3 Dimensions), Version 5.0 [User's Manual]*; ICG. Inc.: Minneapolis, MN, USA, 2016.
41. Richard, T.; Dagrain, F.; Poyol, E.; Detournay, E. Rock strength determination from scratch tests. *Eng. Geol.* **2012**, *147–148*, 91–100. [CrossRef]
42. Wei, J.; Liao, H.; Wang, H.; Chen, J.; Li, N.; Liang, H.; Liu, C.; Zhang, D.; Teng, Z. Experimental investigation on the dynamic tensile characteristics of conglomerate based on 3D SHPB system. *J. Pet. Sci. Eng.* **2022**, *213*, 110350. [CrossRef]
43. Saha, A.; Wiercigroch, M.; Jankowski, K.; Wahi, P.; Stefański, A. Investigation of two different friction models from the perspective of friction-induced vibrations. *Tribol. Int.* **2015**, *90*, 185–197. [CrossRef]
44. Saha, A.; Wahi, P.; Wiercigroch, M.; Stefański, A. A modified LuGre friction model for an accurate prediction of friction force in the pure sliding regime. *Int. J. Non-Linear Mech.* **2016**, *80*, 122–131. [CrossRef]

45. Zhang, H.; Ni, H.; Wang, Z.; Huang, B.; Liu, S.; Xu, X.; Liu, C. Discrete element modeling and simulation study on cutting rock behavior under spring-mass-damper system loading. *J. Pet. Sci. Eng.* **2022**, *209*, 109872. [CrossRef]
46. He, Q.; Li, Y.; Li, D.; Zhang, C. Microcrack fracturing of coal specimens under quasi-static combined compression-shear loading. *J. Rock Mech. Geotech. Eng.* **2020**, *12*, 1014–1026. [CrossRef]
47. Kyllingstad, A.; Halsey, G.W. SPE Annual Technical Conference and Exhibition, Dallas. In *A Study of Slip/stick Motion of the Bit, SPE 16659*; Society of Petroleum Engineers: Dallas, TX, USA, 1987.
48. Richard, T.; Germay, C.; Detournay, E. A simplified model to explore the root cause of stick-slip vibrations in drilling systems with drag bits. *J. Sound Vib.* **2007**, *305*, 432–456. [CrossRef]
49. Germay, C.; Denoël, V.; Detournay, E. Multiple Mode Analysis of the Self-Excited Vibrations of Rotary Drilling Systems. *J. Sound Vib.* **2009**, *325*, 362–381. [CrossRef]
50. Brett, J.F. The genesis of torsional drillstring vibrations. *SPE Drill. Eng.* **1992**, *7*, 168–174. [CrossRef]
51. Liu, S.; Ni, H.; Wang, X.; Wang, Y.; Wang, P. IADC/SPE Asia Pacific Drilling Technology Conference and Exhibition. In *Rock-Breaking Mechanism Study of Axial and Torsional Impact Hammer and Its Application in Deep Wells*; Society of Petroleum Engineers: Bangkok, Thailand, 2018.
52. Zhang, H.; Ni, H.; Liu, S.; Huang, B.; Liang, H. An Axial and Torsional Percussion Hammer with Considerable Potential to Increase the Drilling Speed of High-Temperature Geothermal Wells. In *IOP Conference Series: Earth and Environmental Science*; IOP Publishing: Wuhan, China, 2020; p. 012131.

processes

Article

Analysis of Bottomhole Rock Stress in Deep-Well Drilling Considering Thermal-Hydro-Mechanical Coupling

Bin Yang and Honglin Xu *

School of Petroleum and Natural Gas Engineering, Chongqing University of Science and Technology, Chongqing 401331, China
* Correspondence: xuhlaca1986_jy@163.com

Abstract: Drilling is a key step in the exploitation of deep oil and gas resources. In order to clarify the stress state of the rocks and the mechanism of rock breakage in deep-well drilling, a thermal-hydro-mechanical coupling model for deep-well drilling was established, and the effects of drilling on the temperature, pressure, and stress in the formation were studied. Furthermore, the effects of the formation parameters and wellbore parameters on the bottomhole stress were analyzed. The results revealed that after the formation was drilled, the temperatures in different horizontal in situ stress directions were not significantly different, but the difference in the pore pressure between the maximum and minimum horizontal stress directions was large. The average effective stress at the bottom of the hole was the smallest, and in some areas, it was tensile stress. For deep-well drilling, as the formation pressure increased, the in situ stress increased, and the permeability decreased, leading to greater average effective stress of the bottomhole rock. As a result, it was harder to break the rock, and the drilling efficiency decreased. Reducing the wellbore pressure and wellbore temperature is conducive to forming tensile stress near the borehole axis in the bottomhole, causing tensile damage. The average effective stress of the formation near the shoulder of the drill bit was compressive stress, and it is advisable to take advantage of the rock shear failure characteristics to improve the drilling efficiency in this area. The results of this study can help us to understand the stress state of the bottomhole rocks and the mechanism of rock breakage and can provide a reference for the optimization of drilling tools and drilling parameters in deep-well drilling.

Keywords: thermal-hydro-mechanical coupling; bottomhole stress field; deep-well drilling; finite element analysis; rock breakage mechanism

Citation: Yang, B.; Xu, H. Analysis of Bottomhole Rock Stress in Deep-Well Drilling Considering Thermal-Hydro-Mechanical Coupling. *Processes* **2023**, *11*, 683. https://doi.org/10.3390/pr11030683

Academic Editor: Yidong Cai

Received: 8 February 2023
Revised: 18 February 2023
Accepted: 21 February 2023
Published: 23 February 2023

1. Introduction

With rapid economic and social development, there has been an increasing demand for energy. With the background that shallow oil and gas resources are depleted, energy resources from the deep part of the earth have become an important alternative for increasing oil and gas reserves and production [1,2]. The deep oil and gas reserves in the world are rich, and efficient development of deep oil and gas resources is of great significance to the development of the global economy [3,4]. Drilling is a key step in deep oil and gas exploitation. It not only plays a role in confirming the oil and gas reserves in deep formations but also provides oil and gas migration channels for later deep oil and gas extraction [5]. However, due to the complex lithology, high strength, strong abrasiveness, and poor drillability of deep formations [6–9], as well as the significant coupled effect of multiple physical fields during drilling [10], deep-well drilling faces the challenges of low drilling speed and efficiency, which greatly restrict the exploitation of deep oil and gas resources.

In order to increase the rock-breaking efficiency and rate of penetration (ROP) of deep-well drilling, a variety of drill bits have been developed, such as polycrystalline diamond compact (PDC) bits, composite bits, and dual-diameter PDC bits [11–15]. New drilling techniques have also been developed, such as high-pressure water jet drilling, electro-pulse drilling, particle jet impact drilling, and laser drilling [9,16–19]. In addition to the development of new tools and drilling techniques, analysis of the stress state of bottomhole rocks during drilling can also help clarify the mechanism of bottomhole rock breakage during drilling, which can provide references for optimizing drilling tools and drilling parameters. In this field, Peng et al. [20] used the finite element method to simulate formation stress state before drilling, during drilling, and after drilling of the formation. They found that there was stress concentration in the bottom area near the wellbore wall, which led to poor drillability of the rocks in this area. Regarding the high-pressure water jet drilling process, Li et al. [21] studied bottomhole rock stress under high-pressure water jet drilling using finite element and finite volume methods. Their results showed that the maximum principal stress of bottomhole rocks increased as the jet velocity increased. Furthermore, Wang et al. studied the stress release mechanism of deep bottomhole rock using ultra-high-pressure water jet slotting [16]. Their results suggest that the stress release was due to the combined action of three drivers, namely, horizontal stress release, stress concentration in the area away from the cutting plane, and increased pore pressure caused by rock mass expansion. Under the condition of underbalanced drilling, Li et al. [22] used the finite element method to study the stress state of bottomhole rocks, and their results indicate that the mechanical properties of the rocks near the bottomhole were affected by a number of factors, including the in situ stress, pore pressure, well diameter, and fluid column pressure. Compared to a permeable borehole wall, rocks are easier to break under impermeable borehole conditions. In view of the influence of the bit structure, Heydarshahy and Karekal [23] carried out finite element simulations to study the stress distribution characteristics under different drill bit shapes. It was found that increasing the height of the cone-shaped bit led to an increase in the high-stress area of bottomhole rocks. Shen and Peng [24] analyzed the bottomhole stress field when a dual-diameter PDC bit was used. Their results showed that there was significant equivalent stress unloading, which improved the ROP.

The above studies focused on bottomhole rock stress under different drilling techniques and different drill bit shapes but did not consider the multi-physical coupling effects of the bottomhole rocks during drilling. Actually, after the formation is drilled, because some of the rocks are replaced by the drilling fluid, the stress, temperature, and pressure of the bottomhole rock changes accordingly, and these changes impact the stress state of bottomhole rocks. This process is a complex thermal-hydro-mechanical coupling process. In order to analyze the influence of the coupling effect on bottomhole rock stress, Warren and Smith [25] established a hydro-mechanical coupling model for studying the formation stress near the bottom of a borehole and analyzed the distribution of the mean bottomhole stress under different over-balance pressures. They concluded that a high over-balance pressure was not conducive to improving the rock-breaking efficiency. Chang et al. [26] analyzed the formation stress under different bottomhole pressure differences using numerical simulation methods. Their results showed that reducing the wellbore pressure could promote bottomhole rock breakage. Chen et al. [27] used the finite difference method to analyze the formation stress near the bottom of the wellbore by considering the hydro-mechanical coupling effect during the drilling process, and they found that the rock mass expansion at the bottom of the wellbore led to a decrease in the pore pressure, thereby affecting the drilling efficiency. Hu et al. [28] carried out finite element simulations to study the dynamic variations in bottomhole rock stress and pore pressure, considering hydro-mechanical coupling. Then, Hu et al. [29] analyzed the dynamic variations in the stress field under balanced drilling conditions, and their results showed that the variations in bottomhole stress were caused by the changes in the pore pressure. In addition, Hu et al. [13,14] designed a dual-diameter PDC bit and studied the bottomhole stress release mechanism of the drill bit under the influence of coupled hydro-mechanical conditions. Their results showed that the stress concentration area at the shoulder of the reaming bit was significantly reduced, and a large stress unloading area formed at the intersection of the reaming bit and the pilot bit. Moreover, some scholars have studied the comprehensive influence of the thermal-hydro-mechanical coupling effect. For instance, Li et al. [30] studied the stress of bottomhole rock in underbalanced drilling using the finite element method. Their results suggest that both the wellbore pressure and temperature can affect bottomhole stress. Moreover, Zhang et al. [31] established an analytical model for calculating bottomhole pressure and the thermally induced near-wellbore stress. They found that when the wellbore pressure changed from overbalanced to underbalanced, the stress state of the bottomhole rocks changed from compressive stress to tensile stress. In addition, Zhang et al. [32] also analyzed bottomhole stress distribution and its influencing factors in underbalanced drilling by considering thermal-hydro-mechanical coupling.

Based on the above analysis, it can be seen that most previous studies analyzed the bottomhole rock stress while neglecting the coupling effect or only considering hydro-mechanical coupling. A few studies have considered the influence of thermal-hydro-mechanical coupling in underbalanced drilling conditions. Although underbalanced drilling is beneficial to improving the drilling efficiency, its safety is still controversial, and the related risks are high. In deep-well drilling practices, in order to ensure drilling safety, overbalanced drilling is generally used. In overbalanced drilling, the wellbore fluid under positive pressure difference will penetrate the formation. Due to the difference between the temperature and pressure of the wellbore fluid and the original formation, bottomhole rocks will be directly affected by the thermal-hydro-mechanical coupling effect. The objective of this study was to analyze the stress of bottomhole rocks during deep-well drilling by considering the thermal-hydro-mechanical coupling effect. First, a thermal-hydro-mechanical coupling model was established. Then, numerical simulations were carried out to reveal the dynamic evolution of the temperature, pressure, and stress of the bottomhole rocks. Furthermore, the mechanisms by which the formation pressure, in situ stress, formation permeability, wellbore pressure, and wellbore temperature influence bottomhole rock stress were analyzed. The results of this study can improve our understanding of bottomhole stress and the mechanism of rock breakage; moreover, the results can be used to analyze wellbore stability by combing the rock strength criterion, which is of great significance for improving the safety and efficiency of deep-well drilling.

2. Thermal-Hydro-Mechanical Coupling Model for Deep-Well Drilling

2.1. Assumptions

In order to simplify the model, the following assumptions were made.

(1) The formation is continuous, homogeneous, isotropic, and meets the small deformation condition.
(2) The fluid seepage in the formation is single-phase flow.
(3) The variations in the wellbore temperature and pressure were neglected.
(4) The effect of the water jet on bottomhole rock stress was neglected.
(5) The influence of the thermal radiation on the wellbore temperature was neglected.

2.2. Field Equations

2.2.1. Stress Field

For porous elastic materials, when the external load, pore pressure, and temperature are considered, the constitutive equation of the porous media is [33]

$$\sigma_{ij} = 2G\varepsilon_{ij} + \frac{2G\nu}{1-2\nu}\varepsilon_v\delta_{ij} - \alpha p\delta_{ij} - \beta_s K(T - T_0)\delta_{ij} \tag{1}$$

where

$$G = \frac{E}{2(1+\nu)}, \tag{2}$$

$$K = \frac{2G(1+\nu)}{3(1-2\nu)}, \tag{3}$$

and

$$\alpha = \frac{3(\nu_u - \nu)}{B(1-2\nu)(1+\nu_u)}. \tag{4}$$

In Equations (1)–(4), σ_{ij} is the second-order stress tensor component (MPa). ε_{ij} is the dimensionless second-order strain tensor component. ε_v is the dimensionless volume strain; δ_{ij} is the Kronecker symbol, which is dimensionless. G is the shear modulus (GPa). E is the Young's modulus (GPa). K is the bulk modulus (GPa); ν is the Poisson's ratio. ν_u is the undrained Poisson's ratio, which is dimensionless. B is Skempton's coefficient, which is dimensionless. α is the dimensionless Biot coefficient. p is the pore pressure (MPa). β_s is the volumetric expansion coefficient of the rock (1/°C). T is the formation temperature (°C). T_0 is the initial formation temperature (°C).

By substituting Equation (1) into the momentum conservation equation and combining it with the deformation coordination equation, Equation (5) can be obtained.

$$Gu_{i,jj} + \frac{G}{1-2\nu}u_{j,ji} - \alpha p_{,i} - \beta_s K T_{,i} + F_i = 0, \tag{5}$$

where u is the displacement (m), and F_i is the body load component (MPa).

Next, Equation (5) is written in tensor notation. Without considering the body load, the governing equation of the stress field, considering thermal-hydro-mechanical coupling, is

$$G\nabla^2 \boldsymbol{u} + \left(\frac{G}{1-2\nu}\right)\nabla(\nabla \cdot \boldsymbol{u}) - \alpha\nabla p - \beta_s K \nabla T = 0. \tag{6}$$

2.2.2. Seepage Field

Considering the thermal-hydro-mechanical coupling effect, changes in the volumetric strain, pore pressure, and temperature of the formation will lead to changes in the fluid content in the pores as follows [34]

$$\zeta = \alpha(\nabla \cdot \boldsymbol{u}) + \frac{p}{M} - \beta_{sf}(T - T_0), \tag{7}$$

where

$$M = \frac{BK}{\alpha(1-\alpha B)}, \tag{8}$$

and

$$\beta_{sf} = (\alpha - \phi)\beta_s + \phi\beta_f. \tag{9}$$

In Equations (7)–(9), ζ is the change in the fluid content in a unit volume of rock (m^3). M is the Biot's modulus (1/MPa). ϕ is the formation porosity, which is dimensionless, and β_f is the coefficient of volumetric expansion of the fluid (1/°C).

Because the formation is considered to be a closed system, there is no external mass exchange. Therefore, the mass flux of the fluid is equal to the rate of change in the mass of the fluid. According to the law of conservation of mass, the process can be expressed as follows:

$$\frac{\partial\left(\rho_f \zeta\right)}{\partial t} + \nabla \cdot \left(\rho_f v_f\right) = 0, \tag{10}$$

where

$$v_f = -\frac{k}{\mu}\nabla p. \tag{11}$$

In Equations (10) and (11), ρ_f is the fluid density (kg/m^3). v_f is fluid seepage velocity (m/s). k is the formation permeability (m^2). μ is the fluid viscosity (mPa·s).

When the fluid density is constant, the governing equation of the seepage field, considering thermal-hydro-mechanical coupling, is

$$\alpha\frac{\partial}{\partial t}(\nabla\cdot\boldsymbol{u})+\frac{1}{M}\frac{\partial p}{\partial t}-\left[(\alpha-\phi)\beta_s+\phi\beta_f\right]\frac{\partial T}{\partial t}+\nabla\cdot\left(-\frac{k}{\mu}\nabla p\right)=0. \tag{12}$$

2.2.3. Temperature Field

Considering the thermal-hydro-mechanical coupling condition, according to the second law of thermodynamics, deformation of an object, changes in the pore pressure, and changes in the temperature will all lead to a change in the system's entropy. The entropy change can be expressed as follows:

$$S=-\beta_s K(\nabla\cdot\boldsymbol{u})-\phi\beta_f p+\rho_m c_m\frac{T-T_0}{T_0}, \tag{13}$$

where

$$\rho_m c_m=(1-\phi)\rho_s C_s+\phi\rho_f C_f. \tag{14}$$

In Equations (13) and (14), S is the change in entropy (J/(K·m^3)). C_s is the specific heat capacity of the rock (J/(kg·K)). C_f is the specific heat capacity of the fluid (J/(kg·K)). ρ_s is the density of the solid phase (kg/m^3).

Based on the relationship between the change in entropy, temperature, and energy, the change in energy caused by the change in entropy can be expressed as follows:

$$Q=S\cdot T_0=-\beta_s K T_0(\nabla\cdot\boldsymbol{u})-\phi\beta_f T_0 p+\rho_m c_m(T-T_0), \tag{15}$$

where Q is the change in energy (J/m^3).

The heat transfer in the formation includes thermal conduction and convection. Based on Fourier's law of thermal conduction, the heat flux in the formation is as follows [35]:

$$\boldsymbol{q}=-\lambda_m\nabla(T-T_0)+\rho_f C_f\boldsymbol{v}_f(T-T_0), \tag{16}$$

where

$$\lambda_m=(1-\phi)\lambda_s+\phi\lambda_f. \tag{17}$$

In Equations (16) and (17), $\boldsymbol{q}$ is the heat flux (J/m). λ_s is the thermal conductivity of the rock (J/(m·K·s)). λ_f is the thermal conductivity of the fluid (J/(m·K·s)).

Since the system satisfies the law of energy conservation, the sum of the rate of change in the energy and the heat flux is zero, that is,

$$\frac{\partial Q}{\partial t}+\nabla\cdot\boldsymbol{q}=0. \tag{18}$$

By substituting Equations (15) and (16) into Equation (18), the governing equation of the temperature field of the rock, considering thermal-hydro-mechanical coupling, can be obtained as follows:

$$-\beta_s K T_0\frac{\partial}{\partial t}(\nabla\cdot\boldsymbol{u})-\phi\beta_f T_0\frac{\partial p}{\partial t}+\rho_m c_m\frac{\partial T}{\partial t}+\nabla\cdot\left[-\lambda_m\nabla(T-T_0)+\rho_f C_f\boldsymbol{v}_f(T-T_0)\right]=0. \tag{19}$$

2.3. Initial and Boundary Conditions

Due to the symmetry of the formation conditions, a 1/4 formation model was established to reduce the amount of numerical computations. The side length of the formation was 2 m, and the borehole radius was 0.111 m. The geometric models before and after the drilling of the borehole in the formation are shown in Figure 1a,b. It should be noted that the numbers in the figure represent the different surface of the geometric model. Before the formation is drilled, the formation is in a state of equilibrium under in situ stress; the initial temperature of the formation is equal to the original formation temperature; and the initial formation pressure is equal to the original formation pressure. Therefore, the initial conditions of the whole system are as follows:

$$\begin{cases} u = 0 \\ p = p_i \\ T = T_i \end{cases} \quad (\text{when } t = 0), \tag{20}$$

where p_i is the original formation pressure (MPa), and T_i is the original formation temperature (°C).

After the formation is drilled, some rocks are replaced by drilling fluid. At this time, the boundary conditions in the system are determined by the wellbore parameters:

$$\begin{cases} u = f(p_w) \\ p = p_w \\ T = T_w \end{cases} \quad (\text{On } \Gamma_{10} \text{ when } t > 0), \tag{21}$$

where p_w is the wellbore pressure (MPa), and T_w is the wellbore temperature (°C).

For the outer boundary of the system, the formation displacement, temperature, and pressure remain stable. Therefore, the internal and external boundary conditions of the system are as follows:

$$\begin{cases} u = 0 \\ p = p_i \\ T = T_i \end{cases} \quad (\text{On } \Gamma_4 \cup \Gamma_5 \text{ when } t > 0). \tag{22}$$

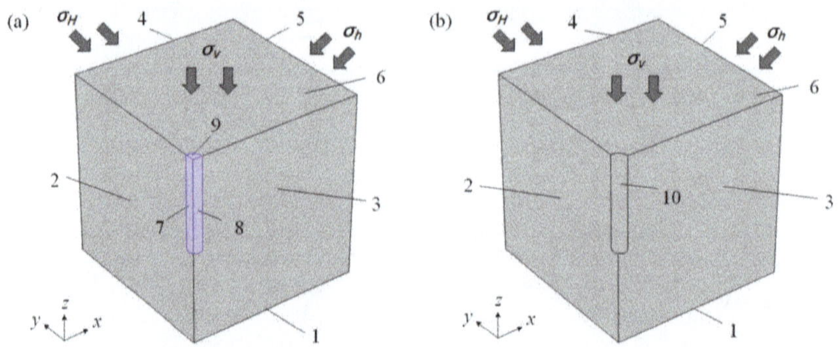

Figure 1. Geometric models of the formation before and after the drilling of the wellbore. (**a**) Before drilling; (**b**) After drilling.

2.4. Model Solving and Validation

Since the contour of the drill bit is complex, and the model involves thermal-hydro-mechanical coupling, the finite element method was used to solve the proposed model, which was carried out in two steps. The first step was the in situ stress balance process, in which, the boundary conditions of surfaces #1, #2, #3, #7, and #8 in Figure 1a were set as the roller support. The maximum horizontal in situ stress was applied on surface #4, and the minimum horizontal in situ stress was applied on surface #5. In addition, vertical stress was applied on surfaces #6 and #9. In this way, the in situ stress distribution before drilling was simulated, which provided the prestress for subsequent drilling and eliminated the influence of the strain generated by the in situ stress on the formation deformation around the well. The second step was the simulation after the formation was drilled. In this step, the geometric elements in the blue area in Figure 1 were deleted from the model to simulate the formation after drilling. The initial conditions and internal and external boundary conditions of the model were set according to Equations (20)–(22). Next, the fully coupled solution method was used to calculate the transient changes in the formation stress, temperature, and pressure after drilling. Positive values represented compressive stress, and negative values represented tensile stress. After obtaining the stress components, the following equation was used to describe the stress components in the cylindrical coordinate system:

$$\begin{bmatrix} \sigma_r & \tau_{r\theta} & \tau_{rz} \\ \tau_{r\theta} & \sigma_\theta & \tau_{\theta z} \\ \tau_{rz} & \tau_{\theta z} & \sigma_z \end{bmatrix} = \begin{bmatrix} \cos\theta & \sin\theta & 0 \\ -\sin\theta & \cos\theta & 0 \\ 0 & 0 & 1 \end{bmatrix} \begin{bmatrix} \sigma_x & \tau_{xy} & \tau_{zx} \\ \tau_{xy} & \sigma_y & \tau_{yz} \\ \tau_{zx} & \tau_{yz} & \sigma_z \end{bmatrix} \begin{bmatrix} \cos\theta & \sin\theta & 0 \\ -\sin\theta & \cos\theta & 0 \\ 0 & 0 & 1 \end{bmatrix}^{\mathrm{T}}, \quad (23)$$

where σ_r, σ_θ, σ_z, $\tau_{r\theta}$, τ_{rz}, and $\tau_{\theta z}$ are the stress components in the cylindrical coordinate system (MPa), and θ is the well circumference angle (°).

After obtaining the stress components in the cylindrical coordinate system, the three principal stresses of the bottomhole rocks were obtained:

$$\begin{cases} \sigma_i = \sigma_r \\ \sigma_j = \frac{1}{2}(\sigma_\theta + \sigma_z) + \frac{1}{2}\left[(\sigma_\theta - \sigma_z)^2 + 4\tau_{\theta z}^2\right]^{\frac{1}{2}} \\ \sigma_k = \frac{1}{2}(\sigma_\theta + \sigma_z) - \frac{1}{2}\left[(\sigma_\theta - \sigma_z)^2 + 4\tau_{\theta z}^2\right]^{\frac{1}{2}} \end{cases}, \quad (24)$$

where σ_i, σ_j, and σ_k are the three principal stresses (MPa).

Furthermore, based on the effective stress expression, the average effective stress of the bottomhole rocks was obtained:

$$\overline{\sigma} = \frac{1}{3}(\sigma_i + \sigma_j + \sigma_k) - \alpha p, \quad (25)$$

where $\overline{\sigma}$ is the average effective stress of the bottomhole rocks after drilling (MPa).

In order to verify the accuracy of the model, the bottomhole stress model proposed by Warren et al. [25] was used to calculate the bottomhole rock stress using the parameters listed in Table 1. The results of Warren et al.'s model are shown in Figure 2a. Hu et al. [28] also employed Warren et al.'s model to calculate the average effective stress of bottomhole rocks after drilling, using the finite element method. Their results are shown in Figure 2b. The results of the model proposed in this study using the same parameters are shown in Figure 2c. It can be seen that the results of the proposed model are consistent with those of the two previous studies. It should be noted that the influence of temperature was not considered in the above analysis. In order to further verify the thermal-hydro-mechanical coupling process, the parameters in Table 2 were used to calculate the temperature, pressure, and stress distribution along the radial direction of the borehole after the formation was drilled, using the finite element method. The numerical simulation results were compared with the analytical results [36] (Figure 3). It can be seen that the numerical simulation results obtained in this study are in good agreement with the analytical solution, which further demonstrates the accuracy of the model proposed in this study. Therefore, the proposed model can be used to study bottomhole stress in deep-well drilling, considering the thermal-hydro-mechanical coupling effect.

Table 1. Basic parameters of the Warren et al. model.

No.	Parameters	Value	Units
1	Well depth, h	3048	m
2	Borehole radius, r_w	0.111	m
3	Drilling fluid density, ρ_f	1050	kg/m^3
4	Vertical in situ stress, σ_v	68.95	MPa
5	Wellbore pressure, p_w	32.41	MPa
6	Formation pressure, p_0	32.41	MPa
7	Maximum horizontal in situ stress, σ_H	48.26	MPa
8	Minimum horizontal in situ stress, σ_h	48.26	MPa
9	Young's modulus of the rock, E	13,789.5	MPa
10	Poisson's ratio of the rock, ν	0.25	-
11	Formation permeability, k	1	mD
12	Formation porosity, ϕ	0.15	-
13	Rock density, ρ_s	2262	kg/m^3
14	Rock compression coefficient, c_s	2.697×10^{-5}	MPa^{-1}
15	Drill bit profile code	IADC5-3-7bit	-

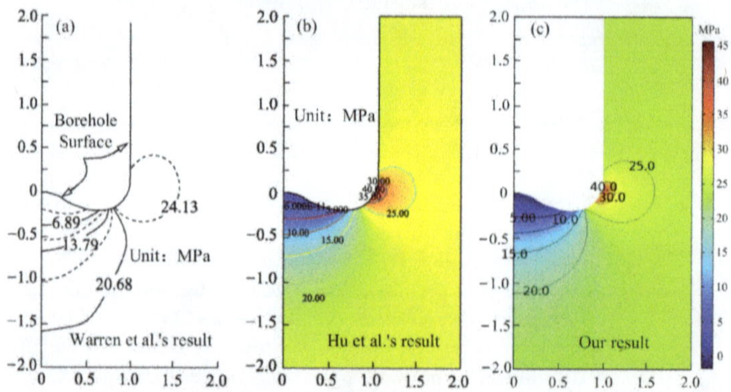

Figure 2. Average effective stress of the bottomhole rock after the formation was drilled. (**a**) Warren et al.'s result; (**b**) Hu et al.'s result; (**c**) Our result.

Table 2. Basic parameters of the thermal-fluid-solid coupling model.

No.	Parameters	Value	Units
	Material parameters		
1	Poisson's ratio of the rock(drained), ν	0.291	-
2	Young's modulus of the rock, E	26.3	GPa
3	Poisson's ratio of the rock(undrained), ν_u	0.45	-
4	Skempton's coefficient of the rock, B	0.85	-
5	Formation permeability, k	1	mD
6	Formation fluid viscosity, μ	5	mPa·s
7	Formation porosity, ϕ	0.1	-

Table 2. *Cont.*

No.	Parameters	Value	Units
8	Rock density, ρ_s	2680	kg/m^3
9	Formation fluid density, ρ_f	1000	kg/m^3
10	Specific heat capacity of the rock, C_s	890	J/(kg·K)
11	Specific heat capacity of the fluid, C_f	2510	J/(kg·K)
12	Thermal conductivity of the rock, λ_s	46	J/(m·K·s)
13	Thermal conductivity of the fluid, λ_f	26	J/(m·K·s)
14	Coefficient of volumetric expansion of the rock, β_s	5×10^{-5}	1/°C
15	Coefficient of volumetric expansion of the fluid, β_f	2×10^{-4}	1/°C
	Formation parameters		
16	Original formation temperature, T_0	130	°C
17	Original formation pressure, p_0	50	MPa
18	Maximum horizontal in situ stress, σ_H	132	MPa
19	Minimum horizontal in situ stress, σ_h	110	MPa
20	Vertical in situ stress, σ_v	144	MPa
	Wellbore parameters		
21	Borehole radius, r_w	0.111	m
22	Wellbore pressure, p_w	54	MPa
23	Wellbore temperature, T_w	70	°C

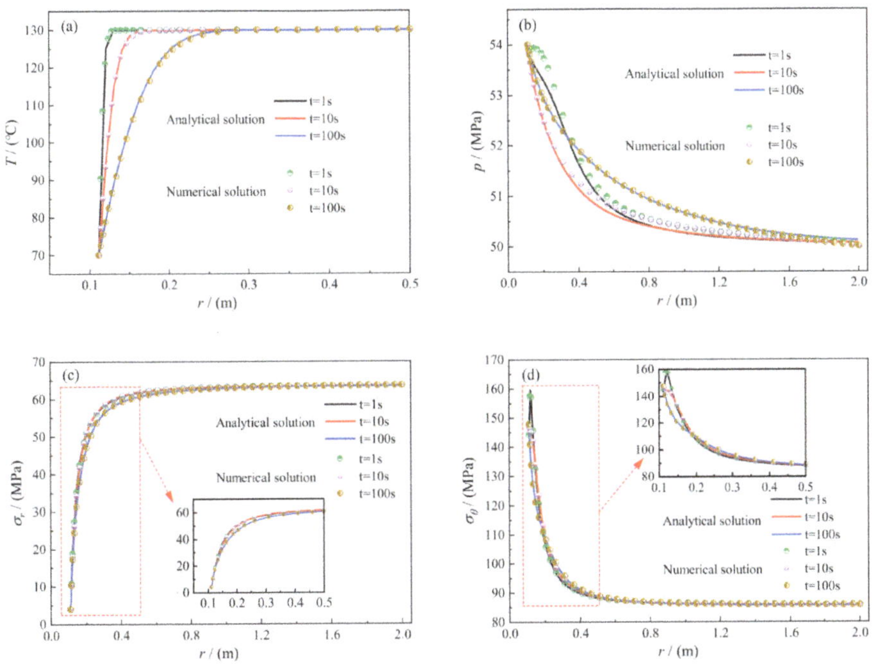

Figure 3. Radial distributions of the temperature, pressure, and stress after drilling. (**a**) Temperature; (**b**) Pore pressure; (**c**) Radial stress; (**d**) Circumferential stress.

3. Distribution of Formation Temperature, Pressure, and Stress due to Drilling

Understanding the distribution and evolution of the formation temperature, pressure, and stress under the action of thermal-hydro-mechanical coupling is the basis for analyzing bottomhole rock stress and studying the mechanism of rock breakage. Therefore, based on the model proposed in this study and the parameters in Table 2, the distributions of the formation temperature, pressure, and average effective stress along the directions of the maximum and minimum horizontal in situ stresses were calculated. The results are shown in Figures 4–6.

Figure 4a–d show the formation temperature in the direction of the maximum horizontal in situ stress at different drilling stages, and Figure 4e–h show the formation temperature in the direction of the minimum horizontal in situ stress at different drilling stages. It can be seen that for the homogeneous and isotropic formation modelled in this study, the temperatures around the wellbore along the directions of the maximum and minimum horizontal crustal stresses are approximately the same; that is, the change in the formation temperature caused by the deformation of the formation around the wellbore and the change in the pore pressure are small, and the formation temperature is mainly controlled by thermal conduction and thermal convection. After the formation is drilled, under the cooling effect of the drilling fluid, the temperatures of the wellbore wall and bottomhole are the same as that of the drilling fluid. Then, under the joint action of thermal conduction and convection, the low temperature zone gradually increases and expands to the deep part of the formation.

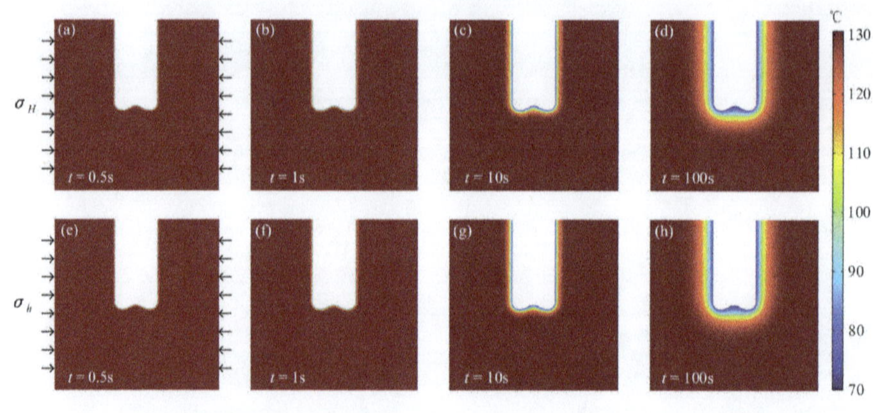

Figure 4. Formation temperature around the wellbore after drilling. (**a**) t = 0.5 s; (**b**) t = 1 s; (**c**) t = 10 s; (**d**) t = 100 s; (**e**) t = 0.5 s; (**f**) t = 1 s; (**g**) t = 10 s; (**h**) t = 100 s.

Figure 5a–d show the formation pressure in the direction of the maximum horizontal in situ stress, and Figure 5e–h show the formation pressure in the direction of the minimum horizontal in situ stress. It can be seen that shortly after the start of drilling (t = 0.5 s and t = 1 s), the formation pressures along the directions of the maximum and minimum horizontal in situ stresses are greatly different. This mainly occurs on the wellbore wall, and the difference near the bottom of the well is small. When the tangent plane of the stratum is along the direction of the maximum horizontal in situ stress, the pore pressure near the wellbore wall is the lowest. However, when the tangent plane of the stratum is along the direction of the minimum horizontal in situ stress, the pore pressure near the wellbore wall is the highest. At t = 0.5 s, the maximum difference in the pore pressure between the two directions is about 10 MPa. The main reason for this difference is that the wellbore along the direction of the maximum horizontal in situ stress is under tension, and the volume of the pores of formation rock increases under the tensile stress, which in turn leads to a decrease in the pore pressure around the wellbore. Conversely, the wellbore along the direction of the minimum horizontal in situ stress is subjected to compressive stress, and the formation is further compressed, which leads to an increase in the pore pressure around the wellbore. With the progression of drilling, the high-pressure area around the wellbore gradually spreads toward the low-pressure area; the minimum pore pressure along the direction of the maximum horizontal in situ stress gradually increases, whereas the maximum pore pressure along the direction of the minimum horizontal in situ stress gradually decreases. After a certain period of time, the pore pressures in the two directions become approximately equal.

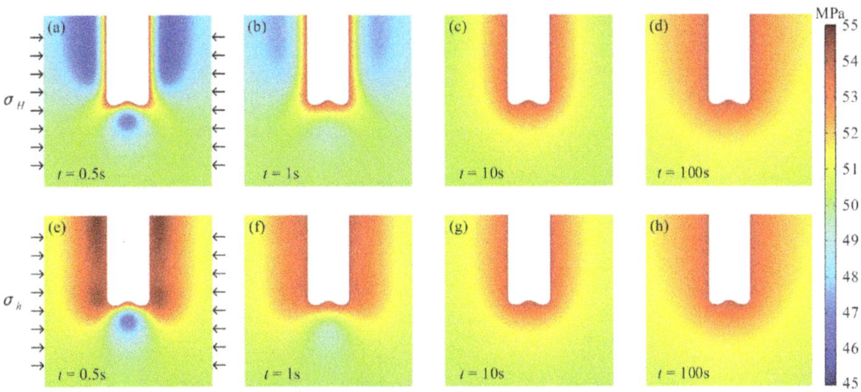

Figure 5. Formation pressure around the wellbore after drilling. (**a**) t = 0.5 s; (**b**) t = 1 s; (**c**) t = 10 s; (**d**) t = 100 s; (**e**) t = 0.5 s; (**f**) t = 1 s; (**g**) t = 10 s; (**h**) t = 100 s.

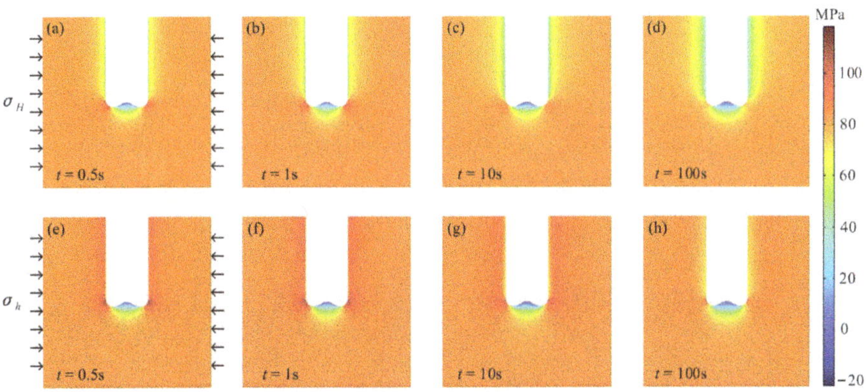

Figure 6. Average effective stress of the formation around the wellbore after drilling. (**a**) t = 0.5 s; (**b**) t = 1 s; (**c**) t = 10 s; (**d**) t = 100 s; (**e**) t = 0.5 s; (**f**) t = 1 s; (**g**) t = 10 s; (**h**) t = 100 s.

Figure 6a–d show the average effective stress in the direction of the maximum horizontal in situ stress, and Figure 6e–h show the average effective stress in the direction of the minimum horizontal in situ stress. The results show that when the maximum horizontal in situ stress and the minimum horizontal in situ stress are not the same, the average effective stresses of the formation around the wellbore in the directions of the maximum and minimum horizontal in situ stresses after drilling are also different, and the average effective stress along the direction of the minimum horizontal in situ stress is larger than that along the direction of the maximum horizontal in situ stress. It can also be seen that the average effective stress near the shoulder of the drill bit is the largest, while the average effective stress near the centerline of the bit is the smallest. Moreover, the average effective stress is tensile stress in some areas, and the minimum average effective stress is located on the bottom of the wellbore. Starting from the centerline of the drill bit, the average effective stress at the bottomhole initially increases and then decreases along the contour of the drill bit. With the progression of drilling, the maximum average effective stress gradually decreases, and the minimum average effective stress remains stable. Moreover, the bottomhole area under tension increases slightly.

4. Factors Affecting Bottomhole Rock Stress under the Thermal-Hydro-Mechanical Coupling Effect

The influences of the formation parameters and wellbore parameters on bottomhole rock stress in deep-well drilling were investigated using the proposed model. Under each set of conditions, the parameter of interest was set to different values, while the other parameters were held constant. The average effective stress of the bottomhole rocks under different formation pressures, in situ stress, formation permeability, wellbore pressures, and wellbore temperature values were calculated. In order to facilitate the analysis, the average effective stress was calculated at three locations, namely, the drill bit contour parallel to the direction of the maximum horizontal in situ stress; the contour parallel to the direction of the minimum horizontal in situ stress; and the axis of the wellbore (Figure 7). The results are described in Sections 4.1–4.5.

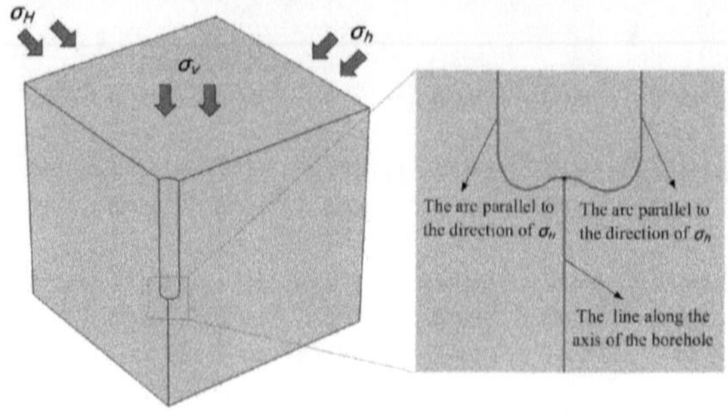

Figure 7. Points of data extraction in the analysis of the bottomhole stress parameters.

4.1. Formation Pressure

The formation pressure is very important to the drilling process. The safe density window of the drilling fluid is determined based on the formation pressure. In addition, the formation pressure directly affects bottomhole rock stress during the drilling process. In order to understand the influence of the formation pressure on bottomhole rock stress, the average effective stress was calculated at the above three locations under different formation pressure values (40, 50, 60, and 70 MPa). The results are shown in Figure 8. It should be noted that the overbalance pressure was set to 4 MPa under the four formation pressure conditions, and the result was the average effective stress at 1 s after the formation had been drilled. It can be seen from Figure 8a that the average effective stress at the bottom of the wellbore was the smallest, and the average effective stress was tensile stress in some areas. The average effective stress was the largest near the shoulder of the drill bit, and there were large fluctuations in this area. In summary, the smaller the formation pressure was, the greater the average effective stress near the shoulder of the drill bit was. After the wellbore was formed, the average effective stresses of the wellbore walls parallel to the directions of the maximum and minimum horizontal in situ stresses were different. Under formation pressures of 40, 50, 60, and 70 MPa, the average effective stress of the wellbore wall parallel to the direction of the maximum horizontal in situ stress was 42.2, 33.8, 25.3, and 16.9 MPa, respectively, and the average effective stress of the wellbore wall parallel to the direction of the minimum horizontal in situ stress was 74.5, 66.0, 57.6, and 49.1 MPa, respectively. That is, the higher the formation pressure was, the lower were the average effective stresses of the wellbore wall parallel to the directions of maximum and minimum horizontal in situ stresses. It can be seen from Figure 8b that after the formation was drilled, the average effective stress at the bottomhole was the smallest. For formation pressures of 40, 50, 60, and 70 MPa, the average effective stress of the bottomhole rock near the borehole axis was −19.2, −18.3, −17.4, and −16.5 MPa, respectively, indicating that a larger formation pressure did not lead to a smaller average effective stress at the bottom of the wellbore. The increase in the average effective stress at the bottom of the hole with increasing formation pressure may have been caused by the high fluid column pressure at the bottomhole under high formation pressure conditions. With increasing depth, the average effective stress initially increased and then became stable. In the region of the formation far away from the bottom of the wellbore, due to the very limited influence of the drilling, the average effective stress was controlled by the in situ stress and the original formation pressure. Therefore, the higher the formation pressure in this region was, the smaller the average effective stress was.

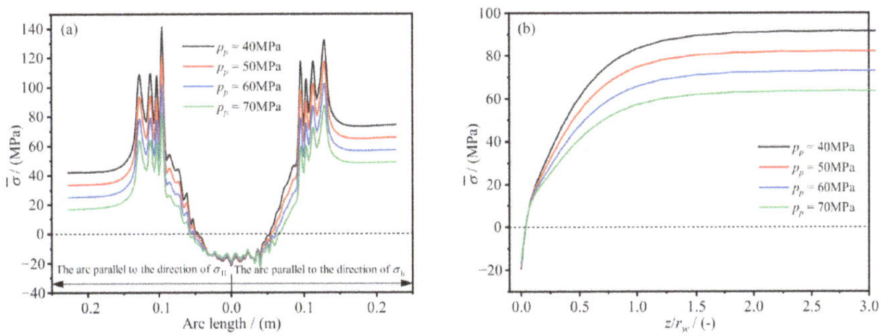

Figure 8. Average effective stress of the bottomhole rocks under different formation pressure conditions. (**a**) The arc parallel to the direction of σ_H and σ_h; (**b**) The line along the axis of the borehole.

4.2. In Situ Stress

In situ stress directly affects bottomhole rock stress. In order to understand the influence of the in situ stress on the bottomhole rock stress in deep-well drilling, the maximum horizontal in situ stress was adjusted, and the average effective stress was calculated at the three locations under maximum horizontal in situ stress to minimal horizontal in situ stress ratios of 1.0, 1.2, 1.4, and 1.6, while holding the minimum horizontal in situ stress in Table 2 constant. The results are shown in Figure 9. It can be seen from Figure 9a that the in situ stress had a large influence on the bottomhole rock stress. When the maximum and minimum horizontal in situ stresses were equal, for the same arc length, the average effective stress of the wellbore parallel to the direction of the maximum horizontal in situ stress was the same as that parallel to the direction of the minimum horizontal in situ stress. When the maximum and minimum horizontal in situ stresses were not equal, the difference in the average effective stresses at the two locations was small near the center of the bottomhole. With increasing arc length, the difference in the average effective stress increased accordingly. In the bottomhole area from the center to the shoulder of the drill bit, the higher the in situ stress ratio was, the higher the average effective stress at the bottom of the hole was. However, on the wellbore wall, the average effective stresses parallel to the maximum and minimum horizontal in situ stresses exhibited different change trends. When the wellbore wall was parallel to the direction of the maximum horizontal in situ stress, the average effective stress decreased with increasing in situ stress ratio. For σ_H/σ_h ratios of 1.0, 1.2, 1.4, and 1.6, the average effective stress was 43.7, 33.8, 23.7, and 13.7 MPa, respectively. However, when the wellbore wall was parallel to the direction of the minimum horizontal in situ stress, the average effective stress increased with increasing in situ stress ratio. For σ_H/σ_h ratios of 1.0, 1.2, 1.4, and 1.6, the average effective stress was

43.7, 66.0, 88.3, and 110.5 MPa, respectively. It can be seen from Figure 9b that as the σ_H/σ_h ratio increased, the average effective stress at the bottom of the hole increased. For σ_H/σ_h ratios of 1.0, 1.2, 1.4, and 1.6, the average effective stress of the bottomhole rock near the borehole axis was −19.3, −18.3, −17.3, and −16.4 MPa, respectively. With increasing depth, the average effective stress of the formation initially increased and then became stable. In summary, the higher the in situ stress ratio was, the higher the average effective stress of the bottomhole rock was. Since the rock strength increases with increasing effective stress, the bottomhole rocks become more difficult to break under a high in situ stress, which leads to a low drilling efficiency.

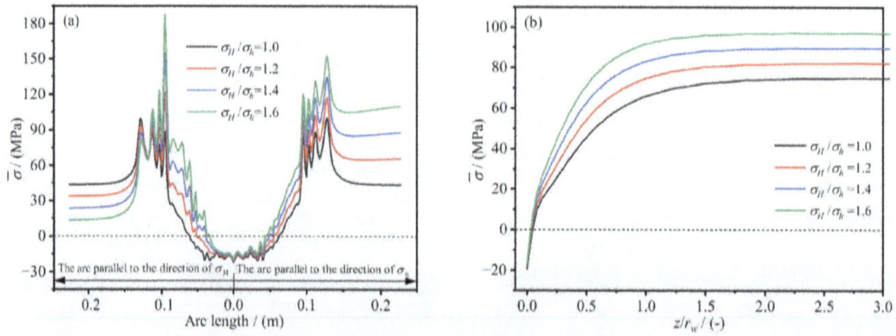

Figure 9. Average effective stress of the bottomhole rocks under different in situ stress ratio conditions. (**a**) The arc parallel to the direction of σ_H and σ_h; (**b**) The line along the axis of the borehole.

4.3. Formation Permeability

After the formation is drilled, the wellbore drilling fluid filtrate penetrates the formation due to the overbalance pressure, which leads to changes in the formation pressure and temperature at the bottom of the hole. Thus, the penetration process is closely related to the formation permeability. In order to analyze the impact of permeability on bottomhole rock stress, the average effective stress of the bottomhole rocks was calculated under formation permeabilities of 10, 1, 0.1, and 0.01 mD at 1 s after the formation had been drilled. The results are shown in Figure 10. It can be seen that the formation permeability had a significant influence on the average effective stress of the bottomhole rock near the borehole axis. The lower the permeability was, the smaller the average effective stress was, and the smaller the area under tension was. At the center of the bottomhole along the borehole axis, for formation permeabilities of 10, 1, 0.1, and 0.01 mD, the average effective stress was −18.5, −18.3, −14.9, and −2.4 MPa, respectively. In the area near the shoulder of the drill bit, the difference in the average effective stress under the different permeabilities was small. As the formation depth increased, the average effective stress of the bottomhole rock initially increased and then became stable. The lower the formation permeability was, the smaller the bottomhole area disturbed by the drilling was. In the undisturbed area, the average effective stress of the rocks was the same, and it was not affected by the formation permeability. The higher the permeability was, the easier it was for the drilling fluid to penetrate the formation. Therefore, the range of the high-pressure area at the bottomhole was large when the permeability was high. Moreover, a high permeability resulted in more drilling fluid penetrating the formation; therefore, the drilling fluid had a more significant cooling effect on the bottomhole rock under the effect of thermal convection. For the above two reasons, the average effective stress of the bottomhole rocks was smaller under a low formation permeability. In terms of deep-well drilling, the greater the formation depth is, the higher the degree of formation compaction is; that is, the lower the formation permeability is, the greater the average effective stress of the formation is, and the harder it is to break the bottomhole rocks, which results in a slower ROP.

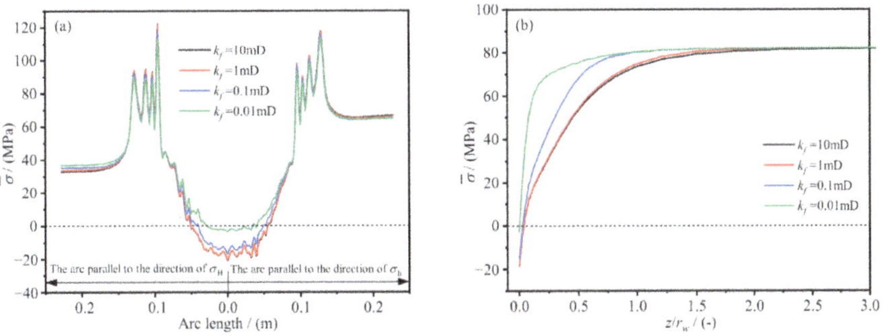

Figure 10. Average effective stress of the bottomhole rocks under different formation permeability conditions. (**a**) The arc parallel to the direction of σ_H and σ_h; (**b**) The line along the axis of the borehole.

4.4. Wellbore Pressure

In order to study the influence of wellbore pressure on bottomhole rock stress, the average effective stress of the bottomhole rocks at 1 s after the formation had been drilled was calculated, under wellbore pressures of 40, 45, 50, and 55 MPa. The results are shown in Figure 11. It can be seen that in the bottomhole area near the borehole axis, the lower the wellbore pressure was, the smaller the average effective stress of the bottomhole rocks was. For wellbore pressures of 40, 45, 50, and 55 MPa, the average effective stress was −19.9, −19.3, −18.8, and −18.2 MPa, respectively. However, in the bottomhole area near the shoulder of the drill bit and the wellbore wall, the average effective stress increased with decreasing wellbore pressure. Since the maximum and minimum horizontal in situ stresses were not the same, the average effective stresses on the wellbore walls parallel to the directions of the maximum and minimum horizontal in situ stresses were also different. Specifically, the average effective stress on the wellbore wall parallel to the direction of the maximum horizontal in situ stress was significantly lower than that parallel to the direction of the minimum horizontal in situ stress. With increasing formation depth, the average effective stress of the bottomhole rocks initially increased and then became stable. In the area of the formation near the bottomhole, the average effective stress decreased with decreasing wellbore pressure. With increasing formation depth, when z/r_w was, for example, greater than 0.25, the average effective stress increased slightly with decreasing wellbore pressure. In summary, decreasing the wellbore pressure is conducive to generating tensile stress in the bottomhole formation near the borehole axis, causing tensile damage to this area. However, the average effective compressive stress of the formation near the shoulder of the drill bit is large. Therefore, the recommendation is to generate shear failure in the formation near the shoulder of the drill bit so as to improve the drilling efficiency.

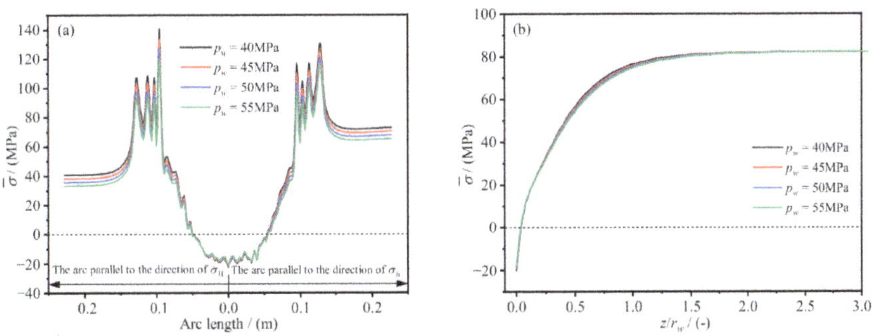

Figure 11. Average effective stress of the bottomhole rocks under different wellbore pressure conditions. (**a**) The arc parallel to the direction of σ_H and σ_h; (**b**) The line along the axis of the borehole.

4.5. Wellbore Temperature

In deep-well drilling, due to the presence of a geothermal gradient, a deep formation generally has a high temperature. Under the cooling effect of the drilling fluid, the rock at the bottom of the wellbore will contract, thereby generating thermal stress. In order to understand the influence of wellbore temperature on bottomhole rock stress, the average effective stress was calculated at three locations in the bottomhole under wellbore temperatures of 130 °C, 110 °C, 90 °C, and 70 °C. The results are shown in Figure 12. The results show that with decreasing wellbore temperature, the average effective stress of the bottomhole rocks decreased. In the bottomhole area near the borehole axis, the average effective stress was the smallest. For wellbore temperatures of 130 °C, 110 °C, 90 °C, and 70 °C, the average effective stress at this location was 1.0, −5.4, −11.8, and −18.3 MPa, respectively. In the bottomhole area between the borehole axis and the shoulder of the drill bit, the average effective stress gradually increased and exhibited fluctuations near the shoulder. Furthermore, between the drill bit shoulder and the wellbore wall, the average effective stress initially decreased and then became stable. With increasing formation depth, the average effective stress initially increased and then became stable. The difference in the average effective stress at different wellbore temperatures gradually decreased with increasing formation depth, which was mainly due to the shorter drilling duration and the limited variation in the formation temperature at the bottomhole. In summary, a lower wellbore temperature is beneficial to reducing the average effective stress of the bottomhole formation and accelerating bottomhole rock breakage. Therefore, under the premise that the drilling fluid meets the relevant requirements, it is recommended to keep the wellbore temperature as low as possible so as to improve the drilling efficiency.

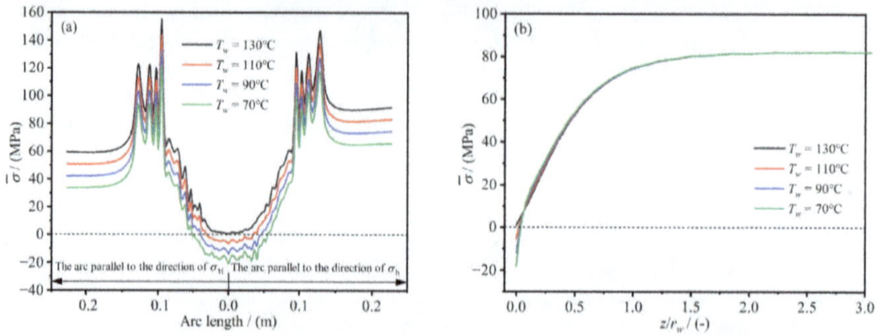

Figure 12. Average effective stress of the bottomhole rocks under different wellbore temperatures. (**a**) The arc parallel to the direction of σ_H and σ_h; (**b**) The line along the axis of the borehole.

5. Conclusions

In this study, based on the laws of momentum conservation, mass conservation, and energy conservation, a thermal-hydro-mechanical coupling model for deep-well drilling was established. The finite element method was used to simulate the distribution and evolution characteristics of the formation temperature, pressure, and stress due to drilling. Finally, the thermal-hydro-mechanical coupling model was used to analyze the influences of the formation pressure, in situ stress, formation permeability, wellbore pressure, and wellbore temperature on the average effective stress of the bottomhole rock. The main conclusions of this study are as follows:

(1) The formation temperature around the wellbore was controlled by thermal conduction and convection. The formation temperature around the wellbore wall was approximately the same in the different in situ stress directions. With the progression of drilling, the low-temperature zone gradually extended deeper within the formation. In the initial stage of drilling, the pore pressure near the wellbore wall along the direction of the maximum horizontal in situ stress was the smallest, and the pore

pressure along the direction of the minimum horizontal in situ stress direction was the largest. With the progression of drilling, the difference in the pore pressure in the two directions gradually decreased.

(2) After the wellbore was drilled, the average effective stress of the formation near the shoulder of the drill bit was the largest, and the average effective stress of the formation near the axis of the drill bit was the smallest. With the progression of drilling, the maximum average effective stress gradually decreased, while the minimum average effective stress remained stable. The bottomhole area under tension increased slightly.

(3) In the bottomhole area near the borehole axis, the average effective stress increased with increasing formation pressure. The higher the in situ stress ratio was, the larger the average effective stress of the bottomhole rocks was. For the wellbore walls, the variation in the average effective stress was different in the different in situ stress directions. The average effective stress in the direction parallel to the maximum horizontal in situ stress decreased as the in situ stress ratio increased, whereas the average effective stress in the direction parallel to the minimum horizontal in situ stress increased as the in situ stress ratio increased.

(4) The formation permeability had a significant effect on the average effective stress of the bottomhole rock near the borehole axis. The lower the formation permeability was, the smaller the average effective stress was, and the smaller the area under tension was. As the wellbore pressure decreased, the average effective stress of the bottomhole rocks near the borehole axis decreased. However, in the bottomhole area between the drill bit shoulder and the wellbore wall, the average effective stress increased with decreasing wellbore pressure. Reducing the wellbore temperature is beneficial to reducing the average effective stress of the bottomhole formation, thereby promoting bottomhole rock breakage.

(5) With increasing formation depth, the average effective stress of the bottomhole rocks initially increased and then became stable. The formation permeability and wellbore temperature had greater influence on the average effective stress of the formation near the bottomhole area. However, the formation pressure and in situ stress had greater influence on the average effective stress of the formation farther away from the bottomhole area. In addition, the influence of the wellbore pressure on the average effective stress of the formation below the bottomhole area was minimal.

(6) The present work focuses on the thermal-hydro-mechanical coupling effect on bottomhole rock stress in deep-well drilling, and the proposed model can be used to calculate the bottomhole stress, pore pressure and temperature during drilling. However, the model also exhibits some limitations. For example, the model neglected the influence of water jet, dynamic wellbore pressure, dynamic wellbore temperature, and multiphase flow in formation, and those aspects need to be improved in future work.

Author Contributions: Conceptualization, B.Y.; Methodology, B.Y. and H.X.; Validation, B.Y.; Formal analysis, B.Y. and H.X.; Investigation, B.Y. and H.X.; Writing—original draft, B.Y. and H.X.; Writing—review & editing, B.Y. and H.X.; Funding acquisition, B.Y. and H.X. All authors have read and agreed to the published version of the manuscript.

Funding: This work was supported by the Fund of the Natural Science Foundation of Chongqing, China (Grant No. CSTB2022NSCQ-MSX0989) and the Sichuan Province Regional Innovation Cooperation Project (Grant No. 2022YFQ0012).

Data Availability Statement: The data presented in this study are available on request from the corresponding author.

Conflicts of Interest: The authors declare no conflict of interest.

References

1. Wang, H.G.; Huang, H.C.; Bi, W.X.; Ji, G.D.; Zhou, B.; Zhuo, L.B. Deep and ultra-deep oil and gas well drilling technologies: Progress and prospect. *Nat. Gas Ind. B* **2022**, *9*, 141–157. [CrossRef]
2. Su, Y.N.; Lu, B.P.; Liu, Y.S. Status and research suggestions on the drilling and completion technologies for onshore deep and ultra deep wells in China. *Oil Drill. Prod. Technol.* **2020**, *42*, 527–542.
3. Zou, C.N.; Zhai, G.M.; Zhang, G.Y.; Wang, H.J.; Zhang, G.S.; Li, J.Z.; Wang, Z.M.; Wen, Z.X.; Ma, F.; Liang, Y.B.; et al. Formation, distribution, potential and prediction of global conventional and unconventional hydrocarbon resources. *Pet. Explor. Dev.* **2015**, *42*, 14–28. [CrossRef]
4. Vedachalam, N.; Srinivasalu, S.; Rajendran, G.; Ramadass, G.A.; Atmanand, M.A. Review of unconventional hydrocarbon resources in major energy consuming countries and efforts in realizing natural gas hydrates as a future source of energy. *J. Nat. Gas Sci. Eng.* **2015**, *26*, 163–175. [CrossRef]
5. Li, G.S.; Song, X.Z.; Tian, S.C. Intelligent drilling technology research status and development trends. *Pet. Drill. Tech.* **2020**, *48*, 1–8.
6. Li, G.; Yang, M.; Meng, Y.F.; Liu, H.; Han, L.; Zhou, F.; Zhang, H. The assessment of correlation between rock drillability and mechanical properties in the laboratory and in the field under different pressure conditions. *J. Nat. Gas Sci. Eng.* **2016**, *30*, 405–413. [CrossRef]
7. Fang, T.C.; Ren, F.S.; Liu, H.X.; Zhang, Y.; Cheng, J.X. Progress and development of particle jet drilling speed-increasing technology and rock-breaking mechanism for deep well. *J. Pet. Explor. Prod. Technol.* **2022**, *12*, 1697–1708. [CrossRef]
8. Peng, N.; Ma, T.S.; Zhu, G.S.; Su, Q. Anti-drilling ability of Ziliujing conglomerate formation in Western Sichuan Basin of China. *Petroleum* **2022**, *in press*. [CrossRef]
9. Li, C.P.; Duan, L.C.; Wu, L.J.; Tan, S.C.; Zheng, J.; Chikhotkin, V. Experimental and numerical analyses of electro-pulse rock-breaking drilling. *J. Nat. Gas Sci. Eng.* **2020**, *77*, 103263. [CrossRef]
10. Jia, S.P.; Wen, C.X.; Deng, F.C.; Yan, C.L.; Xiao, Z.Q. Coupled THM modelling of wellbore stability with drilling unloading, fluid flow, and thermal effects considered. *Math. Probl. Eng.* **2019**, *2019*, 1–20. [CrossRef]
11. Tian, J.L.; Li, J.R.; Cheng, W.M.; Zhu, Z.; Yang, L.; Yang, Y.; Zhang, T.J. Working mechanism and rock-breaking characteristics of coring drill bit. *J. Pet. Sci. Eng.* **2018**, *162*, 348–357. [CrossRef]
12. Yang, Y.X.; Yang, Y.; Liu, X.M.; Huang, K.L.; Ren, H.T. Optimized design and application of a directional reaming-while-drilling polycrystalline diamond compact bit. *Eng. Failure Anal.* **2019**, *105*, 699–707. [CrossRef]
13. Hu, H.G.; Guan, Z.C.; Zhang, B.; Xu, Y.Q.; Liu, Y.W.; Wang, B. Structure design of weight-on-bit self-adjusting PDC bit based on stress field analysis and experiment evaluation. *J. Pet. Sci. Eng.* **2021**, *196*, 107692. [CrossRef]
14. Hu, H.G.; Guan, Z.C.; Wang, B.; Lu, B.P.; Chen, W.Q.; Xu, Y.Q.; Liu, Y.W. Research on weight-on-bit self-adjusting dual-diameter pdc bit design and effect evaluation utilizing stress-releasing effect of rock at bottomhole. *Arabian J. Sci. Eng.* **2021**, *46*, 6925–6937. [CrossRef]
15. Ma, T.S.; Chen, P.; Zhao, J. Overview on vertical and directional drilling technologies for the exploration and exploitation of deep petroleum resources. *Geomech. Geophys. Geo-Energy Geo-Resour.* **2016**, *2*, 365–395. [CrossRef]
16. Wang, H.J.; Liao, H.L.; Wei, J.; Liu, J.S.; Niu, W.L.; Liu, Y.W.; Guan, Z.C.; Sellami, H.; Latham, J.P. Stress release mechanism of deep bottom hole rock by ultra-high-pressure water jet slotting. *Pet. Sci.* **2022**, *in press*. [CrossRef]
17. Cao, S.R.; Ge, Z.L.; Zhang, D.; Zhou, Z.; Lu, Y.Y.; Zhao, H.Y. An experimental study of ultra-high pressure water jet-induced fracture mechanisms and pore size evolution in reservoir rocks. *Int. J. Rock Mech. Min.* **2022**, *150*, 104995. [CrossRef]
18. Ren, F.S.; Fang, T.C.; Cheng, X.Z. Theoretical modeling and experimental study of rock-breaking depth in particle jet impact drilling process. *J. Pet. Sci. Eng.* **2019**, *183*, 106419. [CrossRef]
19. Chen, K.; Huang, Z.Q.; Deng, R.; Zhang, W.L.; Kang, M.Q.; Ma, Y.C.; SHI, M.J.; Yan, J. Research on the temperature and stress fields of elliptical laser irradiated sandstone, and drilling with the elliptical laser-assisted mechanical bit. *J. Pet. Sci. Eng.* **2022**, *211*, 110147. [CrossRef]
20. Peng, Y.; Shen, Z.H.; Fan, S.H. Finite element model for analyzing stress field of bottom hole based on excavating method. *Acta Pet. Sin.* **2006**, *27*, 133–137.
21. Li, G.; Chang, D.; Shen, Z.; Huang, Z.; Tian, S.; Shi, H.; Song, X. Study of the bottom-hole rock stress field under water jet impact. *Energy Sources Part A* **2016**, *38*, 164–173. [CrossRef]
22. Li, W.; Yan, T.; Bi, X.L.; Sun, S.H. Study of Rock Stress State Near the Bottom to Underbalange Drilling. *Sci. Technol. Eng.* **2012**, *12*, 5139–5143.
23. Heydarshahy, S.A.; Karekal, S. Influences of bit profiles on possible fracture modes. *Pet. Explor. Dev.* **2017**, *44*, 667–674. [CrossRef]
24. Shen, Z.; Peng, Y. Analysis of bottom-hole stress field of two-stage PDC bit. *J. China Univ. Pet. Ed. Nat. Sci.* **2006**, *30*, 62–67.
25. Warren, T.M.; Smith, M.B. Bottomhole stress factors affecting drilling rate at depth. *J. Pet. Technol.* **1985**, *37*, 1523–1533. [CrossRef]
26. Chang, D.; Li, G.; Shen, Z.; Huang, Z.; Tian, S.; Shi, H.; Song, X. A study on the effect of bottom-hole differential pressure on the rock stress field. *Energy Sources Part A* **2014**, *36*, 275–283. [CrossRef]
27. Chen, P.J.; Miska, S.; Yu, M.J.; Ozbayoglu, E. Poroelastic modeling of cutting bottom-hole rock–Part I: Stress state of bottom-hole rock. *J. Pet. Sci. Eng.* **2020**, *189*, 107014. [CrossRef]
28. Hu, H.G.; Guan, Z.C.; Xu, Y.Q.; Han, C.; Liu, Y.W.; Liang, D.Y.; Lu, B.P. Bottom-hole stress analysis of ultra-deep wells based on theory of poroelastic mechanics. *J. China Univ. Pet. Ed. Nat. Sci.* **2020**, *44*, 52–61.

29. Hu, H.G.; Guan, Z.C.; Shor, R.; Xu, Y.Q.; Han, C.; Liu, Y.W.; Lu, B.P. Dynamic response and strength failure analysis of bottomhole under balanced drilling condition. *J. Pet. Sci. Eng.* **2020**, *194*, 107561. [CrossRef]
30. Li, S.B.; Dou, T.W.; Dong, D.R.; Zhang, H.J.; Wang, M.; Liu, T.E. Stress state of bottom-hole rocks in underbalanced drilling. *Acta Pet. Sin.* **2011**, *32*, 329–334. [CrossRef]
31. Zhang, H.; Zhang, H.Y.; Guo, B.Y.; Gang, M.H. Analytical and numerical modeling reveals the mechanism of rock failure in gas UBD. *J. Nat. Gas Sci. Eng.* **2012**, *4*, 29–34. [CrossRef]
32. Zhang, R.; Li, G.S.; Tian, S.C. Stress distribution and its influencing factors of bottom-hole rock in underbalanced drilling. *J. Cent. South Univ.* **2018**, *25*, 1766–1773. [CrossRef]
33. Wang, Y.L.; Dusseault, M.B. A coupled conductive–convective thermo-poroelastic solution and implications for wellbore stability. *J. Pet. Sci. Eng.* **2003**, *38*, 187–198. [CrossRef]
34. Coussy, O. *Mechanics and Physics of Porous Solids*; John Wiley & Sons: Oxford, UK, 2011; pp. 78–80.
35. Salimzadeh, S.; Paluszny, A.; Nick, H.M.; Zimmerman, R.W. A three-dimensional coupled thermo-hydro-mechanical model for deformable fractured geothermal systems. *Geothermics* **2018**, *71*, 212–224. [CrossRef]
36. Abousleiman, Y.; Ekbote, S. Solutions for the inclined borehole in a porothermoelastic transversely isotropic medium. *J. Appl. Mech.* **2005**, *72*, 102–114. [CrossRef]

processes

MDPI

Article

A Method of Reducing Friction and Improving the Penetration Rate by Safely Vibrating the Drill-String at Surface

Yuan Long [1,2,3], Xueying Wang [1,2,3,*], Peng Wang [4] and Feifei Zhang [1,2,3]

1 School of Petroleum Engineering, Yangtze University, Wuhan 430100, China
2 National Engineering Research Center for Oil & Gas Drilling and Completion Technology, Yangtze University, Wuhan 430100, China
3 Hubei Key Laboratory of Oil and Gas Drilling and Production Engineering, Yangtze University, Wuhan 430100, China
4 School of Petroleum Engineering, China University of Petroleum (East China), Qingdao 266580, China
* Correspondence: wangxy@yangtzeu.edu.cn

Abstract: Drill-string axial vibration at the surface technology is proposed to reduce the friction between the drill-string and the borehole wall, and to improve load transfer efficiency, the rate of penetration (ROP), and the extended-reach limit of a horizontal well. An analytical framework utilizing the "soft-string" model is constructed. The results obtained from numerical simulations reveal that during the slide drilling operation, the drill-string experiences an axial stick–slip motion, and the weight on bit (WOB) undergoes periodic oscillations. The conventional calibration method of the WOB in the weight indicator gauge is not applicable when the ROP is low. After applying axial vibration on the drill-string at the surface, the WOB increases and becomes smooth because of a release of friction. The amplitude and frequency of the exciting force are the main factors affecting surface vibration effectiveness. There is an optimal frequency for a given case (10 Hz in this paper). This means that the conventional manual pick-up and slack-off by drillers with a high amplitude and a low frequency has little effect on friction reduction. In addition, the conventional method can bring in high risk because of its high root mean square (RMS) acceleration. Safety evaluation results indicate that the drill-string is in a safe state under most of the exciting parameters. The results verify the feasibility and advantages of the proposed technology, and lay a solid theoretical foundation for its application in real drilling applications.

Keywords: petroleum drilling; drill-string mechanics; friction reduction; vibration

Citation: Long, Y.; Wang, X.; Wang, P.; Zhang, F. A Method of Reducing Friction and Improving the Penetration Rate by Safely Vibrating the Drill-String at Surface. *Processes* **2023**, *11*, 1242. https://doi.org/10.3390/pr11041242

Academic Editors: Tianshou Ma and Yuqiang Xu

Received: 19 March 2023
Revised: 7 April 2023
Accepted: 14 April 2023
Published: 17 April 2023

1. Introduction

Drilling complex structural wells (such as directional, horizontal, and extended reach wells) is essential for the successful exploitation of petroleum and natural gas reserves. These types of wells ensure optimal formation exposure and increase the chances of hitting the pay zone. Moreover, these wells are suitable for future stimulation techniques, which can significantly improve well production efficiency and final recovery rates. Directional drilling realized by combining a positive displacement motor (PDM) and measurement while drilling (MWD), which is known as slide drilling, is an effective well trajectory control method and is widely used worldwide. This technology has its drawbacks, as the lack of rotation in the drill-string creates significant friction with the borehole wall. As a result, the transfer of axial load to the bit decreases, leading to a reduction in both the rate of penetration (ROP) and the extended-reach limit. Meanwhile, the stick–slip phenomenon of the drill-string is further intensified by the frequent transform between static and dynamic friction force. Reducing the amount of resistance between the drill-string and the borehole wall and enhancing the effectiveness of load transfer during directional drilling are essential objectives with significant importance in petroleum drilling that have been given attention for several years [1].

Two main approaches are typically employed in friction reduction research, namely reducing the normal contact force and decreasing the coefficient of friction. To decrease the normal contact force, one can optimize the well track and make use of a lightweight drill pipe. To reduce the friction coefficient, it is possible to develop a high-performance lubricant, utilize a cylindrical roller sub, and incorporate a nonrotating protective joint. Adding vibrators to the drill-string was first proposed by Roper in 1983 as a means of reducing friction between the drill-string and the borehole wall. Although categorized as passive methods of friction reduction, these techniques have limited effectiveness [2]. In the past, the concept of utilizing a vibrator for petroleum drilling technology was not extensively researched. However, in recent years, various technology service companies have started conducting application studies on this idea and concentrated their efforts towards enhancing the vibrator's development [3,4]. Axial vibration friction reduction tools were first developed and widely used, represented by Agitator of National Oilwell Varco [5,6]. Then lateral vibration friction reduction tools were developed, represented by Xcite. The research concluded that axial vibration tools provide significantly more effective friction reduction in horizontal wells [7]. Axial and lateral vibration friction reduction patterns are realized by adding downhole tools in the drill-string, which will consume hydraulic energy. Moreover, drillers need to trip out and reassemble the makeup of the string when the well depth goes to a deeper extent. Contrary to the aforementioned vibration technologies, the automatic surface rocking technology utilizes the top drive to rotate the drill-string to the right limit and then to the left limit by a predetermined amount, without relying on any downhole tool. This innovative technique is exemplified by the "Slider" [8] and "DSCS" [9] systems. However, it is sometimes difficult to control the real-time toolface orientation because of dynamic reactive torque from bit–rock interaction.

After weighing the pros and cons of the aforementioned vibration methods, our proposition is to apply axial vibration on the drill-string at the surface. This will reduce the friction between the drill-string and the borehole wall while enhancing the efficiency of load transfer. The excitation means may be displacement or force. Compared to other exciting manners (such as the axial vibration tool Agitator, the lateral vibration tool Xcite, and the torsional vibration system Slider), the technology of drill-string axial vibration at the surface proposed in this paper has the following advantages: Firstly, no drilling fluid energy is needed for vibrating the drill-string, which make it feasible to use and suited for a deep well. Secondly, exciting parameters are adjustable at any time according to the real-time drilling situation through simple surface tools. Thirdly, key seating and pressure differential sticking can also be overcome by applying a higher amplitude and lower frequency excitation.

Scholars have developed many torque and drag calculation models in the last decades, which can be divided into soft-string models [10–14] and stiff-string models [15–18]. However, these models can only be used for calculating the drag while tripping in or tripping out, and the torque while rotating drilling. When the drill-string is vibrated axially at the surface, the drill-string elements are frequently in the states of forward sliding, resting, and backward sliding, which makes the above models not applicable. Meanwhile, dynamic models [19] can be used to study the motion and force characteristics along the whole drill-string, and the load transfer efficiency can be obtained based on the deformation of the drill-string. However, these dynamic models usually focus on the rotating drilling mode, and neglect the axial friction or assume it as linear viscous damping. The study of friction reduction and load transfer in drill-strings under vibrational conditions is significantly lagging behind the advancements made in friction reduction technologies through the applications of vibrating drill-strings. Several scholars have devised models for determining the load transfer of a drill-string subjected to vibration [20–23]. However, the friction model in these models is the Coulomb model that does not consider the viscous effect of friction. The research on the mechanism of friction reduction by applying vibration can provide a reference for choosing a friction model [24]. The friction reduction mechanisms of vibration mainly include: (a) the static friction transforms to dynamic friction due to axial and torsional vibrations [6]; (b) during the cycle of vibration, the direction of dynamic friction is altered by both axial and torsional vibrations, leading to a reduction in the mean friction force [25,26]; and (c) the lateral vibration decreases the normal contact force periodically [27]. Based on these mechanisms, under vibration conditions, the dynamic friction models that take pre-slip displacement into account are better choices to simulate drill-string dynamics.

In this paper, the load transfer characteristics of the drill-string while being vibrated axially at the surface is studied. Firstly, friction is incorporated based on the pre-slip displacement of the drill-string to accurately calculate the friction distribution of the whole drill-string during its axial vibration. The dynamic friction model is used to capture the change in coefficient and a basic model is derived according to the equilibrium of forces. The established numerical model is solved using a finite difference method. Subsequently, evaluations of the load transfer characteristics of the drill-string under both displacement and force excitations are conducted and compared with other types of vibration patterns. The main objective of this study is to demonstrate the viability and benefits of a new technique, which involves axial vibration at the surface to reduce friction. By comparing its load transfer properties with other types of vibration, we aim to establish a strong theoretical basis for the application of this technology in actual drilling.

2. Model

2.1. Assumptions

The drill-string undergoes a complex motion (rolling, revolution, and swirling) and deformation (bending, sinusoidal buckling, and helical buckling) due to the intricate forces exerted on it while drilling underground. In the process of directional drilling, the drill-string remains stationary and experiences significant resistance and drag from the borehole wall. As a result, both rotational and lateral movements are negligible, while axial movement becomes the primary motion. The resistance and drag also decrease the likelihood of buckling occurring, particularly in horizontal sections [28]. The main focus of this study is to examine how axial excitation impacts the load transfer in the directional drilling process of a structurally complex well. Therefore, the following assumptions are made:

(a) Assuming there is no clearance between the drill-string and the borehole wall, the centerline of the drill-string perfectly aligns with the centerline of the borehole, and the drill-string maintains consistent contact with the borehole wall.

(b) Each vibrator is seen as a length of drill-string, and its exciting force is sinusoidal.

(c) Friction and viscous damping of drilling fluid are the two main types of damping forces acting on the drill-string, while mechanical resistance caused by key seating and differential pressure sticking are excluded from consideration.

(d) The analysis focuses on the axial dynamic impact of the drill-string, disregarding the cross-sectional forces of shear and bending moment.

2.2. Equilibrium Equation

To derive the dynamic equation for the vibration process of a drill-string, a small drill-string element ds is selected and analyzed. The element is considered in natural curvilinear coordinates ($\vec{e}_t$, $\vec{e}_n$, $\vec{e}_b$), and is subjected to various forces, such as internal tension force $\vec{T}$, normal contact force $\vec{F}$, friction force F_f, damping force, and buoyant weight of drilling (as displayed in Figure 1). By balancing these forces, we can arrive at an equation that describes the dynamic equilibrium condition [29]:

$$\begin{aligned}\vec{T}(s+ds,t) - \vec{T}(s,t) + \vec{F}\left(s+\tfrac{ds}{2},t\right)ds - [F_f(s,t) + c(s+\tfrac{ds}{2},t)\tfrac{\partial u(s+\frac{ds}{2},t)}{\partial t}]ds\,\vec{e}_t \\ +g_s\left(s+\tfrac{ds}{2}\right)ds\,\vec{e}_g = \rho(s+\tfrac{ds}{2})A(s+\tfrac{ds}{2})ds\tfrac{\partial^2 u(s+\frac{ds}{2},t)}{\partial t^2}\vec{e}_t\end{aligned} \quad (1)$$

where $T_t(s,t)$, $T_n(s,t)$, and $T_b(s,t)$ are internal forces acting on cross-section in $\vec{e}_t$, $\vec{e}_n$, and $\vec{e}_b$ direction, respectively; F_n, F_b are the lateral contact force acting on drill-string in principal normal $\vec{e}_n$ and binormal direction $\vec{e}_b$, respectively.

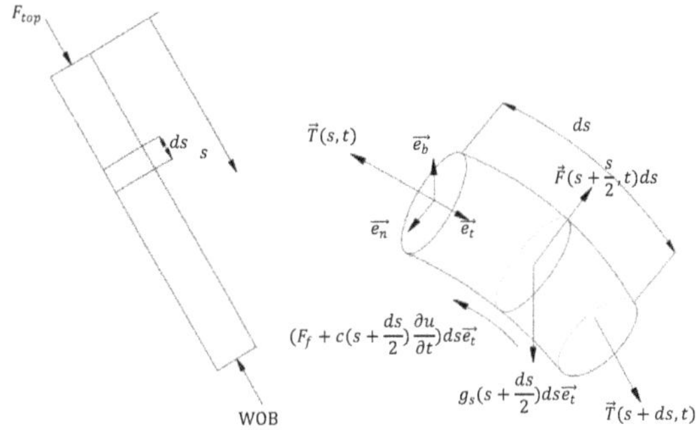

Figure 1. Forces acting on a drill-string element.

The following simpler equilibrium equation can be derived from the Taylor expansion in (s, t) of variables by omitting higher order terms:

$$\frac{\partial \vec{T}(s,t)}{\partial s} + \vec{F}(s,t) - [F_f(s,t) + c(s,t)\frac{\partial u}{\partial t}]\vec{e}_t + g_s(s)\vec{e}_g = \rho(s)A(s)\frac{\partial^2 u(s,t)}{\partial t^2}\vec{e}_t \quad (2)$$

Separating the internal force $\vec{T}$ and distributed lateral contact force $\vec{F}$ into components in natural curvilinear coordinates ($\vec{e}_t$, $\vec{e}_n$, $\vec{e}_b$):

$$\vec{T}(s,t) = T_t(s,t)\vec{e}_t + T_n(s,t)\vec{e}_n + T_b(s,t)\vec{e}_b \quad (3)$$

$$\vec{F}(s,t) = F_n(s,t)\vec{e}_n + F_b(s,t)\vec{e}_b \quad (4)$$

where $T_t(s,t)$,$T_n(s,t)$, and $T_b(s,t)$ are internal forces acting on cross-section in $\vec{e}_t$, $\vec{e}_n$, and $\vec{e}_b$ direction, respectively; F_n, F_b are the lateral contact forces acting on drill-string in principal normal $\vec{e}_n$ and binormal direction $\vec{e}_b$, respectively.

The Frenet–Serret formulas can be used to describe the centerline of the borehole [15]:

$$\begin{aligned} \vec{e}_g \cdot \vec{e}_t &= \cos\bar{\alpha} \\ \vec{e}_g \cdot \vec{e}_n &= \tfrac{k_\alpha}{k_b}\sin\bar{\alpha} \\ \vec{e}_g \cdot \vec{e}_b &= -\tfrac{k_\varphi}{k_b}(\sin\bar{\alpha})^2 \\ \vec{e}_g &= -\cos\bar{\alpha}\,\vec{e}_t + \tfrac{k_\alpha}{k_b}\sin\bar{\alpha}\,\vec{e}_n + \tfrac{k_\varphi}{k_b}\sin^2\bar{\alpha}\,\vec{e}_b \end{aligned} \quad (5)$$

where α, φ, and $\bar{\alpha}$ are inclination angle, azimuth angle, and mean inclination angle, respectively; k_α, k_φ, and k_b are rate of change of inclination angle, rate of change of azimuth angle, and total bending curvature, respectively.

The following scalar equations can be obtained by substituting Equations (3)–(5) into Equation (2):

In $\vec{e}_t$ direction:

$$\frac{\partial T_t}{\partial s} - F_f(s,t) - c\frac{\partial u}{\partial t} + g_s\cos\bar{\alpha} = \rho A\frac{\partial^2 u}{\partial t^2} \quad (6)$$

In $\vec{e}_n$ direction:

$$F_n + T_t k_b + g_s \frac{k_\alpha}{k_b} \sin\alpha = 0 \tag{7}$$

In $\vec{e}_b$ direction:

$$F_b + g_s \frac{k_\varphi}{k_b} (\sin\alpha)^2 = 0 \tag{8}$$

The internal tension force T_t in axial direction for a rod can be written in the following form:

$$T_t = EA\frac{\partial u}{\partial s} \tag{9}$$

where E is the elastic (Young's) modulus of drill-string.

The distributed lateral contact force between drill-string and borehole wall can be obtained by combining Equations (7) and (8):

$$F = \sqrt{F_n^2 + F_b^2} = [(Tk_b + g_s \frac{k_\alpha}{k_b} \sin\alpha)^2 + (g_s \frac{k_\varphi}{k_b} (\sin\alpha)^2)^2]^{\frac{1}{2}} \tag{10}$$

The friction force F_f is calculated by Dahl friction model [30], which has the following form:

$$\begin{cases} \frac{dz}{dt} = v_r \cdot \left[1 - \frac{K}{F_c} \cdot z \cdot sgn(v_r)\right]^i \\ F_f = K_t \cdot z \end{cases} \tag{11}$$

where F_f is the axial friction force; F_c is the Coulomb friction force, $F_c = \mu_d \cdot F$, μ_d is the dynamic sliding friction coefficient; K is the tangential stiffness of contact surfaces; i is the parameter that determines the shape of strain–stress curve, in general $i = 1$; t is the time; $z(t)$ is the offset displacement of asperities at time t; v_r is the relative velocity of drill-string.

2.3. Boundary and Initial Conditions

2.3.1. Boundary Condition

(a) Surface boundary

During the drilling process, drillers aim to control the hook load at surface in order to adjust the actual weight on bit (WOB) to the desired value. This is achieved by ensuring that the hook load, which is the difference between the buoyant weight of the drill-string in the axial direction and the friction force and nominal WOB, is equivalent to the desired value. The request for the driller to maintain a constant hook load results in the friction force and nominal WOB being viewed as constant values. This, however, is subject to change due to the heterogeneity of rocks and the stick–slip motion of the drill-string, which causes the friction force and nominal WOB to transform interchangeably. The zero WOB is determined by running downward the drill-string at constant speed and setting the pointer on the weight indicator to zero before the bit reaches bottom of the well. However, the ROP may be very small in drilling processes, and partial drill-string may be in static friction state. In this case, the real WOB may be very small. With that in mind, we adopt static friction of drill-string to calculate hook load. The hook load can be determined through an analysis of the drill-string under static conditions where the frictional coefficient is equal to the static friction coefficient and the nominal WOB remains constant throughout subsequent calculations.

$$F_{top} = \left(G_t - WOB - F_{f,s}\right) + F_a \sin(\omega t) \tag{12}$$

where F_{top} is the hook load; G_t is the axial component of gravity of the whole drill-string; $F_{f,s}$ is the total static friction of drill-string; F_a is the amplitude of exciting force; ω is the circular frequency of the excitation.

(b) Bit boundary

Generally, a vibrator is typically only required when the ROP is low. Whether the ROP is high or low, the drill-string movement will inevitably experience a stick–slip phenomenon during the slide drilling process, which will cause fluctuations in WOB. Varying WOB will result in varying ROP, which in turn affects the WOB. In 1970, Amoco petroleum exploration and development corporation proposed binary drilling rate model according to a great field of data, and used different bit weight exponent and rotate speed exponent in different hardness stratum. When the stratum changes from soft to hard, bit weight exponent changes from small to big, and rotate speed exponent changes from big to small, and the variation trend of penetration rate changes from fast to slow. The drilling rate model has the following form [31]:

$$ROP = K \cdot WOB^a \cdot n^b = K' \cdot WOB^a \tag{13}$$

where n is bit rotate speed; a and b are WOB exponent and bit rotate speed exponent, respectively.

2.3.2. Initial Conditions

If we suppose that the drill-string starts from rest, with zero initial velocity. The initial position of the drill bit is set to zero, while the initial position of the remaining section of the drill-string can be determined through static force equilibrium calculations.

$$\begin{gathered} \left.\frac{\partial u}{\partial t}\right|_{t=0} = 0 \\ u|_{t=0} = \Phi(T_t),\ u(L)|_{t=0} = 0 \end{gathered} \tag{14}$$

where Φ is a mapping function from the initial tension distribution to initial displacement distribution of the drill-string.

2.4. Solution Method

Essentially, the model described involves the propagation of elastic waves. To solve it, a finite difference method has been selected in this section. The governing differential equation has been dispersed using a central difference scheme. After this, the differential equations are converted into algebraic equations and solved using MATLAB programing.

To begin with, the drill-string is segmented into N sections, spanning from the top to the bottom of the well. Difference grids are formed while setting time step τ and space step h. Subsequently, the node displacement of the drill-string is represented as $u(p,k)$ where p and k represent the position and time, respectively, and are abbreviated as $u(p{\cdot}h, k{\cdot}\tau)$. There is nonlinear term $\frac{\partial(EA\frac{\partial u}{\partial s})}{\partial s}$ in the governing differential equation. In order to deal with this nonlinear preferably, the governing differential equation is integrated for random node p in interval $\left[\left(p-\frac{1}{2}\right)h, \left(p+\frac{1}{2}\right)h\right]$. We can obtain the following Equation [29]:

$$\begin{gathered} \left(EA\frac{\partial u}{\partial s}\right)^{p+\frac{1}{2}} - \left(EA\frac{\partial u}{\partial s}\right)^{p-\frac{1}{2}} - \frac{u(p,k+1)-u(p,k-1)}{2\tau}\int_{(p-\frac{1}{2})h}^{(p+\frac{1}{2})h} c ds + \int_{(p-\frac{1}{2})h}^{(p+\frac{1}{2})h} g_s \cos\bar{\alpha} ds \\ -\mathrm{sgn}(u(p,k)-u(p,k-1))\int_{(p-\frac{1}{2})h}^{(p+\frac{1}{2})h} F\mu ds = \frac{u(p,k+1)-2u(p,k)+u(p,k-1)}{\tau^2}\int_{(p-\frac{1}{2})h}^{(p+\frac{1}{2})h} \rho A ds \end{gathered} \tag{15}$$

where sgn() is the sign function; h is space size; τ is time step.

We denote:

$$T^{p+\frac{1}{2}} = (EA\frac{\partial u}{\partial s})^{p+\frac{1}{2}} = EA\frac{u(p+1,k)-u(p,k)}{h} \tag{16}$$

$$T^{p-\frac{1}{2}} = (EA\frac{\partial u}{\partial s})^{p-\frac{1}{2}} = EA\frac{u(p,k)-u(p-1,k)}{h} \tag{17}$$

$$F_f(p,k) = \text{sgn}(u(p,k) - u(p,k-1)) \int_{(p-\frac{1}{2})h}^{(p+\frac{1}{2})h} F\mu ds \tag{18}$$

$$C^p = \int_{(p-\frac{1}{2})h}^{(p+\frac{1}{2})h} cds \tag{19}$$

$$A^p = \int_{(p-\frac{1}{2})h}^{(p+\frac{1}{2})h} \rho A ds \tag{20}$$

$$g^p = \int_{(p-\frac{1}{2})h}^{(p+\frac{1}{2})h} g_s \cos\bar{\alpha} ds \tag{21}$$

Substitute Equations (16)–(21) into (15), we can obtain the recurrence algorithm of drill-string axial displacement:

$$u(p,k+1) = \frac{2\tau^2}{2A^p + C^p\tau}\left[T^{p+\frac{1}{2}} - T^{p-\frac{1}{2}} - F_f(p,k) + g^p + \frac{2A^p}{\tau^2}u(p,k) + (\frac{C^p\tau - 2A^p}{2\tau^2})u(p,k-1)\right] \tag{22}$$

Difference solution format (22) is explicit. If we know the displacement of two former time steps, we can calculate drill-string displacement at any time step, and then the drilling parameters, such as axial tension force and WOB, can be obtained.

In order to achieve convergence of the solution, there must be a specific relationship between the time step τ and the space step h. By utilizing the Fourier error analysis method, it has been proven that the difference scheme of Equation (22) is only stable under certain conditions. This condition can be expressed as the convergence condition:

$$\tau \leq \frac{h}{\sqrt{\frac{E}{\rho}}} \tag{23}$$

3. Results and Discussion

3.1. Advantages of Surface Vibrating Technology

Nominal WOB equals the vector sum of the hook load F_{top}, the gravity of the drill-string, the real WOB, and the static friction between the drill-string and the borehole in the axial direction of the drill-string. A nominal constant WOB is applied and we expect to obtain a constant real WOB and ROP in conventional slide drilling. However, the real WOB and ROP are variational because of the existence of friction between the drill-string and the borehole wall and the stick–slip motion of the drill-string. We choose a representative horizontal well to analyze its load transfer characteristics. The well profile is shown in Figure 2. The well depth is 4200 m, and the kickoff point and landing point are 1290 m and 2190 m, respectively, and the build-up rate is 0.0524 rad/30 m. The drill-string is made up of ∅127 mm drill pipe, whose inner diameter, density, and elasticity modulus are 108.6 mm, 7850 kg/cm^3, and 210 GPa, respectively. Other calculating parameters are as follows: the static and dynamic frictional coefficient are equal to 0.3 and 0.25 (static coefficient is always 1.2 times of dynamic coefficient [6]), respectively. The time step is 0.00005 s and the space step is 5 m. The exchange critical velocity of static friction and dynamic friction is 0.01 m/s. A binary model of the ROP and the WOB is adopted (see Equation (13)), in which WOB_{cri} = 20 KN, a = 1, and K' = 5×10^{-5}. Under such circumstances, the static friction and dynamic friction of the whole drill-string are 173 kN and 144 kN, respectively. Therefore, the maximal friction released by vibration equals 29 kN. The increase in the WOB also includes the exciting force and inertia force of the drill-string.

Figure 3 shows the variations in the real WOB and ROP with a different nominal WOB. From Figure 3, although the nominal WOB applied at the surface is constant, the real WOB and ROP at the bottom are fluctuant because of the static–dynamic friction switch and stick–slip. Moreover, the real WOB transmitted to the bit increases with the increase in the nominal WOB. The amplitude and frequency of real WOB fluctuation decrease with the increase in the nominal WOB because of the continuous and steady motion of the drill-string under a high nominal WOB.

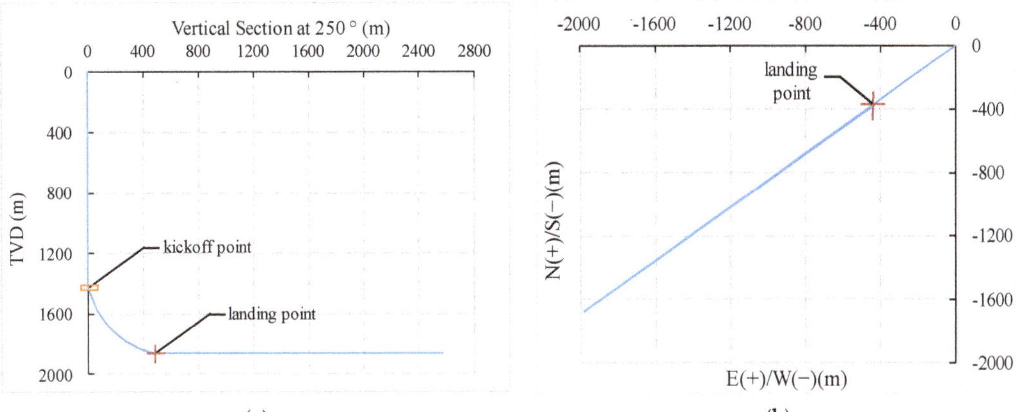

(a) (b)

Figure 2. Well profile of a horizontal well. (**a**) Vertical projection view. (**b**) Horizontal projection view.

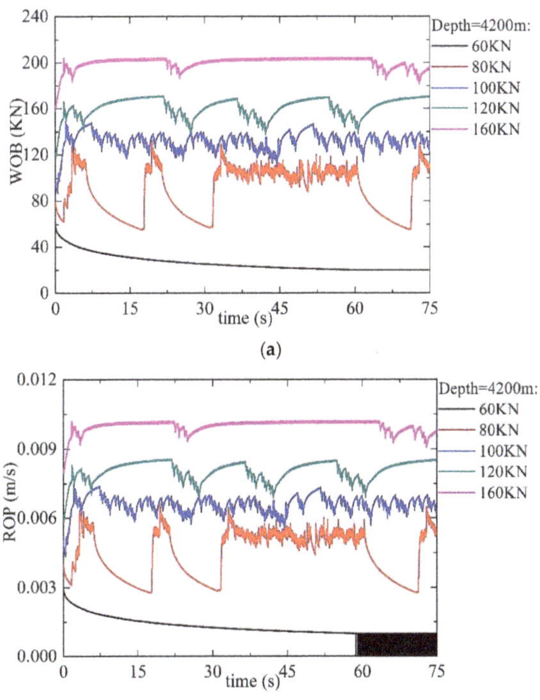

(a)

(b)

Figure 3. Variations in real WOB and ROP with different nominal WOB. (**a**) Real WOB; (**b**) ROP.

Even though under a certain nominal WOB, the load transfer efficiency decreases in a well with a longer horizontal section and higher friction coefficients. As can be seen in Figures 4 and 5, the amplitude of the downhole WOB fluctuation increases and the frequency decreases with the increase in the horizontal section length and friction coefficients when the nominal WOB equals 100 kN, which means that the load transfer becomes worse.

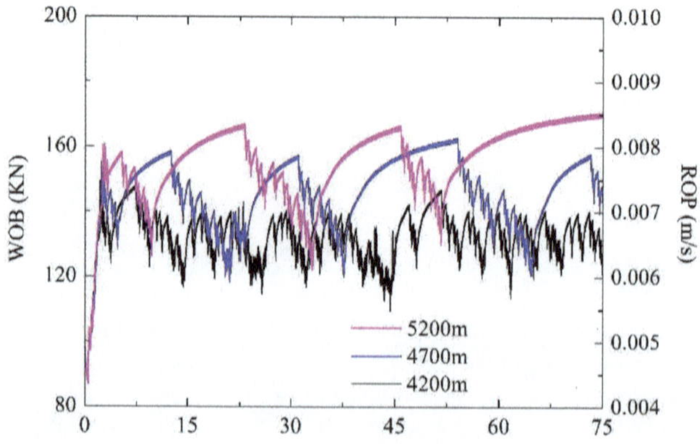

Figure 4. The effect of horizontal section length on WOB.

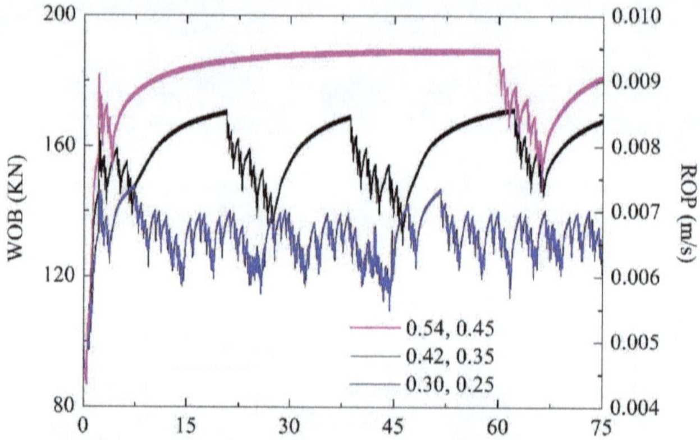

Figure 5. The effect of friction coefficients on WOB and ROP.

As can be seen from Figure 3, the WOB fluctuates severely under a lower WOB, and the drill-string cannot move continuously under 60 kN WOB. The reason for this is the stick–slip phenomenon of the drill-string caused by the static–dynamic friction switch between the drill-string and the borehole wall. If we apply an axial exciting force with an amplitude of 20 kN and a frequency of 10 Hz on the drill-string at the surface, will the load transfer be improved? Figure 6 shows the WOB changes during typical slide drilling and surface vibration drilling. From Figure 6, the real WOB and ROP improve significantly after applying vibration at the surface, and the WOB curve becomes smoother, which is beneficial for the drilling operation.

3.2. *Influence Factor Analysis*

According to the analysis of Section 3.1, we know that the real WOB and ROP improve significantly after applying vibration at the surface, which means that the drill-string axial vibration at the surface technology is effective. However, the questions as to whether it is efficient and how to set the vibrating parameters should be answered. In the application process of this technology, we can only adjust the amplitude and the frequency of the exciting force to control the effectiveness of load transfer. Figure 7 shows the influences of the amplitude and the frequency of the exciting forces on the WOB and ROP. As can be seen from Figure 7, the WOB becomes smooth and fluctuates less with the increase in the amplitude or frequency of the exciting force. The stick–slip phenomenon is suppressed significantly. The WOB and ROP are sensitive to the exciting frequency at a low exciting amplitude, but they are rarely affected at a larger exciting force. Therefore, better load transfer results can be obtained at a higher exciting force amplitude even though the exciting frequency changes a lot in drilling.

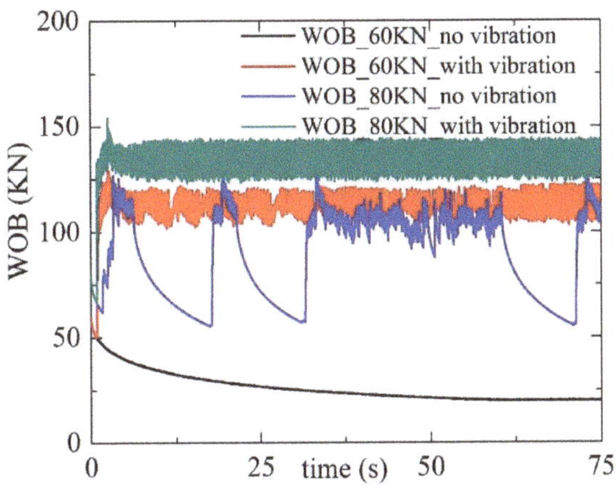

Figure 6. WOB changes during surface vibration drilling.

3.3. *Safety Evaluation of Surface Vibrated Drill-String*

The introduction of vibration in drilling can reduce the friction between the drill-string and the wellbore wall and improve the load transfer efficiency, and a higher exciting amplitude or frequency is good for improving friction reduction and load transfer effectiveness. However, vibration of the drill-string will simultaneously increase the internal stress and the failure risk of the drill-string. Therefore, the increase in the exciting amplitude and frequency must be restricted by the dynamic safety of the drill-string. The methods for evaluating the drill-string's safety mainly include the safety factor method and the real-time monitoring of downhole tools. Drill stem design and operation [32] indicates that overload or fatigue are the two main reasons for drill-string failure. In the process of drill-string design, the safety factors in the stretch and torsional direction are 1.0 to 1.3, which guarantee the real stress of the drill-string is lower than the allowable stress.

However, the safety factor method does not consider the effect of induced vibration on drill-string safety. A large number of statistical results of drill-string failure indicate that drill-string vibration will result in a dramatic change in the drill-string stress on the premise that the drill-string stress is lower than the allowable stress, which will influence the safety of the drill-string. Based on the above considerations, several petroleum technology service corporations have developed underground vibration monitoring systems (such as the VSS system of Baker Hughes, MVC system of Schlumberger, and TVM system of Weatherford; the installation methods of these systems are different) to control the vibration and decrease the dynamic stress level of the drill-string by adjusting the drilling parameters in the drilling process. We adopted the method of Baker Hughes (see Table 1 [33]) to evaluate the safety of the drill-string axial vibration at the surface technology proposed in this paper.

Table 1. Vibration level classification (Baker Hughes INTEQ company).

Grade	Vibration Level in Axial Direction RMS(g)	Color Marking	Evaluation Result
0	[0.0, 0.5)	green	safe
1	[0.5, 1.0)		
2	[1.0, 2.0)	yellow	caution
3	[2.0, 3.0)		
4	[3.0, 5.0)	red	danger
5	[5.0, 8.0)		
6	[8.0, 15.0)		
7	[15.0, ∞)		

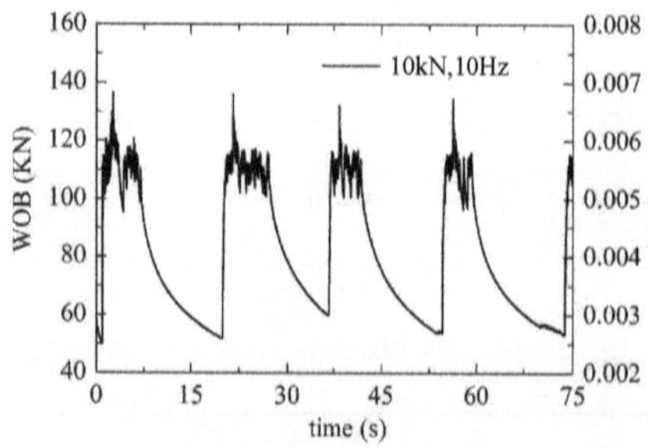

(**a**)

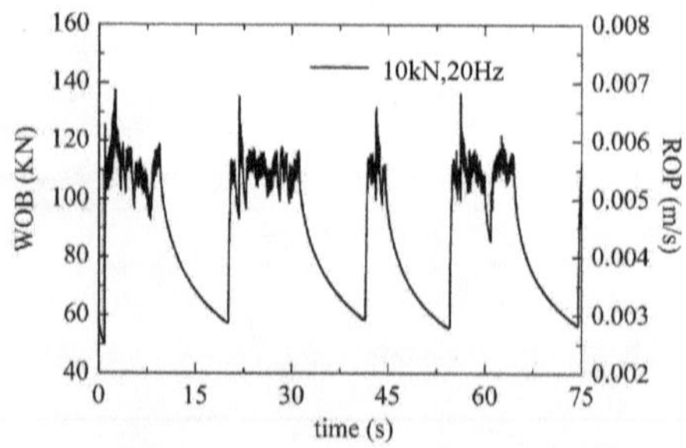

(**b**)

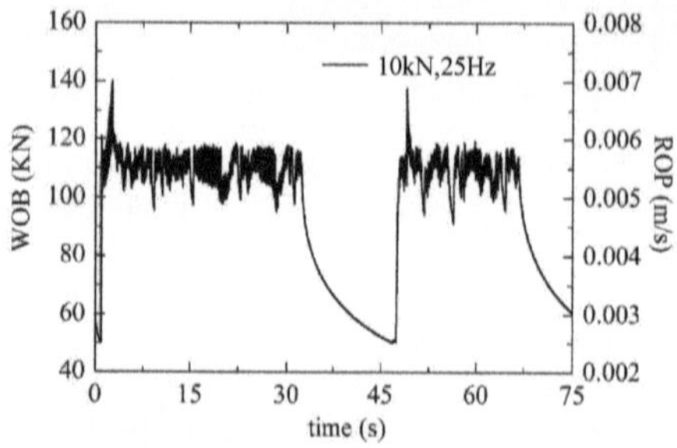

(**c**)

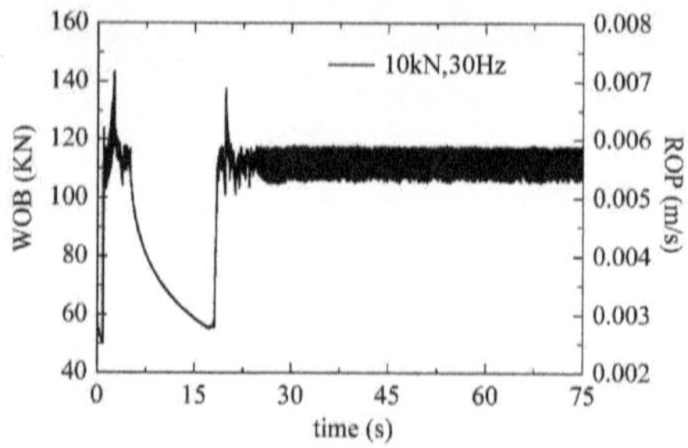

(**d**)

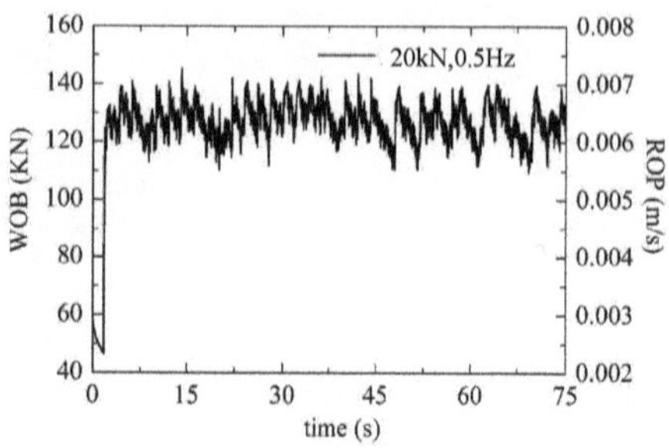

(**e**)

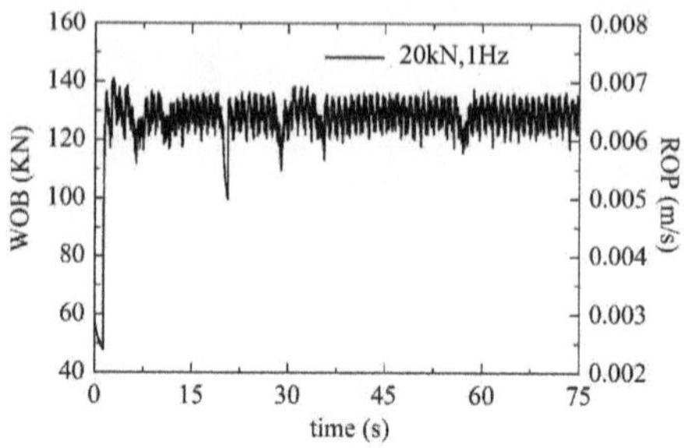

(**f**)

Figure 7. *Cont.*

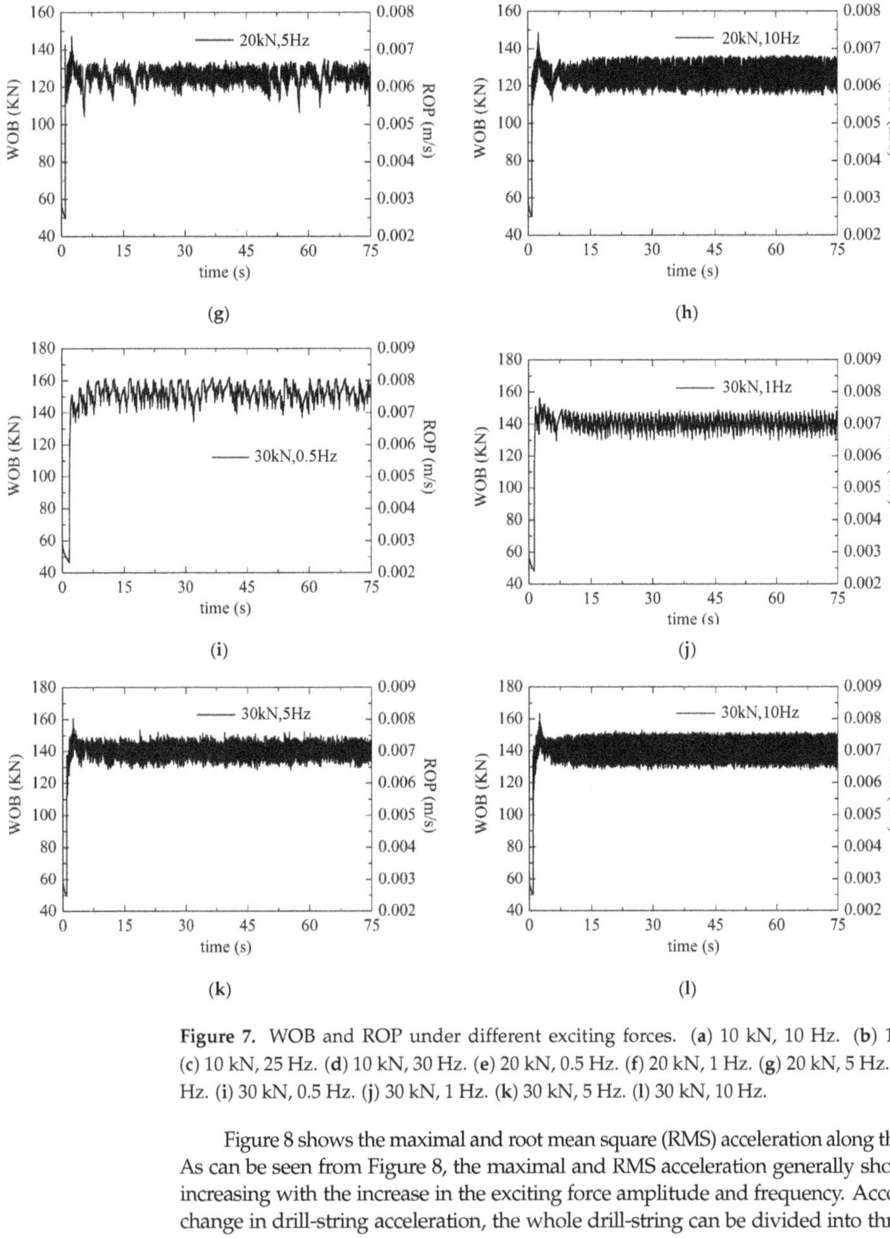

Figure 7. WOB and ROP under different exciting forces. (**a**) 10 kN, 10 Hz. (**b**) 10 kN, 20 Hz. (**c**) 10 kN, 25 Hz. (**d**) 10 kN, 30 Hz. (**e**) 20 kN, 0.5 Hz. (**f**) 20 kN, 1 Hz. (**g**) 20 kN, 5 Hz. (**h**) 20 kN, 10 Hz. (**i**) 30 kN, 0.5 Hz. (**j**) 30 kN, 1 Hz. (**k**) 30 kN, 5 Hz. (**l**) 30 kN, 10 Hz.

Figure 8 shows the maximal and root mean square (RMS) acceleration along the drill-string. As can be seen from Figure 8, the maximal and RMS acceleration generally show a trend of increasing with the increase in the exciting force amplitude and frequency. According to the change in drill-string acceleration, the whole drill-string can be divided into three segments by the kickoff point (depth = 1290 m) and landing point (depth = 2190 m). In the case that the exciting force amplitude equals 10 kN, the maximal acceleration of the building up section and the horizontal section is larger than that of the vertical section, while the RMS acceleration is close. The maximal acceleration and RMS acceleration show a lower value under 10 Hz and 30 Hz. From Figure 7, we can see the load transfer under 10~25 Hz is far from 30 Hz. Therefore, a higher frequency should be applied in order to obtain a better load transfer effect and lower acceleration. In the case that the exciting force amplitude equals 20 kN and 30 kN, the maximal acceleration of the building up section and the horizontal section is larger than that of the vertical section, while the RMS acceleration shows the opposite trend. For maximal acceleration, a lower value is obtained at 5 Hz and 10 Hz and the highest value is obtained at 0.5 Hz in the vertical section, while a lower value is obtained at 10 Hz to 30 Hz and the highest value is obtained at 0.5 Hz to 5 Hz in the horizontal section. For RMS acceleration, a lower value is obtained at 0.5 Hz to 5 Hz and the highest value is obtained at 20 Hz to 30 Hz in the vertical section, while a lower value is obtained at 10 Hz to 30 Hz and the highest value is obtained at 0.5 Hz to 5 Hz in the horizontal well section. Therefore, we can see that the RMS acceleration shows an opposite trend in the vertical and horizontal sections. When the frequency equals 10 Hz, the differences in the RMS between the different well sections are small and the RMS acceleration of the whole drill-string has a lower value. Therefore, 10 Hz is the optimum value for drill-string safety under the conditions set by the given parameters.

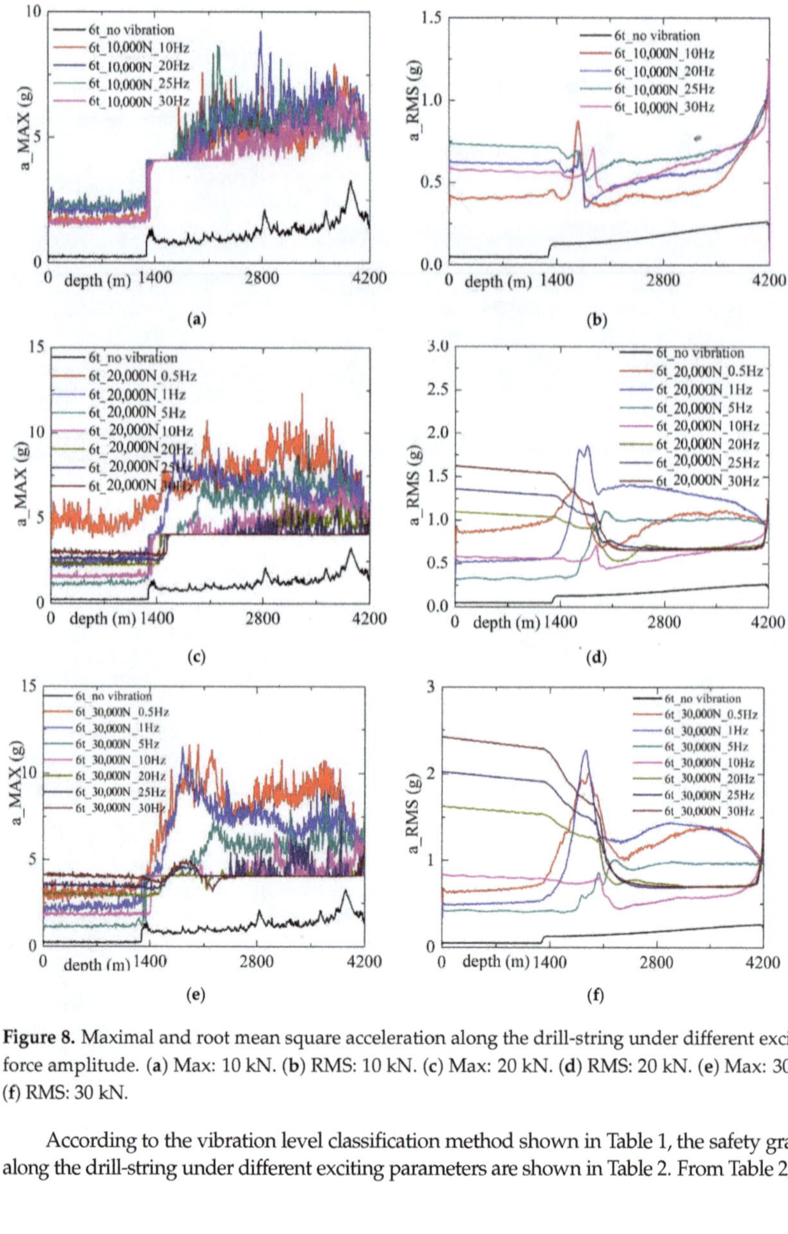

Figure 8. Maximal and root mean square acceleration along the drill-string under different exciting force amplitude. (**a**) Max: 10 kN. (**b**) RMS: 10 kN. (**c**) Max: 20 kN. (**d**) RMS: 20 kN. (**e**) Max: 30 kN. (**f**) RMS: 30 kN.

According to the vibration level classification method shown in Table 1, the safety grades along the drill-string under different exciting parameters are shown in Table 2. From Table 2, the drill-string near bit is all in the caution (two grade) state, which means that its RMS acceleration ranges from 1 g to 2 g. Only when the amplitude of the exciting force equals 30 kN and the frequency is lower than 1 Hz or more than 25 Hz, is it certain that the well section is in the caution (three grade) state, which means that its RMS acceleration ranges from 2 g to 3 g. Caution (two grade) should be avoided, and caution (three grade) is dangerous for longstanding drilling. According to Table 2, 5 Hz to 10 Hz is the optimum frequency range for drill-string axial vibration at the surface technology under the given parameters.

Table 2. Safety evaluation results of drill-string while being axially vibrated at surface.

Frequency/Hz Exciting Force/kN	0	0.5	1	5	10	20	25	30
100	0~4200				0~1600 ~1750 ~3530 ~4170 ~4200	0~1750 ~2450 ~4150 ~4200	0~4150 ~4200	0~1920 ~2330 ~4170 ~4200
200	0~4200	0~1230 ~1755 ~2780 ~4095 ~4180 ~4200	0~1520 ~4110 ~4180 ~4200	0~1660 ~1940 ~3970 ~4175 ~4200	0~1920 ~2330 ~4170 ~4200	0~1415 ~4180 ~4200	0~1865 ~4180 ~4200	0~1895 ~4180 ~4200
300	0~4200	0~1455 ~1890 ~1915 ~4200	0~555 ~1620 ~1790 ~1915 ~4200	0~1775 ~2220 ~2285 ~4160 ~4200	0~2210 ~2555 ~4160 ~4200	0~2075 ~4175 ~4200	0~275 ~2130 ~4165 ~4200	0~1570 ~2155 ~4165 ~4200

Safe (zero grade)	Safe (one grade)	Caution (two grade)	Caution (three grade)

4. Conclusions

This work proposed a method to reduce friction and improve the ROP of slide drilling. The method applies axial vibration on the drill-string at the surface. A mathematical vibration model was established to simulate the load transfer in the drill-string, and this paper discussed the advantages, influencing factors, and feasibility of the drill-string axial vibration at the surface technology.

Numerical modeling results show that the motion of the drill-string in the conventional drilling process has stick–slip features, and the WOB is fluctuant with certain frequencies. If the friction force equals the weight of the upper drilling string, no weight can be transmitted to the bit. The conventional method of calibrating the WOB using a weight indicator gauge is not suitable when the ROP is slow. The proposed technology of applying axial vibration to the drill-string at the surface has been proven to be highly effective. This is supported by the significant increase in the WOB resulting from the release of friction after the application of the vibration. The amplitude and frequency of the exciting forces are the main influence factors of the effectiveness of the application of this technology and the safety of the drill-string. The horizontal well section exhibits a high RMS acceleration at low frequencies, whereas the vertical well section demonstrates a high RMS acceleration at high frequencies. There is an optimal frequency for a given case (10 Hz in this paper). This means that the conventional manual pick-up and slack-off at the surface with a high amplitude and low frequency has little friction reduction effect. Safety evaluation results indicate that the drill-string is in a safe state under most of the exciting parameters.

Some simplification and issues should be considered in future research, such as:

(a) Utilizing the soft-string model in the mathematical model accommodates accurate results for wells with less curvature on site. However, it may lead to significant discrepancies in high curvature intervals.

(b) During the drilling process, the drill-string is composed of joints and centralizers. The borehole wall is not uniform in shape and the gap between the drill-string and the borehole wall cannot be disregarded. These factors will inevitably impact the principles of friction reduction and load transfer. Therefore, the actual conditions should be considered in future research.

Author Contributions: Y.L. conceived and designed the program, and X.W. supervised the study and debugged the program. P.W. assisted in the development of the algorithm. F.Z. analyzed the computed results and wrote the paper. All authors have read and agreed to the published version of the manuscript.

Funding: This paper is sponsored by the Hubei Key Laboratory of Oil and Gas Drilling and Production Engineering (Yangtze University) (Grant No. YQZC202208) (X.W.), and Hubei Provincial Science and Technology Agency (Grant No. 2019CFA093) (Z.F.).

Data Availability Statement: The data used to support the findings of this study are available from the corresponding author upon request.

Acknowledgments: The authors also gratefully acknowledge the financial support of the CNPC Bohai Drilling Corporation.

Conflicts of Interest: The authors declare no conflict of interest.

Nomenclature

Roman symbols:	
a, b	weight on bit exponent and rotate speed exponent, respectively
A	cross-section area of drill-string, m^2
c	drilling fluid drag, $N\ s/m^2$
D	inner diameter of drill-string, m
E	Elastic (Young's) modulus of drill-string, Pa
$\vec{e}_g$	vector of submerged drill-string weight
$\vec{e}_t, \vec{e}_n, \vec{e}_b$	unit base vectors in natural curvilinear system
f	frequency of vibration, Hz
F, F_n, F_b	normal contact force and its components in $\vec{e}_n$ and $\vec{e}_b$ direction, N
F_c	Coulomb friction force, N
F_a	amplitude of exciting force, N
F_f	axial friction force, N
$F_{f,s}$	total static friction of drill-string, N
F_{top}	hook load, N
g_s	linear buoyant weight of drill-string, N/m
$g(x)$	constraint condition
G_t	axial component of gravity of the whole drill-string, N
h	space step, m
i	parameter determines the shape of strain–stress curve
k_α	rate of change of deviation angle, and, respectively, rad/m
k_φ	rate of change of azimuth angle, rad/m
k_b	total bending curvature, rad/m
K_t	tangential stiffness of contact surfaces, N/m
L	length of drill-string, m
n	rotate speed of drill-string, 2π rad/s
ROP	rate of penetration, m/s
s	well depth, m
t	time, s
T_{max}	strength of drill-string, N
T, T_t, T_n, T_b	internal tension force and its components in $\vec{e}_t, \vec{e}_n, \vec{e}_b$ direction, N
u	axial displacement of drill-string, m
v_r	relative velocity of drill-string, m/s
WOB	weight on bit, N
z(t)	offset displacement of asperities at time t, m

Greek symbols:

α	deviation angle, rad
φ	azimuth angle, rad
τ	time step, s
Φ	initial displacement distribution of drill-string, m
$\overline{\alpha}$	mean deviation angle, rad
ρ	density of drill-string, kg/m^3
μ	instantaneous friction coefficient in $\vec{e}_t$ direction
μ_s	static friction coefficient
μ_d	dynamic friction coefficient
ω	circular frequency of the excitation, Hz

References

1. Gao, D.; Tan, C.; Tang, H. Limit analysis of extended reach drilling in South China Sea. *Pet. Sci.* **2009**, *6*, 166–171. [CrossRef]
2. Roper, W.F.; Dellinger, T.B. Reduction of the Frictional Coefficient in a Borehole by the Use of Vibration. US Patent 4384625, 24 May 1983.
3. Maidla, E.; Haci, M.; Jones, S.; Cluchey, M.; Alexander, M.; Warren, T. Field Proof of the New Sliding Technology for Directional Drilling. In Proceedings of the SPE/IADC Drilling Conference, Amsterdam, The Netherlands, 23–25 February 2005.
4. Jones, S.; Feddema, C.; Sugiura, J.; Lightey, J. A New Friction Reduction Tool with Axial Oscillation Increases Drilling Performance: Field-Testing with Multiple Vibration Sensors in One Drill String. In Proceedings of the IADC/SPE Drilling Conference and Exhibition, Fort Worth, TX, USA, 1–3 March 2016.
5. Ali, A.A.; Barton, S.; Mohanna, A. Unique axial oscillation tool enhances performance of directional tools in extended reach applications. In Proceedings of the Brasil Offshore Conference and Exhibition, Macae, Brazil, 14–17 June 2011.
6. Skyles, L.; Amiraslani, Y.; Wilhoit, J. Converting Static Friction to Kinetic Friction to Drill Further and Faster in Directional Holes. In Proceedings of the IADC/SPE Drilling Conference and Exhibition, San Diego, CA, USA, 6–8 March 2012.
7. Gee, R.; Hanley, C.; Hussain, R.; Canuel, L.; Martinez, J. Axial Oscillation Tool vs. Lateral Vibration Tools for Friction Reduction-What's the Best Way to Shake the Pipe? In Proceedings of the SPE/IADC Drilling Conference and Exhibition, London, UK, 17–19 March 2015.
8. Eric, M.; Marc, H.; Daniel, W. Case history summary: Horizontal drilling performance improvement due to torque rocking on 800 horizontal land wells drilled for unconventional gas resources. In Proceedings of the SPE Annual Technical Conference and Exhibition, New Orleans, Louisiana, 4–7 October 2009.
9. Gillan, C.; Boone, S.; Kostiuk, G.; Schlembach, C.; Pinto, J.; LeBlanc, M. Applying precision drill pipe rotation and oscillation to slide drilling problems. In Proceedings of the SPE/IADC Drilling Conference, Amsterdam, The Netherlands, 17–19 March 2009.
10. Sheppard, M.C.; Wick, C.; Burgess, T. Designing well paths to reduce drag and torque. *SPE Drill. Eng. New Orleans* **1987**, *2*, 344–350. [CrossRef]
11. Maidla, E.E.; Wojtanowicz, A.K. Field Comparison of 2D and 3D Methods for the Borehole Friction Evaluation in Directional Wells. In Proceedings of the 62nd Annual Technical Conference and Exhibition, Dallas, TX, USA, 27–30 September 1987.
12. Paslay, P.R. Stress Analysis of Drillstrings. In Proceedings of the University of Tulsa Centennial Petroleum Engineering Symposium, Tulsa, Oklahoma, 29–31 August 1994.
13. Aadnoy, B.S.; Andersen, K. Design of oil wells using analytical friction models. *J. Pet. Sci. Eng.* **2001**, *32*, 53–71. [CrossRef]
14. Johancsik, C.A.; Friesen, D.B.; Dawson, R. Torque and drag in directional wells-prediction and measurement. *J. Pet. Tecknol.* **1984**, *36*, 987–992. [CrossRef]
15. Ho, H.S. An Improved Modeling Program for Computing the Torque and Drag in Directional and Deep Wells. In Proceedings of the SPE Annual Technical Conference and Exhibition, Houston, TX, USA, 2–5 October 1988.
16. Adewuya, O.A.; Pham, S.V. A robust torque and drag analysis approach for well planning and drillstring design. *J. Pet. Technol.* **1998**, *50*, 1–16.
17. Rezmer-Cooper, I.; Chau, M.; Hendricks, A.; Woodfine, M.; Stacey, B.; Downton, N. Field Data Supports the Use of Stiffness and Tortuosity in Solving Complex Well Design Problems. In Proceedings of the IADC/SPE Drilling Conference, Amsterdam, Holland, 9–11 March 1999. IADC/SPE 52819.
18. Zifeng, L.; Xisheng, L.; Daqian, Z. A Steady Tension-torque Model for Drillstring in Horizontal Wells. *Pet. Drill. Technol.* **1993**, *20*, 1–6.
19. Dykstra. *Nonlinear Drill String Dynamics*; University of Tulsa: Tulsa, OK, USA, 1996.
20. Pabon, J.; Wicks, N.; Chang, Y.; Dow, B.; Harmer, R. Modeling Transient Vibrations While Drilling Using a Finite Rigid Body Approach. In Proceedings of the SPE Deepwater Drilling and Completions Conference, Galveston, TX, USA, 5–6 October 2010.
21. Wicks, N.; Pabon, J.; Auzerais, F.; Kats, R.; Godfrey, M.; Chang, Y.; Zheng, A. Modeling of Axial Vibrations to Allow Intervention in Extended Reach Wells. In Proceedings of the SPE Deepwater Drilling and Completions Conference, Galveston, TX, USA, 20–21 June 2012.
22. Ritto, T.G.; Escalante, M.R.; Rosales, M.B. Drill-string horizontal dynamics with uncertainty on the frictional force. *J. Sound Vib.* **2013**, *332*, 145–153. [CrossRef]
23. Wang, X.M.; Chen, P.; Ma, T.S.; Liu, Y. Modeling and experimental investigations on the drag reduction performance of an axial oscillation tool. *J. Nat. Gas Sci. Eng.* **2017**, *39*, 118–132. [CrossRef]

24. Wang, P.; Ni, H.; Wang, R. Experimental investigation of the effect of in-plane vibrations on friction for different materials. *Tribol. Int.* **2016**, *99*, 237–247. [CrossRef]
25. Gutowski, P.; Leus, M. The effect of longitudinal tangential vibrations on friction and driving forces in sliding motion. *Tribol. Int.* **2012**, *55*, 108–118. [CrossRef]
26. Gutowski, P.; Leus, M. Computational model for friction force estimation in sliding motion at transverse tangential vibrations of elastic contact support. *Tribol. Int.* **2015**, *90*, 455–462. [CrossRef]
27. Tolstoi, D.M.; Borisova, G.A.; Grigorova, S.R. Friction regulation by perpendicular oscillation. *Sov. Phys. Dokl.* **1973**, *17*, 80–86.
28. Gao, B.K.; Gao, D.L. Possibility of drill stem buckling in inclined hole. *Oil Drill. Prod. Technol.* **1995**, *17*, 6–11.
29. Wang, P.; Ni, H.J.; Wang, X.Y.; Wang, R.; Lu, S. Research on the characteristics of earthworm-like vibration drilling. *J. Pet. Sci. Eng.* **2018**, *160*, 60–71. [CrossRef]
30. Dahl, P.R. Solid friction damping of mechanical vibrations. *AIAA J.* **1976**, *14*, 1675–1682. [CrossRef]
31. Yan, T. *Theory and Practice of Optimizing Drilling Parameters*; Harbin Institute of Technology Press: Harbin, China, 1994.
32. TH HILL. *Standard DS-1: Drill Stem Design and Operation*; TH HILL Associates: Houston, TX, USA, 2004.
33. Hu, Y.B.; Di, Q.F.; Zou, H.Y. The New Developments of Monitoring Technology and Researches on Drill String Dynamics. *Pereoleum Drill. Technol.* **2006**, *34*, 7–10.

processes

MDPI

Article

Stick–Slip Characteristics of Drill Strings and the Related Drilling Parameters Optimization

Chao Wang [1], Wenbo Chen [1], Zhe Wu [2,*], Jun Li [2] and Gonghui Liu [3]

1 School of Mechanical Engineering, Yangtze University, Jingzhou 434023, China; cw_paper@163.com (C.W.); cwb_paper@163.com (W.C.)
2 College of Petroleum Engineering, China University of Petroleum-Beijing, Beijing 102249, China; lijun446@vip.163.com
3 Beijing University of Technology, Beijing 100124, China; lgh_1029@163.com
* Correspondence: zhewu_paper@163.com

Abstract: To eliminate or reduce stick–slip vibration in torsional vibration of the drilling string and improve the rate of penetration (ROP), a stick–slip vibration model of the drilling string considering the ROP was established based on the multidimensional torsional vibration model of the drilling string. The model was verified by simulation analysis. The characteristics of the drilling string stick–slip vibration in the three stages of stationary, slip, and stick were analyzed. This paper investigated the influence of rotary torque, rotary speed, and weight on bit (WOB) on stick–slip vibrations in the drill string. Based on this, the relationship between the drilling parameters and ROP was established. Drilling parameter optimization was completed for soft, medium-hard, and hard formations. Results showed that appropriately increasing torque and decreasing WOB can reduce or even eliminate stick–slip vibrations in the drill string and increase the ROP. The parameter optimization increased the ROP by 11.5% for the soft formation, 13.7% for the medium-hard formation, and 14.3% for the hard formation. The established drill string stick–slip vibration model provides theoretical guidance for optimizing drilling parameters in different formations.

Keywords: drill string; stick–slip vibration; drilling parameter; optimization

Citation: Wang, C.; Chen, W.; Wu, Z.; Li, J.; Liu, G. Stick–Slip Characteristics of Drill Strings and the Related Drilling Parameters Optimization. *Processes* **2023**, *11*, 2783. https://doi.org/10.3390/pr11092783

Academic Editors: Tianshou Ma and Yuqiang Xu

Received: 26 August 2023
Revised: 8 September 2023
Accepted: 14 September 2023
Published: 18 September 2023

1. Introduction

Nowadays, human demand for petroleum resources is increasingly strong. Exploration and drilling/production of oil/gas resources are gradually moving deeper, and formation structures are more complex. The number of complex structured wells such as deep, ultra-deep, and extended-reach wells is increasing, making drilling engineering operations more difficult. During drilling, drill strings vibrate due to formation friction, lithology, and pressure, potentially causing drilling accidents such as premature bit failure, drill string damage, and low ROP.

Drill string vibrations are typically categorized as axial vibration, transverse vibration, and torsional vibration. The coupled vibrations lead to complex behaviors such as bit jumping, stick–slip vibration, and vortex motion. Among these complex behaviors, stick–slip vibration is most harmful. It induces cyclical stress and strain fluctuations in the drill string, accelerating fatigue failure and severely impacting drill string and bit life. Severe stick–slip vibration also causes intense vibration of the rig and derrick, damaging surface equipment and greatly reducing drilling efficiency [1].

Therefore, to improve drilling efficiency and protect equipment, drill string stick–slip vibration characteristics must be analyzed and proper measures taken to reduce or eliminate stick–slip vibration. Researchers discovered stick–slip phenomena in drill strings and conducted in-depth research as early as the 1980s [2]. Many studies on eliminating stick–slip vibration in drill string torsional vibration have laid the foundation for this research. In terms of modeling and analysis, references [2–5] established drill string stick–slip vibration dynamics models considering various coupling effects, analytically obtained modal characteristics of the complex coupled systems, and laid theoretical foundations for studying the dynamics mechanisms of stick–slip vibrations in drill strings. Reference [6] obtained an analytical expression of the velocity-weakening friction law through a toothed-bit model, revealing the role of bit–rock interactions in the generation of stick–slip vibrations. Reference [7] analyzed the combined torsional stick–slip and lateral whirling vibrations through simplified models, explaining the dynamics mechanisms between different vibration modes. Regarding control strategies, reference [8] proposed an active damping system based on feedback control that effectively suppresses self-excited torsional vibrations of the drill string during drilling by adjusting drive system parameters, expanding the non-vibrating rotary speed range of the drill string. This provided new ideas and references for eliminating nonlinear stick–slip vibrations during drill string rotation. Reference [9] proposed robust control strategies based on model error compensation techniques and constructed cascade and decentralized control schemes. Reference [10] adopted dynamic sliding mode control, established discontinuous torsional dynamics models and dual discontinuous surfaces, achieved rotary speed control of the oilwell drill string, and effectively suppressed self-excited stick–slip vibrations. This provided an important control strategy reference for eliminating stick–slip vibrations in drill strings. Reference [11] designed an improved OSKIL mechanism called D-OSKIL by using axial load as an additional control variable to suppress drill string limit cycles during drilling. This control law rendered the closed-loop system globally asymptotically stable. Simulations verified that stick–slip vibrations can be effectively eliminated without redesigning rotary speed control. Regarding parameter optimization, reference [12] established an oilwell drilling dynamics model, analyzed the self-excited vibration issue at the bottom hole assembly, and proposed key drilling parameter selection guidance to avoid drill string torsional vibrations. This provided new insights into parameter design for eliminating nonlinear stick–slip vibrations during drill string rotation. Reference [13] studied the effects of damping, active control, and interface parameter optimization on stick–slip vibrations. Reference [14] pointed out that bit interface conditions are key factors affecting vibrations. In terms of mechanism theory, reference [15] laid the foundation for drill string torsional vibration research by proposing and validating the stick–slip vibration theory. Reference [16] proposed a new perspective wherein velocity weakening is a system response rather than an intrinsic characteristic. Reference [17] revealed the mechanism of normal and tangential vibrations in stick–slip limit cycles through a two-degrees-of-freedom model. Regarding coupled vibrations, reference [18] avoided self-excited vibrations to suppress stick–slip vibrations. Reference [19] analyzed drill string torsional vibration issues caused by derrick stick–slip motion through establishing a stick–slip vibration model and proposed avoiding large-amplitude vibrations through rotary speed control strategies. Reference [20] proposed a new robust active controller based on the fuzzy sliding mode approach to suppress stick–slip vibration in drill strings and maintain the angular velocity of drill components at desired values. Reference [21] established an axial–lateral–torsion coupling nonlinear model for drill string vibration in ultra-HPHT curved wells considering wellbore constraints, bit–rock interaction, and mud properties. A finite element method was used for the numerical solution. Reference [22] developed a modified integral resonant control scheme with tracking to suppress stick–slip vibrations and achieve the desired drilling velocity in drill strings, showing better performance than sliding mode control. Reference [23] designed an H observer-based controller to estimate and suppress high-frequency stick–slip vibrations in a 10-DOF drill string model, showing better performance than LQG control in handling unstructured perturbations. Reference [24] introduced a new model for analyzing the Anti-Stick–Slip Tool at the drill string's end. This model overcomes the limitations of previous analyses, utilizing two degrees of freedom in the non-activated state and three degrees upon activation.

In summary, the above studies have laid theoretical foundations for this study's drill string stick–slip vibration elimination. However, there has been less systematic discussion on the stick–slip mechanism and its influencing factors. What is more, less work has focused on combining a stick–slip vibration elimination model and ROP model in order to ensure high ROP and less stick–slip at the same time, especially when drilling in different formations.

This paper establishes and validates a drill string stick–slip vibration behavior model based on the drill string torsional vibration model. It reveals characteristics of stick–slip vibrations in different positions along the drill strings and analyzes influencing factors based on the simulation results. With one-time integration and two-time integration, the stick–slip angular velocity and angular displacement along the drill strings can be achieved, which can further reveal the stick–slip mechanization and characteristics. To reduce stick–slip and, in turn, increase ROP, this paper carries out drilling parameter optimization for soft, medium-hard, and hard formations by combining an ROP calculation model and stick–slip model. The optimization method generates preferred drilling parameter ranges applicable to different formations, improving mechanical ROP under relatively stable torsional vibrations. This has very important guiding significance for on-site drilling operation safety.

2. Drill String Stick–Slip Vibration and Drilling Parameter Optimization

2.1. The Multidimensional Drill String Torsional Vibration Model

In actual drilling, surface-applied torque on the rotary drives continuous rotation of the drill pipe, drill collar, drill bit, and other tools. Meanwhile, under the drill bit's WOB, the bit cuts rock to achieve penetration. Based on extensive research worldwide, reasonable simplification and assumptions were made for drill string torsional vibrations considering differences in the mechanical properties of various drill pipes and drill collars. Thus, a multidimensional drill string torsional vibration model was constructed.

This model views the rotary table, m drill pipes, n drill collars, drill bit, etc., as lumped masses; other tools are viewed as springs with torsional stiffness and viscous damping. Under actual conditions, the following assumptions are made:

(1) The research object is a vertical well;
(2) The drill string is simplified into lumped masses including the rotary table, m drill pipes, n drill collars, and the bit;
(3) BHA is equivalent to springs and viscous damping;
(4) Overall stick–slip behavior of the drill string is approximated by stick–slip at the bit.

Figure 1 shows the multidimensional drill string torsional-vibration-model-based rotary table, drill pipe 1…n, drill collars 1…n, and drill bit. J_r, J_{p1}, J_{pm}, J_{c1}, J_{cl}, and J_b represent the rotational inertias of the rotary table, drill pipe 1, drill pipe n, drill collar 1, drill collar n, and drill bit, respectively, kg·m^2. $\ddot{\varphi}_r$, $\ddot{\varphi}_{p1}$, $\ddot{\varphi}_{pm}$, $\ddot{\varphi}_{c1}$, $\ddot{\varphi}_{cn}$, and $\ddot{\varphi}_b$ represent the angular accelerations of the rotary table, drill pipe 1, drill pipe m, drill collar 1, drill collar n, and drill bit, respectively, rad/s^2. $\dot{\varphi}_r$, $\dot{\varphi}_{p1}$, $\dot{\varphi}_{pm}$, $\dot{\varphi}_{c1}$, $\dot{\varphi}_{cn}$, and $\dot{\varphi}_b$ represent the angular velocities of the rotary table, drill pipe 1, drill pipe m, drill collar 1, drill collar n, and drill bit, respectively, rad/s. φ_r, φ_{p1}, φ_{pm}, φ_{c1}, and φ_{cn} represent the angular displacements of the rotary table, drill pipe 1, drill pipe m, drill collar 1, and drill collar n, respectively, rad. c_r, c_{rp}, c_{p1}, $c_{pm\text{-}1}$, c_{pc}, c_{c1}, $c_{cn\text{-}1}$, c_{cb}, and c_b represent the damping coefficients of the rotary table, between the rotary table and drill pipe, between drill pipes, between the drill pipe and collar, between drill collars, between the drill collar and bit, and of the drill bit, respectively, (N·m·s)/rad. T_m, T_e, T_{ar}, T_f, T_{ab}, T_{fb}, T_{sb}, and T_{cb} represent the torque of the rotary table, torque transferred to the drill bit, viscous torque of the rotary table, friction torque of the drill bit, viscous torque of the drill bit, dry friction torque of the drill bit, maximum static friction torque, and Coulomb friction torque, respectively, N·m. k_{rp}, k_{p1}, $k_{pm\text{-}1}$, k_{pc}, k_{c1}, $k_{cn\text{-}1}$, and k_{cb} represent the stiffness coefficients between the rotary table and drill pipe, between drill pipes, between the drill pipe and collar, between drill collars, and between the drill collar and bit, respectively, N·m/rad. D_v represents the boundary layer thickness, rad/s. R_b represents the drill bit radius, m. W_{ob} represents the weight on bit, N.

μ_{cb}, μ_{sb}, and γ_b represent the Coulomb friction coefficient, static friction coefficient, and Stribeck constant, respectively. T_m can be obtained by the following formula:

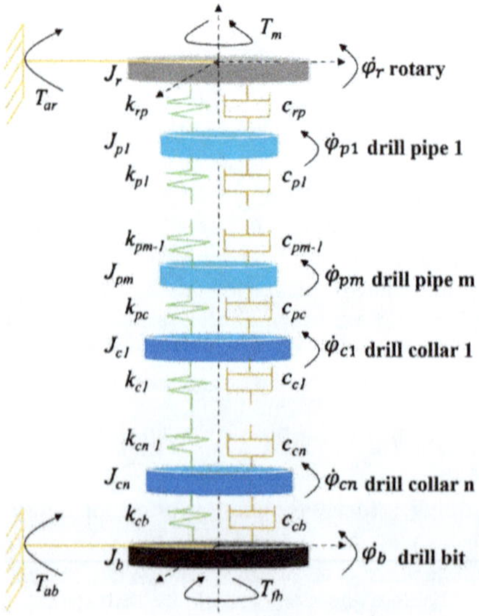

Figure 1. The multidimensional drill string torsional vibration model.

$$T_m = \frac{30P}{\pi n}\eta \tag{1}$$

where P is the rotary table power, W; η is the transmission efficiency, dimensionless; n is the rotation speed of rotary, rad/s. The fluid sticking torque Tar at the rotary table is obtained by the following formula:

$$T_{ar} = -c_r\dot{\varphi}_r \tag{2}$$

where c_r is the rotary table viscous damping coefficient, (N·m·s)/rad. The drill bit is mainly under two torques: the torque T_e transferred from the drill collar to the bit and the friction torque T_f at the bit. The friction torque T_f includes the fluid sticking torque T_{ab} and the dry friction torque T_{fb} between the bit and the rock, which can be expressed as:

$$T_f = \begin{cases} R_b W_{ob}\left[\mu_{cb} + (\mu_{sb} - \mu_{cb})e^{\gamma_b\dot{\varphi}_b}\right] - c_b\dot{\varphi}_b, & \dot{\varphi}_b \le -D_v \\ -f_e, & |\dot{\varphi}_b| < D_v and |T_e| < T_{sb} \\ f_s(-sgn(f_e)), & |\dot{\varphi}_b| < D_v and |T_e| \ge T_{sb} \\ -R_b W_{ob}\left[\mu_{cb} + (\mu_{sb} - \mu_{cb})e^{-\gamma_b\dot{\varphi}_b}\right] - c_b\dot{\varphi}_b, & \dot{\varphi}_b \ge D_v \end{cases} \tag{3}$$

Here, the maximum static friction torque can be obtained by:

$$T_{sb} = \mu_{sb} R_b W_{ob} \tag{4}$$

where μ_{sb} is the drill bit coefficient of maximum static friction, dimensionless; R_b is the bit radius, m; W_{ob} is the WOB, N. Force analysis was performed on the rotary table, m drill pipes, n drill collars, and drill bit separately. Then, based on the law of rotation, the following dynamics equilibrium equations were established [25,26]:

$$\left\{\begin{array}{l} J_r\ddot{\varphi}_r + c_{rp}\left(\dot{\varphi}_r - \dot{\varphi}_{p1}\right) + k_{rp}\left(\varphi_r - \varphi_{p1}\right) = T_m + T_{ar} \\ J_{p1}\ddot{\varphi}_{p1} + c_{p1}\left(\dot{\varphi}_{p1} - \dot{\varphi}_{p2}\right) + k_{p1}\left(\varphi_{p1} - \varphi_{p2}\right) \\ \qquad = c_{rp}\left(\dot{\varphi}_r - \dot{\varphi}_{p1}\right) + k_{rp}\left(\varphi_r - \varphi_{p1}\right) \\ \cdot \\ J_{pm}\ddot{\varphi}_{pm} + c_{pc}\left(\dot{\varphi}_{pm} - \dot{\varphi}_{c1}\right) + k_{pc}\left(\varphi_{pm} - \varphi_{c1}\right) \\ = c_{pm-1}\left(\dot{\varphi}_{pm-1} - \dot{\varphi}_{pm}\right) + k_{pm-1}\left(\varphi_{pm-1} - \varphi_{pm}\right) \\ J_{c1}\ddot{\varphi}_{c1} + c_{c1}\left(\dot{\varphi}_{c1} - \dot{\varphi}_{c2}\right) + k_{c1}\left(\varphi_{c1} - \varphi_{c2}\right) \\ \qquad = c_{pc}\left(\dot{\varphi}_{pm} - \dot{\varphi}_{c1}\right) + k_{pc}\left(\varphi_{pm} - \varphi_{c1}\right) \\ \cdot \\ J_{cn}\ddot{\varphi}_{cn} + c_{cb}\left(\dot{\varphi}_{cn} - \dot{\varphi}_b\right) + k_{cb}\left(\varphi_{cn} - \varphi_b\right) \\ \quad = c_{cn-1}\left(\dot{\varphi}_{cn-1} - \dot{\varphi}_{cn}\right) + k_{cn-1}\left(\varphi_{cn-1} - \varphi_{cn}\right) \\ J_b\ddot{\varphi}_b = c_{cb}\left(\dot{\varphi}_{cn} - \dot{\varphi}_b\right) + k_{cb}\left(\varphi_{cn} - \varphi_b\right) + T_f \end{array}\right. \tag{5}$$

2.2. Drilling Parameters Optimization Based on Drill String Stick–Slip Torsional Vibration

The relationship between drilling parameters and ROP is established. With the goals of vibration reduction/elimination and increased ROP, optimization of drilling parameters such as torque and WOB is completed. This can eliminate stick–slip vibrations and increase ROP simultaneously. This study uses the drill string torsional vibration model and ROP equations to establish the relationship between torque, WOB, and ROP. Combining measured field data and geological data can yield the ROP. The ROP equation is as follows [27]:

$$v_{pc} = K_R(W - M)n^{\lambda}\frac{1}{1 + C_2h}C_pC_h \tag{6}$$

Here, v_{pc} is the ROP, m/h; W is the WOB, kN; n is the rotate speed, r/min; K_R is the formation drillability coefficient; M is the threshold weight, kN; λ is the rotate speed index; C_2 is the tooth wear coefficient; C_p is the pressure difference influence coefficient; C_h is the water purification coefficient; h is the tooth wear.

Figure 2 is a block diagram of drill string stick–slip vibration drilling parameter optimization. Rotary table torque T_m and WOB W are input. First, the angular rate w of the drill bit can be obtained through the stick–slip vibration model to determine vibration intensity. Second, the rotate speed n of the drill bit can be obtained through the stick–slip vibration model. By combining with WOB W, the ROP v can be obtained. Finally, with the goals of eliminating stick–slip vibration and increasing ROP, the optimal solutions for WOB and rotary table torque can be obtained.

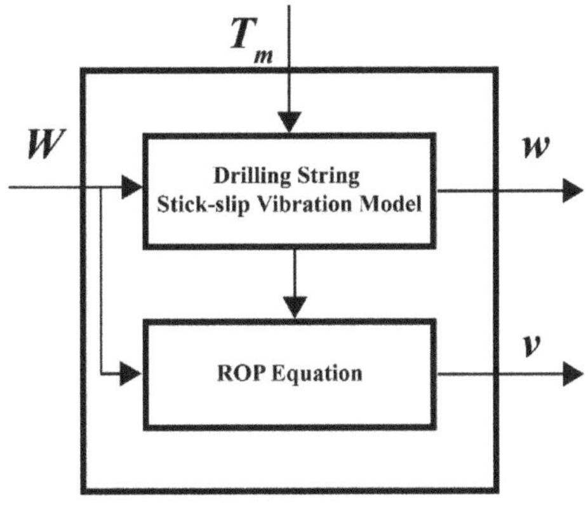

Figure 2. Block diagram of drilling string stick–slip vibration drilling parameter optimization.

3. Stick–Slip Model Validation and Influencing Factors Analysis

3.1. Validation of Stick–Slip Vibration Model

Based on [18], the basic simulation parameters of the drill string stick–slip vibration model are set as shown in Table 1. The simulation time is from 0 to 100 s.

Table 1. Model simulation parameter table.

Parameter Description	Parameter Symbol	Parameter Value
turning inertia of the rotary	J_r	930 kg·m^2
turning inertia of the drill pipe	J_p	2782.25 kg·m^2
turning inertia of the drill collar	J_c	750 kg·m^2
turning inertia of the drill bit	J_b	471.97 kg·m^2
equivalent stiffness coefficient between the rotary and drill pipe	k_{rp}	698.06 N·m/rad
equivalent stiffness coefficient between the drill pipe and drill collar	k_{pc}	1080 N·m/rad
equivalent stiffness coefficient between the drill collar and drill bit	k_{cb}	907.48 N·m/rad
damping coefficient of rotary	c_r	425 N·m/rad
equivalent damping coefficient between the rotary and drill pipe	c_{rp}	139.61 N·m/rad
equivalent damping coefficient between the drill pipe and drill collar	c_{pc}	190 N·m/rad
equivalent damping coefficient between the drill collar and drill bit	c_{cb}	181.49 N·m/rad
damping coefficient of drill bit	c_b	50 N·m/rad
coefficient of coulomb frication	μ_{cb}	0.5
coefficient of static friction	μ_{sb}	0.8
constant of Stribeck	γ_b	0.9
thickness of border stratum	D_v	0.000001 rad/s
radius of drill bit	R_b	0.155 m
WOB applied to the drill bit	W_{ob}	97,347 N
rotating torque	T_m	9400 Nm

The angular velocity results of the drill string stick–slip vibration simulation in this study are shown in Figure 3. The subscripts r, p, l, and b of angular velocity represent the rotary table, drill pipe, drill collar, and drill bit, respectively.

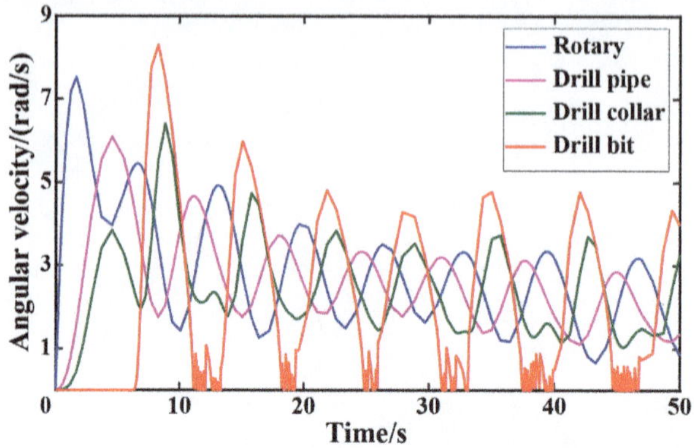

Figure 3. Simulation block diagram of drill string torsional vibration mode.

The angular velocity simulation results of drill string stick–slip vibration in this study have, overall, similar regular patterns to the simulation results in [18], with slight differences in details. The drill bit angular velocity in the stick section has slight fluctuations, while it is completely 0 in [18]. The variation regular patterns of the rotary table, drill pipe, and drill collar are identical, exhibiting angular velocities generated in sequence with decreasing initial wave crests. Therefore, the drill string stick–slip vibration simulation research in this study is reliable and effective.

3.2. Analysis of Stick–Slip Vibration Characteristics

Based on the established drill string stick–slip vibration model simulation, the time domain variations of drill bit angular displacement, angular velocity, and angular acceleration at 9400 N·m rotary table torque are shown in Figure 4. For a clear description of typical stick–slip patterns, an image was drawn over 30 s.

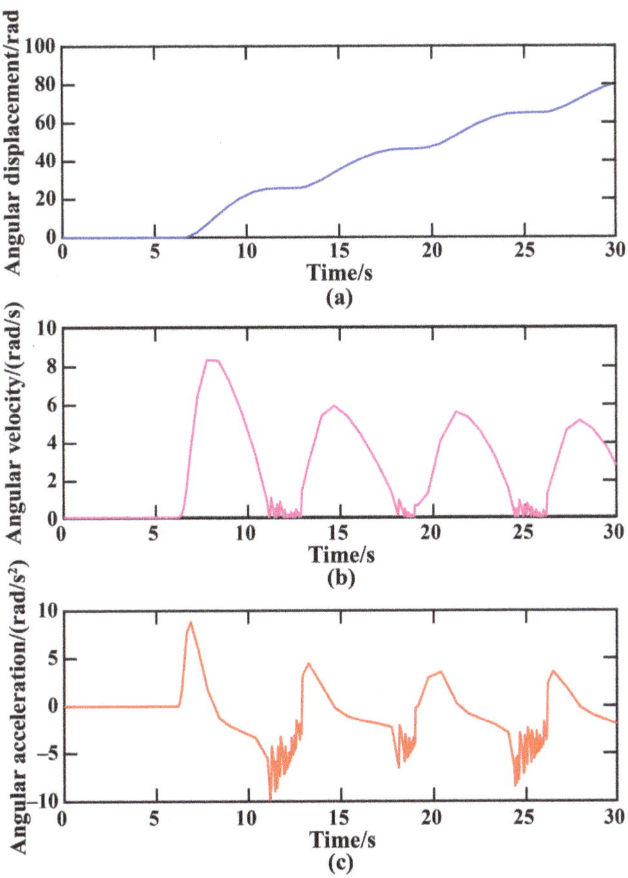

Figure 4. Time domain diagram of drill bit angular displacement, angular velocity, and angular acceleration. (**a**) drill bit angular displacement; (**b**) angular velocity; (**c**) angular acceleration.

Figure 4 shows the time domain graphs of drill bit angular displacement, angular velocity, and angular acceleration. When applying a 9.4 kN·m rotary table torque, the drill bit experiences three stages: stationary, slip, and stick. From 0 to 6.2 s, the drill bit is in the stationary stage with 0 angular displacement, velocity, and acceleration, as shown in Figure 4a–c. The drill bit needs to overcome the stiction torque. It can be seen from 6.2 to 11.2 s in Figure 4a–c that the drill bit is in the slip stage, which can be divided into three periods. Early slip period: The accumulated torque in the drill string is suddenly released, and the angular acceleration of the drill bit begins to increase dramatically. When the torque is fully released, the angular acceleration reaches its maximum, so the angular velocity increases rapidly. Middle slip period: The drill bit is then mainly affected by the formation friction. The angular acceleration starts to decrease. When the drill string torque balances the formation friction torque, the angular acceleration is 0, and the angular velocity reaches its maximum, so the angular displacement increases fastest. Late slip period: Due to the formation friction, the angular acceleration starts to increase in the opposite direction, as shown in Figure 4c. The angular velocity quickly decreases to 0, so the angular displacement is also 0. From 11.2 to 12.9 s, when entering the stick stage, the drill bit angular velocity experiences slight changes of small magnitude belonging to minor fluctuations, as shown in Figure 4b. The angular displacement barely changes, as shown in Figure 4a. After breaking through the stationary stage, continuous and stable slip and stick stages occur, forming a stable stick–slip behavior in the drill string.

When the drill string exhibits stick–slip behavior, the angular velocity response of the rotary table, drill pipe, drill collar, and drill bit in the multidimensional drill string torsional vibration model is analyzed. Figure 5 shows the time domain graphs of the angular velocities of the four lumped masses.

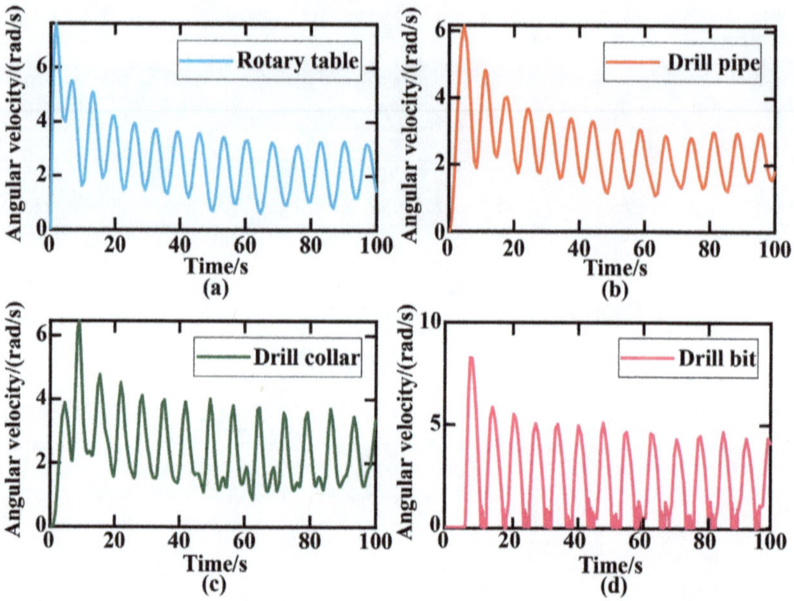

Figure 5. Time diagram of drill tool angular velocity during stick–slip vibration of drill string. (**a**) rotary table; (**b**) drill pipe; (**c**) drill collar; (**d**) drill bit.

It can be seen from Figure 5 that when the drill bit in Figure 5d exhibits stick–slip behavior, its angular velocity shows periodical fluctuations. The rotary table in Figure 5a, drill pipe in Figure 5b, and drill collar in Figure 5c also exhibit periodical fluctuations accordingly. When the drilling rig overcomes the stationary stage, the rotary table, drill pipe, drill collar, and drill bit generate angular velocities in sequence. So, there is a displacement lag between the drill bit and rotary table. Continuing to study their angular displacement regular patterns, the time domain variations of angular displacement are as follows.

It can be seen from Figure 6 that the rotary in Figure 6a starts rotating at 1 s, the drill pipe in Figure 6b at 2 s, the drill collar in Figure 6c at 2.7 s, and the drill bit in Figure 6d at 7 s. This is because the drill bit is also under the dry friction torque between the bit and th rock. At this time, the drill bit angular displacement already lags behind the rotary table by 36.5 rad.

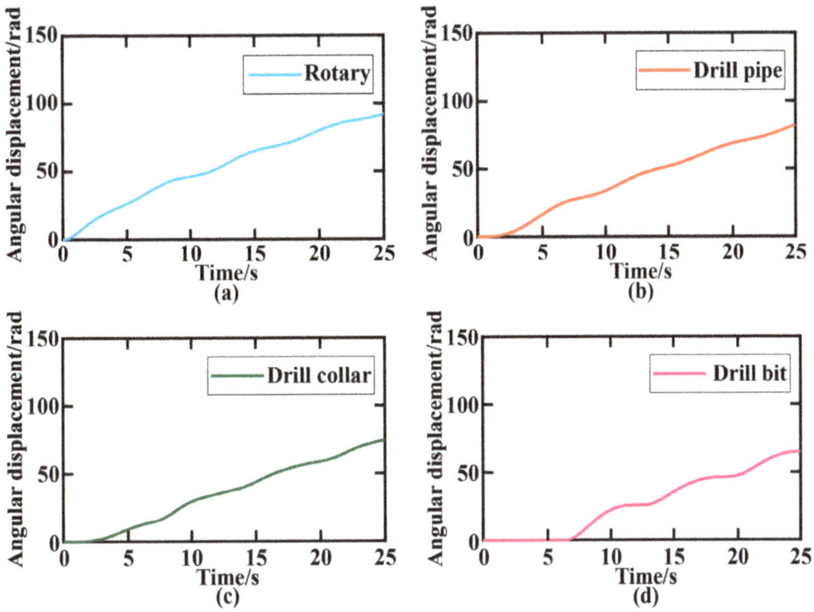

Figure 6. Time diagram of drill tool angular displacement during stick–slip vibration. (**a**) rotary; (**b**) drill pipe; (**c**) drill collar; (**d**) drill bit.

3.3. Analysis of Influencing Factors

As the rotary table torque increases, the drill string exhibits five different phenomena: the drill bit always sticking, minor stick–slip, typical stick–slip, stick–slip stability, and no stick–slip stability. To study the specific impact of rotary table torque on the typical stick–slip characteristics of the drill string, the typical stick–slip section is selected for further study. Based on example simulations, rotary table torques are set to 9000 N·m, 9200 N·m, 9300 N·m, 9400 N·m, and 9600 N·m, respectively. WOB is set to 97,347 N during the calculation. The time domain variations of drill bit angular velocity are obtained as follows.

Setting the rotary table speeds to 1 rad/s, 3 rad/s, 5 rad/s, and 7 rad/s, respectively, WOB is 97,347 N, and the torque is 9400 N·m during the calculation. The time domain variations of drill bit angular velocity are obtained as follows.

Taking the 9400 N·m rotary table torque, resulting in stick–slip behavior as a baseline, WOB is set to 97,000 N, 99,000 N, 100,000 N, and 110,000 N, respectively. The time domain variations of drill bit angular velocity are obtained as follows.

Analyzing the relatively stable stick–slip behavior of the drill string from 50 s to 100 s, shown in Figures 7–9, it can be seen that when the drill string exhibits stick–slip, the larger the rotary table torque, rotary speed, or WOB, the shorter the stick–slip period and stick duration. The drill bit angular velocity departs from the zero point sooner. The peak angular velocity of the drill bit becomes larger and larger. The angular velocity of the drill bit is greater. Moreover, the ratio of drill bit to rotary table angular velocity decreases as rotary table torque, speed, or WOB increases. The ratio is 1 without stick–slip.

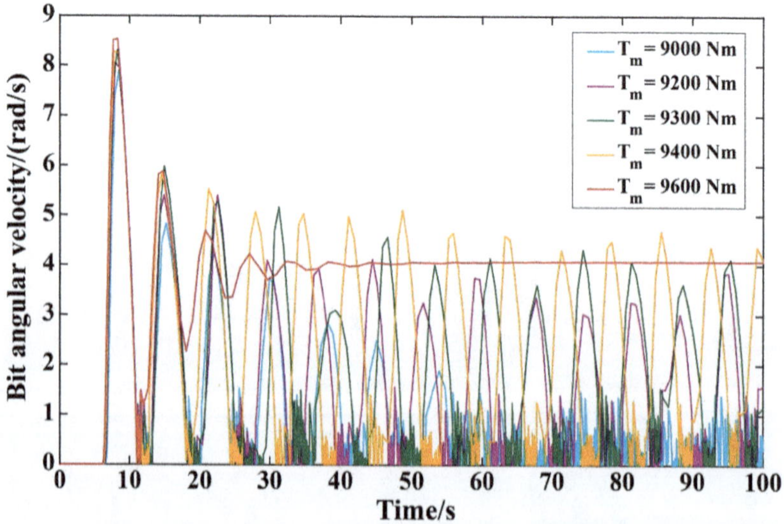

Figure 7. Time domain diagram of drill bit angular velocity with different turntable torque.

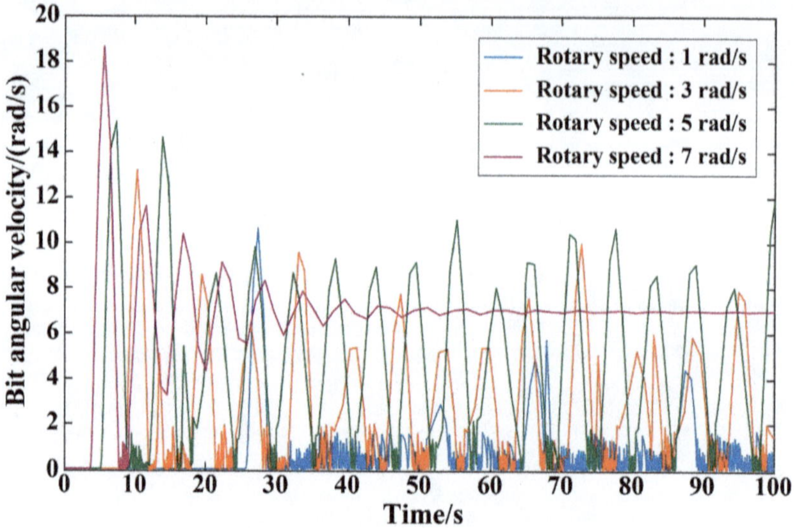

Figure 8. Time domain diagram of drill bit angular velocity at different turntable speeds.

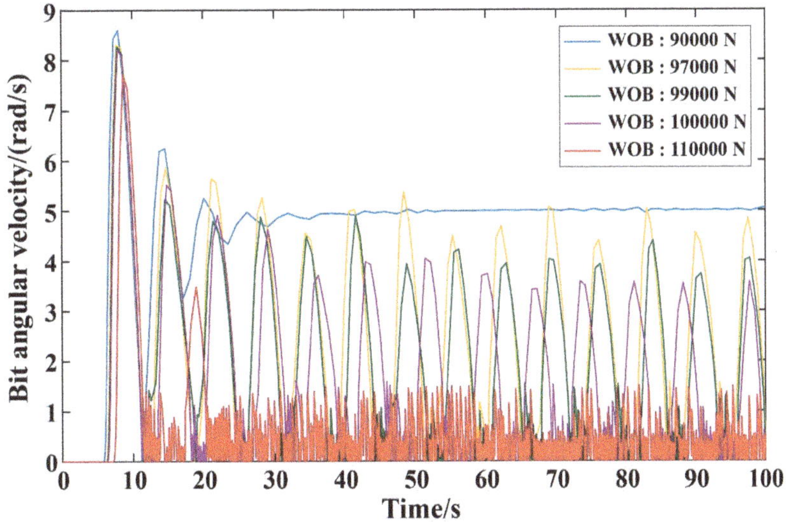

Figure 9. Time domain diagram of bit angular velocity at different weight-on-bit.

4. Case Study of the Stick–Slip-Model-Based Drilling Parameters Optimization

4.1. Drilling Parameter Optimization for Soft, Medium-Hard, and Hard Formations

The drillability of soft formation rating is set as 4.19, with the main rocks being mudstone, sandstone, and pebbled sandstone. Based on the rock friction coefficient table, the dynamic friction coefficient is 0.15, and maximum static friction coefficient is 0.4. The medium-hard formation simulates siliceous limestone with a rock drillability rating of 7. Its dynamic friction coefficient is 0.35, and maximum static friction coefficient is 0.6. In drilling parameter optimization for the hard formation, quartzite is simulated with a rock drillability rating of 11. Its dynamic friction coefficient is 0.45, and maximum static friction coefficient is 0.7. Parameters required for the ROP of the three formations are shown in Table 2.

Table 2. Parameter table of ROP in 3 kinds of formations.

Parameter Description	Parameter Symbol	Soft Formation	Medium-Hard Formation	Hard Formation
threshold weight	M	10 kN	20 kN	30 kN
rotate speed index	λ	0.68	0.68	0.68
tooth wear coefficient	C_2	3.26	3.82	4.68
pressure difference influence coefficient	C_p	1	1	1
water purification coefficient	C_h	1	1	1
tooth wear	h	0.7	0.75	0.8
formation drillability coefficient	K_R	0.0038	0.0020	0.0008

In the soft formation, the torque is 3 kN·m, and WOB is 93 kN when drill string torsional vibration occurs. Taking these as initial values, equal intervals are divided. Torque is increased by 0.3 kN·m each time, and WOB is decreased by 7 kN each time. In the medium-hard formation, the initial torque is 5 kN·m, and WOB is 93 kN during drill string torsional vibration. Torque is increased by 0.3 kN·m each time, and WOB is decreased by 4 kN each time. In the hard formation, the initial torque is 6 kN·m, and WOB is 93 kN during drill string torsional vibration. Torque is increased by 0.3 kN·m each time, and WOB is decreased by 4 kN each time. The vibration state and ROP of the drill string under different drilling parameters in the three formations can be obtained. If the drill string is always in stick–slip vibration, the ROP fluctuates periodically and is marked as stick–slip. If stick–slip vibration disappears after a period, the ROP is obtained, in m/h, as shown in Tables 3–5.

Table 3. Table of optimization results of drilling parameters for softer formations.

WOB (kN) \ Torque (kN·m)	3	3.3	3.6	3.9	4.2	4.5
93	stick–slip	stick–slip	stick–slip	stick–slip	1.44	1.56
86	stick–slip	stick–slip	stick–slip	1.26	1.36	1.46
79	stick–slip	stick–slip	stick–slip	1.18	1.27	1.36
72	stick–slip	stick–slip	1.01	1.09	1.17	1.25
65	stick–slip	0.84	0.92	1.00	1.07	1.14
58	0.69	0.76	0.83	0.89	0.96	1.01

Table 4. Optimal result table of drilling parameters for medium-hard formations.

WOB (kN) \ Torque (kN·m)	5.0	5.3	5.6	5.9	6.2	6.5
93	stick–slip	stick–slip	stick–slip	stick–slip	0.57	0.61
89	stick–slip	stick–slip	stick–slip	stick–slip	0.56	0.60
85	stick–slip	stick–slip	stick–slip	0.50	0.54	0.58
81	stick–slip	stick–slip	0.46	0.49	0.53	0.56
77	stick–slip	0.41	0.44	0.48	0.51	0.54
73	0.36	0.40	0.43	0.46	0.49	0.52

Table 5. Table of optimization results of drilling parameters for hard formations.

WOB (kN) \ Torque (kN·m)	6.0	6.3	6.6	6.9	7.2	7.5
93	stick–slip	stick–slip	stick–slip	stick–slip	0.159	0.171
89	stick–slip	stick–slip	stick–slip	0.144	0.157	0.168
85	stick–slip	stick–slip	stick–slip	0.142	0.153	0.163
81	stick–slip	stick–slip	0.128	0.138	0.148	0.157
77	stick–slip	0.115	0.124	0.133	0.142	0.150
73	0.102	0.111	0.119	0.127	0.135	0.142

It can be seen from Table 3 that in the soft formation, with the goals of eliminating drill string stick–slip vibration and having an ROP greater than 1.40 m/h, there are three solutions in the table. When torque is increased to above 4.2 kN and WOB can be reduced to 86 kN, the ROPs are 1.44 m/h, 1.46 m/h, and 1.56 m/h, respectively. From Table 4, in the medium-hard formation, with the goals of eliminating drill string stick–slip vibration and having an ROP greater than or equal to 0.57 m/h, there are four solutions in the table. When torque is increased to above 6.2 kN and WOB can be reduced to 85 kN, the ROPs are 0.57 m/h, 0.58 m/h, 0.60 m/h, and 0.61 m/h, respectively. From Table 5 it can be seen that in the hard formation, with the goals of eliminating stick–slip vibration and having an ROP greater than or equal to 0.159 m/h, there are four solutions in the table. When torque is increased to above 7.2 kN and WOB can be reduced to 85 kN, the ROPs are 0.159 m/h, 0.163 m/h, 0.168 m/h, and 0.171 m/h, respectively.

The soft formation has three optimization solutions; the medium-hard and hard formations have four. The ROP time domain graphs are compared in Figures 10–12.

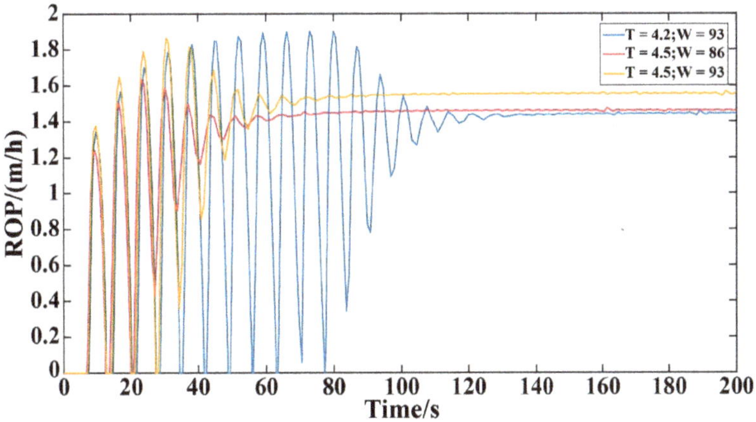

Figure 10. Time domain diagram of drilling speed optimization for soft formations.

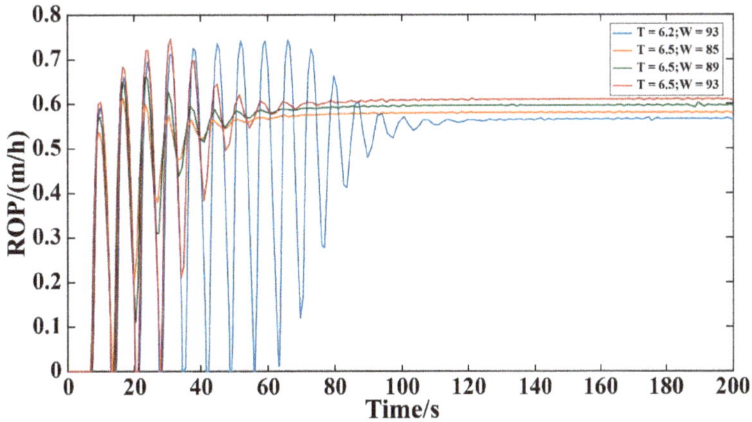

Figure 11. Time domain diagram of drilling speed optimization for medium-hard formations.

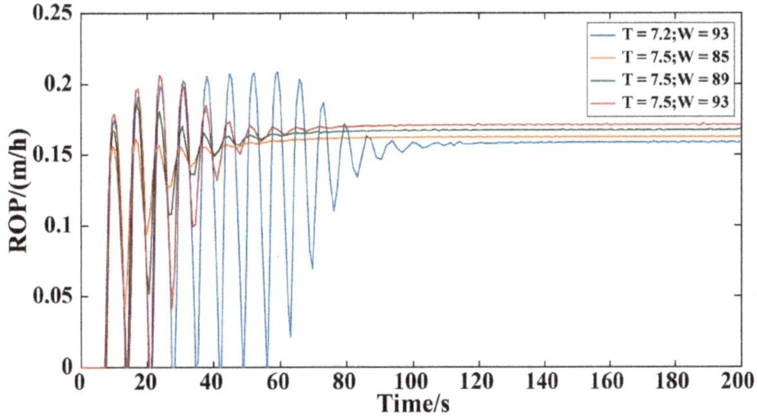

Figure 12. Time domain diagram of mechanical drilling rate optimization for drilling parameters in hard formations.

4.2. Comparison of the Optimized Results

In order to minimize the adjustments, among the three solutions for the soft formation, the WOB of 93 kN and torque of 4.2 kN·m were selected as the optimal drilling parameters for this formation. Among the four solutions for the medium-hard formation, the WOB of 93 kN and torque of 6.2 kN·m were selected as the optimal drilling parameters. Among the four solutions for the hard formation, the WOB of 93 kN and torque of 7.2 kN·m were selected as the optimal drilling parameters. The drill bit vibration effects and ROP without and with optimization of the drilling parameters for the three formations were then compared, as shown in Figures 13–15.

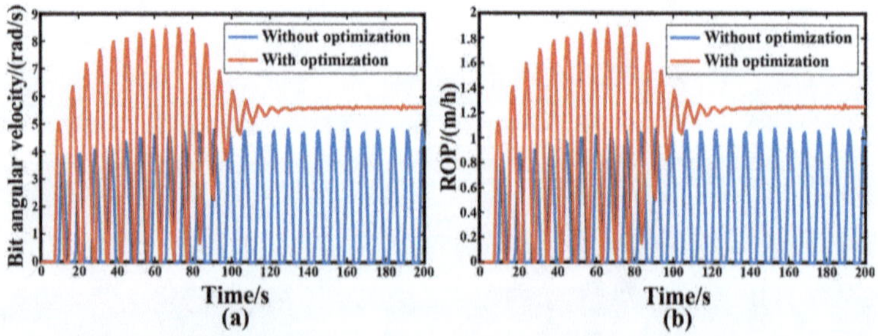

Figure 13. Diagram of bit angular velocity optimization of drilling parameters in soft formation. (**a**) bit angular velocity; (**b**) ROP.

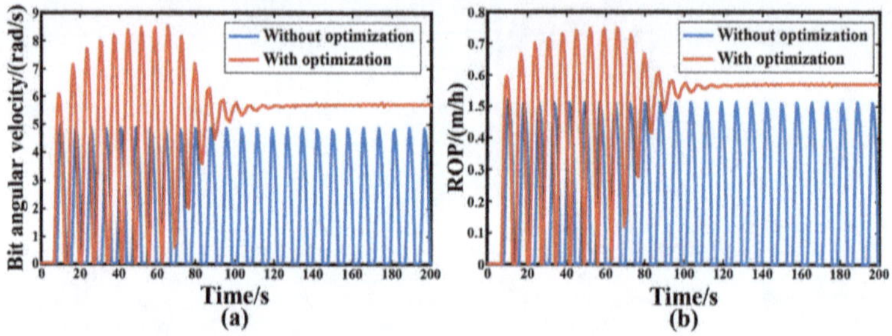

Figure 14. Diagram of bit angular velocity optimization of drilling parameters in medium-hard formation. (**a**) bit angular velocity; (**b**) ROP.

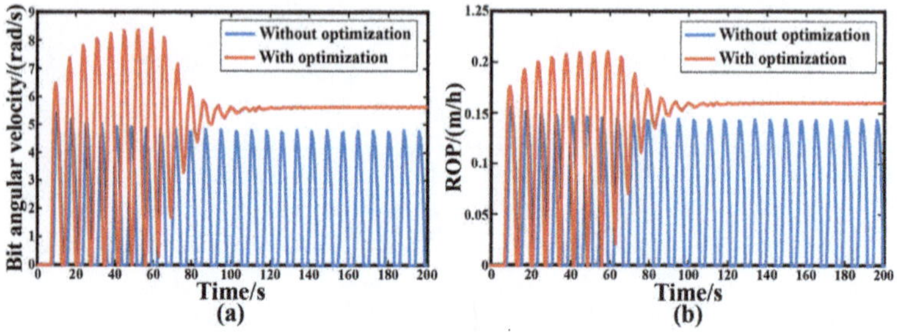

Figure 15. Diagram of bit angular velocity optimization of drilling parameters in hard formation. (**a**) bit angular velocity; (**b**) ROP.

Figure 13 shows that in the soft formation, without optimization of the drilling parameters, the drill string experienced steady stick–slip vibration. The peak value of the drill bit angular velocity was about 4.8 rad/s, as shown in Figure 13a, and the ROP fluctuated periodically with a peak value of approximately 1.3 m/h, as shown in Figure 13b. With optimization of the drilling parameters, the drill string experienced 125 s of stick–slip vibration. During this stage, the maximum drill bit angular velocity was about 8.4 rad/s, as shown in Figure 13a, and the maximum ROP was about 1.9 m/h, as shown in Figure 13b. After that, stick–slip vibration no longer occurred. The drill bit angular velocity was about 5.6 rad/s, and the ROP was about 1.45 m/h. This indicates that through optimization of the drilling parameters, after 125 s, the drill string no longer experienced stick–slip vibration. The drill bit angular velocity increased by 16.7%, and the ROP increased by 11.5%.

Figure 14 shows that in the medium-hard formation, without optimization of the drilling parameters, the drill string experienced steady stick–slip vibration. The peak value of the drill bit angular velocity was about 4.9 rad/s, as shown in Figure 14a, and the ROP fluctuated periodically with a peak value of approximately 0.51 m/h, as shown in Figure 14b. With optimization of the drilling parameters, the drill string experienced 110 s of stick–slip vibration. During this stage, the maximum drill bit angular velocity was about 8.2 rad/s, as shown in Figure 14a, and the maximum ROP was about 0.74 m/h, as shown in Figure 14b. After that, stick–slip vibration no longer occurred. The drill bit angular velocity was about 5.6 rad/s, and the ROP was about 0.57 m/h. This indicates that through optimization of the drilling parameters, after 110 s, the drill string no longer experienced stick–slip vibration. The drill bit angular velocity increased by 14.3%, and the ROP increased by 13.7%.

Figure 15 shows that in the hard formation, without optimization of the drilling parameters, the drill string experienced steady stick–slip vibration. The peak value of the drill bit angular velocity was about 4.8 rad/s, as shown in Figure 15a, and the ROP fluctuated periodically with a peak value of approximately 0.14 m/h, as shown in Figure 15b. With optimization of the drilling parameters, the drill string experienced 105 s of stick–slip vibration. During this stage, the maximum drill bit angular velocity was about 8.3 rad/s, as shown in Figure 15a, and the maximum ROP was about 0.21 m/h, as shown in Figure 15b. After that, stick–slip vibration no longer occurred. The drill bit angular velocity was about 5.6 rad/s, and the ROP was about 0.16 m/h. This indicates that through optimization of the drilling parameters, after 105 s, the drill string no longer experienced stick–slip vibration. The drill bit angular velocity increased by 16.7%, and the ROP increased by 14.3%.

5. Conclusions

Based on the drill string torsional vibration model, this paper established and validated a drill string stick–slip vibration condition model and revealed the characteristics of drill string stick–slip vibration. It was found that the drill bit periodically changes between stationary, slip, and stick stages, and the angular velocity in the slip stage is much greater than that in the stick stage. When stick–slip vibration occurred in the drill string, there was significant displacement lag and torque fluctuation between the drill bit and rotary. The analysis showed that the higher the rotary torque, rotate speed, and WOB, the shorter the period and the greater the amplitude of the stick–slip vibration. On this basis, the relationship between torque, WOB, and ROP was established, and drilling parameter optimization was completed for soft, medium-hard, and hard formations. The results showed that appropriately increasing torque and reducing WOB can effectively reduce or even eliminate stick–slip vibration of the drill string and improve the ROP. Parameter optimization in the soft formation improved ROP by 11.5%, in the medium-hard formation by 13.7%, and in the hard formation by 14.3%. This provides theoretical guidance for field operations and has important engineering application value.

Author Contributions: Conceptualization, C.W. and Z.W.; methodology, C.W.; software, Z.W.; validation, J.L. and G.L.; formal analysis, C.W.; investigation, Z.W.; resources, Z.W.; data curation, Z.W.; writing—original draft preparation, C.W.; writing—review and editing, W.C.; visualization, W.C.; supervision, J.L.; project administration, C.W.; funding acquisition, C.W. All authors have read and agreed to the published version of the manuscript.

Funding: This research was funded by The Major Scientific Research Instrument Development Program of National Natural Science Foundation of China grant number No. 52227804 and National Natural Science Foundation of China grant number No. 52304001.

Data Availability Statement: Data is unavailable due to privacy or ethical restrictions.

Conflicts of Interest: The authors declare no conflict of interest.

References

1. Yigit, A.; Christoforou, A. Coupled torsional and bending vibrations of actively controlled drillstrings. *J. Sound Vib.* **2000**, *234*, 67–83. [CrossRef]
2. Eronini, I.; Somerton, W.; Auslander, D. A dynamic model for rotary rock drilling. *J. Energy Resour. Technol.* **1982**, *104*, 108–120. [CrossRef]
3. Richard, T.; Germay, C.; Detournay, E. Self-excited stick–slip oscillations of drill bits. *Comptes Rendus Mécanique* **2004**, *332*, 619–626. [CrossRef]
4. Besselink, B.; Van De Wouw, N.; Nijmeijer, H. A semi-analytical study of stick-slip oscillations in drilling systems. *J. Comput. Nonlinear Dyn.* **2011**, *6*, 021006. [CrossRef]
5. Khulief, Y.; Al-Sulaiman, F.; Bashmal, S. Vibration analysis of drillstrings with self-excited stick–slip oscillations. *J. Sound Vib.* **2007**, *299*, 540–558. [CrossRef]
6. Germay, C.; Van De Wouw, N.; Nijmeijer, H.; Sepulchre, R. Nonlinear Drillstring Dynamics Analysis. *SIAM J. Appl. Dyn. Syst.* **2009**, *8*, 527–553. [CrossRef]
7. Leine, R.; Van Campen, D.; Keultjes, W. Stick-slip whirl interaction in drillstring dynamics. *J. Vib. Acoust.* **2002**, *124*, 209–220. [CrossRef]
8. Jansen, J.D.; Van Den Steen, L. Active damping of self-excited torsional vibrations in oil well drillstrings. *J. Sound Vib.* **1995**, *179*, 647–668. [CrossRef]
9. Puebla, H.; Alvarez-Ramirez, J. Suppression of stick-slip in drillstrings: A control approach based on modeling error compensation. *J. Sound Vib.* **2008**, *310*, 881–901. [CrossRef]
10. Navarro-Lopez, E.M.; Cortes, D. Sliding-Mode Control of a Multi-DOF Oilwell Drillstring with Stick-Slip Oscillations. In Proceedings of the 2007 American Control Conference, New York, NY, USA, 9–13 July 2007.
11. Canudas-De-Wit, C.; Rubio, F.R.; Corchero, M.A. D-OSKIL: A New Mechanism for Controlling Stick-Slip Oscillations in Oil Well Drillstrings. *IEEE Trans. Control Syst. Technol.* **2008**, *16*, 1177–1191. [CrossRef]
12. Navarro-López, E.M.; Cortés, D. Avoiding harmful oscillations in a drillstring through dynamical analysis. *J. Sound Vib.* **2007**, *307*, 152–171. [CrossRef]
13. Zamanian, M.; Khadem, S.E.; Ghazavi, M. Stick-slip oscillations of drag bits by considering damping of drilling mud and active damping system. *J. Pet. Sci. Eng.* **2007**, *59*, 289–299. [CrossRef]
14. Yigit, A.S.; Christoforou, A.P. Stick-slip and bit-bounce interaction in oil-well drillstrings. *J. Energy Resour. Technol.* **2006**, *128*, 268–274. [CrossRef]
15. Lin, Y.-Q.; Wang, Y.-H. Stick-slip vibration of drill strings. *J. Eng. Ind.* **1991**, *113*, 38–43. [CrossRef]
16. Richard, T.; Germay, C.; Detournay, E. A simplified model to explore the root cause of stick–slip vibrations in drilling systems with drag bits. *J. Sound Vib.* **2007**, *305*, 432–456. [CrossRef]
17. Richard, T.; Detournay, E. Stick–slip motion in a friction oscillator with normal and tangential mode coupling. *Comptes Rendus De L'Académie Des Sci.-Ser. IIB-Mech.* **2000**, *328*, 671–678. [CrossRef]
18. Navarro-López, E.M.; Licéaga-Castro, E. Non-desired transitions and sliding-mode control of a multi-DOF mechanical system with stick-slip oscillations. *Chaos Solitons Fractals* **2009**, *41*, 2035–2044. [CrossRef]
19. Kyllingstad, Å.; Halsey, G. A study of slip/stick motion of the bit. *SPE Drill. Eng.* **1988**, *3*, 369–373. [CrossRef]
20. Krama, A.; Gharib, M.; Refaat, S.S.; Sassi, S. Hybrid fuzzy sliding mode for stick-slip suppression in drill string systems. *J. Mech. Sci. Technol.* **2022**, *36*, 1089–1102. [CrossRef]
21. Liu, J.; Wang, J.; Guo, X.; Dai, L.; Zhang, C.; Zhu, H. Investigation on axial-lateral-torsion nonlinear coupling vibration model and stick-slip characteristics of drilling string in ultra-HPHT curved wells. *Appl. Math. Model.* **2022**, *107*, 182–206. [CrossRef]
22. MacLean, J.D.; Vaziri, V.; Aphale, S.S.; Wiercigroch, M. Suppressing stick–slip oscillations in drill-strings by Modified Integral Resonant Control. *Int. J. Mech. Sci.* **2022**, *228*, 107425. [CrossRef]
23. Riane, R.; Doghmane, M.Z.; Kidouche, M.; Djezzar, S. Observer-Based H Controller Design for High Frequency Stick-Slip Vibrations Mitigation in Drill-String of Rotary Drilling Systems. *Vibration* **2022**, *5*, 264–289. [CrossRef]
24. Taraghi Osguei, A.; Mohammad Alizadeh, B.; Dobakhti, A. Two-three degree of freedom model for Anti Stick-Slip Tool of Drill-string. *Amirkabir J. Mech. Eng.* **2023**, *55*, 4.
25. Long, Y.; Wang, X.; Wang, P.; Zhang, F. A Method of Reducing Friction and Improving the Penetration Rate by Safely Vibrating the Drill-String at Surface. *Processes* **2023**, *11*, 1242. [CrossRef]

26. Rill, G.; Schuderer, M. A Second Order Dynamic Friction Model Compared to Commercial Stick-Slip Models. *Modelling* **2023**, *4*, 366–381. [CrossRef]
27. Guan, Z.; Chen, T.; Liao, H. *Theory and Technology of Drilling Engineering*; Springer: Berlin, Germany, 2021.

processes MDPI

Article

Fluid-Loss Control Technology: From Laboratory to Well Field

Shamil Islamov [1,*], Ravil Islamov [2], Grigory Shelukhov [1], Anar Sharifov [3], Radel Sultanbekov [4], Rustem Ismakov [5], Akhtyam Agliullin [6] and Radmir Ganiev [6]

1 Department of Petroleum Engineering, Saint Petersburg Mining University, 199106 Saint Petersburg, Russia; gsheluhov@mail.ru
2 Department of Technological Development, RusWellGroup JSC, 115409 Moscow, Russia; rrislamov@rosatom.ru
3 Department of Product Development, Nedra—New Digital Resources for Assets, 190000 Saint Petersburg, Russia; anarshar@mail.ru
4 Resource Management Centre, Gazpromneft Marine Bunker, 199106 Saint Petersburg, Russia; radelsultanbekov@mail.ru
5 Department of Drilling Engineering, Ufa State Petroleum Technological University, 450064 Ufa, Russia; ismakovrustem@gmail.com
6 World-Class Scientific Center, Ufa State Petroleum Technological University, 450064 Ufa, Russia; dir@et.center (A.A.); radmirganiev@mail.ru (R.G.)
* Correspondence: sh.islamoff@gmail.com

Abstract: Effective fluid-loss control in oil wells is a critical concern for the oil industry, particularly given the substantial reserves situated in carbonate reservoirs globally. The prevalence of such reservoirs is expected to rise with the slow depletion of hydrocarbons, intensifying the need to address challenges related to deteriorating reservoir properties post well-killing operations. This deterioration results in significant annual losses in hydrocarbon production at major oil enterprises, impacting key performance indicators. To tackle this issue, this study focuses on enhancing well-killing technology efficiency in carbonate reservoirs with abnormally low formation pressures. To address this issue, the authors propose the development of new blocking compositions that prevent the fluid loss of treatment fluids by the productive reservoir. The research tasks include a comprehensive analysis of global experience in well-killing technology; the development of blocking compositions; an investigation of their physico-chemical, rheological, and filtration properties; and an evaluation of their effectiveness in complicated conditions. The technology's application in the oil and gas condensate fields of the Volga-Ural province showcases its practical implementation. This study provides valuable insights and solutions for improved fluid-loss control in carbonate reservoirs, ultimately enhancing well performance and hydrocarbon recovery.

Keywords: well-killing technology; workover; hydrophobic emulsion composition; fluid-loss; abnormally low reservoir pressure; fractured carbonate reservoir; complicated conditions

Citation: Islamov, S.; Islamov, R.; Shelukhov, G.; Sharifov, A.; Sultanbekov, R.; Ismakov, R.; Agliullin, A.; Ganiev, R. Fluid-Loss Control Technology: From Laboratory to Well Field. *Processes* **2024**, *12*, 114. https://doi.org/10.3390/pr12010114

Academic Editors: Tianshou Ma and Yuqiang Xu

Received: 24 November 2023
Revised: 26 December 2023
Accepted: 28 December 2023
Published: 2 January 2024

1. Introduction

The modern oil and gas industry is confronted with significant technological challenges associated with well killing before workover operations, especially in fields with carbonate reservoirs and abnormally low reservoir pressures. This issue has gained particular relevance, as 40% to 60% of the world's oil reserves are concentrated in carbonate reservoirs [1–5].

Fracturing is typical for carbonate reservoirs. Fractures in the rock can form as a result of tectonic processes during the formation of geological faults and folds in carbonate reservoirs. Such fractures are called natural. Fractures can also have technogenic origins, formed as a result of hydraulic fracturing and hydrochloric acid treatment [6,7].

Both natural and technogenic fracturing enhance the well productivity coefficient, ensuring an increase in the production rate of oil wells. Conversely, fractures can pose additional problems linked to breakthrough into an aquifer or into an overlying gas cap, attracting water and gas to the wells. They can also contribute to the fluid loss of technological fluid by the formation during workover operations [8,9].

It is anticipated that the proportion of such reservoirs will increase due to expanded geological and geophysical explorations. Equally important to note is that many developing fields with carbonate reservoirs face challenges related to the deterioration of reservoir properties after well killing before workover operations, resulting in significant losses in hydrocarbon production [10–12].

This article presents a study aimed at improving the efficiency of well-killing technology before workover operations in fields with fractured carbonate reservoirs and abnormally low reservoir pressures [13–15]. The focus of the work is on the development of blocking compositions capable of preventing complete fluid losses, thereby enhancing well operating conditions.

In practice, two main physical principles are employed to reduce fluid losses: increasing the viscosity of the blocking technological fluid and plugging the pores and filtration channels with solid particles [16–18]. A literature review revealed that the most preferable blocking compositions are structured hydrocarbon systems, such as inverse emulsions, which may additionally contain acid-soluble solid additives [19].

In the post-repair period, acid treatments may be required to remove the filler. However, when using gentle emulsion compositions, there is a possibility of creating a blocking screen with insufficient strength to retain the gas. An alternative in this case may be the use of gel-forming compositions [20] with fillers. One way to reduce the density and filtration of reverse emulsions and polymer solutions is by filling them with aluminosilicate microspheres with a concentration of up to 30% wt. [21].

The results of numerous observations of well-killing operations with various types of solutions enable their arrangement according to the decreasing negative impact on the bottomhole formation zone. The order is as follows: reverse emulsion; formation waters + surfactants; formation water + surfactant + polyacrylamide; and aqueous solutions of calcium or sodium chloride salts + surfactant and without surfactant. Field data from the results of well killing with reverse emulsions indicate the preservation of the filtration properties of the bottomhole formation zone in the post-repair period [22].

Regarding the characteristics of reverse emulsions, it is important to note their rheological properties and stability. These characteristics are influenced by the ratio of hydrocarbons to aqueous phases and their composition, as well as the type and concentration of the emulsifier reagent, methodology, and preparation procedure. In addition to sedimentation stability, stability under dynamic conditions must also be taken into account to justify the choice of composition. To maintain aggregative stability, it is necessary that inverse emulsions have a certain ultimate shear stress. In this case, the performance characteristics are determined based on the initial shear deformation. Inverse emulsions, especially highly concentrated ones, can be destroyed under various flow conditions, once again emphasizing the importance of their stability under dynamic conditions.

Thus, active research and development of existing and promising reverse emulsions for use as process fluids should play an important role in the introduction of new technologies [23].

One of the common trends in well killing is the use of viscoelastic compositions, polymers "cross-linked" with polyvalent metals, mainly the polysaccharide base. This ensures reliable shielding of the bottomhole formation zone from intensive penetration of water filtrates. At the same time, these technologies have a number of disadvantages, one of which is the complex sequence of technological operations and the complexity of their implementation [1].

Another disadvantage of polymer blocking compositions is the need to use destructors.

Destructors can be used in two ways: together with the technological fluid during workover or individually after workover operations.

The first method relies on the mixing of the destructor with the entire volume of the technological fluid, its complete destruction, and, consequently, a high permeability recovery coefficient. To effectively combat solution fluid loss, the study [24] describes polyglycolic acid as a moderator. The disadvantage of this method is the impossibility of slowing down the liquefaction of the technological fluid for more than 2 days [25].

The second method does not depend on the time of work and provides fluid-loss control throughout the entire workover [26]. However, the adsorption properties of polymers and the difference in rheological properties of technological fluids and destructors should be considered. This makes destruction impossible, and productivity is partially lost due to a decrease in the permeability coefficient [27].

When carrying out the well-killing process by operating two or more objects with high-permeability hydraulic fractures [28], or with a high degree of cavernousness of the productive reservoir and abnormally low reservoir pressure, in order to combat fluid loss and showings of gas, oil, and water, the common practice is the introduction of dispersed filtration reducers into the blocking composition. The most suitable reducer for this is particularly fractionated chalk [29]. According to research consensus, the average size of dispersed particles in technological fluids should be one third of the average pore size of the productive reservoir, with their concentration being at least 5% vol. Solid particles can be represented by NaCl, $Ca(OH)_2$, or $CaCO_3$. Water-soluble polymers are used as stabilizers: starch, hydroxypropyl guar, hydroxyethyl cellulose, etc. [30,31].

To prevent long-term blockage of formation pores, calcium carbonate ($CaCO_3$) is usually used, as carbonate particles can be removed by acid treatment [32]. Unfortunately, the rate of diffusion of the acid to the plugging particles is slow due to the thickened carbonate solution. Even after the acid diffuses through the thickened liquid, most of it is wasted on the dolomite formation before the particles are dissolved.

The paper [33] describes an experience where an emulsion suspension hydrophobic composition with a wide range of fillers is used for horizontal open holes.

From a technological point of view, an interesting method for well killing [34] consists of lowering polymer rods, in particular made of (composed of/containing) methylcellulose, which additionally include a crystalline gas-releasing agent: calcium carbide and sodium nitrite. This ensures rapid gelation of well water opposite the perforation interval, simultaneously simplifying and reducing the cost of the well-killing process. However, the dissolution time of the methylcellulose rod is long, which leads to a slower formation of gel-forming composition for well killing [35].

The analysis of the methods used in the oil and gas industry to increase the blocking properties of the process fluid has shown the prospects of using fillers with a multi-fractional composition. This occurs because solid filler particles that are too small can penetrate deeply into the formation and lead to a partial loss of its permeability, and larger filler particles, in turn, do not provide a reliable barrier preventing the technological fluid losses by the productive formation.

This work includes a review of global experience [36–45], the development of blocking compositions, the study of their physico-chemical, rheological, and filtration properties, as well as the modeling of well killing and oil well development processes. The research results will contribute to enhancing the efficiency of well-killing technology and reducing production losses in fields with carbonate reservoirs and abnormally low reservoir pressures.

Thus, this work is of significant importance for the oil and gas industry and has the potential to enhance hydrocarbon recovery conditions in fields with complex reservoirs and pressure conditions.

2. Materials and Methods

This chapter is dedicated to describing the laboratory equipment and methodologies used to study properties of the developed blocking composition to substantiate the well-killing technology for oil wells before workover operations in fractured carbonate reservoirs. Laboratory tests were carried out according to the program presented in Figure 1, which includes stages for determining the physico-chemical, rheological, and filtration properties of the technological fluid compositions.

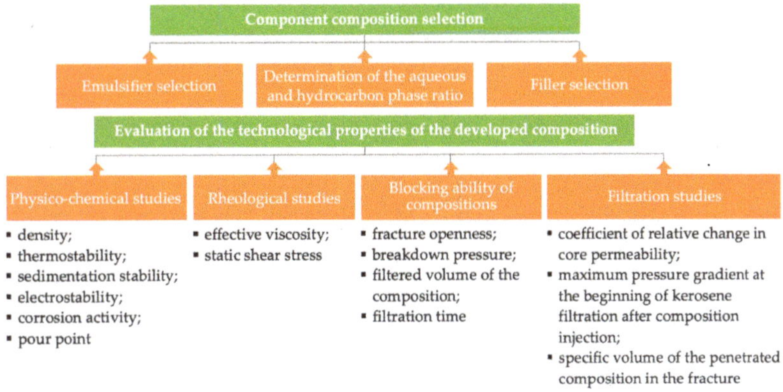

Figure 1. Laboratory test program.

2.1. Preparation of the Blocking Composition BHES-MC

The following chemical reagents were used for the preparation of the developed blocking hydrophobic emulsion solution with marble chips (BHEC-MC):

1. Emulsifier "Yalan-E-2" brand B2—a mixture of reaction products of amines and amino alcohols with fatty acids, produced according to TS 2458-012-22657427-2000 with amendment 1.
2. Oil prepared in accordance with GOST R 51858-2002.
3. Water phase mineralizer—calcium chloride, produced in accordance with GOST 450-77.
4. Fresh water.
5. Mineral filler—multi-fraction marble chips, produced according to TS 5716-001-56390243-2015 (fraction size and loading depend on the fracturing and fluid-loss capacity of the bottomhole formation zone).

The preparation of the blocking composition BHES-MC in laboratory conditions was carried out as follows: emulsifier "Yalan-E-2" brand B2 was dosed into the hydrocarbon phase (oil or diesel fuel), after which the resulting mixture was stirred using the laboratory overhead stirrer for 5 min at a frequency of 1000 rpm. Then, water, mineralized with calcium chloride or sodium chloride, was gradually introduced into the resultant substance equal, in amount, to the total volume of oil and emulsifier. After that, the mixing time was 5 min at 1000 rpm. Then, the remaining mineralized water was introduced in portions over 2 min at a stirrer rotation speed of 1000 rpm, followed by stirring for 13 min at 1500 rpm. After 15 min from the start of composition preparation, marble chips were added in portions, and the resulting mixture was stirred for 10 min at 1500 rpm.

Overall, the total preparation time for the blocking composition BHES-MC was 35 min.

The sequence of adding components of the developed blocking composition BHES-MC is schematically depicted in Figure 2.

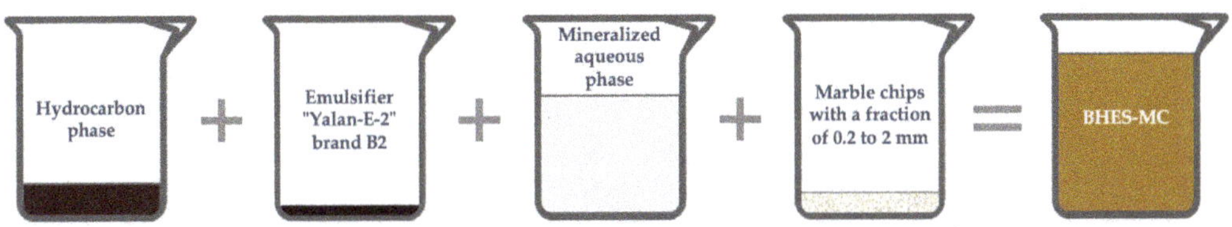

Figure 2. Blocking agent, BHES-MC, preparation procedure.

The prepared blocking composition BHES-MC is a stable, homogeneous, thickened liquid of light brown color. The components of the BHES-MC composition are available, inexpensive, and non-toxic reagents.

The preparation of the blocking composition BHES-MC is possible both on the basis of production support and in the field using mobile and stationary units and installations.

2.2. Determination of the Physico-Chemical Properties of Blocking Compositions BHES and BHES-MC

2.2.1. Determination of Density

The density of individual components constituting the BHES composition was measured using the DE 40 densitometer from Mettler Toledo (Zürich, Switzerland) at the standard temperature 20 °C. The principle of determining density with this device involved measuring the period of oscillation of a glass U-shaped tube filled with several milliliters of the test liquid. Density measurement was performed automatically with a precision of 0.0001 g/cm^3.

The density measurement process consisted of several stages:

1. The mass of the empty pycnometer was determined using precise scales, such as the XP 204 analytical balances from Mettler Toledo (Zürich, Switzerland) with a division value of 0.0001 g.
2. Distilled water was poured into the pycnometer up to the mark.
3. The mass of the pycnometer filled with distilled water was measured.
4. The water was poured out of the pycnometer, and its contents were dried in an oven.
5. The pycnometer was filled up to the mark with the tested technological fluid, and its mass was determined on the scales.
6. The density of the tested technological fluid (ρ_{tf}) was determined using Equation (1):

$$\rho_{tf} = \frac{m_2 - m}{m_1 - m} \cdot \rho_{dw}, \tag{1}$$

where m is the mass of the pycnometer, g; m_1 is the mass of the pycnometer with water, g; m_2 is the mass of the pycnometer with the tested technological fluid, g; and ρ_{dw} is the density of distilled water, g/cm^3.

The density measurement of the tested compositions was also conducted using hydrometers according to the methodology [46] at 20 °C. This is due to the density measurement of the technological fluid in field conditions before injection into the well being typically performed using a hydrometer. A 450 mL sample of the tested composition was poured into a 500 mL graduated cylinder volume at standard temperature. Subsequently, a hydrometer was placed in the cylinder, and readings were taken after one minute.

2.2.2. Thermal Stability Assessment

Determination of the thermal stability of the investigated compositions was conducted as follows: the prepared composition was placed in a thermostat, for example, in the VO 400 incubator by Memmert GmbH (Büchenbach, Germany), with a set temperature of 90 °C, and maintained for 7 days. Simultaneously, control samples were kept for 7 days at standard conditions. Every 24 h, the condition of the sample and the amount of separated hydrocarbons or the aqueous phase were visually assessed.

The criterion for evaluating the thermal stability of the investigated compositions was the absence of phase separation, clouding, color change, or sedimentation.

2.2.3. Sedimentation Stability Assessment

Evaluating sedimentation stability is crucial for determining the dispersed system's capacity to sustain an equilibrium state throughout the volume of the dispersion medium, as well as to identify the tendency of the composition towards phase separation.

Experiments to determine sedimentation stability were conducted according to the following procedure. The investigated compositions were poured into glass test tubes.

Then, the first half of the prepared samples were placed in an oven with the average reservoir temperature of the Volga-Ural oil and gas province (VUOGP) (37 °C), while the other half were kept at standard temperature for 7 days. The condition of the samples was visually assessed (photographed) after 1 h, 24 h, and 7 days.

In order to more reliably evaluate sedimentation stability, additional studies were conducted to determine the indicator of change in the density of the compositions by volume after thermostating. After the specified period (7 days), a sample of the investigated composition was taken from the upper and lower parts of the test tube to measure its density using a pycnometer or hydrometer (to simulate field conditions).

Sedimentation stability was calculated using Equation (2):

$$S_s = \frac{\rho_2}{\rho_1} \cdot 100, \tag{2}$$

where S_s is the sedimentation stability, %; ρ_1 is the density of the upper part of the sample of the investigated composition, kg/m^3; and ρ_2 is the density of the lower part of the sample of the investigated composition, kg/m^3.

Thus, the investigated composition was considered sedimentationally stable; the difference in densities between the lower layer and the upper layer did not exceed 20 kg/m^3, which was a change of 1.5%. In this case, the sedimentation stability range was found to lie within the 98.5% to 100% range.

2.2.4. Determination of Electrical Stability

Electrical stability is another parameter indirectly characterizing the stability of emulsions. This parameter describes the emulsion's resistance to the coalescence of dispersed phase particles and subsequent separation. The method for determining electrical stability involved measuring the voltage corresponding to the moment of emulsion destruction. In the present study, the TEE-01Ts device from LLC Analytica (Moscow, Russia) was used, with a measurement range from 0 to 750 V.

The device operates on the principle of measuring the electrical resistance of the liquid when a constant current of a certain magnitude passes through it. Thus, the breakdown voltage is determined, and its magnitude characterizes the electrical stability of the emulsion—the higher the breakdown voltage, the higher the stability of the emulsion.

The electrical stability indicator for reservoir temperatures up to 80 °C should not be less than 300 V.

It is important to note that the experience of using emulsions indicates that the absolute values of the electrical stability of the initial compositions stabilized by various emulsifiers may not correlate with the values of technological parameters such as thermal stability.

2.2.5. Determination of Corrosion Activity

The study of corrosion activity of blocking compositions involved exposing metal samples, made from steel grade St. 20, to the liquid medium of the acidizing composition for a specified period. Before testing, the surfaces of the metal plates were prepared according to [47].

The experimental studies were conducted at a temperature of 37 °C, within the limit not exceeding ±2 °C. No stirring of the liquid medium was applied. The corrosion rate was calculated based on the mass loss of the metal samples, and the exposure time of the metal samples in the test compositions was 24 h.

The corrosion rate V_c (g/m^2 × hour) was calculated using Equation (3):

$$V_c = \frac{m_2 - m_1}{S \cdot t}, \tag{3}$$

where m_1 and m_2 are the masses of the metal plates before and after testing, respectively, g; S is the surface area of the metal plate, m^2; and t is the test duration in hours.

The corrosion rate of the investigated compositions should not exceed the established standard (0.10–0.12 mm/year). In this case, their use is considered acceptable for conducting downhole operations in accordance with the requirements [47].

2.2.6. Determination of Pour Point

To determine the pour point, the method described in [48] was used. The test tube with a sample of the test composition was placed in the cooling thermostat Proline RP 890 by LAUDA GmbH (Lauda-Königshofen, Germany), where it was cooled down to −45 °C. The flowability of the investigated compositions was assessed visually every 3 °C.

2.2.7. Rheological Studies

To conduct rheological studies for determining the effective viscosity of the investigated compositions, a rotational automated viscometer from Messgerate Medingen GmbH (Ottendorf-Okrilla, Germany)—Rheotest RN 4.1, presented in Figure 3, was used.

Figure 3. Rotary viscometer Rheotest RN 4.1.

The methodology for conducting these studies involved constructing the initial rheological curves of the compositions and determining their static shear stress. The studies were conducted at a temperature of 37 °C.

The controlled shear rate mode was used to obtain the flow curve, representing the shear stress dependence on the shear rate with a gradual increase in the shear rate from 0 to 300 s^{-1}.

For a more informative analysis of the obtained results of flow curves for the investigated compositions, the dependence of effective viscosity (μ_{eff}) on the shear rate was plotted.

The procedure for determining the static shear stress was as follows:

1. Set the viscometer to the "constant shear rate" mode.
2. Pour the test liquid into the cylinder with a volume of 45 cm^3.
3. Set the number of revolutions of the cylinder D = 200 rpm.
4. Mix the test composition for 10 s.

5. After 1 and 10 min of rest of the composition, at a cylinder rotation speed of 3 rpm, readings were taken at which the structure of the test composition began to break down.

A high value of effective viscosity indicates good blocking capabilities of the compositions, contributing to the reduction of the penetration radius of technological fluids into the bottomhole formation zone. This is because an increase in effective viscosity leads to enhanced structural strength of the emulsion. During well development, such a reduction in the penetration radius of the liquid zone can lead to an increase in the permeability recovery coefficient in this zone.

Static shear stress is an indicator of the strength of the internal structure of well-killing fluids formed during the resting period. Its value is determined by the stress that is required to be exceeded in a fluid at rest to break its internal structure and set it in motion. The higher value of the static shear stress in well-killing fluids, the lower the probability of fluid loss of the composition in the reservoir zone is, especially when well killing with reservoir pressure gradients significantly below the potential capabilities of the well-killing fluids in terms of providing the required repression on the reservoir.

2.3. Determination of Interfacial Tension at the "Oil with Emulsifier—Aqueous Solution of Calcium Chloride" Boundary

To measure the interfacial tension of oil with varying emulsifier concentrations at the boundary with an aqueous solution of calcium chloride with a density of 1140 kg/m^3, the EasyDrop tensiometer from KRÜSS GmbH (Hamburg, Germany) was used.

The methodology for using the tensiometer and this software package involved determining the interfacial tension based on the shape and size of the oil droplet containing the emulsifier.

For each type of emulsifier, the dependence of the interfacial tension at the "oil with emulsifier—aqueous solution of calcium chloride" boundary with a density of 1140 kg/m^3 on the emulsifier concentration (at three points—0.05, 0.1, and 0.5%) was determined under standard conditions. Each data point was determined five times, after which the minimum and maximum values were discarded. Subsequently, the average interfacial tension value was calculated for each emulsifier concentration [49].

2.4. Study of the Blocking Ability of the BHES-MC Composition in Modeling a Fractured Reservoir

Studies to assess the blocking ability of compositions in modeling a fractured reservoir were conducted using the HPHT 500 filter press from FANN (Houston, TX, USA). This device is designed to measure the fluid loss characteristics of the studied compositions in the bottomhole formation zone, and it can also be applied to predict the potential for filter cake formation.

For laboratory investigations of the blocking ability of compositions concerning fractures of various degrees of openness, the design of the filter press was modified. A metal disk, simulating a fracture with different degrees of openness (0.1, 0.5, 1, 1.5, 2, 3, and 5 mm), was installed in the lower part of the cell (Figure 4). The lower part of the disk was machined in a "U"-shaped form to engage the entire working length of the fracture and allow unimpeded discharge of the blocking composition from the cylinder (Figure 5a). An enlarged 5 mm diameter exit hole was created in the lower plug of the cell to prevent its closure by the solid phase of the studied blocking agent (Figure 5b).

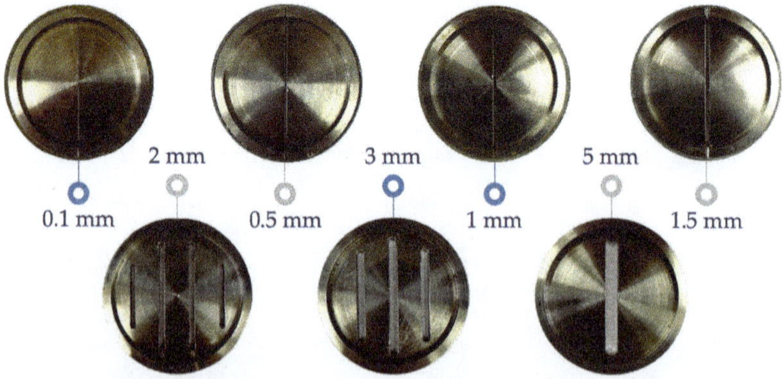

Figure 4. Metal disks simulating a fracture with varying degrees of openness.

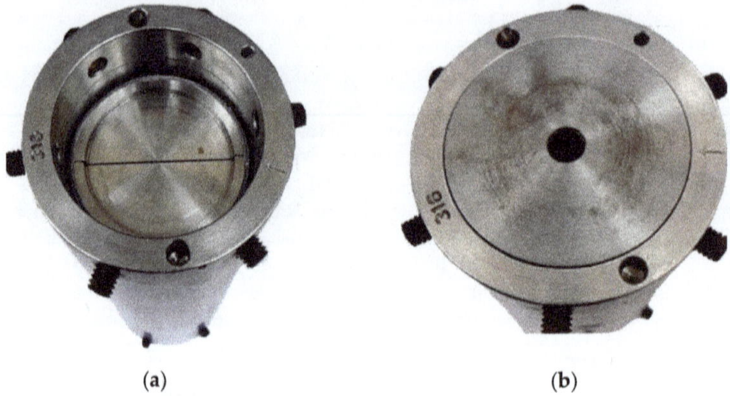

Figure 5. (**a**) Lower part of the filter press cell (exit point of the blocking composition from the crack); (**b**) Bottom plug of filter press (exit point of the blocking composition from the filter press).

The schematic diagram of the modified filter press construction is presented in Figure 6. The sequence of actions during the experiment consisted of the following stages:

1. In the lower part of the filter press, a metal disk with a crack of a specified degree of openness was placed between two rubber gaskets.
2. The lower plug was installed, and its outlet hole was closed.
3. The test composition was poured into the cell with a volume of 450 cm^3.
4. The filter press cell was closed with the upper plug, after which all clamping screws were tightened, and the top valve was closed.
5. The cell was placed in a thermo-jacket with a given temperature (37 °C) and left for the complete heating of the cell and the test composition (1–2 h).
6. The gas line from the nitrogen cylinder was connected and fixed to the upper valve.
7. The valve in the lower plug was opened.
8. A measuring capacity was placed under the cell to fix the volume of the filtered liquid.
9. The required pressure (1 MPa) was created in the gas line.
10. Along with the opening of the top valve, the stopwatch was started.
11. The test composition was maintained at each pressure (from 1 to 8 MPa, with a step of 1 MPa) for 5 min or until the filtration of liquid through the fracture ceased.
12. The volume of the filtered test composition was recorded for a certain period.

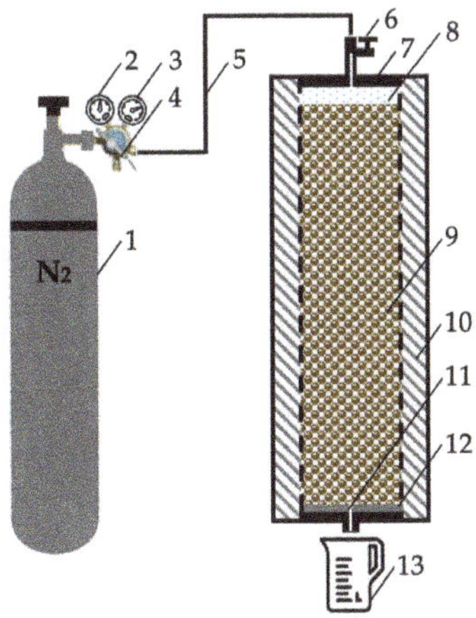

Figure 6. Schematic diagram of the stand for assessing the blocking ability of technological fluids during the well-killing process in a fractured reservoir: 1—gas cylinder; 2—regulator; 3—pressure gauge (pressure in the cylinder); 4—pressure gauge (pressure in the line); 5—gas line (gas supply); 6—top valve (open/close); 7—upper plug; 8—gas; 9—blocking composition; 10—thermo-jacket; 11—fracture; 12—metal disk with a fracture; 13—measuring capacity.

The main experimental conditions are summarized in Table 1.

Table 1. Experimental conditions on the filter press.

Parameter Name	Value
Simulated pressure drops, MPa	1–8
Temperature, °C	37
Fracture width, mm	0.1–5
Volume of blocking composition filtered, cm^3	450
Experiment duration, min	40

The studies used a minimum fracture width of 0.1 mm, corresponding to the minimum sizes of natural microfractures according to data [50] presented in Figure 7. A literature review indicated that in carbonate reservoirs of the VUOGP, microfractures are widely developed, with widths less than 0.5 mm. Among them, fractures developed along stylolites stand out, with widths not exceeding 1–2 mm [1].

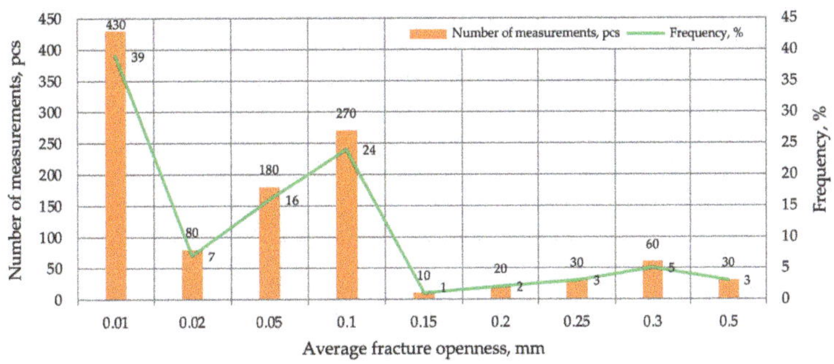

Figure 7. Assessment of fracture openness based on the analysis of 1110 carbonate core samples from one field of the VUOGP.

2.5. Conducting Filtration Tests of the Blocking Composition BHES-MC on Natural Core Samples with a Fracture

2.5.1. Preparation for Filtration Tests

For laboratory filtration tests, samples of natural core material from the VUOGP, measuring 6.45 cm in length and 3 cm in diameter, were prepared in accordance with regulatory documents [51–54], featuring artificially simulated fractures.

During filtration tests, three fracture widths were modeled: 0.1, 0.5, and 1.5 mm. A fracture width of 0.1 mm was simulated using two aluminum foil shims, each 0.1 mm thick and 5 mm wide, placed on the inner side of the natural core sample (Figure 8). Fracture widths of 0.5 and 1.5 mm were modeled using specially prepared metal plates of the corresponding thickness and 5 mm width, positioned on the inner side of the core.

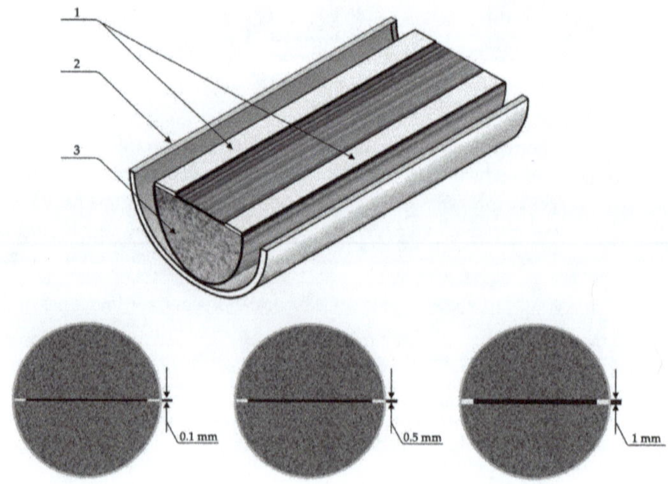

Figure 8. Appearance of a natural core in cross-section with fractures of varying degrees of width: 1—aluminum foil shims for 0.1 mm fractures or metal plates for 0.5 and 1.5 mm fractures; 2—plastic heat-shrink tubing; 3—natural carbonate core sample.

To conduct filtration studies with high-viscosity fluids containing a filler, modifications were made to the design of the standard core holder: tubes with an increased diameter of the flow passage (4.5 mm) were used in the "cylinder-reservoir—core holder—measuring capacity" system. Steel rings were placed before and after the core, facilitating unimpeded flushing of the core face even in the presence of a solid phase with a large particle size in the fluid. This core holder design also permits testing cross-linked polymer systems [10] without the risk of clogging the filtration tubes (Figure 9).

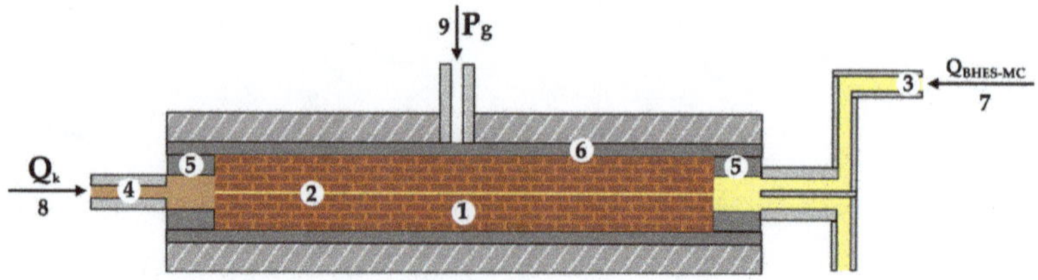

Figure 9. Core holder scheme: 1—natural core sample; 2—fracture; 3—blocking composition BHES-MC; 4—kerosene; 5—buffer metal rings; 6—rubber collar for core compression; 7—filtration direction of the composition BHES-MC; 8—filtration direction of kerosene; 9—fluid injection direction for creating compression pressure.

Prepared core samples were vacuum-saturated with kerosene and placed in the core holder (Figure 10) of the filtration unit FDES-645 (Formation Damage Evaluation System) by Coretest Systems (Morgan Hill, CA, USA). The tests were conducted under thermobaric conditions characteristic of the VUOGP.

Figure 10. Core holder and oven of the filtration unit FDES-645.

Figure 11 shows a schematic diagram of the operation of stand for conducting filtration tests in modeling the processes of well killing and oil well development using developed blocking compositions.

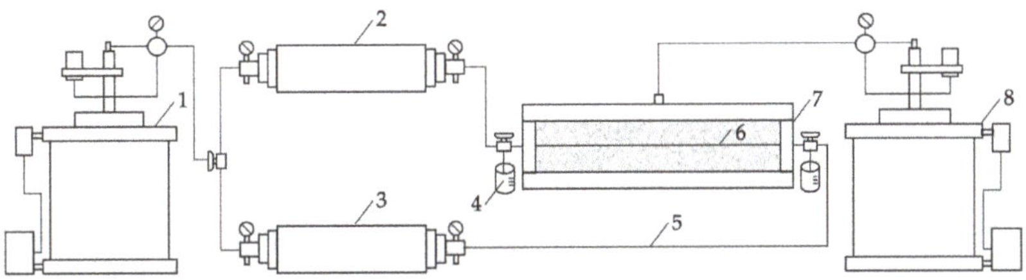

Figure 11. Schematic diagram of the stand for conducting filtration tests in the simulation of the well-killing processes and oil well development: 1—injection pump; 2—cylinder with kerosene; 3—cylinder with a developed blocking composition; 4—measuring capacity; 5—alloy steel tubes; 6—fracture; 7—core holder; 8—compression system.

2.5.2. Methodology for Conducting Filtration Tests

Assessment of the filtration properties' effectiveness of the investigated compositions is based on determining the permeability changes in reservoir models (cores) with fractures of varying degrees of width (0.1, 0.5, and 1.5 mm). The main objective of the research was to compare compositions "with" and "without" fillers in modeling the processes of well killing and oil well development.

Filtration tests were conducted using the FDES-645 unit in conditions closely approximating the average reservoir conditions of the VUOGP:

- repression during well-killing modeling—1 MPa (5% of reservoir pressure);
- reservoir pressure—20 MPa;

- average reservoir temperature of the VUOGP—37 °C.

The filtration rate (flow rate) value was set at 10 cm^3/min. Kerosene with a viscosity of 0.8 MPas was used as the saturating fluid.

The repression value of 1 MPa during well-killing modeling reproduced the condition of creating a liquid column of hydrostatic pressure at the wellhead in accordance with the requirements.

The result of the laboratory test is the determination of

- the volume of blocking composition filtered into the core at given pressure drop values during well-killing modeling;
- initial shear pressure gradient of the blocking composition in the fracture during well-development modeling;
- the coefficient of relative permeability changes in the core samples (with fractures) by kerosene after modeling well killing and well development processes.

The injection and filtration direction of the investigated compositions in the core corresponded to the direction of fluid and technological fluid movement in oil wells. Accordingly, direct filtration described the process of oil inflow from the reservoir into the well, i.e., the well development process. Reverse filtration modeled the well-killing process, which, in turn, involved creating a constant pressure drop.

Laboratory filtration tests were carried out in four stages:

1. Determination of the initial core permeability for kerosene. Prepared samples of natural core with a fracture saturated with kerosene were placed in a core holder FDES-645, where thermobaric conditions were created, closely approximating the averaged reservoir conditions of the VUOGP. Subsequently, filtration through the kerosene-saturated core was conducted. The initial phase permeability of the core to kerosene was measured in constant flow conditions until the pressure gradient stabilized at reservoir temperature. The filtration direction was "direct."
2. Modeling the process of closing the interval of perforations in the reservoir zone of an oil well with the investigated composition. In the constant pressure drop conditions, the investigated composition was injected into the core in the "reverse" filtration direction for 30 min, and the volume of the injected composition was recorded. Afterward, the core was left at rest for 1 h.
3. Modeling the process of treating the bottomhole formation zone with a hydrochloric acid composition. This stage of work was performed only for cases where the injection of the acid composition led to the formation damage of the bottomhole zone, resulting in a corresponding reduction in the coefficient of relative permeability change of the core after well development. The hydrochloric acid composition was injection at a constant volumetric flow rate of 0.5 cm^3/min. The amount of injected hydrochloric acid composition was set at 6 cm^3.
4. Modeling the process of oil well operation after the well killing and development operations. After the core had been at rest for a specified period, the restoration of kerosene filtration in the "direct" direction was carried out in constant flow conditions until the pressure gradient stabilized. The initial shear pressure gradient of the investigated composition in the core with kerosene was recorded.

The processing of the results involved the following steps:

- Pressure gradients were determined before and after the injection of the investigated composition into the core, based on which phase permeability coefficients for kerosene were calculated before and after injection.
- The initial shear pressure gradient of the investigated composition in the core with kerosene was recorded during the modeling of the well development process.
- The coefficient of relative change in the permeability of the core after its treatment with the investigated composition was calculated according to the following Equation (4):

$$\bar{k} = \frac{k_2 - k_1}{k_1} \cdot 100, \tag{4}$$

where k—permeability change coefficient, %; k_1—permeability of the core sample before the well-killing process, m^2; k_2—permeability of the core sample after the development process, m^2.

A positive value of the coefficient of relative permeability change indicated an increase in the permeability of the core sample after the well-killing process, while a negative value indicated a decrease in permeability.

The main controlled parameter during the experiments was the change in pressure drop, based on which the change in kerosene mobility as a result of pumping the investigated composition was determined.

In accordance with the key requirements imposed on fluids for well killing before workover operation in fractured carbonate reservoirs, a comprehensive research program on the technological properties of created blocking compositions was developed. The studies of blocking compositions, including physico-chemical, rheological, and filtration characteristics, were carried out using modern high-precision equipment.

To evaluate the blocking ability of the developed compositions, the author-designed stand was used, created for modeling a fractured reservoir using a high-pressure filter press. This equipment has been modernized and adapted for more precise investigation of the blocking screen quality.

Experiments to evaluate the filtration properties of the compositions were carried out on a special filtration unit in accordance with the developed methodology. They were based on the use of natural core samples with artificially simulated fractures, bringing the conditions as close as possible to reservoir conditions.

These results highlight not only the comprehensiveness of the research program but also the importance of the methods used for a more precise understanding of the properties of the developed blocking compositions in the context of fractured carbonate reservoirs.

3. Results

Based on the analysis of literary sources, the main requirements for blocking compositions can be formulated as follows:

- manufacturability in preparation, storage, and use;
- accessibility and low cost of composition components;
- non-toxicity of applied reagents;
- thermal and sedimentation stability in a wide range of reservoir temperatures;
- adjustable density and viscosity;
- low corrosive activity;
- effective blocking of pores and filtration channels;
- ability to prevent fluid loss of technological fluids into the productive formation;
- possibility of destruction;
- insignificant penetration into the bottomhole formation zone and maximum preservation of its reservoir properties.

The developed blocking composition is a thermally stable reverse oil-in-water emulsion stabilized by an emulsifier reagent, which is a mixture of reaction products of amines and amino alcohols with fatty acids. To enhance the blocking ability, a multi-fraction mineral filler based on calcium carbonate (marble chips) is introduced into the emulsion composition. The developed composition is conditionally named BHES (Blocking Hydrophobic Emulsion Solution), and when used with marble chips, it is called BHES-MC. After preparation, these compositions are stable, homogeneous, thickened liquids of light brown color with adjustable viscosity and density values.

The initial components used in formulating the developed emulsion technological fluids were as follows: hydrocarbon phase—light oil from the VUOGP; aqueous phase—aqueous solutions of calcium chloride; surfactant—emulsifiers of various brands and manufacturers; filler—marble chips. The components of the BHES-MC composition are available, cost-effective, and non-toxic reagents of domestic production.

The primary task in developing formulations for blocking compositions was the selection of the type and optimal concentration of the emulsifier reagent, which influences the thermal and aggregative stability of emulsions.

As a result of measuring interfacial tension at the interface "oil with emulsifier—aqueous solution of calcium chloride," the three (out of 20 investigated) most effective emulsifiers were selected (based on the lowest interfacial tension values at a surfactant concentration of 0.5% wt.) under the conditional numbers: E-09, E-14, and E-20. Subsequently, based on the selected emulsifiers, the value of their optimal concentration in oil was assessed in the range from 0.05 to 5% wt. The obtained results (Figure 12) showed that the critical micelle concentration (CMC) range is within the emulsifier content in oil from 0.5 to 1.5% wt. In comparison with other types of emulsifiers, their application, under equal conditions, will contribute to the creation of an emulsion with greater thermal and aggregative stability. This is due to the fact that low interfacial tension at the phase boundary is one of the most important conditions for obtaining a stable emulsion.

In selecting the optimal ratio of initial reagents for the developed emulsion blocking compositions, three selected emulsifier brands were used with concentrations of 0.5, 1, and 1.5% wt. During the series of laboratory experiments, the following ratios of aqueous and hydrocarbon phases in the emulsion were chosen: 70/30, 80/20, and 90/10% wt., respectively. Thermal and aggregate stability at 90 °C were determined for all prepared compositions.

The results of the thermal stability tests are shown in Figure 13. The volumes of the separated hydrocarbon phase in the emulsion solution after 7 days of thermostating are presented in the respective cells. All prepared compositions are thermally and aggregatively stable at 90 °C, indicating their potential application as blocking compositions for well killing under elevated reservoir temperatures (up to 90 °C).

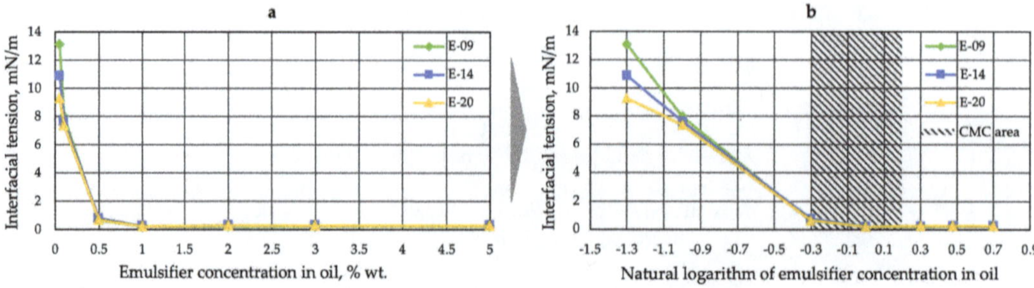

Figure 12. Results of measuring interfacial tension at the interface "oil with emulsifier—aqueous solution of calcium chloride" as a function of emulsifier concentration in oil: (**a**)—dependence of interfacial tension on emulsifier concentration in oil; (**b**)—dependence of interfacial tension on the natural logarithm of emulsifier concentration in oil.

Based on the results of the thermal stability tests at 90 °C, the optimal concentration and ratio of aqueous and hydrocarbon phases were determined for each emulsifier. A concentration of 1.5% wt. was selected for all three emulsifier types (considering the possibility of demulsifier content in oil, temperature increase, specific conditions of emulsion preparation in the field, etc.). With an aqueous-to-hydrophobic phase ratio of 80/20 and 90/10% wt., all emulsion compositions are stable. Variations in the phase ratio within these limits can be used to regulate the viscosity and density of the emulsion composition from less viscous (80/20% wt.) to more viscous (90/10% wt.).

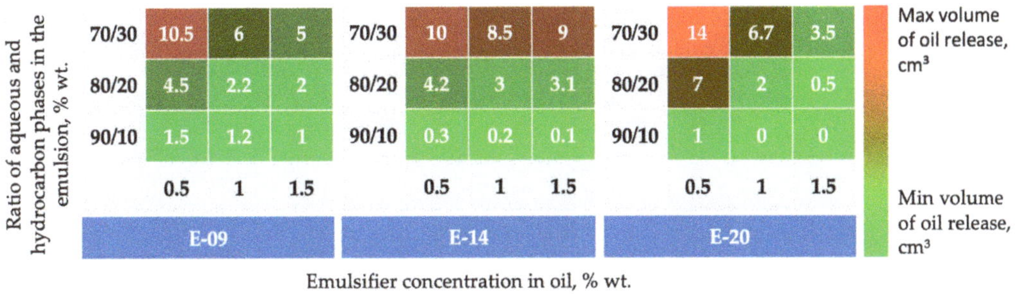

Figure 13. Thermal stability of emulsion compositions at different phase ratios at 90 °C after 7 days.

Based on the conducted research, it can be concluded that when using the emulsifier E-09 in the emulsion composition, the minimum amount of the hydrocarbon phase is separated compared to other emulsifier brands (E-14 and E-20).

Thus, for the preparation of BHES in further research, the emulsifier E-09 ("Yalan-E-2" brand B2) was used. This emulsifier is produced according to TS 2458-012-22657427-2000 and is a mixture of reaction products of amines and amino alcohols with fatty acids.

As a result, the following formulation for the BHES composition was recommended, % wt.: emulsifier "Yalan-E-2" brand B2—1.5; oil—18.5; aqueous solution of calcium chloride—80.

The use of the blocking composition BHES in its pure form (without a filler) does not prevent its fluid loss by the productive formation during well killing in conditions of fractured carbonate reservoirs and abnormally low reservoir pressures. Therefore, the next stage of research involved selecting the type and optimal concentration of the filler. To enhance the blocking ability of the BHES composition in practice, various types of plugging fillers are used to create a low-permeability blocking screen. An analysis of global experience in well killing with fluid-loss control has shown that the most promising fillers are marble chips, chalk, halite, and fibrous additives.

The assessment of the blocking ability of the compositions in modeling the fractured reservoir was carried out using a high-pressure filter press on the author's developed stand, as described in the previous chapter. The studies covered the most common range of fracture width from 0.1 to 1.5 mm, corresponding to the data from numerous laboratory studies conducted using carbonate core materials.

During laboratory experiments conducted on the developed stand, it was established that for reliable blocking of fractures with a width of up to 0.5 mm, the most effective approach is to use MC-500 brand marble chips in the composition of BHES (BHES-MC-0.5). The distribution of the fractional composition ranges from 0 to 0.5 mm (Figure 14). At the same time, the most optimal concentration of marble chips in the BHES composition for this fracture width is 5% wt. It was found that at this concentration, a blocking screen is formed at the entrance to the fracture, withstanding a pressure drop of up to 8 MPa. At a lower content of marble chips in the BHES composition, breakthrough of the blocking screen occurs at a pressure drop of less than 1 MPa.

With an increase in fracture width to 1.5 mm, it is recommended to add to the BHES a mixture of marble chips MC-500 and MC-1000/1500 (BHES-MC-1.5), taken at the optimal concentration of 3% by mass for both brands, in a 1:1 ratio. The distribution of the fractional composition of this mixture ranges from 0 to 1.5 mm (Figure 14). In this case, the BHES-MC-1.5 composition contributes to the formation of a blocking screen at the entrance to the fracture, withstanding a pressure drop of up to 8 MPa.

As a result of the conducted research, the mechanism of forming a durable and impermeable blocking screen by the BHES-MC composition was established. This mechanism involves creating a hydrodynamic barrier at the entrance to the fracture in the carbonate reservoir through the correctly selected fractional composition of the filler, covering the entire range of fracture openness (from 0 to 1.5 mm). This allows withstanding high pressure drops (more than 8 MPa). The blocking screen is formed due to the arch effect of the multi-fraction marble chips. Thus, high particle compaction is achieved under the influence of frictional forces (Figure 15).

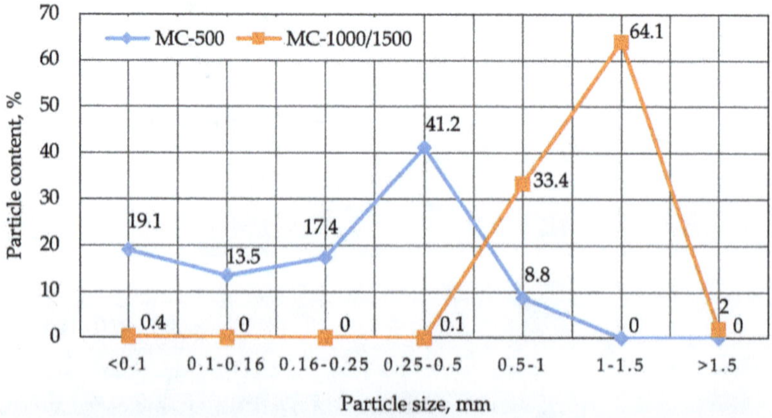

Figure 14. Distribution of the fractional composition of the filler (marble chips) in the BHES.

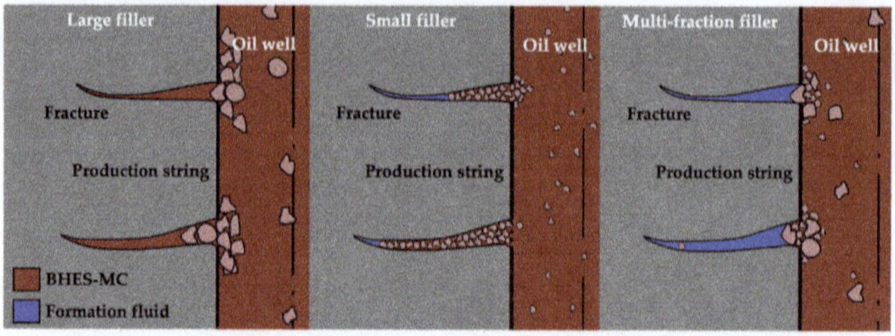

Figure 15. Mechanism of forming a blocking screen at the entrance to a fracture in the bottomhole formation zone when using fillers with various fractional compositions in process fluids.

The correct selection of the fractional composition of marble chips is important to reduce the penetration of small filler particles into the formation, which can lead to a decline in permeability and, consequently, complications during well development and production startup. On the other hand, larger filler particles are incapable of forming an impermeable blocking screen, which leads to the penetration of process fluid filtrate into the bottomhole formation zone (Figure 15).

The assessment of the sedimentation stability of BHES-MC is important to determine the ability of the dispersed system to maintain an equilibrium state throughout the volume of the dispersion medium and to identify the stability of the composition of a given phase separation. The sedimentation stability of BHES-MC was determined based on the changes in the density of the upper and lower layers of the composition after thermostating at 37 °C. After a specified time interval (7 days), a sample of the investigated composition was taken from the upper and lower parts of the test tube to measure its density using a pycnometer. The composition was considered sedimentationally stable if the difference in density between the lower and upper layers did not exceed 20 kg/m^3, representing a change of 1.5%. Thus, the sedimentation stability range was within 98.5 and 100%. In Figure 16, a dependence on changes in the sedimentation stability of the blocking composition BHES-MC is observed based on the ratio of the aqueous/hydrocarbon phases and the concentration of the filler. It was found that at an aqueous and hydrocarbon phases ratio of 80/20 and 90/10% wt., respectively, all compositions have similar sedimentation stability, indicating that the 80/20% wt. ratio is the most optimal. At this ratio, the solid particles of marble chips in the BHES composition are suspended due to the increased density and viscosity of the composition. Reducing the aqueous phase content in BHES to less than 80% wt. leads to a decrease in the density and viscosity of the solution, which impairs its sedimentation stability.

As a result, the main physico-chemical properties of the developed blocking composition BHES-MC were determined and are presented in Table 2. This composition meets the requirements imposed on technological fluids, indicating its applicability in the well-killing process in conditions of fractured carbonate reservoirs and abnormally low reservoir pressure, aiming to prevent the fluid loss of technological fluids and preserve the filtration characteristics of the bottomhole formation zone.

The study of the influence of the developed blocking compositions on the filtration characteristics of carbonate reservoirs was conducted on a specialized filtration unit, described in the previous chapter, using a modified core holder design that allows modeling fractures of different degrees of openness (0.1, 0.5, and 1.5 mm).

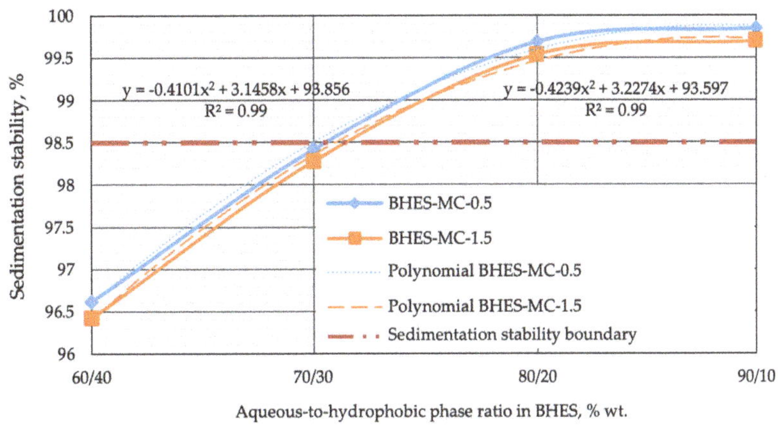

Figure 16. Dependence of sedimentation stability of BHES-MC on the ratio of aqueous/ hydrocarbon phases and filler concentration.

Table 2. The main physico-chemical properties of blocking compositions.

Parameter Name		BHES	BHES-MC-0.5	BHES-MC-1.5
Density at 20 °C, kg/m^3		1221	1287	1300
Thermal stability at 37 °C for 7 days		stable	stable	stable
Sedimentation stability after thermostating for 7 days, %		–	99.7	99.5
Electrical stability, V		367	355	351
Corrosion rate of steel grade St.20, mm/year		0.052	0.055	0.056
Pour point, °C		−21	−21	−21
Effective viscosity (D = 300 rpm), mPa·s		382	463	467
Static shear stress (D = 3 rpm), Pa	after 1 min	25.1	18.2	18.5
	after 10 min	36.6	39.4	42.1

The final results of the conducted filtration tests modeling the well-killing process and oil well development are presented in Table 3.

Table 3. Results of filtration tests of blocking compositions at a constant pressure drop (repression) of 1 MPa.

Fracture Width, mm	Composition Name	Coefficient of Relative Change in Core Permeability, %	Maximum Pressure Gradient at the Beginning of Kerosene Filtration after Composition Injection, MPa/m	Specific Volume of the Penetrated Composition in the Fracture, cm^3
0.1	BHES	−32	5.49	0.3
	BHES-MC-0.5	−2	5.07	<0.2
0.5	BHES	−93	20.9	14
	BHES-MC-0.5	−4	2.20	<0.2
1.5	BHES	−61	0.30	300
	BHES-MC-0.5	−56	0.27	300
	BHES-MC-1.5	−9	0.13	<0.2

The conducted filtration tests have shown that for conditions of a fractured carbonate reservoir, in order to prevent the fluid loss of technological fluid and maintain the filtration characteristics of the bottomhole formation zone, the use of a mineral filler (marble chips) is recommended. It is advisable to use a wider range of filler particle size with the increasing degree of natural fracture openness. Thus, with a moderate fracture openness of up to 0.5 mm, the use of the BHES-MC-0.5 is recommended. For a greater fracture openness of up to 1.5 mm, the use of the BHES-MC-1.5 is recommended. These compositions allow for the maximum preservation of the core's filtration characteristics by forming a fine-dispersed crust at its entrance from various-sized marble chips, which prevents further filtration into the fracture of the blocking composition BHES and the aqueous solution of inorganic salts. In addition to preserving filtration properties, a significant indicator of the effectiveness of the applied well-killing fluids is the low value of the pressure gradient at the beginning of kerosene filtration after well killing. This parameter characterizes the depression in the "well-bottomhole formation zone" system that needs to be created during well development to ensure the influx of reservoir fluids.

Negative values of the relative permeability change coefficient indicate the penetration of marble chips into the fracture until the formation of an impermeable blocking screen, which, in turn, leads to a deterioration of permeability. The restoration of permeability can be achieved by treating the bottomhole formation zone with a hydrochloric acid composition (12% HCl), as confirmed by the results of the filtration experiment presented in Figure 17.

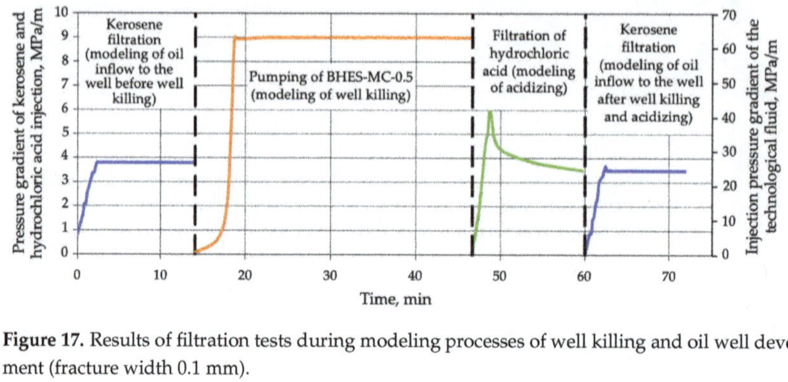

Figure 17. Results of filtration tests during modeling processes of well killing and oil well development (fracture width 0.1 mm).

Based on the presented results, it can be concluded that conducting acid treatment after the well-killing process using the BHES-MC-0.5 composition allows the following:

- improvement of the efficiency of well development by dissolving marble chips in the BHES composition and within the fracture (the maximum pressure gradient during kerosene injection after acid treatment reached 3.7 MPa/m);
- restoration and enhancement of the filtration characteristics of the bottomhole formation zone due to the impact on the fracture and matrix of the carbonate reservoir (the relative permeability change coefficient of the core was +8%).

Field tests of the blocking composition BHES-MC were carried out at the oil and gas condensate field in conditions of a fractured carbonate reservoir, abnormally low reservoir pressure, high gas factor, and long open-ended horizontal wells.

During the field test, the technology for preparing the blocking composition BHES-MC was also tested in field conditions. The following reagents and materials were used in the preparation of the blocking composition BHES-MC:

- mineralized water (saline solution) with the required density (according to the work plan)—aqueous solution of calcium or sodium chloride—78%;
- hydrocarbon phase (oil, gas condensate, diesel fuel)—10%;
- emulsifier reagent—2%;
- marble chips (ground or crushed marble), produced according to TS 5716-001-56390243-2015:
  - ground fractionated marble MC-500—3% wt.;
  - crushed marble MC-1000/1500—3% wt.

During the field testing, the following equipment was used:

- pump truck—1 unit;
- oilfield tanker—1 unit;
- on-board truck;
- 30 m^3 capacity tank—1 unit;
- crane—1 unit.

The process of preparing the blocking composition BHES-MC in field conditions, accompanying the well-killing process, and its discharge is presented in Figure 18.

Figure 18. The process of preparing the blocking composition BHES-MC in field conditions, accompanied by pilot test and well discharge.

The results of well killing using the blocking composition BHES-MC are presented in Table 4.

Analysis of the results of field tests of the BHES-MC technology on one of the oil and gas condensate fields with a carbonate reservoir showed that during the well-killing process, there is an increase in pressure when pumping the blocking composition into the bottomhole formation zone compared to a similar well treatment with a traditional emulsion composition without colmatant. This indicates the formation of a blocking screen in the fractures, contributing to preventing the fluid loss of technological fluid (Figure 19). The pressure at the pump truck at the end of the well-killing process was 12 MPa, which is 2 MPa higher compared to the previous well killing operation of the same well with an emulsion without a filler. This may indicate the formation of a filter cake of marble chips in the bottomhole formation zone.

Table 4. Results of well killing using blocking composition BHES-MC.

Parameter Name	Value
Bottomhole pressure, MPa	9.12
Vertical depth of the formation roof, m	1834
Calculated density of well killing fluid, kg/m^3	510
Current depth, m	2661
Open hole interval, m	2037–2661
Volume of blocking composition with colmatant, m^3	32
Volume of gas blocking screen (emulsion without colmatant), m^3	18
Pressure at the unit upon completion of well killing, MPa	12
Fluid level in the well after technical settling (pipe/annular), m	578/584
Presence of gas at the outlet after discharge	no
Transfer of the well to the workover team	yes

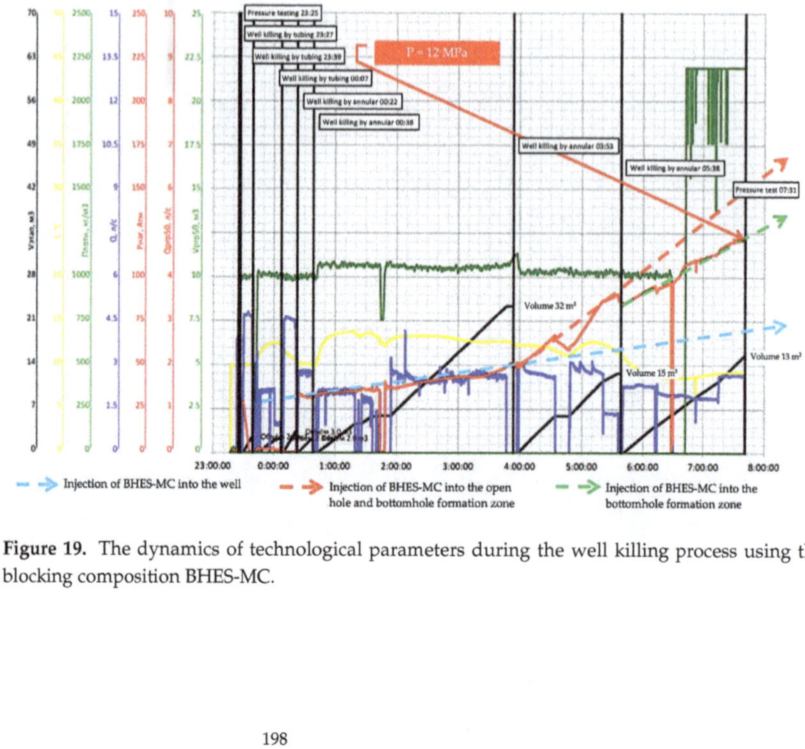

Figure 19. The dynamics of technological parameters during the well killing process using the blocking composition BHES-MC.

4. Discussion

In this section, the results of conducted investigations aimed at improving the efficiency of well-killing technology before workover operations in conditions of fractured carbonate reservoirs with abnormally low reservoir pressure are detailed and discussed. The discussion of the results is presented in the context of previous research and in consideration of the working hypotheses.

The obtained blocking composition, BHES-MC, possesses several key properties: high thermal stability up to 90 °C, sedimentation stability for 7 days at an average reservoir temperature (37 °C), adjustable density and viscosity over a wide range, and low corrosion activity.

The influence of the ratio of aqueous and hydrophobic phases, as well as the filler concentration, on the sedimentation stability of the blocking composition has been considered. At an aqueous-to-hydrophobic phase ratio of 80/20 and 90/10% wt., the composition demonstrates sedimentation stability, making it promising for use as a blocking technological fluid in the well killing before workover operations.

The mechanism of forming the blocking screen BHES-MC is described in detail, including the creation of a hydrodynamic barrier using different fractions of marble chips with fracture widths up to 1.5 mm. Such a screen can withstand pressure drops up to 8 MPa, which is extremely important in abnormally low reservoir pressure conditions.

In the context of filtration tests, the importance of the blocking composition BHES-MC in maintaining the filtration characteristics of reservoir rocks is emphasized, preventing the penetration of technological fluid into natural rocks. Particular attention is given to the possibility of restoring filtration characteristics after well killing, which is achieved using hydrochloric acid treatment.

The results of pilot tests of the BHES-MC technology on one of the oil and gas condensate fields in the VUOGP with a carbonate reservoir showed a pressure increase of 2 MPa when injecting the blocking composition into the bottomhole formation zone compared to a similar well treatment with a traditional emulsion composition without colmatant. This indicates the formation of a blocking screen in the fractures, contributing to preventing the fluid loss of the technological fluid.

The technical and economic analysis of the well-killing efficiency in complicated conditions using the blocking composition BHES-MC demonstrated a cumulative economic impact of 600 000.00 RUB (5555.00 USD) per well. Over a 4-year period for one of the oil and gas condensate fields, this amounted to a final accumulated net present value (NPV) of 80 million RUB (888,888.00 USD). The main factor of this economic effect is the prevented damage from oil production losses by reducing the number of well-killing cycles to 1, reducing the well development period, expediting its commissioning, and maintaining oil production rates.

The results obtained in this work provide a basis for further research and the development of innovative methods for well killing and workover operations, adapted to various reservoir conditions, ensuring long-term stability and operational effectiveness of oil wells.

5. Patents

Blocking hydrophobic-emulsion solution with marble chips. Available online: https://www1.fips.ru/ofpstorage/Doc/IZPM/RUNWC1/000/000/002/736/671/%D0%98%D0%97-02736671-00001/document.pdf (accessed on 23 October 2023).

Blocking biopolymer composition. Available online: https://www1.fips.ru/ofpstorage/Doc/IZPM/RUNWC1/000/000/002/757/626/%D0%98%D0%97-02757626-00001/document.pdf (accessed on 23 October 2023).

Program for selection and calculation of main parameters of well killing fluids during underground repair. Available online: https://new.fips.ru/ofpstorage/Doc/PrEVM/RUNWPR/000/002/020/615/706/2020615706-00001/document.pdf (accessed on 15 October 2023).

Program for selecting the fractional composition of marble chips for blocking hydrocarbon fluid for an oil well killing in a fractured reservoir. Available online: https://new.fips.ru

/ofpstorage/Doc/PrEVM/RUNWPR/000/002/020/616/170/2020616170-00001/document.pdf (accessed on 23 October 2023).

Database of process fluids for workover operations. Available online: https://new.fips.ru/ofpstorage/Doc/PrEVM/RUNWDB/000/002/022/621/227/2022621227-00001/document.pdf (accessed on 23 October 2023).

Database of modern technologies for killing oil and gas wells. Available online: https://new.fips.ru/ofpstorage/Doc/PrEVM/RUNWDB/000/002/022/621/272/2022621272-00001/document.pdf (accessed on 23 October 2023).

Author Contributions: Conceptualization, S.I., R.I. (Ravil Islamov), R.I. (Rustem Ismakov) and A.S.; methodology, S.I.; software, S.I.; validation, R.I. (Ravil Islamov), R.I. (Rustem Ismakov) and R.G.; formal analysis, G.S.; investigation, S.I. and A.S.; resources, R.I. (Ravil Islamov), R.I. (Rustem Ismakov) and A.A.; data curation, R.G.; writing—original draft preparation, S.I., R.I. (Ravil Islamov) and R.I. (Rustem Ismakov); writing—review and editing, G.S., A.S. and R.S.; visualization, R.S.; supervision, A.S.; project administration, S.I. All authors have read and agreed to the published version of the manuscript.

Funding: This research received no external funding.

Data Availability Statement: Data are contained within the article.

Acknowledgments: The authors thank Saint Petersburg Mining University for enabling the laboratory experiments. The investigations were carried out using the equipment of the Scientific Center "Arctic" of Saint Petersburg Mining University.

Conflicts of Interest: Author Ravil Islamov was employed by the company RusWellGroup JSC. The remaining authors declare that the research was conducted in the absence of any commercial or financial relationships that could be construed as a potential conflict of interest.

Abbreviations

BHES	Blocking hydrophobic emulsion solution
BHES-MC	Blocking hydrophobic emulsion solution with marble chips
GOST	Russian government standard
LLC	Limited liability company
TS	Technical specification
VUOGP	Volga-Ural oil and gas province

References

1. Al Manthari, M.; Al Malki, Z.; Kumar, R.; Ali, A.; Al Zaabi, A.; Das, A.; Al Sinani, H.; Al Dhayabi, S.; Ferdiansyah, E.; Al Harthi, N.; et al. Propped Fracturing in Tight Carbonates Reservoir—A Case Study. In Proceedings of the SPE International Hydraulic Fracturing Technology Conference and Exhibition, Muscat, Sultanate of Oman, 12–14 September 2023; pp. 1–10.
2. Golf-Rakht, T.D. *Fundamentals of Fractured Reservoirs Engineering*; Elsevier Publishing: Amsterdam, The Netherlands, 1982.
3. Rylance, M.; Sherishorin, V. Process and Lessons Learned Using Hollow Glass Spheres for Drilling Depleted Fields. In Proceedings of the SPE/IATMI Asia Pacific Oil & Gas Conference and Exhibition, Jakarta, Indonesia, 10–12 October 2023; pp. 1–10.
4. Nikitin, M.N.; Petukhov, A.V. Gel-forming composition on sodium silicate base for water shutoff in complex structure fractured reservoir conditions. *Pet. Eng.* **2011**, *5*, 143–154.
5. Wang, W.; Chang, S.; Gizzatov, A.; Thomas, G. Novel Sustainable Nanofluids from Waste Plastics for Oil Recovery Displacement and Decarbonization. In Proceedings of the ADIPEC, Abu Dhabi, United Arab Emirates, 2–5 October 2023; pp. 1–10.
6. Nikitin, M.N. Substantiation of Technology for Oil Recovery Increase from Deposits with High Viscosity Oil in Fractured-Porous Reservoirs with Application of Gel-Forming Composition on the Basis of Sodium Silicate. Ph.D. Thesis, Saint Petersburg Mining University, Saint Petersburg, Russia, 2012.
7. Fuller, M.J. An Innovative Approach to Gel Breakers for Hydraulic Fracturing. In Proceedings of the SPE International Conference and Exhibition on Formation Damage Control, Lafayette, LA, USA, 23–24 February 2016; pp. 1–16.
8. Alekseev, A.A. Heading for carbonates: Developing carbonate reservoirs at Gazprom Neft. *Sib. Neft* **2017**, *138*, 28–35.
9. Dandekar, A.Y. *Petroleum Reservoir Rock and Fluid Properties*; CRC Press: Boca Raton, FL, USA, 2013.
10. Foxenberg, W.E.; Ali, S.A.; Ke, M. Effects of Completion Fluid Loss on Well Productivity. In Proceedings of the SPE Formation Damage Control Symposium, Lafayette, LA, USA, 14–15 February 1996; pp. 1–16.

11. Savari, S.; Whitfill, D.; Walker, J. Acid-Soluble Lost Circulation Material for Use in Large, Naturally Fractured Formations and Reservoirs. In Proceedings of the SPE Middle East Oil & Gas Show and Conference, Manama, Kingdom of Bahrain, 6–9 March 2017; pp. 1–8.
12. Reiss, L.H. *The Reservoir Engineering Aspects of Fractured Formations*; TECHNIP Editions: Paris, France, 1980.
13. Blocking Hydrophobic-Emulsion Solution with Marble Chips. Available online: https://www1.fips.ru/ofpstorage/Doc/IZPM/RUNWC1/000/000/002/736/671/%D0%98%D0%97-02736671-00001/document.pdf (accessed on 23 October 2023).
14. Blocking Biopolymer Composition. Available online: https://www1.fips.ru/ofpstorage/Doc/IZPM/RUNWC1/000/000/002/757/626/%D0%98%D0%97-02757626-00001/document.pdf (accessed on 23 October 2023).
15. Sergeev, V.; Tanimoto, K.; Abe, M. Innovative Emulsion-Suspension Systems Based on Nanoparticles for Drilling and Well Workover Operation. In Proceedings of the Abu Dhabi International Petroleum Exhibition & Conference, Abu Dhabi, United Arab Emirates, 1–12 November 2019; pp. 1–12.
16. Bridges, K.L.; Berry, S.L. *Treatment of Completion/Workover Fluids to Remove Particulates Efficiently: Advances in Filtration and Separation Technology*; Gulf Publishing Co.: Houston, TX, USA, 1991.
17. Gizzatov, A.; Mashat, A.; Kosynkin, D.; Alhazza, N.; Kmetz, A.; Eichmann, S.; Abdel-Fattah, A. Nanofluid of Petroleum Sulfonate Nanocapsules for Enhanced Oil Recovery in High-Temperature and High-Salinity Reservoirs. *Energy Fuels* **2019**, *33*, 11567–11573. [CrossRef]
18. Savari, S.; Whitfill, D. Managing Losses in Naturally Fractured Formations: Sometimes Nano is too Small. In Proceedings of the SPE/IADC Drilling Conference and Exhibition, London, UK, 17–19 March 2015; pp. 1–10.
19. Savchenkov, S.A. The research of obtaining master alloys magnesium-gadolinium process by the method of metallothermic recovery. *Tsvetnye Met.* **2019**, *5*, 33–39. [CrossRef]
20. Chesser, B.G.; Nelson, G.F. Applications of weighted acid-soluble workover fluids. *JPT* **1979**, *31*, 36–41. [CrossRef]
21. Jafarpour, H.; Moghadasi, J.; Khormali, A.; Petrakov, D.; Ashena, R. Increasing the stimulation efficiency of heterogeneous carbonate reservoirs by developing a multi-bached acid system. *J. Pet. Sci. Eng.* **2019**, *172*, 50–59. [CrossRef]
22. Lutfullin, A.; Khusainov, R.; Manurov, I.; Ovchinnikov, K.; Malyavko, E. Conducting Multi-Stage Acid Hydraulic Fracturing in Carbonate Formations with Subsequent Intervals Production Efficiency Monitoring. In Proceedings of the SPE Russian Petroleum Technology Conference, Abu Dhabi, United Arab Emirates, 22–24 October 2019; pp. 1–12.
23. Kosolapova, S.M.; Smal, M.S.; Rudko, V.A.; Pyagay, I.N. A new approach for synthesizing fatty acid esters from linoleic-type vegetable oil. *Processes* **2023**, *11*, 1534. [CrossRef]
24. Raupov, I.; Rogachev, M.; Sytnik, J. Design of a polymer composition for the conformance control in heterogeneous reservoirs. *Energies* **2023**, *16*, 515. [CrossRef]
25. Podoprigora, D.; Byazrov, R.; Sytnik, J. The comprehensive overview of large-volume surfactant slugs injection for enhancing oil recovery: Status and the Outlook. *Energies* **2022**, *15*, 8300. [CrossRef]
26. Scheuerman, R.F. Guidelines for Using HEC Polymers for Viscosifying Solids-Free Completion and Workover Brines. *J. Pet. Technol.* **1983**, *35*, 306–314. [CrossRef]
27. McNeely, W.H.; Kang, K.S. *Industrial Gums, Polysaccharides and Their Derivatives*; Academic Press: New York, NY, USA, 1973.
28. Rylance, M.; Hoq, A. Opportunity in Adversity: A Paradigm Shift in Chemical Stimulation. In Proceedings of the SPE Russian Petroleum Technology Conference, Moscow, Russia, 16–18 October 2017; pp. 1–12.
29. Song, H.; Gizzatov, A.; Wang, W.; Thomas, G.; Mashat, A.; Abdel-Fattah, A. Wettability-Based Retention of Encapsulated Petroleum Sulfonate in Carbonate Rocks at High-Temperature and High-Salinity Conditions. *Energy Fuels* **2022**, *36*, 8210–8215. [CrossRef]
30. Caenn, R.; Darley, H.C.H.; Gray, G.R. *Composition and Properties of Drilling and Completion Fluids*; Gulf Professional Publishing: Houston, TX, USA, 2011.
31. Binks, B.P. *Modern Aspects of Emulsion Science Technology*; Royal Society of Chemistry: Cambridge, UK, 1998.
32. Sharikov, F.Y.; Rudko, V.A.; Smyshlyaeva, K.I. Oxidation thermolysis kinetics of asphaltenes with various chemical prehistory. *Thermochim. Acta* **2023**, *726*, 179550. [CrossRef]
33. Lu, S.; Foxenberg, E. Double-Emulsion Spacer Design for Highly Efficient Invert Emulsion Displacement. In Proceedings of the SPE International Symposium and Exhibition on Formation Damage Control, Lafayette, LA, USA, 15–17 February 2012; pp. 1–18.
34. Gomaa, A.M.; Nino-Penaloza, A.; McCartney, E.; Mayor, J. Engineering Solid Particulate Diverter to Control Fracture Complexity: Experimental Study. In Proceedings of the SPE Hydraulic Fracturing Technology Conference, The Woodlands, TX, USA, 9–11 February 2016; pp. 1–19.
35. Beloglazov, I.; Morenov, V.; Leusheva, E.; Gudmestad, O.V. Modeling of heavy-oil flow with regard to their rheological properties. *Energies* **2021**, *14*, 359. [CrossRef]
36. Savchenkov, S.A.; Bazhin, V.G. Research on the process of gadolinium recovery from the melt of salts on formation of Mg—Zn—Gd master alloys for manufacturing of magnesium and aluminium special-purpose alloys. *Non-Ferr. Met.* **2020**, *1*, 35–40. [CrossRef]
37. Ilyushin, Y.V.; Kapostey, E.I. Developing a comprehensive mathematical model for aluminium production in a soderberg electrolyser. *Energies* **2023**, *16*, 6313. [CrossRef]
38. Ilyushin, Y.V. Development of a process control system for the production of high-paraffin oil. *Energies* **2022**, *15*, 6462. [CrossRef]
39. Beloglazov, I.; Morenov, V.; Leusheva, E. Flow modeling of high-viscosity fluids in pipeline infrastructure of oil and gas enterprises. *Egypt. J. Pet.* **2021**, *30*, 43–51. [CrossRef]

40. Asadulagi, M.M.; Ioskov, G.V. Simulation of the control system for hydrodynamic process with random disturbances. In *Topical Issues of Rational Use of Natural Resources*; Taylor & Francis: London, UK, 2019; pp. 399–405.
41. Asadulagi, M.M.; Ioskov, G.V.; Tronina, E.V. Synthesis of lumped and distributed controllers for control system of hydrodynamic process. In Proceedings of the International Multi-Conference on Industrial Engineering and Modern Technologies, Institute of Electrical and Electronics Engineers, Vladivostok, Russia, 1–4 October 2019; pp. 399–405.
42. Efimov, I.; Smyshlyaeva, K.; Povarov, V.; Buzyreva, E.; Zhitkov, N.; Vovk, M.; Rudko, V. UNIFAC residual marine fuels stability prediction from NMR and elemental analysis of SARA components. *Fuel* **2023**, *352*, 129014. [CrossRef]
43. Fetisov, V.; Davardoost, H.; Mogylevets, V. Technological aspects of methane–hydrogen mixture transportation through operating gas pipelines considering industrial and fire safety. *Fire* **2023**, *6*, 409. [CrossRef]
44. Fetisov, V.; Shalygin, A.V.; Modestova, S.A.; Tyan, V.K.; Shao, C. Development of a numerical method for calculating a gas supply system during a period of change in thermal loads. *Energies* **2023**, *16*, 60. [CrossRef]
45. Savchenkov, S.; Beloglazov, I. Features of the process obtaining of Mg-Zn-Y master alloy by the metallothermic recovery method of yttrium fluoride melt. *Crystals* **2022**, *12*, 771. [CrossRef]
46. *GOST 3900-85*; Petroleum and Petroleum Products. Methods for Determination of Density. Standards Publishing House: Moscow, Russia, 1985.
47. *GOST 9.502-82*; Unified System of Corrosion and Ageing Protection. Inhibitors of Metals Corrosion for Aqueous Systems. Methods of Corrosion Tests. Standards Publishing House: Moscow, Russia, 1982.
48. *GOST 20287-91*; Petroleum Products. Methods of Test for Flow Point and Pour Point. Standards Publishing House: Moscow, Russia, 1991.
49. *GOST 29232-91*; Anionic and Non-Ionic Surface Active Agents. Determination of the Critical Micellization Concentration. Method by Measuring Surface Tension with a Plate, Sturrup or Ring. Standards Publishing House: Moscow, Russia, 1991.
50. *GOST 12536-2014*; Soils. Methods of Laboratory Granulometric (Grain-Size) and Microaggregate Distribution. Standards Publishing House: Moscow, Russia, 2014.
51. *GOST 26450.0-85*; Rocks. General Requirements for Sampling and Sample Preparation for Determination of Collecting Properties. Standards Publishing House: Moscow, Russia, 1985.
52. *GOST 26450.1-85*; Rocks. Method for Determination of Open Porosity Coefficient by Fluid Saturation. Standards Publishing House: Moscow, Russia, 1985.
53. *GOST 26450.2-85*; Rocks. Method for Determination of Absolute Gas Permeability Coefficient by Stationary and Non-Stationary Filtration. Standards Publishing House: Moscow, Russia, 1985.
54. *GOST 39-235-89*; Oil. Method for Determining Phase Permeabilities in Laboratory Conditions with Joint Stationary Filtration. Standards Publishing House: Moscow, Russia, 1989.

processes

MDPI

Article

The Application of Breakthrough Pressure in the Evaluation of the Sealing Ability of Cement–Casing Interface and Cement Matrix in Underground Gas-Storage Wells

Yan Yang *, Lukuan Li, Wenyan Yu, Yan Zhou, Kuanliang Zhu and Bin Yuan

Drilling & Production Technology Research Institute of Jidong Oilfield, Petrochinay, Tangshan 063000, China; lulk@petrochina.com.cn (L.L.); jdzc_yuwy@petrochina.com.cn (W.Y.); jdzc_zhouyan@petrochina.com.cn (Y.Z.); zkl@petrochina.com.cn (K.Z.); 202199010013@swpu.edu.cn (B.Y.)

* Correspondence: jdzc_yangyan@petrochina.com.cn

Abstract: This paper proposes an evaluation system for the sealing ability of cement–casing interface and cement matrix, which was developed based on a permeability testing device and a model that predicts the breakthrough pressure of cement sheath matrix and interfacial transition zone (*ITZ*). It was found that the breakthrough pressure of *ITZ* was much smaller than that of cement matrix. Moreover, compared with water-based drilling fluid, oil-based one led to lower breakthrough pressure of *ITZ* even after the flushing treatment. Meanwhile, latex, resin and elastic materials enabled a substantial rise in the breakthrough pressure of cement matrix. However, compared with the latex, resin and elastic cement, the expansive cement had higher breakthrough pressure of *ITZ*, indicating an improvement on the interface sealing ability. Additionally, a small enlargement rate of the hole diameter and long effective bond were able to prevent gas storage wells from leakage.

Keywords: underground gas storage; breakthrough pressure; cement sheath; cement–casing interface; interfacial transition zone

Citation: Yang, Y.; Li, L.; Yu, W.; Zhou, Y.; Zhu, K.; Yuan, B. The Application of Breakthrough Pressure in the Evaluation of the Sealing Ability of Cement–Casing Interface and Cement Matrix in Underground Gas-Storage Wells. *Processes* **2022**, *10*, 620. https://doi.org/10.3390/pr10040620

Academic Editor: Xiaoyan Ji

Received: 10 February 2022
Accepted: 21 March 2022
Published: 22 March 2022

1. Introduction

"To cut greenhouse gas emissions by 80–100% by 2050 or sooner" is the goal of Carbon Neutral Cities Alliance [1] and has promoted the development of the natural-gas industry. As a critical component of the natural-gas supply and distribution chain, underground gas storage (UGS) is playing an increasingly important role in this sector [2]. Maintaining the wellbore sealing integrity of UGS wells is a key issue concerning the safety of the staff, the surrounding environment and the UGS itself [3]. The sealing failure of the cement sheath results in sustained casing pressure (SCP), which is a common phenomenon that raises potential risk concerns [4].

The quality of the resultant annular cement sheath is crucial for sealing integrity [5]. The quality of the cement sheath is evaluated not only by the compressive strength, bond strength, and permeability [6], but also by the sealing ability of the interface of the cement casing and cement formation. Yang, Kuru [7] found that leakage is most likely to occur near the interface, either the one between the cement and casing or the cement and formation. Therefore, interfaces play an essential role in the sealing integrity of the cement sheath. An important component of the cement sheath is the interfacial transition zone (*ITZ*) between the casing (and formation) and cement [5], which significantly affects the sealing integrity of the cement sheath. In the present study, the sealing ability of the cement–casing interface and cement matrix was investigated.

Numerous methods have been used to assess the interface's sealing integrity of the cement sheath, mainly including the interface's bonding strength, gas channeling, scanning electron microscope (SEM), and μ-CT [5,7–9]. For instance, Yang [7] employed SEM and μ-CT to investigate the effects of the environmental cement-drying shrinkage and the expansion agent on the sealing ability of the interface. However, the interface's bonding strength cannot be used to evaluate the leakage capacity of channels such as micro-cracks and large pores. In other words, a high interfacial bonding strength may still accompany serious SCP. The channeling pressure of the gas-channeling test is time dependent and is thus unable to quantitatively represent the sealing ability of the cement sheath. SEM and μ-CT are not suitable for field applications due to the cost, complex operating procedure and strong professionalism required. Underground gas-storage cement is in need of an evaluation system with convenient operation, high applicability and low cost.

Considering its wide and successful application in reservoir caprocks [10], the breakthrough pressure was selected to assess the sealing ability of the cement–casing interface and is generally described by the Laplace law [11]:

$$p_c = \frac{2\sigma\cos\theta}{r} \tag{1}$$

where σ is the interfacial tension, N/m; θ is the contact angle, °; and r is the radius, m. Breakthrough pressure is a vital parameter for evaluating the effectiveness of the cement sheath on gas sealing. One of the major reasons accounting for SCP is gas leakage due to the differential pressure of the cap rock and cement sheath. High breakthrough pressure of the cement sheath ensures better sealing integrity and lower leakage risk. The porosity of the matrix and the *ITZ* of the cement sheath is close to that of rock, whose breakthrough pressure can be used to determine the sealing ability of the cement sheath.

In this work, a model was built to predict the breakthrough pressure of the *ITZ* of the cement–casing interface and cement matrix, based on which a measuring method was presented. The accuracy and reliability of the new evaluation system was verified. The effects of oil- and water-based drilling fluids and flushing fluids on the sealing ability of the cement sheath were also investigated by breakthrough pressure. The system was finally applied to an underground gas-storage well.

2. Interfacial Transition Zone

The *ITZ* is a typical micro-meso structure with a thickness of 40–100 μm [12–14]. The properties of the *ITZ* differ from that of the cement matrix [15]. The radial micro-cracks, micro-annuli and macro-porosity that develop in the *ITZ* of the cement–casing interface [5,16,17] result in the permeability reaching 600 mD or even higher [7,18,19], which is higher than that of the cement matrix (usually less than 0.01 mD). Therefore, the *ITZ* is a potential leakage channel for natural gas in underground gas storage.

The *ITZ* of the cement sheath has two parts, including the casing–cement interface (first interface) and the formation–cement interface (second interface) (Figure 1). However, the pore structure and permeability are different between the two transition zones [7]. Besides, the cement sheath is a porous media with three permeabilities: the *ITZ* is highly permeable while the cement matrix is a low-permeability medium. This study investigated the breakthrough pressure of the *ITZ* of the casing–cement interface and matrix.

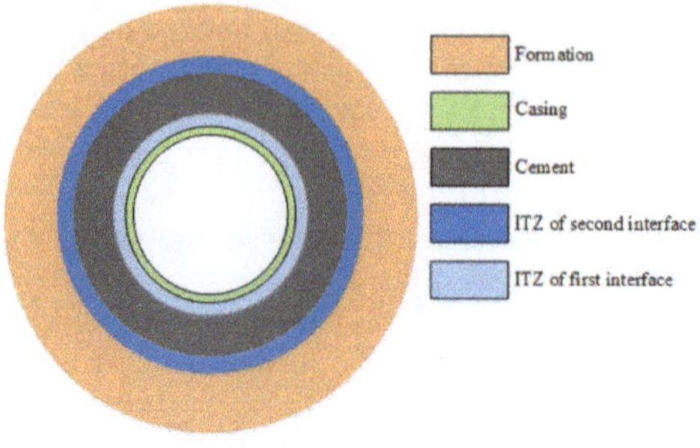

Figure 1. Schematic diagram of the transition zones in cement sheath.

3. Breakthrough-Pressure-Prediction Model

3.1. Breakthrough Pressure of Matrix

The breakthrough pressure of the cement matrix is higher than that of the *ITZ* and can be described as follows [20]:

$$p_{mbt} = \frac{\sigma_g \cos\theta_g}{\tau\sqrt{\psi}}\sqrt{\frac{\phi_m}{k_m}} \tag{2}$$

where p_{mbt} is the breakthrough pressure, Pa; τ is the tortuosity, m/m; σ_g is the interfacial tension of gas, N/m; θ_g is the contact angle of gas, °; k_m is the permeability of the cement matrix, m^2; ψ is the pore-shape factor, 2; and ϕ_m is the porosity of the cement matrix, %.

The porosity of the cement sheath depends on the water-cement ratio and the hydration degree. The porosity of the cement matrix can be determined by the following equation [21]:

$$\phi_m = \frac{r_{w/c} - 0.36\alpha_{max}}{r_{w/c} + 0.32} \tag{3}$$

where α_{max} is the max hydration degree, %. $r_{w/c}$ is the water-cement ratio, %.

The hydration degree of a cement slurry mainly depends on the water-cement ratio, curing conditions and time. After the cementing construction of the gas storage, it takes so long to start gas injection that the cement will be fully hydrated. As the specimens had been cured for a long time, the maximum hydration degree of the cement slurry was applied in this model and was calculated by the following equation [22]:

$$\alpha_{max} = \frac{1.031 \cdot r_{w/c}}{r_{w/c} + 0.194} \tag{4}$$

The tortuosity is the geometric feature of the pore structure and is only closely related to the porosity of the cement stone. Koichi, Tetsuya [23] found that the tortuosity of cement stone decreases as the porosity in the cement increases, and its changing is expressed mathematically as:

$$\tau = -1.5\tanh[8.0(\phi - 0.25)] + 2.5 \tag{5}$$

3.2. Breakthrough Pressure of ITZ

According to the seepage characteristics, the high permeability of the *ITZ* and the low permeability of the matrix have the following relationship [24]:

$$k_T R_1^2 = k_{ITZ}(2R_1 - h)h + k_m(R_1 - h)^2 \tag{6}$$

where R_1 is the radius of the cement sheath, m; h is the thickness of the *ITZ*, m, which is 7×10^{-5} in this paper; k_T, k_m, k_{ITZ} is the permeability of the cement–casing interface, matrix and *ITZ*, respectively.

When $h << R_1$, Equation (6) can be written as:

$$k_{ITZ} = \frac{(k_T - k_m)R_1}{2h} \tag{7}$$

The porosity and permeability of the *ITZ* can be calculated by the following equation [25]:

$$k_{ITZ} = \frac{\phi_{ITZ}^2}{50B^2\mu}\left[1 - \ln\left(\frac{\phi_{ITZ}}{1-\phi_{ITZ}}\right) \cdot (1 - \phi_{ITZ})\right]^2 \tag{8}$$

where ϕ_{ITZ} is the porosity of *ITZ*; B represents the Raleigh–Ritz pore-size-distribution constant in HYMOSTRUC and is 3.6×10^7 m^{-1}; μ is the viscosity, Pa·s.

The breakthrough pressure of the *ITZ* is presented as follows:

$$p_{ITZbt} = \frac{\sigma_g \cos\theta_g}{\tau\sqrt{\psi}}\sqrt{\frac{\phi_{ITZ}}{k_{ITZ}}} \tag{9}$$

4. Model Verification

4.1. Contact Angle

The neat-cement and expansion-cement slurries with a water-cement ratio of 0.44 were used to verify the accuracy of the breakthrough-pressure-prediction model. These cement slurries were cured in water (90 °C, 21 MPa) for seven days. The contact angles of the cements, as indicated in Figure 2, were measured and the averages of the measurements are shown in Table 1. The addition of expansion agents increased the contact angle, suggesting the hydrophobic nature of these agents.

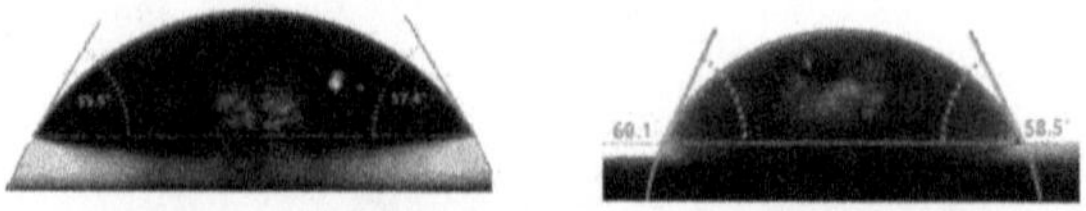

Figure 2. Contact angles of two cement slurries. (**a**) neat cement; (**b**) expansion cement.

Table 1. Composition of the prepared cement samples.

Sample	Constituents (Additive to Cement Ratio) (wt. %)						
	Cement	Expensive Agent	Stabilizer	Drag Reducer	Filtrate Reducer	De-Foamer	Retarder
Neat cement	100	0	4	0.75	3.5	1	3
Expansive cement		4					

The breakthrough-pressure-testing device is composed of a nitrogen source, intermediate container, core holder, simulation casing mold, cement sheath, and flowmeter, etc. The essential component is the simulation casing mold, which has a certain slope inside with a value satisfying Equation (10). The purpose of this design is to prevent the pressure difference between the two ends of the cement column from exceeding the bonding strength of the cement, which will lead to micro-annuli.

$$\frac{D_2 - D_1}{2L} = 0.017 \quad (10)$$

4.2. Verification Method

Step 1: Measure the gas permeability k_m of the cement matrix by the gas-permeability method.
Step 2: Use Equation (2) to calculate the breakthrough pressure of the cement matrix (p_{mbt}).
Step 3: Measure the gas permeability k_T of the *ITZ* and use Equation (7) to calculate the permeability of the *ITZ* (k_{ITZ}).
Step 4: Use Equation (9) to calculate the breakthrough pressure of the *ITZ* p_{ITZbt}.
Step 5: Put the cured specimen of the *ITZ*-testing mold (Figure 4) into the core holder and apply a confining pressure that is double the matrix breakthrough pressure (p_{mbt}).
Step 6: Apply the p_{ITZbt} of the *ITZ* to the specimen as a test of the pressure difference. If no bubbles appear after waiting for 10 h, then increase the pressure by 0.1 MPa. Repeat this step until bubbles appear and record the pressure difference.

The permeability of the total mold and cement matrix are shown in Table 2 and the verified results are shown in Figure 3. It was found that the errors of the breakthrough pressure of the cement matrix and the *ITZ* of the neat-cement-slurry system were 3.85% and 4.17%, respectively, and those of the expansive-cement-slurry system were 2.44% and 6.74%, respectively, indicating accurate predictions of the breakthrough pressure of the *ITZ* by this model. The breakthrough pressure of the *ITZ* of the expansion cement was 185.4% higher than that of the neat cement. Expansion agents were added to the cement to reduce or avoid the shrinkage during cement-slurry setting, contributing to minimized gap or cracks of the *ITZ* [7,26]. This means a reduced permeability of the *ITZ* and the resultant increase in the breakthrough pressure.

Table 2. Verified results of breakthrough pressure.

Cement Slurries	k_m ($\times 10^{-19}$ m^2)	k_t ($\times 10^{-19}$ m^2)	θ (°)
Neat cement	5.82	15.10	56.45
Expansive cement	2.33	2.71	59.30

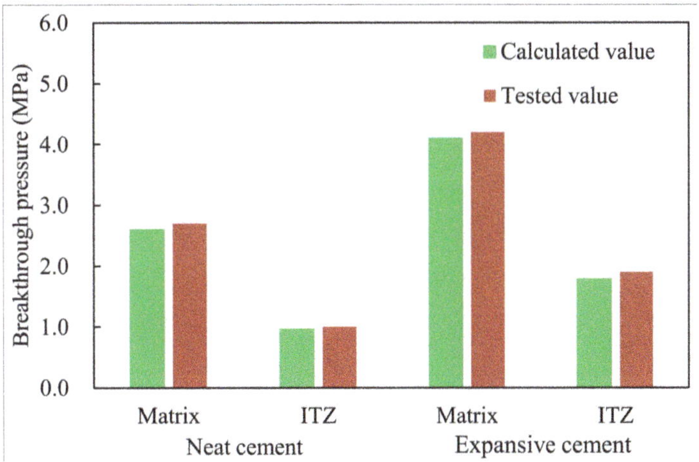

Figure 3. Experimentally verified results of breakthrough pressure of matrix and *ITZ*.

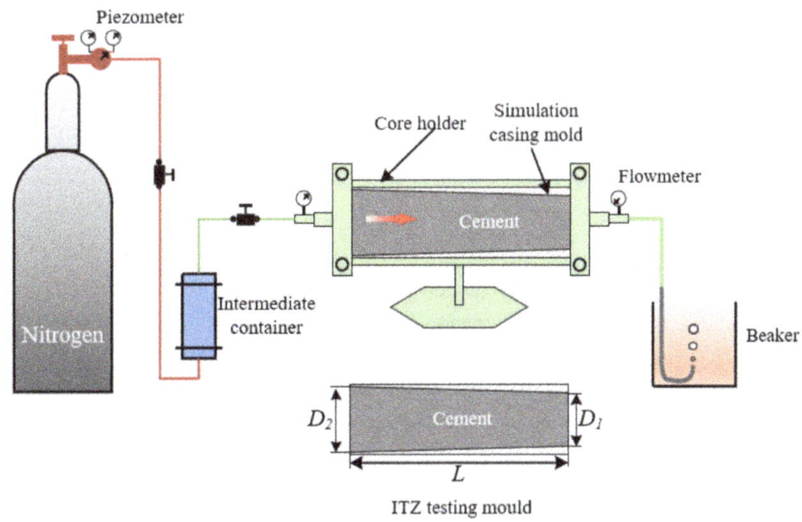

Figure 4. Schematic diagram of breakthrough-pressure-testing device.

5. Factors Influencing Breakthrough Pressure

5.1. Drilling Fluid

Since drilling fluid and flushing fluid mainly act on the casing–cement interface, this study only analyzed the breakthrough pressure of the *ITZ*. To determine the effect of water-based and oil-based drilling fluids on the breakthrough pressure of the casing–cement interface, the steel tubes were immersed in two drilling fluids for 24 h, and then flushed with flushing fluid. The effect of the treatment on the breakthrough pressure of the cement–casing interface was analyzed, and the results are shown in Figure 5. Compared with the oil-based drilling fluid, the water-based fluid led to a much higher breakthrough pressure of the treated steel tube, which was further augmented by the flushing-liquid treatment. The oil-based-drilling-fluid treatment left an oil film on the surface of the steel tube, which could not be completely removed by the flushing-fluid treatment. The oil film changed the hydrophilicity of the surface of the casing: the contact angles of the steel treated by oil-based and water-based drilling fluid were 86.5° and 68.1°, respectively, which were reduced to 65.6° and 53.3° by flushing. According to Equation (10), the degree of the contact angle negatively correlates with the breakthrough pressure.

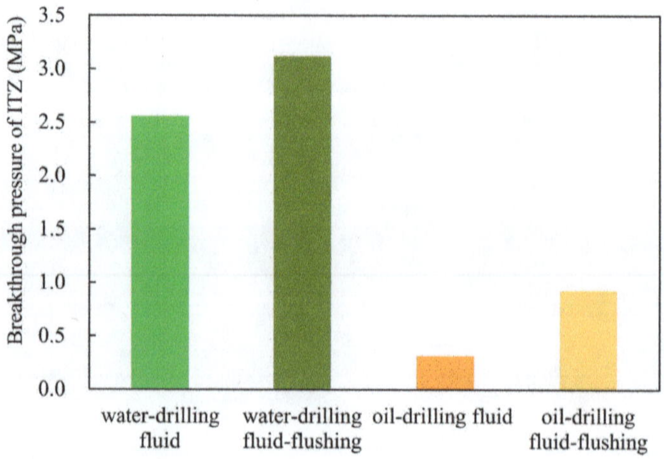

Figure 5. Impact of drilling fluids on the breakthrough pressure of *ITZ*. Drilling fluid and flushing fluid are from the construction site.

5.2. Cement Slurries

The breakthrough pressure of the matrix and the *ITZ* of the expansive-cement, elastic-cement, latex-cement and resin-cement slurries were analyzed and the results are shown in Figures 6 and 7. The different types of cement ranked in descending order of the breakthrough pressure of the matrix are as follows: latex cement, resin cement, expansive cement, elastic cement and neat cement, with the pressure of the first two types much higher than that of the rest. First of all, latex and resin have a film-forming effect, which can greatly reduce the permeability of the cement matrix [27,28]. Additionally, latex and resin change the hydrophilicity of the cement and increase the contact angle, which raises the matrix breakthrough pressure. In descending order of the breakthrough pressure of the *ITZ* are expansive cement, latex cement, resin cement, elastic cement, and neat cement (Figure 7). Expansion agents reduce the micro-gap caused by cement shrinkage, and latex and resin improve the bonding effect of the interface, both of which cause a dramatic increase in the breakthrough pressure of the *ITZ*. The breakthrough pressures of elastic cement, latex cement and expansive cement were 17.7%, 64.5% and 85.4%, respectively, which were higher than that of the neat slurry. Compared with swellable materials and latex, elastic materials have less influence on the breakthrough pressure. Figure 7 shows that the breakthrough pressure of the matrix remained constantly high while the breakthrough pressure of the *ITZ* varied. The expansion material had a better effect than latex and resin on increasing the breakthrough pressure of the *ITZ*.

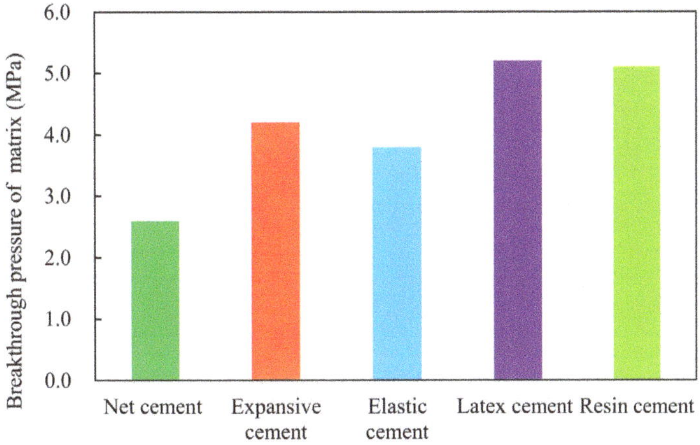

Figure 6. Impact of cement slurries on the breakthrough pressure of matrix.

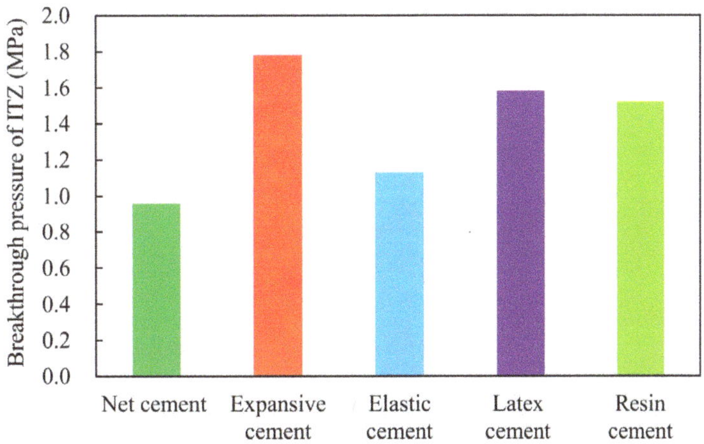

Figure 7. Impact of cement slurries on the breakthrough pressure of *ITZ*.

6. Application

We applied the evaluation system to an underground gas-storage well in Sichuan Basin, the basic information of which is shown in Table 3. The expansive elastic-cement slurry was used for cementing the gas-storage well, and its performance is shown in Table 4. The leakage rate of the *ITZ* was predicted by Equation (11). The result in Figure 8 shows that the breakthrough pressure is negatively correlated with the leakage rate.

$$\mathrm{Q} = \begin{cases} \frac{\pi(R_w^3 - R_c^2 R_w)(P_o - \rho_w g H)(k_T - k_m)}{2hL_c\mu} \, p_{ITZbt} < (P_o - \rho_w g H) \\ \left[\frac{R_w}{2h}(k_T - k_m) + k_m\right] \frac{\pi(R_w^2 - R_c^2)(P_o - \rho_w g H)}{L_c\mu} p_{\mathrm{mbt}} < (P_o - \rho_w g H) \end{cases} \quad (11)$$

where P_o is the formation pressure, MPa; ρ_w is the density of the formation water, kg/m^3; g is the gravity acceleration, m/s^2; H is the depth of the top of caprock, m; R_w is the radius of the casing, m; L_c is the thickness of the caprock, m.

Table 3. Basic information of an underground gas storage.

Parameter	Depth (m)	Caprock (m)	Formation Pressure (MPa)	Interfacial Tension ($\times 10^{-3}$ N/m)	Contact Angle (°)	Viscosity ($\times 10^{-3}$ Pa·s)
Value	2310	40	12.6	34.1	35	0.02

Table 4. Performance parameter of expansive-elastic cement.

Parameter	k_m ($\times 10^{-19}$ m^2)	k_t ($\times 10^{-19}$ m^2)	p_{mbt} (MPa)	p_{ITZbt} (MPa)	Bonding Strength (MPa)
Value	2.42	2.84	4.85	1.80	4.30

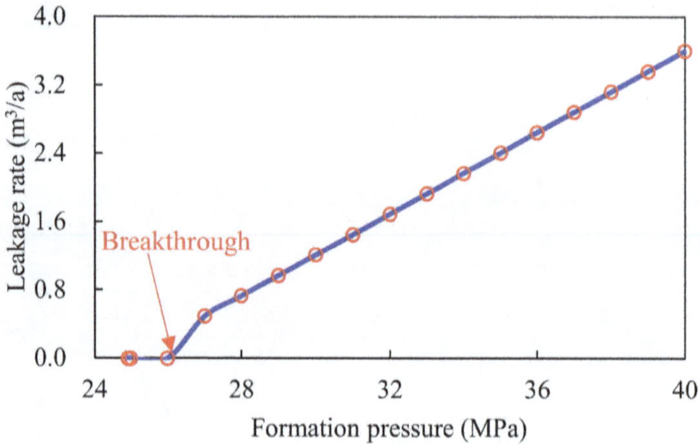

Figure 8. The relationship between leakage rate and formation pressure.

When gas was first injected, the formation-pore pressure was too low to leak (Figure 8). However, the pore pressure continued rising with the continuous injection of natural gas into the formation. When the pore pressure of the formation reached 27.0 MPa, the difference between the pore pressure and hydrostatic column one (2.1 MPa) exceeded the breakthrough pressure (1.8 MPa) of the *ITZ*, resulting in leakage. When the pressure difference was only slightly higher than the breakthrough pressure of the cement matrix (4.85 MPa), the increase in the leakage rate was not significant, because the contribution of the matrix to the rate is extremely low. When the pore pressure increased to 40.0 MPa, the leakage rate of the matrix was only 0.006 m^3/a. When the pressure difference outran the breakthrough pressure, the leakage rate rose linearly with the increase in pore pressure. When the pore pressure increased by 33.3% from 30.0 to 40.0 MPa, the leakage rate climbed by 197.1%.

With poor cementing quality, the effective bonding length of the cement–casing-interface sheath greatly impacted the leakage rate, as shown in Figure 9. As the effective bonding length increased, the drop in the leakage rate was initially dramatic and then gradually slowed down. The leakage rate dropped by 400.0% as the effective bonding length increased from 4 m to 20 m, but only reduced by 66.5% when the length rose from 24 m to 40 m. The cement sheath with such performance ensured that significant leakage could be avoided by an effective bond between the cement–casing interface longer than 24 m.

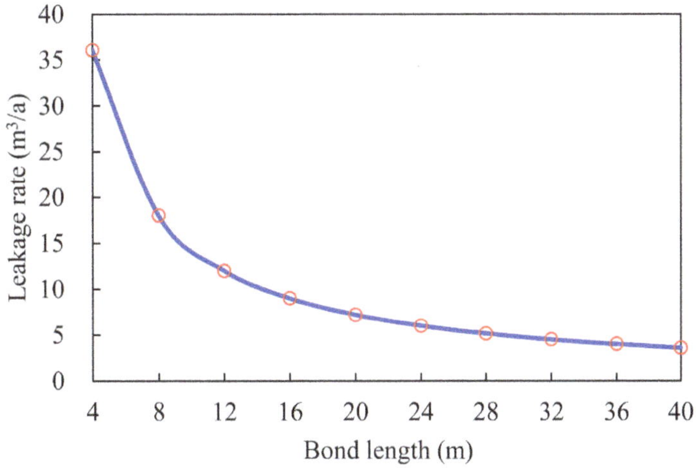

Figure 9. The influence of the effective bonding length of cement sheath on the leakage rate.

As the viscosity of fluid in the formation increased, the leakage rate decreased rapidly and then gradually slowed down (Figure 10). Viscosity had a comparable effect on the leakage rate as bonding length. Viscosity negatively correlated with the leakage rate, which reduced from 7.2 to 0.9 m^3/a as the viscosity was increased from 0.01 to 0.08 mPa·s. Natural gas contains more impurities under formation conditions [29]. The heterogeneity of the gas caused high viscosity, which decelerated the leakage.

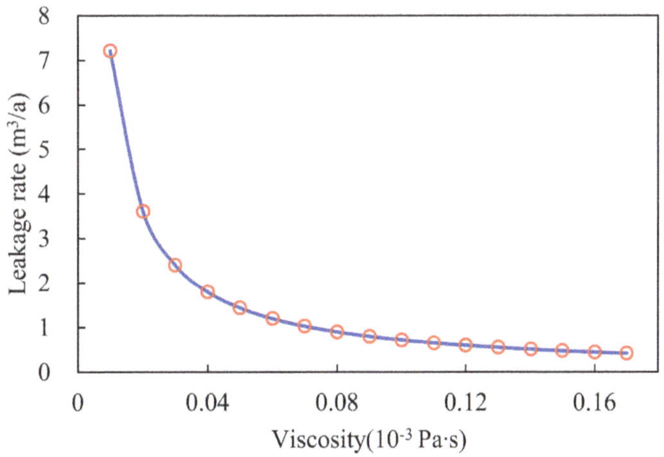

Figure 10. The effect of the viscosity of formation fluids on the leakage rate.

The borehole diameter positively correlated with the leakage rate (Figure 11). An increase in the borehole diameter raised the contact area between the cement sheath and the casing, as well as the *ITZ*. When the diameter of the drill bit was constant, a larger diameter of the wellbore led to a greater expansion rate, which increased by 88% as the hole diameter was enlarged by 14%. The expansion rate of the borehole should be controlled during the drilling.

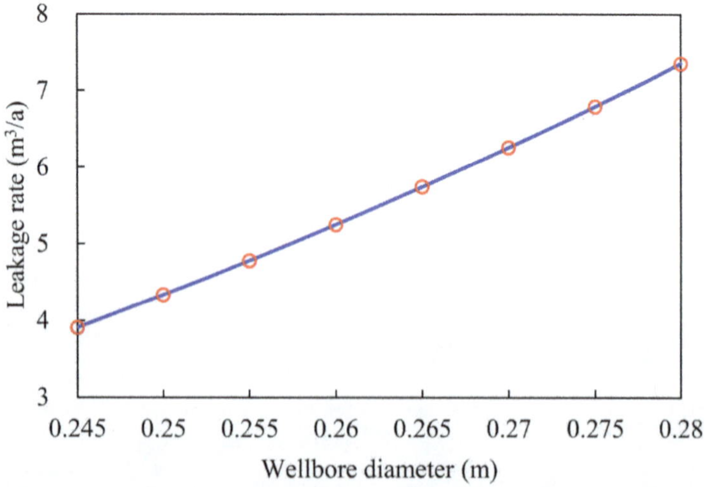

Figure 11. The wellbore diameter effect on the leakage rate.

7. Conclusions

In the present study, an evaluation system used the breakthrough-pressure model to determine the sealing ability of the cement–casing interface and cement matrix in underground gas-storage wells. The quantitative evaluation generated by this system allows the rational selection of flushing fluids and cement slurries. We employed the system to evaluate the effect of the drilling fluid, flushing fluid, and sealing ability of the cement slurries on breakthrough pressure, and an analysis of the leakage of the cement–casing-interface was performed in an underground gas-storage well. The following are the obtained results:

1. Compared with water-based drilling fluid, oil-based fluid decreased the breakthrough pressure of the *ITZ* by changing the contact angle. The pressure could be increased by fluid flushing.
2. The breakthrough pressure of the matrix of latex and resin cement was higher than that of expansive cement, whereas the breakthrough pressure of the *ITZ* of latex and resin was lower than that of expansive cement. Expansive cement had a better performance in interface sealing.
3. The *ITZ* had a dominant role in the leakage of the cement–casing interface. Continuous injection of natural gas increased the pore pressure of the formation, leading to an increase of leakage in the cement–casing-interface sheath.
4. A longer effective bonding length, higher gas viscosity and smaller enlargement rate of the hole diameter led to a smaller leakage rate of the cement sheath-casing.

Author Contributions: Writing-Original draft preparation, Y.Y.; writing-reviewing, and editing, L.L.; methodology, W.Y.; data curation, Y.Z.; data statistics, K.Z.; validation, B.Y. All authors have read and agreed to the published version of the manuscript.

Funding: This research was funded by Feasibility study on gas-storage engineering of Nanpu 1–29 and Baogu 2 blocks in Jidong Oilfield grant number KF2019A02.

Institutional Review Board Statement: Not applicable.

Informed Consent Statement: Not applicable.

Data Availability Statement: Not applicable.

Acknowledgments: The authors acknowledge the Feasibility study on gas-storage engineering of Nanpu 1–29 and Baogu 2 blocks in Jidong Oilfield (KF2019A02). Without their support, this work would not have been possible.

Conflicts of Interest: The authors declare no conflict of interest. The authors declare that they have no known competing financial interests or personal relationships that could have appeared to influence the work reported in this paper.

References

1. Becker, S.; Bouzdine-Chameeva, T.; Jaegler, A. The carbon neutrality principle: A case study in the French spirits sector. *J. Clean. Prod.* **2020**, *274*, 122739. [CrossRef] [PubMed]
2. Zhao, L.; Yan, Y.; Wang, P.; Yan, X. A risk analysis model for underground gas storage well integrity failure. *J. Loss Prev. Process Ind.* **2019**, *62*, 103951. [CrossRef]
3. Zhang, H.; Shen, R.; Yuan, G.; Ba, Z.; Hu, Y. Cement sheath integrity analysis of underground gas storage well based on elastoplastic theory. *J. Pet. Sci. Eng.* **2017**, *159*, 818–829. [CrossRef]
4. Benge, G. Improving wellbore seal integrity in CO_2 injection wells. *Energy Procedia* **2009**, *1*, 3523–3529. [CrossRef]
5. Torsæter, M.; Todorovic, J.; Lavrov, A. Structure and debonding at cement–steel and cement–rock interfaces: Effect of geometry and materials. *Constr. Build. Mater.* **2015**, *96*, 164–171. [CrossRef]
6. Nelson, E.B.; Guillot, D. *Well Cementing*, 2nd ed.; Schlumberger: Sugar Land, TX, USA, 2006.
7. Yang, X.; Kuru, E.; Gingras, M.; Iremonger, S.; Chase, P.; Lin, Z. Characterization of the microstructure of the cement/casing interface using esem and micro-CT scan techniques. *SPE J.* **2021**, *26*, 1131–1143. [CrossRef]
8. Lavrov, A.; Panduro, E.A.C.; Torsæter, M. Synchrotron Study of Cement Hydration: Towards Computed Tomography Analysis of Interfacial Transition Zone. *Energy Procedia* **2017**, *114*, 5109–5117. [CrossRef]
9. Zeng, Y.; Lu, P.; Zhou, S.; Sang, L.; Liu, R.; Tao, Q.; Qian, T. A new prediction model for hydrostatic pressure reduction of anti-gas channeling cement slurry based on large-scale physical modeling experiments. *J. Pet. Sci. Eng.* **2018**, *172*, 259–268. [CrossRef]
10. Xu, L.; Ye, W.; Chen, Y.; Chen, B.; Cui, Y. A new approach for determination of gas breakthrough in saturated materials with low permeability. *Eng. Geol.* **2018**, *241*, 121–131. [CrossRef]
11. Song, J.; Zhang, D. Comprehensive Review of Caprock-Sealing Mechanisms for Geologic Carbon Sequestration. *Environ. Sci. Technol.* **2012**, *47*, 9–22. [CrossRef]
12. He, S.; Li, Z.; Yang, E.-H. Quantitative characterization of anisotropic properties of the interfacial transition zone (*ITZ*) between microfiber and cement paste. *Cem. Concr. Res.* **2019**, *122*, 136–146. [CrossRef]
13. Hussin, A.; Poole, C. Petrography evidence of the interfacial transition zone (*ITZ*) in the normal strength concrete containing granitic and limestone aggregates. *Constr. Build. Mater.* **2011**, *25*, 2298–2303. [CrossRef]
14. Gao, Y.; De Schutter, G.; Ye, G.; Huang, H.; Tan, Z.; Wu, K. Characterization of *ITZ* in ternary blended cementitious composites: Experiment and simulation. *Constr. Build. Mater.* **2013**, *41*, 742–750. [CrossRef]
15. Scrivener, K.L.; Crumbie, A.K.; Laugesen, P. The Interfacial Transition Zone (*ITZ*) Between Cement Paste and Aggregate in Concrete. *Interface Sci.* **2004**, *12*, 411–421. [CrossRef]
16. Bentur, A.; Diamond, S.; Mindess, S. The microstructure of the steel fibre-cement interface. *J. Mater. Sci.* **1985**, *20*, 3610–3620. [CrossRef]
17. Neubauer, C.; Jennings, H.; Garboczi, E. A three-phase model of the elastic and shrinkage properties of mortar. *Adv. Cem. Based Mater.* **1996**, *4*, 6–20. [CrossRef]
18. Rusch, D.W.; Sabins, F.; Aslakson, J. Microannulus Leaks Repaired with Pressure-Activated Sealant. In Proceedings of the SPE Eastern Regional Meeting, Charleston, WV, USA, 15–17 September 2004.
19. Bachu, S.; Bennion, D.B. Experimental assessment of brine and/or CO_2 leakage through well cements at reservoir conditions. *Int. J. Greenh. Gas Control* **2009**, *3*, 494–501. [CrossRef]
20. Wen, M.; Liu, Z.; Gao, X.; Li, Y. Asymptotic Model of Breakthrough Pressure in Partially Saturated Porous Media with Microvisualization Step-by-Step Breakthrough Experiments. *Energy Fuels* **2020**, *35*, 1594–1604. [CrossRef]
21. Tennis, P.D.; Jennings, H.M. A model for two types of calcium silicate hydrate in the microstructure of Portland cement pastes. *Cem. Concr. Res.* **2000**, *30*, 855–863. [CrossRef]
22. Mills, R.H. Factors influencing cessation of hydration in water cured cement pastes. In *Transportation Research Information Services;* Highway Research Board: Washington, DC, USA, 1966.
23. Koichi, M.; Tetsuya, I.; Toshiharu, K. Multi-scale modeling of concrete performance -Integrated material and structural mechanics_decrypted. *J. Adv. Concr. Technol.* **2003**, *1*, 91–126.
24. He, G.; Tang, H. *Reservoir Physical (Second)*; Petroleum Industry Press: Beijing, China, 2016.
25. Chaube, R.; Kishi, T.; Maekawa, K. *Modelling of Concrete Performance: Hydration, Microstructure and Mass Transport*; CRC Press: London, UK, 1999.
26. Goboncan, V.C.; Dillenbeck, R.L. Real-time cement expansion/shrinkage testing under downhole conditions for enhanced annular isolation. In Proceedings of the SPE/IADC Drilling Conference, Amsterdam, The Netherlands, 19–21 February 2003.

27. Xu, B.; Yuan, B.; Wang, Y.; Zeng, S.; Yang, Y. Nanosilica-latex reduction carbonation-induced degradation in cement of CO_2 geological storage wells. *J. Nat. Gas Sci. Eng.* **2019**, *65*, 237–247. [CrossRef]
28. Yang, Y.; Yuan, B.; Wang, Y.; Zhang, S.; Zhu, L. Carbonation resistance cement for CO_2 storage and injection wells. *J. Pet. Sci. Eng.* **2016**, *146*, 883–889. [CrossRef]
29. Jarrahian, A.; Aghel, B.; Heidaryan, E. On the viscosity of natural gas. *Fuel* **2015**, *150*, 609–618. [CrossRef]

processes

MDPI

Article

Theoretical Analysis of the Micro Annulus of an Oil-Well Cement Sheath Formed via Cooling under Acid-Fracturing Conditions

Donghua Su [1,2], Xuning Wu [1,2,3], Zaoyuan Li [1,2,*], Sheng Huang [1,2,*], Jin Li [1,2], Jinfei Sun [1,4] and Guanyi Zheng [1,2]

1 State Key Laboratory of Oil and Gas Reservoir Geology and Exploitation, Southwest Petroleum University, Chengdu 610500, China; swpusdh@stu.swpu.edu.cn (D.S.); 201911000102@stu.swpu.edu (X.W.); 202011000112@stu.swpu.edu.cn (J.L.); 201511000095@stu.swpu.edu.cn (J.S.); 202199010049@swpu.edu.cn (G.Z.)
2 Petroleum Engineering School, Southwest Petroleum University, Chengdu 610500, China
3 Institute of Subsurface Energy Systems, Clausthal University of Technology, 38678 Clausthal-Zellerfeld, Germany
4 School of Sciences, Southwest Petroleum University, Chengdu 610500, China
* Correspondence: swpilzy@swpu.edu.cn (Z.L.); swpuhs@swpu.edu.cn (S.H.)

Abstract: The plastic deformation and interface micro annulus of oil-well cement during acid fracturing are key reasons for the failure of the wellbore seal and sustained casing pressure. However, most of the existing research ignores the influence of the wellbore cooling effect during acid fracturing, owing to which, the design conditions in the theoretical analysis may be inconsistent with those of the actual wellbore, and the calculation results may be biased. In this study, a novel elastoplastic mechanical model of the cement sheath was established. This model can analyze the yield state of the cement sheath under the influence of three-dimensional principal stress and consider the effect of the differential temperature stress on the interface debonding of the cement sheath from the beginning to the end of acid fracturing. Moreover, the generation mechanism and development law of the interface micro annulus were clarified. The findings indicated that the influence of the intermediate principal stress cannot be ignored; otherwise, the elastoplastic analysis results of the cement sheath may be conservative. During acid fracturing, the casing–cement sheath interface is influenced by the differential temperature stress, and the interface is debonded; however, a micro annulus is not generated. The debonding of the cement sheath–formation interface and micro annulus occurs only when the cement sheath is completely plastic. After acid fracturing, the interface micro annulus is likely to be generated at the casing–cement sheath interface, and the presence of the differential temperature stress may increase the formation risk of the interface micro annulus. The research results can provide theoretical guidance for the prediction of oil-well cement sheath interface seals under acid-fracturing conditions.

Keywords: micro annulus; oil-well cement; HPHT wells; acid fracturing; differential temperature stress; elastic-plastic analysis

Citation: Su, D.; Wu, X.; Li, Z.; Huang, S.; Li, J.; Sun, J.; Zheng, G. Theoretical Analysis of the Micro Annulus of an Oil-Well Cement Sheath Formed via Cooling under Acid-Fracturing Conditions. *Processes* **2022**, *10*, 966. https://doi.org/10.3390/pr10050966

Academic Editors: Tianshou Ma and Yuqiang Xu

Received: 6 April 2022
Accepted: 7 May 2022
Published: 11 May 2022

1. Introduction

Cementing, as a key process in the construction of oil and gas wells [1], is aimed at injecting cement slurry into the annular between the casing and formation and ensuring its solidification within a specified time to form a cement sheath [2,3]. To avoid wellbore integrity issues such as wellbore sealing failure, formation fluid channeling, and casing damage, the cement sheath must maintain adequate isolation under the load induced by the wellbore construction operation [4–6]. Notably, the failure of cement sheath sealing during the well construction cycle has not been completely clarified [7]. For example, in an oil field in western China, the high-temperature and high-pressure gas well is often subjected to sustained casing pressure (SCP) after acid fracturing, which critically threatens the safety of the ground personnel, equipment, and environment [8,9]. Early studies highlighted that the plastic yield of the cement sheath under high pressure in a wellbore and the generation of the interface micro annulus are key causes of SCP [10–12].

Extensive research has been performed on the micro annulus at the cement sheath interface. Goodwin and Crook [13] considered a full-size casing and cement sheath test device to experimentally analyze the failure mode of the cement sheath. The authors reported that under steam–flood injection conditions, the cement sheath interface maybe lose its sealing and annular control of gas or water. Andrade et al. [14–16] applied a cyclic temperature load to the casing–cement sheath–formation combination and noted that the interface of the cement sheath is influenced by the temperature and debonded. Zeng et al. [17] established a wellbore simulation device and applied cyclic loads to the cement sheath to simulate the failure of the cement sheath under large-scale hydraulic fracturing conditions. Under cyclic loading, the cement sheath is expected to undergo cumulative plastic deformation, and an interface micro annulus is expected to be formed. Li et al. [18] conducted equivalent physical experiments based on a self-developed wellbore simulation device and observed that the cement sheath enters a plastic state under high internal casing pressure conditions, and interface debond occurs during the unloading stage, eventually leading to gas channeling.

In addition to experimental methods, several scholars have established theoretical models to predict whether the cement sheath generates a micro annulus under wellbore loads. Mueller et al. [19] used the finite element method to analyze the micro annulus of the cement sheath and noted that temperature changes lead to the bonding failure of the interface. Chu et al. [20] used the Mohr–Coulomb criterion as the yield criterion to establish an elastoplastic mechanical model of the cement sheath, and based on Jackson's experimental parameters, evaluated the size of the micro annulus generated at the cement sheath interface during the unloading stage. Dusseault et al. [21] and Taleghani et al. [22] also highlighted that the cement sheath may produce an interface micro annulus under cycling temperature and pressure conditions. Through theoretical analysis of the failure mode of the cement sheath of gas storage wells, Zhang et al. [23] indicated that the interface debonding and micro annulus generation occurs when the interface tensile stress is greater than the bonding strength during the unloading stage. Chen et al. [24] noted that when the cement sheath completely enters the plastic state, a larger micro annulus may be generated when the internal casing pressure decreases. They recommended the limitation of the phenomenon of the wellbore cement sheath completely entering the plastic state. Zhao et al. [25] reported that the decrease in the temperature in the wellbore during fracturing influences the stress distribution of the cement sheath and may lead to tensile failure of the cement sheath.

According to the abovementioned studies, a cement sheath is expected to undergo plastic deformation under the influence of the internal casing pressure and generate a micro annulus during the unloading stage. In addition, temperature changes are expected to adversely influence the cement sheath interface bonding. In the actual wellbore operation, as shown in Figure 1, the temperature and pressure change during the acid-fracturing are the most severe. Acid-fracturing refers to the process of artificially creating fractures in the formation by pumping the acid-fracturing fluid into the formation, which can effectively improve the seepage characteristics of the formation. Due to the pumping of the acid fracturing fluid, the temperature in the wellbore will decrease continuously and the wellbore pressure will increase rapidly. At this time, the cement sheath will withstand both the wellbore cooling effect and high internal casing pressure. However, in existing research, the actual conditions of "high-pressure expansion" and "cooling contraction" of the cement sheath during acid-fracturing have been neglected; in particular, the influence of the cooling effect in the acid-fracturing process on the cement sheath interface and micro annulus. Therefore, the theoretical analysis results may deviate from the actual wellbore.

Notably, the oil and gas industry lack an elastoplastic mechanical model of oil-well cement that can consider the acid-fracturing conditions and integrate the effects of the wellbore pressure loads and cooling phenomena.

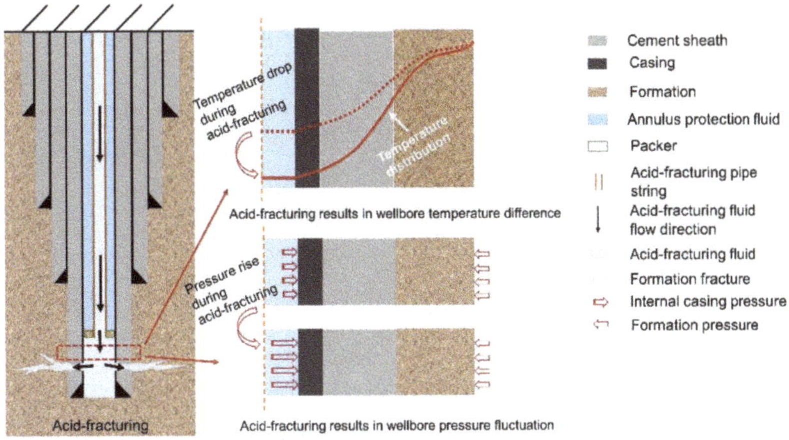

Figure 1. Schematic diagram of acid-fracturing and the change rule of the temperature and pressure during acid-fracturing.

Considering these aspects, this study was aimed at establishing an elastoplastic model that can describe the mechanical integrity of the cement sheath under acid-fracturing conditions. The proposed model can analyze the yield state of the cement sheath under the influence of three-dimensional principal stress and consider the effect of the differential temperature stress generated by the wellbore cooling on the cement sheath interface debonding and micro annulus generation from the beginning to the end of acid fracturing. The accuracy and effectiveness of the elastoplastic model were validated through a comparison with the elastoplastic analysis results of cement sheaths associated with Chu and Jackson [20]. In addition, based on the wellbore conditions of a high-temperature and high-pressure gas well, the generation mechanism and development law of the cement sheath micro annulus during and after the acid-fracturing process were clarified. The proposed model was expected to effectively restore the stress state of the cement sheath under acid-fracturing conditions and ensure that the analysis results of the cement sheath interface debonding and micro annulus more accurately represented the state of actual wellbores.

2. Model Establishment

2.1. Mechanical Model of Casing–Cement Sheath-Formation

To establish the elastoplastic model of the cement sheath, the following assumptions are implemented [20]:

1. The casing and formation are elastomers, the cement sheath is an elastoplastic material, and the yield condition of the cement sheath satisfies the twin-shear unified strength theory;
2. The casing, cement sheath, and formation correspond to homogeneous and isotropic materials;
3. The casing is centered, and the cementing quality is good.

Tensile stress and compressive stress are defined as normal and negative stresses, respectively. Because the radial stress at the inner wall of the cement sheath is always greater than that at the outer wall, the inner wall of the cement sheath yields first as the internal casing pressure increases during acid fracturing. With the gradual expansion of the plastic zone of the cement sheath, the cement sheath can be divided into a plastic zone and an elastic zone. When the pressure reaches a critical value, the cement sheath enters the plastic state. The case in which the casing, formation, and cement sheath are in the elastoplastic state is illustrated in Figure 2.

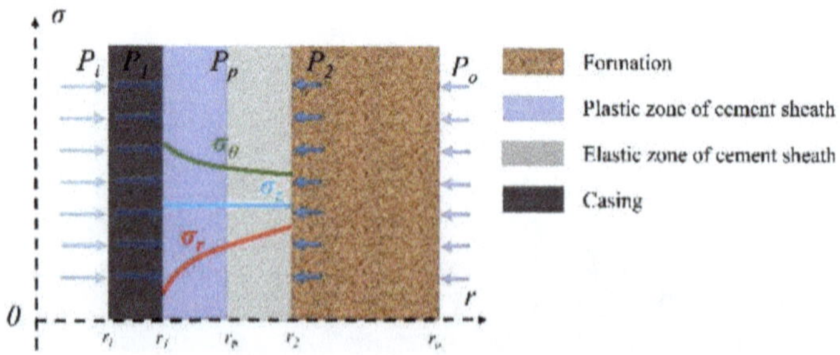

Figure 2. Schematic of casing–elastoplastic cement sheath-formation combination.

where σ_r, σ_θ, and σ_z represent the radial, circumferential, and axial stresses (MPa), respectively. P_i, P_1, P_p, P_2, and P_o represent the internal casing pressure, casing–plastic cement sheath interface contact pressure, plastic cement sheath–elastic cement sheath interface contact pressure, elastic cement sheath-formation interface contact pressure, and formation pressure (MPa), respectively. r_i, r_1, r_p, r_2, and r_o represent the radius of the casing inner wall, plastic cement sheath inner wall (casing outer wall), plastic–elastic interface of cement sheath, elastic cement sheath outer wall (formation inner wall), and formation outer wall (mm), respectively. The 0 represents the origin of coordinates. r represents the distance from the center of the wellbore (mm). σ represents the magnitude of stress (MPa).

2.1.1. Stress and Displacement Analysis of the Plastic Cement Sheath

The cement sheath is divided into the plastic and elastic zones, and the zones are separately discussed. The plastic zone cement sheath is represented as the area pertaining to $r_1 \le r \le r_p$ in Figure 2. Because the cement sheath bears radial, circumferential, and axial stresses in the wellbore, the twin-shear unified strength theory considering the three-dimensional principal stress is selected to define the yield criterion for the plastic zone of the cement sheath [26]. The relevant theoretical equations are

$$\begin{cases} F = \frac{1}{\alpha}\sigma_1 - \frac{1}{1+b}(b\sigma_2 + \sigma_3) = \sigma_c & \left(\sigma_2 \le \frac{\sigma_1+\alpha\sigma_3}{1+\alpha}\right) \\ F\prime = \frac{1}{\alpha(1+b)}(\sigma_1 + b\sigma_2) - \sigma_3 = \sigma_c & \left(\sigma_2 \ge \frac{\sigma_1+\alpha\sigma_3}{1+\alpha}\right) \end{cases} \quad (1)$$

where σ_1, σ_2, and σ_3 represent the maximum, intermediate, and minimum principal stresses (MPa), respectively. σ_c represents the yield strength (MPa). α and b represent the tensile-compression strength ratio (MPa) of the material and weight coefficient used to reflect the influence of the intermediate principal stress, respectively. b can be defined as

$$b = \frac{1+\alpha-B}{B-1} \quad (2)$$

where B denotes the tensile-shear strength ratio (MPa).

When the wellbore cement sheath bears the load from the formation and casing, radial and circumferential compressive stresses are distributed over the cement sheath (it may also be tensile stress in the circumferential direction), with the radial compressive stress being significantly larger than the circumferential stress. Because $\sigma_1 \ge \sigma_2 \ge \sigma_3$, it can be considered that $\sigma_1 = \sigma_\theta > (\text{or} <)\ 0$, $\sigma_2 = \sigma_z = (\sigma_r + \sigma_\theta)/2$, and $\sigma_3 = \sigma_r < 0$. Moreover, the tensile-compression strength ratio of cement materials must be less than 1, so $(\sigma_r + \sigma_\theta)/2 = \sigma_2 \leq (\sigma_\theta + \alpha\sigma_r)/(1 + \alpha)$, which is consistent with the first expression in Equation (1), and the yield condition can be defined as

$$\frac{2+2b-\alpha b}{2\alpha(1+b)}\sigma_\theta - \frac{b+2}{2(1+b)}\sigma_r = \sigma_c \tag{3}$$

Furthermore, the stress components of the cement sheath in the radial and circumferential satisfy the balance equation:

$$\frac{d\sigma_r}{dr} + \frac{\sigma_r - \sigma_\theta}{r} = 0 \tag{4}$$

Considering Equations (3) and (4) and the boundary conditions $r = r_1$ and $\sigma_r = -p_1$, the stress distribution equations for the plastic zone of the cement sheath can be solved:

$$\begin{cases} \sigma_r = \frac{\alpha}{1-\alpha}\sigma_c - \left(\frac{r_1}{r}\right)^{\frac{2+2b-2b\alpha-2\alpha}{2+2b-b\alpha}}\left[p_1 + \frac{\alpha\sigma_c}{1-\alpha}\right] \\ \sigma_\theta = \frac{\alpha}{1-\alpha}\sigma_c - \left(\frac{2\alpha+b\alpha}{2+2b-b\alpha}\right)\left(\frac{r_1}{r}\right)^{\frac{2+2b-2b\alpha-2\alpha}{2+2b-b\alpha}}\left[p_1 + \frac{\alpha\sigma_c}{1-\alpha}\right] \end{cases} \tag{5}$$

Considering the boundary conditions $r = r_p$ and $\sigma_r = -p_1$, the contact pressure of the elastoplastic interface of the cement sheath can be determined:

$$p_p = \left(\frac{r_1}{r_p}\right)^{\frac{2+2b-2b\alpha-2\alpha}{2+2b-b\alpha}}\left[p_1 + \frac{\alpha\sigma_c}{1-\alpha}\right] - \frac{\alpha}{1-\alpha}\sigma_c \tag{6}$$

Neglecting the plastic cement sheath volumetric strain according to the law of volume elasticity

$$\varepsilon_r + \varepsilon_\theta = \frac{(1+v)(1-2v)}{E}(\sigma_r + \sigma_\theta) \tag{7}$$

where ε_r and ε_θ represent the radial and circumferential strains, respectively. E represents the Young's modulus (MPa). v represents the Poisson's ratio.

According to the geometry equation,

$$\begin{cases} \varepsilon_r = du/dr \\ \varepsilon_\theta = u/r \end{cases} \tag{8}$$

where u represents the displacement (mm).

By substituting Equations (5) and (8) into Equation (7), the displacement of the plastic zone u_{cp} of the cement sheath can be determined as

$$u_{cp} = \frac{(1+v_c)(1-2v_c)}{E_c}\left[\left(\frac{\alpha\sigma_c}{1-\alpha}\right)r - \left(p_1 + \frac{\alpha\sigma_c}{1-\alpha}\right){r_1}^{\frac{2+2b-2b\alpha-2\alpha}{2+2b-b\alpha}} r^{\left(\frac{2\alpha+b\alpha}{2+2b-b\alpha}\right)}\right] + \frac{K}{r} \tag{9}$$

where u_{cp} denotes the displacement of the plastic cement sheath (mm). E_c represents the Young's modulus of the cement sheath (MPa). v_c represents the Poisson's ratio of the cement sheath. K is the integral constant, which is an unknown quantity, and r corresponds to any radial position in the plastic zone (mm). The displacement at the outer boundary of the plastic zone of the cement sheath can be defined as:

$$u_{cpi} = -\frac{(1+v_c)(1-2v_c)}{E_c}p_1 r_1 + \frac{K}{r_1} \tag{10}$$

$$u_{cpo} = \frac{(1+v_c)(1-2v_c)}{E_c}\left[\frac{\alpha\sigma_c}{1-\alpha}r_p - \left(p_1 + \frac{\alpha\sigma_c}{1-\alpha}\right){r_1}^{\frac{2+2b-2\alpha b-2\alpha}{2+2b-b\alpha}} r_p^{\frac{2\alpha+b\alpha}{2+2b-b\alpha}}\right] + \frac{K}{r_p} \tag{11}$$

where u_{cpi} and u_{cpo} denote the displacements of the inner and outer walls of the cement sheath plastic zone (mm), respectively.

2.1.2. Stress and Displacement Analysis of the Elastic Cement Sheath

The elastic zone of the cement sheath can be evaluated considering the Lame formula [27] for elastic mechanics. The elastic zone in Figure 2 refers to the area corresponding to $r_p \le r \le r_2$. The pressure of the elastic inner wall and outer wall is p_p and p_2, respectively. In this manner, the expressions for the radial stress σ_{rc} and circumferential stress $\sigma_{\theta c}$ of the cement sheath can be derived:

$$\begin{cases} \sigma_{rc} = \frac{r_2{}^2(p_2-p_p)}{r_2{}^2-r_p{}^2} + \frac{r_p{}^2 p_p - r_2{}^2 p_2}{r_2{}^2-r_p{}^2} \\ \sigma_{\theta c} = \frac{r_2{}^2(p_p-p_2)}{r_2{}^2-r_p{}^2} + \frac{r_p{}^2 p_p - r_2{}^2 p_2}{r_2{}^2-r_p{}^2} \end{cases} \tag{12}$$

In addition, the radial stress at the boundary and elastic—plastic interface in the elastic zone satisfies the yield criterion. Thus, the equation for the contact pressure p_2 of the cement sheath-formation interface can be defined as

$$p_2 = \frac{1}{r_2^2(2+2b-\alpha b)}\left\{\left[(1-\alpha+b-\alpha b)r_p^2 + (1+\alpha+b)r_2^2\right]p_p - \alpha(1+b)\left(r_2^2-r_p^2\right)\sigma_c\right\} \tag{13}$$

The radial displacement at the inner and outer boundaries of the elastic zone of the cement sheath can be determined through the theoretical displacement formula of the thick-walled cylinder [27]:

$$u_{cei} = \frac{(1-2\nu_c)r_p^3 + (1+v_c)r_p r_2^2}{E_c\left(r_2^2-r_p^2\right)}p_p - \frac{(2-\nu_c)r_p r_2^2}{E_c\left(r_2^2-r_p^2\right)}p_2 \tag{14}$$

$$u_{ceo} = \frac{(2-\nu_c)r_p^2 r_2}{E_c\left(r_2^2-r_p^2\right)}p_p - \frac{(1-2\nu_c)r_2^3 + (1+v_c)r_p^2 r_2}{E_c\left(r_2^2-r_p^2\right)}p_2 \tag{15}$$

where u_{cei} and u_{ceo} denote the boundary displacement of the inner and outer elastic zone (mm), respectively.

2.1.3. Continuity Condition and Model Solution

Because the casing, cement sheath, and formation are in close contact, under the action of the casing pressure and formation pressure, the displacement of the outer wall of the casing (u_{so}) is equal to that of the inner wall of the cement sheath, and the displacement of the outer wall of the cement sheath is equal to that of the inner wall of the formation (u_{fi}). The displacements at the elastic–plastic boundary of the cement sheath are also equal [20]. This state corresponds to the continuity condition of the casing–elastic–plastic cement sheath-formation combination, which can be characterized by the following equations:

$$\begin{cases} u_{so} = u_{cpi} \\ u_{cpo} = u_{cei} \\ u_{ceo} = u_{fi} \end{cases} \tag{16}$$

where:

$$u_{so} = \frac{1}{E_s\left(r_1^2-r_i^2\right)}\left[(1-2\nu_s)\left(r_i^2 p_i - r_1^2 p_1\right)r_1 + (1+\nu_s)r_i^2 r_1^2(p_i-p_1)\frac{1}{r_1}\right] \tag{17}$$

$$u_{fi} = \frac{1}{E_f\left(r_o^2-r_2^2\right)}\left[\left(1-2\nu_f\right)\left(r_2^2 p_2 - r_o^2 p_o\right)r_2 + \left(1+\nu_f\right)r_2^2 r_o^2(p_2-p_o)\frac{1}{r_2}\right] \tag{18}$$

where E_s and E_f denote the Young's modulus of the casing and the formation (MPa), respectively, and v_f denotes the Poisson's ratio of the casing and the formation, respectively.

By combining Equation (16) with Equations (6) and (13) and substituting the values of the casing, cement sheath, and formation sizes, the mechanical properties, internal casing pressure, and formation pressure, as well as the unknowns p_1, p_p, p_2, r_p, and K, can be determined. Thus, the stress and displacement at any position inside the casing–elastic–plastic cement sheath-formation combination can be determined.

2.2. Differential Temperature Stress Model of Casing–Cement Sheath-Formation

The differential temperature stress of the casing, cement sheath, and formation can be solved using elastic mechanics [28]. The corresponding model and solution method are presented in Appendix A; in cases 1–3, the cement sheath is in the elastic state, elastic–plastic state, and plastic state, respectively.

2.3. Combined Stress Calculation Model of Casing–Cement Sheath-Formation

Under the combined influence of the wellbore temperature change and acid fracturing, the casing–cement sheath-formation combination simultaneously bears the differential temperature stress and pressure load, and the superposition of the two loads correspond to the combined stress:

$$\begin{cases} \sum\sigma_r = \sigma_r + \sigma_r^t \\ \sum\sigma_\theta = \sigma_\theta + \sigma_\theta^t \end{cases} \tag{19}$$

where σ and σ^t represent the stresses caused by the wellbore pressure and the temperature difference of the wellbore (MPa), respectively. Subscript r and θ denote radial and circumferential directions, respectively.

During acid fracturing, the wellbore load is rapidly transmitted to the bottom of the well. As construction progresses, the temperature at the bottom of the well gradually decreases. Therefore, in the calculation, first, the elastic–plastic interval of the cement sheath and interface contact pressure must be analyzed according to the wellbore load. Subsequently, the differential temperature stress calculation methods must be identified according to the elastic–plastic interface position. Finally, the stress and displacement must be superimposed under the influence of the fracturing load. The combined stress and displacement of the casing–cement sheath-formation combination can thus be determined.

2.4. Calculation of Micro Annulus during Acid Fracturing

During acid fracturing, the cement sheath is simultaneously subjected to high pressure and cooling effects. When the temperature decreases, the cement sheath undergoes "chilling shrinkage", which may lead to the generation of radial tensile stress at the interface. If the cement sheath is in a plastic state, a micro annulus may be generated. Therefore, a calculation method for the micro annulus in the acid-fracturing process is established, considering the casing–cement sheath interface as an example.

Considering the assumption specified in Section 2.1, the highest casing pressure is defined as p_{im}, and the pressure values of the inner wall, elastic–plastic interface, and outer wall of the cement sheath are ${p^p}_{1m}$, ${p^p}_p$, and ${p^p}_{2m}$, respectively. During acid fracturing, the pressure in the casing remains constant. Changes in the size of the casing–cement sheath-formation combination and position of the elastic–plastic interface of the cement sheath under the above-mentioned wellbore load conditions are identified.

The displacement of the casing inner wall, u_{sim}, under the action of the internal casing pressure and the formation pressure can be expressed as:

$$u_{sim} = \frac{1+v_s}{E_s}\frac{r_1^2 r_i + (1-2v_s)r_i^3}{r_1^2 - r_i^2}p_{im} - \frac{1+v_s}{E_s}\frac{2(1-v_s)r_1^2 r_i}{r_1^2 - r_i^2}p_{1m}^p \tag{20}$$

The displacements of the inner wall of the casing, inner and outer walls of the cement sheath, and the elastic–plastic boundary have been specified in Section 2.1. The displacement of the formation outer wall u_{fom} under the acid-fracturing load is expressed as

$$u_{fom} = \frac{1+v_f}{E_f}\frac{2\left(1-v_f\right)r_2^2 r_o}{r_o^2 - r_2^2}p_{2m}^p - \frac{1+v_f}{E_f}\frac{r_2^2 r_o + \left(1-2v_f\right)r_o^3}{r_o^2 - r_2^2}p_o \tag{21}$$

By setting the wellbore size as the initial wellbore size, the method described in Section 2.2 is used to calculate the differential temperature stress. Before acid fracturing is initiated, the temperature in the wellbore is constant, which also represents the formation temperature. The differential temperature stress is zero. As the acid fracturing is initiated, the temperature in the wellbore gradually decreases. At the moment when the wellbore temperature is at its lowest, the temperatures of the inner and outer walls of the casing are recorded as t_{1m} and t_{2m}, respectively. The temperatures of the inner and outer walls of the formation are recorded as t_{3m} and t_{4m}, respectively, and the temperature at the elastic-plastic interface of the cement sheath is defined as t_{pm}.

When the wellbore temperature is minimized, the cement sheath interface exhibits sufficient bonding strength. Moreover, the casing–cement sheath-formation combination is in close contact, and thus, the displacement continuity condition is satisfied. According to Appendix A, the radial stress at the inner and outer walls of the cement sheath and elastic–plastic interface can be calculated:

$$\begin{cases} \sigma_{1m}^t = -\dfrac{E_{cp}C_{1cp}'}{(1+v_{cp})(1-2v_{cp})} + \dfrac{E_{cp}C_{2cp}'}{(1+v_{cp})r_1^2} \\ \sigma_p^t = \dfrac{E_{cp}\alpha_{cp}}{(1-v_{cp})r_p^2}\displaystyle\int_{r_1}^{r_p} trdr - \dfrac{E_{cp}C_{1cp}'}{(1+v_{cp})(1-2v_{cp})} + \dfrac{E_{cp}C_{2cp}'}{(1+v_{cp})r_p^2} \\ \sigma_{2m}^t = \dfrac{E_{ce}\alpha_{ce}}{(1-v_{ce})r_2^2}\displaystyle\int_{r_p}^{r_2} trdr - \dfrac{E_{ce}C_{1ce}'}{(1+v_{ce})(1-2v_{ce})} + \dfrac{E_{cp}C_{2ce}'}{(1+v_{ce})r_2^2} \end{cases} \tag{22}$$

where ${\sigma^t}_{1m}$ and ${\sigma^t}_{2m}$ denote the stresses of the cement sheath inner and outer walls (MPa) caused by the temperature difference of the wellbore, respectively. ${\sigma^t}_p$ denotes the differential temperature stress at the elastic–plastic cement sheath interface (MPa).

To ensure that the cement sheath does not exhibit interface bonding failure, the combined stress at the inner and outer walls of the cement sheath must satisfy

$$\begin{cases} \sigma_{1m}^t + \sigma_{1m}^p \geq \sigma_{1b} \\ \sigma_{2m}^t + \sigma_{2m}^p \geq \sigma_{2b} \end{cases} \tag{23}$$

where ${\sigma^p}_{1m}$ and ${\sigma^p}_{2m}$ denote the stresses of the cement sheath inner and outer walls (MPa), respectively, with ${p^p}_{1m} = -{\sigma^p}_{1m}$ and ${p^p}_{2m} = -{\sigma^p}_{2m}$; and σ_{1b} and σ_{2b} denote the bonding strength of the cement sheath inner and outer walls (MPa), respectively.

If the inner and outer walls of the cement sheath are subjected to tensile stress that is less than the interface bonding strength, the cement sheath can exhibit adequate interface sealing performance. If the tensile stress is greater than the interface bonding strength, a micro annulus may be generated.

For example, for the inner wall of the cement sheath, when the tensile stress at the interface is greater than the bonding strength of the casing and the cement sheath, the casing and cement sheath are debonded, and the contact pressure between the two entities becomes zero. At this time, the casing bears the load p_{im} associated with the acid-fracturing and contact pressure p_{1m} (=0), and the displacement of the casing outer wall (${u^p}_{som}$) can be expressed as

$$u_{som}^p = \frac{1+v_s}{E_s}\frac{2(1-v_s)r_i^2 r_1}{r_1^2 - r_i^2}p_{im} \tag{24}$$

When the casing and the cement sheath are debonded, the inner and outer walls of the casing are free to deform. Because rapid heat transfer occurs inside the casing, it can be considered that no temperature difference exists between the inner and outer walls. Therefore, the differential temperature stress caused by the temperature difference acting on the casing disappears, and the displacement caused by the differential temperature stress is zero.

After the interface is debonded, the pressure on the cement sheath inner wall is zero. The pressure on the inner wall of the cement sheath decreases from p_{1m} to zero. This process corresponds to elastic unloading. The radial displacement of the cement sheath inner wall under the fracturing load ($u^p{}_{cim}$) is the sum of the boundary displacement in the plastic zone ($u^p{}_{cpim}$) and inner wall displacement at the time of debonding:

$$u^p_{cim} = u^p_{cpim} + \frac{1+v_c}{E_c}\frac{r_1 r_2^2 + (1-2v_c)r_1^3}{r_2^2 - r_1^2}\left(-p^p_{1m}\right) - \frac{1+v_c}{E_c}\frac{2(1-v_c)r_1 r_2^2}{r_2^2 - r_1^2}\left(p^p_{2m}{}' - p^p_{2m}\right) \quad (25)$$

where $p_{2m}{}'$ represents the contact pressure of the cement sheath outer wall when the contact pressure of the cement sheath inner wall is zero in the acid-fracturing stage (MPa).

In addition, the inner wall of the cement sheath exhibits displacement $u^t{}_{ci}$ caused by the differential temperature stress. Owing to the interface debonding, the radial differential temperature stress at the inner wall of the cement sheath is zero. However, the outer wall of the cement sheath is still in close contact with the ground, and thus, under the temperature difference, a differential temperature stress exists inside the cement sheath.

Because the cement sheath undergoes elastic unloading, in the analysis of the differential temperature stress, the cement sheath and formation are considered a combined body. At this time, taking the displacement at the cement sheath-formation interface is equal as the continuity condition; taking the temperature difference stress between the inner wall of the cement sheath and the outer wall of the formation is zero as the boundary condition. Equation (26) is established to calculate the radial differential temperature stress at the cement sheath-formation interface:

$$\begin{cases} -\dfrac{E_c\alpha_c}{(1-v_c)r_2^2}\displaystyle\int_{r_1}^{r_2} trdr + \dfrac{E_c C_1{}'}{(1+v_c)(1-2v_c)} - \dfrac{E_c C_2{}'}{(1+v_c)r_2^2} - \dfrac{E_f C_1{}''}{\left(1+v_f\right)\left(1-2v_f\right)} + \dfrac{E_f C_2{}''}{\left(1+v_f\right)r_2^2} = 0 \\ \left(\dfrac{1+v_c}{1-v_c}\right)\dfrac{\alpha_c}{r_2}\displaystyle\int_{r_1}^{r_2} trdr + C_1{}' r_2 + \dfrac{C_2{}'}{r_2} - C_1{}'' r_2 - \dfrac{C_2{}''}{r_2} = 0 \\ \dfrac{E_c C_1{}'}{(1+v_c)(1-2v_c)} - \dfrac{E_c C_2{}'}{(1+v_c)r_1^2} = 0 \\ -\dfrac{E_f\alpha_f}{\left(1-v_f\right)r_0^2}\displaystyle\int_{r_2}^{r_0} trdr + \dfrac{E_f C_1{}''}{\left(1+v_f\right)\left(1-2v_f\right)} - \dfrac{E_f C_2{}''}{\left(1+v_f\right)r_0^2} = 0 \end{cases} \quad (26)$$

$C_1{}'$, $C_2{}'$, $C_1{}''$, and $C_2{}''$ are solved and substituted into the displacement equations of the differential temperature stress. Subsequently, the displacement, $u^t{}_{cim}$, of the cement sheath inner wall can be determined. The displacement when the cement sheath inner wall is debonded from the interface during the acid-fracturing process (u_{cim}) is

$$u_{cim} = u^p_{cim} + u^t_{cim} \quad (27)$$

Therefore, the micro annulus at the inner wall of the cement sheath (d_m) can be calculated using the following equation:

$$d_m = u_{cim} - u_{som} \quad (28)$$

where u_{som} represents the displacement when the casing outer wall is debonded from the interface during the acid-fracturing process (mm).

2.5. Calculation of Micro Annulus after Acid Fracturing

In contrast to the wellbore seal failure mechanism in which the cement sheath is affected by cooling during acid fracturing, thereby producing the micro annulus, after the acid fracturing, the temperature in the wellbore is relatively static and does not increase rapidly. However, the internal casing pressure rapidly decreases, and the contact pressure acting on the inner and outer walls of the cement sheath also decreases. The interface of the cement sheath may be affected by the pressure in the wellbore, resulting in the risk of debonding.

It is assumed that the combined stress in the acid-fracturing stage does not generate a micro annulus in the cement sheath. At the end of fracturing, the casing pressure decreases to p_{in}, and the pressures of the inner and outer walls of the cement sheath are $p^p{}_{1n}$ and $p^p{}_{2n}$, respectively. The calculation method is described in Appendix B. The differential temperature stress is solved like that described in the previous section: Under the highest internal casing pressure, the size of the casing–cement sheath-formation combination is determined, and the differential temperature stress under the temperature difference condition in the wellbore at the end of the fracturing is calculated considering this size. Assuming that the interface does not debond, the contact pressure caused by the temperature difference between the inner and outer walls of the cement sheath can be expressed as $p^t{}_{1n}$ and $p^t{}_{2n}$. The solution method is described in Appendix A. The combined contact pressure of the inner and outer walls of the cement sheath is

$$\begin{cases} p_{1n} = p^t_{1n} + p^p_{1n} \\ p_{2n} = p^t_{2n} + p^p_{2n} \end{cases} \tag{29}$$

where p_{1n} and p_{2n} represent the combined contact pressure of the inner and outer walls of the cement sheath (MPa), respectively.

When the combined contact pressure is negative and positive, respectively, this means that the corresponding combined stress is compressive stress and tensile stress. When the tensile stress as the combined stress is greater than the interface bonding strength, the cement sheath produces a micro annulus. The schematic diagram is shown in Figure 3.

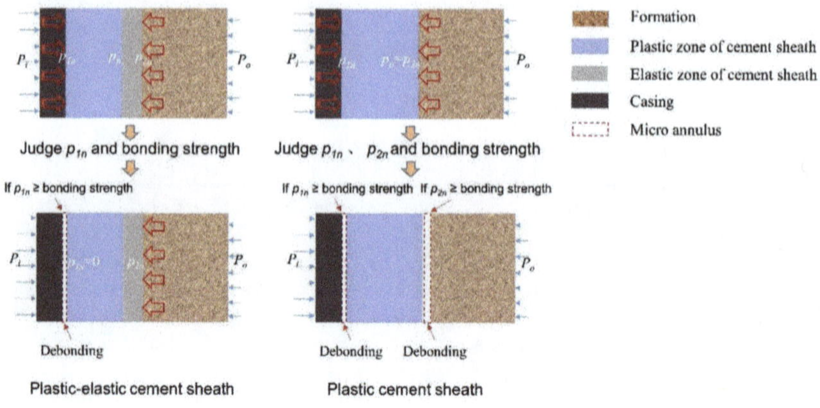

Figure 3. Schematic diagram of the micro annulus generation.

For example, if the casing and cement sheath interface debonds, the contact pressure between the cement sheath and casing is zero. The displacement of the outer wall of the casing and the inner wall of the cement sheath at the end of fracturing is calculated using the method described in Appendix B. The temperature difference and differential temperature stress of the cement sheath in the wellbore are calculated using the method described in Section 2.2. The following expression can be obtained:

$$u_{cin} = u^p_{cin} + u^t_{cin} \tag{30}$$

where u_{cin}, $u^p{}_{cin}$, and $u^t{}_{cin}$ denote the displacements (mm) of the inner wall of the cement sheath, cement sheath affected by the unloading pressure, and cement sheath affected by the differential temperature stress, at the end of fracturing, respectively.

The size of the cement sheath micro annulus (d_n) at the end of fracturing can be calculated as

$$d_n = u_{cin} - u_{son} \tag{31}$$

where u_{son} denotes the displacement of the outer wall of the casing after acid-fracturing (mm).

3. Case Study

The twin-shear unified strength theory and the Mohr–Coulomb criterion are used as the yield criteria to compare and analyze the plastic change of cement sheath under a given pressure load. The influence of the different yield criteria on the development of the plastic zone of the cement sheath, interface contact pressure, and micro annulus is discussed. Considering the example of a high-pressure deep well, the elastic–plastic change in the cement sheath and generation of micro annulus under the combined influence of the temperature and pressure during acid fracturing are analyzed.

3.1. *Influence of Yield Criterion on Elastic–Plastic Change and Micro Annulus of the Cement Sheath*

Chu et al. [20] used the Mohr–Coulomb criterion as the yield criterion and analyzed the changes in the micro annulus of the cement sheath based on the test data reported by Jackson et al. [29]. In the experiment conducted by Jackson et al., the cement sheath was maintained between the inner and outer casings, and the gas channeling flow rate of the cement sheath was determined by applying gas channeling pressure on the end of the cement sheath and gradually increasing the internal casing pressure. The objective was to identify the damage to the cement sheath or failure of the interface seal. According to the experimental results, when the internal casing pressure increased to 69 MPa, no gas channeling flow was detected, indicating that the cement sheath was not damaged. However, gas channeling occurred during the pressure drop in the casing, likely owing to the failure of the interface seal caused by the micro annulus.

In this study, the abovementioned working conditions and parameters listed in Table 1 are used to calculate the elastic–plastic and micro annulus changes associated with the cement sheath through the Mohr–Coulomb criterion and twin-shear unified strength theory.

Table 1. Parameters in Jackson et al.'s experiment and Chu et al.'s theoretical calculation [20,29].

Inner radius of inner casing (mm)	54.3	Outer radius of inner casing (mm)	63.5
Inner radius of outer casing (mm)	77.39	Outer radius of outer casing (mm)	88.9
Casing internal pressure (MPa)	69	Casing external pressure (MPa)	0
Young's modulus of casing (GPa)	210	Poisson's ratio of casing	0.3
Young's modulus of cement sheath (GPa)	13.8	Poisson's ratio of cement sheath	0.25
Internal friction angle of cement sheath (°)	30	Cement sheath cohesion (MPa)	5.77

3.1.1. Development of Plastic Zone of the Cement Sheath during Loading

The change in the elastic–plastic boundary of the cement sheath when the internal casing pressure increases from 0 to 69 MPa is calculated, as shown in Figure 4.

The calculation results obtained using the Mohr–Coulomb criterion show that when the internal casing pressure increases to 30 MPa, the inner wall of the cement sheath begins to enter a critical state of plastic yield. The elastic–plastic boundary of the cement sheath gradually expands to the outer wall of the cement sheath, until 53.5 MPa, and the cement sheath enters the plastic state. The calculation results based on the twin-shear unified strength theory demonstrate that the cement sheath enters the critical state of plastic yield when the internal casing pressure is 28.1 MPa, and the cement sheath completely enters the plastic state when the internal casing pressure is 51.7 MPa.

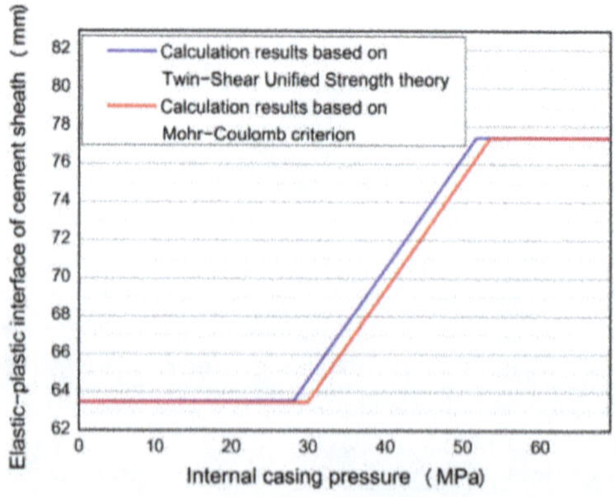

Figure 4. Development of elastic–plastic boundary of cement sheath under different yield criteria.

Therefore, the elastic–plastic analysis results of the cement sheath when the Mohr–Coulomb criterion is used as the yield criterion are conservative [30]. This result occurs because the Mohr–Coulomb criterion ignores the influence of the intermediate principal stress [31].

3.1.2. Contact Pressure of the Cement Sheath

As the internal casing pressure increases, the contact pressure changes at the casing-cement sheath interface and cement sheath-formation interface are calculated, as shown in Figure 5. As the pressure increases, the contact pressure of the cement sheath is always positive. Figure 6 shows the interface stress of the cement sheath when the internal casing pressure decreases from 69 MPa to zero. The compressive stress at the casing–cement sheath interface is higher than that at the cement sheath-formation interface in the initial stage of the decrease of the internal casing pressure. As the internal casing pressure decreases, the difference in the interface stress gradually decreases. As the internal casing pressure further decreases, the casing–cement sheath interface exhibits tensile stress until the pressure decreases to zero. The tensile stress at the casing–cement sheath interface is higher than that of the outer wall. Therefore, the casing–cement sheath interface is highly prone to tensile debonding.

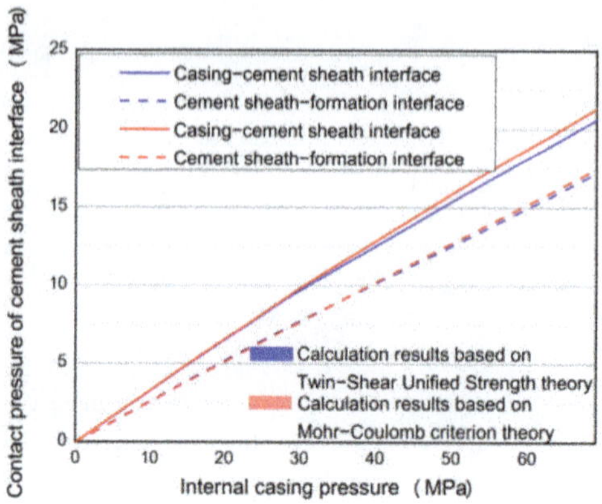

Figure 5. Change in the contact pressure at the cement sheath interface during loading.

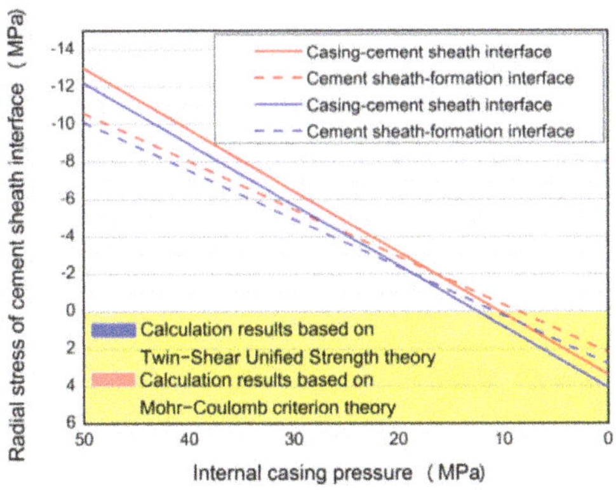

Figure 6. Interface stress change in the cement sheath during unloading.

The results obtained using the Mohr–Coulomb criterion and twin-shear unified strength theory are compared. In the unloading process, the combined stress calculated using the Mohr–Coulomb criterion is greater than that calculated using the twin-shear unified strength theory. However, the tensile stress obtained by the twin-shear unified strength theory under a low internal casing pressure is larger, indicating that the cement sheath interface calculated using this theory is more prone to bonding failure. Moreover, the results calculated using the Mohr–Coulomb criterion are conservative.

3.1.3. Micro Annulus of the Cement Sheath

When the interface bonding strength is 2 MPa, the micro annulus change at the casing—cement sheath interface is calculated under the abovementioned conditions, as shown in Figure 7. In the initial stage, when the internal casing pressure decreases, the interface of the casing and cement sheath are not yet debonded, and the displacement of the two entities is the same. As the internal casing pressure further decreases and the casing–cement sheath interface exhibits tensile stress that is greater than the bonding strength, the casing and cement sheath debond. At this time, the inner wall of the cement sheath suddenly retracts and becomes stationary. The casing continues to retract until the internal casing pressure decreases to zero. This phenomenon occurs because the cement sheath expands under tensile stress. When the interface is debonded, the tensile stress of the cement sheath returns to zero, and the cement sheath shrinks. The casing deforms considerably and continues to deform as the internal casing pressure decreases.

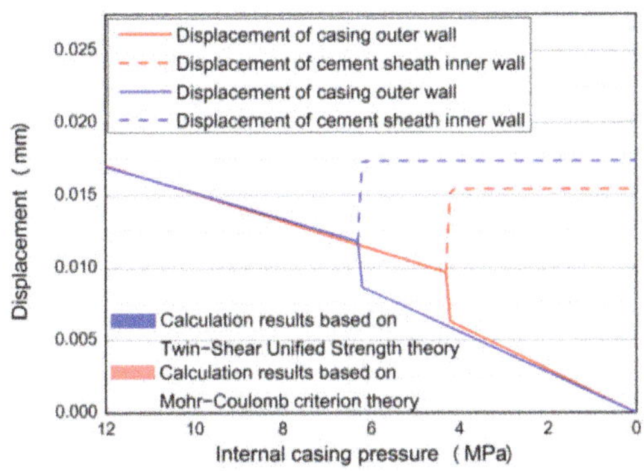

Figure 7. Change in the cement sheath micro annulus in the unloading stage.

In addition, when the twin-shear unified strength theory is used for calculation, the cement sheath generates the micro annulus (0.0174 mm) earlier in the unloading stage, and this micro annulus is larger than that determined using the Mohr–Coulomb criterion (0.0154 mm). The difference between the calculation results and those reported in the literature can be attributed to the different assumptions regarding the interface bonding strength.

3.2. *Analysis of Plastic Deformation and Micro Annulus of Cement Sheath in HTHP Wells during Acid Fracturing*

The twin-shear unified strength theory considers the influence of the intermediate principal stress, thereby avoiding the conservativeness associated with the Mohr–Coulomb criterion in the elastic–plastic analysis of the cement sheath. Therefore, the twin-shear unified strength theory is considered as the yield criterion, and a gas well in western China is considered as an example for calculation. The elastic–plastic change in the cement sheath and generation of micro annulus under changes in the acid-fracturing temperature and pressure are examined.

The well has a five-segment vertical well structure, and the cement sheath 6745 m deep in the acid-fracturing section is considered as the analysis object. The formation pressure at this depth is 115 MPa, the formation temperature is 168.2 °C, the displacement is 6 m^3/min, and the acid-fracturing time is 1 h. The wellbore structure parameters are listed in Table 2.

Table 2. Wellbore structure parameters of the high-pressure gas well.

Segment	Drill Size (mm)	Casing Size (mm)	Casing Thickness (mm)	Casing Shoe Position (m)	Cement Slurry Back to High (m)
1	660.4	508	12.70	200	0
2	444.5	365.13	13.88	4340	0
3	333.4	273.05	13.84	6466	0
4	241.3	201.7	15.12	6250	6700
5	168.3	139.7	12.09	7040	6700

The composition of the cement slurry: Aksu class G cement + 28% ganister sand (SiO2) + 7% micro-silicon + 1.35% channeling agent (FlOK-2) + 4.5% fluid loss agent (FS-23L) + 3% drag reducer (FS-13L) + 3% retarder (HX-13L) + 0.2% defoaming agent (DF-A) + 4.5% industrial salt + water; the density is 1.95 g/cm^3. The physical parameters of the casing, cement sheath, and formation are listed in Table 3.

Table 3. Physical parameters of the casing, cement sheath, and formation.

Parameters	Value	Parameters	Value
Thermal expansion coefficient of casing	1.3×10^{-5} 1/°C	Thermal expansion coefficient of cement sheath	1.5×10^{-5} 1/°C
Thermal expansion coefficient of formation	7×10^{-5} 1/°C	Young's modulus of casing	210,000 MPa
Poisson's ratio of casing	0.3	Young's modulus of cement sheath	13,800 MPa
Poisson's ratio of cement sheath	0.25	Young's modulus of formation	29,500 MPa
Poisson's ratio of formation	0.33	Cement sheath bond strength	4 MPa

It should be noted that the parameters in Tables 2 and 3 are derived from the field engineering design.

3.2.1. Combined Stress Distribution of the Cement Sheath in the Wellbore

The acid-fracturing fluid density is 1.04 g/cm^3, pump pressure is approximately 150 MPa, and friction resistance is approximately 60 MPa. The stress distribution of the cement sheath at the end of the cementing is considered as the initial state, and the elastic–plastic change in the cement sheath during acid fracturing is analyzed, as shown in Figure 8. When the internal casing pressure increases by 30.5 MPa, the cement sheath begins to enter the plastic state. When the internal casing pressure increases to the highest, the radius of the elastic–plastic boundary of the cement sheath is 80.34 mm.

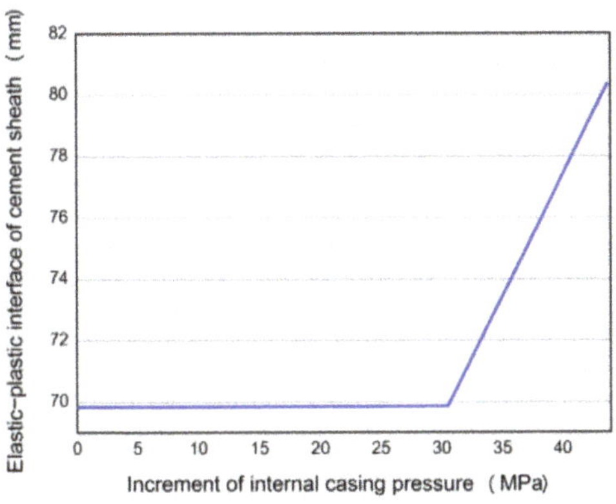

Figure 8. Elastic–plastic interface of the cement sheath during acid fracturing.

The stresses of the elastic–plastic cement sheath associated with the acid-fracturing load and differential temperature stress are added to determine the combined stress distribution of the cement sheath. Considering the temperature of the wellbore in a static state as the initial condition, the Hasan formula [32,33] is used to calculate the temperature change in the wellbore and temperature distribution of the cement sheath at 6745 m under acid-fracturing conditions. The calculation results are shown in Figure 9. During acid fracturing, the temperature of the wellbore continues to decrease. After 1 h, the wellbore temperature decreases to approximately 64.8 °C, and the temperature of the cement sheath gradually decreased owing to acid fracturing. However, the temperature difference between the inner and outer interfaces of the cement sheath is relatively small (~3 °C).

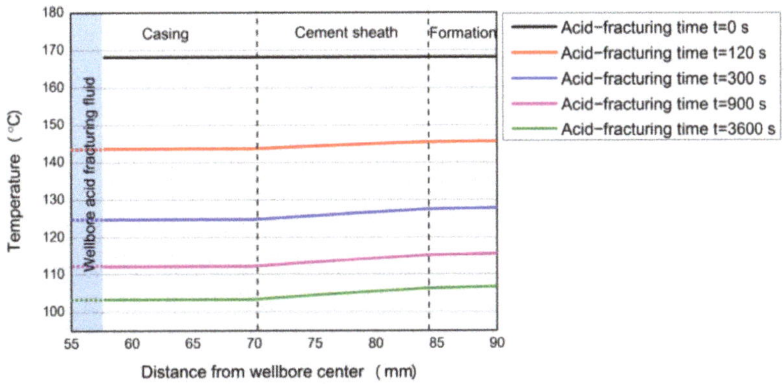

Figure 9. Wellbore temperature and temperature distribution of the cement sheath.

After determining the temperature distribution of the wellbore, the differential temperature stress of the cement sheath is calculated considering the temperature difference between the casing–cement sheath-formation combination and elastic–plastic state of the cement sheath (Case 2 in Appendix A), as shown in Figure 10. The radial differential temperature stress of the cement sheath corresponds to continuous tensile stress. As the construction time increases, the tensile stress gradually increases. The differential temperature stress in the circumferential direction of the cement sheath is dominated by compressive stress. The circumferential differential temperature stress between the plastic and elastic zones of the cement sheath is discontinuous, and the compressive stress in the plastic zone is greater than that in the elastic zone.

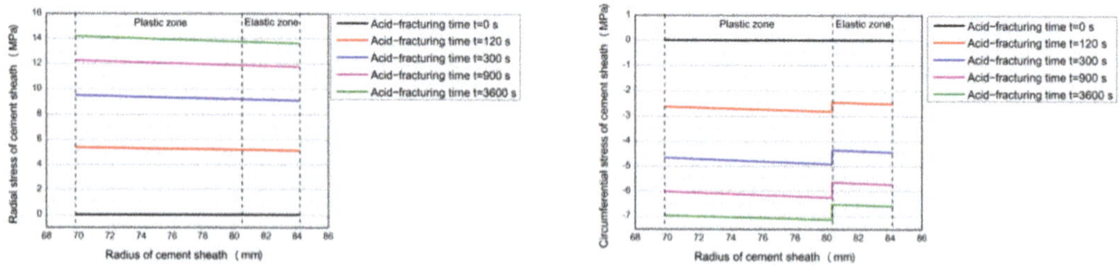

(**a**) (**b**)

Figure 10. Differential temperature stress of cement sheath during acid fracturing. (**a**) Radial differential stress; (**b**) Circumferential differential stress.

Equation (19) can be used to determine the combined stress distribution of the cement sheath, as shown in Figure 11. The radial combined stress of the cement sheath is continuous, and the circumferential combined stress is discontinuous. In the initial stage of acid fracturing, the cement sheath exhibits compressive combined stress in the radial direction and tensile combined stress in the circumferential direction. At this time, if the tensile combined stress in the circumferential direction of the cement sheath exceeds the tensile strength of the cement sheath, the cement sheath may undergo tensile failure, failing the wellbore seal. As the acid fracturing progresses, the circumferential stress of the cement sheath is affected by the differential temperature stress, which gradually transforms into compressive stress, and the risk of tensile failure of the cement sheath decreases. However, the combined stress in the radial direction of the cement sheath is transformed into tensile stress. When the tensile combined stress at the interface of the cement sheath in the plastic zone is greater than the bonding strength, the interface micro annulus may be generated during acid fracturing.

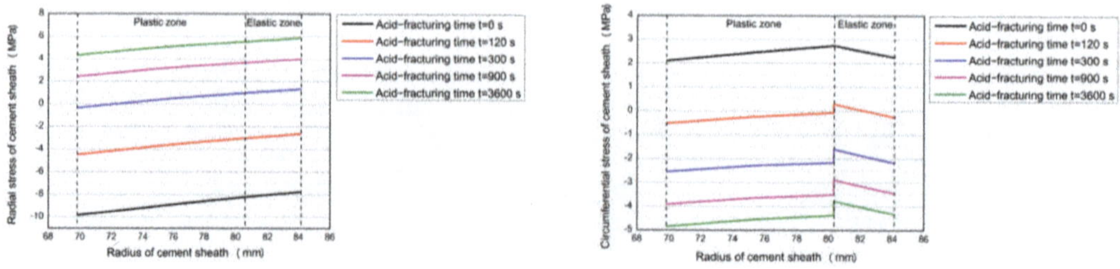

(**a**) (**b**)

Figure 11. Combined stress of cement sheath during acid fracturing. (**a**) Radial combined stress; (**b**) Circumferential combined stress.

3.2.2. Micro Annulus of Cement Sheath during Acid Fracturing

The method described in Section 2.4 is used to analyze the influence of the cooling effect on the cement sheath micro annulus during acid fracturing. As can be seen from Figure 8, the cement sheath does not completely enter the plastic state, and thus, although the tensile stress at the outer wall of the cement sheath is higher, the cement sheath-formation interface does not debond.

Analysis of the casing–cement sheath interface indicates that the cement sheath undergoes plastic deformation, and the interface is influenced by the combined effect of the wellbore load and temperature difference, which generates tensile stress. Considering the combined stress at the acid-fracturing time of 3600 s, the casing–cement sheath interface displacement before the interface debonding is calculated, and the results find that the interface displacement is -0.015 mm (the displacement to the center of the wellbore is defined as negative), indicating that the cement sheath "shrinks" in the wellbore under the influence of acid-fracturing load and temperature. Moreover, the tensile combined stress at the interface exceeds the bonding strength, and the risk of debonding at the interface exists.

Continue to calculate the displacement of the casing and cement sheath after debonding, at this time, the contact stress at the casing–cement sheath interface can be considered zero. The calculation results show that owing to the high internal casing pressure, the casing is displaced (0.05 mm) in the direction of the formation. The cement sheath is affected by the formation pressure and differential temperature stress, and its inner wall is displaced (0.002 mm) in the direction of the formation. Because the displacement of the outer wall of the casing is considerably greater than that of the inner wall of the cement sheath, the casing and cement sheath remain in close contact. However, because the stress at the interface satisfies the debonding conditions, the casing–cement sheath interface is considered to debond. Meanwhile, the debonded casing is influenced by the internal pressure and expands, resulting in close contact between the casing and cement sheath.

In addition, the tensile combined stress of the cement sheath-formation interface is higher than that of the casing–cement sheath interface. To analyze whether the cement sheath-formation interface debonds and produces a micro annulus during acid fracturing, we assume that the cement sheath has a lower yield strength and a higher temperature difference between the inner and outer walls. In this scenario, the cement sheath can fully enter the plastic state, which increases the risk of micro annulus generation.

The calculation finds that the displacement of the cement sheath-formation interface before debonding under the assumed conditions is -0.032 mm, indicating that the cement sheath is in a compressed state. Under the effect of the differential temperature stress, the cement sheath-formation interface is subjected to tensile stress, the interface debonds, and a 0.004 mm micro annulus is generated between the outer wall of the cement sheath and formation.

Overall, during acid fracturing, the cement sheath is expected to undergo plastic deformation, and a radial tensile combined stress is expected to be distributed. The tensile stress at the cement sheath-formation interface is greater than that at the casing–cement sheath interface. According to the calculations based on wellbore conditions, although interface debonding occurs at the casing–cement sheath interface, the micro annulus is not generated under the influence of the internal casing pressure. When the cement sheath completely enters the plastic state, interface debonding may occur at the cement sheath-formation interface, and the interface micro annulus may be generated, owing to the temperature difference of the wellbore.

3.2.3. Micro Annulus of Cement Sheath after Acid Fracturing

The method described in Section 2.5 is used to analyze the generation of the cement sheath micro annulus after acid fracturing, based on the original wellbore working conditions. According to existing research, the casing–cement sheath interface undergoes debonding during acid fracturing; however, because of the internal casing pressure, the casing and cement sheath remain in close contact. As described in this section, the micro-annulus size of the cement sheath interface can be obtained by calculating the internal casing pressure under the condition of interface debonding, and the displacement of the outer wall of the casing is less than that of the inner wall of the cement sheath. In the calculation, it is considered that the rate of decrease in the internal casing pressure is considerably greater than the rate of increase in the wellbore temperature. Therefore, the influence of wellbore temperature changes on the differential temperature stress is ignored (the differential temperature stress remains constant).

Figure 12 shows the displacement of the casing–cement sheath interface and changes in the size of the micro annulus after acid fracturing. The casing–cement sheath interface first shrinks into the wellbore as the internal casing pressure decreases. Next, when the displacement of the outer wall of the casing is equal to that of the inner wall, the casing-cement sheath interface enters a critical state in which the micro annulus is generated. As the internal casing pressure decreases, the displacement of the outer wall of the casing returns to the initial state, and the contact pressure between the casing and the cement sheath is zero. However, the cement sheath-formation combination is influenced by the differential temperature stress and produces displacement along the direction of the wellbore with the formation of a micro annulus sized 0.00196 mm.

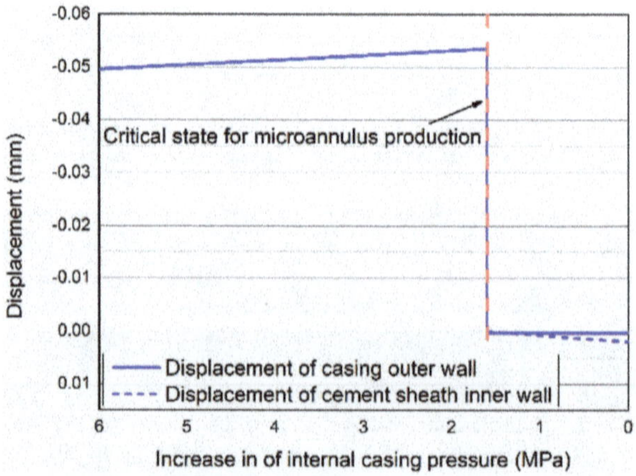

Figure 12. Displacement and micro annulus of casing–cement sheath interface.

To compare the influence of the differential temperature stress on the cement sheath micro annulus, the cement sheath interfaces contact pressure during acid fracturing was calculated based on the same wellbore working conditions while ignoring the differential temperature stress. As shown in Figure 13, when the differential temperature stress is ignored, the contact pressure of the cement sheath interface during acid fracturing corresponds to compressive stress. After the acid fracturing, when the internal casing pressure decreases, the contact pressure of the cement sheath interface transformed to tensile stress that is less than the interface bonding strength. Therefore, the cement sheath interface will not debond.

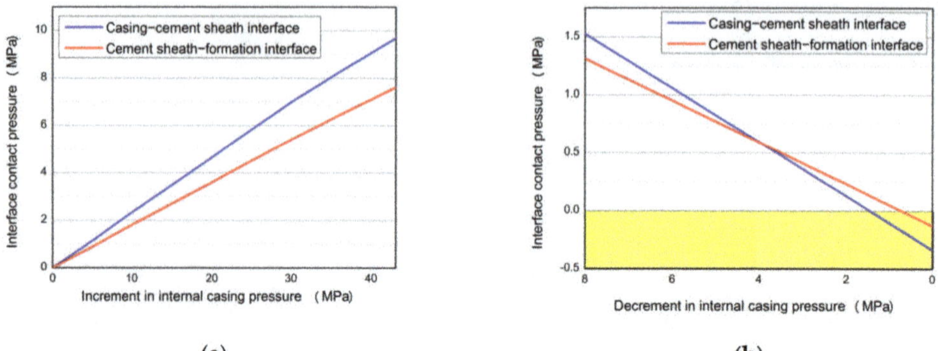

Figure 13. Contact pressure of wellbore cement sheath interface determined without considering the differential temperature stress. (**a**) Acid fracturing stage; (**b**) End of acid fracturing (unloading stage).

Therefore, in the analysis of the cement sheath interface debonding and micro annulus formation, it is necessary to consider the effect of the differential temperature stress; otherwise, the analysis results would be conservative, and the risk associated with the interface debonding and micro annulus formation may not be captured.

4. Discussion

In this study, the micro annulus caused by the wellbore cooling effect was analyzed; the study found that the debonding of the cement sheath interface will occur due to the differential temperature stress during acid-fracturing. Meanwhile, after acid fracturing, the cement sheath will also be affected by the comprehensive stress, and there was a risk of micro annulus formation. Among them, it should be noted that, under the wellbore parameters and conditions in the study, it was found that the elastic–plastic cement sheath will only cause interface debonding, but will not generate micro annulus in the acid fracturing stage. This did not mean that there was no risk of cement sheath interface sealing failure in the actual wellbore acid-fracturing process. The reason was that the debonded cement sheath interface may break into the acid fracturing fluid, causing cracks (or micro annulus) to propagate along the interface. The fluid load applied to the debonding interface of the cement sheath was ignored in our study and this is an aspect that remains to be studied.

5. Conclusions

To address the interface debonding problem of the cement sheath of high-temperature and high-pressure wells under acid fracturing, the cement sheath yield and micro annulus are examined. The plastic deformation of the cement sheath from the beginning to the end of acid fracturing and the mechanism of micro annulus generation are analyzed. The influence of the physical parameters of the cement sheath is discussed, and the following conclusions are derived:

(1) The elastic–plastic mechanical model of cement sheath is established based on the twin-shear unified strength theory. According to the comparative analysis of the theoretical and experimental results reported by Chu et al. and Jackson et al., the proposed model can consider the influence of the intermediate principal stress on the yield state of the cement sheath, and avoid the conservativeness of the plastic analysis results of the cement sheath based on the Mohr-Coulomb criterion.

(2) Considering the wellbore load and temperature difference during acid fracturing, calculation models of the combined stress of the casing–cement sheath-formation combination and interface micro annulus during and after acid fracturing are established. The models are used to determine the combined stress and displacement of the cement sheath from the beginning to the end of acid fracturing. The mechanism and development law of the micro annulus between the casing–cement sheath interface and cement sheath-formation interface are analyzed.

(3) During acid fracturing, if the cement sheath is not fully yielded, the casing–cement sheath interface is influenced by the differential temperature stress, and the interface debonds. However, under the influence of the internal casing pressure, the casing–cement sheath interface remains in close contact, and no micro annulus is generated. If the cement sheath completely enters the plastic state, the cement sheath-formation interface is more susceptible to differential temperature stress, and interface debonding and micro annulus occurs.

(4) After acid fracturing, the micro annulus is more likely to be formed at the casing-cement sheath interface, owing to the decrease in the internal casing pressure. In this stage, the influence of the temperature difference of the cement sheath must be considered; otherwise, the analysis results would be conservative, and the risks associated with interface debonding and micro annulus formation may not be captured.

Author Contributions: Methodology, modeling and validation: D.S., X.W. and S.H.; investigation: J.L. and G.Z.; formal analysis: D.S. and J.S.; resources, supervision: Z.L.; writing—original draft preparation: D.S.; writing—review and editing: X.W. All authors have read and agreed to the published version of the manuscript.

Funding: The authors gratefully acknowledge the financial support provided by the Science and Technology Cooperation Project of the CNPC-SWPU Innovation Alliance (No. 2020CX040000); the National Natural Science Foundation of China (No. 52004231); Sichuan Science and Technology Program (No. 2020JDTD0019).

Institutional Review Board Statement: Not applicable.

Informed Consent Statement: Not applicable.

Data Availability Statement: The data that support the findings of this study are available from the corresponding author upon reasonable request.

Conflicts of Interest: The authors declare no conflict of interest.

Appendix A. Differential Temperature Stress Model of Casing—Cement Sheath-Format during Acid Fracturing

The differential temperature stress and displacement equations of the thick-walled cylinder can be expressed as [28]:

$$\begin{cases} \sigma_r^t = -\frac{E\alpha}{(1-v)r^2}\int_{R_i}^{r} trdr + C_3 - \frac{C_4}{r^2} \\ \sigma_\theta^t = \frac{E\alpha}{(1-v)r^2}\int_{R_i}^{r} trdr + C_3 + \frac{C_4}{r^2} \\ \sigma_z^t = -\frac{E\alpha t}{1-v} + 2vC_3 \end{cases}, \quad \text{(A1)}$$

$$u_t = \left(\frac{1+v}{1-v}\right)\frac{\alpha}{r}\int_{R_i}^{r} trdr + C_1 r + \frac{C_2}{r}, \quad \text{(A2)}$$

where α denotes the coefficient of linear expansion (1/°C), t represents the temperature difference (°C), and ${\sigma^t}_r$, ${\sigma^t}_\theta$, and ${\sigma^t}_z$ represent the radial, circumferential, and axial stresses caused by the temperature difference (MPa), respectively. E represents the Young's modulus (MPa). v represents the Poisson's ratio. r represents radius of any point in the thick-walled cylinder (mm). R_i represents the inner wall radius of the thick-walled cylinder (mm). Furthermore, u_t denotes the displacement caused by the differential temperature stress (mm) and C_1, C_2, C_3, and C_4 are defined as

$$C_3 = \frac{EC_1}{(1+\nu)(1-2\nu)}; C_4 = \frac{EC_2}{1+\nu}. \quad \text{(A3)}$$

Case 1: The cement sheath is in the elastic state

The differential temperature stress for the casing, cement sheath, and formation is described in Appendix B of the existing study [34].

Case 2: The cement sheath is in the elastic–plastic state

In this state, based on the boundary conditions and continuity conditions in Case 1, we re-assume that the casing–elastic cement sheath-formation combination is the casing-plastic cement sheath–elastic cement sheath-formation combination. The interface of the plastic and elastic cement sheath satisfies the continuous radial stress and displacement, as shown in Equation (A4):

$$\sigma_{rcp}^t |_{r=r_p} = \sigma_{rce}^t |_{r=r_p} ; u_{cp}^t |_{r=r_p} = u_{ce}^t |_{r=r_p}, \quad \text{(A4)}$$

where ${\sigma^t}_{rcp}|_{r=rp}$ and ${\sigma^t}_{rce}|_{r=rp}$ denote the radial differential temperature stress of elastic and plastic cement sheaths at the elastic—plastic interface (MPa), respectively. ${u^t}_{cp}|_{r=rp}$

and $u^t{}_{ce}|_{r=rp}$ denote the displacement of the elastic and plastic cement sheaths at the elastic–plastic interface, caused by the temperature difference (mm), respectively.

The differential temperature stress can be solved by the following equations:

$$\begin{cases} -\frac{E_s\alpha_s}{(1-v_s)r_1^2}\int_{r_i}^{r_1} trdr + \frac{E_sC_1}{(1+v_s)(1-2v_s)} - \frac{E_sC_2}{(1+v_s)r_1^2} - \frac{E_{cp}C_{1cp}{}'}{(1+v_{cp})(1-2v_{cp})} + \frac{E_{cp}C_{2cp}{}'}{(1+v_{cp})r_1^2} = 0 \\ -\frac{E_{cp}\alpha_{cp}}{(1-v_{cp})r_p^2}\int_{r_1}^{r_p} trdr + \frac{E_{cp}C_{1cp}{}'}{(1+v_{cp})(1-2v_{cp})} - \frac{E_{cp}C_{2cp}{}'}{(1+v_{cp})r_p^2} - \frac{E_{ce}C_{1ce}{}'}{(1+v_{ce})(1-2v_{ce})} + \frac{E_{ce}C_{2ce}{}'}{(1+v_{ce})r_p^2} = 0 \\ -\frac{E_{ce}\alpha_{ce}}{(1-v_{ce})r_2^2}\int_{r_p}^{r_2} trdr + \frac{E_{ce}C_{1ce}{}'}{(1+v_{ce})(1-2v_{ce})} - \frac{E_{ce}C_{2ce}{}'}{(1+v_{ce})r_2^2} - \frac{E_fC_1{}''}{\left(1+v_f\right)\left(1-2v_f\right)} + \frac{E_fC_2{}''}{\left(1+v_f\right)r_2^2} = 0 \\ \frac{E_sC_1}{(1+v_s)(1-2v_s)} - \frac{E_sC_2}{(1+v_s)r_i^2} = 0 \\ -\frac{E_f\alpha_f}{\left(1-v_f\right)r_o^2}\int_{r_2}^{r_o} trdr + C_3{}''\frac{E_fC_1{}''}{\left(1+v_f\right)\left(1-2v_f\right)} - \frac{E_fC_2{}''}{\left(1+v_f\right)r_o^2} = 0 \\ \left(\frac{1+v_s}{1-v_s}\right)\frac{\alpha_s}{r_1}\int_{r_i}^{r_1} trdr + C_1r_1 + \frac{C_2}{r_1} - C_{1cp}{}'r_1 - \frac{C_{2cp}{}'}{r_1} = 0 \\ \left(\frac{1+v_{cp}}{1-v_{cp}}\right)\frac{\alpha_{cp}}{r_p}\int_{r_1}^{r_p} trdr + C_{1cp}{}'r_p + \frac{C_{2cp}{}'}{r_p} - C_{1ce}{}'r_p - \frac{C_{2ce}{}'}{r_p} = 0 \\ \left(\frac{1+v_{ce}}{1-v_{ce}}\right)\frac{\alpha_{ce}}{r_2}\int_{r_p}^{r_2} trdr + C_{1ce}{}'r_2 + \frac{C_{2ce}{}'}{r_2} - C_1{}''r_2 - \frac{C_2{}''}{r_2} = 0 \end{cases} \quad \text{(A5)}$$

where E_{cp} and E_{ce} denote Young's modulus of the cement sheath in the plastic and elastic zones (MPa), respectively; v_{cp} and v_{ce} denote Poisson's ratio of the cement sheath in plastic and elastic zones, respectively; α_s, α_f, α_{cp}, and α_{ce} denote thermal expansivity of the casing, the formation, and the cement sheath in plastic and elastic zones (1/°C); $C_{1cp}{}'$, $C_{2cp}{}'$, $C_{1ce}{}'$, and $C_{2ce}{}'$ are the intermediate parameters.

In this study, it is considered that Young's modulus and Poisson's ratio of the cement sheath in the plastic zone are the same as those in the elastic zone.

Case 3: The cement sheath is in the plastic state

The method for solving the differential temperature stress and displacement of the casing—cement sheath-formation combination is the same as that in Case 1. The difference is that the sizes and mechanical parameters of the plastic cement sheath must be appropriately substituted for calculations.

Appendix B. Displacement and Contact Pressure of Cement Sheath Associated with Wellbore Unloading

In the wellbore unloading stage, the micro annulus of the cement sheath can be obtained by calculating the difference between the casing–cement sheath interface displacement (or cement sheath-formation interface) after debonding, specifically [20]:

The radial displacement at the outer wall of the casing can be characterized by the following equation:

$$u_{son}^P = \frac{(2-v_s)r_i^2r_1}{E_s(r_1^2-r_i^2)}p_{in} - \frac{(1-2v_s)r_1^3+(1+v_s)r_i^2r_1}{E_s(r_1^2-r_i^2)}p_{1n}^P, \quad \text{(A6)}$$

where u_{son}^{P} represents the displacement of the outer wall of the casing caused by the internal casing pressure (mm). p_{in} represents the internal casing pressure during unloading (MPa). p_{1n}^{P} represents the contact pressure at the casing—cement sheath interface during unloading (MPa).

which can be rewritten as

$$u_{son}^{P} = f_1 p_{in} - f_2 p_{1n}^{P}. \tag{A7}$$

Similarly, the radial displacement of the inner wall of the formation is:

$$u_{fin}^{P} = \frac{(1-2v_f)r_2^3 + (1+v_f)r_2 r_o^2}{E_f(r_o^2 - r_2^2)} p_{2n}^{p} - \frac{(2-v_f)r_2 r_o^2}{E_f(r_o^2 - r_2^2)} p_o, \tag{A8}$$

where u_{fin}^{P} represents the displacement of the inner wall of the formation (mm). p_{2n}^{P} represents the contact pressure at the cement sheath-formation interface during unloading (MPa).

which can be rewritten as

$$u_{fin}^{p} = f_7 p_{2n}^{p} - f_8 p_o. \tag{A9}$$

In the unloading stage, the radial displacement of the cement sheath inner wall is the sum of the displacement of the cement sheath inner wall during loading and radial displacement of the inner wall when the internal casing pressure decrease. The corresponding equations are

$$u_{cin}^{p} = u_{cpim}^{p} + u_{cir}^{p} \tag{A10}$$

$$u_{cin}^{p} = u_{cpim}^{p} + \frac{(1-2v_c)r_1^3 + (1+v)r_1 r_2^2}{E_c(r_2^2 - r_1^2)}(p_{1n}^{p} - p_{1m}^{p}) - \frac{(2-v_c)r_1 r_2^2}{E_c(r_2^2 - r_1^2)}(p_{2n}^{p} - p_{2m}^{p}), \tag{A11}$$

where u_{cin}^{P}, u_{cpim}^{P}, and u_{cir}^{P} denoted the radial displacements of the inner wall of the cement sheath during unloading and loading and when the internal casing pressure decreased (mm), respectively.

The following expression can be obtained:

$$u_{cin}^{p} = u_{cpim}^{p} + f_3(p_{1n}^{p} - p_{1m}^{p}) - f_4(p_{2n}^{p} - p_{2m}^{p}). \tag{A12}$$

Similarly, the radial displacement of the outer wall of the cement sheath during unloading can be expressed as

$$u_{con}^{p} = u_{ceom}^{p} + \frac{(2-v_c)r_1^2 r_2}{E_c(r_2^2 - r_1^2)}(p_{1n}^{p} - p_{1m}^{p}) - \frac{(1-2v_c)r_2^3 + (1-v_c)r_1^2 r_2}{E_c(r_2^2 - r_1^2)}(p_{2n}^{p} - p_{2m}^{p}), \tag{A13}$$

where u_{con}^{P} and u_{ceom}^{P} denote the radial displacements of the outer wall of the cement sheath during unloading and loading (mm), respectively.

The following expression can be obtained:

$$u_{con}^{p} = u_{ceom}^{p} + f_5(p_{1n}^{p} - p_{1m}^{p}) - f_6(p_{2n}^{p} - p_{2m}^{p}). \tag{A14}$$

It is assumed that in the unloading process, the inner wall of the casing and cement sheath have adequate bonding strength to prevent the interface from being debonded. In this case, the casing–cement sheath-formation combination satisfies the continuity condition of radial displacement:

$$\begin{cases} u_{son}^{p} = u_{cin}^{p} \\ u_{con}^{p} = u_{fin}^{p} \end{cases}. \tag{A15}$$

By substituting Equations (A11), (A12), (A13), and (A14) into Equation (A15), the contact pressure of the inner and outer walls of the cement sheath before the interface bonding can be calculated:

$$p_{1n}^{p} = \frac{f_1(f_6+f_7)p_{in} + f_4 f_8 p_o - (f_6+f_7)u'_{cpi} + f_4 u'_{ceo}}{(f_2+f_3)(f_6+f_7) - f_4 f_5}, \quad \text{(A16)}$$

$$p_{2n}^{p} = \frac{f_1 f_5 p_{in} + f_8(f_2+f_3)p_o - f_5 u'_{cpi} + (f_2+f_3)u'_{ceo}}{(f_2+f_3)(f_6+f_7) - f_4 f_5}, \quad \text{(A17)}$$

where

$$u'_{cpi} = u_{cpim}^{p} - f_3 p_{1m}^{p} + f_4 p_{2m}^{p}$$
$$u'_{ceo} = u_{ceom}^{p} - f_5 p_{1m}^{p} + f_6 p_{2m}^{p}$$

For example, in the case of the casing–cement sheath interface, the casing after the interface debonding bears only the internal casing pressure, and the radial displacement of the casing outer wall is

$$u_{son}^{p} = \frac{(2-v_s)r_i^2 r_o}{E_s(r_1^2 - r_i^2)} p_{in}. \quad \text{(A18)}$$

After the interface is debonded, the contact pressure of the inner wall of the cement sheath is zero. The radial displacement of the inner wall of the cement sheath during unloading can be expressed as follows:

$$u_{cin}^{p} = u_{cpim}^{p} + \frac{(1-2v_c)r_1^3 + (1-v_c)r_1 r_2^2}{E_c(r_2^2 - r_1^2)}(-p_{1m}^{p}) - \frac{(2-v_c)r_1 r_2^2}{E_c(r_2^2 - r_1^2)}(p_{2n}^{p} - p_{2m}^{p}). \quad \text{(A19)}$$

References

1. Zhang, C.; Cai, J.; Xu, H.; Cheng, X.; Guo, X. Mechanical properties and mechanism of wollastonite fibers reinforced oil well cement. Constr. *Build. Mater.* **2020**, *260*, 120461. [CrossRef]
2. Cheng, X.; Chen, Z.; Gu, T.; Zeng, L.; Yao, L.; Chen, Z.; Huang, K.; Zhang, Z.; Zhang, C.; Liu, K.; et al. Study on the dynamic and static mechanical properties of microsphere rubber powder reinforced oil well cement composites. *Constr. Build. Mater.* **2021**, *309*, 125145. [CrossRef]
3. Zhang, C.; Li, Y.; Cheng, X.; Liang, S.; Guo, X.; Zhao, H.; Song, Y. Effects of plasma-treated rock asphalt on the mechanical properties and microstructure of oil-well cement. *Constr. Build. Mater.* **2018**, *186*, 163–173. [CrossRef]
4. Xiang, H.; Han, G.; Ma, G.; Zhu, Z.; Zhu, L.; Peng, L. Pressure Transient Analysis and Transient Inflow Performance Relationship of Multiple-Fractured Horizontal Wells in Naturally Fractured Reservoirs by a Trilinear Flow Model. *ACS Omega* **2021**, *6*, 19222–19232. [CrossRef] [PubMed]
5. Xu, H.; Ma, T.; Peng, N.; Yang, B. Influences of fracturing fluid injection on mechanical integrity of cement sheath under four failure modes. *Energies* **2018**, *11*, 3534. [CrossRef]
6. Zhang, X.; Bi, Z.; Wang, L.; Guo, Y.; Yang, C.; Yang, G. Shakedown analysis on the integrity of cement sheath under deep and large-scale multi-section hydraulic fracturing. *J. Pet. Sci. Eng.* **2022**, *208*, 109619. [CrossRef]
7. Chen, Y.; Peng, X.; Yu, H. Mechanical performance experiments on rock and cement, casing residual stress evaluation in the thermal recovery well based on thermal-structure coupling. *Energy Explor. Exploit.* **2017**, *35*, 591–608. [CrossRef]
8. Guo, H.; Wang, G.; Wang, Z. New Practices for Cement Integrity Evaluation in the Complex Environment of Xinjiang Oil Field. In Proceedings of the SPE Asia Pacific Oil and Gas Conference and Exhibition, Perth, Australia, 22–24 October 2012; Society of Petroleum Engineers: Richardson, TX, USA, 2012.
9. Liu, H.; Cao, L.; Xie, J.; Yang, X.; Zeng, N.; Zhang, X.; Chen, F. Research and practice of full life cycle well integrity in HTHP well. In Proceedings of the Tarim Oilfield, International Petroleum Technology Conference, Beijing, China, 26–28 March 2019; Society of Petroleum Engineers: Richardson, TX, USA, 2019.
10. Davies, R.J.; Almond, S.; Ward, R.S.; Jackson, R.B.; Adams, C.; Worrall, F.; Herringshaw, L.G.; Gluyas, J.G.; Whitehead, M.A. Oil and gas wells and their integrity: Implications for shale and unconventional resource exploitation. *Mar. Pet. Geol.* **2014**, *56*, 239–254. [CrossRef]
11. Guo, Y.; Li, X.; Feng, S.; Zhang, C.; Liu, R.; Zhang, Z.; Li, T.; Guo, P.; Wang, R.; Taoutaou, S.; et al. Cementing practices to solve well integrity challenges of ultra deep high temperature wells in western China. In Proceedings of the Abu Dhabi International Petroleum Exhibition & Conference, Abu Dhabi, United Arab Emirates, 7–10 November 2016; Society of Petroleum Engineers: Richardson, TX, USA, 2016.

12. Oyarhossein, M.; Dusseault, M.B. Wellbore Stress Changes and Microannulus Development Because of Cement Shrinkage. In Proceedings of the 49th US Rock Mechanics/Geomechanics Symposium, San Francisco, CA, USA, 29 June–1 July 2015; American Rock Mechanics Association: Alexandria, VA, USA, 2015.
13. Goodwin K., J.; Crook R., J. Cement Sheath Stress Failure. In Proceedings of the SPE Annual Technical Conference and Exhibition, New Orleans, USA, 23–26 September 1993; Society of Petroleum Engineers: Richardson, TX, USA, 1993.
14. Albawi, A.; De Andrade, J. Experimental set-up for testing cement sheath integrity in Arctic wells. In Proceedings of the Offshore Technology Conference, Houston, TX, USA, 5–8 May 2014.
15. De Andrade, J.; Torsaeter, M.; Todorovic, J.; Opedal, N.; Stroisz, A.; Vrålstad, T. Influence of casing centralization on cement sheath integrity during thermal cycling. In Proceedings of the IADC/SPE Drilling Conference and Exhibition, Fort Worth, TX, USA, 4 March 2014; Society of Petroleum Engineers: Richardson, TX, USA, 2014; p. 10.
16. Shadravan, A.; Schuber, J.; Amani, M.; Teodori, C. HPHT cement sheath integrity evaluation method for unconventional wells. In Proceedings of the SPE International Conference on Health, Safety, and Environment, Long Beach, CA, USA, 17–19 March 2014.
17. Zeng, Y.; Liu, R.; Li, X.; Zhou, S.; Tao, Q.; Lu, P. Cement sheath sealing integrity evaluation under cyclic loading using large-scale sealing evaluation equipment for complex subsurface settings. *J. Pet. Sci. Eng.* **2019**, *176*, 811–820. [CrossRef]
18. Li, Z.; Zhang, K.; Guo, X.; Liu, J.; Cheng, X.; Du, J. Study of the failure mechanisms of a cement sheath based on an equivalent physical experiment. *J. Nat. Gas Sci. Eng.* **2016**, *31*, 331–339. [CrossRef]
19. Mueller, D.T.; GoBoncan, V.; Dillenbeck, R.L.; Heinold, T. characterizing casing-cement-formation interactions under stress conditions: Impact on long-term zonal isolation. In Proceedings of the SPE Annual Technical Conference and Exhibition, Houston, TX, USA, 26–29 September 2004.
20. Chu, W.; Shen, J.; Yang, Y.; Li, Y.; Gao, D. Calculation of micro-annulus size in casing-cement sheath-formation system under continuous internal casing pressure change. *Pet. Explor. Dev.* **2015**, *42*, 414–421. [CrossRef]
21. Dusseault, M.B.; Gray, M.N.; Nawrocki, P.A. Why oilwells leak: Cement behavior and long-term consequences. In Proceedings of the SPE International Oil and Gas Conference and Exhibition, Beijing, China, 7–10 November 2000; Society of Petroleum Engineers: Richardson, TX, USA, 2000.
22. Taleghani, A.D.; Klimenko, D. An Analytical Solution for Microannulus Cracks Developed Around a Wellbore. *J. Energy Resour. Technol.* **2015**, *137*, 062901. [CrossRef]
23. Zhang, H.; Shen, R.; Yuan, G.; Ba, Z.; Hu, Y. Cement sheath integrity analysis of underground gas storage well based on elastoplastic theory. *J. Petrol. Sci. Eng.* **2017**, *159*, 818–829. [CrossRef]
24. Chen, Z.; Dai, C.; Liao, M. Analyses of Mechanical Conditions and Affecting Factors for Forming Micro-Annuli in a Casing-Cement-Formation System. In Proceedings of the 52nd U.S. Rock Mechanics/Geomechanics Symposium, Seattle, WA, USA, 17–20 June 2018; American Rock Mechanics Association: Alexandria, VA, USA, 2018.
25. Zhao, C.; Li, J.; Liu, G.; Zhang, X. Analysis of the influence of cement sheath failure on sustained casing pressure in shale gas wells. *J. Nat. Gas Sci. Eng.* **2019**, *66*, 244–254. [CrossRef]
26. Yu, M.-H.; Kolupaev, V.; Li, Y.-M.; Li, J.-C. Advances in Unified Strength Theory and its Generalization. *Procedia Eng.* **2011**, *10*, 2508–2513. [CrossRef]
27. Amenzade, Y.A. *Theory of Elasticity*; Mir Publishers: Moscow, Russia, 1979.
28. Boles, M.A. *Thermodynamics: An Engineering Approach*; McGraw-Hill Higher Education: New York, USA, 2008.
29. Jackson, P.B.; Murphey, C.E. Effect of casing pressure on gas flow through a sheath of set cement. In Proceedings of the SPE/IADC Drilling Conference, Amsterdam, The Netherlands, 22 February 1993; Society of Petroleum Engineers: Richardson, TX, USA, 1993.
30. Huang, J.; Zhao, M.; Du, X.; Dai, F.; Ma, C.; Liu, J. An Elasto-Plastic Damage Model for Rocks Based on a New Nonlinear Strength Criterion. *Rock Mech. Rock Eng.* **2018**, *51*, 1413–1429. [CrossRef]
31. Chen, S.; Bao, W.; Jin, S. Twin τ2 Strength Theory and Its Application to Concrete Material. *J. Univ. Hydraul. Electr. Eng./Yichang* **2003**, *25*, 504–506.
32. Hasan, A.R.; Kabir, C.S.; Wang, X. A Robust Steady-State Model for Flowing-Fluid Temperature in Complex Wells. *SPE Prod. Oper.* **2009**, *24*, 269–276. [CrossRef]
33. Hasan, A.R.; Kabir, C.S.; Wang, X. Wellbore Two-Phase Flow and Heat Transfer During Transient Testing. *SPE J.* **1998**, *3*, 174–180. [CrossRef]
34. Su, D.; Li, Z.; Huang, S.; Wu, X.; Li, J.; Xue, Y. Experiment and failure mechanism of cement sheath integrity under development and production conditions based on a mechanical equivalent theory. *Energy Sci. Eng.* **2021**, *9*, 2400–2422. [CrossRef]

processes

MDPI

Article

Mechanism Analysis and Potential Solutions for Casing Deformation of Shale GAS Fracturing Wells in Sichuan Basin

Bihua Xu [1], Shuo Yang [1], Bin Yuan [1,*], Lu Ma [2] and Leding Wang [2]

[1] State Key Laboratory of Oil and Gas Reservoir Geology and Exploitation, Southwest Petroleum University, Chengdu 610500, China
[2] Sichuan Shale Gas Project Management Department, CNPC Bohai Drilling Engineering Company Limited, Chengdu 610057, China
* Correspondence: 202199010013@swpu.edu.cn

Abstract: Casing deformation caused by fault and fracture sliding derived from fracturing has attracted growing attention. Casing deformation frequently occurs during the hydraulic fracturing process in the Sichuan Basin. Although its mechanism has been intensively studied, this issue is becoming increasingly severe and demands immediate solutions, especially in the shale gas blocks of Changning-Weiyuan-Zhaotong. The present study summarizes and analyzes the research progress relevant to casing deformation based on the existing literature. It is shown that the casing deformation rate of the deflection point on the shale gas horizontal well is much higher than that of other places and that shear deformation is the dominant form. The main factors influencing the casing deformation of shale gas horizontal wells include weakened strength of the collapsing casing, geological factors, cement, cement quality sheath, fracturing engineering factor, etc. We propose to reduce casing deformation by optimizing well trajectory, improving casing strength and cementing quality or optimizing fracturing operation. In addition, a hierarchical relationship between the influencing factors is also provided. However, the mechanisms of some forms of casing deformation need to be further studied, and the casing deformation in shale gas exploitation must be solved urgently.

Keywords: casing deformation; hydraulic fracturing; shale gas horizontal well; Sichuan Basin; countermeasures

Citation: Xu, B.; Yang, S.; Yuan, B.; Ma, L.; Wang, L. Mechanism Analysis and Potential Solutions for Casing Deformation of Shale GAS Fracturing Wells in Sichuan Basin. *Processes* **2022**, *10*, 1711. https://doi.org/10.3390/pr10091711

Academic Editors: Tianshou Ma, Yuqiang Xu and Albert Ratner

Received: 14 June 2022
Accepted: 20 August 2022
Published: 27 August 2022

1. Introduction

The Sichuan Basin is a typical shale gas production base in China. There are many famous shale gas blocks, such as the well-known Jiaoshiba [1–3], Changning-Weiyuan [3–5], and Fushun-Yongchuan blocks [6]. The permeability and porosity of the shale matrices are very low [7], but their natural fractures and bedding planes make them worth exploiting commercially [8,9]. However, the conventional depressurization production used in sandstone and carbonate reservoirs is no longer suitable for shale gas [8,9]. To achieve commercial exploitation, long horizontal wells combined with massive multistage hydraulic fracturing technology are employed to communicate natural fractures and bedding planes to form a fracture network [10–12].

The longer horizontal segment of a well passes through a reservoir with different mechanical properties, which results in a non-uniform stress distribution on the horizontal segment [13]. Furthermore, multistage fracturing operations cause drastic changes in the pressure and temperature of the casing [14,15]. The casing of a horizontal segment is easily deformed and damaged in a complex mechanical environment [16,17]. The casing deformation renders it difficult for bridge plugs to pass the deformation section, which significantly impacts subsequent simulation operations and production efficiency, even leading to the well's retirement before the fracture operation's completion [18–20].

Casing deformation during fracturing is a common issue in shale gas horizontal wells in the Sichuan Basin, and especially in the Changning-Weiyuan shale gas block. Although shale gas development is a mature technology in China, casing deformation causes delays in its exploration. To explain why its occurrence happens in shale reservoirs, researchers have investigated factors that may lead to casing deformation, including casing strength reduction, geological and fracturing engineering factors, etc. [14,21–25]. Hence, there is a need to research and analyze recent progress and developments in the study of casing deformation in shale reservoirs.

This paper studies the development of casing deformation concerning developments and problems arising from casing deformation in order to provide technical support for shale gas production. The first part of the research focuses on casing deformation characteristics. The third section analyzes the internal relationship between each influencing factor and casing deformation. Both historic and newer methods proposed for studying casing deformation are analyzed. In the third section, the internal relationship between each influencing factor and casing deformation is analyzed. In the fourth section, the countermeasures to casing deformation are summarized. Finally, casing deformation in Lu 203H60-3 well was analyzed, and future development will be prospected.

2. Casing Deformation Characteristics

2.1. Frequent Occurrence

Statistics show that casing deformation is a widespread and prominent problem in Sichuan Basin (Figure 1). By March 2022, 993 horizontal wells were fractured in the Changning-Weiyuan-Zhaotong shale gas block, and the casing deformation occurred in 269 Wells; the deformation rate is 26.93%. The casing deformation ratio of the Changning, Weiyuan, and Zhaotong shale gas wells reached 21.51% (77/358), 48.75% (136/279), and 24.65% (53/215), respectively. The casing deformation ratio was only 18.75% (3/16) in the Yongchuan shale gas block. Only 800 wells had casing deformation in the Fuling shale gas block [18]. These data indicate that casing deformation is more serious and frequent in the Changning and Weiyuan shale gas blocks than at Yongchuan and Fuling.

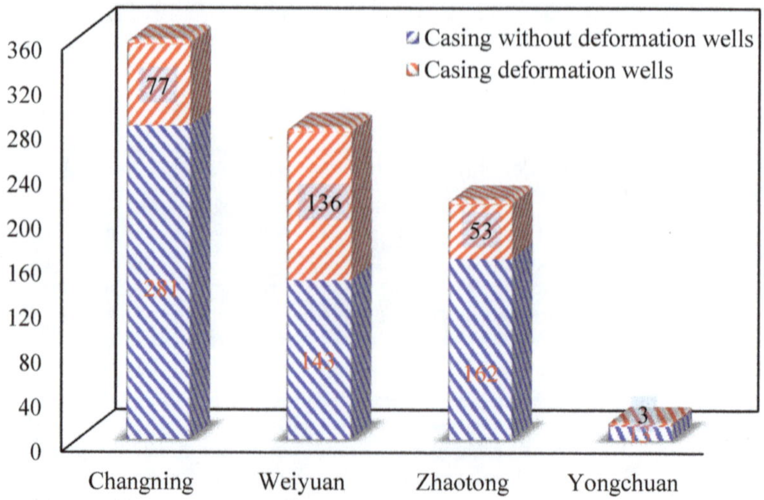

Figure 1. Statistics on casing deformation in some shale gas fields of Sichuan Basin.

2.2. *Distribution Characteristics of Deformation*

According to the statistics on casing deformation positions, the number occurring at target "A" is higher than 47.0% in the Changning-Weiyuan [18,19] and 43.8% in the Weiyuan-Rongxian shale gas areas (Figure 2). In Changning-Weiyuan, 65.0% of deformation points were located near the heel and 29.0% between the heel and toe. Figure 2 shows that others occurred at the toe [20]. The following aspects summarize the distribution characteristics of casing deformation:

(1) The deformation points increased from the toe to the heel of a horizontal well.
(2) The number of casing deformation points within 200 m of the target "A" was much higher than others.
(3) Most of the casing deformations occurred after the operation of several fractured segments or during the drilling of the bridge plug.
(4) The damage extent of deformation points increased with the fracturing operation time.
(5) The number of deformation points near natural fractures and faults was much higher than others.

Figure 2. Distribution of casing deformation points of shale gas horizontal well in Sichuan Basin.

2.3. *Deformation Morphology*

The statistical results show that shear deformation and collapsing deformation are the primary forms of casing deformation [26–28]. Three-dimensional logging imaging of casing in the Changning-Weiyuan area shows that most casing deformations are shear deformations [16,21]. Shear deformation was induced by fracture/fault slip and nonuniform loads on casing during hydraulic fracturing [22,23]. However, a stereotypical test of shale gas wells in the Fuling area indicated that the casing deformation was dominated by collapsing deformation [18]. This may be due to the fact that the geological structure of the Fuling area is more stable, and there is no fracturing-induced fracture/fault slip. In addition to shear and collapsing deformation, there was also a small amount of bending deformation and axial "S"-shaped deformation owing to casing "hanging" during fracturing in Sichuan Basin. These two deformations are mainly due to the well trajectory and casing strength.

3. Casing Deformation Influence Factors

3.1. *Casing Collapsing Strength Reduction*

The burial depth of shale reservoirs in the Sichuan Basin ranges from 2000 m to 6000 m [6], and the reservoir temperatures range from 50 °C to 130 °C [24]. Furthermore, the casing collapsing strength is reduced by worn, bending, temperature, and perforation. [25]. The casing collapsing strength is reduced by wearing, bending, temperature, and perforation.

3.1.1. Bending Reducing Casing Collapse Strength

The trajectory of a horizontal well cannot remain horizontal in a horizontal segment, and there are many curved sections. Furthermore, the casing in the building angle section is also bending. A dogleg angle will increase the casing bending stress and weaken the casing collapsing strength. It was found in experiments when the dogleg angle was 7°/30 m. The casing collapsing strength decreased from 73.8 MPa to 61.2 MPa for a TP110-grade casing [26]. The curvature also increased the wear of the casing and the eccentricity, which are additional factors that affect the casing deformation. The bending of the casing explains the reason for the high deformation rate of the casing near the "A" target point.

3.1.2. Temperature Reducing Casing Collapse Strength

High temperature or dramatic temperature fluctuations will reduce the casing collapsing strength [20,27]. The casing collapse strength decreases at high temperatures; the reduction law can be described by Equation (1) [28]. When the casing is in a 100 °C environment, its strength decreases by 4.32%. Due to the effect of thermal expansion and cold contraction, a rapid temperature decrease causes axial stress in the casing, thus reducing its collapsing strength [24,29]. For the type-TP110 casing, the casing strength decreases by 14% when the temperature is lowered by 70 °C. For TP140, the strength is reduced by 10% [29]. Sudden reductions in temperature also lead to annulus fluid shrinkage, increasing the casing stress [14]. Therefore, Kaldal, Jonsson [27] think that the influence of temperature on the casing collapsing strength was considered one of the main factors. However, this view has not been proven.

$$\sigma_{Temp} = [1 - 0.00054(T - 20)]\sigma_{20}. \quad (1)$$

where σ_{Temp} is the casing collapsing strength at high temperature, MPa; T is the temperature, °C; σ_{20} is the casing collapsing strength at 20 °C.

3.1.3. Wore Reducing Casing Collapse Strength

During drilling and hydraulic fracturing operations, the drilling strings, coiled tubing, and other tools easily rub against the casing in curved sections, resulting in wearing on the casing [25]. According to field caliper logging data, the casing wear rate is as high as 12.0%. Stress concentration occurs after casing wear, and the wear depth increases with the dogleg [22]. A worn casing is more susceptible to deformation under the coupling of temperature and nonuniform stress [30], especially at target "A" [19]. The bending of the wellbore trajectory causes casing wear, so casing bending and casing wear will coincide. Therefore, the combined effect of bending and wear is an essential factor that induces the deformation of the casing.

3.1.4. Perforation Reducing Casing Collapse Strength

A perforation will destroy the integrity of the casing structure and cause stress concentrations at the perforation holes. At the same time, perforations will cause the casing to crack. A casing with multicluster perforation was placed in complicated stress conditions during fracturing operations [31]. However, Zhao [32] and Xi, Li [19] found that casing deformation did not occur at the perforation position even though casing deformation is the most serious at Changning-Weiyuan.

3.2. Geological Factors

3.2.1. In-Situ Stress

The in-situ stresses in three directions are rarely equal, so the casing is always under nonuniform stress. When the horizontal segment of a shale gas well is under nonuniform in-situ stress, the casing is subjected to shear stress from the in-situ stress [33]. Furthermore, multistage hydraulic fracturing changes the in-situ stress, increasing the heterogeneity of the in-situ stress and thus making it easier for the casing to collapse [31]. Stress shadows will form during multistage hydraulic fracturing as the in-situ stress gradually increases from toe to heel. Therefore, casing collapse failure often occurs quickly under high in-situ stress, especially at target "A" [34]. Table 1 lists the in-situ stresses of typical blocks in the Sichuan Basin. Changning—Weiyuan is an anticlinal—synclinal slope zone and extrusion structure, and the difference in horizontal crustal stress is significant, while Fuling Jiaoshiba is a tension-extrusion anticline structure, and the contrast of horizontal crustal stress is slight. This is one of the reasons why the casing deformation rate of the Changning-Weiyuan block is much higher than that of the Fuling block.

Table 1. Comparison of in-situ stress in typical blocks of Sichuan Basin.

Well No.	Maximum Horizontal Principal Stress (MPa)	Minimum Horizontal Principal Stress (MPa)	Horizontal Stress Difference (MPa)	Coefficient of Horizontal Stress Difference
Wei 202	70.0	54.0	16.0	0.296
Wei 204	88.3	69.6	18.7	0.269
Ning 201	57.0	44.6	12.4	0.278
Jiao 1	54.0	49.0	5.0	0.102

3.2.2. Fracture/Fault Slip and Lithological Interface

Many faults, natural fractures, lithologic interfaces, and beddings in shale reservoirs are basic conditions for commercial exploitation [17]. However, during multistage hydraulic fracturing, the in-situ stress balance near the wellbore is destroyed owing to a large amount of injected fracturing fluid. Then, the faults, natural fractures, lithologic interfaces, and beddings are prone to slipping. This can cause casing shear deformation [20,23]. The non-uniform distribution of natural fractures in space and the heterogeneity of in-situ stress are other factors that lead to slippage [35]. A statistical analysis of the geological conditions of deformation points using seismic and logging data showed that about 61.7% of the total casing deformation points were related to fractures/faults and lithologic interfaces/bedding [18,21]. In a numerical simulation, Guo [36] found that casing deformation dramatically increases with the fracture/fault slip distance. About 52.38% of the casing deformation points were located in the lithologic interface area, according to statistics by Chen, Shi [21] and Xi, Li [18]. The casing deformation locations in different blocks are shown in Figure 3. Statistics show that 80% of casing deformation risks in Luzhou block are located at the natural fractures and bedding interface, and nearly 50% of casing deformation sections in Weiyuan block fractures developed. Therefore, fracture/fault slip and lithological interfaces are significant causes of casing deformation in Changning-Weiyuan-Rongxian-Luzhou shale blocks [19,37].

3.2.3. Microseism

Multistage hydraulic fracturing can induce microseisms [38]. For example, hydraulic fracturing triggered earthquakes in western Canada, and the most significant moment of magnitude was 3.9 [39]. Changning-Weiyuan-Rongxian-Luzhou is located in a geologically active area. However, the geological structures of Fuling and Yongchuan are relatively stable. During massive multistage hydraulic fracturing, microseism is easily triggered, thus increasing the slip distance of fractures/faults. Figure 4 shows the microseism and impacts before and after fracturing in different blocks. Microseism is the inducing factor of fracture/fault slip, and the real cause of casing deformation is still fracture/fault slip. Therefore, deformation in Changning-Weiyuan-Rongxian-Luzhou is more serious than in Fuling and Yongchuan.

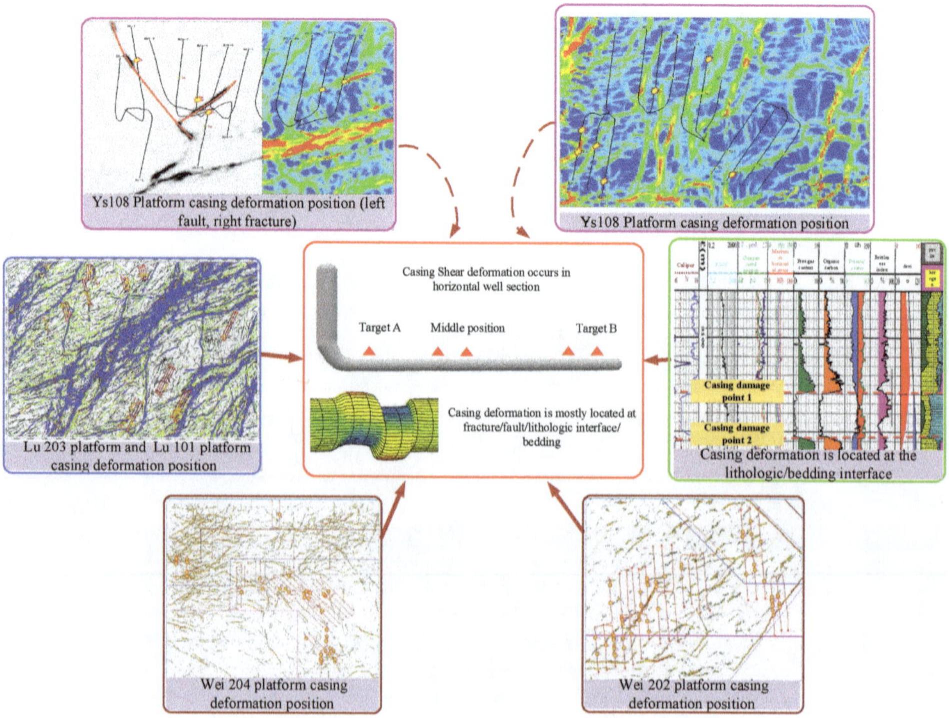

Figure 3. Location of casing deformation in different blocks.

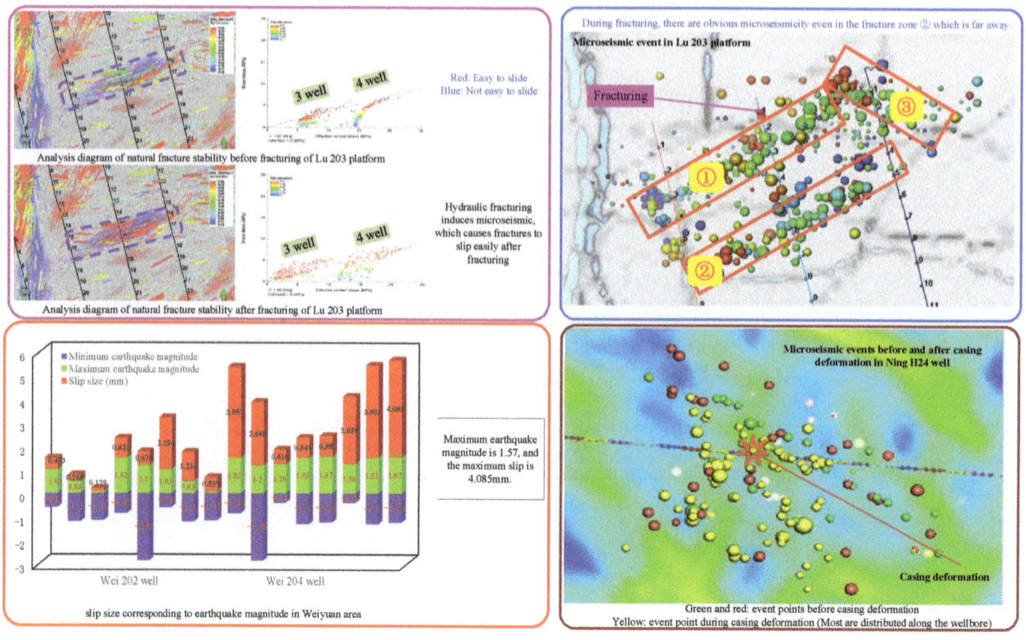

Figure 4. Microseismic events and their effects before and after fracturing in different blocks.

3.2.4. Shale Swelling

In the process of producing shale gas by fracturing, the water and ions from fracturing fluid enter the shale, causing the shale to absorb water and expand and changing the stress distribution of the casing [37]. Some studies have shown that the casing stress is 500 MPa when there is no shale expansion, while it increases to 1100 MPa when the shale expansion rate is 0.4% [40]. Figure 5 shows the impact of shale swelling on fractures and casing. Due to hydraulic fracturing, many artificial fractures are formed, and some natural fractures are connected, resulting in a significant increase in the contact area between fracturing fluid and shale. This may be one of the reasons why the number of casing deformations near the natural fracture is much more than in other locations.

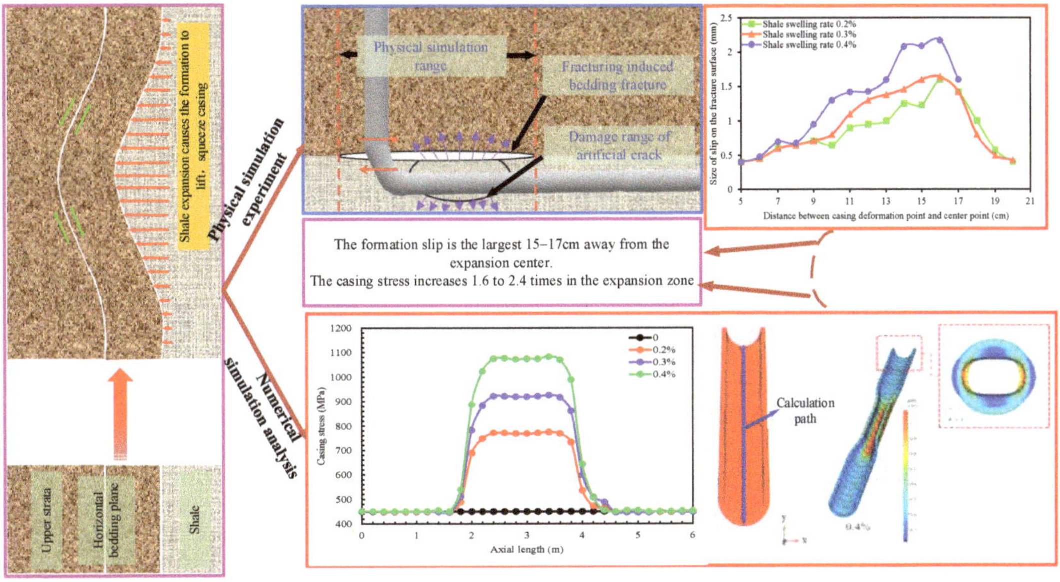

Figure 5. Effects of shale swelling on fractures and casing (numerical simulation from Li [41]).

3.3. Casing Eccentricity

Casing weight, the collapse of the well wall, and improper positioning of the centralizer resulting in casing eccentricity easily occur in the buildup and horizontal segments in cementing horizontal shale gas wells [42]. Casing eccentricity leads to a heterogeneous distribution of the annulus velocity of the cement slurry during cementing, resulting in rate efficiency at narrow gaps. This easily leads to cement sheath voids and channeling [43]. Figure 6 shows that when the centering degree of the casing is 67%, the outer extrusion stress of the casing is more significant, and the cement sheath is damaged at the thinnest part. An eccentric casing is easily deformed owing to the mechanical-thermal coupling effect [40].

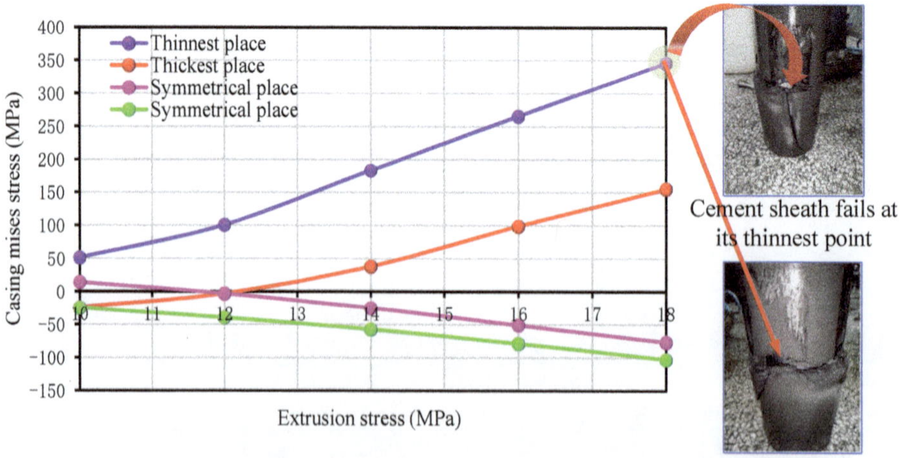

Figure 6. Casing mises stress distribution when 67% casing is centered.

3.4. *Cement Quality and Cement Sheath Properties*

A horizontal segment is long, and it is not difficult to achieve good cement quality. Poor cement quality is one factor that causes casing deformation in shale gas horizontal wells [44]. Poor cement quality is characterized by the poor quality of cementation, cement sheath voids, and channeling in the annulus [19,33]. Poor cement quality combined with the asymmetry of the fracturing area results in more serious casing deformation. Furthermore, the mechanical properties and thickness of the cement sheath also affect the casing deformation [19,45]. Figure 7, an experiment conducted in our laboratory, shows the effects of casing center degree, cement elastic modulus, cement gas penetration length, and missing cement thickness on casing circumferential stress.

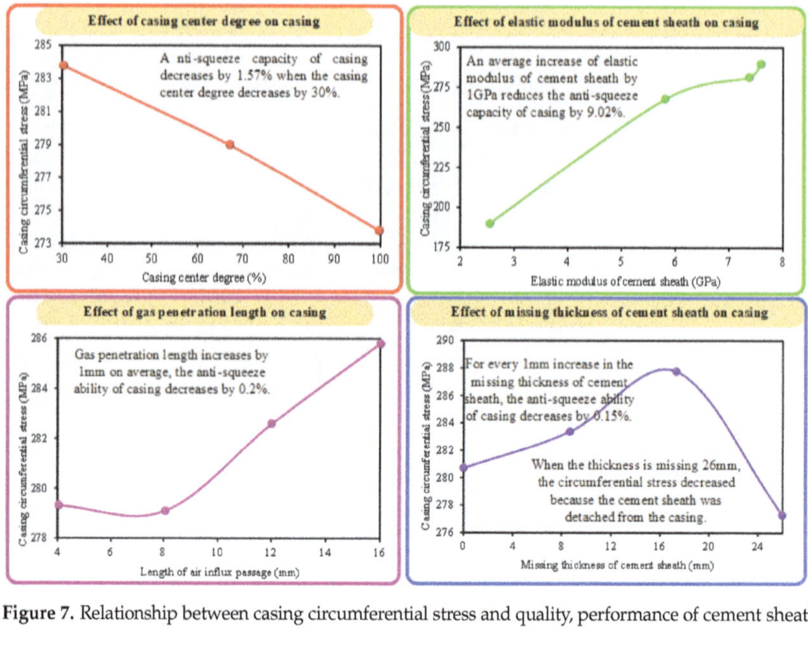

Figure 7. Relationship between casing circumferential stress and quality, performance of cement sheath.

3.4.1. Cement Sheath Voids and Channeling

Channeling and voids frequently occur in shale gas horizontal wells owing to eccentricity and low rate efficiency. The casing is under nonuniform stress, and when the cement sheath contains voids, a phenomenon similar to "annular pressure buildup" appears [14,33]. A cement void amplifies the nonuniform stress on the casing, thus increasing the casing shear stress, which causes casing shear deformation.

The geometry of cement sheath voids also affects the casing stress since it affects the stress distribution on the casing [14,44]. Research results showed that the casing stress increases and decreases with the void angle, reaching its maximum value when the void angle is 45° [40]. The injection of a large amount of cooling fracturing fluid into the casing resulted in a significant reduction in the internal pressure of the cement sheath voids. The loss in pressure can be expressed as follows [14]:

$$p_{loss} = \left(\frac{\Delta T \alpha}{B_N} \right) \tag{2}$$

where B_N is the fluid compressibility, m^2/N; α is the fluid thermal expansion coefficient, $1/°C$; and ΔT is the reduction in temperature, °C.

The casing partly lacks external support when the fluid pressure is deficient in the voids. Under the combined effect of high internal pressure and asymmetrical external supporting force, plastic deformation will occur on a casing as the stress concentration increases. Near the target "A", cement sheath voids occur easily owing to the curved wellbore trajectory and sudden temperature changes. Therefore, filling the annulus with a cement slurry to ensure cementing quality is one of the necessary conditions for reducing casing deformation.

3.4.2. Properties of Cement Sheath

The mechanical properties of a cement sheath (compressive strength, bonding strength, Young's modulus, Poisson's ratio, etc.) affect the stress on the casing. With a decrease in Young's modulus of a cement sheath, the maximum casing stress decreases sharply [45,46]. For a TP140-grade casing, its safety coefficient improved from 0.98 to 1.2 when Young's modulus of the cement sheath was reduced from 10 GPa to 5 GPa [20]. Shale is generally hard and brittle and has a high Young's modulus [47]. Suppose Young's modulus of a cement sheath is close to that of formation (high Young's modulus). In that case, the in-situ stress is more easily and efficiently transmitted to the casing, resulting in deformation.

A cement sheath with a low Young's modulus and high Poisson's ratio can significantly reduce the radial and tangential stresses of the cement sheath, thereby promoting integrity in the cement sheath [46]. Improving the integrity of a cement sheath can reduce local stress on the casing and fluid channeling in the annulus, thus reducing the volume of bound fluid in the annulus [14]. Moreover, an integrated cement sheath can reduce the casing deformation lowering the temperature and pressure changes of the bound fluid in the annulus.

3.4.3. Cement Sheath Thickness

Increasing the thickness of the cement sheath involves adding more stress-absorbing materials between the formation and casing, thereby reducing the effect of in-situ stress on the casing. The casing stress decreased slightly with an increase in the cement sheath thickness during faults/natural fracture slippage [20]. Increasing the thickness of the cement sheath involves adding more stress-absorbing materials between the formation and casing, thereby reducing the effect of in-situ stress on the casing. However, the thickness of the cement ring has a limited effect on casing deformation compared to other influencing factors [48]. Only considering increasing the thickness of the cement sheath cannot solve the problem of casing deformation.

3.5. Alternating Temperature and Pressure

There are alternating internal pressure and temperature during the multistage fracturing, which affects the casing deformation in two ways. On the one hand, the coupling of alternating temperature and pressure imposes alternating stress on the casing [30]. Under the alternating stress, the cement sheath is easily peeled from the casing to form a micro gap [14]. On the other hand, they will break down the integrity of the cement sheath sealing, thus forming a micro-annulus in the casing-cement formation [49], increasing the amount of annular fluid. If the alternate stress exceeds the yield strength of the casing, the yield strength of the casing decreases with the increase in the number of alternations [18]. The relationship between temperature and safety factor of casing triaxial in the Changning-Weiyuan area is analyzed, and Wei202 well is taken as an example. Under a small temperature range, the casing compressive strength is affected to a certain extent (Figure 8). The effect of the cement-sheath-sealing integrity on the casing deformation is similar to that of a cement sheath void on a local load. The casing at target "A", with the lowest collapsing strength under the most significant number of cyclic loads, was more prone to deformation. Therefore, the abrupt temperature change during fracturing was considered one of the main factors resulting in casing deformation [50].

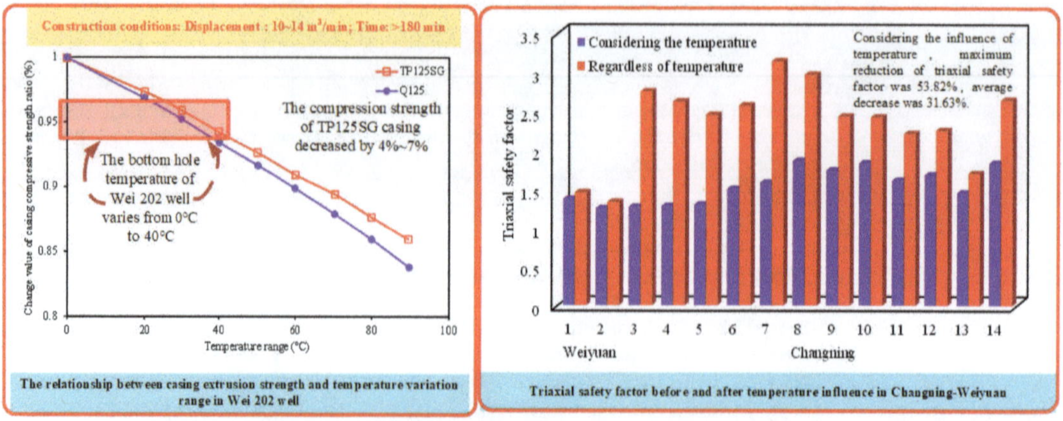

Figure 8. Relationship between temperature and casing strength.

3.6. Hydraulic Fracturing Parameters

3.6.1. Injection Rate

One of the characteristics of multistage hydraulic fracturing is that each stage injects thousands of cubic meters of liquid at a pump rate of over 10 m^3/min at a high pump pressure [51]. Figure 9 shows the fracturing injection rate of nine deformed casing Wells in Weiyuan, with an average maximum injection rate of 14 m^3/min. When there is bound fluid in the annulus, the temperature of the bound fluid within the cement sheath void will continue to decrease as the fracturing fluid is continuously injected. Then, the pressure of the cement-sheath void segments drops sharply [24]. With an increase in the rate, the maximum temperature difference increases continuously from heel to toe [24,29]. The casing stress increases with the injection rate. In addition, a higher injection rate can increase the influence of stress accumulation [37]. Therefore, the deformation risk of the casing increases with the injection rate [31].

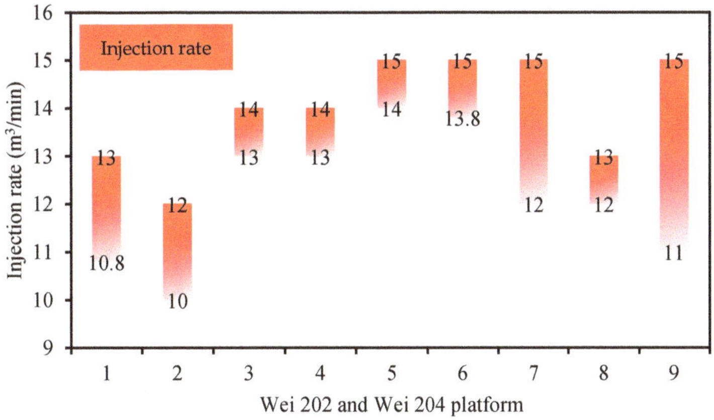

Figure 9. Fracturing injection rates of nine casing deformation wells in the Weiyuan area.

3.6.2. Injection Pressure

High injection pressure is needed during multistage hydraulic fracturing to break up shale reservoirs to form fracture networks. Figures 10 and 11 show the fracturing injection pressure of some casing deformation Wells in the Changning-Weiyuan area. The average injection pressure was 76.2 MPa on Chang 201 platform and 81.7 MPa on Wei 204. When the casing is under nonuniform loading, it is challenging to avoid casing deformation when improving its grade. As a cement sheath is integrated, the casing's internal pressure has a negligible effect on the casing stress. For example, even if the casing internal pressure reaches 110 MPa, the maximum casing stress is only 291.2 MPa [19,31]. However, once the cement sheath void is broken or the casing is under nonuniform loading, the casing stress will rapidly increase with the casing's internal pressure [19,20]. Calculations showed that when the internal pressure is 95 MPa, the casing stress could reach 1000 MPa (the yield strength of TP140 is 965 MPa) [20].

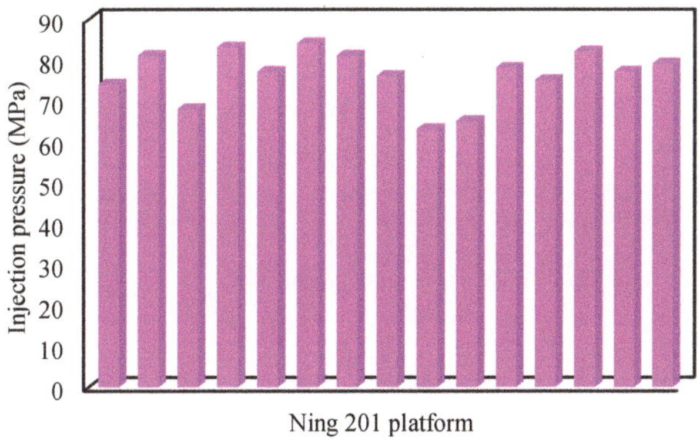

Figure 10. Fracturing injection pressure of 15 deformed casing wells in the Changning 201 platform.

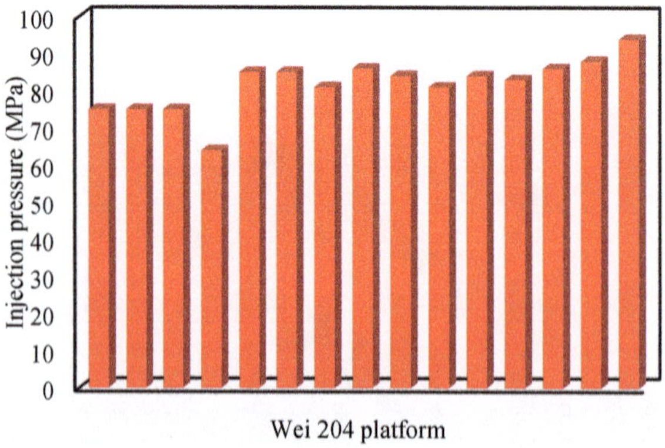

Figure 11. Fracturing injection pressure of 15 deformed casing wells in the Weiyuan 204 platform.

4. Analysis of Relationship between Influence Factors

Two or more influencing factors are required to deform the casing. Based on the frequency of casing deformation caused by these influencing factors, we have summarized the effect degree of these influencing factors, as shown in Table 2. The fracture/fault slip and microseisms are the strong influence factors and the leading cause of casing deformation. They can cause casing deformation. Casing wear, temperature, non-uniform in-situ stress, cement sheath voids and channeling, alternating temperature and pressure, injection rate, and injection pressure are medium influence factors. Only weaker influence factors cannot make casing deformation. However, if the casing has been affected by other strong or medium influence factors, weaker factors will make the casing deformation more serious.

Table 2. Effect degree of influence factors.

Influence Factors		Effect Degree of Influence Factors		
		Strong	Medium	Weaker
Casing collapsing strength reduction	Casing bending		√	
	Temperature		√	
	Casing wear		√	
	Perforation			√
Geological factors	Non-uniform in-situ stress		√	
	Fracture/fault slip and lithological interface	√		
	Microseism	√		
	Shale swelling	√		
Cement quality and cement sheath properties	Casing eccentricity		√	
	Cement sheath voids and channeling		√	
	Properties of cement sheath			√
	Cement sheath thickness			√
Fracturing engineering factors	Alternating temperature and pressure		√	
	Injection rate		√	
	Injection pressure		√	

However, the influence factors of casing deformation are not independent. The relationship between various influence factors is shown in Figures 12 and 13. The wellbore trajectory and gravity cause bending in the casing, leading to casing wear and forming a narrow eccentric annulus. Bending and casing wear lead to the collapsing strength of the casing reduces. Then, poor cement quality is inevitable due to narrow eccentric annulus, resulting in a cement sheath with voids and channels. When the cement sheath has voids and channeling, there is bound fluid in the annulus. A large amount of cool fracturing fluid injected through the casing leads to fluid pressure in the bound fluid, reducing due abrupt reduction in temperature. Therefore, there will be a significant differential pressure between the inside and outside of the casing and local loading; deformation easily happens to the casing. If the casing collapsing strength has been reduced by perforations and high temperature, deformation will likely occur under the above-complicated conditions. Furthermore, a large amount of fracturing fluid entering the formation will unbalance the in-situ stress and increase its heterogeneity. In addition, multistage fracturing induces slips in the faults and natural fractures. It also causes microseisms, which increases the slip distance in unstable formations.

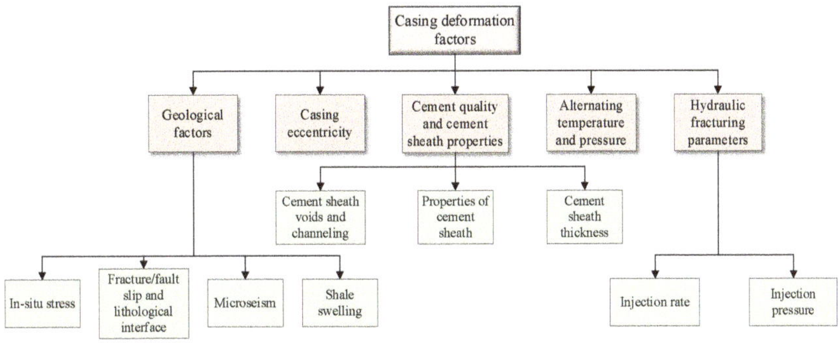

Figure 12. The relationship between the influencing factors.

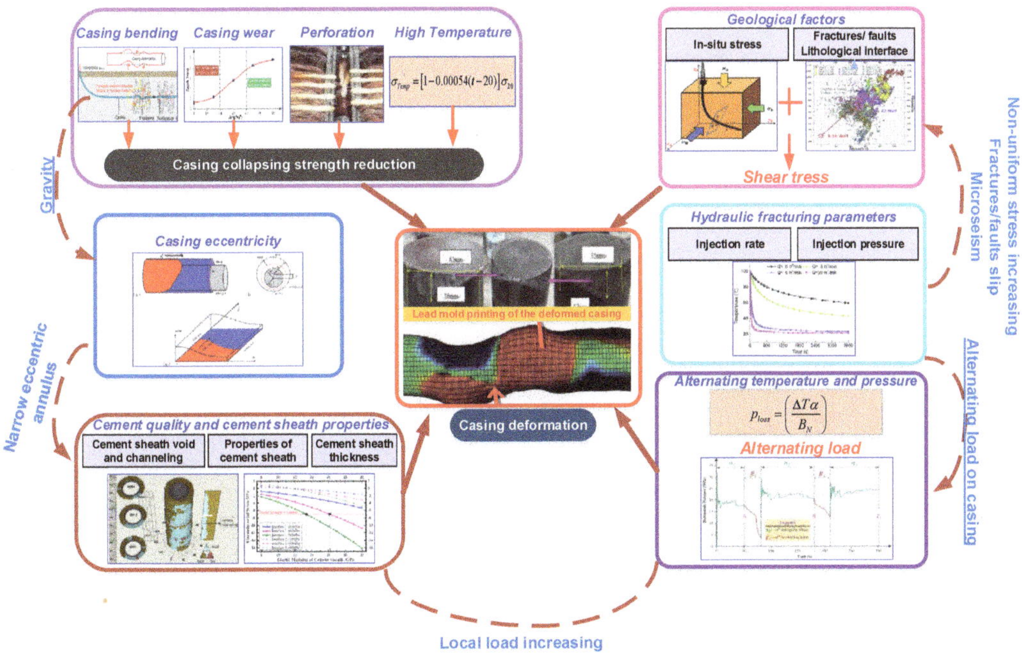

Figure 13. Schematic diagram of the relationship between various influencing factors. Some pictures in this figure from Yu [52], Xi, Li [19], Lian, Yu [31], and Liu, Gao [48].

5. Countermeasures

In studying casing deformation mechanisms and countermeasures, the researchers proposed a series of methods to prevent casing deformation, including well trajectory optimization, improving the cement quality, fracturing construction parameters optimization, casing material selection, preventing shale expansion, etc.

5.1. Optimization of Well Trajectory

Although about 61.7% of the total casing deformation points were related to fractures/faults and lithologic interfaces/bedding [18,21], these can be avoided by optimizing the wellbore trajectory design. Before optimization of the wellbore trajectory, the seismic data should be strengthened to determine the location and size of the fault to avoid shear deformation owing to the fault slip [53]. During well location selection and well trajectory design, the trajectory of the horizontal segment should avoid lithological interfaces, natural fractures, and fault development areas such as "ridges" or "valley bottoms" with severe geological structures or drill along the fracture belt [19,20]. In this way, the risk of right-angle shear casings caused by faults can be avoided, reducing casing deformation. Optimizing the wellbore trajectory can also reduce the narrow gap and casing wear caused by bending. Then, the cement quality can be improved to reduce the channeling and cement voids. In addition, the stress concentration of the casing caused by the bending stress can be reduced [45].

5.2. Optimization of Cement Sheath Properties and Cementing Parameters

In high-in-situ stress shale reservoirs, cement properties cause casing deformation [14]. Xi, Li [19] and [45] suggested that a high strength and low Young's modulus can be adopted for shale gas well cementing. Young's modulus of the cement sheath can be decreased to the maximum extent while keeping the strength is not or slightly reduced. Thus, the ability of the cement to maintain its sealing integrity can be promoted. The injection of high-viscosity fluid instead of cement was recommended since this provide space for slippage and can change the nonuniform load on the casing into a uniform load [45]. However, this method is difficult to construct, and it is difficult to find this material. The research shows that the young's modulus of foamed cement can be reduced to less than 2000 MPa [54], so foamed cement can effectively restrain casing deformation [41]. Reasonable placement of centralizers and floating casing cementing technology were adopted to improve the eccentricity of the casing to reduce channeling and voids [44]. Then, the effect of the sealed fluid pressure drop owing to the temperature drop was weakened. Yan, Zou [14] suggested that rotating the casing string during cementing can improve the cement quality and prevent cement voids' formation. In a word, shale gas horizontal wells require a high-strength and low-modulus cement slurry, and the displacement efficiency must be ensured during cementing to sure cement slurry fill the entire annulus.

5.3. Improvement Casing Strength

Maintaining the steel grade of the casing and improving its thickness, or maintaining the wall thickness of the casing and improving its steel grade, can improve the casing's collapsing resistance to reduce casing deformation [33]. In addition, it is recommended that the casing be externally thickened while maintaining the same size to reduce the impact of fracturing tools [20]. If there is a stress concentration or fracture/fault slip, reducing the casing deformation can be achieved by increasing the casing grade or wall thickness [20]. However, increasing the flexural strength by simply increasing steel grade and wall thickness cannot radically solve the failure of axial S-shaped casing deformation [31]. Moreover, increasing the casing grade or wall thickness makes it challenging to run casing. Therefore, reducing the deformation during fracturing by optimizing the casing should consider the difficulty, and other preventive measures should be considered.

5.4. Optimization of Hydraulic Fracturing Parameters

Hydraulic fracturing parameters contribute to casing deformation; thus, these parameters must be optimized. Fluid pressure in fault fractures should be controlled and reduced to reduce fault fracture activity [55], avoiding casing shear deformation. A technical scheme using a large-diameter bridge plug without drilling or a full-diameter and infinite-stage solubility ball seat seal can reduce the number of operations, reducing the casing deformation [56]. In sections with poor cement quality, the fracturing pressure should be controlled to avoid the formation of axial fractures of the cement sheath [53]. Dual or multicard packers and long-rubber-barrel packers can be installed to separate fracture zones in wells where natural fractures and beddings develop [21]. Xi, Li [19] proposed warm fracturing fluids to minimize pressure drops inside voids as an innovative strategy. However, this is difficult to achieve.

5.5. Optimization of Shale Inhibitor

Shale expansion affects casing deformation [41] significantly, and shale expansion occurs in every area in contact with fracturing fluid. Therefore, it is necessary to provide a reasonable plan to restrain shale expansion from avoiding casing deformation. High salt content is used to solve the problem of shale swelling, but it can affect the environment and fracturing fluid rheological characteristics [57]. Wang [58] proposed guar gum fracturing fluid to enhance pore connectivity, inhibit shale expansion, and improve flowback efficiency. This method can reduce the risk of casing deformation and improve the efficiency of the fracturing operation. We summarize the corresponding countermeasures for each influencing factor, as shown in Table 3. Figure 14 shows a flowchart of casing deformation countermeasures.

Table 3. The countermeasures for influence factors.

Influence Factors		Countermeasures
Casing collapsing strength reduction	Casing bending Temperature Casing wear Perforation	Well trajectory optimization Optimization casing strength Casing strength and well trajectory optimization Perforation parameters optimization
Geological factors	Non-uniform in-situ stress Fracture/fault slip and lithological interface Microseism Shale swelling	Casing strength optimization Well trajectory optimization Hydraulic fracturing parameters and well trajectory optimization Shale inhibitor optimization
Cement quality and cement sheath properties	Casing eccentricity Cement sheath voids and channeling Properties of cement sheath Cement sheath thickness	Well trajectory optimization Cementing parameters optimization Cement sheath properties optimization
Fracturing engineering factors	Alternating temperature and pressure Injection rate Injection pressure	Hydraulic fracturing parameters optimization

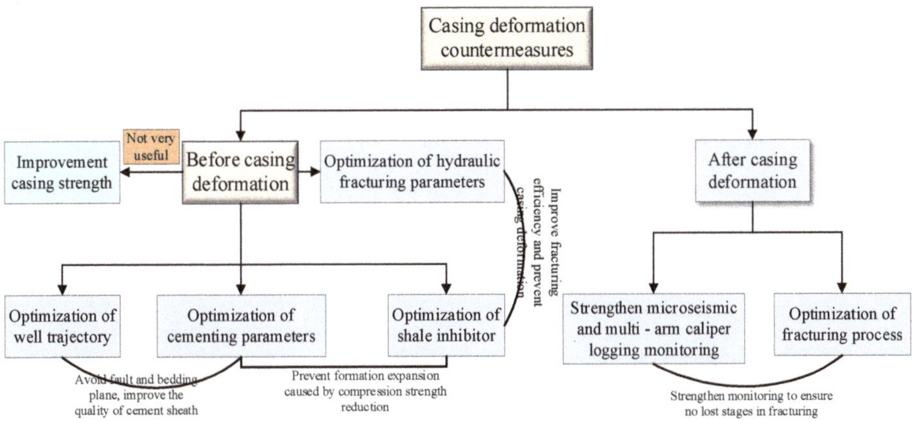

Figure 14. Countermeasures before and after casing deformation.

6. Analyze the Cause of Casing Deformation in Lu 203H60-3 Well

By February 2022, casing deformation in Luzhou 203 platform reached 48.1%. The casing deformation depth of well Lu 203H60-3 was determined through multi-arm well diameter and pumping resistance depth, and formation lithology data and construction parameters determined the cause of deformation. The multi-arm well diameter showed casing deformation at 3938.12 m, 3944.99 m, and 4012.77 m, with a maximum of 39.97 mm. While pumping 90 mm plugs to 3939.65 m and 3941.32 m in stage 22, encountered resistance. They were pumping 98 mm plug-in stage 8 encountered resistance at 4746.72 m. Figure 15 shows the well trajectory and casing deformation distribution. The five casing deformation positions measured by pumping resistance and multi-arm well diameter were identified as the same casing deformation point since 4000 m was located at the interface of the two zones, and the dogleg degree reached this point at 8.9°/30 m. A large dogleg degree specifically influences the anti-squeeze strength of the casing. In the process of multistage hydraulic fracturing, the sliding of the bedding interface and the constant change of temperature and pressure will significantly reduce the anti-squeeze strength of the casing, resulting in casing deformation. The second casing deformation location was identified at 4746.72 m. It was located near the interface between the first and second zones, and fractures developed. Acoustic cementing showed good cementing quality, but the casing deformation still occurred since the bedding and fractures during hydraulic fracturing were sliding to damage the cement sheath. The pressure difference between the casing and the casing is 80~90 MPa.

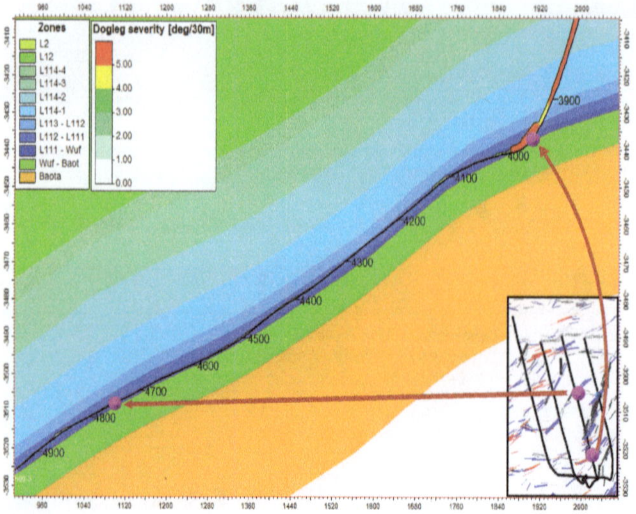

Figure 15. Comparison of casing deformation position, horizon and dogleg degree of well Lu 203H60-3.

7. Expectation

As mentioned above, many researches have been carried out on the mechanism of casing deformation, but some are not thorough enough. For example, some scholars believe that temperature is the main factor affecting the casing's compressive strength, but this view has not been proven. Perforation can destroy the integrity of the casing, but the deformation point of the casing is far from the perforation point. Currently, only field data support this idea, and the reason is not known. There are few studies on casing deformation caused by shale swelling and lifting formation, but shale water absorption and swelling occur in every stage of hydraulic fracturing. There are few countermeasures against these factors, most of which are based on numerical simulation and lack of physical experimental verification. On the other hand, the future is a time of both opportunities and challenges. In particular, the industry is moving toward big data. For example, when the influence of multiple factors makes it difficult to determine the primary mechanism of casing deformation, sensitivity analysis can make the ranking method more reliable and enable the recommendation of countermeasures [22,59,60]. Shale welling will raise the formation. Abbaszadeh Shahri [61] and Ghaderi [62] proposed that 3D models can be built to map the soil and predict future events so that they can accurately locate the casing deformation point. As mentioned earlier, hydraulic fracturing contributes to casing deformation. However, fracture extension is a complex process. Jin [59] proposes to monitor cracks based on DAS signals to avoid the uncertainty of traditional monitoring methods, and Huang [60] proposes a series of geological models to predict the fracture network. Abbaszadeh Shahri [63] proposes a state-of-the-art method (ARDCW) to integrate multiple models for visual estimation and exhibit superior capabilities. Then applying this technique to the petroleum engineering industry may have positive aspects. In short, the opportunities are enormous, and using artificial intelligence and big data analytics could lead to a dramatic shift in the oil and gas industry.

8. Conclusions

This work analyzed the associated developments in casing deformation. The author researched casing deformation characteristics, critical issues affecting casing deformation, and the relationships between various influence factors. In addition, effective countermeasures to casing deformation were highlighted. Based on our research, the following conclusions can be drawn:

(1) Casing deformation frequently occurs in shale gas development in the Sichuan Basin. The probability of casing deformation is the largest at target "A", and shear deformation is dominant, especially in the Changning-Weiyuan-Luzhou, since the area's geological structure is poor owing to natural fracture/fault and lithological interface development, and they are easily induced to slip. Furthermore, the natural fracture/fault and lithological interface are the main factors leading to casing deformation. Analysis shows that the casing deformation of the Lu 203H60-3 well is mainly caused by the bedding interface and fracture sliding.

(2) Although significant progress has been made in research on the mechanism of casing deformation in shale gas horizontal wells during hydraulic fracturing, at present, casing deformation cannot be sufficiently resolved, implying that the present understanding of its mechanism has not yet reached a significant level of maturity.

(3) Considering the effects of stress concentration and running casing, reducing casing deformation by increasing the wall thickness and grade of the casing is not optimal. By contrast, well trajectory optimization, cementing optimization, hydraulic fracturing parameter optimization, and shale inhibitor optimization are more desirable and operational.

(4) It is challenging to prevent casing deformation only by one preventive measure, so combining multiple measures is necessary to compensate for each other's shortcomings.

(5) Experiments on physical models can be used to verify the addition of nonuniform stress by cement voids on casings. At present, the results of research into casing deformation are based on numerical simulations and lack experimental research. If numerical simulations are combined with experimental research, more accurate research results should be obtained. The development of big data and artificial intelligence will provide new directions for casing deformation prevention.

Author Contributions: Methodology: B.X.; Writing-reviewing and editing: S.Y.; Writing-Original draft preparation: B.Y.; Data statistics and Language check: L.M.; Data curation: L.W. All authors have read and agreed to the published version of the manuscript.

Funding: The authors would like to acknowledge that this work was completed through financial support from the Natural Science Foundation National Science, Technology Major Project of China (2016ZX05002004-012).

Conflicts of Interest: The authors declare no conflict of interest.

References

1. Wang, Y.; Xu, S.; Hao, F.; Lu, Y.; Shu, Z.; Yan, D.; Lu, Y. Geochemical and petrographic characteristics of Wufeng-Longmaxi shales, Jiaoshiba area, southwest China: Implications for organic matter differential accumulation. *Mar. Pet. Geol.* **2019**, *102*, 138–154. [CrossRef]
2. Gao, J.; Zhang, J.-K.; He, S.; Zhao, J.-X.; He, Z.-L.; Wo, Y.-J.; Feng, Y.-X.; Li, W. Overpressure generation and evolution in Lower Paleozoic gas shales of the Jiaoshiba region, China: Implications for shale gas accumulation. *Mar. Pet. Geol.* **2019**, *102*, 844–859. [CrossRef]
3. Feng, Z.; Dong, D.; Tian, J.; Wu, W.; Cai, Y.; Shi, Z.; Peng, W. Geochemical characteristics of Lower Silurian shale gas in the Changning-Zhaotong exploration blocks, southern periphery of the Sichuan Basin. *J. Pet. Sci. Eng.* **2019**, *174*, 281–290. [CrossRef]
4. Xie, J. Rapid shale gas development accelerated by the progress in key technologies: A case study of the Changning–Weiyuan National Shale Gas Demonstration Zone. *Nat. Gas Ind. B* **2018**, *5*, 283–292. [CrossRef]
5. Wang, Q.; Wang, T.; Liu, W.; Zhang, J.; Feng, Q.; Lu, H.; Peng, P.A. Relationships among composition, porosity and permeability of Longmaxi Shale reservoir in the Weiyuan Block, Sichuan Basin, China. *Mar. Pet. Geol.* **2019**, *102*, 33–47. [CrossRef]
6. Zhai, G.-Y.; Wang, Y.-F.; Zhou, Z.; Yu, S.-F.; Chen, X.-L.; Zhang, Y.-X. Exploration and research progress of shale gas in China. *China Geol.* **2018**, *1*, 257–272. [CrossRef]
7. Guo, T.; Zhang, S.; Zou, Y.; Xiao, B. Numerical simulation of hydraulic fracture propagation in shale gas reservoir. *J. Nat. Gas Sci. Eng.* **2015**, *26*, 847–856. [CrossRef]
8. Yuan, B.; Wang, Y.; Zeng, S. Effect of slick water on permeability of shale gas reservoirs. *J. Energy Res. Technol.* **2018**, *140*, 112901–112907. [CrossRef]
9. Chen, Z.; Liao, X.; Zhao, X.; Dou, X.; Zhu, L. Performance of horizontal wells with fracture networks in shale gas formation. *J. Pet. Sci. Eng.* **2015**, *133*, 646–664. [CrossRef]
10. Yuan, B.; Zheng, D.; Moghanloo, R.G.; Wang, K. A novel integrated workflow for evaluation, optimization, and production predication in shale plays. *Int. J. Coal Geol.* **2017**, *180*, 18–28. [CrossRef]
11. Liu, Y.; Chen, C.; Ma, T.; Zhu, G.; Peng, N.; Zhang, X. Experimental Investigation on the Initiation of Hydraulic Fractures from a Simulated Wellbore in Laminated Shale. *Lithosphere* **2021**, *2021*, 4152918. [CrossRef]
12. Abbaszadeh Shahri, A.; Rezaei, F.; Mehdizadeh Farsad, S.; Mehdizadeh Farsad, K.; Panaei, N. Investigation of Engineering Geology Parameters in Creation of Hydraulic Fracturing in Order to Enhance Oil Recovery from Oil Reservoir Using Finite Element Method. *Sci. Q. J. Geosci.* **2015**, *24*, 3–9. [CrossRef]
13. Zhou, D.; Yang, H.; Jianjun, L.; Wang, X.; Wang, H.; Xue, Y. A test method for the in-situ stress of salt rock. *Oil Gas Storage Transp.* **2017**, *36*, 1385–1390. [CrossRef]
14. Yan, W.; Zou, L.; Li, H.; Deng, J.; Ge, H.; Wang, H. Investigation of casing deformation during hydraulic fracturing in high geo-stress shale gas play. *J. Pet. Sci. Eng.* **2017**, *150*, 22–29. [CrossRef]
15. Wang, W.; Zhang, H.; Liu, P.; Li, Z.; Ni, W.; Uechi, H.; Matsumura, T. A finite element method approach to the temperature distribution in the inner casing of a steam turbine in a combined cycle power plant. *Appl. Therm. Eng.* **2016**, *105*, 18–27. [CrossRef]
16. Liao, S.; Yu, S.; Yi, S.; Bo, Z.; Wang, L.; Lei, Y. Research and field tests of staged fracturing technology for casing deformation section in horizonta shale gas wells. *Nat. Gas Ind.* **2017**, *37*, 40–45. [CrossRef]
17. Qian, B.; Yin, C.; Li, Y.; Xu, B.; Qin, G. Diannostics of caing deformation in multistage hydraulic fracturing stimulation in lower Silurian Marine shale play in Southwestern China. In Proceedings of the Unconventional Resources Technology Conference, San Antonio, TX, USA, 20–22 July 2015; pp. 1–119.
18. Xi, Y.; Li, J.; Liu, G.; Zeng, Y.; Li, J. Ooverview of casing deformation in multistage fracturing of shale gas horizontal wells. *Spec. Oil Gas Reserv.* **2019**, *26*, 1–6. [CrossRef]
19. Xi, Y.; Li, J.; Liu, G.; Cha, C.; Fu, Y. Numerical investigation for different casing deformation reasons in Weiyuan-Changning shale gas field during multistage hydraulic fracturing. *J. Pet. Sci. Eng.* **2018**, *163*, 691–702. [CrossRef]
20. Guo, X.; Li, J.; Liu, G.; Xi, Y.; Zeng, Y.; He, M.; Yan, H. Numerical simulation of casing deformation during volume fracturing of horizontal shale gas wells. *J. Pet. Sci. Eng.* **2019**, *172*, 731–742. [CrossRef]
21. Chen, Z.; Shi, L.; Xiang, D. Mechanism of casing deformation in the Changning–Weiyuan national shale gas demonstration area and countermeasures. *Nat. Gas Ind. B* **2017**, *4*, 1–6. [CrossRef]
22. Xu, Z.; Yin, F.; Han, L.; Yang, S.; Wu, X. Mechanism study of casing deformation in multistage hydraulic fracturing shale reservoir. In Proceedings of the SPE/AAPG/SEG Unconventional Resources Technology Conference, Houston, TX, USA, 23–25 July 2018.
23. Yin, F.; Xiao, Y.; Han, L.; Wu, X. Quantifying the induced fracture slip and casing deformation in hydraulically fracturing shale gas wells. *J. Nat. Gas Sci. Eng.* **2018**, *60*, 103–111. [CrossRef]
24. Xi, Y.; Liu, G.; Li, J.; Zha, C.; Chao, W.; Liu, M. Calculation of wellbore temperature and analysis of its effect on casing strength during volume fracturing in shale gas well. *Fault Block Oil Gas Field* **2017**, *24*, 561–564. [CrossRef]

25. Wang, J.; Rao, X.; Han, L.; Li, Y.; Zhou, l.; Lu, X. Selection criteria and evaluation technology for casing used in shale gas wells in China. In Proceedings of the SPE Asia Pacific Unconventional Resources Conference and Exhibition, Brisbane, Australia, 9 November 2015; p. 10.
26. Jia, Z.; Liu, S.; Geng, Y.; Wang, M.; Ji, B. Theoretical and experimental study on the collapse strength of casing in curved well section. *China Pet. Mach.* **2018**, *46*, 100–105. [CrossRef]
27. Kaldal, G.S.; Jonsson, M.T.; Palsson, H.; Karlsdottir, S.N. Structural modeling of the casings in high temperature geothermal wells. *Geothermics* **2015**, *55*, 126–137. [CrossRef]
28. Wang, Z.; Ma, Z. Effect by thermal well temperature on casing properties and calculation method for pretension. *Steel Pipe* **2007**, *36*, 24–27. [CrossRef]
29. Yin, H.; Zhang, Y. A quantitative evaluation method for the effect of temperature on casing collapsing strength: A case study of large-scale hydraulic fracturing in shale gas horizontal wells. *Nat. Gas Ind.* **2016**, *36*, 73–77. [CrossRef]
30. Song, X.; Li, J.; Liu, G.; Xi, Y.; Lian, W.; Guo, X. Casing wear and stress distribution under coupling effects of temperature and pressure. *Fault Block Oil Gas Field* **2018**, *25*, 670–674. [CrossRef]
31. Lian, Z.; Yu, H.; Lin, T.; Guo, J. A study on casing deformation failure during multi-stage hydraulic fracturing for the stimulated reservoir volume of horizontal shale wells. *J. Nat. Gas Sci. Eng.* **2015**, *23*, 538–546. [CrossRef]
32. Zhao, C.; Li, J.; Liu, G.; Zhang, H.; Wang, C.; Ren, K.; Zhang, X. The casing deformation during shale gas hydraulic fracturing: Why it is so serious in weiyuan-changning, China? In Proceedings of the SPE Trinidad and Tobago Section Energy Resources Conference, Port of Spain, Trinidad and Tobago, 25–26 June 2018.
33. Liu, K.; Gao, D.; Wang, Y.; Liu, Y. Effects of local load on shale gas well casing deformation. *Nat. Gas Ind.* **2016**, *11*, 76–82. [CrossRef]
34. Guohua, W.; Zhengmao, C.; Jiyou, X.; Keli, Y. Study on the Effect of Non-Uniformity Load and Casing Eccentricity on the Casing Strength. *Energy Procedia* **2012**, *14*, 285–291. [CrossRef]
35. Liu, W.; Tao, C.; Wan, Y.; Chi, X.; LI, Y.; Lin, H.; Dend, J. Numerical analysis of casing deformation during massive hydraulic fracturing of horizontal wells in a tight-oil reservoir. *Pet. Sci. Bull.* **2017**, *2*, 466–477. [CrossRef]
36. Guo, X.; Li, J.; Liu, G.; Lian, W.; Zeng, Y.; Tao, Q.; Song, X. Shale experiment and numerical investigation of casing deformation during volume fracturing. *Arab. J. Geosci.* **2018**, *11*, 723. [CrossRef]
37. Li, H.; Li, Z.; Li, G.; Yu, H.; Jiang, Z.; He, L.; Guo, B.; Dong, M. Casing deformation mechanisms of horizontal wells in Weirong shale gas field during multistage hydraulic fracturing. *J. Nat. Gas Sci. Eng.* **2020**, *84*, 103646. [CrossRef]
38. He, Y.-Y.; Zhang, B.-P.; Duan, Y.-T.; Xue, C.-J.; Yan, X.; He, C.; Hu, T.-Y. Numerical simulation of surface and downhole deformation induced by hydraulic fracturing. *Appl. Geophys.* **2014**, *11*, 63–72. [CrossRef]
39. Bao, X.; Eaton, D.W. Fault activation by hydraulic fracturing in western Canada. *Science* **2016**, *354*, 1406–1409. [CrossRef]
40. Xi, Y.; Li, J.; Liu, G.; Zha, C.; QIin, X.; Yan, P. Impact analysis of cement sheath shape on casing stress under transient mechanical-thermal coupling effect. *Fault Block Oil Gas Field* **2017**, *24*, 700–704. [CrossRef]
41. Li, Z.; Li, H.; Li, G.; Yu, H.; Jiang, Z.; Liu, H.; Hu, S.; Tang, B. The influence of shale swelling on casing deformation during hydraulic fracturing. *J. Pet. Sci. Eng.* **2021**, *205*, 108844. [CrossRef]
42. Panjwani, S.; McDaniel, J.; Nikolaou, M. Improvement of zonal isolation in horizontal shale gas wells: A data-driven model-based approach. *J. Nat. Gas Sci. Eng.* **2017**, *47*, 101–113. [CrossRef]
43. Choueiri, G.H.; Tavoularis, S. Experimental investigation of flow development and gap vortex street in an eccentric annular channel. Part 1. Overview of the flow structure. *J. Fluid Mech.* **2014**, *752*, 521–542. [CrossRef]
44. Jiang, K.; Li, Q.; Chen, Y.; Guo, X.; Fu, Y.; Li, J. Influence of cementing quality on casing failures in horizontal shale gas wells. *Nat. Gas Ind.* **2015**, *35*, 77–82.
45. Yin, F.; Han, L.; Yang, S.; Deng, Y.; He, Y.; Wu, X. Casing deformation from fracture slip in hydraulic fracturing. *J. Pet. Sci. Eng.* **2018**, *166*, 235–241. [CrossRef]
46. Liu, S.; Li, D.; Yuan, J.; Qi, F.; Shen, J.; Guo, M. Cement sheath integrity of shale gas wells: A case study from the Sichuan Basin. *Nat. Gas Ind. B* **2018**, *5*, 22–28. [CrossRef]
47. Shi, X.; He, Z.; Long, S.; Peng, Y.; Li, D.; Jiang, S. Loading rate effect on the mechanical behavior of brittle longmaxi shale in nanoindentation. *Int. J. Hydrog. Energy* **2019**, *44*, 6481–6490. [CrossRef]
48. Liu, K.; Gao, D.; Taleghani, A.D. Analysis on integrity of cement sheath in the vertical section of wells during hydraulic fracturing. *J. Pet. Sci. Eng.* **2018**, *168*, 370–379. [CrossRef]
49. Liu, K.; Gao, D.; Wang, Y.; Yang, Y. Effect of local loads on shale gas well integrity during hydraulic fracturing process. *J. Nat. Gas Sci. Eng.* **2017**, *37*, 291–302. [CrossRef]
50. Zhang, B.; Guan, Z.; Hasan, A.R.; Lu, N.; Wang, Q.; Xu, Y.; Zhang, Q.; Liu, Y. Development and design of new casing to mitigate trapped annular pressure caused by thermal expansion in oil and gas wells. *Appl. Therm. Eng.* **2017**, *118*, 292–298. [CrossRef]
51. Wu, Y.; Cheng, L.; Huang, S.; Jia, P.; Zhang, J.; Lan, X.; Huang, H. A practical method for production data analysis from multistage fractured horizontal wells in shale gas reservoirs. *Fuel* **2016**, *186*, 821–829. [CrossRef]
52. Yu, H.; Dahi Taleghani, A.; Lian, Z. On how pumping hesitations may improve complexity of hydraulic fractures, a simulation study. *Fuel* **2019**, *249*, 294–308. [CrossRef]
53. Zhang, H.; Chen, Z.; Shi, l.; Xiang, D.; Wang, S.; Wang, Q.; Zhou, L. Mechanism of how fluid passage formed and application in sichuan shale gas casing deformation analysis. *Drill. Prod. Technol.* **2018**, *41*, 8–11. [CrossRef]

54. Velayati, A.; Roostaei, M.; Rasoolimanesh, R.; Soleymani, M.; Fattahpour, V. Colloidal gas aphron (CGA) based foam cement system. *Pet. Explor. Dev.* **2019**, *46*, 1281–1287. [CrossRef]
55. Zhang, P.; He, Y.; Liu, Z.; Tong, H.; Deng, C.; Ren, X.; Zhang, H.; Li, Y.; Qu, L.; Fu, Q.; et al. Shear compression deformation test and deformation prevention practice of casing in shale gas horizontal wells. *Nat. Gas Ind. B* **2021**, *8*, 514–522. [CrossRef]
56. Li, L.; Wang, G.; Lian, Z.; Zhang, L.; Mei, J.; He, Y. Deformation mechanism of horizontal shale gas well production casing and its engineering solution: A case study on the Huangjinba Block of the Zhaotong National Shale Gas Demonstration Zone. *Nat. Gas Ind. B* **2018**, *5*, 261–269. [CrossRef]
57. Quainoo, A.K.; Negash, B.M.; Bavoh, C.B.; Ganat, T.O.; Tackie-Otoo, B.N. A perspective on the potential application of bio-inhibitors for shale stabilization during drilling and hydraulic fracturing processes. *J. Nat. Gas Sci. Eng.* **2020**, *79*, 103380. [CrossRef]
58. Wang, B.; Liu, B.; Yang, J.; Bai, L.; Li, S. Compatibility characteristics of fracturing fluid and shale oil reservoir: A case study of the first member of Qingshankou Formation, northern Songliao Basin, Northeast China. *J. Pet. Sci. Eng.* **2022**, *211*, 110161. [CrossRef]
59. Jin, G.; Roy, B. Hydraulic-fracture geometry characterization using low-frequency DAS signal. *Lead. Edge* **2017**, *36*, 975–980. [CrossRef]
60. Huang, J.; Safari, R.; Mutlu, U.; Burns, K.; Geldmacher, I.; McClure, M.; Jackson, S. Natural-Hydraulic Fracture Interaction: Microseismic Observations and Geomechanical Predictions. In Proceedings of the SPE/AAPG/SEG Unconventional Resources Technology Conference, Denver, CO, USA, 25–27 August 2014. [CrossRef]
61. Abbaszadeh Shahri, A.; Kheiri, A.; Hamzeh, A. Subsurface Topographic Modeling Using Geospatial and Data Driven Algorithm. *ISPRS Int. J. Geo-Inf.* **2021**, *10*, 341. [CrossRef]
62. Ghaderi, A.; Abbaszadeh Shahri, A.; Larsson, S. A visualized hybrid intelligent model to delineate Swedish fine-grained soil layers using clay sensitivity. *CATENA* **2022**, *214*, 106289. [CrossRef]
63. Abbaszadeh Shahri, A.; Shan, C.; Larsson, S. A Novel Approach to Uncertainty Quantification in Groundwater Table Modeling by Automated Predictive Deep Learning. *Nat. Resour. Res.* **2022**, *31*, 1351–1373. [CrossRef]

processes

MDPI

Article

Wellbore Temperature and Pressure Calculation of Offshore Gas Well Based on Gas–Liquid Separated Flow Model

Jun Jing [1,2,*], Hongbin Shan [1], Xiaohua Zhu [1,3], Yixiang Huangpu [1] and Yang Tian [1]

1 College of Mechanical and Electrical Engineering, Southwest Petroleum University, Chengdu 610500, China
2 State Key Laboratory of Oil and Gas Reservoir Geology and Exploitation, Southwest Petroleum University, Chengdu 610500, China
3 Oil and Gas Equipment Technology Sharing and Service Platform of Sichuan Province, Chengdu 610500, China
* Correspondence: 201999010045@swpu.edu.cn

Abstract: Compared with land wells, the production environment and reservoir depth of offshore oil and gas wells are more complex and shallower. Further, HPHT production fluid there will produce strong temperature and pressure disturbance that affects the wellbore, which easily generates wellbore safety problems, such as wellhead growth and leakage caused by the incompatible deformation of casing and cement sheath. Therefore, obtaining an accurate wellbore temperature and pressure field is the key to implementing a wellbore safety assessment. Based on the gas–liquid two-phase separated method, this paper established an improved calculation model of wellbore temperature and pressure field for offshore HPHT wells. This model also takes into account the heat transfer environment characteristics of "formation-seawater-air" and the influence of well structure. Compared with the measured data of the case well, the error of temperature and pressure calculation results of the improved model are only 0.87% and 2.46%. Further, its calculation accuracy is greatly improved compared to that of the traditional gas–liquid homogeneous flow calculation model. Based on this model, the influencing factors of wellbore temperature and pressure in offshore gas wells are analyzed. The results show that forced convection heat exchange between seawater–air and wellbore is stronger than that between wellbore and formation. Reducing the gas–liquid ratio of the product can effectively reduce wellbore temperature and increase wellbore pressure. The gas production has a significant impact on the wellbore temperature. When the gas production rises from $10 \times 10^4 \cdot m^3/d$ to $60 \times 10^4 \cdot m^3/d$, the wellhead temperature rises from 63 °C to 99 °C. However, due to the mutual influence of friction pressure drop and hydrostatic pressure drop, wellbore pressure increases first and then decreases with the increase in gas production. The improved model can provide a more accurate estimate of the time to reach the rated wellhead temperature. Meanwhile, this model displays accurate theoretical support for the rational formulation of the production plan after the well opening, so as to avoid excessive restrictions on the initial production rate.

Keywords: well temperature; well pressure; HPHT; multiple annulus temperature; gas–liquid two phase flow; separated flow

Citation: Jing, J.; Shan, H.; Zhu, X.; Huangpu, Y.; Tian, Y. Wellbore Temperature and Pressure Calculation of Offshore Gas Well Based on Gas–Liquid Separated Flow Model. *Processes* **2022**, *10*, 2043. https://doi.org/10.3390/pr10102043

Academic Editors: Tianshou Ma, Yuqiang Xu and Angelo Lucia

Received: 15 September 2022
Accepted: 7 October 2022
Published: 10 October 2022

1. Introduction

The offshore oil and gas field will become the main arena in which to improve domestic oil and gas production capacity and ensure national energy security. However, the harsh and complex offshore environment and operating conditions are the realistic challenges that offshore oil and gas development and production must face. The complex geological genesis of offshore oil and gas results in some HPHT reservoirs. For example, there are HPHT wells with an absolute reservoir temperature of more than 200 °C and a formation pressure coefficient up to 2.3 in Yingqiong Basin and the Ledong area [1,2]. Compared with Tarim Basin, where reservoir temperature and pressure are comparable despite buried depths of more than 7500 m [3,4], the buried depth of offshore oil and gas is usually only about 3000 m. The HPHT fluid in offshore oil and gas reservoirs will reach the wellhead in a shorter time, causing a stronger thermal-pressure coupling effect on the entire wellbore, which leads to difficulties in operation and production. Further there are even safety problems such as wellhead growth, well leakage, and well integrity damage [5].

In 1962, Ramey established a temperature distribution model for single-phase fluid flowing in the wellbore, which transformed heat transfer in the wellbore into steady state heat transfer and formation heat transfer into unsteady state heat transfer. This model is still widely used today [6]. Shiu and Beggs developed semi-empirical expressions for the relaxation distance parameters in the Ramey model based on the analysis of field data and gas and fluid flow characteristics, which improved the utility of the model [7]. Hasan and Kabir transformed single-phase flow into two-phase flow based on the Ramey model and established a semi-empirical analytical solution model based on gas–liquid two-phase flow by using the calculation method of homogeneous-phase flow [8]. Spindler obtained the exact analytical solution of the Hasan model by establishing dimensionless intermediate variables, which improved the accuracy of calculation [9]. Wu and Xu established the pressure and temperature coupling model of HPHT gas wells under single-phase gas flow and analyzed the influence of geothermal gradient on wellbore temperature field [10,11]. Martínez found that the change of fluid viscosity would have a significant impact on the fluid flow pattern and pressure drop, and the change of wellbore temperature and pressure would also have a great impact on the wellbore fluid flow pattern, and then change the heat-transfer ability of the wellbore fluid [12]. Based on this, Yin and Gao, combined with the correction of compressibility and gas–liquid mixing density, established a pseudo single-phase flow wellbore-temperature field model considering the change of wellbore fluid flow pattern [13,14]. Hou studied the influence of gas–liquid two-phase flow pattern change on fluid heat-transfer ability through experiments [15]. According to the change of fluid flow characteristics after well shut-in, Song established a transient model of the temperature field of ocean wellbore under the condition of well shut-in and studied the heat-transfer mechanism during well shut-in [16]. Ma simulated the heat-transfer experiment of gas–liquid two-phase flow and gave the calculation formula of convective heat transfer between seawater and string [17]. Sivaramkrishnan developed a simulation algorithm for an unsteady state non-isothermal two-phase wellbore model to predict the downward flow of a wet steam based on a drift-flux model. However, the frictional multiplier of gas–liquid two-phase flow is not considered [18].

The above studies mainly focus on the influence of physical parameters of production fluid and the change of external heat-transfer environment on wellbore temperature and pressure. There are few studies on gas–liquid two-phase flow in wellbore, and these studies usually use the homogeneous flow method, treating gas–liquid two-phase flow as quasi-single flow, in order to calculate wellbore temperature and pressure. The homogenous flow model assumes that the two phases are well mixed and move with identical velocities. However, the separated flow model assumes the two phases to flow separately and share a definite and continuous interface. This method is more in line with the actual situation of oil and gas production. Therefore, this paper establishes a new gas–liquid two-phase flow wellbore temperature and pressure mathematical model by using the calculation method of gas–liquid-phase flow. Further, this model also takes into account the heat-transfer environment characteristics of "formation-seawater-air" and the influence of well structure.

2. Calculation Model of Wellbore Temperature Field

Figure 1 shows the wellbore structure. According to different wellbore structures, regions are divided, and a heat-transfer model between the wellbore and the external environment is established. The temperature field and pressure-field coupling should be calculated first, followed by multiple annular temperatures in semi-steady mode. Due to fluid migration from the bottom-up along the tubing, fluid temperature is significantly higher than environment temperature in the same plane, resulting in heat transfer between them. Before establishing the calculation model, the following assumptions must be made:

1. Assume that gas–liquid two-phase fluid flows in a one-dimensional steady direction in the tubing. The fluid flow is turbulent flow, and there is no phase change.
2. Tubing is concentric with the riser, casing, formation, and cement sheath. Further, the cement sheath is well-bonded to the formation and casing.
3. The temperature profile of the formation and seawater is approximately linear. Further, the air temperature is considered the same as the temperature of the sea level.
4. When the fluid flows in the tubing, only the heat-transfer process occurs, and no energy exchange with the outside world occurs.
5. Radial heat transfer occurs between the fluid in the tubing and the environment, and axial heat transfer in the flow direction is ignored.

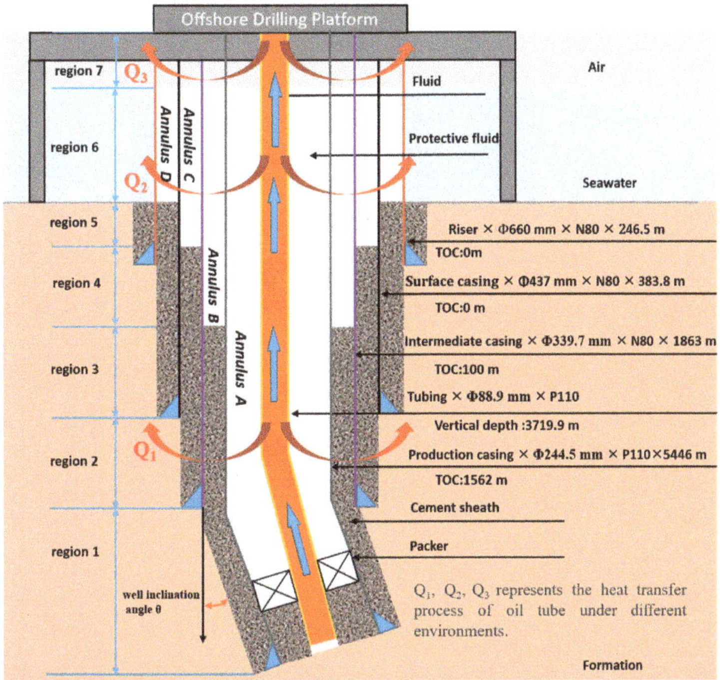

Figure 1. Schematic of wellbore structure.

2.1. Tubing Heat-Transfer Model

Figure 2 depicts wellbore heat transfer. Taking a micro-segment of the tubing for example, the arrow indicates the flow direction. According to the law of energy conservation, the energy expression of this micro-segment is obtained:

$$-Q = \Delta H + mg\cos\theta\Delta h + \frac{m\Delta v^2}{2} \tag{1}$$

where H is the fluid of enthalpy, J; g is gravitational acceleration, m/s^2; v is fluid velocity, m/s; and Q is heat loss in a micro-segment, J.

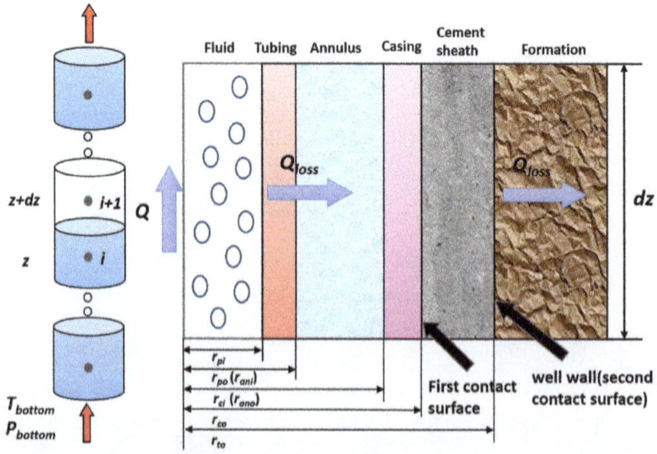

Figure 2. Schematic of wellbore heat transfer.

The differential formula can be obtained by deriving Equation (2):

$$-\frac{dq}{dz} = \frac{dh}{dz} + g\cos\theta + v\frac{dv}{dz} \tag{2}$$

where q is the heat loss of the micro-segment in unit time, $\mathrm{J/(m \cdot s)}$; h is the specific enthalpy of gas, J/kg; and θ is the well inclination angle, °.

Fluid heat flowing into the tubing (Q_{in}) equals the sum of heat flowing out of the micro-segment per unit time (Q_{out}) and radial heat loss (Q_{loss}).

$$Q_{in} = Q_{out} + Q_{loss} \tag{3}$$

Fluid flows axially from bottom to wellhead, so the loss of the wellbore micro-segment in unit time—namely, heat transmitted from fluid to well wall—is [19,20]:

$$\frac{dq}{dz} = \frac{2\pi r_{po} U_T \left(T_f - T_{to}\right)}{\omega} \tag{4}$$

where r_{po} is the outer diameter of the tubing, m; U_T is the total heat-transfer coefficient of the wellbore, $\mathrm{J/(s \cdot m^2 \cdot {}^\circ C)}$; T_f is the fluid temperature of the tubing, °C; T_{to} is wellbore temperature, °C; and ω is mass flow rate, kg/s.

The total thermal resistance of radial heat transfer in the wellbore is composed of the thermal resistance of fluid in tubing (R_f), string thermal resistance (R_p and R_c), annulus fluid thermal resistance (R_{an}), and cement sheath thermal resistance (R_{cem}) in a series. The specific calculation equation of the total heat transfer coefficient of the wellbore, according to the heat-transfer principle of thick wall cylinder, is [21]:

$$U_T == \frac{1}{2\pi r_{po}} R_T^{-1} = \begin{bmatrix} \frac{r_{po}}{r_{pi} h_f} + \frac{r_{po}\ln\left(\frac{r_{po}}{r_{pi}}\right)}{k_p} + \frac{r_{po}\ln\left(\frac{r_{to}}{r_{kco}}\right)}{k_{cem}} + \\ \sum\limits_{k=1}^{k} \frac{1}{h_{kanc}+h_{kanr}} + \sum\limits_{k=1}^{k} \frac{r_{po}\ln\left(\frac{r_{kco}}{r_{kci}}\right)}{k_{ck}} \end{bmatrix}^{-1} \tag{5}$$

where R_T is the total heat-transfer coefficient of the wellbore, $\mathrm{J/(s \cdot m^2 \cdot {}^\circ C)}$; k is the number of annulus in the wellbore; h_{kanc} is the convective heat-transfer coefficient of the kth annulus, $\mathrm{J/(s \cdot m^2 \cdot {}^\circ C)}$; h_{kanr} is the radiative heat-transfer coefficient of the kth annulus, $\mathrm{J/(s \cdot m^2 \cdot {}^\circ C)}$; r_{kco} is the outer radius of the outer casing of the kth annulus, m; r_{kci} is the inner radius of the outer casing of the kth annulus, m; and k_{ck} is the thermal conductivity of the outer casing of the kth annulus, $J/(s \cdot m^2 \cdot °C)$.

The convection heat-transfer coefficient of the production fluid is [22]:

$$\begin{aligned} h_L &= 0.0135 Re_L^{0.8} Pr_L^{0.33} \left(\frac{k_L}{r_{pi}}\right)\left(\frac{\mu_L}{\mu_w}\right)^{0.14} \\ h_{TP} &= h_L F_p \left[1 + 0.55\left(\frac{x}{1-x}\right)^{0.1}\left(\frac{1-F_p}{F_p}\right)^{0.4}\left(\frac{Pr_g}{Pr_L}\right)^{0.25}\left(\frac{\mu_L}{\mu_g}\right)^{0.25} I^{0.25}\right] \end{aligned} \tag{6}$$

where h_L is the liquid convection heat-transfer coefficient of the tubing, $J/(s \cdot m^2 \cdot °C)$; k_L is the thermal conductivity of the tubing liquid, $J/(s \cdot m \cdot °C)$; Re_L is the in situ liquid-phase Reynolds number of the tubing liquid, dimensionless; Pr is the Prandtl constant, dimensionless; μ_w is the dynamic viscosity of the production casing wall, $Pa \cdot s$; and F_p is the fluidity factor, dimensionless.

The calculation formula of the in situ Reynolds number of the tubing liquid is:

$$Re_L = \frac{2\omega_L}{\pi r_{pi} \mu_L \sqrt{1-\varphi_g}} \tag{7}$$

The calculation formula of the fluidity factor is:

$$F_p = (1-\varphi_g) + \varphi_g F_s^2 \tag{8}$$

where φ_g is gas holdup, dimensionless; F_s is the shape factor, dimensionless.

The shape factor calculation formula is:

$$F_s = \left[\frac{2}{\pi}\arctan\left(\sqrt{\frac{\rho_g (v_g - v_L)^2}{2 g r_{pi}(\rho_L - \rho_g)}}\right)\right] \tag{9}$$

The calculation formula of the inclination coefficient is:

$$I = 1 + \frac{4(\rho_L - \rho_g) g r_{pi}^2}{\sigma_p}|\cos\theta| \tag{10}$$

where σ_p is surface tension, N/m.

Heat transfer from well-wall to formation in a wellbore micro-segment:

$$\frac{dq'}{dz} = \frac{2\pi k_d (T_{to} - T_d)}{f(t_D)} \tag{11}$$

where $f(t_d)$ is the dimensionless time function of formation, dimensionless; k_d is the thermal conductivity of the formation, $J/(s \cdot m \cdot °C)$; and T_d is the formation temperature at any depth of well, °C.

According to the heat-transfer model and the transient heat-transfer function improved by Ramey and Hasan in the steady heat-transfer process, the dimensionless time function of the formation can be obtained as follows [23]:

$$\begin{cases} t_D = \lambda_d \frac{t}{r_{to}^2} \\ f(t_D) = 1.128\sqrt{t_D}(1 - 0.3\sqrt{t_D}) & t_D \le 1.5 \\ f(t_D) = (0.4063 + 0.5 ln t_D)\left(1 + \frac{0.6}{t_D}\right) & t_D > 1.5 \end{cases} \tag{12}$$

where λ_D is the thermal diffusion coefficient of the formation, m^2/s; t_D is dimensionless time, dimensionless; and t is the production time, s.

When drilling and completing a well on an offshore platform, the seawater section of the mud line is run into a riser to isolate the seawater. Below the mudline, the wellbore temperature gradually decreases with the decrease in formation depth. Above the mud line, the wellbore temperature is affected by the seawater temperature, which drops from the sea level temperature to the temperature at the mud line (the mudline temperature is 0–4 °C).

The heat-transfer process of the riser section wellbore is similar to the heat-transfer model below the mudline, including the heat transfer of the oil casing wall, the heat transfer between the casing wall and the riser wall, and the convective heat transfer between the riser wall and the seawater.

Heat transfer from well wall to seawater in a wellbore micro-segment:

$$\frac{dq'}{dz} = 2\pi r_{po} h_{sea}(T_{to} - T_{sea}) \tag{13}$$

where h_{sea} is the convective heat-transfer coefficient between seawater and casing wall, $\mathrm{J/(s \cdot m^2 \cdot {}^\circ C)}$; T_{sea} is the formation temperature at any depth of well, °C.

The convective heat-transfer coefficient of seawater:

$$h_{sea} = \frac{Nu_{sea} k_{sea}}{2r_{ro}} \tag{14}$$

where Nu_{sea} is the Nusselt coefficient of seawater, dimensionless; k_{sea} is the thermal conductivity of seawater, $\mathrm{J/(s \cdot m \cdot {}^\circ C)}$; and r_{ro} is the outer diameter of the riser, m.

The Nusselt relation of the riser under different seawater flow states is as follows [17]:

$$\begin{array}{ll} Nu_{sea} = 0.02183\mathrm{Re}^{0.9434} & \mathrm{Pr}^{1/3}\ \mathrm{Re} \le 2000 \\ Nu_{sea} = 35.7273\mathrm{Re}^{0.0191} & \mathrm{Pr}^{1/3}\ 2000 < \mathrm{Re} \le 10000 \\ Nu_{sea} = 19.1161\mathrm{Re}^{0.0647} & \mathrm{Pr}^{0.4}\ 10000 < \mathrm{Re} \end{array} \tag{15}$$

Between the sea level and the offshore platform, the convective heat transfer between the casing and the external air environment is not negligible. The heat transfer in the air section is similar to that in the seawater section, except that the external heat-transfer medium is different. Therefore, the seawater heat-transfer model is still adopted, and the convective heat-transfer coefficient of seawater can be changed into air.

Heat transfer from well wall to air in a wellbore micro-segment:

$$\frac{dq'}{dz} = 2\pi r_{po} h_{air}(T_{to} - T_{air}) \tag{16}$$

where h_{air} is the convective heat-transfer coefficient between air and casing wall, $\mathrm{J/(s \cdot m^2 \cdot {}^\circ C)}$; T_{air} is the air temperature, °C.

The air convective heat-transfer coefficient:

$$h_{air} = \frac{Nu_{air} k_{air}}{2r_{ro}} \tag{17}$$

where Nu_{air} is the Nusselt coefficient of the air, dimensionless; k_{air} is the thermal conductivity of air, $\mathrm{J/(s \cdot m \cdot {}^\circ C)}$.

The Nusselt relation of the riser under different air-flow states is as follows [24]:

$$\begin{array}{l} \text{Laminar flow :} \\ Nu_{air} = 1 + 0.3\left[32^{0.5}(Gr_h)^{-0.25}\left(\frac{h}{d}\right)^{0.909}\right]\ 10^4 \le Ra_h \le 4 \times 10^9 \\ \text{Transitional flow : } 1.08 \times 10^4 \le Gr_d \le 6.9 \times 10 \\ Nu_{air} = 2.9Gr_d^{1/12}(Gr_h Pr)^{1/4}\ 9.88 \times 10^7 \le Ra_h \le 2.7 \times 10^9 \\ Nu_{air} = 0.47Gr_d^{-1/12}(Gr_h Pr)^{1/4}\ 2.7 \times 10^9 < Ra_h < 2.95 \times 10^{10} \\ \text{Turbulence :} \\ Nu_{air} = 0.13Ra_h^{1/3} \\ Nu_{air} = 0.582 \times 10^{-5} Ra_h^{0.675}\ 70 \le h/d \le 136 \end{array} \tag{18}$$

where:

$$\begin{aligned} Gr_d &= \frac{g\alpha_{air}\Delta T_r d^3}{\gamma_{air}^2} \\ Gr_h &= \frac{g\alpha_{air}\Delta T_r h^3}{\gamma_{air}^2} \\ Ra_h &= Gr_h Pr(Pr_{air} = 0.71) \end{aligned} \tag{19}$$

where Ra_h is the Rayleigh number of the air, dimensionless.

Radial heat in and out of the micro-segment per unit time can be considered equal for the second contact surface.

$$\frac{dq'}{dz} = \frac{dq}{dz} \tag{20}$$

Combined with Equations (4), (12), (13), (16) and (20), the well wall temperature in different heat-transfer environments is:

$$\begin{aligned} \text{Formation}: \quad T_{\text{to}} &= \frac{f(t_D)r_{po}U_T T_f + k_d T_d}{f(t_D)r_{po}U_T + k_d} \\ \text{Seawater}: \quad T_{\text{to}} &= \frac{U_T T_f + h_{air}\omega T_{air}}{U_T + h_{air}\omega} \\ \text{Air}: \quad T_{\text{to}} &= \frac{U_T T_f + h_{sea}\omega T_{sea}}{U_T + h_{sea}\omega} \end{aligned} \tag{21}$$

The relationship between specific enthalpy and temperature and pressure is obtained from a specific heat capacity at a constant pressure and the Joule–Thomson coefficient, according to the first law of thermodynamics:

$$\frac{dh}{dz} = -C_J C_P \frac{dP_f}{dz} + C_p \frac{dT_f}{dz} \tag{22}$$

where C_J is the Joule–Thomson coefficient, K/Pa; C_p is the specific heat capacity of the fluid at constant pressure, $\text{J}/(\text{kg}\cdot\text{K})$.

The pressure gradient formula of gas–liquid two-phase flow is:

$$\frac{dp}{dz} = -\rho_f g\cos\theta - f\frac{\rho_f {v_f}^2}{4r_{pi}} - \rho_f v_f \frac{dv_f}{dz} \tag{23}$$

where ρ_f is the mixed fluid density, kg/m^3; f is the friction coefficient, dimensionless; and r_{pi} is the tubing radius, m.

The gas-phase flow rate in the tubing is:

$$\begin{aligned} v_g &= B_f v_{st} = 1.27 \times 10^{-9}\frac{Q_{st} Z T_f}{r_{pi}^2 P_f} \\ B_f &= 3.447 \times 10^{-4}\frac{Z T_f}{P_f} \end{aligned} \tag{24}$$

where B_f is the volume coefficient of natural gas, K/MPa; v_{st} is the gas-phase flow rate under standard state, m/s; Z is the deviation factor of natural gas, dimensionless; P_f is the pressure of the fluid acting on the tubing, MPa; and Q_{st} is the gas flow under standard state, m^3/d.

The liquid-phase flow rate in the tubing is:

$$v_l = \frac{Q_l}{86400\pi r_{pi}^2} \tag{25}$$

where v_l is the liquid flow rate, m/s; Q_l is the liquid flow of the tubing, m^3/d.

The actual velocity of gas–liquid two-phase is [25]:

$$v_f = v_g + v_l \tag{26}$$

The gas density in the tubing is:

$$\rho_g = \frac{P_f M_f}{ZRT_f} = 3484.28\frac{P_f S_g}{ZT_f} \tag{27}$$

where s_g is the gas relative density of the tubing.

By the volume-weighted calculation of gas–liquid two-phase density, the mixed fluid density in the tubing is:

$$\rho_f = \rho_g \varphi_g + \rho_l(1 - \varphi_g) \tag{28}$$

where ρ_l is the liquid density of the tubing, kg/m^3.

Liquid holdup reflects the ratio of liquid to production fluid on the section of tubing. The calculation formula is:

$$\varphi_g = \frac{v_g}{C_o v_f + v_{gm}} \tag{29}$$

where C_o is a distribution parameter, dimensionless; v_{gm} is the drift speed, m/s.

The calculation formula of the distribution parameters is:

$$C_o = \lambda\left[1 + \left(\frac{v_L}{v_g}\right)^{(\rho_g/\rho_L)^{0.1}}\right] \tag{30}$$

The calculation formula of drift speed is [22]:

$$v_{gm} = 2.9\left[\frac{2gr_{pi}\sigma_p(1+\cos\theta)(\rho_L - \rho_g)}{\rho_L^2}\right]^{0.25}(1.22 + 1.22\sin\theta)^{p_{atm}/p_{sys}} \tag{31}$$

The calculation formula of gas volume flow fraction λ is:

$$\lambda = \frac{v_g}{v_g + v_L} \tag{32}$$

Bring Equations (4), (21) and (23) into Equation (2) to obtain:

$$\beta\left(T_f - T_d\right) = C_p\frac{dT_f}{dz} + g\cos\theta - C_p C_J\frac{dP_f}{dz} + v_f\frac{dv_f}{dz} \tag{33}$$

Fluid temperature at the outlet of a micro-segment in formation is obtained by solving the first-order differential Equation (33):

$$\begin{aligned} T_{fo} &= T_{do} + e^{-\Delta z\beta/C_p}\left(T_{fi} - T_{di}\right) + \tfrac{1}{\beta}\left(1 - e^{-\Delta z\beta/C_p}\right) \\ &\quad \left(C_p C_J\tfrac{P_{fo}-P_{fi}}{\Delta z} - v_f\tfrac{v_{fo}-v_{fi}}{\Delta z} - g\cos\theta + C_p G_h\cos\theta\right) \\ \beta &= \frac{2\pi r_{po}U_T}{\omega\left(r_{po}U_T f(t_D) + k_d\right)} \end{aligned} \tag{34}$$

Fluid temperature at the outlet of a micro-segment in seawater is obtained using the same solution process:

$$\begin{aligned} T_{fo} &= T_{seao} + e^{-\Delta z\beta/C_p}\left(T_{fi} - T_{di}\right) + \tfrac{1}{\beta}\left(1 - e^{-\Delta z\beta/C_p}\right) \\ &\quad \left(C_p C_J\tfrac{P_{fo}-P_{fi}}{\Delta z} - v_f\tfrac{v_{fo}-v_{fi}}{\Delta z} - g\cos\theta + C_p G_h\cos\theta\right) \\ \beta &= \frac{2\pi r_{po}U_T h_{sea}}{\omega(U_T + h_{sea})} \end{aligned} \tag{35}$$

Similarly, fluid temperature at the outlet of a micro-segment in seawater is obtained by:

$$\begin{array}{l} T_{fo} = T_{airo} + e^{-\Delta z\beta/C_p}\left(T_{fi} - T_{airi}\right) + \frac{1}{\beta}\left(1 - e^{-\Delta z\beta/C_p}\right) \\ \left(C_p C_J \frac{P_{fo}-P_{fi}}{\Delta z} - v_f \frac{v_{fo}-v_{fi}}{\Delta z} - g\cos\theta\right) \\ \beta = \frac{2\pi r_{po} U_T h_{air}}{\omega(U_T + h_{air})} \end{array} \quad (36)$$

The wellbore is divided into countless micro-segments, and the outlet temperature of each micro-segment is calculated as the inlet temperature of the next section, which is successively calculated until the wellhead fluid temperature is obtained, and finally, the tubing temperature profile is obtained.

2.2. Wellbore Annulus Temperature Calculation

Figure 3a illustrates the single-layer annulus heat transfer. The heat flowing into the annulus Q_{ano} is equal to the heat flowing out of the annulus per unit of time. Further, the single-layer annulus temperature can be calculated by [26]:

$$T_{an1} = \frac{2\pi T_f U_f r_{po} + \beta\omega T_d}{2\pi U_f r_{po} + \beta\omega} \quad (37)$$

where:

$$U_f = \frac{r_{po}}{r_{pi} h_f} + \frac{r_{po}\ln\left(\frac{r_{po}}{r_{pi}}\right)}{k_p} \quad (38)$$

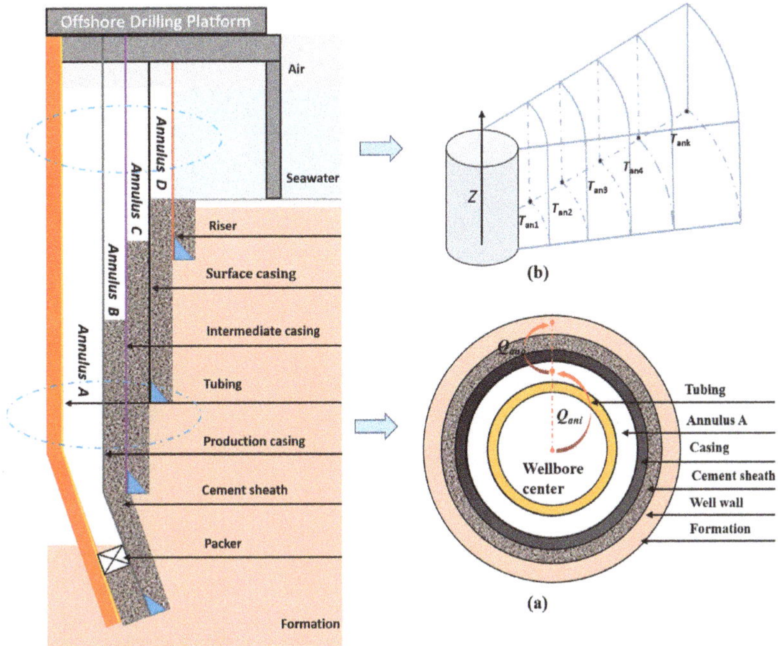

Figure 3. Schematic of multiple annulus heat transfer. (**a**) Schematic of single annulus heat transfer. (**b**) Schematic of multiple annulus heat transfer.

Figure 3b depicts the progress of dividing each annulus into a radial mesh. Establish a coordinate system, and the direction of fluid flow is the positive direction of the z-axis.

According to Fourier's law of thermal conductivity, calculate the annulus temperature. Then, the heat transfer of the kth annulus can be expressed as:

$$2\pi k_{an} r_{ank} h_{anck} \frac{T_{ank-1}^{n+1} - T_{ank}^{n+1}}{\Delta r_{ank}} + 2\pi k_{an} r_{ank+1} h_{anck+1} \frac{T_{ank}^{n+1} - T_{ank+1}^{n+1}}{\Delta r_{ank+1}} = C_{pan} \rho_{an} \pi \Delta r_{ank}^2 h_{anck} \frac{T_{ank}^{n+1} - T_{ank}^{n}}{\Delta t} \tag{39}$$

Where k_{an} is thermal conductivity of fluid in the annulus, $J/(s \cdot m \cdot {}^\circ C)$; Δr_{ank} is the difference between the outer surface radius of the kth annulus and the outer surface radius of the $(k + 1)$th annulus, m.

3. Wellbore Pressure Field Calculation Model

In the process of offshore HPHT gas-well exploitation, a small amount of liquid phase such as water and oil and natural gas are extracted to the wellhead at the same time, so the calculation model of separated flow pressure drop can be adopted. Different from the homogeneous flow model, it considers the slip at the gas–liquid interface.

The pressure drop of a flow pipeline mainly consists of three parts: hydrostatic pressure drop caused by the axial flow of pipeline fluid, pressure drop caused by friction of the pipe wall and gas–liquid interface, and pressure drop caused by acceleration. The contribution of each of these sections to the total pressure drop depends on the flow mode, fluid characteristics, pipe orientation, and pipe diameter.

The hydrostatic pressure drop of gas–liquid two-phase flow is [27]:

$$\left(\frac{dP}{dz}\right)_h = \rho_m g \sin\theta \tag{40}$$

The pressure drop caused by the acceleration of gas–liquid two-phase flow is:

$$\left(\frac{dP}{dz}\right)_a = \rho_m^2 v_f^2 \frac{d}{dz}\left(\frac{x^2}{\varphi_g \rho_g} + \frac{(1-E)^2(1-x)^2}{\beta \rho_l} + \frac{E^2(1-x)^2}{(1-\varphi_g-\beta)\rho_l}\right) \tag{41}$$

β is the correction coefficient of gas holdup, and the calculation formula is:

$$\beta = 1 - \varphi_g - \frac{\varphi_g E(1-x)\rho_g}{x\rho_l} \tag{42}$$

E is the liquid entrainment fraction. Due to the relative motion between the gas and liquid phases, a special phenomenon called entrainment can be observed. The entrainment process is characterized by tiny droplets flowing into a rapidly moving gas phase at the center. Liquid entrainment is due to the obvious shear phenomenon at the gas–liquid interface, which depends on the fluid flow rate, pipe diameter and direction, and the surface tension at the gas–liquid interface. The liquid entrainment fraction E is defined as the ratio of the mass flow rate of droplets entering the gas phase to the total mass flow rate of the liquid phase.

The calculation formula of liquid entrainment fraction is:

$$\frac{E}{1-E} = 0.003 We_c^{1.8} \left(\frac{v_g}{\sqrt{gd}}\right)^{-0.92} Re_g^{-1.24} \left(\frac{\rho_L}{\rho_c}\right)^{0.38} \left(\frac{\mu_L}{\mu_g}\right)^{0.9} \tag{43}$$

where We_c is the Weber number, dimensionless; ρ_c is the fluid core density, kg/m^3.

The calculation formula of the Weber number is:

$$We_c = \frac{\rho v_g^2 d}{\sigma_P} \tag{44}$$

The fluid core density calculation formula is:

$$\rho_c = \rho_g\left[1 + E\left(\frac{v_L}{v_g}\right)\left(\frac{\rho_L}{\rho_g}\right)\right] \tag{45}$$

In this paper, the friction pressure drop is calculated by selecting two-phase frictional multiplier Φ_{lo}^2, which is summarized by Ghajar and Bhagwat through a large number of experiments. The pressure drop caused by interface friction for gas–liquid two-phase flow is:

$$\begin{aligned}\left(\frac{dP}{dz}\right)_f &= \Phi_{LO}^2\left(\frac{dP}{dz}\right)_{LO}\\ \left(\frac{dP}{dz}\right)_{LO} &= \frac{f_l\rho_m^2 v_f^2}{r_{pi}\rho_l}\end{aligned} \tag{46}$$

The calculation formula of two-phase frictional multiplier Φ_{lo}^2 is:

$$\Phi_{lo}^2 = \left\{(1-x)^{0.33}\left[1 + B_1 x\left(Y^2 - 1\right)\right] + B_2 Y^2 x^3\right\}\left(1 + B_3(1-x)^2\right) \tag{47}$$

where B_1, B_2, B_3, Π_1, Π_2, Π_3, ζ, Bo, and Y are the calculation parameters and correction coefficient of two-phase frictional multiplier Φ_{lo}^2(for the specific formula, please refer to the literature [21,25]).

For laminar flow and turbulent flow in a circular pipe, this calculation method can be used to calculate the friction coefficient of single-phase string [28]:

$$f = 2\left[\left(\frac{8}{\mathrm{Re}}\right)^{12} + \frac{1}{(a+b)^{1.5}}\right]^{\frac{1}{12}} \tag{48}$$

where:

$$\begin{aligned} a &= \left\{2.457\ln\left[\frac{1}{(7/\mathrm{Re})^{0.9} + (0.135e/r_{pi})}\right]\right\}^{16}\\ b &= \left(\frac{37530}{\mathrm{Re}}\right)^{16}\end{aligned} \tag{49}$$

where e is the absolute roughness of the tubing wall, m.

4. Example Calculation

4.1. Model Verification

In order to ensure the reliability of the results of the calculation model established in this paper, a gas well in the China sea is used for an example. This case well is a directional well. The basic parameters are shown in Table 1 and Figure 1, and the numerical calculation process is shown in Figure 4.

Table 1. Basic parameters of case well.

Parameter	Value	Parameter	Value
Formation temperature	157 °C	Cement sheath thermal conductivity	$0.958\ \mathrm{J/(s\cdot m\cdot °C)}$
Formation pressure	36 MPa	Formation thermal conductivity	$1.73\ \mathrm{J/(s\cdot m\cdot °C)}$
Mudline temperature	0.0428 °C/m	Annulus liquid thermal conductivity	$0.66\ \mathrm{J/(s\cdot m\cdot °C)}$
Seawater temperature gradient	0.195 °C/m	Annulus liquid density	$1250\ \mathrm{kg/m^3}$
Air temperature	26 °C	Annulus liquid specific heat capacity	$4050\ \mathrm{J/(kg\cdot °C)}$
Formation thermal diffusivity	$1.17\times10^{-6}\ \mathrm{m^2/s}$	Annulus fluid viscosity	$0.051\ \mathrm{mPa\cdot s}$
Gas production	$31.5\times10^4\ \mathrm{m^3/d}$	Tubing thermal conductivity	$43\ \mathrm{J/(s\cdot m\cdot °C)}$
Water production	$30\ \mathrm{m^3/d}$	Casing thermal conductivity	$48\ \mathrm{J/(s\cdot m\cdot °C)}$
Oil production	$140\ \mathrm{m^3/d}$	Well inclination angle	58.78°
Production time	84 h	Seawater depth	120 m

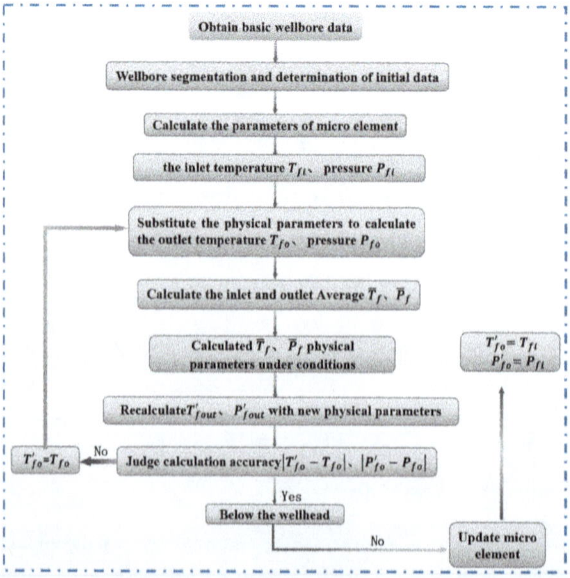

Figure 4. The mathematical model calculation process.

Figure 5 shows the comparison of the wellbore temperature and pressure field results of the case wells, which were obtained by different calculation methods. It can be seen from the figure that the results obtained by the two calculation methods are relatively close, and the variation of the wellbore temperature and pressure curve is the same. However, the wellbore data obtained by the homogeneous flow calculation are larger than that of the separated flow calculation. The wellhead temperature and pressure are selected to compare with the measured data, as shown in Figure 6. The comparison between the calculated results of homogeneous flow and the measured data shows that there is an error of 2.33% in the wellhead temperature and an error of 17.45% in the wellhead pressure. Compared with the measured data, the calculated results of the separated flow in this paper show an error of 0.87% in wellhead temperature and only 2.46% in wellhead pressure. This shows that the calculation model established in this paper has higher accuracy and is more consistent with the measured data, which greatly improves the accuracy of wellbore temperature and pressure.

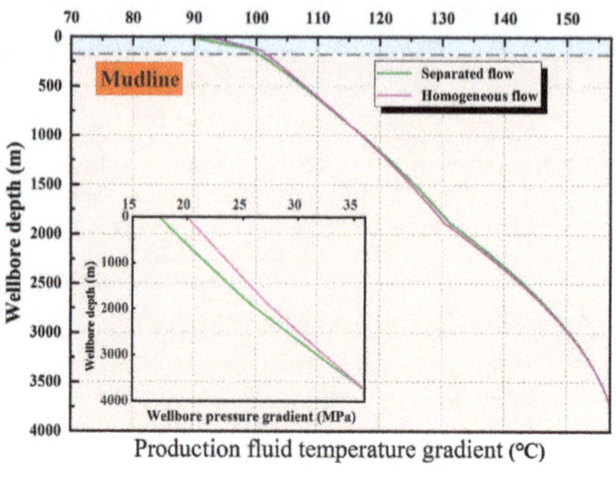

Figure 5. Wellbore temperature pressure comparison.

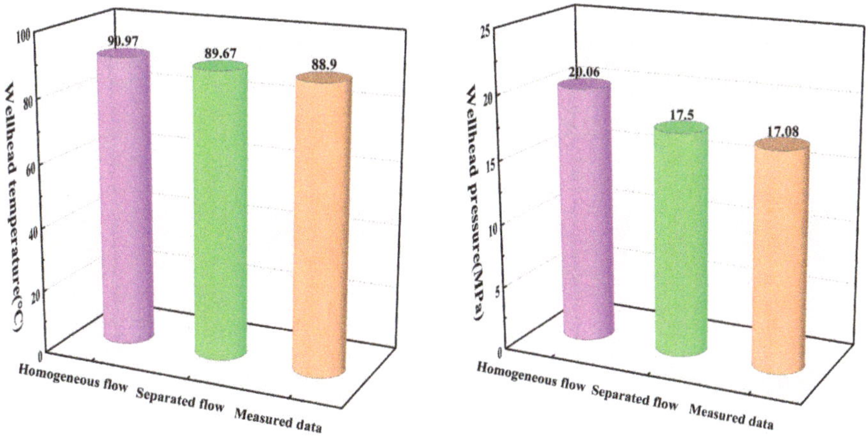

(a) (b)

Figure 6. Comparison between calculation results and measured data. (**a**) Wellhead temperature. (**b**) Wellhead pressure.

4.2. Wellbore Temperature Field

Figure 7 shows the wellbore temperature curve obtained by using the calculation model in this paper. We can see from the drawings: (1) in the formation section, the fluid temperature in the tubing decreases slowly with the decrease in well depth; (2) in the seawater–air section, forced convection heat exchange between seawater–air and wellbore is stronger than that between wellbore and formation. Further, production fluid temperature decreases significantly faster. The annulus temperature of the formation section declines gradually with depth decreases. However, there is a sharp change near the mudline, and the annulus temperature drops rapidly. After entering the seawater section completely, the temperature of each annulus gradually decreases with the decrease in seawater depth, and the temperatures between annulus temperatures are close to each other. The maximum temperature difference is only about 7 °C. As a result, the impact of the seawater section and air section on wellbore temperature cannot be ignored for offshore wells.

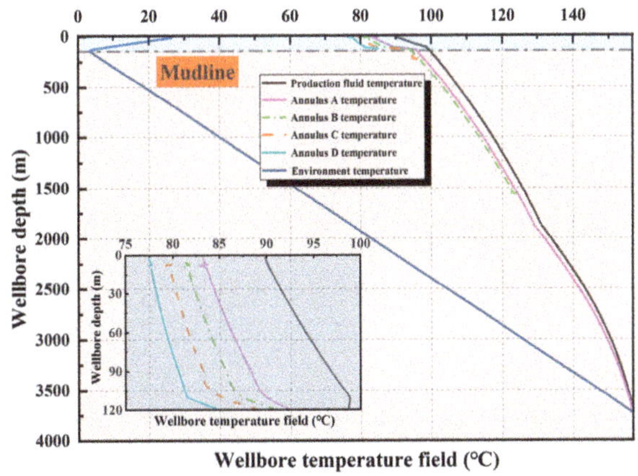

Figure 7. Wellbore temperature field.

4.3. Gas–Liquid Ratio Production Fluid

Figure 8 depicts the influence of gas–liquid ratio on wellbore temperature and pressure under the conditions of gas production of $31.5 \times 10^4 \cdot m^3/d$, production time of 84 h, and the same oil–water ratio of production fluid (14:3). As can be seen from the figure, since the specific heat capacity of liquid phase is much larger than that of gas phase, the wellbore temperature becomes increasingly high with the increase in liquid oil–water content in the mixed fluid. However, the increase in liquid content will increase the density of gas–liquid mixture, which makes hydrostatic pressure drop and friction pressure drop increase. As a result, the wellbore pressure drop decreases continuously, showing a trend of wellbore pressure decreasing with the decrease in the gas–liquid ratio (the increase in liquid content). It can also be seen from Equation (33) that wellbore temperature and pressure change in opposite trends. Therefore, by controlling the gas–liquid ratio reasonably, the wellbore pressure can be adjusted without changing the production process, so as to avoid the risk of increasing casing failure due to the high wellbore pressure.

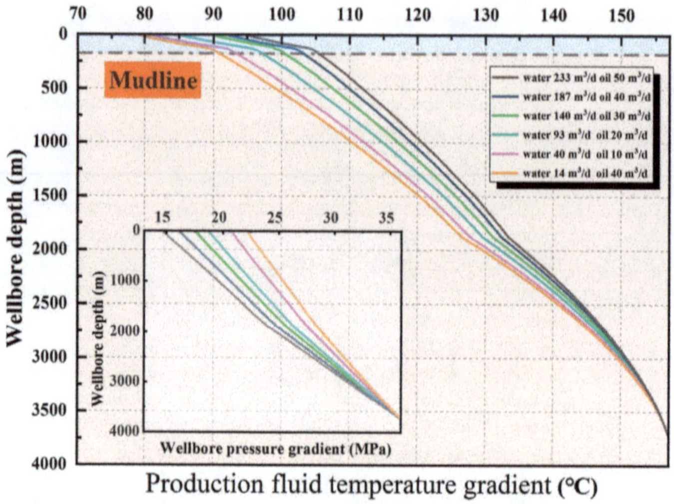

Figure 8. The influence of gas–liquid ratio on wellbore temperature and pressure.

4.4. Production Conditions

Figure 9 shows the influence of gas production on wellbore temperature and pressure. Only gas production is changed, and other conditions are consistent with the case well. As can be seen from the figure, the gas production has a significant impact on the wellbore temperature. When the gas production rises from $10 \times 10^4 \cdot m^3/d$ to $60 \times 10^4 \cdot m^3/d$, the wellhead temperature rises from 63 °C to 99 °C, which is increased by 1.6 times. However, the wellbore pressure increases first and then decreases with the increase in production because of the mutual influence of friction pressure drop and hydrostatic pressure drop. Figure 10 depicts the pressure drop in the micro-segment of the wellhead. Since the accelerated pressure drop is very small, only the friction pressure drop and hydrostatic pressure drop are shown. As can be seen from the figure, when the gas production is small, the liquid phase in the wellbore occupies a relatively high proportion, and the mixed fluid density is relatively large, resulting in a rapid hydrostatic pressure drop, and it accounts for a large proportion of the total pressure drop. However, with the increase in gas production, the fluid mixing density decreases, and the hydrostatic pressure drop slows down. At the same time, the friction at the gas–liquid interface gradually increases with the increase in the gas phase, leading to a significant decrease in the wellbore friction pressure drop, and the final wellbore pressure increases first and then decreases.

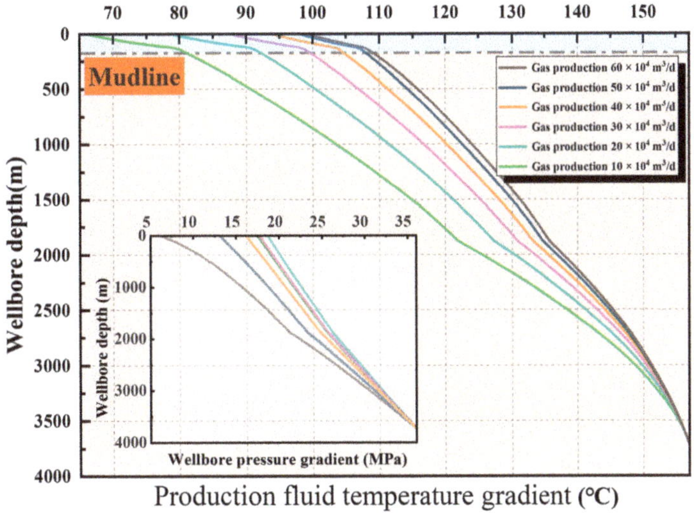

Figure 9. The influence of gas production on wellbore temperature and pressure.

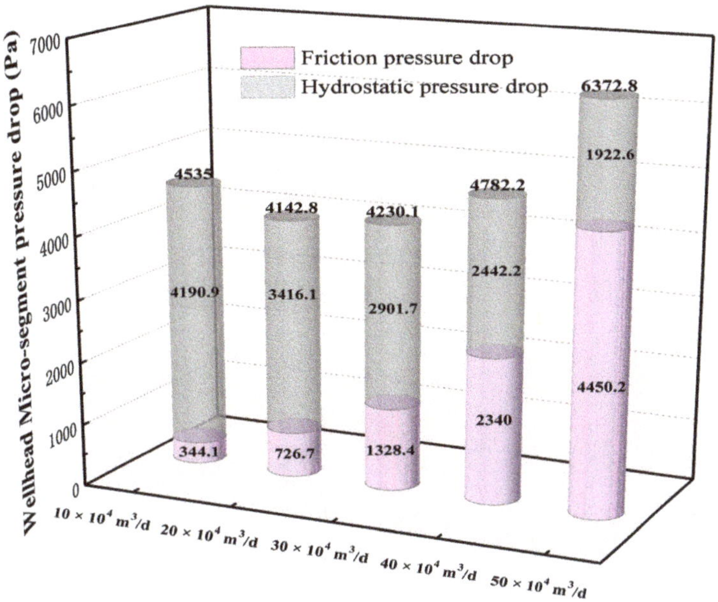

Figure 10. The pressure drop in the micro-segment of the wellhead.

Figure 11 shows the influence of production time on wellbore temperature and pressure under the production conditions of the case well. It can be seen from the figure that the wellhead temperature increased from 91.6 °C to 96.4 °C during the period of 10 d to 100 d, while the wellhead temperature only increased by 1.8 °C during the period of 101 d to 300 d. This indicates that wellbore temperature tends to be stable with the increase in production time. The wellbore pressure does not change with the increase in production time in the 10~300 d production range.

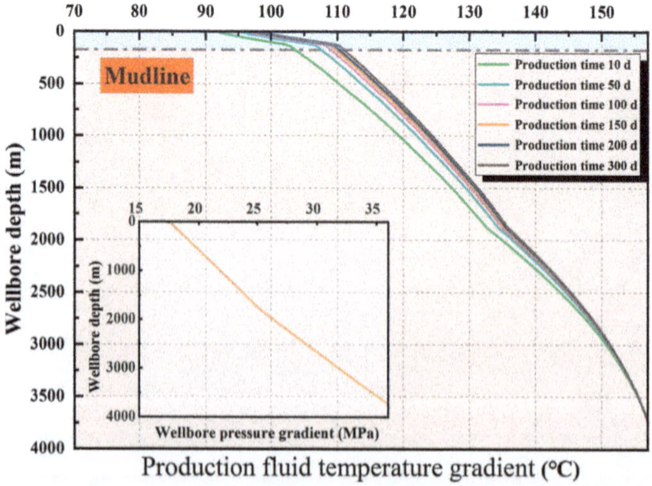

Figure 11. The influence of production time on wellbore temperature and pressure.

Figure 12 compares the variation trend of the wellhead temperature changing with the production time that was obtained by the homogeneous method model and the separated method. As can be seen from the figure, it takes 28 days for the wellhead temperature to reach 95 °C in the model based on homogeneous phase flow, which is much faster than 48 days in the separated model. Therefore, the improved model can provide a more accurate estimate of the time taken to reach the rated wellhead temperature and accurate theoretical support for the rational formulation of the production plan after the well opening, so as to avoid excessive restrictions on the initial production rate.

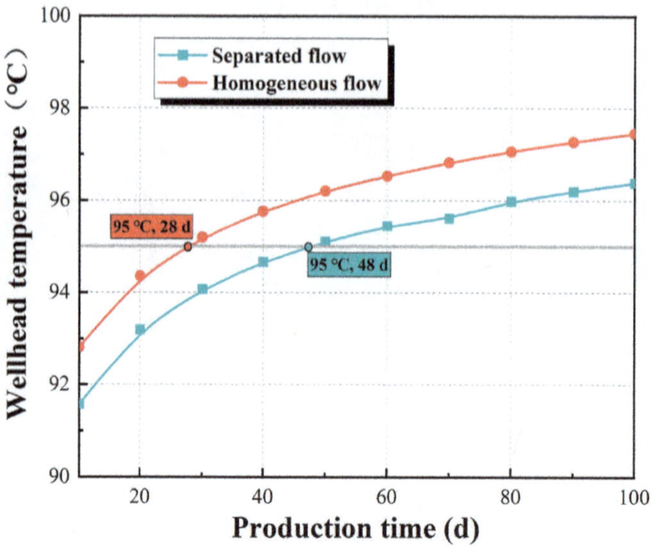

Figure 12. The influence of production time on wellhead temperature.

5. Conclusions

Based on the gas–liquid two-phase separated method, an improved calculation model of wellbore temperature and pressure for gas–liquid two-phase flow in offshore gas wells is established in this paper. Further, the influencing factors of wellbore temperature and pressure are analyzed. The following conclusions are obtained:

1. Although the results of homogeneous flow and separated phase flow have the same trend, the errors of wellhead temperature and pressure obtained by homogeneous flow calculation are large compared with measured data, which are 2.33% and 17.45%. However, the errors of the improved model are only 0.87% and 2.46%, which greatly improves the accuracy of wellbore temperature and pressure calculation.
2. Forced convection heat exchange between seawater–air and wellbore is stronger than that between wellbore and formation. As a result, the temperature drop rate of the tubing in the seawater and air sections is significantly accelerated. Therefore, the influence of the seawater and air sections on the wellbore temperature of offshore gas wells should be fully considered.
3. Since the specific heat capacity of liquid phase is much larger than that of gas phase, the wellbore temperature becomes increasingly high as the gas–liquid ratio of the product decreases (liquid-phase content increases). However, it also leads to the increase in hydrostatic pressure and friction pressure loss as the mixed fluid density rises, causing the wellbore pressure decrease. Based on this trend, the liquid content of the product should be controlled in a reasonable range in order to avoid the risk of casing failure due to high wellbore pressure.
4. Gas production is the key factor affecting wellbore temperature. When the gas production rises from $10 \times 10^4 \cdot m^3/d$ to $60 \times 10^4 \cdot m^3/d$, the wellhead temperature rises from 63 °C to 99 °C. However, due to the mutual influence of friction pressure drop and hydrostatic pressure drop, wellbore pressure increases first and then decreases with the increase in gas production. It takes 28 days for the wellhead temperature to reach 95 °C in the model based on homogeneous-phase flow, which is much faster than 48 days in the separated model. Therefore, the improved model can provide a more accurate estimate of the time to reach the rated wellhead temperature and accurate theoretical support for the rational formulation of the production plan after the well opening, so as to avoid excessive restrictions on the initial production rate.

Author Contributions: Conceptualization, J.J. and H.S.; methodology, J.J.; software, H.S.; validation, X.Z.; formal analysis, H.S., Y.H. and Y.T.; investigation, H.S.; resources, J.J.; data curation, J.J., H.S., X.Z., Y.H. and Y.T.; writing—original draft preparation, H.S.; writing—review and editing, J.J.; visualization, H.S.; supervision, X.Z.; project administration, J.J.; funding acquisition, J.J. All authors have read and agreed to the published version of the manuscript.

Funding: This research was funded by the Youth Program of the National Natural Science Foundation of China, grant number 52004230; the Open Fund of the State Key Laboratory of Oil and Gas Reservoir Geology and Exploitation (Southwest Petroleum University), grant number PLN201921; and the scientific research starting project of SWPU, grant number No.2019QHZ008.

Data Availability Statement: Not applicable.

Conflicts of Interest: The authors declare no conflict of interest.

Abbreviations

HPHT	High temperature and high pressure
TOC	Top of cement

References

1. Luo, M.; Zhu, H.Y.; Liu, Q.Y.; Wang, Z.W.; Li, Y.J.; Han, C.; Zhang, C. A V-cutter PDC bit suitable for ultra-HTHP plastic mudstones. *Nat. Gas Ind.* **2021**, *41*, 97–106.
2. Xu, M.G.; Fan, C.W.; Zhang, D.N.; Hu, G.W.; Tan, J.C. Formation condition and hydrocarbon accumulation model in Ledong 01 Gas Reservoir of super high temperature and high pressure in the Yinggehai Basin. *Nat. Gas Ind.* **2021**, *41*, 43–53.

3. Chen, H.; Yang, C.; Datta-Gupta, A.; Zhang, J.; Chen, L.; Liu, L.; Chen, B.; Cui, X.; Shi, F.; Bahar, A. Fracture inference and optimal well placement using a multiscale history matching in a HPHT tight gas reservoir, Tarim Basin, China. *Upstream Oil Gas Technol.* **2020**, *2*, 100002. [CrossRef]
4. Wang, Y.H.; Hao, Y.Z.; Li, F.; Lu, F.F.; Deng, K.H.; Lin, Y.H. Failure mechanism of wellbore integrity for HTHP gas wells, Shunnan block, Tarim Basin. *Nat. Gas Explor. Dev.* **2019**, *42*, 86–89.
5. Xie, H.; Zhang, G.W.; Liu, X.Q.; Chen, G.M.; Chang, Y.J. Mechanism of Christmas tree rise in offshore oil and gas wells and the related calculation method. *Nat. Gas Ind.* **2013**, *33*, 80–84.
6. Ramey, H.J. Wellbore Heat Transmission. *J. Pet. Technol.* **1962**, *4*, 427–435. [CrossRef]
7. Shiu, K.C.; Begges, H. Predicting Temperatures in Flowing Oil Wells. *J. Energy Resour. Technol.* **1980**, *102*, 2–11. [CrossRef]
8. Hasan, A.R.; Kabir, C.S. Heat transfer during two phase flow in wellbores. In Proceedings of the 66th Annual Technical Conference and ExhiMion of the Society of Petroleum Engineers, Dallas, TX, USA, 6–9 October 1991.
9. Spindler, R. Analytical Models for Wellbore-Temperature Distribution. *SPE J.* **2011**, *1*, 125–133. [CrossRef]
10. Wu, Z.; Xu, J.; Wang, X.; Chen, K.; Li, X.; Zhao, X. Predicting Temperature and Pressure in High-Temperature–High-Pressure Gas Wells. *Petrol. Sci. Technol.* **2011**, *29*, 132–148. [CrossRef]
11. Xu, J.; Hu, J.; Wu, Z.; Wang, S.; Qi, B. Prediction of Temperature and Pressure Distribution in HTHP Injection Gas Wells with Thermal Effect of Wellbore. *Petrol. Sci. Technol.* **2013**, *31*, 1423–1438. [CrossRef]
12. Martínez-Cuenca, R.; Mondragón, R.; Hernández, L.; Segarra, C.; Jarque, J.C.; Hibiki, T.; Juliá, J.E. Forced-convective heat-transfer coefficient and pressure drop of water-based nanofluids in a horizontal pipe. *Appl. Therm. Eng.* **2016**, *98*, 841–849. [CrossRef]
13. Yin, B.; Li, X.; Qi, M. Study of Fluid Phase Behavior and Pressure Calculation Methods of a Condensate Gas Well. *Petrol. Sci. Technol.* **2013**, *31*, 1595–1607. [CrossRef]
14. Gao, Y.; Cui, Y.; Xu, B.; Sun, B.; Zhao, X.; Li, H.; Chen, L. Two phase flow heat transfer analysis at different flow patterns in the wellbore. *Appl. Therm. Eng.* **2017**, *117*, 544–552. [CrossRef]
15. Hou, Z.; Yan, T.; Li, Z.; Feng, J.; Sun, S.; Yuan, Y. Temperature prediction of two phase flow in wellbore using modified heat transfer model: An experimental analysis. *Appl. Therm. Eng.* **2019**, *149*, 54–61. [CrossRef]
16. Song, X.C.; Wei, L.G.; He, L.; Feng, J.; Sun, S.; Gou, Y.B. Transient temperature field model of shut-in offshore wells. *J. China Univ. Pet.* **2013**, *37*, 94–99.
17. Ma, Y.Q. *Study on Law of Gas-Liquid Two-Phase Flow Heat Transfer in Deep-Water Wellbore*; China University of Petroleum: Qingdao, China, 2010.
18. Sivaramkrishnan, K.; Huang, B.; Jana, A.K. Predicting wellbore dynamics in a steam-assisted gravity drainage system: Numeric and semi-analytic model, and validation. *Appl. Therm. Eng.* **2015**, *91*, 679–686. [CrossRef]
19. Hasan, A.R.; Jang, M. An analytic model for computing the countercurrent flow of heat in tubing and annulus system and its application: Jet pump. *J. Petrol. Sci. Eng.* **2021**, *203*, 108492. [CrossRef]
20. Zhang, Z.; Wang, H. Effect of thermal expansion annulus pressure on cement sheath mechanical integrity in HPHT gas wells. *Appl. Therm. Eng.* **2017**, *118*, 600–611. [CrossRef]
21. Hasan, A.R.; Lzgec, B.; Kabir, C.S. Sustained production by managing annular-pressure buildup. *SPE J.* **2010**, *25*, 195–202.
22. Ghajar, A.J. *Single-and Two-Phase Flow Pressure Drop and Heat Transfer in Tubes*; Springer: Berlin/Heidelberg, Germany, 2021.
23. Izgec, B.; Hasan, A.R.; Lin, D.; Kabir, C.S. Flow-rate estimation from wellhead-pressure and temperature data. In Proceedings of the SPE Annual Technical Conference and Exhibition (ATCE 2008), Denver, CO, USA, 21–24 September 2008; Volume 25, pp. 31–39.
24. Popiel, C.O. Free Convection Heat Transfer from Vertical Slender Cylinders: A Review. *Heat Transf. Eng.* **2008**, *29*, 521–536. [CrossRef]
25. Hasan, A.R.; Kabir, C.S. A Simple Model for Annular Two-Phase Flow in Wellbores. In Proceedings of the Annual Technical Conference and Exhibition of the Society of Petroleum Engineers 2005 (ATCE 2005), Dallas, TX, USA, 9–12 October 2005; pp. 168–175.
26. Zhang, Z.; Wang, H. Sealed annulus thermal expansion pressure mechanical calculation method and application among multiple packers in HPHT gas wells. *J. Nat. Gas Sci. Eng.* **2016**, *31*, 692–702. [CrossRef]
27. Bhagwat, S.M.; Ghajar, A.J. Experimental investigation of non-boiling gas-liquid two phase flow in upward inclined pipes. *Exp. Therm. Fluid Sci.* **2016**, *79*, 301–318. [CrossRef]
28. Bhagwat, S.M. *Experimental Measurements and Modeling of Void Fraction and Pressure Drop in Upward and Downward Inclined Non-Boiling Gas-Liquid Two Phase Flow*; Oklahoma State University: Stillwater, OK, USA, 2015.

processes

Article

Effects of Make-Up Torque on the Sealability of Sphere-Type Premium Connection for Tubing and Casing Strings

Bin Yang [1], Honglin Xu [1,*], Shilin Xiang [2], Zhi Zhang [2], Kanhua Su [1] and Yan Yang [2]

[1] School of Petroleum and Natural Gas Engineering, Chongqing University of Science & Technology, Chongqing 401331, China
[2] State Key Laboratory of Oil & Gas Reservoir Geology and Exploitation, Southwest Petroleum University, Chengdu 610500, China
* Correspondence: xuhlaca1986_jy@163.com

Abstract: The present investigations on sealability evaluation for tubing and casing premium connections depend on the FEM with testing. This paper proposed a theoretical model to evaluate the sealability of a sphere-type premium connection based on make-up torque, which combines Hertz contact pressure and the von Mises yield criterion for calculating elastic–plastic contact pressure distribution on sealing interface and adopts the gas sealing criterion obtained from Murtagian's experimental results for deducing gas sealing capacity. With the proposed model, the effects of additional make-up torque from the sealing interface on the sealing contact pressure distribution and key sealability parameters, including contact width, yield width, average contact pressure and gas sealing capacity, were analyzed and compared. The results show that additional make-up torque from the sealing interface closely influenced sealability parameters' variation and gas sealing capacity. The gas sealing index based on the sealing contact energy theory should be recommended for sealability evaluation other than average contact pressure on the sealing interface. For improving gas sealability, make-up torque should be controlled accurately for ensuring enough average contact pressure and contact width but a proper yield width, and a lager sphere radius should be selected for reducing the risk of yield sticking.

Keywords: sphere-type premium connection; sealability evaluation; elastic–plastic contact; contact pressure; make-up torque

Citation: Yang, B.; Xu, H.; Xiang, S.; Zhang, Z.; Su, K.; Yang, Y. Effects of Make-Up Torque on the Sealability of Sphere-Type Premium Connection for Tubing and Casing Strings. *Processes* **2023**, *11*, 256. https://doi.org/10.3390/pr11010256

Academic Editor: Yidong Cai

Received: 19 December 2022
Revised: 3 January 2023
Accepted: 10 January 2023
Published: 12 January 2023

1. Introduction

Premium connection has become one of the key technologies for wellbore integrity and safety in harsh serving wells, such as high-pressure high-temperature (HPHT) gas wells, thermal recovery wells and shale gas wells [1,2]. Differently from the API casing and tubing connection that only depends on thread interference contact sealing and a high-quality thread compound filled in the leakage path between the pin and coupling threads to meet a lower requirement of gas sealability, premium connection adopts a special metal-to-metal seal structure to attain high gas leakage resistance and a torque shoulder to control make-up torque and provide additional sealability, and it also usually adopts American Petroleum Institute (API) buttress threads or modified buttress threads for high joint strength, shown in Figure 1.

The general metal-to-metal seal structures for a premium connection can be mainly divided into four types, namely cone-to-cone seal, sphere-to-cone seal, sphere-to-cylinder seal and sphere-to-sphere seal; the latter three are collectively referred to as sphere-type seals, shown in Figure 2. Compared with a cone-to-cone seal, a sphere-type seal is a typical non-coordinated contact static seal, which tends to have higher contact pressure distribution and also yield more easily on a sealing interface. A certain yield width on the sealing interface can be conducive to blocking the micro leakage channel and forming an excellent gas seal rapidly, while a relevant larger yield width can easily lead to yield bonding and damage on the sealing surface. Consequently, accurate and efficient acquirement of the elastic–plastic contact pressure distribution on the sealing interface gives great importance to sealability evaluation for sphere-type premium connections.

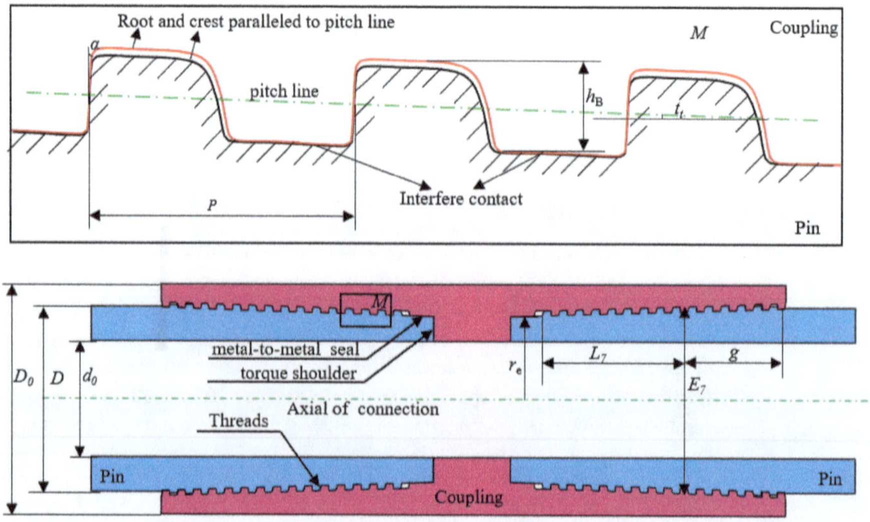

Figure 1. The schematic diagram of a premium connection.

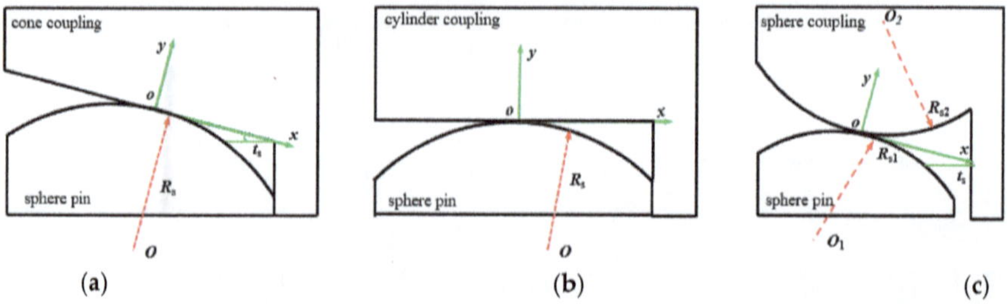

Figure 2. Metal-to-metal seal structures of a sphere-type premium connection, the sphere-to-cone type is for (**a**), the sphere-to-cylinder type is for (**b**) and the sphere-to-sphere is for (**c**).

Up to now, the sealability of premium connections was investigated in many previous references, whose methods can mainly be divided into three categories. The first category is based on experimental testing. Payne and Schwind proposed an international standard for casing/tubing connection testing [3]. Based on those testing procedures, a lot of leak-proof tests have been carried out. And new sealing structures have been developed for premium connections. Salzano et al. investigated the sealing mechanism and structural integrity of a premium connection under mechanical and pressure loads and exposed to extremely low temperature in the experiment [4]. Hamilton et al. presented adopting ultrasonic techniques to accurately examine seal surface contact pressure in premium connections [5]. Ernens et al. adopted an experimental setup and a stochastic numerical sealing model to describe the mechanisms of sealing metal-to-metal seals [6]. Keita et al. investigated the influence of grease on high-pressure gas tightness by metal-to-metal seals of premium connections [7]. There is no doubt that implementing a full-scale experiment is the most direct and effective mothed for sealability evaluation of a premium connection; however, it is expensive and time-consuming. The second category is the theoretical method, which is on the basis of macroscopic or microscopic contact mechanics. The former gets the sealing contact pressure distribution first and then uses average sealing contact pressure or a sealing index integrated by sealing width and sealing pressure to evaluate the sealability [8], while the latter can calculate the gas leakage rate of a premium connection directly [9]. Although the theoretical method is convenient for usage, the stringent assumptions of the elastic material and contact pressure make its application limited. The third method is the finite element method (FEM), which is an effective way to deal with material nonlinearity, contact nonlinearity and complex load conditions for premium connections and has been taken as the most popular and necessary tool for new product development [10–13]. Chen et al. investigated the sealing mechanism of tubing and casing premium threaded connections under complex loads [14]. Kim et al. analyzed the effects of stabbing flank angle, upper stabbing flank corner radius on the von Mises stress of premium connection and presented the design criteria [15]. Zhang et al. adopted the viscoelastic finite element model to predict the relaxation of contact pressure on the premium connections' sealing surface versus time under different temperatures [16]. Yu et al. have analyzed the effect of energy dissipation on the sealing surface of premium connections with a microslip shear layer model [17]. Dou et al. have carried out a sealing ability simulation for a premium connection based on ISO 13679 CAL IV tests with FEM [18]. Generally, a preset interference between the sealing interfaces is made to produce sealing contact pressure. However, the preset interference on the sealing interface in FEM models differs mostly from the actual produced interference because of improper operation during make-up torque, thus leading to inaccurate simulated results of sealing contact pressure. As a result, it is hard to evaluate the sealability exactly and design the sealing parameters optimally. Xu and Yang have theoretically investigated the effect of make-up torque on the joint strength of premium connections but not focused on its sealability [19]. In fact, the sealing contact pressure distribution depends closely on the real additional make-up torque from the sealing interface, so we can calculate the sealing contact pressure on the basis of the real make-up toque curve.

In this paper, a theoretical model for evaluating the sealability of a sphere-type premium connection is firstly proposed based on the additional make-up torque from the sealing interface, which combines Hertz contact pressure and the von Mises yield criterion for calculating elastic–plastic contact pressure distribution on the sealing interface and adopts the gas sealing criterion obtained from Murtagian's experimental results [20] for deducing gas sealing capacity. Then, based on the proposed model, the effects of additional make-up torque from the sealing interface on the contact pressure distribution and key sealing parameters, including contact width, yield width, average contact pressure and gas sealing capacity for sphere-type premium connection are analyzed and also compared for three kinds of sphere-type premium connections. In addition, some measures for ensuring the gas sealability of sphere-type premium connections for tubing and casing strings are proposed; those results can provide a useful guidance for sealability evaluation, parameter design and make-up torque control for sphere-type premium connections.

2. Model Development

The sealing structure of a sphere-type premium connection belongs to a typical non-coordinated radial interference contact seal. According to the equivalent principle of contact mechanics [21], sphere-to-sphere contact can be regarded as sphere-of-equivalent-radius-to-cone contact, and sphere-to-cylinder contact can also be taken as sphere-to-cone with zero taper. Consequently, sphere-to-cone contact was selected as an archetype to develop the sealability evaluation model for a sphere-type premium connection. Figure 3 shows the sealing formation process for a sphere-to-cone contact premium connection. An initial circular line contact lies between the sphere pin and the cone coupling when the sealing surface starts to contact at the point of O. With make-up torque, normal interference is produced between the sphere pin and cone coupling, thus leading to normal contact pressure of $p_{sN}(x)$ on the contact interface. When make-up torque operation completes, a circumferential sealing interface with a certain contact width of $2w_s$ and yield width of $2w_y$

can form to meet the required gas sealing capacity for the premium connection. Figure 3a,b show, respectively, the initial sealing contact and final sealing contact for a sphere-to-cone premium connection.

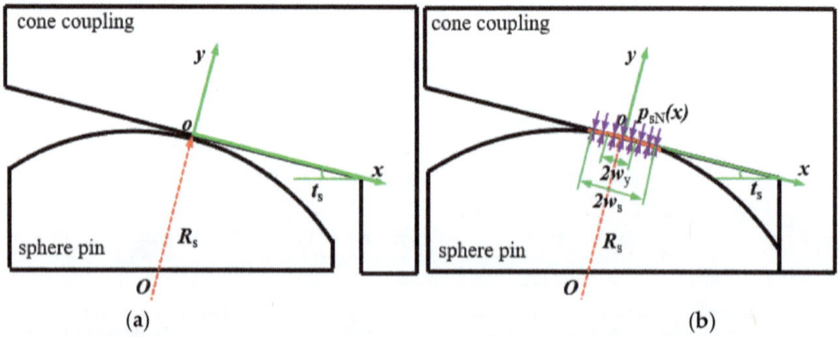

Figure 3. Schematic of sealing formation process for a sphere-to-cone contact premium connection, initial sealing contact is for (**a**) and final sealing contact is for (**b**).

To evaluate the sealability of a premium connection, the first key is to acquire the elastic–plastic contact pressure distribution $p_{sN}(x)$ on the sealing interface and to get the main sealability parameters, including average contact pressure, contact width and yield width on the sealing interface, as well as gas sealing capacity for the sphere-type premium connection. During practical making-up of torque for the premium connection, the contact pressure on the sealing interface brings out additional make-up torque, which can be easily accessed by the make-up torque curve. Consequently, it is more accurate and convenient or calculating the $p_{sN}(x)$ based on the real make-up torque data.

2.1. The Model of Additional Make-Up Torque from Sealing Interface for Sphere-Type Premium Connection

Figure 4 shows the typical curve between make-up torque and turns for premium connection. It can be easily found that the final make-up torque T_D involves mainly four parts, namely the leading torque T_A, the thread interference torque T_{ti}, the additional make-up torque from sealing interface T_{se} and the additional make-up torque from shoulder T_{sh}, as follows [22]:

$$T_D = T_A + T_{ti} + T_{se} + T_{sh} \tag{1}$$

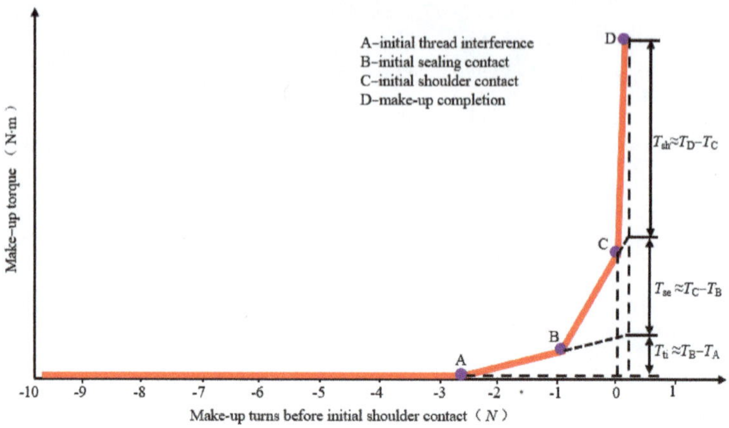

Figure 4. The typical curve between make-up torque and turns for a premium connection.

Additional make-up torque from sealing interface T_{se} can be relatively accurately calculated by the torque at the initial contact on sealing interface T_B and that on shoulder T_C.

$$T_{se} = T_C - T_B \tag{2}$$

To acquire the elastic–plastic contact pressure distribution on a sealing interface, a model for additional make-up torque from the sealing interface should be established first. We regard the real sealing interface as a circular cone surface with its generatrix length of approximately $2w_s$, as in Figure 3b. Obviously, an additional axial pre-tightening force will generate on the sealing interface and lead to additional thread torque T_{set}. Meanwhile, additional friction torque on sealing interface T_{sef} will also be produced. Consequently, the total additional make-up torque from the sealing interface is the sum of the additional thread torque and additional friction torque.

$$T_{se} = T_{set} + T_{sef} \tag{3}$$

2.1.1. Additional Thread Torque from Sealing Interface T_{set}

The relevant structure parameters for a premium connection are shown in Figure 1. As cone taper t_s and actual contact width $2w_s$ are both very small in practice, sealing radius r_s can be regarded as equal, and the additional axial pre-tightening force F_{se} from the sealing interface is as follows:

$$F_{se} = \int_{-w_s}^{w_s} p_{sN}(x) \times 2\pi r_s dx \times \frac{t_s}{\sqrt{4+t_s^2}} \tag{4}$$

Further, according to the rotary shoulder connection with Farr's formula [23], the additional thread torque from the sealing interface can be calculated by:

$$T_{set} = \frac{F_{se}}{1000}\left(\frac{P}{2\pi} + \frac{\mu_t R_t}{\cos\alpha}\right) \tag{5}$$

In Equation (5), the equivalent moment arm for thread friction torque is as follows:

$$R_t = \frac{2E_7 + (g - L_7)t_t}{4} \tag{6}$$

2.1.2. Additional Friction Torque from Sealing Interface T_{sef}

Additional friction torque from the sealing interface can be easily obtained by the following integration:

$$T_{sef} = \frac{1}{1000}\int_{-w_s}^{w_s} \mu_s p_{sN}(x) \times 2\pi r_s^2 dx \tag{7}$$

2.1.3. Total Additional Make-Up Torque from Sealing Interface T_{se}

By combining Equation (3) to Equation (7), the total additional make-up torque from the sealing interface can be obtained as follows:

$$T_{se} = \left\{\frac{2\pi r_s t_s}{1000\sqrt{4+t_s^2}}\left[\frac{P}{2\pi} + \frac{\mu_t}{\cos\alpha}\left(\frac{2E_7 + (g-L_7)t_t}{4}\right)\right] + \frac{2\pi\mu_s r_s^2}{1000}\right\}\int_{-w_s}^{w_s} p_{sN}(x)dx \tag{8}$$

In Equation (8), we let:

$$\zeta = \frac{2\pi r_s t_s}{1000\sqrt{4+t_s^2}}\left[\frac{P}{2\pi} + \frac{\mu_t}{\cos\alpha}\left(\frac{2E_7 + (g-L_7)t_t}{4}\right)\right] + \frac{2\pi\mu_s r_s^2}{1000}$$

and then the additional make-up torque from the sealing interface can be further written as:

$$T_{se} = \zeta \int_{-w_s}^{w_s} p_{sN}(x)dx \tag{9}$$

When the thread parameters, sealing radius, cone taper and friction coefficient for the premium connection are all constant, the coefficient of ζ is also unchangeable. Apparently, it can be seen from Equation (9) that T_{se} is very directly relative to the elastic–plastic contact pressure distribution p_{sN} (x). Consequently, when T_{se} is known, p_{sN} (x) can be calculated with Equation (9).

2.2. The Model of Elastic–Plastic Contact Pressure on Sealing Interface for Sphere-Type Premium Connection

2.2.1. Hertz Elastic Contact Pressure

According to the Hertz theory of contact mechanics [21], the elastic contact pressure on a sealing interface for a sphere-type premium connection with sphere radius R_s and equivalent elastic modulus E^* can be expressed as follows:

$$p_{sN}(x) = \frac{E^*}{2R_s}\sqrt{w_s^2 - x^2}\ (-w_s \leq x \leq w_s, 0 < 2w_s << R_s) \tag{10}$$

For the plane strain or axisymmetric problems, the equivalent elastic modulus E^* can be calculated by the following equation:

$$\frac{1}{E^*} = \frac{1-\nu_p^2}{E_p} + \frac{1-\nu_c^2}{E_c} \tag{11}$$

For contact of sphere with radius R_{s1} to sphere with radius R_{s2}, the sphere radius equals equivalent sphere radius R_{se}:

$$R_s = R_{se} = \frac{R_{s1}R_{s2}}{R_{s1} + R_{s2}} \tag{12}$$

2.2.2. Initial Yield Condition on Sealing Interface

Based on Hertz contact pressure state and the von Mises yield criterion, Green deduced the critical half contact width w_s^c with initial yield on the sealing interface for a rigid plane loaded by an elastic–plastic cylinder, as follows [24]:

$$w_s^c = 2R_s\frac{C_1 S_y}{E^*} \tag{13}$$

In Equation (13), C_1S_y denotes yield strength of the sealing contact pair, and the coefficient of C_1 can be calculated with Poisson's ratio ν as follows:

$$C_1 = \begin{cases} \frac{1}{\sqrt{1+4(\nu-1)\nu}} & (\nu \leq 0.1938) \\ 1.164 + 2.975\nu - 2.906\nu^2 & (\nu > 0.1938) \end{cases} \tag{14}$$

Obviously, for contact between different materials, C_1S_y equals to:

$$C_1S_y = \min\left[C_{1p}S_{yp}, C_{1c}S_{yc}\right] \tag{15}$$

2.2.3. Elastic–Plastic Contact Pressure on Sealing Interface

For sphere-type premium connections, after the initial yield occurs at the maximum contact pressure point, the sealing interface turns into an elastic–plastic contact state along with the continued make-up operation. In the plastic contact zone, we assume that the yield width is $2w_y$ and the contact pressure still equals that when initial yield occurs. In another elastic contact zone, the contact pressure equals to the Hertz contact pressure. By combing Equation (10) with Equation (13), the elastic–plastic contact pressure on the sealing interface can be described as follows [25]:

$$p_{\mathrm{sN}}(x) = \begin{cases} C_1 S_{\mathrm{y}} & 0 \le |x| \le w_{\mathrm{y}} \\ \frac{E^*}{2R_{\mathrm{s}}}\sqrt{w_{\mathrm{s}}^2 - x^2} & w_{\mathrm{y}} \le |x| \le |w_{\mathrm{s}}| \end{cases} \tag{16}$$

By submitting Equation (16) into Equation (9), the additional make-up torque from the sealing interface can be further written as:

$$\begin{aligned} T_{\mathrm{se}} &= \zeta \int_{-w_{\mathrm{s}}}^{w_{\mathrm{s}}} p_{\mathrm{sN}}(x)dx = \zeta \int_{-w_{\mathrm{y}}}^{w_{\mathrm{y}}} C_1 S_{\mathrm{y}} dx + 2\zeta \int_{w_{\mathrm{y}}}^{w_{\mathrm{s}}} \frac{E^*}{2R_{\mathrm{s}}}\sqrt{w_{\mathrm{s}}^2 - x^2}dx \\ &= 2\zeta C_1 S_{\mathrm{y}} w_{\mathrm{y}} + \frac{\zeta E^*}{R_{\mathrm{s}}}\left[\frac{\pi w_{\mathrm{s}}^2}{4} - \frac{1}{2}w_{\mathrm{y}}\sqrt{w_{\mathrm{s}}^2 - w_{\mathrm{y}}^2} - \frac{w_{\mathrm{s}}^2}{2}\arcsin\frac{w_{\mathrm{y}}}{w_{\mathrm{s}}}\right] \end{aligned} \tag{17}$$

Now we discuss Equation (17) in the following three cases:

(a) When $w_{\mathrm{y}} = 0$ and $w_{\mathrm{s}} < w_{\mathrm{s}}^{\mathrm{c}}$, the contact between the sealing interface is completely elastic. On this occasion, the additional make-up torque from the sealing interface $T_{\mathrm{se}}^{\mathrm{e}}$ is as follows:

$$T_{\mathrm{se}}^{\mathrm{e}} = \frac{\zeta \pi E^* w_{\mathrm{s}}^2}{4R_{\mathrm{s}}} \tag{18}$$

(b) When $w_{\mathrm{y}} = 0$ and $w_{\mathrm{s}} = w_{\mathrm{s}}^{\mathrm{c}}$, initial yield occurs on the sealing interface. On this occasion, the additional make-up torque from the sealing interface $T_{\mathrm{se}}^{\mathrm{c}}$ is as follows:

$$T_{\mathrm{se}}^{\mathrm{c}} = \frac{\zeta \pi R_{\mathrm{s}} (C_1 S_{\mathrm{y}})^2}{E^*} \tag{19}$$

(c) When $w_{\mathrm{y}} > 0$, the contact between the sealing interface is under an elastic–plastic contact state. The outer boundary condition in the plastic contact zone can be obtained from Equation (16) as follows:

$$\frac{E^*}{2R_{\mathrm{s}}}\sqrt{w_{\mathrm{s}}^2 - w_{\mathrm{y}}^2} = C_1 S_{\mathrm{y}} \text{ or } w_{\mathrm{s}}^2 = \left(\frac{2R_{\mathrm{s}} C_1 S_{\mathrm{y}}}{E^*}\right)^2 + w_{\mathrm{y}}^2 \tag{20}$$

Then, by submitting Equation (20) into Equation (17), we can get the additional make-up torque from sealing interface $T_{\mathrm{se}}^{\mathrm{ep}}$ under this condition:

$$\begin{aligned} T_{\mathrm{se}}^{\mathrm{ep}} &= 2\zeta C_1 S_{\mathrm{y}} w_{\mathrm{y}} + \frac{\zeta E^*}{R_{\mathrm{s}}}\left[\frac{\pi w_{\mathrm{s}}^2}{4} - \frac{1}{2}w_{\mathrm{y}}\frac{2R_{\mathrm{s}} C_1 S_{\mathrm{y}}}{E^*} - \frac{w_{\mathrm{s}}^2}{2}\arcsin\frac{w_{\mathrm{y}}}{w_{\mathrm{s}}}\right] \\ &= \zeta C_1 S_{\mathrm{y}} w_{\mathrm{y}} + \frac{\zeta E^*}{2R_{\mathrm{s}}}\left[\left(\frac{2R_{\mathrm{s}} C_1 S_{\mathrm{y}}}{E^*}\right)^2 + w_{\mathrm{y}}^2\right]\left[\frac{\pi}{2} - \arcsin\frac{w_{\mathrm{y}}}{\sqrt{\left(\frac{2R_{\mathrm{s}} C_1 S_{\mathrm{y}}}{E^*}\right)^2 + w_{\mathrm{y}}^2}}\right] \\ &= f(w_{\mathrm{y}}) \end{aligned} \tag{21}$$

It can be seen from Equation (21) that when relevant parameters are constant, $T_{\mathrm{se}}^{\mathrm{ep}}$ can be expressed as a function of w_{y}. Consequently, if $T_{\mathrm{se}}^{\mathrm{ep}}$ is known, w_{y} can be easily obtained by a trial method. Then, the half-contact width on the sealing interface w_{s} can also be further calculated according to Equation (20). It can be concluded comprehensively that when the actual additional make-up torque from the sealing interface after make-up operation T_{se} is known, if $T_{\mathrm{se}} < T_{\mathrm{se}}^{\mathrm{c}}$, then $w_{\mathrm{y}} = 0$; if $T_{\mathrm{se}} = T_{\mathrm{se}}^{\mathrm{c}}$, initial yield occurs; and if $T_{\mathrm{se}} > T_{\mathrm{se}}^{\mathrm{c}}$, w_{y} can be determined by the following implicit function:

$$w_{\mathrm{y}} = f^{-1}(T_{\mathrm{se}}^{\mathrm{ep}}) \tag{22}$$

By combining Equations (18)–(22), the half-contact width on the sealing interface w_s can be expressed as follows:

$$w_s = \begin{cases} \sqrt{\frac{4R_s T_{se}}{\zeta \pi E^*}} & (T_{se} < T_{se}^c) \\ 2R_s \frac{C_1 S_y}{E^*} & (T_{se} = T_{se}^c) \\ \sqrt{\left(\frac{2R_s C_1 S_y}{E^*}\right)^2 + w_y^2} & (T_{se} > T_{se}^c) \end{cases} \tag{23}$$

2.3. *Average Contact Pressure on Sealing Interface*

The average contact pressure on sealing interface p_{ave} can be calculated from Equation (16):

$$\begin{aligned} p_{ave} &= \frac{1}{2w_s}\int_{-w_s}^{w_s} p_{sN}(x)dx \\ &= \frac{1}{2w_s}\left[C_1 S_y w_y + \frac{E^*}{R_s}\frac{w_s^2}{2}\left(\frac{\pi}{2} - \arcsin\frac{w_y}{w_s}\right)\right] \end{aligned} \tag{24}$$

2.4. *Gas Sealing Capacity for Sphere-Type Premium Connection Base on Sealing Contact Energy Theory*

The sealing contact energy theory holds that the integral of macro-sealing stress multiplied by sealing length can represent the sealability of a metal-to-metal seal, which is called the gas sealing index. According to the research from Murtagian's paper [20], the gas sealing index W_a for a premium connection in HPHT gas wells can be expressed as:

$$W_a = \int_{-w_s}^{w_s} p_{sN}^{1.4}(x)dx \tag{25}$$

Murtagian et al. also proposed that when the gas pressure needing to be sealed p_g and the atmospheric pressure p_a are both known, a premium connection has reliable gas sealability only when W_a in Equation (25) exceeds a critical value W_{ac} in Equation (26):

$$W_{ac} = 103.6 \times \left(\frac{p_g}{p_a}\right)^{0.838} \quad \left(\text{mm} \cdot \text{MPa}^{1.4}\right) \tag{26}$$

However, this evaluation criterion is too conservative because the gas leakage rate considered for reliable sealing in Murtagian's experiment was 0.025 $\text{cm}^3/15$ min, while the permitted limit gas leakage rate for a casing and tubing connection is 0.9 $\text{cm}^3/15$ min in ISO 13679. Considering this fact and the existing failure cases of tubing premium connections in HPHT gas wells, Xie et al. further proposed a practical critical value of W_{ac} [26]:

$$W_{ac} = 10 \times \left(\frac{p_g}{p_a}\right)^{0.838} \tag{27}$$

When the sealing contact pressure distribution is known, by combining Equation (16), Equation (25) and Equation (27), the gas sealing capacity of a sphere-type premium connection in HPHT gas wells can be obtained as:

$$p_{gm} = p_a\left[\frac{\int_{-w_s}^{w_s} p_{sN}^{1.4}(x)dx}{10}\right]^{\frac{1}{0.838}} = p_a\left[\frac{2w_y C_1 S_y + 2\int_{w_y}^{w_s}\left(\frac{E^*}{2R_s}\sqrt{w_s^2 - x^2}\right)^{1.4} dx}{10}\right]^{\frac{1}{0.838}} \tag{28}$$

3. Results and Discussion

According to the previous theoretical analysis, the contact pressure distribution on the sealing interface for a sphere-type premium connection of a certain specification is directly related to the additional make-up torque from the sealing interface. Consequently, to provide theoretical guidance on make-up torque control and sealing parameter design to ensure gas sealability for a sphere-type premium connection in tubing and casing strings, by example analysis, the effects of additional make-up torque from the sealing interface on the contact pressure distribution and key sealing parameters, including contact width, yield width, average contact pressure and gas sealing capacity, were analyzed and also compared for three kinds of sphere-type premium connections. This paper takes P110 steel, 127 mm outer diameter casing with API buttress threads and different sphere-type premium connections as an example for analysis. The basic calculation parameters are shown in Table 1.

Table 1. The basic parameters for sphere-type premium connection.

Symbol	Value	Unit	Symbol	Value	Unit
E_p	206,000	MPa	t_s	0.1	mm/mm
ν_p	0.28	dimensionless	P	5.08	mm
E_c	100,000	MPa	α	3	°
ν_c	0.32	dimensionless	R_s	100	mm
S_{yp}	758.42	MPa	R_{s1}	100	mm
S_{yc}	250	MPa	R_{s2}	100	mm
E_7	125.83	mm	r_s	59.40	mm
g	50.39	mm	μ_s	0.08	dimensionless
L_7	45.17	mm	μ_t	0.08	dimensionless
t_t	0.0625	mm/mm			

3.1. Contact Pressure Distribution on Sealing Interface

According to the previous theoretical analysis, the contact pressure distribution on the sealing interface for a sphere-type premium connection of a certain specification is directly related to the additional make-up torque from the sealing interface. Sphere-to-sphere contact can be regarded as a sphere-of-equivalent-radius-to-cone contact, and sphere-to-cylinder contact can be also taken as the sphere-to-cone with zero taper. Consequently, we firstly took the sphere-to-cone seal with sphere radius of 100 mm and cone taper on diameter of 0.1 as an example to investigate the effects of additional make-up torque from the sealing interface on the contact pressure distribution on the sealing interface. According to Equation (19), we firstly calculated the additional make-up torque from the sealing interface when initial yield T_{se}^{c} was 1644 N·m. Then, we analyzed the contact pressure distribution when the actual additional make-up torque from the sealing interface T_{se} was 1000 N·m, 1644 N·m and 3000 N·m, respectively. The results are shown in Figure 5.

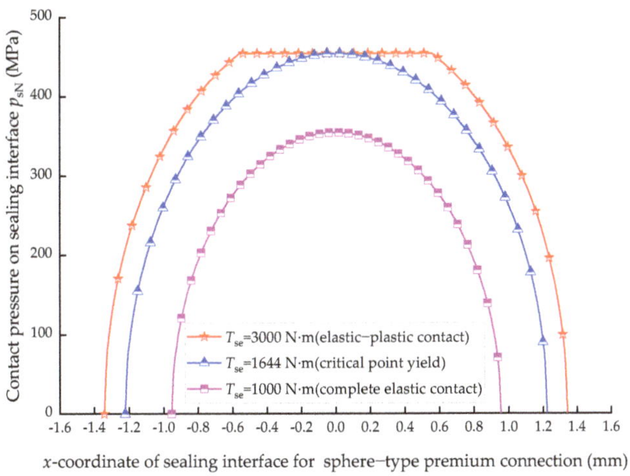

Figure 5. Contact pressure distribution on sealing interfere for sphere-type premium connection under different additional make-up torque from sealing interface.

Figure 5 shows that the contact pressure on the sealing interface was symmetrically distributed with respect to the normal direction of the initial contact point and in a shape of a downward parabola or a parabolic cup. If $T_{se} < T_{se}^{c} = 1644$ N·m, the sealing interface was under a complete elastic contact state; if $T_{se} = T_{se}^{c} = 1644$ N·m, initial yield occurred on the sealing interface and the yield point was located at the initial contact point; if $T_{se} > T_{se}^{c} = 1644$ N·m, the sealing interface was under an elastic–plastic contact state and yield width was 1.11 mm. The average contact pressure was, respectively, 278.5 MPa, 357.0 MPa and 379.8 MPa when T_{se} was 1000 N·m, 1644 N·m and 3000 N·m. In general, with an increase of T_{se}, the contact width, yield width and average contact pressure increased gradually. Therefore, sufficient T_{se} should be attained after the make-up operation, ensuring excellent gas sealability for a premium connection.

3.2. Sealability Parameters for Sphere-Type Premium Connection

In this section, we also take the sphere-to-cone seal with sphere radius of 100 mm and cone taper on diameter of 0.1 as an example to investigate the effects of the additional make-up torque from the sealing interface on the contact width, yield width, average contact pressure on sealing interface and gas sealing capacity of a premium connection.

3.2.1. Contact Width and Yield Width on Sealing Interface

Contact width and yield width are both important parameters that reflect the gas sealability for a premium connection, and contact width should be as large as possible to increase the leakage resistance. For a sphere-type premium connection, the usual higher contact pressure may easily cause the sealing interface to yield and block off leak passage quickly, thus giving enough gas sealing capacity. However, the yield of the sealing surface also easily leads to sticking and stress relaxation, so the yield width of a sphere-type connection should be controlled to be less than 1–2 mm in general. In this paper, contact width $2w_s$ and yield width $2w_y$ were both calculated and compared under different additional make-up torque from sealing interface T_{se}, and the results are shown in Figure 6.

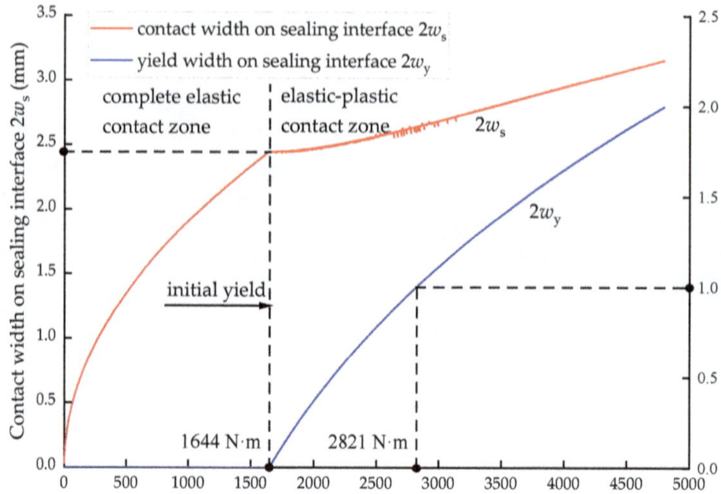

Figure 6. Effects of additional make-up torque from sealing interface on contact width and yield width on sealing interface for sphere-type premium connection.

It can be seen from Figure 6 that if $T_{se} < T_{se}^{c} = 1644$ N·m, the sealing interface was under a complete elastic contact state and contact width showed parabolic increases with the increase of T_{se} while the yield width was zero. If $T_{se} = T_{se}^{c} = 1644$ N·m, initial yield occurred and the critical contact width was 2.45 mm. If $T_{se} > T_{se}^{c} = 1644$ N·m, the sealing interface was under an elastic–plastic contact state. With the increase of T_{se}, the contact width increased slower and slower, and it finally increased approximately linearly, while yield contact showed a parabolic increase and it gradually approached the contact width. In the example, to make the yield width less than 1 mm, the additional make-up torque from sealing interface T_{se} should be controlled within 2821 N·m.

3.2.2. Average Contact Pressure on Sealing Interface and Gas Sealing Capacity of Premium Connection

The general sealing criterion for a metal-to-metal seal is that the sealing contact pressure should exceed the fluid pressure inside the string, which has also been taken as the early design criterion for tubing and casing connections. However, this criterion has been established based on the assumption of a complete smooth sealing surface and it does not conform to the actual rough contact and sealing. The sealing contact energy theory adopts the gas sealing index W_a to reflect the leakage resistance and takes a critical gas sealing index W_{ac} from laboratory testing to denote reliable sealing, which builds a bridge between macro-contact pressure and micro-leakage on the sealing interface. Consequently, the average contact pressure and gas sealing capacity were both calculated and compared under different additional make-up torque from sealing interface T_{se}, and the results are shown in Figure 7.

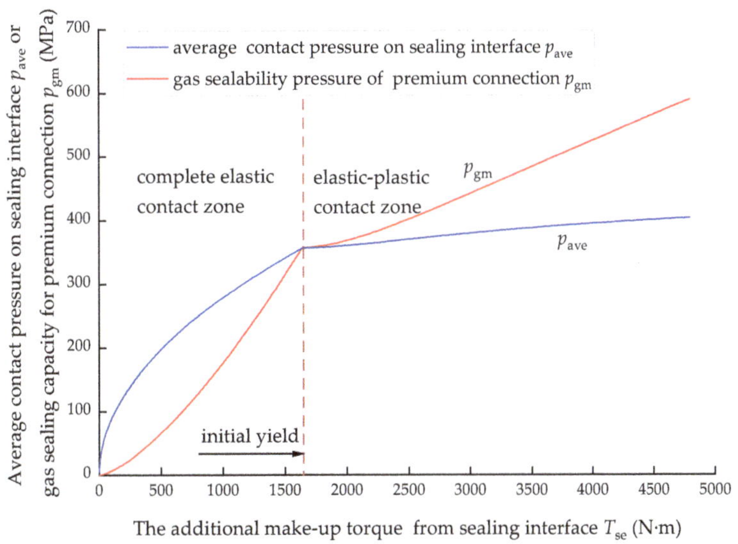

Figure 7. Effects of additional make-up torque from sealing interface on average contact pressure on sealing interface and gas sealing capacity for sphere-type premium connection.

It can be seen from Figure 7 that in the complete elastic contact zone, the average contact pressure on sealing interface p_{ave} and gas sealing capacity for premium connection p_{gm} both showed a parabolic increase with the increase of additional make-up torque from sealing interface T_{se}, although p_{gm} was smaller than p_{ave} but gradually approached it and nearly equal to p_{ave} at the initial yield point. Later, the sealing interface turned into an elastic–plastic contact state. With the increase of T_{se}, p_{gm} increased quickly and gradually exceeded p_{ave}. This was because the plastic deformation occurred and it blocked off the leakage passage quickly, which led to nearly invariable p_{ave} and larger and larger p_{gm}. Consequently, the adoption of p_{ave} will overestimate the sealability of a sphere-type premium connection in a complete elastic contact zone while underestimating it in an elastic–plastic contact zone. Therefore, we recommend the sealing contact energy theory.

3.3. Sealability Parameters' Comparison for Three Kinds of Sphere-Type Premium Connections

To provide theoretical guidance on sealing parameters' design and make-up torque control for sphere-type premium connections, a comparison analysis for three types of premium connections were also conducted. Considering the different additional make-up torque from the sealing interface, the contact width, yield width, average contact pressure and gas sealing capacity for three types of premium connections, including sphere-to-cone seal, sphere-to-cylinder seal and sphere-to-sphere seal, were calculated, and the results are shown in Figures 8–11, respectively.

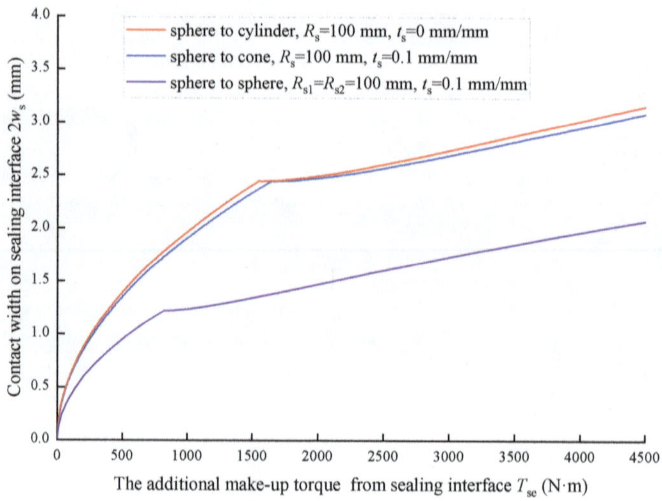

Figure 8. Effects of additional make-up torque from sealing interface on contact width on sealing interface for three kinds of sphere-type premium connections.

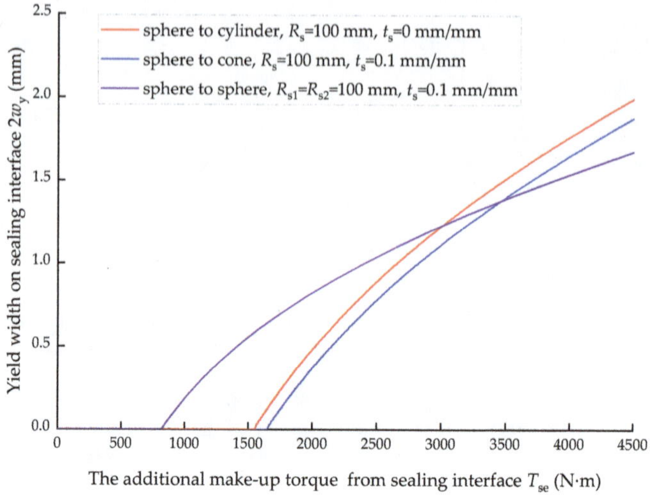

Figure 9. Effects of additional make-up torque from sealing interface on yield width on sealing interface for three kinds of sphere-type premium connections.

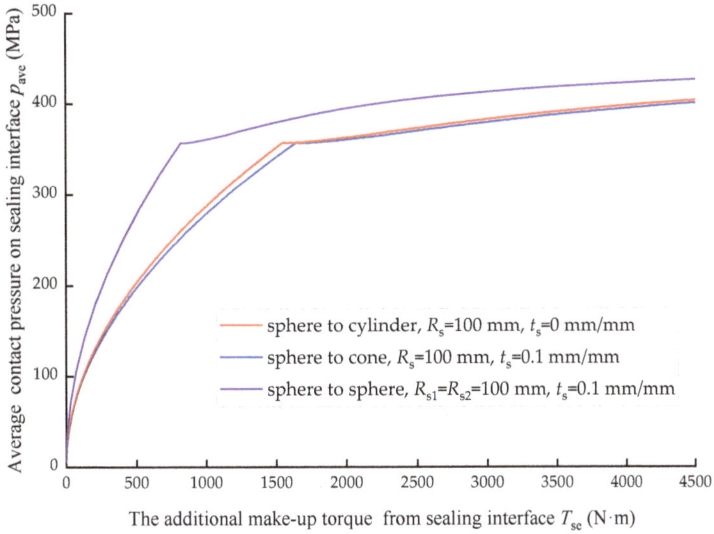

Figure 10. Effects of additional make-up torque from sealing interface on average contact pressure on sealing interface for three kinds of sphere-type premium connections.

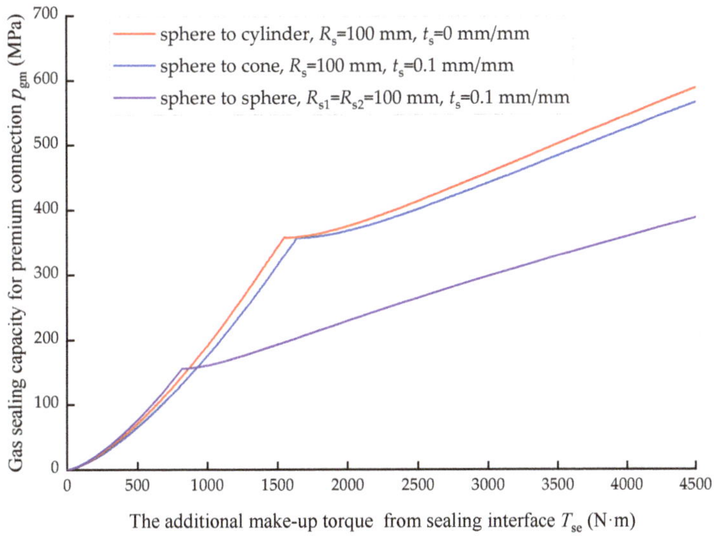

Figure 11. Effects of additional make-up torque from sealing interface on gas sealing capacity for three kinds of sphere-type premium connections.

It can be comprehensively seen from Figures 8–11 that the sealing capacity for a sphere-to-cylinder premium connection was slightly larger than that for a sphere-to-cone premium connection under the same additional make-up torque from sealing interface T_{se}. This was because a fraction of T_{se} was adopted by the thread for the sphere-to-cone premium connection. The value of T_{se} for a sphere-to-sphere premium connection when initial yield occurred was obviously lower than that of sphere-to-cone and sphere-to-cylinder premium connections (see Figure 9). In the complete elastic zone, the gas sealing capacity of each sphere-type premium connection was nearly equal, and this was because the sphere-to-cone and sphere-to-cylinder premium connections had larger contact width and smaller average contact pressure while the sphere-to-sphere premium connection had smaller contact width and larger average contact pressure (see Figures 8 and 10), which produced fair sealing contact energy for the three types of premium connections. In the elastic–plastic zone, although the average contact pressures for three types of premium connections were relatively close to each other, the contact width for sphere-to-sphere was obviously smaller than that of sphere-to-cone and sphere-to-cylinder premium connections, thus leading to obvious lower gas sealing capacity for the sphere-to-sphere premium connection. Consequently, a proper larger sphere radius should be selected for a sphere-type premium connection, which can reduce the risk of yield sticking on the sealing interface and also improve the gas sealability of the connection.

4. Conclusions

(1) Differently from the widely used FEM method with testing, a theoretical model for evaluating the sealability of a sphere-type premium connection was proposed based on the additional make-up torque from the sealing interface, which combined Hertz contact pressure and the von Mises yield criterion for calculating elastic–plastic contact pressure distribution on the sealing interface and adopted the gas sealing criterion obtained from Murtagian's experimental results for deducing gas sealing capacity.

(2) The sensitivity analysis on sealability parameters indicated that the additional make-up torque from sealing interface T_{se} closely influenced sealability parameters' variation and gas sealing capacity:

- The contact pressure on the sealing interface was symmetrically distributed with respect to the normal direction of the initial contact point and in a shape of a downward parabola or a parabolic cup. The corresponding T_{se}^{c} when initial yield occurred divided the contact state into complete elastic contact and elastic–plastic contact.
- Under a complete elastic contact state, the contact width showed parabolic increases with increase of T_{se} while the yield width was zero; under an elastic–plastic contact state, with the increase of T_{se}, the contact width increased slower and slower and it finally increased approximately linearly, while the yield contact showed a parabolic increase and it gradually approached the contact width.
- Under a complete elastic contact state, average contact pressure on sealing interface and gas sealing capacity both showed a parabolic increase with the increase of T_{se}, although the latter was smaller but gradually approaching the former and nearly equal to it at the initial yield point; under an elastic–plastic contact state, with the increase of T_{se}, the gas sealing capacity increased quickly and exceeded average contact pressure on the sealing interface.
- The gas sealing capacity of each sphere-type premium connection was nearly equal in the complete elastic zone, but that of sphere-to-sphere was obviously smaller than that of sphere-to-cone and sphere-to-cylinder premium connections in the elastic–plastic zone.

(3) For ensuring the gas sealability of a sphere-type premium connection for tubing and casing strings, the gas sealing index based on the sealing contact energy theory should be recommended for sealability evaluation alongside average contact pressure on the sealing interface. Make-up torque is also suggested to be controlled accurately to ensure enough average contact pressure and contact width but a proper yield width, and a lager sphere radius should be selected to reduce the risk of yield sticking.

(4) This paper mainly adopted macroscopic contact mechanics combined with previous experimental results to establish the model, which could not accurately descript the actual microcosmic contact state between sealing interfaces and may influence the accuracy of the model to some extent, so the model should be improved by considering microcosmic contact mechanics on the sealing interface, and corresponding full-scale experimental testing should be carried out in later work.

Author Contributions: Conceptualization, B.Y. and H.X.; Methodology, B.Y., H.X. and Z.Z.; Formal analysis, H.X.; Investigation, B.Y., H.X., S.X. and K.S.; Writing—original draft, H.X.; Writing—review & editing, B.Y.; Supervision, Z.Z.; Funding acquisition, H.X., K.S. and Y.Y. All authors have read and agreed to the published version of the manuscript.

Funding: This research was funded by the National Natural Science Foundation of China (51804060, 51974052), the Open Fund (PLN2022-23) of State Key Laboratory of Oil and Gas Reservoir Geology and Exploitation (Southwest Petroleum University), the Natural Science Foundation of Chongqing, China (CSTB2022NSCQ-MSX0989) and Sichuan Province Regional Innovation Cooperation Project (2022YFQ0012).

Data Availability Statement: The data used to support the findings of this study are available from the corresponding author upon request.

Conflicts of Interest: The authors declare no conflict of interest.

Nomenclature

T_A	leading torque, N·m
T_B	make-up torque value at the initial contact on sealing interface, N·m
T_C	make-up torque value at the initial contact on shoulder interface, N·m
T_D	final make-up torque for premium connection, N·m
T_{ti}	thread interference torque, N·m
T_{se}	additional make-up torque from sealing interface, N·m
T_{sh}	additional make-up torque from shoulder, N·m
T_{set}	additional thread torque from sealing interface, N·m
T_{sef}	additional friction torque from sealing interface, N·m
p_{sN}	normal contact pressure on sealing interface, MPa
F_{se}	additional axial pre-tightening force from sealing interface, N
r_s	sealing radius, mm
t_s	cone taper on diameter, mm/mm
w_s	half contact width on sealing interface, mm
w_y	half yield width on sealing surface, mm
w_s^c	half contact width with initial yield on sealing interface, mm
P	thread pitch, mm
μ_t	friction coefficient on thread surface, dimensionless
μ_s	friction coefficient on sealing interface, dimensionless
R_t	equivalent moment arm for thread friction torque, mm
α	loading angle for thread tooth, °
E_7	thread pitch diameter, mm
g	intact thread length, mm
L_7	complete thread length, mm
t_t	thread taper on diameter, mm/mm
ζ	the coefficient of additional make-up torque from sealing interface, 10^{-3} mm^2
E^*	equivalent elastic modulus for sealing contact pair, MPa
R_s	sphere radius, mm
R_{se}	equivalent sphere radius for sphere-to-sphere contact, mm
R_{s1}, R_{s2}	two sphere radii, respectively, for sphere-to-sphere contact, mm
E_p, E_c	elastic modulus for sphere pin and cone coupling, respectively, MPa
ν_p, ν_c	Poisson's ratio for sphere pin and cone coupling, respectively, dimensionless
ν	Poisson's ratio of material, dimensionless
C_1	yield strength coefficient of the softer material for sealing contact pair, dimensionless
S_y	yield strength of the softer material for sealing contact pair, MPa
C_{1p}, C_{1c}	Yield strength coefficient of sphere pin and cone coupling, respectively, dimensionless
S_{yp}, S_{yc}	yield strength of sphere pin and cone coupling, respectively, MPa
T_{se}^e	additional make-up torque from sealing interface with complete elastic contact, N·m
T_{se}^c	additional make-up torque from sealing interface when initial yield occurs on the sealing interface, N·m
T_{se}^{ep}	additional make-up torque from sealing interface when sealing interface is under elastic–plastic contact state, N·m

p_{ave}	average contact pressure on sealing interface, MPa
W_a	gas sealing index for premium connection, mm·MPa$^{1.4}$
W_{ac}	critical gas sealing index, mm·MPa$^{1.4}$
p_a	atmospheric pressure, MPa
p_g	gas pressure needing to be sealed, MPa
p_{gm}	gas sealing capacity of sphere-type premium connection, MPa

References

1. Sugino, M.; Nakamura, K.; Yamaguchi, S.; Daly, D.; Briquet, G.; Verger, E. Development of an innovative high-performance premium threaded connection for OCTG. In Proceedings of the Offshore Technology Conference, Houston, TX, USA, 3–6 May 2010; pp. 1–9.
2. Zara, E.M.; Grittini, S.J.; Rodriguez Jordan, G.; Perello, L.; Sanders, M. Premium Connections Fatigue Assessment Methodology Fit for Multi-stage Hydraulic Fracturing Operations. In Proceedings of the SPE Annual Technical Conference and Exhibition, Dubai, United Arab Emirates, 21–23 September 2021; pp. 1–12.
3. Payne, M.L.; Schwind, B.E. A new international standard for casing/tubing connection testing. In Proceedings of the SPE/IADC drilling conference, Amsterdam, The Netherland, 9–11 March 1999; pp. 1–12.
4. Salzano, T.; Racca, S.; Ferrara, G.; Nutricato, G.; Repetto, C.; Ricci Maccarini, G.; Santi, N.; Zambetti, R.; Cordero, J.; Giraldo, L.; et al. Proceedings of the Low Temperature Testing of OCTG Premium Connections for CCS Application. In Proceedings of the OMC Med Energy Conference and Exhibition, Ravenna, Italy, 28–30 September 2021; pp. 1–12.
5. Hamilton, K.A.; Wagg, B.; Roth, T. Using ultrasonic techniques to accurately examine seal surface contact stress in premium connections. In Proceedings of the SPE Annual Technical Conference and Exhibition, Anaheim, CA, USA, 11–14 November 2007; pp. 1–14.
6. Ernens, D.; Pérez–Ràfols, F.; Hoecke, D.V.; Roijmans, R.F.; van Riet, E.J.; Vande Voorde, J.B.; Almqvist, A.; Bas de Rooij, M.; Roggeband, S.M.; van Haaften, W.M.; et al. On the sealability of metal-to-metal seals with application to premium casing and tubing connections. *SPE Drill. Complet.* **2019**, *34*, 382–396. [CrossRef]
7. Inose, K.; Sugino, M.; Goto, K. Influence of grease on high-pressure gas tightness by metal-to-metal seals of premium threaded connections. *Tribol. Online* **2016**, *11*, 227–234. [CrossRef]
8. Xu, H.L.; Yang, B. A Quantitative Model to Calculate Gas Sealing Capacity and Design Sealing Parameters for Premium Connection. *Math. Probl. Eng.* **2020**, *2020*, 9074381. [CrossRef]
9. Xu, Z.Q.; Yan, X.Z.; Yang, X.J.; Yin, X.K.; Wang, M.D.; Zheng, X.Y. Application of micro-leakage mechanism for evaluating the sealing performance of non-API casing connections. *Acta Pet. Sin.* **2014**, *35*, 963–971.
10. Chen, F.; Di, Q.F.; Wang, W.C.; Zhu, W.P.; Wang, M.J.; Shen, Z.J. The reliability evaluation of threaded connections in challenging drilling by three-dimensional finite element analysis. In *Advanced Materials Research*; Trans Tech Publications Ltd.: Bäch, Switzerland, 2013; Volume 690, pp. 2831–2839.
11. Hilbert, L.B.; Kalil, I.A. Evaluation of premium threaded connections using finite-element analysis and full-scale testing. In Proceedings of the IADC/SPE Drilling Conference, New Orleans, LA, USA, 18–21 February 1992; pp. 563–580.
12. Takano, J.; Yamaguchi, M.; Kunishige, H. Development of premium connection "KSBEAR" for withstanding high compression, high external pressure, and severe bending. *Kawasaki Steel Tech. Rep.* **2002**, *47*, 14–22.
13. Zhang, M.L.; Zhang, Q.; Cheng, L.; Liao, R.Q.; Feng, J. Study on the performances of two kinds of special tubing connector. *Chem. Eng. Trans.* **2015**, *46*, 1063–1068.
14. Chen, W.; Di, Q.; Zhang, H.; Chen, F.; Wang, W. The sealing mechanism of tubing and casing premium threaded connections under complex loads. *J. Pet. Sci. Eng.* **2018**, *171*, 724–730. [CrossRef]
15. Kim, B.; Huang, J.; Yoon, J.Y. Design optimization of OCTG premium connection system based on the effect of stabbing flank angle. *J. Mech. Sci. Technol.* **2021**, *35*, 545–561. [CrossRef]
16. Zhang, Y.; Lian, Z.; Zhou, M.; Lin, T. Viscoelastic behavior of a casing material and its utilization in premium connections in high-temperature gas wells. *Adv. Mech. Eng.* **2018**, *10*, 1–8. [CrossRef]
17. Yu, Y.; Qu, Z.; Dou, Y.; Cao, Y. Analysis of Energy Dissipation on the Sealing Surface of Premium Connection Based on a Microslip Shear Layer Model. *Energies* **2022**, *15*, 8400. [CrossRef]
18. Dou, Y.; Li, Y.; Cao, Y.; Yu, Y.; Zhang, J.; Zhang, L. FE simulation of sealing ability for premium connection based on ISO 13679 CAL IV tests. *Int. J. Struct. Integr.* **2020**, *12*, 138–148. [CrossRef]
19. Xu, H.L.; Yang, B.; Zhang, Z.; Shi, T.H. Special considerations to calculate joint strength of premium connections. *J. Pet. Sci. Eng.* **2019**, *182*, 106295. [CrossRef]
20. Murtagian, G.R.; Fanelli, V.; Villasante, J.A.; Johnson, D.H.; Ernst, H.A. Sealability of stationary metal-to-metal seals. *J. Ribologi.* **2004**, *126*, 591–596. [CrossRef]
21. Johnson, K.L. *Contact Mechanics*; Cambridge University Press: Cambridge, UK, 1987; pp. 103–119.
22. Xu, H.L.; Shi, T.H.; Zhang, Z. Theoretical analysis on makeup torque in tubing and casing premium threaded connections. *J. Southwest Pet. Univ. Sci. Technol. Ed.* **2014**, *36*, 160–168.
23. American Petroleum Institute. *Recommended Practice for Drill Stem Design and Operating Limits*; API RP 7G 16 (E).; API Publishing Services: Washington, DC, USA, 1998; pp. 133–135.

24. Green, I. Poisson ratio effects and critical valus in spherical and cylindrical Hertzian contacts. *Appl. Mech. Eng.* **2005**, *10*, 451–462.
25. Xu, H.L.; Yang, B.; Zhang, Z.; Shi, T.H. Elastic-plastic contact stress analysis on sphere to cone sealing structure for tubing and casing premium connection. *Chin. J. Appl. Mech.* **2021**, *38*, 1258–1263.
26. Xie, J.; Matthews, C.; Hamilton, A. A study of sealability evaluation criteria for casing connections in thermal wells. In Proceedings of the SPE Canada Heavy Oil Technical Conference, Calgary, AB, Canada, 7–9 June 2016; pp. 1–12.

processes

Article

A Three-Dimensional Analytical Solution of Stress Field in Casing-Cement-Stratum System Considering Initial Stress State

Xiaoyang Wang [1,2], Tingxue Jiang [1,2], Yayun Zhang [1,2], Jun Zhou [1,2], Hecheng Xiao [3,4] and Wenda Li [3,*]

1 State Key Laboratory of Shale Oil and Gas Enrichment Mechanisms and Effective Development, Beijing 102206, China
2 SINOPEC Research Institute of Petroleum Engineering Co., Ltd., Beijing 102206, China
3 Key Laboratory of In-Situ Property-Improving Mining of Ministry of Education, Taiyuan University of Technology, Taiyuan 030024, China
4 College of Mining Engineering, Taiyuan University of Technology, Taiyuan 030024, China
* Correspondence: liwenda@tyut.edu.cn

Abstract: Accurate stress field calculation of the casing-cement-stratum system is crucial for evaluating wellbore integrity. Previous models treated in-situ stress as boundary pressure loads, leading to unrealistic infinite displacements at infinity. This study presents a three-dimensional (3D) analytical solution for the stress field within the casing-cement-stratum system in inclined wells, considering in-situ stress and hydrostatic stress in cement as the initial stress state and taking into account stress components related to the axial direction. Assuming a plane strain condition and superimposing the in-plane plane strain problem, elastic uni-axial stress problem and anti-plane shear problem, a 3D analytical solution is obtained. Comparisons with previous models indicate that the existing model overestimates the absolute values of stress components and failure potential of casing and cement in both 2D and 3D scenarios. The presence of initial stress in cement greatly increases the absolute value of the compressive stress state but decreases the failure potential in cement, which has not been well studied. Additionally, a low Young's modulus and high initial stress state of the cement benefits the cement's integrity since the maximum Mises stress significantly decreases. The new 3D analytical solution can provide a benchmark for 3D numerical simulation and quick assessment for wellbore integrity.

Keywords: casing-cement-stratum system; 3D analytical solution; initial stress state of cement; in-situ stress; inclined well

Citation: Wang, X.; Jiang, T.; Zhang, Y.; Zhou, J.; Xiao, H.; Li, W. A Three-Dimensional Analytical Solution of Stress Field in Casing-Cement-Stratum System Considering Initial Stress State. *Processes* **2023**, *11*, 1164. https://doi.org/10.3390/pr11041164

Academic Editors: Tianshou Ma and Yuqiang Xu

Received: 6 March 2023
Revised: 5 April 2023
Accepted: 7 April 2023
Published: 10 April 2023

1. Introduction

Zonal isolation often fails throughout the life of oil/gas wells due to mechanical failure of casing and cement [1]. Moreover, the oil/gas leaks from the reservoir to the shallow aquifers and abandonment of the well occurs, and environmental and safety concerns are inevitable [2,3]. Hence, precise stress field calculations of the casing-cement-stratum system play a crucial role in evaluating mechanical wellbore integrity.

Aside from numerical modeling works, analytical solutions have been developed for determining the stress field in the casing-cement-stratum system [4–6]. Yin et al. [7] developed a theoretical solution for in-plane stress distribution in the casing-cement-stratum system. Moreover, Yin and Gao [8] derived the in-plane stress field in directional (inclined) cemented wells. Then, in the directional wells [9] and vertical section of the wells [10], the stress field and collapse resistance of multilayer cemented casing were examined. The wellbore integrity of HPHT gas wells [11], carbon sequestration projects [12], well testing and production process [13] have all been analytically analyzed by taking into consideration the thermal effect. Moreover, the impact of casing eccentricity on the failure of the cement sheath was analytically examined [14]. Xia et al. [15] obtained a transient response to the casing-cement-stratum system under dynamic radial tractions. Li et al. [16]

obtained an analytical thermo-poroelastic response of the casing-cement-stratum system under internal hydraulic and thermal perturbation. However, the aforementioned models only obtained the in-plane stress field in vertical and inclined wells, neglecting the 3D stress field, which takes into account stress components with the axial direction. Atkinson and Eftaxiopoulos [17] employed complex potentials to determine the stress field in the in-plane and anti-plane issues for an inclined, cased and cemented wellbore. Vitali et al. [18] combined classical Kirsch and Einstein-Schwartz solutions to obtain an analytical solution for tunnels not aligned with geostatic principal stress directions, which provides the stress and displacement distribution induced by far-field axial shear stresses. Jo [19] derived the 3D stress field for inclined wells under the assumption of generalized plane strain condition and taking thermal effect into account.

However, the majority of models [7–19] considered the in-situ stress that already existed prior to excavation as the boundary pressure load rather than as the initial stress; this results in infinite displacements at infinity in the stratum and significant error for calculating an appropriate stress field [20]. Li et al. [21] considered the in-situ stress as the initial stress state to assess the casing integrity during the hydraulic fracturing of shale gas wells. Wei et al. [22] simulated the factors influencing the cement sheath integrity in hydraulic fractured wells, discovering that injection pressure significantly affects cement debonding. Yan et al. [23] simulated the debonding degree of cementing interfaces during fracturing and found that improving the interfacial bonding strength and reducing the elastic modulus of cement can promote wellbore integrity. Simone et al. [24] developed an analytical solution to calculate the in-plane stress field of the vertical casing-cement-stratum system with uniform in-situ stress, taking into account the drilling, construction, and production phases. To the best of the authors' knowledge, a 3D stress field has not yet been established for the inclined casing-cement-stratum system that treats in-situ stress as the initial stress state.

A further factor that must be overlooked is the cement's initial stress state. According to Saint-Marc et al. [25], attaining long-term cement-sheath integrity depends on the initial condition of stress. Petersen and Ulm [26] examined the impact of early-age stress and pressure developments in the setting of cement on cement failure. When Meng et al. [27] proposed a model to estimate the drop in total interface stress in cement during hydration, they discovered that increasing the cement's initial stress will protect it. For the sake of simplification, the hydrostatic stress of cement is frequently taken to be equal to the initial stress state [28–30].

This work establishes a 3D analytical model of the stress field in the casing-cement-stratum system for inclined wells to assess wellbore integrity. The 3D problem is separated into three parts on the plane strain condition: in-plane plane strain problem, elastic uni-axial stress problem and anti-plane shear problem. Moreover, the in-situ stress in the stratum and the hydrostatic stress in the cement are regarded as the initial stress state. The stress field in the casing-cement-stratum system for inclined wells is then obtained by applying continuity of stress and displacement at the interfaces and boundary conditions. Comparisons with the existing model are made to demonstrate the necessity of treating the in-situ stress as the initial stress. In addition, a sensitivity analysis is conducted to estimate the influence of different factors on the integrity of the casing and cement. In the end, a 3D analytical solution can provide a benchmark for numerical simulation and a rapid assessment of wellbore integrity.

2. Formulation

2.1. Problem Description, Decomposition and Basic Equations

The system of steel casing, cement sheath, and stratum constitutes the near wellbore region after wellbore drilling and completion, as depicted in Figure 1. Assuming perfect bonding at the interfaces of casing-cement and cement-stratum, the continuity of the corresponding stress and displacement components is satisfied. The inner radii of casing, cement and stratum are a, b, and c, respectively. It should be noted that in-situ compressive stresses in the stratum have been applied before well construction, so they represent the initial stress state for the casing-cement-stratum system. Then, the three principle stresses in the stratum at the Cartesian coordinate can be decomposed by the vertical stress component σ_v and unequal horizontal stress components σ_H and σ_h, respectively. Moreover, the gravity of the casing and cement (hydrostatic pressure) has also been exerted before wellbore completion. In this study, the initial stress state in the cement is simulated by the hydraulic pressure of the slurry. Meanwhile, a uniform wellbore pressure p_w acts on the inner surface of the casing. Given that the cross-sectional dimension is trivial compared to the length dimension of the system, this 3D problem can be treated as a plane strain problem [31,32].

Figure 1. The structure of cylindrical casing-cement-stratum system.

The transformation from the Cartesian coordinates to the borehole coordinate is shown in Figure 2, where an inclined or horizontal wellbore in the petroleum industry is considered. We assume that the borehole is inclined to the X-Y-Z far-field principal stress axes. The Cartesian coordinate system coincides with the principal axes of the in-situ far-field stress, which are designated as σ_x, σ_y, σ_z. The borehole coordinate is defined by a local coordinate system x-y-z, with the borehole axes coinciding with the z-axis. In the borehole system, α is an azimuth angle to the Z-axis, and β is a zenith angle toward the Z-axis.

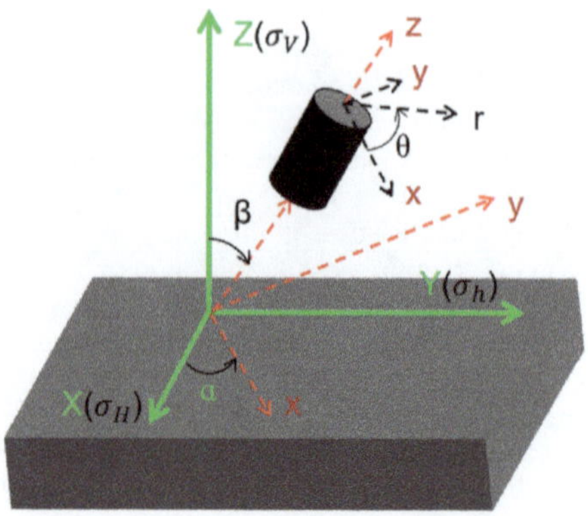

Figure 2. Coordinate transformation of inclined-vertical shaft (modified from Gao et al. [33]).

Then, in the borehole coordinates, the stress tensor reads:

$$\sigma = \begin{bmatrix} \sigma_{xx} & \sigma_{xy} & \sigma_{xz} \\ \sigma_{xy} & \sigma_{yy} & \sigma_{zy} \\ \sigma_{xz} & \sigma_{zy} & \sigma_{zz} \end{bmatrix} = \boldsymbol{L} \begin{bmatrix} \sigma_H & 0 & 0 \\ 0 & \sigma_h & 0 \\ 0 & 0 & \sigma_v \end{bmatrix} \boldsymbol{L}^T \quad (1)$$

where

$$L = \begin{bmatrix} \cos\beta\cos\alpha & -\cos\beta\sin\alpha & \sin\beta \\ \sin\alpha & \cos\alpha & 0 \\ -\sin\beta\cos\alpha & \sin\beta\sin\alpha & \cos\beta \end{bmatrix} \tag{2}$$

To obtain the principal horizontal stress components, a rotation of coordinates around the z-axis is performed. The angle of rotation is given as:

$$\theta_r = \frac{1}{2}\tan^{-1}\frac{2\sigma_{xy}}{\sigma_{xx} - \sigma_{yy}}. \tag{3}$$

Then, the initial in-situ stress in the polar coordinate reads:

$$\sigma^{s}_{rr0} = -P - S\cos(2\theta'),\ \sigma^{s}_{\theta\theta0} = -P + S\cos(2\theta'),\ \tau^{s}_{r\theta0} = S\sin(2\theta'). \tag{4}$$

where $P = (\sigma_{xx} + \sigma_{yy})/2, S = (\sigma_{xx} - \sigma_{yy})/2, \theta' = \theta - \theta_r$. Variables with superscripts containing 'i = ca, ce, s' are related to corresponding casing, cement and stratum, respectively. Moreover, the subscript 0 denotes the initial stress state.

Owing to the linearity of elasticity theory and the in-situ far-field conditions, the problem can be decomposed into three sub-problems: (I) in-plane plane strain problem; (II) elastic uni-axial stress problem; (III) elastic anti-plane shear problem [31–33]. Then, the total solution of the inclined casing-cement-stratum system can be obtained by superposition. In addition, the tensile stress is taken as a positive.

The boundary conditions of three Sub-problems I, II and III can be listed as:

(1) In-plane plane strain Sub-problem I: this sub-problem can be divided into an axisymmetric and asymmetric mode, which is widely applied. Mode one, axisymmetric loading:

$$\begin{aligned} &\sigma^{I(1)}_{rr}\Big|_{r=r_w} = -p_w,\ \tau^{I(1)}_{r\theta}\Big|_{r=r_w} = 0, \\ &\sigma^{ce,I(1)}_{rr0} = \sigma^{ce,I(1)}_{\theta\theta0} = -\sigma^{ce}_{0} = -\rho_{ce}gh, \\ &\sigma^{s,I(1)}_{rr0} = \sigma^{s,I(1)}_{\theta\theta0} = -P, \end{aligned} \tag{5}$$

Mode two, asymmetric loading:

$$\begin{aligned} &\sigma^{I(2)}_{rr}\Big|_{r=r_w} = 0,\ \tau^{I(2)}_{r\theta}\Big|_{r=r_w} = 0, \\ &\sigma^{s,I(2)}_{rr0} = -Scos(2\theta'),\ \sigma^{s,I(2)}_{\theta\theta0} = Scos(2\theta'), \tau^{s,I(2)}_{r\theta0} = Ssin(2\theta'), \end{aligned} \tag{6}$$

(2) Elastic uni-axial stress problem II:

$$\sigma^{ce,II}_{zz0} = -\sigma^{ce}_{0} = -\rho_{ce}gh, \sigma^{s,II}_{zz0} = -\sigma^{s}_{zz0}, \tag{7}$$

(3) Elastic anti-plane shear problem III:

$$\begin{aligned} &\sigma^{ca,III}_{rz}\Big|_{r=r_w} = 0,\ \sigma^{ca,III}_{\theta z}\Big|_{r=r_w} = 0, \\ &\sigma^{s,III}_{rz0} = -(\sigma^{s}_{xz0}\cos\theta + \sigma^{s}_{yz0}\sin\theta), \\ &\sigma^{s,III}_{\theta z0} = (\sigma^{s}_{xz0}\sin\theta - \sigma^{s}_{yz0}\cos\theta), \end{aligned} \tag{8}$$

2.2. *Solution Formulation*

2.2.1. Solution of Mode One in Sub-Problem I

The classical solution for an elastic hollow cylinder subjected to axisymmetric loads is used for the casing, cement, and stratum. Chemical shrinkage during the cement setting process, which can affect the internal stress state in cement, is not considered in this paper since the stress state in the cement is caused by fluid column pressure and the deepest cement sets first. Then, the general solution for the stress and displacement field in increment form is given as [19]:

$$\begin{aligned}
\Delta\sigma_{rr}^{i,I(1)} &= \frac{A^i}{2} + \frac{B^i}{r^2}, \\
\Delta\sigma_{\theta\theta}^{i,I(1)} &= \frac{A^i}{2} - \frac{B^i}{r^2}, \\
u_{rr}^{i,I(1)} &= \frac{1}{2G_i}\left[\left(\frac{1}{2} - \nu_i\right)A^i r - \frac{B^i}{r}\right],
\end{aligned} \tag{9}$$

Then the total stress in Mode one can be written as:

$$\begin{aligned}
\sigma_{rr}^{i,I(1)} &= \sigma_{rr0}^{i} + \Delta\sigma_{rr}^{i,I(1)}, \\
\sigma_{\theta\theta}^{i,I(1)} &= \sigma_{\theta\theta 0}^{i} + \Delta\sigma_{\theta\theta}^{i,I(1)}.
\end{aligned} \tag{10}$$

Due to the infinity of displacement at infinity in the stratum, $A_s = 0$ applies. In contrast, the existing models [19] treat the in-situ stress as boundary pressure loads instead of an initial stress state, thus Equations (9) and (10) can be rewritten as:

$$\begin{aligned}
\sigma_{rr}^{s,I(1)} &= \frac{A^s}{2} + \frac{B^s}{r^2}, \\
\sigma_{\theta\theta}^{i,I(1)} &= \frac{A^s}{2} - \frac{B^s}{r^2},
\end{aligned} \tag{11}$$

In this case, according to $\sigma_{rr}^{s,I(1)}\Big|_{r=\infty} = -P$ in Equation (5), $A^s = -2P$ applies. Consequently, the corresponding ultimate stress field will differ from that of the present model with $A^s = 0$.

The normal traction and displacement continuity and boundary conditions at the casing-cement-stratum interfaces in Mode one are:

$$\begin{aligned}
&\sigma_{rr}^{ca,I(1)}\Big|_{r=R_1} = -P_w, \\
&\sigma_{rr}^{ca,I(1)}\Big|_{r=R_2} = \sigma_{rr}^{ce,I(1)}\Big|_{r=R_2}, \quad \mathbf{u}_{rr}^{ca,I(1)}\Big|_{r=R_2} = \mathbf{u}_{rr}^{ce,I(1)}\Big|_{r=R_2}, \\
&\sigma_{rr}^{ce,I(1)}\Big|_{r=R_3} = \sigma_{rr}^{s,I(1)}\Big|_{r=R_3}, \quad \mathbf{u}_{rr}^{ce,I(1)}\Big|_{r=R_3} = \mathbf{u}_{rr}^{s,I(1)}\Big|_{r=R_3},
\end{aligned} \tag{12}$$

Substituting Equations (9) and (10) into Equation (12), the linear equations can be given as:

$$\boldsymbol{M}^{I(1)} \cdot \mathbf{X}^{I(1)} = \mathbf{N}^{I(1)} \tag{13}$$

where $\boldsymbol{M}^{I(1)}$ is a five-by-five matrix whose non-zero entries are given in Appendix A, and:

$$\begin{aligned}
\mathbf{X}^{I(1)} &= \begin{bmatrix} A^{ca} & B^{ca} & A^{ce} & B^{ce} & B^{s} \end{bmatrix}^T, \\
\mathbf{N}^{I(1)} &= \begin{bmatrix} -P_w & -\sigma_0^{ce} & 0 & \sigma_0^{ce} - P_0 & 0 \end{bmatrix}^T.
\end{aligned} \tag{14}$$

From Equation (13), the array $\mathbf{X}^{I(1)}$ consisting of the unknown coefficients can be solved straightforwardly. Then, the corresponding stress and displacement distributions within the casing-cement-stratum system in Mode one can be yielded. The vertical stress components induced by horizontal stress, assuming plane strain condition, in Mode one are given as:

$$\Delta\sigma_{zz}^{i,I(1)} = v_i\left(\Delta\sigma_{rr}^{i,I(1)} + \Delta\sigma_{\theta\theta}^{i,I(1)}\right) \tag{15}$$

2.2.2. Solution of Mode two in Sub-Problem I

In Mode two, the in-situ stress in the stratum varies with the polar angle θ' in Equation (6). By solving the stress function, the solution of a hollow cylinder subjected to cosinoidal and sinusoidal distributions of normal and shear traction can be given as [20,21]:

$$\begin{cases} \sigma_{rr}^{i,\mathrm{I}(2)} = \sigma_{rr0}^{i,\mathrm{I}(2)} + \Delta\sigma_{rr}^{i,\mathrm{I}(2)} = \sigma_{rr0}^{i,\mathrm{I}(2)} - 2\left(B^i + \frac{3C^i}{r^4} + \frac{2D^i}{r^2}\right)\cos(2\theta'), \\ \sigma_{\theta\theta}^{i,\mathrm{I}(2)} = \sigma_{\theta\theta0}^{i,\mathrm{I}(2)} + \Delta\sigma_{\theta\theta}^{i,\mathrm{I}(2)} = \sigma_{\theta\theta0}^{i,\mathrm{I}(2)} + 2\left(6\mathrm{A}^i r^2 + B^i + \frac{3D^i}{r^4}\right)\cos(2\theta'), \\ \sigma_{r\theta}^{i,\mathrm{I}} = \sigma_{r\theta0}^{i,\mathrm{I}(2)} + \Delta\sigma_{r\theta}^{i,\mathrm{I}(2)} = \sigma_{r\theta0}^{i,\mathrm{I}(2)} + 2\left(3\mathrm{A}^i r^2 + B^i - \frac{C^i}{r^2} - \frac{3D^i}{r^4}\right)\sin(2\theta'), \end{cases} \tag{16}$$

Moreover, the displacement components are given as [18,19]:

$$\begin{cases} u_{rr}^{i,\mathrm{I}(2)} = -\frac{2(1+v_i)}{E_i}(2v_i A^i r^3 + B^i r - (1-v_i)\frac{2C^i}{r} - \frac{D^i}{r^3})\cos(2\theta'), \\ u_{\theta\theta}^{i,\mathrm{I}(2)} = \frac{2(1+v_i)}{E_i}((3-2v_i)A^i r^3 + B^i r - (1-2v_i)\frac{C^i}{r^3} + \frac{D^i}{r^3})sin(2\theta'), \end{cases} \tag{17}$$

Similarly, the vertical stress components in Mode two under the plane strain condition are given as:

$$\Delta\sigma_{\mathrm{zz}}^{\mathrm{i,I}(2)} = v_{\mathrm{i}}(\Delta\sigma_{rr}^{\mathrm{i,I}(2)} + \Delta\sigma_{\theta\theta}^{\mathrm{i,I}(2)}) \tag{18}$$

In Mode two, the initial stress components are zero in the casing and cement, while the initial stress components are non-zero for the stratum. Since the increment of stress and the displacement of the stratum in Equations (16) and (17) at infinity is zero, we obtain:

$$A^{\mathrm{s}} = B^{\mathrm{s}} = 0 \tag{19}$$

The continuity of stresses and displacements at the interfaces and boundary conditions in Equation (6) give the following linear aligns:

$$\boldsymbol{M}^{\mathrm{I}(2)} \cdot \boldsymbol{X}^{\mathrm{I}(2)} = \mathbf{N}^{\mathrm{I}(2)}, \tag{20}$$

where $\boldsymbol{M}^{\mathrm{I}(2)}$ is a 10-by-10 matrix whose non-zero entries are given in Appendix A, and:

$$\begin{aligned} \mathbf{X}^{\mathrm{I}(2)} &= [A^{\mathrm{ca}}, B^{\mathrm{ca}}, C^{\mathrm{ca}}, D^{\mathrm{ca}}, A^{\mathrm{ce}}, B^{\mathrm{ce}}, C^{\mathrm{ce}}, D^{\mathrm{ce}}, C^{\mathrm{s}}, D^{\mathrm{s}}]^{\mathrm{T}}, \\ \mathbf{N}^{\mathrm{I}(2)} &= [0,0,0,0,S/2,S/2,0,0,0,0]^{\mathrm{T}}, \end{aligned} \tag{21}$$

By substituting the unknown coefficients in $\boldsymbol{X}^{\mathrm{I}(2)}$, the incremental and total stress distribution in Mode two of the system can be obtained. In contrast, the existing model [19] treats the in-situ stress as boundary pressure loads instead of an initial stress state; Equation (16) can be rewritten as:

$$\begin{cases} \sigma_r^{i,\mathrm{I}(2)} = -2\left(B^i + \frac{3C^i}{r^4} + \frac{2D^i}{r^2}\right)\cos 2\theta', \\ \sigma_\theta^{i,\mathrm{I}(2)} = 2\left(6A^i r^2 + B^i + \frac{3D^i}{r^4}\right)\cos 2\theta', \\ \tau_{r\theta}^{i,\mathrm{I}(2)} = 2\left(3A^i r^2 + B^i - \frac{C^i}{r^2} - \frac{3D^i}{r^4}\right)\sin 2\theta', \end{cases} \tag{22}$$

According to the boundary conditions at infinity for the stratum in Equation (6), $A^s = 0$, $B^s = S/2$ differing from those in corresponding Equation (19) applies. Correspondingly, the matrix $\boldsymbol{N}^{\mathrm{I}(2)}$ in Equation (21) can be rewritten as:

$$\mathbf{N}_{\mathrm{old}}^{\mathrm{I}(2)} = [0,0,0,0,S/2,S/2,0,0,Sc/2/Gs,Sc/2/Gs]^{\mathrm{T}}, \tag{23}$$

2.2.3. Comparison between the Present and Existing Model

Previous models considered in-situ stress as the boundary pressure load at infinity for the stratum, which led to displacements of the stratum at infinity being infinite, not reflecting the actual situation [20,21]. The present model improves upon this by considering in-situ stress as the initial stress state instead. For instance, the incremental radial displacement in the stratum in Sub-problem I includes the terms in Equations (9) and (17):

$$\begin{cases} u_{rr}^{s,I(1)} = -\frac{(1-2v_s)}{4G_s} A^{s,I(1)} r, \\ u_{rr}^{s,I(2)} = \frac{2(1+v_s)}{E_s} B^{s,I(2)} r\cos(2\theta'), \end{cases} \tag{24}$$

In Equation (24) $A^{s,I(1)} = -2P, B^{s,I(2)} = S/2$ applies in the existing model while $A^{s,I(1)} = B^{s,I(2)} = 0$ applies in the present model. So the displacements at infinity in the stratum will be infinity rather than zero in the existing model. This difference between the previous and present model ensures that the displacement and stress distribution in the stratum are modeled more accurately, and this applies to Sub-problem III.

2.2.4. Solution of Sub-Problem II: Elastic Uni-Axial Stress Problem

Based on the plane strain assumption, vertical stress can be given as:

$$\sigma_{zz}^i = \sigma_{zz0}^i + v_i(\Delta\sigma_{zz}^{i,I(1)} + \Delta\sigma_{zz}^{i,I(2)}) \tag{25}$$

2.2.5. Solution of Sub-Problem III: Anti-Plane Shear Problem

In this problem, the Navier equation of equilibrium is [16]:

$$\nabla^2 u_z^i = 0, \tag{26}$$

where u_z^i is the vertical displacement component. Given that the wellbore is long, $\partial u_i/\partial z = 0$ applies, and we can assume that u_z is a function of the radial coordinate r and polar angle θ. Then, the solution of Equation (26) has the following forms [18]:

$$u_z^i = \sum_{n=1}^{\infty} (A_n^i \sin(n\theta) + B_n^i \cos(n\theta)) r^n + \sum_{n=1}^{\infty} (C_n^i \sin(n\theta) + D_n^i \cos(n\theta)) r^{-n} \tag{27}$$

Using symmetry considerations, it can be suggested that the dependence of the stress upon the polar angle θ in Equation (8) makes $n = 1$. Therefore, the stress components are derived as follows:

$$\begin{aligned} \sigma_{rz}^i &= \sigma_{rz0}^i + \Delta\sigma_{rz}^i = \sigma_{rz0}^i + G_i((A^i - \tfrac{C^i}{r^2})\sin\theta + (B^i - \tfrac{D^i}{r^2})\cos\theta), \\ \sigma_{\theta z}^i &= \sigma_{\theta z0}^i + \Delta\sigma_{\theta z}^i = \sigma_{\theta z0}^i + G_i((A^i + \tfrac{C^i}{r^2})\cos\theta - (B^i + \tfrac{D^i}{r^2})\sin\theta), \end{aligned} \tag{28}$$

According to the boundary and continuity conditions, linear equations can be obtained for the stress distributions induced by the initial stress in Sub-problem III. Additionally, since the displacement of the stratum at infinity must be zero, $A^s = B^s = 0$ applies in Equations (27) and (28). Therefore, the corresponding linear aligns are given as:

$$\begin{aligned} \boldsymbol{K} \cdot \boldsymbol{X}_1^{III} &= \boldsymbol{L}_1, \\ \boldsymbol{K} \cdot \boldsymbol{X}_2^{III} &= \boldsymbol{L}_2, \end{aligned} \tag{29}$$

where

$$\begin{aligned} &\boldsymbol{X}_1^{III} = [A^{ca}, C^{ca}, A^{ce}, C^{ce}, C^s]^T, \boldsymbol{X}_2^{III} = [B^{ca}, D^{ca}, B^{ce}, D^{ce}, D^s]^T, \\ &\boldsymbol{L}_1 = \left[0, 0, S_{yz0}^s, 0, 0\right]^T, \ \boldsymbol{L}_2 = \left[0, 0, S_{xz0}^s, 0, 0\right]^T, \end{aligned} \tag{30}$$

and $\boldsymbol{K}$ is listed in Appendix A.

In contrast to the present model, the existing model assumes that the sum displacement and stress at the interface is continuous. The coefficients in Equations (27) and (28) can be given as $A_{old}^s = S_{yz}/G_s$, $B_{old}^s = S_{xz}/G_s$, according to the initial stress state in Equation (8), following a similar approach as in Sub-problem I. However, this leads to the displacement at infinity of the stratum being infinite due to the term $(A_{old}^s \sin\theta + B_{old}^s \cos\theta)r$, which is not realistic and will be discussed later. Correspondingly, the modified parameters L_1, L_2 in the existing model can be given as:

$$L_{1old} = \left[0, 0, S^{s}_{yz0}, 0, S^{s}_{yz0}c/G_s\right]^T, \ L_{2old} = \left[0, 0, S^{s}_{xz0}, 0, S^{s}_{xz0}c/G_s\right]^T \quad (31)$$

2.3. The 3D Total Stress Field within the Casing-Cement-Stratum System

The ultimate stress components can be obtained as a superposition solution of three Sub-problems: in-plane plane strain, uni-axial stress and anti-plane shear problems. Additionally, the initial stress of cement is considered, which affects the stress field and failure potential. The ultimate stresses can be given as:

$$\begin{aligned}
\sigma^i_{rr} &= \sigma^i_{rr0} + \Delta\sigma^{i,I(1)}_{rr} + \Delta\sigma^{i,I(2)}_{rr}, \\
\sigma^i_{\theta\theta} &= \sigma^i_{\theta\theta 0} + \Delta\sigma^{i,I(1)}_{\theta\theta} + \Delta\sigma^{i,I(2)}_{\theta\theta}, \\
\sigma^i_{r\theta} &= \sigma^i_{r\theta 0} + \Delta\sigma^{i,I(2)}_{r\theta}, \\
\sigma^i_{zz} &= \sigma^i_{zz0} + \nu_i(\Delta\sigma^{i,I(1)}_{zz} + \Delta\sigma^{i,I(2)}_{zz}), \\
\sigma^i_{rz} &= \sigma^i_{rz0} + \Delta\sigma^i_{rz}, \\
\sigma^i_{\theta z} &= \sigma^i_{\theta z0} + \Delta\sigma^i_{\theta z}.
\end{aligned} \quad (32)$$

3. Results and Discussions

3.1. Validation of Present Analytical Model

The purpose of this paper is to introduce a straightforward and comprehensive 3D analytical solution for the stress field of the casing-cement-stratum system, taking into account the initial stress state of cement and the stratum. The input parameter values used in the simulation, listed in Table 1, are obtained from the literature of Liu et al. [20]. The new analytical solution to the present model in Sub-problem I is compared with the Liu model [20], which considers the initial displacement induced by the in-situ stresses in the stratum and ignores the hydrostatic stress in the cement. In contrast, this paper uses the initial stress in the stratum and cement directly to obtain the ultimate stress field, resulting in a more straightforward and understandable approach.

Table 1. Input parameters (Data from Liu et al. [20]).

Component	Casing	Cement	Stratum
	Elastic coefficients and geometry		
Young's modulus E/GPa	210	12	10
Poisson's ratio ν	0.3	0.25	0.2
Inner radius/mm	59.31	69.85	107.95
	Stress state		
Initial vertical stress σ_V/MPa	0	0	−100 (assumed)
Initial horizontal stress σ_H, σ_h/MPa	0	0	−82, −55
Internal pressure p_w/MPa		−34	

The accuracy of the model results in Sub-problem I, which considers the in-situ stress as the initial stress state (present model) or as boundary pressure loads (existing model); this is validated by excellent agreement with existing results and finite element analysis (FEA) [20], as shown in Figure 3. The existing model considers the initial stress in the stratum as the boundary pressure loads, which causes great error compared to the present model. A detailed analysis has been conducted by Liu et al. [20] and Li et al. [21], which is not discussed here.

Furthermore, the existing model leads to an infinite radial displacement at infinity in the stratum, which is clearly unrealistic, as shown in Figure 4. In contrast, in the present model, as the radius increases, the radial displacement approaches zero. The reason lies in the fact that the initial stress in the stratum has been exerted before wellbore completion and cannot be regarded as the boundary pressure loads, which is always assumed in the existing model.

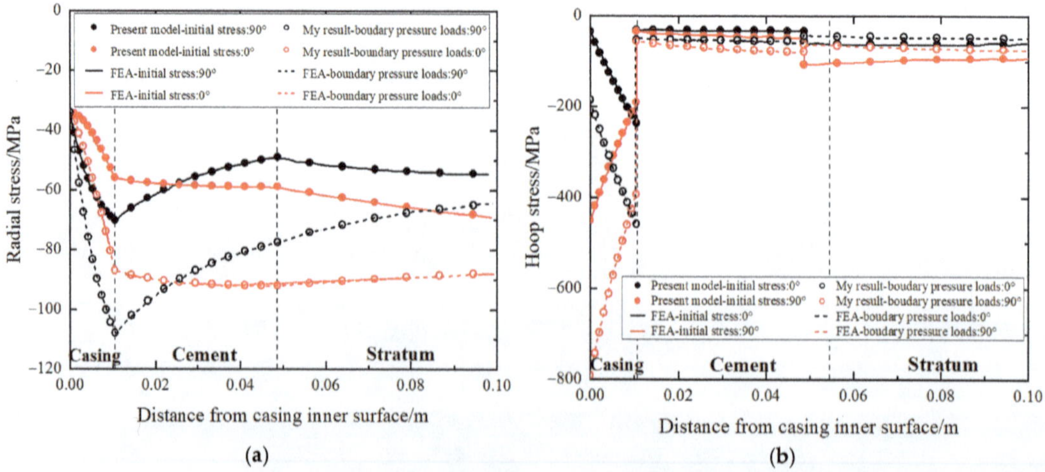

Figure 3. Comparisons of model results in Sub-problem I with existing model results and corresponding FEA [20] (Reproducted with permission from Liu W, *Appl. Math. Mech*): (**a**) radial stress; (**b**) hoop stress. In addition, model results consider in-situ stress as boundary pressure loads, which coincide with corresponding FEA [20] and validate the accuracy of the analytical method. The input parameters are given in Table 1.

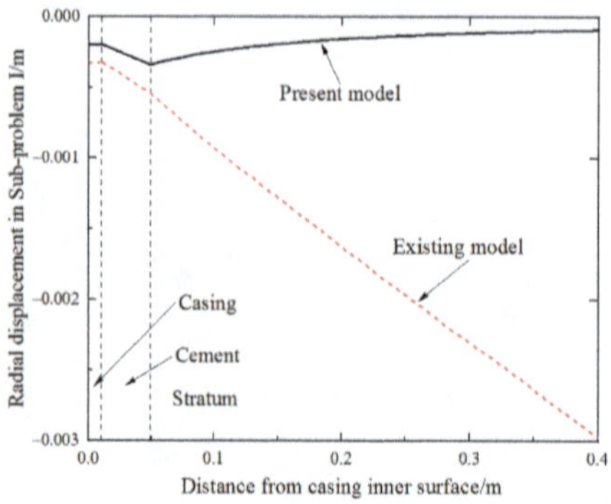

Figure 4. Radial displacement along 0° direction with increasing radius in Sub-problem I. The displacement in stratum increases linearly with increasing radius, which implies unrealistic infinite displacement at infinity.

3.2. Model Comparison between the Present Solution and Existing One in Antiplane Shear Problem III

In the case of an inclined wellbore with a deviation azimuth of 30° and a deviation angle of 60°, the in-situ anti-plane shear stresses in the stratum can be given as $\sigma^s_{xz0} = 16.6$ MPa, $\sigma^s_{yz0} = 5.8$ MPa. The induced anti-plane stresses using both the present and existing analytical models are shown in Figure 5. Similar to Sub-problem I, the existing analytical model leads to significant errors compared to the corresponding present model in Sub-problem III.

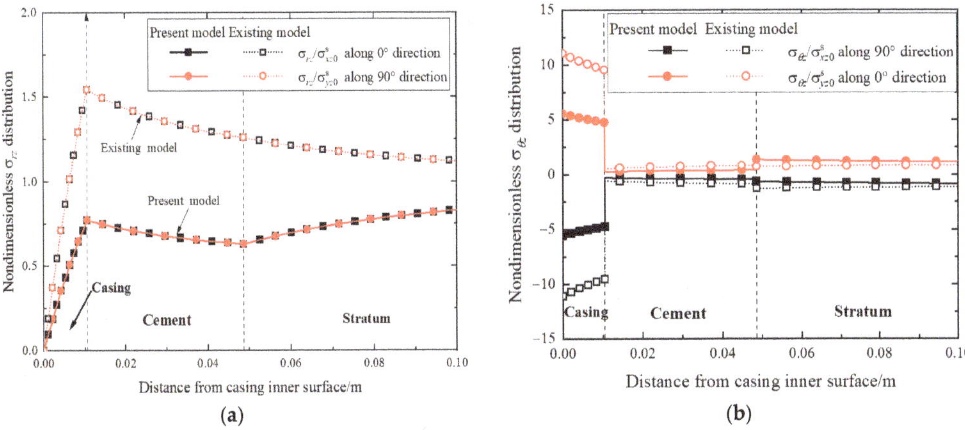

Figure 5. Comparisons of model results of anti-plane stress field in the casing-cement-stratum system with existing model: (**a**) nondimensionless $\sigma_{rz}/\sigma^s_{xz0}$ distribution along 90° direction; (**b**) nondimensionless $\sigma_{\theta z}/\sigma^s_{yz0}$ distribution along 0° direction, which show great differences between them.

Equations (29) and (30) imply that the coefficients $\boldsymbol{X}_1^{\mathrm{III}}, \boldsymbol{X}_2^{\mathrm{III}}$ are proportional, resulting in the same nondimensionless anti-plane shear stress as the $\sigma_{rz}/\sigma^s_{xz0}, \sigma_{rz}/\sigma^s_{yz0}$ distributions when considering the dependence of cosinoidal and sinusoidal anti-plane shear stress in Equation (8). However, the existing model greatly amplifies the induced anti-plane shear stresses $\sigma_{rz}, \sigma_{\theta z}$ compared to the corresponding present model, especially in casing and cement. Therefore, the present analytical model is recommended to accurately calculate the stress field in the casing-cement-stratum system in all Sub-problems. Additionally, the existing model in Sub-problem III causes linearly increasing radial displacement in the stratum in Figure 6, which is also unrealistic at infinity.

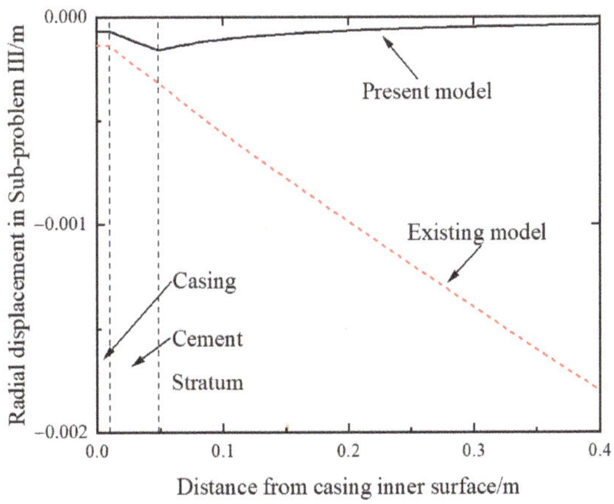

Figure 6. Radial displacement along 0°direction with increasing radius in the anti-plane shear Sub-problem III. The radial displacement in stratum increases linearly with increasing radius, which implies unrealistic infinite displacement at infinity.

Similar to the result of Sub-problem I, a low Young's modulus of cement will decrease the anti-plane shear stress $\sigma_{rz}, \sigma_{\theta z}$ in the casing and cement in both the present and existing models, as shown in Figure 7. This reduction will benefit wellbore integrity. However, Poisson's ratio of cement has little influence on the anti-plane shear stress.

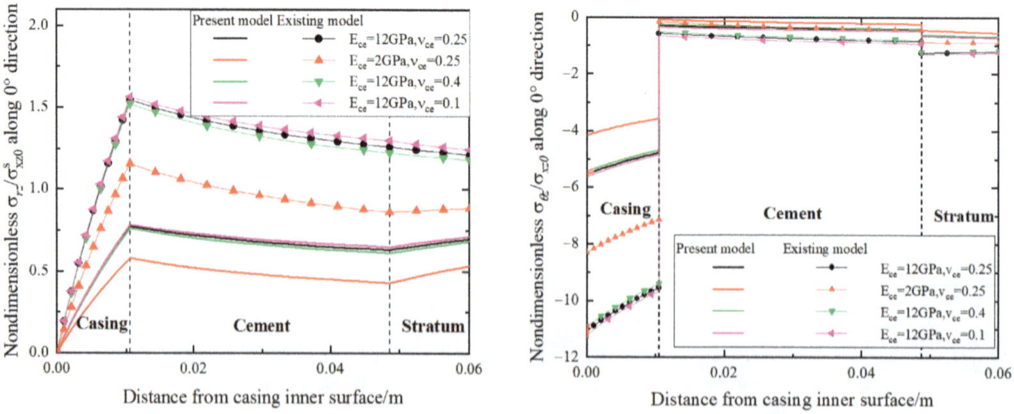

(a) (b)

Figure 7. Influence of Young's modulus and Poisson's ratio of cement on the anti-plane shear stress distribution: (**a**) nondimensionless $\sigma_{rz}/\sigma^s_{xz0}$; (**b**) nondimensionless $\sigma_{\theta z}/\sigma^s_{yz0}$ distribution along 0° direction.

3.3. The Effect of Initial Stress State of Cement on the Stress Field

In previous models [7–21], the initial stress of the cement was often ignored. However, the existence of initial stress in cement has a significant influence on the stress distribution, as shown in Figure 8. Due to the isotropic initial stress state in the cement, the effect of the initial stress state of cement is studied in Mode one of Sub-problem I. The initial stress state in cement promotes the uniformity of radial and hoop stress along the radial direction and increases the absolute values of compressive radial and hoop stresses. Moreover, for the given parameters in this case, the initial stress state in cement has little influence on the stress field in the casing.

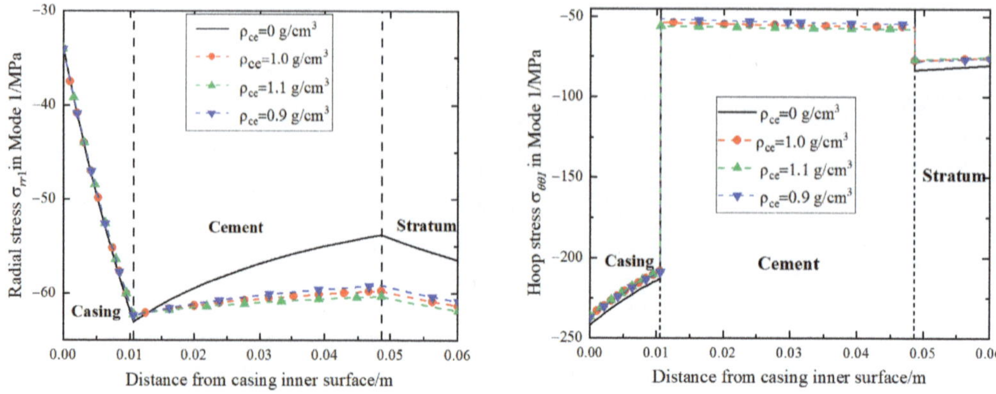

(a) (b)

Figure 8. Influence of the initial stress state of cement on stress field within the casing-cement-stratum system: (**a**) Radial stress; (**b**) Hoop stress in Mode 1 of Sub-problem I. In this case, a vertical well is considered.

The effect of Young's modulus and Poisson's ratio on cement in the stress field in Mode one of Sub-problem I, considering the initial stress state of cement, is studied in Figure 9. Considering the initial stress state of cement, a low Young's modulus also induces low absolute values of radial and hoop stress. Ignoring the initial stress state of cement will amplify the decreasing degree of absolute values in radial and hoop stresses in the casing and cement under low Young's modulus of cement condition.

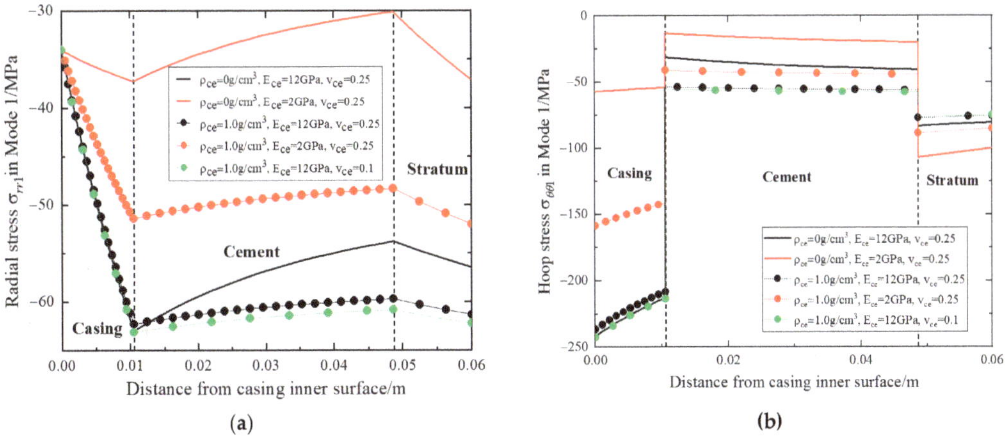

Figure 9. Influence of Young's modulus and Poisson's ratio of cement on stress field with and without the initial stress state of cement: (**a**) Radial stress; (**b**) Hoop stress in Mode 1 of Sub-problem I. In this case, a vertical well is considered.

4. Estimation of the Casing and Cement Integrity

The Mohr-Coulomb failure criterion has been widely used to evaluate cement failure potential, while the von Mises yield criterion is used to assess longitudinal zonal isolation in the cement [34]. In this paper, the 3D von Mises yield criterion is adopted to evaluate the failure potential of casing and cement:

$$\sigma_{\text{Mises}} = \sqrt{\frac{1}{2}\left[(\sigma_r - \sigma_\theta)^2 + (\sigma_r - \sigma_z)^2 + (\sigma_z - \sigma_\theta)^2\right] + 3\tau_{r\theta}^2 + 3\tau_{rz}^2 + 3\tau_{\theta z}^2} \tag{33}$$

Correspondingly, the 2D von Mises yield criterion is simplified as:

$$\sigma_{\text{Mises,2D}} = \sqrt{\frac{1}{2}((\sigma_r - \sigma_\theta)^2 + \sigma_r^2 + \sigma_\theta^2 + 6\tau_{r\theta}^2)} \tag{34}$$

The Mises stress in the 2D and 3D solution at the inner surfaces of casing and cement are evaluated by the present and existing models, as shown in Figure 10. In this case, an inclined wellbore with a deviation azimuth of 30°, a deviation angle of 60° and zero initial stress of cement is considered. The symmetry of von Mises stress in the 2D solution, θ_r, validates the accuracy of the model results in Sub-problem I. In contrast, the existence of an anti-plane shear stress distribution in Sub-problem III breaks the symmetry of the Mises stress distribution, θ_r, in the casing and cement.

Consistent with the law above, the existing model greatly signifies the values of Mises stress in casing and cement both in the 2D and 3D solution. The 3D solutions of the present and existing models differ from the corresponding 2D solutions, which highlights the necessity of establishing the 3D analytical solution to evaluate wellbore integrity. Counter-intuitively, the 2D Mises stress in both the present and existing models is higher than that in the 3D case, particularly for the cement. This suggests that the common 2D solution overestimates the failure potential of cement compared to the 3D model. This could be because ignoring the existence of stress components with axial direction increases the nonuniform degree of stress components.

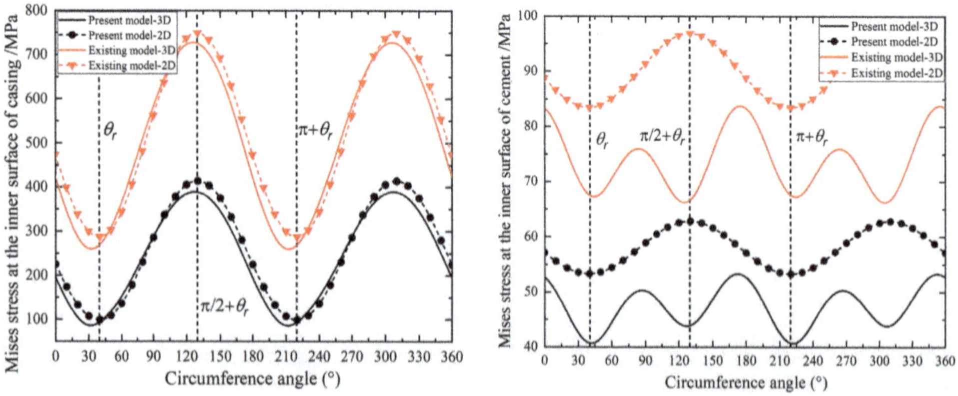

(**a**) (**b**)

Figure 10. The Mises stress distribution in 2D and 3D solution at the inner surfaces of (**a**) casing and (**b**) cement using the present and existing model. Two-dimensional von Mises stress distribution with a changing circumference angle shows symmetry θ_r, while anti-plane shear stress in the 3D model breaks it. In this case, an inclined wellbore with a deviation azimuth of 30° and a deviation angle of 60° is considered.

Figure 11 demonstrates how Young's modulus and the initial stress state of cement affect the maximum Mises stress in casing and cement in the present 3D model. As Young's modulus increases, the maximum Mises stress in the casing, and cement approaches a constant value. Simultaneously, the presence of initial stress in cement significantly decreases the maximum Mises stress in the cement with moderate Young's modulus values of cement. This occurs because the initial hydrostatic compressive stress state in cement moves Mohr's circle of stress in cement away from the failure envelope compared to the case of zero initial stress. Additionally, under a low Young's modulus of cement conditions, the presence of initial stress in cement aggravates the failure potential of the casing, and vice versa.

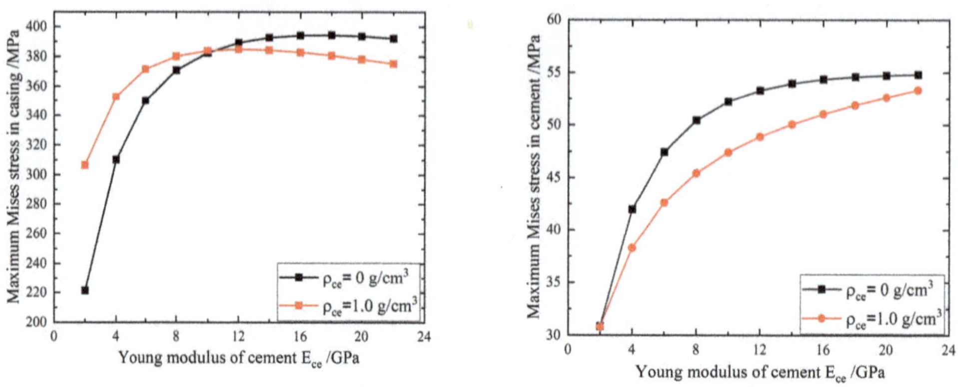

(**a**) (**b**)

Figure 11. The effect of Young's modulus and initial stress of cement on the maximum von Mises stress in (**a**) casing and (**b**) cement in the present 3D model.

Figure 12 illustrates the influence of wellbore pressure on the maximum Mises stress in casing and cement in the present 3D model. As the wellbore pressure increases, the failure potential of the casing decreases while that of the cement increases. Furthermore, the presence of initial stress in cement significantly decreases the increased failure potential of cement with increasing internal wellbore pressure. The initial stress state has little influence on the maximum Mises stress in the casing since radial and hoop stresses hardly change with increasing initial stress in cement for the given specific parameters, as shown in Figure 8.

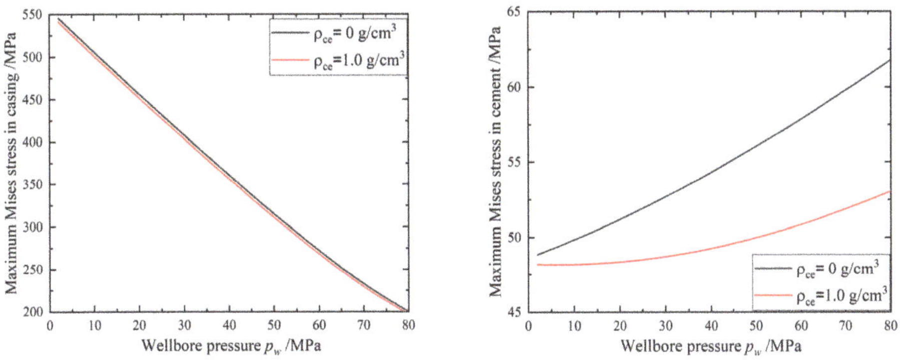

Figure 12. The initial stress state of cement's effect on the maximum von Mises stress in (**a**) casing and (**b**) cement in the present 3D model for the given specific parameters in Table 1.

5. Conclusions

This paper presents a new 3D analytical model for evaluating the stress field in inclined oil/gas wells, considering in-situ stress as the initial stress instead of boundary pressure loads and the existence of the initial stress of cement. The 3D analytical solution is obtained by dividing the model into an in-plane plane strain problem, elastic uni-axial stress problem and anti-plane shear problem. Moreover, the accuracy of the present model is validated by the existing model results and finite element analysis. The following main conclusions can be made:

(1) Compared to the existing model, the present model provides a more accurate stress level in the anti-plane shear problem and eliminates unrealistic infinite displacements at infinity.
(2) The maximum Mises stress in the present 3D model differs from that in the corresponding 2D model, highlighting the importance of considering the uni-axial stress problem and anti-plane shear problem. For the specific conditions studied in this paper, the maximum Mises stress in the cement in the present 3D model is lower than that in the corresponding 2D model, resulting in a more conservative failure potential in the 3D circumstance.
(3) The existence of initial stress in cement has a significant effect on the cement failure potential. For the specific conditions studied in this paper, as the wellbore pressure increases, the failure potential of the casing decreases while that of cement increases. However, the presence of initial stress in cement significantly decreases the increased failure potential of cement with increasing internal wellbore pressure.

Author Contributions: Conceptualization, X.W. and W.L.; methodology, T.J.; software, Y.Z.; validation, Y.Z.; investigation, Y.Z.; resources, J.Z.; writing—original draft preparation, X.W.; writing—review and editing, W.L.; visualization, H.X.; supervision, T.J.; project administration, T.J.; funding acquisition, T.J. All authors have read and agreed to the published version of the manuscript.

Funding: This research was financially supported by the NSFC Basic Research Program on Deep Petroleum Resource Accumulation and Key Engineering Technologies (No. U19B6003), and also the support of the Fundamental Research Program of Shanxi Province, China (No. 20210302124664).

Institutional Review Board Statement: Not applicable.

Informed Consent Statement: Not applicable.

Data Availability Statement: Data is contained in the article.

Acknowledgments: The authors appreciate valuable comments and help from W. Liu at China University of Petroleum, Beijing and X.L. Guo in CNPC Engineering Technology R&D Company Limited, Beijing 102206, China. All individuals included in this section have consented to the acknowledgement.

Conflicts of Interest: The authors declare no conflict of interest.

Appendix A

$$
\begin{aligned}
&M_{1,1}^{I(1)} = 1/2, M_{1,2}^{I(1)} = 1/a^2,\\
&M_{2,1}^{I(1)} = 1/2, M_{2,2}^{I(1)} = 1/b^2, M_{2,3}^{I(1)} = -1/2, M_{2,4}^{I(1)} = -1/b^2,\\
&M_{3,1}^{I(1)} = (1-2\nu_{ca})b/(4G_{ca}), M_{3,2}^{I(1)} = -1/(2G_{ca}b),\\
&M_{3,3}^{I(1)} = -(1-2\nu_{ce})b/(4G_{ce}), M_{3,4}^{I(1)} = 1/(2G_{ce}b),\\
&M_{4,3}^{I(1)} = 1/2, M_{4,4}^{I(1)} = 1/c^2, M_{4,5}^{I(1)} = -1/c^2,\\
&M_{5,3}^{I(1)} = (1-2\nu_{ce})c/(4G_{ce}), M_{5,4}^{I(1)} = -1/(2G_{ce}c),\\
&M_{5,5}^{I(1)} = 1/(2G_{s}c),\\
&M_{1,2}^{I(2)} = 1, M_{1,3}^{I(2)} = 2/a^2, M_{1,4}^{I(2)} = 3/a^4,\\
&M_{2,1}^{I(2)} = 3a^2, M_{2,2}^{I(2)} = 1, M_{2,3}^{I(2)} = -1/a^2, M_{2,4}^{I(2)} = -3/a^4,\\
&M_{3,2}^{I(2)} = 1, M_{3,3}^{I(2)} = 2/b^2, M_{3,4}^{I(2)} = 3/b^4,\\
&M_{3,6}^{I(2)} = -1, M_{3,7}^{I(2)} = -2/b^2, M_{3,8}^{I(2)} = -3/b^4,\\
&M_{4,1}^{I(2)} = 3b^2, M_{4,2}^{I(2)} = 1, M_{4,3}^{I(2)} = -1/b^2, M_{4,4}^{I(2)} = -3/b^4,\\
&M_{4,5}^{I(2)} = -3b^2, M_{4,6}^{I(2)} = -1, M_{4,7}^{I(2)} = 1/b^2, M_{4,8}^{I(2)} = 3/b^4,\\
&M_{5,6}^{I(2)} = 1, M_{5,7}^{I(2)} = 2/c^2, M_{5,8}^{I(2)} = 3/c^4,\\
&M_{5,9}^{I(2)} = -2/c^2, M_{5,10}^{I(2)} = -3/c^4,\\
&M_{6,5}^{I(2)} = 3c^2, M_{6,6}^{I(2)} = 1, M_{6,7}^{I(2)} = -1/c^2, M_{6,8}^{I(2)} = -3/c^4,\\
&M_{6,9}^{I(2)} = 1/c^2, M_{6,10}^{I(2)} = 3/c^4,\\
&M_{7,1}^{I(2)} = 2\nu_{ca}b^3/G_{ca}, M_{7,2}^{I(2)} = b/G_{ca}, M_{7,3}^{I(2)} = -2(1-\nu_{ca})/(bG_{ca}), M_{7,4}^{I(2)} = -1/(b^3G_{ca}),\\
&M_{7,5}^{I(2)} = -2\nu_{cc}b^3/G_{ce}, M_{7,6}^{I(2)} = -b/G_{ce}, M_{7,7}^{I(2)} = 2(1-\nu_{ce})/(bG_{ce}), M_{7,8}^{I(2)} = 1/(b^3G_{ce}),\\
&M_{8,1}^{I(2)} = (3-2\nu_{ca})b^3/G_{ca}, M_{8,2}^{I(2)} = b/G_{ca}, M_{8,3}^{I(2)} = -(1-2\nu_{ca})/(bG_{ca}), M_{8,4}^{I(2)} = 1/(b^3G_{ca}),\\
&M_{8,5}^{I(2)} = -(3-2\nu_{ce})b^3/G_{ce}, M_{8,6}^{I(2)} = -b/G_{ce}, M_{8,7}^{I(2)} = (1-2\nu_{ce})/(bG_{ce}), M_{8,8}^{I(2)} = -1/(b^3G_{ce}),\\
&M_{9,5}^{I(2)} = 2\nu_{ce}c^3/G_{ce}, M_{9,6}^{I(2)} = c/G_{ce}, M_{9,7}^{I(2)} = -2(1-\nu_{ce})/(cG_{ce}), M_{9,8}^{I(2)} = -1/(c^3G_{ce}),\\
&M_{9,9}^{I(2)} = 2(1-\nu_{s})/(cG_{s}), M_{9,10}^{I(2)} = 1/(c^3G_{s}),\\
&M_{10,5}^{I(2)} = (3-2\nu_{ce})c^3/G_{ce}, M_{10,6}^{I(2)} = c/G_{ce}, M_{10,7}^{I(2)} = -(1-2\nu_{ce})/(cG_{ce}), M_{10,8}^{II} = 1/(c^3G_{ce}),\\
&M_{10,9}^{I(2)} = (1-2\nu_{s})/(cG_{s}), M_{10,10}^{I(2)} = -1/(c^3G_{s}),
\end{aligned}
$$

$$
\begin{aligned}
&K_{1,1} = 1, K_{1,2} = -1/a^2,\\
&K_{2,1} = G_{ca}, K_{2,2} = -G_{ca}/b^2, K_{2,3} = -G_{ce}, K_{2,4} = G_{ce}/b^2,\\
&K_{3,3} = G_{ce}, K_{3,4} = -G_{ce}/c^2, K_{3,5} = G_{s}/c^2\\
&K_{4,1} = b, K_{4,2} = 1/b, K_{4,3} = -b, K_{4,4} = -1/b,\\
&K_{5,3} = c, K_{5,4} = 1/c, K_{5,5} = -1/c,
\end{aligned}
$$

References

1. Su, D.; Li, Z.; Huang, S.; Wu, X.; Li, J.; Xue, Y. Experiment and failure mechanism of cement sheath integrity under development and production conditions based on a mechanical equivalent theory. *Energy Sci. Eng.* **2021**, *9*, 2400–2422. [CrossRef]
2. Lian, W.; Li, J.; Tao, Q.; Du, J.; Wang, L.; Xi, Y. Formation mechanism of continuous gas leakage paths in cement sheath during hydraulic fracturing. *Energy Sci. Eng.* **2020**, *8*, 2527–2547. [CrossRef]
3. Dusseault, M.; Jackson, R. Seepage pathway assessment for natural gas to shallow groundwater during well stimulation, in production, and after abandonment. *Environ. Geosci.* **2014**, *21*, 107–126. [CrossRef]
4. Gray, K.E.; Podnos, E.; Becker, E. Finite-element studies of near-wellbore region during cementing operations: Part I. *SPE Drill. Complet.* **2009**, *24*, 127–136. [CrossRef]
5. Wang, W.; Taleghani, A.D. Three-dimensional analysis of cement sheath integrity around Wellbores. *J. Pet. Sci. Eng.* **2014**, *121*, 38–51. [CrossRef]
6. Shen, Z.; Beck, F.E. Three-dimensional modeling of casing and cement sheath behavior in layered, nonhomogeneous formations. In Proceedings of the IADC/SPE Asia Pacific Drilling Technology Conference and Exhibition, Tianjin, China, July 2012.
7. Yin, Y.; Chen, Z.; Li, P. Theoretical solutions of stress distribution in casing-cement and stratum system. *AATA Mech. Sin.* **2006**, *38*, 835–842. (In Chinese)
8. Yin, F.; Gao, D.L. Mechanical analysis and design of casing in directional well under in-situ stresses. *J. Nat. Gas Sci. Eng.* **2014**, *20*, 285–291. [CrossRef]
9. Fang, J.; Wang, Y.; Gao, D. On the collapse resistance of multilayer cemented casing in directional well under anisotropic formation. *J. Nat. Gas Sci. Eng.* **2015**, *26*, 409–418. [CrossRef]
10. Liu, K.; Gao, D.; Taleghani, A.D. Analysis on Integrity of Cement Sheath in the Vertical Section of Wells during Hydraulic Fracturing. *J. Pet. Sci. Eng.* **2018**, *168*, 370–379. [CrossRef]
11. Zhang, L.; Yan, X.; Yang, X.; Zhao, X. Evaluation of wellbore integrity for HTHP gas wells under solid-temperature coupling using a new analytical model. *J. Nat. Gas Sci. Eng.* **2015**, *25*, 347–358. [CrossRef]
12. Li, B.; Guo, B.; Li, H.; Shi, Y. An analytical solution to simulate the effect of cement/formation stiffness on well integrity evaluation in carbon sequestration projects. *J. Nat. Gas Sci. Eng.* **2015**, *27*, 1092–1099. [CrossRef]
13. Wang, Y.; Gao, D.; Fang, J. Assessment of wellbore integrity of offshore drilling in well testing and production. *J. Eng. Mech.* **2016**, *142*, 04016030. [CrossRef]
14. Liu, K.; Gao, D.; Taleghani, A.D. Impact of Casing Eccentricity on Cement Sheath. *Energies* **2018**, *11*, 2557. [CrossRef]
15. Xia, Y.; Jin, Y.; Chen, M.; Lu, Y. Dynamic analysis of a cylindrical casing–cement structure in a poroelastic stratum. *Int. J. Numer. Anal. Methods Geomech.* **2017**, *41*, 1362–1389. [CrossRef]
16. Li, W.; Chen, M.; Jin, Y.; Lu, Y.; Zhang, Y.; Bao, Y.; Yang, R. The Porothermoelastic Response of a Casing-Cement-Stratum System. In Proceedings of the 52nd US Rock Mechanics/Geomechanics Symposium, Seattle, WA, USA, 17–20 June 2018.
17. Atkinson, C.; Eftaxiopoulos, D.A. A plane model for the stress field around an inclined, cased and cemented wellbore. *Int. J. Numer. Anal. Methods Geomech.* **1996**, *20*, 549–569. [CrossRef]
18. Vitali, O.P.M.; Celestino, T.B.; Bobet, A. Analytical solution for tunnels not aligned with geostatic principal stress directions. *Tunn. Undergr. Space Technol.* **2018**, *82*, 394–405. [CrossRef]
19. Jo, H.; Gray, K.E. Mechanical behavior of concentric casing, cement, and formation using analytical and numerical methods. In Proceedings of the 44th US Rock Mechanics Symposium and 5th U.S.-Canada Rock Mechanics Symposium, Salt Lake City, UT, USA, 27–30 June 2010.
20. Liu, W.; Yu, B.; Deng, J. Analytical method for evaluating stress field in casing-cement-formation system of oil/gas wells. *Appl. Math. Mech.* **2017**, *38*, 1273–1294. [CrossRef]
21. Li, J.; Guo, X.; Liu, G.; Liu, S.; Xi, Y. New Analytical Method to Evaluate Casing Integrity during Hydraulic Fracturing of Shale Gas Wells. *Shock. Vib.* **2019**, 4253241. [CrossRef]
22. Wei, S.; Ergun, K.; Jin, Y.; Yang, X. Numerical investigation of the factors affecting the cement sheath integrity in hydraulically fractured wells. *J. Pet. Sci. Eng.* **2022**, *215*, 110582. [CrossRef]
23. Yan, Y.; Guan, Z.; Zhang, B.; Chen, W. Numerical investigation of debonding extent development of cementing interfaces during hydraulic fracturing through perforation cluster. *J. Pet. Sci. Eng.* **2021**, *197*, 107970. [CrossRef]
24. Simone, M.D.; Pereira, F.L.G.; Roehl, D.M. Analytical methodology for wellbore integrity assessment considering casing-cement-formation interaction. *Int. J. Rock Mech. Min. Sci.* **2017**, *94*, 112–122. [CrossRef]
25. Saint-Marc, J.; Garnier, A.; Bois, A.P. Initial State of Stress: The Key to Achieving Long-Term Cement-Sheath Integrity. In Proceedings of the SPE Annual Technical Conference and Exhibition, Seattle, WA, USA, 21–24 September 2008.
26. Petersen, T.; Ulm, F.J. Early-Age Stress and Pressure Developments in a Wellbore Cement Liner: Application to Eccentric Geometries. *J. Appl. Mech.* **2016**, *83*, 091012. [CrossRef]
27. Meng, M.; Frash, L.; Carey, B.; Niu, Z.; Welch, N.J. Predicting Cement Sheath Integrity with Consideration of Initial State of Stress and Thermoporoelastic Effects. *SPE J.* **2020**, *26*, 3505–3528. [CrossRef]
28. Nygaard, R.; Salehi, S.; Lavoie, R.G. Effect of Dynamic Loading on Wellbore Leakage for the Wabamun Area CO_2-Sequestration Project. *J. Can. Pet. Technol.* **2014**, *53*, 69–82. [CrossRef]
29. Frash, L.P.; Carey, J.W. Engineering Prediction of Axial Wellbore Shear Failure Caused by Reservoir Uplift and Subsidence. *SPE J.* **2018**, *23*, 1039–1066. [CrossRef]

30. Bois, A.P.; Garnier, A.; Galdiolo, G.; Laudet, J.B. Use of a mechanistic model to forecast cement-sheath integrity. *SPE Drill. Complet.* **2012**, *27*, 303–314. [CrossRef]
31. Cui, L.; Cheng, A.; Leshchinsky, D.; Abousleiman, Y.; Roegiers, J.C. Stability analysis of an inclined borehole in an isotropic poroelastic medium. In Proceedings of the 35th US Symposium on Rock Mechanics (USRMS), Reno, NV, USA, 5 June 1995.
32. Gao, J.; Lau, H.C.; Sun, J. A Semianalytical Poroelastic Solution to Evaluate the Stability of a Borehole Drilled Through a Porous Medium Saturated with Two Immiscible Fluids. *SPE J.* **2020**, *25*, 2319–2340. [CrossRef]
33. Gao, J.; Lin, H.; Sun, J.; Chen, X.; Wang, H.; Liu, X. A porothermoelastic model considering the dynamic temperature-perturbation boundary effect for borehole stability in fractured porous rocks. *SPE J.* **2022**, *27*, 2491–2509. [CrossRef]
34. Wang, H.; Shi, L.; Zheng, Y.; Zhang, C. Sealing Capacity Analysis of Cement Sheath Based on Combined Mechanical Model. *J. Northeast. Univ. Nat. Sci.* **2020**, *41*, 1334. (In Chinese)

processes

Article

Research on Managed-Pressure Running Casing in Oil and Gas Wells with the Negative Pressure Window

Yuntao Mei [1,2], Huanqiang Yang [1,2,*], Zhuo Zhang [3] and Mengjia Ji [4]

1 College of Petroleum Engineering, Yangtze University, Wuhan 430100, China; 2021720411@yangtzeu.edu.cn
2 National Engineering Research Center for Oil & Gas Drilling and Completion Technology, Yangtze University, Wuhan 430100, China
3 Engineering Technology Department of Western Drilling Company, Karamay 834000, China
4 Sinopec Zhongyuan Oilfield, Puyang 457000, China; 202072342@yangtzeu.edu.cn
* Correspondence: yanghuanqiang@yangtzeu.edu.cn

Abstract: The failure of managed-pressure running casing in oil and gas wells may lead to complex accidents such as overflow or leakage. The technique of using multi-density gradient drilling fluids in wellbores with negative pressure windows (NPWs) is often used to deal with this situation. Therefore, it is vital to analyze the dynamic slurry column structure and calculate the wellbore pressure during casing running. For this issue, the model of transient surge pressure is established during casing running. The calculation equation of the model is proposed, and the calculations of the wellbore pressure are carried out with the exploration of Well LT-X1, located in the Xinjiang oil field. A circulation scheme is designed as follows: Circulate 125 m^3 of drilling fluid with a density of 2.45 g/cm^3 and 155 m^3 of drilling fluid with a density of 2.35 g/cm^3 at a depth of 3560 m. From there, circulate 164 m^3 of drilling fluid with a density of 2.35 g/cm^3 at a depth of 5900 m. Finally, at a depth of 7050 m, circulate 250 m^3 of drilling fluid with a density of 2.30 g/cm^3. The casing running speeds and back-pressure values were designed as follows for the respective well sections: 0–1523 m: 0.160 m/s casing speed, 0 MPa back pressure; 1523–3560 m: 0.160 m/s casing speed, 1.641 MPa back pressure; 3560–5900 m: 0.145 m/s casing speed, 2.427 MPa back pressure; 5900–6674 m: 0.137 m/s casing speed, 4.041 MPa back pressure; 6674–7050 m: 0.124 m/s casing speed, 4.457 MPa back pressure. The results show that optimizing structure of the multi-density gradient drilling fluid with different densities and applying annular back pressure in stages, with the accurate calculation of wellbore pressure, can achieve the goals of leak-proofing and pressure-stabilization. It is concluded that this result may serve as the foundation for managed-pressure running casing technology.

Keywords: negative pressure window; managed-pressure running casing; multi-density gradient drilling fluids; transient surge pressure; leak-proof and pressure-stabilized

Citation: Mei, Y.; Yang, H.; Zhang, Z.; Ji, M. Research on Managed-Pressure Running Casing in Oil and Gas Wells with the Negative Pressure Window. *Processes* **2023**, *11*, 2210. https://doi.org/10.3390/pr11072210

Academic Editors: Tianshou Ma and Yuqiang Xu

Received: 26 June 2023
Revised: 16 July 2023
Accepted: 18 July 2023
Published: 22 July 2023

1. Introduction

Complicated formation pressure systems, such as NPWs in oil and gas wells, are increasing, which may lead to complex accidents. Lost circulation and overflow in the wellbore are two typical accidents that occur when the wellbore pressure is beyond the scope of pore pressure and leakage pressure during the drilling operation [1–3]. To solve this problem, managed-pressure drilling (MPD) continues to rely on the back pressure applied by the surface high-pressure pump at the wellhead to precisely control the annular pressure profile, despite the noise pollution generated by the high-pressure pump [4]. After drilling to the predetermined depth with MPD technology, the next step is always to pull out of hole (POOH) and prepare the next construction called casing running. The calculation of wellbore pressure in the processes of pre-casing preparation, down-casing and drilling fluid circulation are important for developing the casing running plan reasonably and adjusting the back pressure in a timely manner.

Surge pressure, generated during the process of casing running, is an important part of wellbore pressure that is crucial for avoiding complex accidents in wellbores. In various working conditions, extensive studies have been conducted to analyze and gain a comprehensive understanding of surge pressure. The field studies aimed to examine the impact of fluid properties and wellbore geometry on surge and swab pressures [5–9]. The findings of these studies consistently indicate that both surge pressures exhibit a positive correlation with increasing tripping speeds and wellbore depths. These studies only investigated two types of surge pressures, both of which are positively correlated with an increase in tripping speeds and wellbore depths. More recent studies have further confirmed the strong impact of tripping speeds, wellbore geometry, flow regime, and fluid rheology on surge pressures [10–16].

However, the drilling industry greatly relies on surge models and simulators. At present, some models, including steady-state models and transient models, have been proposed to calculate the surge pressure during POOH and casing running processes. Before the 1970s, the models were mainly based on the theory of the steady-state method and usually considered the fluid in the wellbore as a Newtonian model [17–20]. Furthermore, more and more models were developed considering the Bingham model, power-law model and Roberston–Stiff model [21–23]. In recent years, the impact of stabilizer placement on the annular clearance has been taken into account in the prediction of the volatility stress model [24]. Meanwhile, the error of surge pressure calculation results reaches $\pm 3\%$ [25]. These models mainly analyze the influence of drilling fluid density, stabilizer quantity, annular clearance, and drill string eccentricity on the surge pressure. In summary, while surge models and simulators have made significant progress in predicting surge pressure during POOH and casing running processes, there are limitations in capturing the full complexity of fluid behavior and accurately accounting for all factors that influence surge pressure. Further research and advancements are needed to improve the accuracy and reliability of surge models in the drilling industry.

After mid-1970s, with the improvement of computer calculation capabilities, the transient method was applied in the calculation of surge pressure. Fluid compressibility, duct wall expansion, and fluid inertia were considered with the establishment of the surge pressure model [26]. The variation of maximum surge pressure and swabbing pressure at each depth within the wellbore with time were predicted proposed solving the transient surge pressure of each passage using the mixing implicit feature line method, which obtained the numerical solutions of transient surge pressure under various conditions [27]. To acquire the wellbore pressure and ground stress field in the near-wellbore section of deepwater drilling, some scholars proposed a method for analyzing wellbore stability based on the oscillation coupling of transient surge pressure [28]. Considering the liquid–solid two-phase medium, researchers studied the transient surge pressure during tripping operations with the use of the numerical simulation method [29]. Based on the transient method, the effect of overflow, drilling speed and acceleration were analyzed in the calculation of surge pressure [30,31]. Using spectrum analyses, it was found that high-pressure hoses are most susceptible at certain frequencies to deformation, and to the response and surge pressures of the fluid within. This can even lead to irreversible degradation changes in the internal structure of the hose [32]. However, it is important to note that despite the advancements and insights gained from these studies, there are limitations to consider. The basis of their model establishment is limited to a single-density fluid within the wellbore.

The above-established calculation models for the surge pressure are all based on homogeneous fluid within the wellbore and cannot be applied to drilling fluids with multiple density gradients that are usually used in wells with NPW. This paper presents the structure of the slurry column and surge pressure calculation models during the casing running process, considering the drilling fluids with multiple density gradients in the wellbore. Taking Well LT-X1 as an example, the pressure-managed technology for casing running under NPWs is developed.

The formation pressure data of Well LT-X1 are shown in Figure 1. At the depth of 5852 m, and 6152 m in the wellbore, the equivalent densities of pore pressure and leakage pressure are 2.49 g/cm^3 and 2.52 g/cm^3, 2.425 g/cm^3 and 2.445 g/cm^3, respectively. The leakage pressure at the depth of 6152 m is lower than the pore pressure at 5852 m, which is defined as the NPW. So, the reservoir pore pressure cannot be balanced by drilling fluid for which the density is equal to pore pressure. This will result in lost circulation at the bottom of the hole and kick (probably evolving to blowout) at the upper depth of the well. MPD technology was employed in this well at the key well sections with NPWs. After drilling, the drill rods are pulled out of hole, and the fluid in the wellbore will have a loss of circulation. Therefore, drilling fluids with multi-density gradient column structures are needed to balance the reservoir pore pressure.

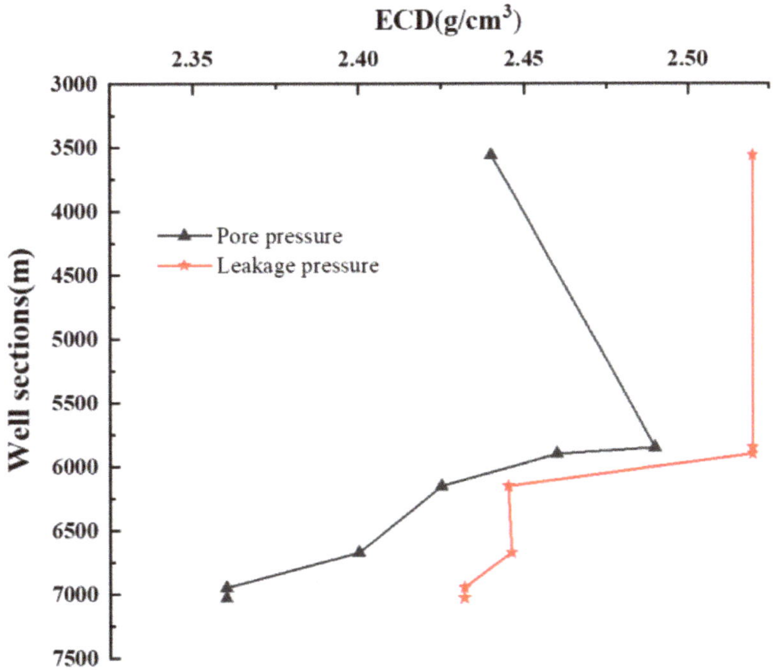

Figure 1. The formation pressure of Well LT-X1 with an NPW.

2. Methodology

2.1. Model Description

As shown in Figure 2a, we assume there is an oil and gas well drilled to a depth of H, which has n sets of pressure systems with the depths of H_1', H_2', H_3', …… , H_n' in an open-hole section, and the corresponding pore pressure and leakage pressure are (ρ_{p1}, ρ_{f1}), (ρ_{p2}, ρ_{f2}), (ρ_{p3}, ρ_{f3}), …… , (ρ_{pn}, ρ_{fn}), respectively. The density of drilling fluid in the wellbore is ρ_0, since the leakage pressure ρ_{f1} at the bottom hole is less than the pore pressure at the upper depth, and the design method of slurry column structures with multi-density gradients is adopted as follows.

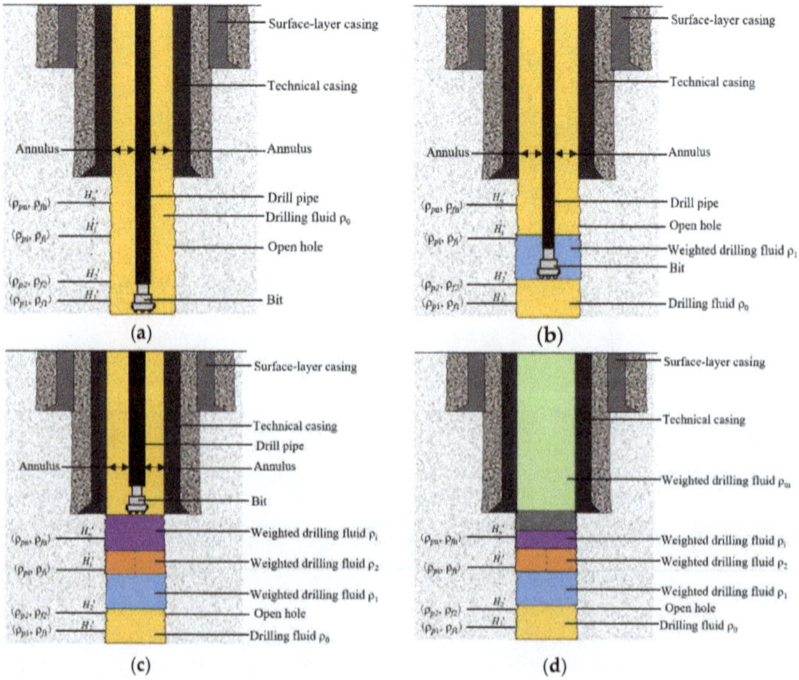

Figure 2. Slurry column structure in the wellbore during tripping. Each drilling fluid density ρ_i and length h_i are designed with the requirements of the formula (**a**) Before tripping; (**b**) Drilling fluid circulation after tripping; (**c**) Drill to a certain height; (**d**) Drilling completed.

(1) To stabilize the upper formation and prevent leakage, a substitution is made by replacing a portion of the bottom-hole drilling fluid ρ_0 with weighted drilling fluid ρ_1 before commencing the tripping-out process.

(2) After raising the drill pipe string to a specific height, the height of the weighted drilling fluid ρ_1 is changed to $h_{1.}$ This adjustment is necessary to stabilize the upper formation and facilitate the circulation of the weighted drilling fluid ρ_2, as shown in Figure 2b.

(3) The design of the height h_i of the tripped-out drill pipe string and the density of the weighted drilling fluid ρ_i is based on the specific pressure systems involved, as shown in Figure 2c.

(4) The hydrostatic column pressure of the multi-density gradient drilling fluid in the wellbore plays contributes entirely to preventing leaks and ensuring pressure-stabilization after all the drill pipe strings are pulled out The density ρ_i and length h_i of the multi-density gradient drilling fluids are designed as (ρ_1, h_1), (ρ_2, h_2), (ρ_3, h_3), ... , (ρ_m, h_m), respectively, as shown in Figure 2d.

$$\rho_{oj} \leq \sum_{j=k}^{m} \rho_i g h_i \leq \rho_{pj} \tag{1}$$

2.2. Model of Slurry Column Structure

In the construction process of this well, step-by-step casing running is employed, accompanied by the circulation of drilling fluids with different densities. Typically, the pipe string, comprising the casing string and the drill pipe string, is run into the well during the casing running procedure, as depicted in Figure 3. As the drilling pipe string is inserted, the fluid volume in the well is continuously drained by the pipe string. Furthermore, once the casing string reaches a certain depth, the drilling fluids in the well must be circulated and discharged using another drilling fluid from the wellhead. Calculating the slurry column structure in the wellbore during casing running poses particular challenges, especially in oil and gas wells with complex wellbore structures. In order to accurately calculate the slurry column structure in the wellbore during casing running, the following calculation methods are established.

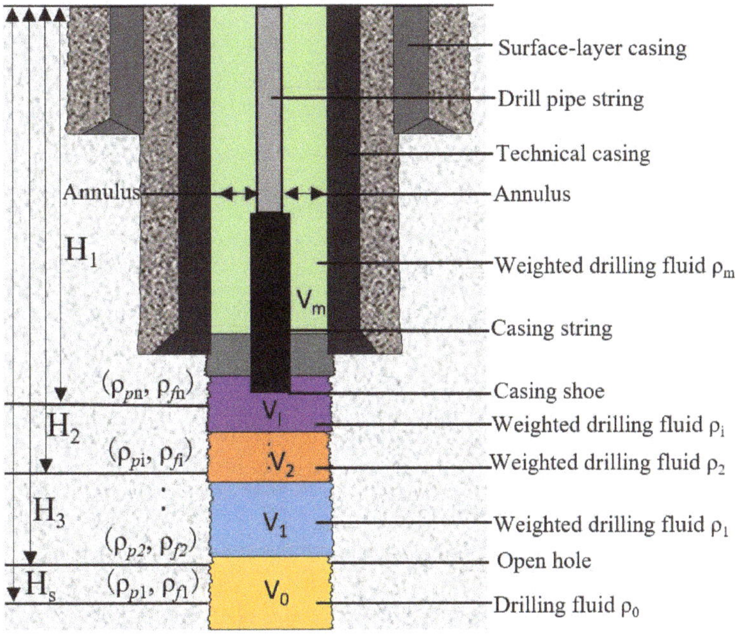

Figure 3. Casing running process diagram.

(1) Assume that the volumes of multi-density gradient drilling fluids in the wellbore from bottom to top are $V_0, V_1, V_2, \ldots \ldots, V_m$, respectively. The lengths of the drill pipe string and the production casing used for cementing are denoted as LD and LC, respectively. The total well depth is given by the sum of LD and LC. The casing running process is divided into s stages, the casing running speeds in each stage are denoted as $v_1, v_2, v_3, \ldots \ldots, v_s$, respectively, and the corresponding well running depths are $H_1, H_2, H_3, \ldots \ldots, H_s$, respectively, as shown in Figure 3.

(2) Calculate the volume Vp of the drill pipe string and production casing in the well at any time and compare it with the sum of the volume V_f of the first k fluids (from top to bottom) in the wellbore, as shown in Equation (2). If $V_p > V_f$, the $m, m-1, m-2, \ldots \ldots, k$ fluids will be replaced out of the wellbore; otherwise, $k = k-1$, and repeat step (2).

$$V_f = \sum_{i=k}^{m} V_i \tag{2}$$

(3) The calculation process of the fluid slurry column structure in the annulus above the casing shoe is complex; it is shown in Figure 4.

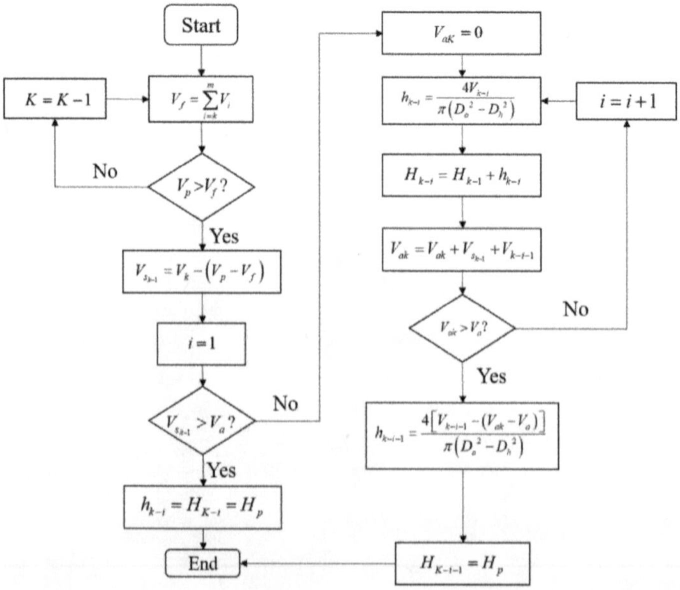

Figure 4. Calculation flow chart of annular fluid slurry column structure during the casing running process.

where V_{Sk-1} is the $k-1$ residual fluid volume in the wellbore, m^3; V_a is the volume of the annulus above the casing shoe, m^3; h_{k-1} is the height of $k-1$ fluid in the annulus, m; H_{k-1} is the top depth of $k-1$ fluid, m; V_{ak} is the sum of the volumes of $k-i-1$ fluids in the annulus, m^3; D_o is the outer diameter of the annulus, m; and D_h is the inner diameter of the annulus, m.

(4) When casing is run into the well, it reaches a depth of Hp in the p stage. At this stage, drilling fluid with a density of ρ_{cp} is injected from the drill pipe string at the wellhead, filling both the drill pipe string and the annular space. Subsequently, casing continues to be run to a depth of H_{p+1} in the $p = p + 1$ stage, and the process described in Equation (2) to (3) is repeated to calculate the slurry column structure in the wellbore during the casing running process.

The calculation of the wellbore fluid column pressure P_h when running casing to the depth of H_p can be formulated as:

$$P_h(H_p) = \sum_{n=0}^{k-i} \rho_n g h_n + \sum_{j=k-i}^{k-1} \rho_j g H_j \tag{3}$$

where ρ_n is the density of the drilling fluid in the n-section of the upper annular space of the casing shoe, g/cm^3; H_n is the length of the drilling fluid column of the n-section of the annular space, m; ρ_j is the density of drilling fluid in the lower j-section of the casing shoe, g/cm^3; and H_j is the length of the drilling fluid column in the lower j-section of the casing shoe, m.

In situations where the drill pipe string does not fully enter the well, the ground equipment is unable to control the annular pressure. The wellbore pressure under such circumstances is the combination of the hydrostatic column pressure P_h and the surge pressure Po. The wellbore pressure P_1 during casing running up to the depth of Hp can be formulated as:

$$P_1 = P_o(v_p) + P_h(H_p) \tag{4}$$

With the complete insertion of the drill pipe string into the well, back pressure can be applied to stabilize the formation. The wellbore pressure during this phase comprises three key elements: the hydrostatic column pressure P_h at the lower and upper ends of the casing shoe, the surge pressure Po induced by casing running, and the back pressure P_b exerted by ground equipment.

The wellbore pressure $P_1{}'$ during casing running to the depth of H_p can be formulated as:

$$P_1{}' = P_0(v_p) + P_h(H_p) + P_b \quad (5)$$

The algorithm of this model has been implemented in Visual Studio platform using the C# programming language to calculate the mid-column structure in the wellbore during the casing running process.

2.3. Model of Wellbore Pressure

During the casing running operation, the wellbore is divided into two parts: *LD* (above the casing shoe) and *LC* (below the casing shoe). This division allows for the analysis and calculation of the wellbore pressure. Furthermore, considering the specific working conditions of casing running operations, two stages of wellbore pressure-control models are developed. These models are utilized to calculate the wellbore pressure during both the casing running process and the circulating drilling fluid process.

2.3.1. Wellbore Pressure during Casing Running

When the DP non-completely enters into the well, the annular pressure cannot be managed by the ground equipment. The wellbore pressure at this time is P_1. The speed of the lower casing within the safe density window range in this formation can be calculated using the following formula:

$$P_1 = P_0(v_p) + P_h(H_p) < P_{Leakage} \quad (6)$$

When the DP completely enters into the well, back pressure can be applied to stabilize the formation. In this case, the wellbore pressure at this time is $P_1{}'$. Strict control of the wellbore pressure within the safe density window is required to ensure the safe lowering of the pressure control equipment into the well. The model analysis diagram is shown in Figure 5.

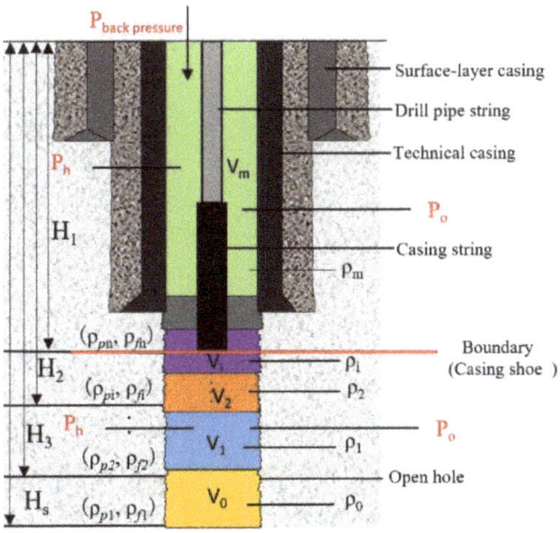

Figure 5. Schematic diagram of wellbore pressure model analysis during casing running.

And the wellbore pressure calculation model is formulated as:

$$P_{1'} = P_0(v_p) + P_h(H_p) + P_b < P_{Leakage} \tag{7}$$

By performing calculations of the hydrostatic column pressure and surge pressure, the correlation between wellbore pressure and casing running speed can be established. This analysis facilitates the determination of the appropriate casing running speed and back-pressure compensation values within the safe density window range.

2.3.2. Surge Pressure

In Formulas (4) and (5), the surge pressure $P_o(v_p)$ plays a crucial role in calculating the wellbore pressure. To account for the pressure changes transmitted to the wellhead and bottom hole, the calculations of surge pressure utilize the transient-flow method based on unstable flow theory.

The normal pressure window casing assumes the presence of a single drilling fluid in the wellbore. However, under the normal pressure window (NPW), there are multiple types of drilling fluids present in the casing wellbore, making the calculation of the friction pressure drop more complex. An accurate calculation of friction pressure drop is crucial for modeling surge pressure. The model of the slurry column structure is based on the principles described in Section 2.2, and the friction pressure drop in the annular space is formulated as:

$$P_f = \sum_{j=k-i}^{k-1} \frac{4K_jH_j}{D_{hy}} (\frac{2n_j+1}{3n_j} \frac{12v}{D_{hy}})^n \tag{8}$$

where K_j *is* the consistency coefficient of the jth fluid, $Pa{\cdot}s^n$; H_j is the return distance of the jth fluid, m; n is the liquidity index of the jth fluid; and D_{hy} is the hydraulic diameter of the annular space, m.

(1) Governing Equation

Based on the principles of mass and momentum conservation, the governing equation for one-dimensional unstable flow in the wellbore can be derived as follows (9) and (10):

$$\begin{cases} v\frac{\partial p_0}{\partial z} + \frac{\partial p_0}{\partial t} + \rho C^2 \frac{\partial v}{\partial z} = 0 \\ \frac{\partial p_0}{\partial z} + \rho \frac{\partial v}{\partial t} + \rho v \frac{\partial v}{\partial z} + p_f = 0 \end{cases} \tag{9}$$

where

$$C = \frac{1}{\sqrt{\rho(\alpha + \beta)}} \tag{10}$$

where Q is the flow rate of drilling fluid, L/s; p is the surge pressure, MPa; C is the propagation velocity of the pressure wave, m/s; A is the cross-sectional area of flow channel, m^2; t is the casing decentralization time, s; Z is the ordinate value along casing, m; p_f is the friction pressure drop, MPa; β is the expansion coefficient of the flow channel; α is the compression coefficient of drilling fluid; and ρ is the density of drilling fluid, g/cm^3.

(2) Grid Division

The characteristic curve method is employed to solve the system of governing equations. Firstly, the characteristic equation of the governing equations (Equation (9)) is established using the modified Lister method. Subsequently, the finite difference method is applied to solve the characteristic equation, yielding the solution for the governing equations.

The characteristic equation of governing equationsis formulated as Formula (11).

$$\begin{cases} \pm dp_0 + \rho C dv + C p_f dt = 0 \\ \frac{dz}{dt} = v \pm C \end{cases} \tag{11}$$

The forward characteristic line, originating from point M with a slope of arctan ($\frac{1}{v+c}$) intersects the backward characteristic line, originating from point N with a slope of arctan ($\frac{1}{v-c}$), at point T. In the case where the selected Δz and Δt are sufficiently small, the short line segments of the characteristic lines MT and NT can be approximated as straight lines, as shown in Figure 6.

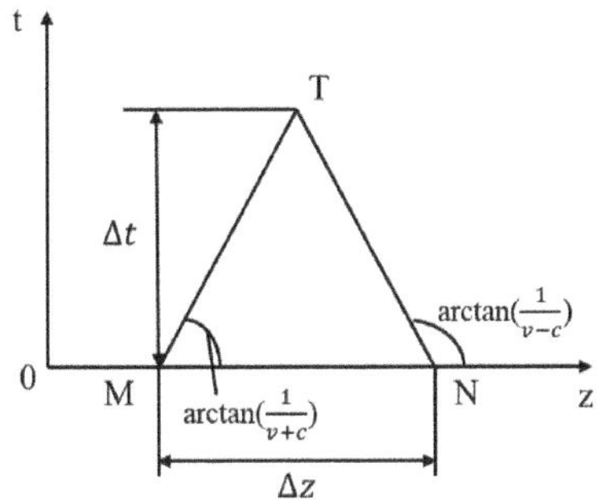

Figure 6. Characteristic diagram.

On the lines MT and NT, the system of equations can be satisfied:

$$\begin{cases} dp_0 + \rho C dv + C p_f dt = 0 \\ -dp_0 + \rho C dv + C p_f dt = 0 \end{cases} \tag{12}$$

The finite difference representation of Equation (12) can be expressed as follows:

$$\begin{cases} p_T - p_M + \rho C(v_T - v_M) + C\Delta t p_{fM} = 0 \\ -p_T + p_N + \rho C(v_T - v_N) + C\Delta t p_{fN} = 0 \end{cases} \tag{13}$$

If P_M, Q_M, P_{fM}, P_N, Q_N, P_{fN} are known, Q_T, P_T can be obtained as follows:

$$Q_T = \frac{p_M - p_N + v(v_M + v_N) - C\Delta t(p_{fM} + p_{fN})}{2\rho C} \tag{14}$$

$$P_T = \frac{p_M + p_N - \rho C(v_M - v_N) - C\Delta t(p_{fM} - p_{fN})}{2} \tag{15}$$

The instantaneous pressure P_T and flow Q_T at any point in the flow channel at any time are obtained.

(3) Boundary Conditions

During actual casing operations, the movement of the casing string within the wellbore induces fluid motion, which can be divided into two distinct flow paths: the annular flow path above the casing shoe and the open-hole flow path below the casing shoe. The surge pressure, resulting from the casing string movement, originates at the casing shoe and gradually attenuates towards both the wellbore and the bottom of the well. The wellhead boundary of the annular flow path is connected to the atmosphere, thereby maintaining a pressure of 0 MPa. At the junction point, which corresponds to the casing shoe, the pressures of both flow paths are equal. The fluid at the casing shoe is displaced by the casing string, causing the fluid within the annular flow path to ascend, while the fluid within the open-hole flow path at the well's bottom descends. Neglecting the permeation of drilling fluid at the well's bottom, the flow rate at the bottom boundary of the open-hole flow path is 0. The sum of the flow rates at the junction point, i.e., the casing shoe, equals the flow rate of casing string displacement. The junction point diagram is shown in Figure 7:

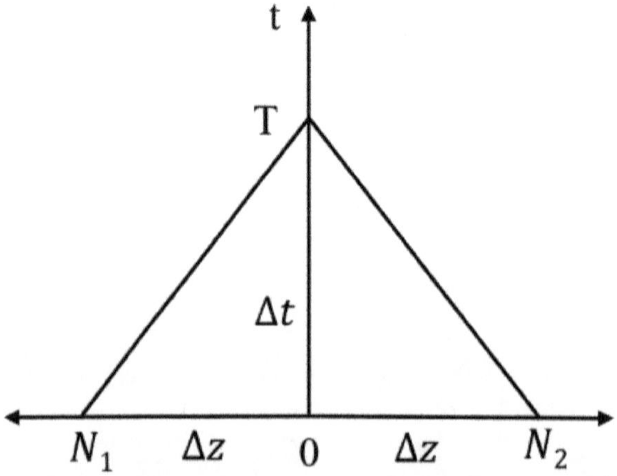

Figure 7. Junction point diagram.

a. The pressure at the wellhead boundary of the annular channel is 0, and the flow rate can be calculated using the equations of the forward characteristic line:

$$\begin{cases} p_T = 0 \\ Q_T = \dfrac{p_M + \rho C_M v_M - C_M \Delta t p_{fM}}{\rho v_M} \end{cases} \tag{16}$$

b. The flow rate at the bottom-hole boundary of the bottom-hole channel is 0, and the pressure can be determined using the equations of the forward characteristic line.

$$\begin{cases} Q_T = 0 \\ p_T = p_M + \frac{\rho C_M}{A_M} Q_M - C_M \Delta t p_{fM} \end{cases} \tag{17}$$

c. The displacement flow at the junction point, i.e., the casing shoe, is equal to the sum of the flow rates in the two channels. The equations be expressed as follows:

$$\begin{cases} Q_P = Q_{T1} + Q_{T2} \\ p_T = p_{N1} + \frac{\rho C_{N1}}{A_{N1}} (Q_{T1} - Q_{N1}) + C_{N1} \Delta t p_{fN1} \\ p_T = p_{N2} + \frac{\rho C_{N2}}{A_{N2}} (Q_{T2} - Q_{N2}) + C_{N2} \Delta t p_{fN2} \end{cases} \tag{18}$$

By combining Equations (14) and (15), the pressure and flow rate at each position in the wellbore during the casing running process can be obtained by substituting the fixed-solution conditions.

2.3.3. Wellbore Pressure during Drilling Fluid Circulation

During the drilling fluid circulation process, the drilling fluid is injected from the casing and returns through the annular space. The wellbore pressure P_2 during this process consists of the hydrostatic column pressure and the friction pressure drop P_f of the drilling fluids in the annular space with varying densities. The model analysis diagram is shown in Figure 8:

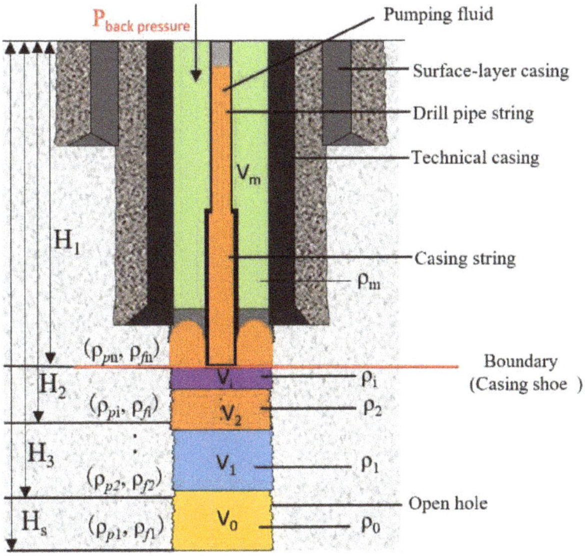

Figure 8. Schematic diagram of wellbore pressure model analysis in the process of circulating drilling fluid.

The formula for calculating the wellbore pressure during the circulating drilling fluid process is expressed as:

$$P_2 = P_h(H_p) + P_f < P_{Leakage} \tag{19}$$

Taking the drilling fluid as a power-law fluid, the friction pressure drop in the annular space is formulated as:

$$P_f = \sum_{j=k-i}^{k-1} \frac{4K_j H_j}{D_{hy}} \left(\frac{2n_j + 1}{3n_j} \frac{12v}{D_{hy}}\right)^n \tag{20}$$

where K_j *is* the consistency coefficient of the jth fluid, Pa·s^n; H_j is the return distance of the jth fluid, m; n is the liquidity index of the jth fluid; D_h is the diameter of the annular space, m; D_{co} is the casing diameter, m; v is the mean flow rate, m/s; D_{hy} is the hydraulic diameter of the annular space, m; and Q is the displacement, L/s.

The mean flow rate is formulated as:

$$v = \frac{4Q}{\pi(D_h^2 - D_{co}^2)} \tag{21}$$

3. Case Study

Well LT-X1 is an exploratory well located in the Xinjiang oil field, situated in the thrust belt of the Homatu anticline and Tugulu anticline in the southern margin of the Junggar Basin. The total depth of the well is 7050 m. To validate the reliability of the model established in this study, a case analysis is conducted on Well LT-X1. The formation pressure of the five layers starting from 5610 m in the open hole of this well is shown in Table 1. The last column of data is the density of drilling fluid circulating during the practical drilling process.

Table 1. Formation pressure of open-hole layers in Well LT-X1.

Serial Number	Well Section (m)	Layer	Pressure Coefficient	Drilling Fluid Density (g/cm³)
				Practical
1	5610–5745	/K1h	2.54–2.49	2.39–2.41
2	5745–6012	/K1h-K1q	2.48–2.45	2.34–2.40
3	6012–6158	K1q	2.45–2.42	2.55–2.59
4	6158–6806	K1q	2.42–2.40	2.56–2.57
5	6806–7050	J3k	2.39–2.34	2.34–2.36

The MPD data indicated that the well had four leaky layers, with equivalent cycle densities (ECD) at depths of 6152 m, 6674 m, 6947 m and 7028 m of 2.445 g/cm^3, 2.446 g/cm^3, 2.432 g/cm^3 and 2.432 g/cm^3. It can be seen that the formation pressure of section 5610–5745 m and section 5745–6012 m is greater than the leakage pressure of 6672–6674 m, 6946–6947 m and 7026–7028 m, which are typical of NPW wells. The pressure system is complex, posing challenges in determining the casing running speed. By simulating the wellbore's slurry column structure and the variation in wellbore pressure during casing running, a drilling fluid circulation scheme is developed for the casing running process, along with recommended casing velocities and pressure control values for different stages.

3.1. *The Slurry Column Structure Inside the Wellbore after Drilling*

Combined with the design model of the wellbore slurry column structure and the formation pressure system of each layer in the open hole of Well LT-X1, Table 2 shows the designed slurry column structure.

Table 2. Slurry column structure in wellbore after managed-pressure tripping.

Well Depth Position (m)	Length(m)	Density (g/cm³)	Fluid Column Pressure (MPa)	Equivalent Density (g/cm³)
1521	1521	2.50	37.352	2.500
3558	2037	2.50	49.957	2.500
5850	2292	2.38	53.513	2.453
5900	50	2.38	1.121	2.452
6150	250	2.35	5.809	2.448
6300	150	2.35	3.412	2.446
6672	372	2.10	7.705	2.427
6946	274	2.10	5.624	2.414
7026	80	2.10	1.669	2.410
7050	24	2.10	0.453	2.409

3.2. *Casing Running Speed*

When the casing is being run to a certain depth, drilling fluid circulation is required. Three different density-gradient drilling fluid circulation schemes were designed based on the calculation method of the slurry column structure in the wellbore during the casing running process and in combination with the formation pressure system, as shown in Table 3.

Table 3. Circulating drilling fluid scheme at different depths of casing running.

Casing Depth (m)	Scheme 1	Scheme 2	Scheme 3
3550	2.43 g/cm^3 + 2.32 g/cm^3	2.45 g/cm^3 + 2.35 g/cm^3	2.47 g/cm^3 + 2.38 g/cm^3
5900	2.32 g/cm^3	2.35 g/cm^3	2.38 g/cm^3
7050	2.32 g/cm^3	2.30 g/cm^3	2.28 g/cm^3

The wellbore pressure is calculated using the model described in Section 2.3.3 for the three proposed schemes, as illustrated in Figure 9. In scheme 1, the equivalent circulating density (ECD) in the section from 6500 to 7100 m exceeded the ECD of the leakage layers. In scheme 3, the ECD in the section from 5900 to 6600 m exceeded the leakage ECD. Scheme 2 had an ECD within the safe density window range, and thus the drilling fluid circulation plan for scheme 2 is selected.

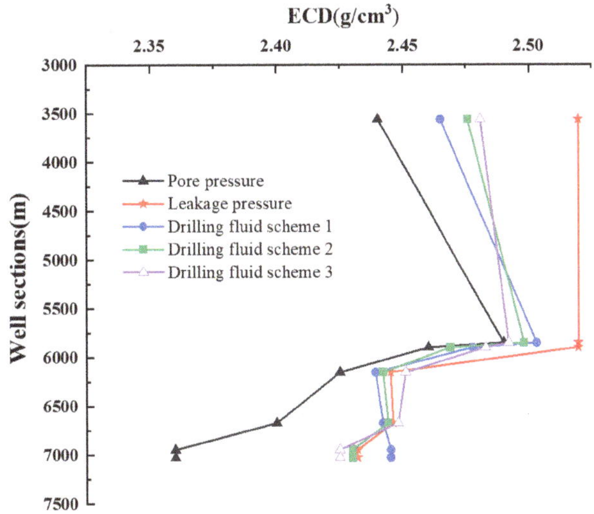

Figure 9. Changes of ECD under different circulating drilling fluid schemes.

Based on the formation pressure system, the density and volume of the circulating drilling fluid for scheme 2 can be calculated. Additionally, the corresponding annular back pressure P_b required during the circulation process is determined. The calculated values are summarized in Table 4:

Table 4. Circulation drilling fluid density when running casing into different positions.

Casing Depth (m)	Circulating Drilling Fluid Density (g/cm^3)	Volume (m^3)	Pressure Control Values (MPa)
3560	2.45	125	0.40–0.10
	2.35	155	1–4.6
5900	2.35	35	4.65
		45	3.90
		40	3.27
		44	2.64
7050	2.30	250	3.60–6.40

According to the calculation model of wellbore pressure during casing running, three schemes of casing speed were designed, as shown in Table 5.

Table 5. Casing speed scheme for casing into different well depth sections.

Casing into the Well Depth Section (m)	Scheme 1 (m/s)	Scheme 2 (m/s)	Scheme 3 (m/s)
0–1521	0.160	0.160	0.160
1521–3558	0.160	0.160	0.160
3558–5900	0.145	0.145	0.145
5900–6672	0.137	0.137	0.137
6672–7050	0.124	0.110	0.137

The wellbore pressure is calculated under the three schemes using the model described in Section 2.3.1, as depicted in Figure 10. Scheme 1 and scheme 2 were found to fall within the safe density window range. However, in scheme 3, the equivalent cycle density (ECD) in the section 6800–7050 m exceeded the ECD of the leakage layer. Running the casing at a low speed can lead to a slower construction progress and increased nonproductive time, thereby raising the overall well budget. Consequently, the casing speed of scheme 1 is selected for the operation.

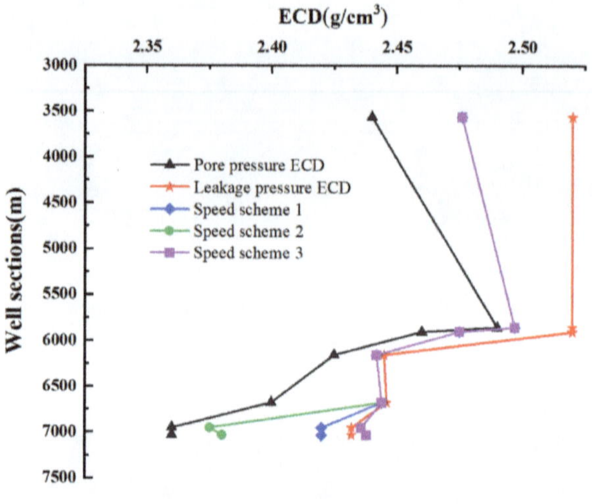

Figure 10. ECD changes of different casing running speed schemes.

Based on the calculation model for surge pressure and wellbore pressure, the surge pressure generated at each casing speed in scheme 1 and the corresponding annular back pressure required during casing running can be determined, as presented in Table 6.

Table 6. The surge pressure, casing running speed and back pressure.

Casing into the Well Section (m)	Surge Pressure (MPa)	Casing Speed (m/s)	Back Pressure (MPa)
0–1523	0.32	0.160	—
1523–3560	0.77	0.160	1.641
3560–5900	1.74	0.145	3.427
5900–6672	1.70	0.137	4.041
6672–7050	1.80	0.124	4.457

The calculation results from Tables 4 and 6 were successfully applied during the casing running operation of Well LT-X1, resulting in a smooth operation without any well leakage or overflow.

4. Discussion

The model studied in this paper still has limitations and uncertainties, which include:

(1) Insufficient validation in practical applications: Although managed-pressure casing running technology under NPWs was developed for Well LT-X1, further validation is required to assess its applicability to other wells and different geological conditions.
(2) Model accuracy and assumptions: The models established in the study may be based on certain assumptions and simplifications, which can impact the accuracy and applicability of the models. Therefore, when using these models for wellbore pressure calculations, it is important to evaluate the accuracy of the models and the assumptions they rely on and make adjustment or corrections as necessary.
(3) Data availability and parameter selection: The accuracy and reliability of wellbore pressure calculation models are limited by the availability of data. In practical applications, there may be instances of missing or incomplete data, which can affect the accuracy of the calculated results. Additionally, the selection and estimation of parameters involved in the models can also impact the results, requiring careful consideration and reasonable estimation.

In conclusion, although the study achieved certain results under specific conditions, further research and practice are needed to overcome these limitations.

5. Conclusions

This paper, using Well LT-X1 as an example, simulated the casing running process and the changes in the slurry column structure of drilling fluids. A circulation scheme is designed as follows: Circulate 125 m^3 of drilling fluid with a density of 2.45 g/cm^3 and 155 m^3 of drilling fluid with a density of 2.35 g/cm^3 at a depth of 3560 m. From there, circulate 164 m^3 of drilling fluid with a density of 2.35 g/cm^3 at a depth of 5900 m. Finally, at a depth of 7050 m, circulate 250 m^3 of drilling fluid with a density of 2.30 g/cm^3. The casing running speeds and back-pressure values were designed as follows for the respective well sections: 0–1523 m: 0.160 m/s casing speed, 0MPa back pressure; 1523–3560 m: 0.160 m/s casing speed, 1.641 MPa back pressure; 3560–5900 m: 0.145 m/s casing speed, 2.427 MPa back pressure; 5900–6674 m: 0.137 m/s casing speed, 4.041 MPa back pressure; 6674–7050 m: 0.124 m/s casing speed, 4.457 MPa back pressure. The results demonstrate that by optimizing the structure of multi-density gradient drilling fluid and applying annular back pressure in stages, along with the accurate calculation of wellbore pressure, the goals of leak-proofing and pressure stabilization can be achieved.

This study provides a foundation for managed-pressure running casing technology under NPWs. The established model for wellbore slurry column structures and wellbore pressure calculations under managed-pressure running casing serves as a theoretical basis for casing running technology under NPWs.

Author Contributions: Investigation, M.J.; Resources, Z.Z.; Writing—original draft, Y.M.; Writing—review & editing, H.Y. All authors have read and agreed to the published version of the manuscript.

Funding: This research was funded by Engineering Technology Research Institute of PetroChina West Drilling Engineering grant number xzgcy2021-055.

Institutional Review Board Statement: Not applicable.

Informed Consent Statement: Not applicable.

Data Availability Statement: Data is unavailable due to privacy or ethical restrictions.

Acknowledgments: The authors wish to express their appreciation to the Engineering Technology Department of Western Drilling Company for their assistance with the experimental study and kindly supplying data.

Conflicts of Interest: The authors declare no conflict of interest.

References

1. Wang, F.; Liu, X.; Jiang, B.; Zhuo, H.; Chen, W.; Chen, Y.; Li, X. Low-loading Pt nanoparticles combined with the atomically dispersed FeN_4 sites supported by Fe_{SA}-N-C for improved activity and stability towards oxygen reduction reaction/hydrogen evolution reaction in acid and alkaline media. *J. Colloid Interface Sci.* **2023**, *635*, 514–523. [CrossRef] [PubMed]
2. Li, Q.; Zhang, C.; Yang, Y.; Ansari, U.; Han, Y.; Li, X.; Cheng, Y. Preliminary experimental investigation on long-term fracture conductivity for evaluating the feasibility and efficiency of fracturing operation in offshore hydrate-bearing sediments. *Ocean. Eng.* **2023**, *281*, 114949. [CrossRef]
3. Li, Q.; Zhao, D.; Yin, J.; Zhou, X.; Li, Y.; Chi, P.; Han, Y.; Ansari, U.; Cheng, Y. Sediment instability caused by gas production from hydrate-bearing sediment in Northern South China Sea by horizontal wellbore: Evolution and mechanism. *Nat. Resour. Res.* **2023**, *32*, 1595–1620. [CrossRef]
4. Kudźma, Z.; Stosiak, M. Reduction of infrasounds in machines with hydrostatic drive. *Acta Bioeng. Biomech.* **2013**, *15*, 51–64. [CrossRef] [PubMed]
5. Clark, E.H. A graphic view of pressure surges and lost circulation. In *Drilling and Production Practices*; American Petroleum Institute: New York, NY, USA, 1956.
6. Moore, P.L. Pressure surges and their effect on hole conditions. *Oil Gas J.* **1965**, *63*, 90.
7. Clark, R.K.; Fontenot, J.E. Field measurements of the effects of drillstring velocity, pump speed, and lost circulation material on downhole pressures. In Proceedings of the Fall Meeting of the Society of Petroleum Engineers of AIME, Houston, TX, USA, 6–9 October 1974; SPE Annual Technical Conference and Exhibition. Society of Petroleum Engineers (SPE): Richardson, TX, USA, 1974. [CrossRef]
8. Lal, M. Surge and swab modeling for dynamic pressures and safe trip velocities. In Proceedings of the IADC/SPE Drilling Conference, New Orleans, LA, USA, 20–23 February 1983; Society of Petroleum Engineers (SPE): Richardson, TX, USA, 1983.
9. Wagner, R.R.; Halal, A.S.; Goodman, M.A. Surge field tests highlight dynamic fluid response. In Proceedings of the SPE/IADC Drilling Conference, Amsterdam, The Netherlands, 22–25 February 1993; Society of Petroleum Engineers (SPE): Richardson, TX, USA, 1993. [CrossRef]
10. Bing, Z.; Kaiji, Z.; Qiji, Y. Equations help calculate surge and swab pressures in inclined wells. *Oil Gas J.* **1995**, *93*, 38.
11. White, W.W.; Zamora, M.; Svoboda, C.F. Downhole measurements of synthetic-based drilling fluid in offshore well quantify dynamic pressure and temperature distributions. *SPE Drill. Complet.* **1997**, *12*, 149–157. [CrossRef]
12. Ward, C.; Andreassen, E. Pressure-while-drilling data improve reservoir drilling performance. *SPE Drill. Complet.* **1998**, *13*, 19–24. [CrossRef]
13. Rudolf, R.L.; Suryanarayana, P.V.R. Field Validation of Swab Effects While Tripping-In the Hole on Deep, High Temperature Wells. In Proceedings of the SPE/IADC Drilling Conference, Dallas, TX, USA, 3–6 March 1998; Society of Petroleum Engineers (SPE): Richardson, TX, USA, 1998. [CrossRef]
14. Isambourg, P.; Bertin, D.L.; Brangetto, M. Field hydraulic tests improve HPHT drilling safety and performance. *SPE Drill. Complet.* **1999**, *14*, 219–227. [CrossRef]
15. Samuel Robello, G.; Sunthankar, A.; McColpin, G.; Bern, P.; Flynn, T. Field validation of transient swab/surge response with PWD data. In Proceedings of the SPE/IADC Drilling Conference, Amsterdam, The Netherlands, 27 February–1 March 2001; Society of Petroleum Engineers (SPE): Richardson, TX, USA, 2001. [CrossRef]
16. Rommetveit, R.; Bjørkevoll, K.S.; Gravdal, J.E.; Goncalves, C.J.C.; Lage, A.C.V.M.; Campos, J.E.A.; Aragão, Á.F.L.; Arcelloni, A.; Ohara, S. Ultradeepwater hydraulics and well-control tests with extensive instrumentation: Field tests and data analysis. *SPE Drill. Complet.* **2005**, *20*, 251–257. [CrossRef]
17. Cannon, G.E. Changes in Hvdrostatic Pressure Due to Withdrawing Drill Pipe from the Hole. In *Drilling and Production Practices*; American Petroleum Institute: New York, NY, USA, 1934.
18. Ormsby, G.S. Calculation and Control of Mud Presures in Drilling and Completion Operations. In *Drilling and Production Practices*; American Petroleum Institute: New York, NY, USA, 1954.
19. Clark, E.H. Bottom-hole pressure surges while running pipe. *Pet. Eng. Int.* **1955**, *27*, B68.
20. Burkhardt, J.A. Wellbore pressure surges produced by pipe movement. *J. Pet. Technol.* **1961**, *13*, 595–605. [CrossRef]
21. Schuh, F.J. Computer makes surge-pressure calculations useful. *Oil Gas J.* **1964**, *31*, 96.
22. Fontenot, J.E.; Clark, R.K. An improved method for calculating swab and surge pressures and circulating pressures in a drilling well. *Soc. Pet. Eng. J.* **1974**, *14*, 451–462. [CrossRef]
23. Haige, W.; Xisheng, L. Study on steady surge pressure for yield-pseudoplastic fluid in a concentric annulus. *Appl. Math. Mech.* **1996**, *17*, 15–23. [CrossRef]
24. Jinchang, W. Analysis of annular fluctuation pressure pattern during casing running operation in Da-niudi Gas Field. *Oil Drill. Prod. Technol.* **2016**, *38*, 36–41. [CrossRef]
25. He, S.; Srivastav, R.; Tang, M.; Ahmed, R. A new simplified surge and swab pressure model for yield-power-law drilling fluids. *J. Nat. Gas Sci. Eng.* **2016**, *28*, 184–192. [CrossRef]
26. Lubinski, A.; Hsu, F.H.; Nolte, K.G. Transient pressure surges due to pipe movement in an oil well. *Rev. Inst. Fr. Pét.* **1977**, *32*, 307–348. [CrossRef]
27. Zhong, B.; Zhou, K.J.; Yuan, Q.J. Equations help calculate surge and swab pressures in inclined well. *Oil Gas J.* **1995**, *18*, 74–77.

28. Zhang, F.; Kang, Y.; Wang, Z.; Miska, S.; Yu, M.; Zamanipour, Z. Real-time wellbore stability evaluation for deepwater drilling during tripping. In Proceedings of the SPE Deepwater Drilling and Completions Conference, Galveston, TX, USA, 14–15 September 2016; Society of Petroleum Engineers (SPE): Richardson, TX, USA, 2016.
29. Pengcheng, W.; Chaoyang, X.; Yingfeng, M.; Hongtao, L. Laws fo bottomhole transient pressure fluctuation during tripping in narrow safety density window formation. *Drill. Prod. Technol.* **2016**, *39*, 22–25.
30. Dong, T. Study on Drilling Overflow Monitoring Method and Application for Pressure-Sensitive Formation. Master's Thesis, Southwest Petroleum University, Chengdu, China, 2018.
31. Junbo, Q. Research on Calculation of Surge and Swab Pressures and Control of Gas Kick in Constant Bottom Hole Managed Pressure Drilling. Ph.D. Thesis, Northeast Petroleum University, Daqing, China, 2019.
32. Karpenko, M. Landing gear failures connected with high-pressure hoses and analysis of trends in aircraft technical problems. *Aviation* **2022**, *26*, 145–152. [CrossRef]

processes

Article

Numerical Simulation Analysis of Wellbore Integrity and Casing Damage in High-Temperature Injection and Production of Shale Oil

Xiaocong Yu [1,2,*], Xueqi Cen [1,3], Changbin Kan [2], Yilin Hu [2], Yanxing Yang [2], Shilin Tao [2], Xinyuan Chen [2], Xiupeng Chen [2] and Zhiqiang Hu [1]

[1] State Key Laboratory of Shale Oil and Gas Enrichment Mechanisms and Effective Development, SINOPEC, Beijing 102206, China; cenxq.syky@sinopec.com (X.C.)
[2] School of Petroleum Engineering, China University of Geosciences, Wuhan 430074, China; kanchangbin@cug.edu.cn (C.K.); shilintao@cug.edu.cn (S.T.); xinyuanchen@cug.edu.cn (X.C.)
[3] Petroleum Exploration and Production Research Institute, SINOPEC, Beijing 102206, China
* Correspondence: yuxc@cug.edu.cn

Abstract: Shale oil represents a relatively new form of unconventional oil and gas resource, and the extensive exploration and development of shale oil resources carry significant implications for China's oil and gas supply and demand dynamics. At present, within the realm of low-maturity shale oil extraction technologies, the reservoir must be subjected to elevated temperatures ranging between 400 to 60 °C. Prolonged exposure of wellbores to such high temperatures can result in a substantial decrease in cement strength, the formation of microcracks due to cement cracking, and damage stemming from thermal stresses on the casing. Casing damage stands out as a prominent factor contributing to wellbore integrity failures and well shutdowns within the context of shale oil development. Given the limited natural energy reservoirs of shale oil formations, it becomes necessary to supplement the reservoir's energy during the development process. Furthermore, shale oil exhibits high viscosity and poor flowability, and conventional water injection methods yield limited efficacy. This situation can induce significant shifts in the stress field and rock mechanical parameters, potentially activating specific formations and complicating the load dynamics on the casing. Consequently, the risk of failure increases. In light of these considerations, this study uses numerical simulations to study the integrity of high-temperature injection and production wellbores in shale oil and aims to encompass a comprehensive evaluation and analysis of the principal factors that influence casing damage, the fluctuations in thermal stress, and the yield strength of various steel grades of casings exposed to alternating stress conditions. Subsequently, this paper developed a model for simulating the temperature and pressure within shale oil and steam injection wellbores to support engineering design analysis. The research results indicate that the application of pre-stress results in a significant increase in stress at the casing pipe head while causing a noticeable decrease in stress within the pipe wall. When N80 casing is used, the entire casing experiences thermal stresses surpassing the casing's yield limit. Stress concentration may arise at both ends of the external seal, potentially leading to casing contraction, shear failure, and, under non-uniform stress conditions, casing bending deformation. The temperature of steam injection significantly influences the temperature field of the casing wall, with stress values experiencing a marked reduction when the steam injection temperature decreases from 350 °C to 200 °C, underscoring the substantial impact of temperature on casing thermal stress. As the steam injection process advances along with injection-production cycles, shear stresses at the interface can exceed the bond strength, resulting in relative slippage between the cement and the casing. The bonding force between the wellbore and the cement primarily depends on the interface's friction, particularly in the context of friction during wellhead lifting. This study endeavors to determine rational injection and production parameters under varying conditions, optimize completion methods, reduce casing damage, and extend the casing's operational life; it aims to offer critical technical support for the safe and efficient development of shale oil resources.

Citation: Yu, X.; Cen, X.; Kan, C.; Hu, Y.; Yang, Y.; Tao, S.; Chen, X.; Chen, X.; Hu, Z. Numerical Simulation Analysis of Wellbore Integrity and Casing Damage in High-Temperature Injection and Production of Shale Oil. *Processes* **2023**, *11*, 3053. https://doi.org/10.3390/pr11113053

Academic Editors: Tianshou Ma and Yuqiang Xu

Received: 27 September 2023
Revised: 15 October 2023
Accepted: 19 October 2023
Published: 24 October 2023

Keywords: shale oil; wellbore integrity; casing damage; injection and production; numerical simulation

1. Introduction

Shale oil is a new type of unconventional hydrocarbon resource, and it has been recognized as a crucial alternative energy source for the 21st century due to its abundant reserves and feasible development prospects. Currently, shale oil production, represented by the Permian Basin, Bakken, and Eagle Ford in North America, has achieved economic viability with an annual output of 2.4×10^8 tons and is on the rise, contributing to energy independence in the United States [1]. China also possesses significant shale oil resources, estimated by the Ministry of Natural Resources to be 1.53×10^9 tons in geological reserves and 3.7×10^8 tons in recoverable resources [2]. Since 2011, key efforts have been made in shale oil development in fields such as Xinjiang Oilfield, Daqing Oilfield, and Jilin Oilfield, resulting in significant progress. Therefore, the large-scale exploration and development of shale oil resources hold great significance for addressing China's oil and gas supply and demand pressures [3,4]. In the early stages of low-maturity shale oil extraction, it is necessary to heat the formation to temperatures ranging from 400 to 600 °C. Prolonged exposure of wellbores to high temperatures can lead to a sharp decrease in cement strength, the development of microcracks in the cement ring, and thermal stress damage to casing and completion tubing, ultimately compromising wellbore integrity and impacting the safe and efficient development of oil and gas resources [5–9]. Based on conventional oil and gas field development experience, the percentage of casing damage increases with the passage of time, especially in the later stages of development, making casing damage a significant reason for well abandonment. For example, as of the end of 2016, Daqing Oilfield had accumulated over 20,000 damaged wells, with a casing damage rate of 21%. By the end of 2017, Bameihe Block in Jianghan Oilfield had 646 damaged wells, with a casing damage rate of 32% [10]. Common causes of casing damage include geological factors, such as structural stress, reservoir creep, sand production, formation subsidence, and fault reactivation, and engineering factors, such as water injection, perforation, acidization, and fracturing, as well as corrosion factors, such as sulfur, carbon dioxide, hydrogen sulfide, and corrosion by formation water and injected water, while casing quality factors include defects in quality, residual stress, and thread sealing. Since deformation issues in casing pipes have been evident in the early stages of shale oil and gas well development, it can be anticipated that wellbore integrity issues will be even more complex during long-term development compared to conventional oil and gas wells.

Therefore, proactive measures, including risk assessment and related contingency plans, need to be developed as soon as possible to minimize losses. Zhang Yonggui [11] established a mechanical model for thermal recovery well casings, studied the calculation methods for thermal stress and extrusion pressure in the model, and fully considered the interaction between the casing and formation in the casing mechanical model. However, the casing stress condition is affected by well depth, and the influence of various factors on casing strength at different well depths still requires further research. Li Jing et al. [12] studied the theoretical solution of thermal stress in the casing–cement–ring-formation coupled system, qualitatively indicating the factors affecting casing thermal stress but not considering the effect of the cement ring's thickness on casing thermal stress variations. Tian Zhonglan et al. [13] have made significant contributions by developing models that enable the assessment and calculation of casing damage resulting from a variety of factors, and these models provide valuable insights into the underlying mechanisms and typical patterns associated with casing damage, as well as the phenomenon of shale slippage. From this, it can be seen that China has conducted extensive research on the structural aspects of wellbore integrity in thermal recovery wells [14–16]. However, there has been relatively limited research on the impact of heating methods for shale oil extraction on the integrity of wellbores, casing, cement, and the geological formations as a whole [17–20].

This research gap is particularly significant due to the inherent limitations of shale oil reservoirs, which necessitate the timely supplementation of subsurface energy during the development process. Furthermore, the high viscosity and poor flow properties of shale oil make conventional water injection methods less effective, resulting in significant changes in subsurface stress fields and rock mechanical parameters [21–27]. This can potentially activate certain geological formations, leading to the complex load conditions experienced by casing pipes and an increased risk of casing failure.

Therefore, this paper conducts a study on the wellbore integrity and casing damage in high-temperature shale oil injection and production wells using numerical simulation methods. The study involves a comprehensive evaluation and analysis of geological factors in the reservoir, single-well completion engineering factors, and steam injection and production process factors. The goal is to clearly identify the primary factors that affect casing damage and to understand the patterns of thermal stress and yield strength variations in casing materials under alternating stress conditions. Subsequently, a thermal and pressure model for shale oil steam injection and production wells is constructed. Advanced simulation software is employed for the numerical calculations and analyses. The research includes engineering design analysis and a temperature–pressure profile analysis, among other aspects of the study. The aim is to determine reasonable injection and production parameters for different operating conditions, achieve optimization of well completion methods, reduce casing damage, and extend the lifespan of casing pipes. This research is intended to provide technical support for the safe and efficient development of shale oil resources.

2. The Primary Causes of Casing Damage in Steam Injection Wells

Currently, many shale oil fields are facing serious casing damage issues in their thermal recovery wells, with the main types of casing damage being contraction, rupture, shear, and corrosion. These casing damage problems have posed significant challenges to the production of shale oil fields. The study of wellbore integrity and casing deformation during shale oil high-temperature injection and production processes can draw insights from the casing damage issues in shale oil recovery. The decline in production in thermal recovery wells, and in severe cases, production shutdown, underscores the importance of understanding the mechanisms of casing damage in shale oil thermal recovery wells. It is only by comprehending the mechanisms of casing damage that appropriate protective and remedial measures can be developed based on the primary causes of casing damage. This is of paramount significance in extending the lifespan of casing and wellbores, stabilizing and increasing thermal recovery production in shale oil fields. The primary causes of casing damage in steam injection wells are shown in Figure 1.

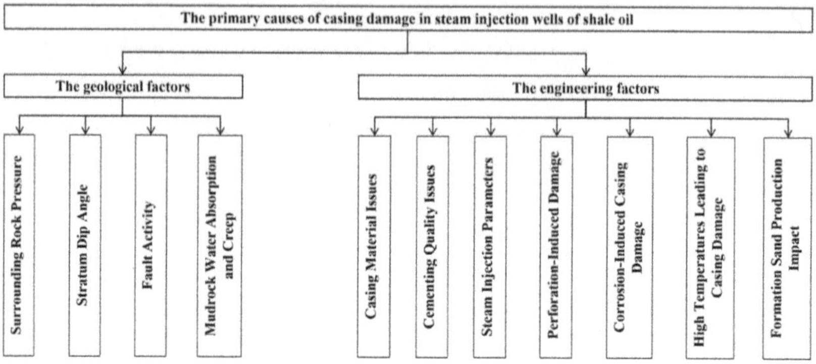

Figure 1. The primary causes of casing damage in steam injection wells.

2.1. The Geological Factors

(1) Influence of Surrounding Rock Pressure

After drilling, an annular space is formed in the surrounding rock, disrupting the original equilibrium. When the stress at the stress concentration point reaches the yield limit of the surrounding rock, plastic deformation or formation rupture occurs. This deformation and rupture are constrained by the casing and the outer cement ring. Additionally, the casing experiences deformation and damage due to the reaction force from the surrounding rock.

(2) Influence of Stratum Dip Angle

Many of the currently developed oil fields are characterized by anticlinal structural reservoirs, resulting in strata having a certain dip angle. Weak structural planes in the strata may weaken during oil and gas production and enhanced recovery measures. When the dip angle of the strata exceeds the internal friction angle of the rocks, gravity can cause sliding in weak rock layers.

(3) Fault Activity

During the oil field development process, changes in subsurface pressure due to crustal movement, earthquakes, high-pressure injection, and other factors can alter rock mechanics properties and subsurface stress. This can induce the reactivation of pre-existing faults, especially after water injection, intensifying damage to the casing. Some oil fields in China, such as Daqing and Jilin, have experienced casing damage concentrated near faults, which is a consequence of fault reactivation. Once the faults are reactivated, it can lead to extensive casing damage near fault zones, often characterized by casing shear failure, and the depth of these damages is generally consistent.

(4) Mudrock Water Absorption and Creep

Mudrock is an unstable rock type, and its mechanical properties and stress state change when the temperature rises or water enters the formation. This leads to the displacement, deformation, and expansion of mudrock, increasing external loads on the casing. When the casing's compressive strength is lower than the external load, it can result in casing compression deformation or even shear failure. Due to steam intrusion into the formation and water ingress into the mudrock, the elastic modulus decreases, reducing the stratum's bearing capacity and increasing the compressive stress on the casing.

2.2. The Engineering Factors

(1) Casing Material Issues

The choice of casing material plays a crucial role in its resistance to deformation. By improving the steel grade and wall thickness and using thermal recovery casing, some progress can be made. However, the design and control of casing integrity measures still may not fully meet on-site requirements. Different steel grades have varying resistance to deformation and strength.

(2) Cementing Quality Issues

Cementing is a critical process before drilling completion and has a direct impact on the well's lifespan and subsequent injection and production operations. The cementing quality can be influenced by various factors such as irregularities in the wellbore, well inclination, non-compliance with cementing standards, inadequate displacement of drilling mud, improper slurry density, mud cake issues, cleanliness of the wellbore and casing exterior before cementing, and inappropriate tension loads on the casing after cementing. These factors can adversely affect the cementing quality, which in turn affects the casing's integrity and lifespan.

(3) Steam Injection Parameters

Many oil fields employ steam injection for shale oil recovery. High-temperature, high-pressure steam injection into the wellbore can have detrimental effects on the casing's integrity. Steam temperatures can exceed the maximum allowable temperatures for certain casing grades, causing a decrease in yield strength and elastic modulus. Additionally, the tension generated on the casing due to thermal expansion and contraction, coupled with axial stress under continuous high temperatures, can result in fatigue cracks and compression deformations, ultimately leading to casing damage. Common casing thread types have temperature limits below 300 °C, which, when exceeded, can cause radial deformation of the threads that may lead to leaks and uncoupling issues.

(4) Perforation-Induced Damage

Casing damage can also be attributed to well-perforation activities. The main reasons include outer cement ring rupture and even casing rupture, or significant perforation depth errors or misfires, which are especially critical for secondary enhanced recovery wells or those with thin interlayers. If thin layers of shale or mudstone are unintentionally penetrated, water intrusion and expansion of these formations can lead to changes in subsurface stress, ultimately resulting in casing damage. An inappropriate perforation density can also impact casing strength. In low-permeability sandstone reservoirs, high-density perforation completion or cyclic high-pressure from long-term water injection or reservoir acidizing can lead to casing damage.

(5) Corrosion-Induced Casing Damage

Casing damage can include corrosion. Casing corrosion can have severe consequences, including perforations and multiple leak points. Corrosion can accelerate casing fatigue, causing early deformation and damage. It is a complex issue in oil and gas wells, requiring comprehensive technical measures for mitigation.

(6) High Temperatures Leading to Casing Damage

High temperatures in the wellbore can adversely affect casing integrity in two main ways: reducing casing strength and increasing thermal stress within the casing. These two factors, along with other contributing factors, can significantly increase the likelihood of casing damage.

(7) Formation Sand Production Impact

The formation and production of sand can affect casing integrity. Analyzing cavity shapes based on the principles of the Provencher natural balance arch theory and considering the interaction between the casing and the formation, a formula for calculating the axial force on the casing during sand production was derived. This can help determine the relationship between casing instability failure and sand production volume. Different steel grades and wall thicknesses of casing can impact casing failure during sand production. In highly sand-prone oil fields, it is advisable to use high-strength, thick-walled casing sections, particularly in critical areas, to prevent casing damage.

3. Casing Stress Analysis

Before steam injection, the casing primarily experiences the following loads: longitudinal axial loads, internal and external pressure loads, frictional loads, and bending loads. These loads stem from the casing's own weight, the buoyancy of drilling mud, internal pressure from drilling fluids, formation stress, frictional forces, and wellbore deviation. After steam injection, the casing undergoes significant temperature changes due to the heat from the injected steam. The steel material expands when heated but is constrained by the wellbore and cement sheath. As a result, substantial thermal stresses develop within the casing. These wells are fully cased and cemented, with packers used at the top of the reservoir. Perforation completion is employed. Steam is delivered through a thermal insulation column to the section below the packer, then through the annular space and perforation casing to heat the viscous crude oil reservoir [28–30]. During this phase, significant temperature differentials occur, leading to substantial thermal stresses. Our analysis comprehensively considers the radial, circumferential, and axial stress variations in the casing string throughout the steam injection process. After oil well production, the casing may experience increased external pressure from formation compaction, which is a significant factor leading to casing damage. Formations containing salt rocks, shale, and mudstone can exhibit viscoelastic deformation under the influence of original in situ stresses. Post-cementing, the casing must withstand the gradually increasing pressure due to wellbore contraction. Over an extended period, this pressure stabilizes and is referred to as the casing's geomechanical stress extrusion force. As viscoelastic solutions and elastic solutions are consistent for stable solutions of the same problem, seeking the final stable extrusion force can be accomplished using the methods of elastic mechanics. The stress model under in situ stress conditions is shown in Figure 2.

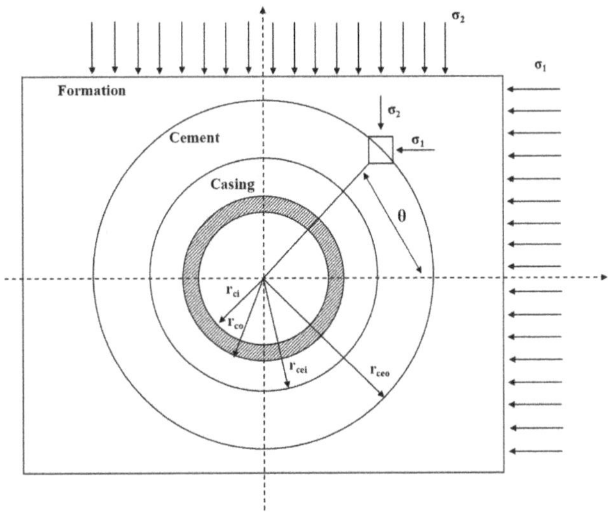

Figure 2. The stress model under in situ stress conditions.

This stress state can be represented in polar coordinates using the average and deviatoric stress in order to consider the stress state as a superposition of uniform and non-uniform forces. Therefore, solving the casing's stress under non-uniform stress conditions can be transformed into solving the casing's stress separately under σ_{ave} and under the action of the non-uniform forces $\sigma \cos 2\theta$ and $\sigma \sin 2\theta$. After superimposing these results, we can obtain the stress state of the casing under non-uniform stress. The final result can be expressed approximately as the external extrusion stress on the casing under non-uniform stress conditions.

$$\begin{cases} \sigma_{ave} = \frac{1}{2}(\sigma_1 + \sigma_2) \\ \sigma_{dev} = \frac{1}{2}(\sigma_1 - \sigma_2) \\ \sigma_r(r,\theta) = \sigma_{ave} + \sigma_{dve} \cos 2\theta \\ \tau_{r\theta}(r,\theta) = \sigma_{dve} \sin 2\theta \end{cases} \quad (1)$$

where σ_{ave} is the average stress, in MPa. σ_{dve} is the deviatoric stress, in MPa. σ_1 is the maximum horizontal stress, in MPa. σ_2 is the maximum horizontal stress, in MPa. θ is the principal direction, in °C.

The extrusion load acting on the casing can be calculated using the formula:

$$P = \rho g H \times 10^{-3} \quad (2)$$

where P is the extrusion load, in MPa. ρ is the density of the annular mud or cement slurry, in g/cm^3. g is the acceleration due to gravity, in m/s^2. H is the depth at which the casing is located, in m.

The axial load on the casing is typically derived from the gravitational force due to the casing's self-weight and the buoyant force exerted by the wellbore fluid. Additionally, the casing's exposure to the bending stress should also be considered. As the majority of the steam injection wells are vertical wells, the influence of the bending loads on the casing can be disregarded. The initial expression for the axial load is obtained by adding the equations for the casing's self-weight and the buoyant force of the wellbore fluid. Assuming that when the casing is first installed, the pressure inside and outside the wellhead is zero. The additional axial force caused by the change in internal and external pressures in the casing can be calculated using the Lame formula, as described in Equation (3).

$$\Delta F_P = \frac{2 \times 10^{-6} \mu_c \sum_{j=1}^{N} \left(\Delta p_i A_{ij} - \Delta p_o A_o\right)\left(L_j / A_j\right)}{\sum_{j=1}^{N} \left(L_j / A_j\right)} \tag{3}$$

where ΔF_P is the additional axial load, in MPa. μ_c is the Poisson's ratio of the casing. Δp_i is the increase in the internal wellhead pressure of the casing, in MPa. Δp_o is the increase in the outer wellhead pressure of the casing, in MPa. A_{ij} is the inner cross-sectional area for the jth casing section, in mm^2. L_j is the length of the jth casing section, in m. A_j is the cross-sectional area for the jth casing section, in mm^2. N is the number of casing sections calculated in the casing string.

High-temperature steam places the casing in a high-temperature environment downhole, and as the casing is typically made of steel, it undergoes thermal expansion when exposed to heat. However, the casing is constrained by the cement annulus and reservoir, preventing it from expanding freely. As a result, an additional stress induced by the temperature difference, known as thermal stress, is generated in the casing. The generation of thermal stress in the casing is primarily influenced by factors such as temperature, material properties, the properties of the cement annulus restraining the casing, and the characteristics of the reservoir. The thermal stress issue of vertical thermal recovery well casings is the foundation for calculating casing thermal stress. When deriving the expressions for thermal stress in the radial, circumferential, and axial directions, it is considered that the casing has not reached its own yield strength. The components of thermal stress are as follows:

$$\begin{cases} \sigma_\tau = \left[-\frac{a_c E_c T}{2(1-\mu_c)} + \frac{E_c C_{cl}}{(1+\mu_c)(1-2\mu_c)}\right]\left(1 - \frac{r_c^2}{r^2}\right) \\ \sigma_\theta = \left[-\frac{a_c E_c T}{2(1-\mu_c)} + \frac{E_c C_{cl}}{(1+\mu_c)(1-2\mu_c)}\right]\left(1 + \frac{r_c^2}{r^2}\right) \\ \sigma_z = -\frac{a_c E_c T}{1-\mu_c} + \frac{2\mu_c E_c C_{cl}}{(1+\mu_c)(1-2\mu_c)} \\ C_{cl} = \frac{a_c T\,(1-2\mu_c)\left[r_{co}^2 - r_{ci}^2\right]\left(\frac{E_c}{E_f} - \frac{1+\mu_c}{1+\mu_f}\right)}{2\,(1-\mu_c)\left[\frac{E_c}{E_f}\frac{r_{co}^2 - r_{ci}^2}{1+\mu_c} + \frac{(1-2\mu_c)\left(r_{co}^2 - r_{ci}^2\right)}{1+\mu_f}\right]} \end{cases} \tag{4}$$

where σ_τ is the radial casing thermal stress, in MPa. σ_θ is the circumferential casing thermal stress, in MPa. σ_z is the axial casing thermal stress, in MPa. a_c is the coefficient of thermal expansion of casing, in 1/°C. T is the temperature change, in °C. E_c is the elastic modulus of the casing, in GPa. C_{cl} is the calculation parameter. r_{ci} is the casing inner diameter, in m. r_{co} is the casing outer diameter, in m. E_f is the elastic modulus of drilling fluid, in GPa. μ_f is the Poisson's ratio of the drilling fluid.

During the steam injection process, the injected hot steam causes the casing to expand due to the high temperature. The casing experiences significant thermal stresses internally due to the restraining effect of the cement sheath. If the thermal stresses exceed the casing material's yield strength, the casing will yield and undergo plastic deformation. During repeated heating cycles, plastic deformation accumulates, gradually reducing the structural strength of the casing and eventually leading to failure. Failure modes can include necking, buckling, and fracture. For metallic materials, the Mises yield criterion is generally employed. When the thermal stress level in the casing exceeds its yield strength, the casing steel undergoes plastic yielding. This is expressed using the Mises yield criterion, represented in terms of principal stresses:

$$\sigma_s = \sqrt{\frac{1}{2}\left[(\sigma_1 - \sigma_2)^2 + (\sigma_1 - \sigma_3)^2 + (\sigma_2 - \sigma_3)^2\right]} \quad (5)$$

where σ_s is the Mises equivalent stress, in MPa. σ_1, σ_2, σ_3 are the three principal stresses, in MPa.

4. Numerical Simulation Analysis

4.1. Theoretical Basis of Finite Element Analysis

As the steam injection process continues, the casing is highly susceptible to damage, which can even lead to the abandonment of the oil well, severely impacting reservoir development and the economic benefits of the oil field. Therefore, understanding the distribution of temperature, stress, and strain in the steam injection well casing during production is of utmost importance. It is difficult to simulate the actual conditions of a steam injection wellbore in a laboratory setting. Therefore, the commonly used approach is to employ finite element simulations with thermal-stress coupling. This involves establishing a wellbore-cement annulus-formation combination and using the principles of elastic mechanics for cylindrical structures to create temperature, stress, and strain models for the steam injection well casing. The goal is to identify the causes of wellbore failure and propose effective preventive measures for casing damage in steam injection wells, ultimately extending the casing's service life and improving production efficiency.

4.2. Assumptions of Finite Element Model

(1) The cement bond is well-established on both sides, and the casing–cement–formation system is tightly connected, with the tubing, casing, and cement annulus co-axially forming a single unit.
(2) The well cementing condition is good, and the influence of cementing quality is not considered.
(3) The initial in situ stress is undisturbed during steam injection.
(4) The casing has consistent specifications, no wall thickness variations and there is no slippage between the casing and cement annulus or between the cement annulus and the formation.
(5) The parameters for steam injection at the entrance of the simulated well section are identical to those at the surface wellhead, and heat loss during steam injection is neglected.
(6) The impact of perforation on the steam injection well is not considered.
(7) The geological formation in the well section is considered infinite, with the exception of temperature changes near the wellbore, where the temperature remains constant in the rest of the formation.
(8) Material anisotropy is not considered, and it is assumed that the properties of the formation, cement annulus, and casing materials are isotropic and homogeneous.
(9) When simulating the wellbore temperature field under a certain temperature, it is assumed that the temperature distribution represents a quasi-steady state condition reached by the wellbore.

4.3. Physical Model

The numerical simulation analysis for casing damage in this study utilizes an indirect coupling approach. Initially, transient thermal analysis using ANSYS finite element analysis is employed to determine the temperature field distribution along the radial direction of the wellbore and the temperatures on the inner and outer walls of the casing. Subsequently, an indirect coupling method is used to import the temperature field of the wellbore into the structural field to obtain the thermal stresses on the inner and outer walls of the casing. The fundamental parameters of the case well and the thermal loss parameters of the surface pipeline are listed in Table 1.

Table 1. Basic parameters.

Parameters	Value	Parameters	Value
Drilled Depth	1016.0 m	Boiler Outlet Discharge Rate	10 t/h
Casing Specifications	177.8 mm	Wellhead Temperature	336 °C
Casing Depth	1015.3 m	Boiler Outlet Dryness Fraction	75.6%
Artificial Bottom	1010.0 m	Wellhead Pressure	14.15 MPa
Oil Buffer Distance	4.75 m	Pipeline Inner Diameter	0.1 m
Casing Repair Distance	5.07 m	Pipeline Material Thermal Conductivity	57.0 W/(m·°C)
Tubing Specifications	114.3 mm	Pipeline Outer Diameter	0.108 m
Tubing Depth	873.14 m	Insulation Material Thermal Conductivity	0.2 W/(m·°C)
Boiler Outlet Temperature	343.4 °C	Pipeline Length	1000 m
Design Steam Injection Volume	1500 m^3	Ambient Temperature	10 °C
Boiler Outlet Pressure	15.3 MPa	Pipeline Surface Emissivity	0.85
Measured Steam Injection Volume	1060 m^3	Average Wind Speed	3 m/s
Insulation Layer Thickness	0.07 m	Calculation Step Length	200 m

The selected wellbore segment for the numerical simulation study is located at a depth of 800 m, with a 50 m interval. Based on the given relationships, the simulated formation temperature at 800 m is 52 °C. The horizontal principal stresses in the horizontal direction are 20 MPa, the overburden pressure is 13 MPa, the cement sheath pressure is 11 MPa, and the casing self-weight load is 50 MPa. The in situ stress in the model is applied as an initial stress state to better represent real geological conditions. The model employs a structured grid division approach, with local refinements in the mesh for the casing and cement sheath and gradual mesh refinement from the outside to the inside for the formation. This results in a total of 20,560 elements and 116,481 nodes, as shown in Figure 3.

The material parameters for the casing, formation, and cement sheath are shown in Table 2.

Table 2. Basic parameters of selected material.

Material	Density (kg/m^3)	Young's Modulus (MPa)	Poisson's Ratio	Coefficient of Thermal Expansion (1/K)	Thermal Conductivity (W/(m·K))
N80 Casing	7850	146	0.3	12	43.27
TP100H Casing	7850	180	0.3	12	43.27
TP110 Casing	7850	180	0.3	12	43.27
TP130TT Casing	7850	165	0.3	12	43.27
Cement Sheath	1900	7.5	0.18	10.5	0.81
Formation	2390	25	0.2	10.3	3.44

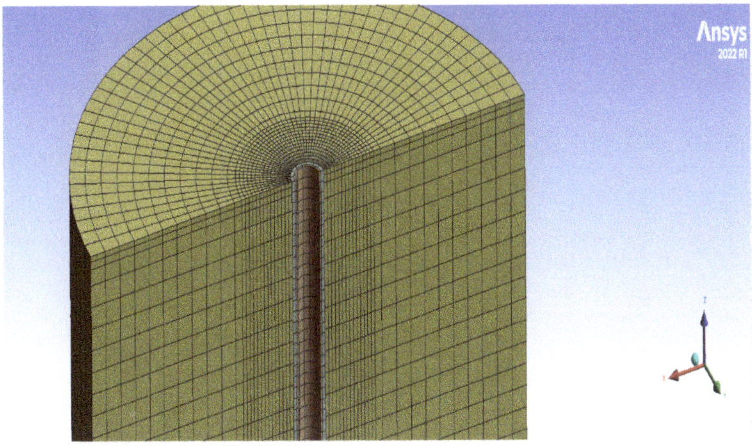

Figure 3. Model mesh division.

Applying stress and displacement to the model as boundary conditions, as shown in Figure 4. A represents the radial pressure of 20 MPa; B represents the axial pressure of 13 MPa from the formation; C represents the axial pressure of 11 MPa from the cement sheath; D represents the axial pressure of 50 MPa from the casing; E represents the outer boundary of the formation and is completely fixed (constraint); F represents the displacement constraint on the cutting plane; and G and H represent the axial displacement constraints on the upper and lower surfaces of the model.

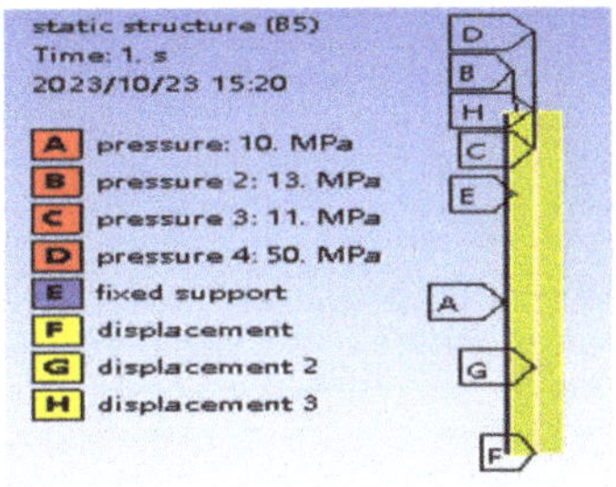

Figure 4. The model as boundary conditions.

In order to address the issue of the excessive axial thermal stresses generated in the casing during steam injection in thermal recovery wells and prevent casing damage, a pre-stressed casing completion technique is employed. By applying a certain level of pre-stress, some of the thermal stresses can be reduced, thereby preventing casing damage. Some casings experience thermal stresses during steam injection that exceed the yield strength of the casing, resulting in plastic deformation and, ultimately, casing failure, which poses a safety risk to the wellbore. Numerical simulation setup: Two analysis steps are used. In the first step, the casing–cement interface is defined as smooth, and pre-tensile stress is applied to the casing to simulate pre-stressing the casing and injecting back cement. In the second analysis step, the casing–cement interface is changed to bonded to simulate the solidification of the cement, bonding between the inner wall of the casing and the cement sheath, and then applying the thermal load. Currently, there is limited research on the calculation methods for reasonable pre-stress values in various oil fields in China. The methods for calculating the pre-stress values mentioned in manuals and textbooks typically involve using the thermal expansion coefficient of steel to calculate the elongation of the casing under thermal conditions and then deriving the corresponding reduction in pre-stress, referred to as uniaxial pre-stress. When applying uniaxial pre-stress, the maximum thermal stress produced in the casing under thermal conditions is calculated as follows:

$$\sigma_{\max} = \lambda E \Delta T \tag{6}$$

where $\sigma_{\max}$ is the maximum thermal stress experienced by the casing, in MPa. λ is the coefficient of linear expansion for the metal, K^{-1}. E is the elastic modulus of the steel material, in MPa. ΔT is the increase in temperature, in °C.

The minimum pre-stress required for the casing string is calculated as follows:

$$\Delta\sigma = \sigma_{\max} - \sigma_s \tag{7}$$

where $\Delta\sigma$ is the minimum pre-stress required for the casing, in MPa. σ_s is the minimum yield strength of the casing material, in MPa.

The minimum pre-tension force required for the casing string is calculated as follows:

$$F = \Delta\sigma S \tag{8}$$

where F is the pre-tension force for the casing, in kN.

Based on theoretical calculations and field data, the following pre-stress loads are selected for the study, as shown in Table 3 below.

Table 3. Pre-stress and corresponding loads.

Applied Pre-Stress (KN)	Corresponding Loads (MPa)
300	45.7
320	48.8
445	67.8
500	76.2
750	114.3
850	129.5
1000	152.4

First, apply the corresponding pre-tension loads to the model individually, with no frictional relative slip allowed between the cement ring and the casing. Next, export the deformations produced in three axes at different nodes under different pre-stress conditions as text files. These files will serve as input for simulating the casing's state under pre-tension conditions. Simulate the changes in Mises stress in the casing with different pre-stress values applied. The distribution patterns of the casing Mises equivalent stress at the pipe head and the pipe wall are inconsistent. The trend of stress at the pipe head and within the pipe wall with changing pre-stress is illustrated in the following Figures 5 and 6.

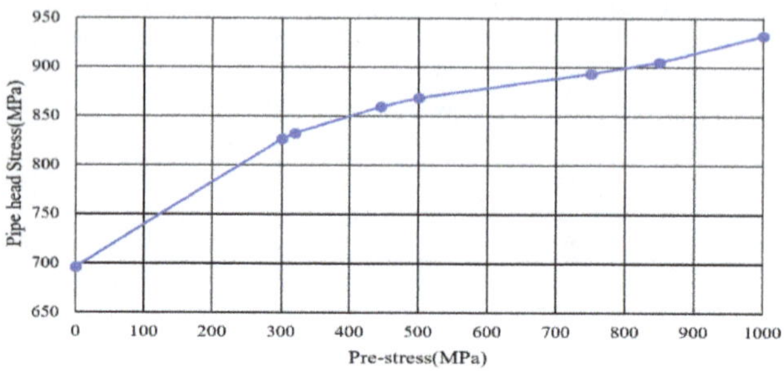

Figure 5. The trend of stress at the pipe head with changing pre-stress.

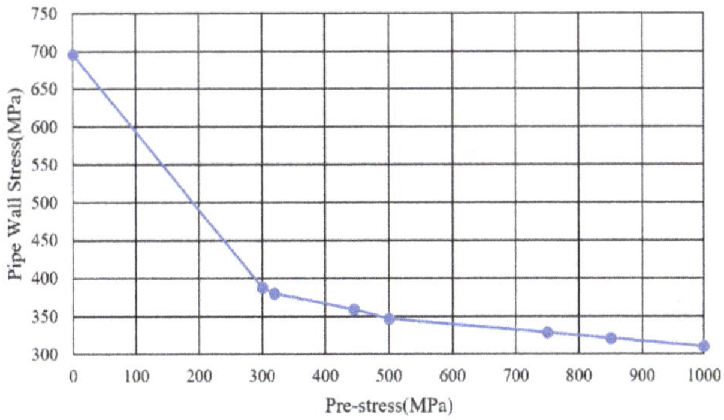

Figure 6. The trend of stress at the pipe wall with changing pre-stress.

From the figures, it can be observed that applying pre-stress leads to a significant increase in the stress at the casing pipe head and a noticeable decrease in stress within the pipe wall. As the pre-tension force increases, the impact on the pipe wall thermal stress gradually diminishes. Without applying pre-stress, the maximum Mises stress in the casing reaches 696 MPa. However, after applying pre-stress levels of 300, 500, 750, 850, and 1000 kN, the actual reduction in thermal stress in the casing due to the sealing of the inner wall by steam injection is significant. In the absence of pre-stress, the casing segment experiences relatively high overall stress, with an inner wall stress of 697 MPa. After applying a 500 kN pre-stress, the casing experiences the maximum equivalent stress of 870 MPa at the wellhead. The application of pre-stress can significantly reduce the thermal stress experienced by the casing wall, but it increases the stress at the wellhead.

5. Analysis of Influencing Factors

5.1. Impact of Different Cement Properties on Casing Damage Patterns

To overcome the challenges of developing high-temperature-resistant elastic cement slurry systems, a series of high-temperature-resistant additive materials were studied in the laboratory. This led to the formulation of high-temperature-resistant elastic cement slurry systems, and their performance was evaluated to provide technical support for cementing operations in high-temperature and complex well environments. Using G-grade cement as the base material, high-temperature elastic additives were incorporated to improve the mechanical properties of the cement rings, as shown in Table 4 below.

Table 4. Cement properties.

NO.	Elastic Agent Content %	Elastic Modulus (GPa)
1	8%	5
2	5%	7.5
3	3%	10
4	2%	15
5	1%	20

Through experimental research, the types and quantities of additives, as well as their influence on the properties of oil well cement rings, were determined to simulate the material properties of the cement rings used. Based on adjusting the content of additives in the elastic cement, changes in the mechanical properties of the cement rings were achieved. Simulations were conducted to examine the thermal stress in the casing for cement rings with different Young's moduli. The continuous increase in the elastic modulus of the cement rings results in a gradual increase in the maximum Mises stress in the casing. However, this increase is not very significant. Although it may not lead to casing yield, it could potentially cause plastic deformation in the cement ring–casing interface, affecting the bond between the cement and the casing. The Numerical simulation results are shown in Table 5.

Table 5. Numerical simulation results.

NO.	Elastic Modulus (GPa)	Equivalent Stress (MPa)	Deformation (mm)	Plastic Strain
1	5	690	3.18	0.008
2	7.5	691	3.27	0.0096
3	10	692	3.33	0.0097
4	15	696	3.30	0.0101
5	20	728	3.59	0.012

Regressing the axial deformation data over the entire wellbore, the axial elongation of the entire wellbore is obtained when the wellbore is unconstrained, as shown in Figure 7.

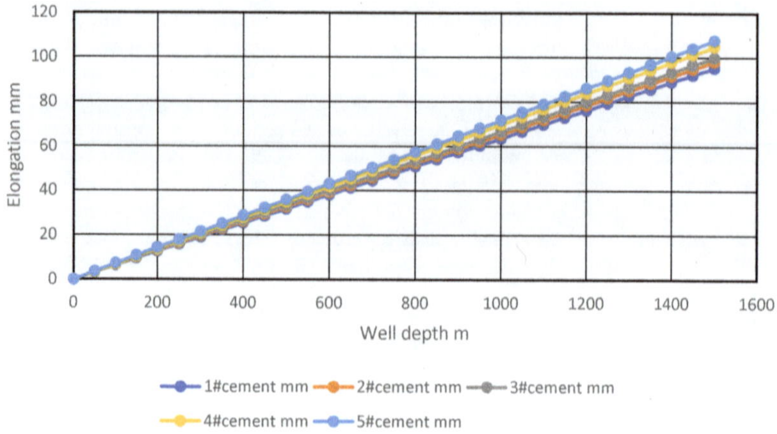

Figure 7. The relationship between the elastic modulus of cement and overall wellbore axial deformation.

As the elastic modulus of the cement ring continues to increase, the maximum Mises stress in the casing gradually increases; however, the change is not very significant. With the increasing elastic modulus of the cement ring, both the overall maximum deformation and the deformation range increase. As the elastic modulus of the cement ring increases, the model's plastic strain also begins to increase, indicating more significant plastic deformation in the cement ring and the formation. However, the strength of the cement and its bond with the casing plays a significant role in their impact on the casing. In practical field applications, selecting appropriate cement properties can provide some protection to the casing; however, the effect is limited.

5.2. Simulation Analysis of the Influence of External Loads on Casing

At a depth of 800 m below the packer, the Mises stress levels in the casing above and below the packer are significantly different. This disparity is due to the packer's function of isolating the annulus above and below it. The packer experiences horizontal stress due to the thermal compression of the casing. Near and especially below the packer, there is a concentration of local stress, and the average casing stress can reach over 750 MPa, which exceeds the yield strength of the casing and makes it prone to damage. If a packer is not used or if it fails, the overall casing experiences relatively high stress levels and is subjected to very high radial and axial stresses, making the casing more susceptible to damage. At a depth of 800 m in the wellbore, a 2 m rigidly constrained section with packers is installed.

Local circumferential pressure is applied to simulate the squeezing of the casing wall at the packer location. The simulation results are obtained, as shown in Figure 8.

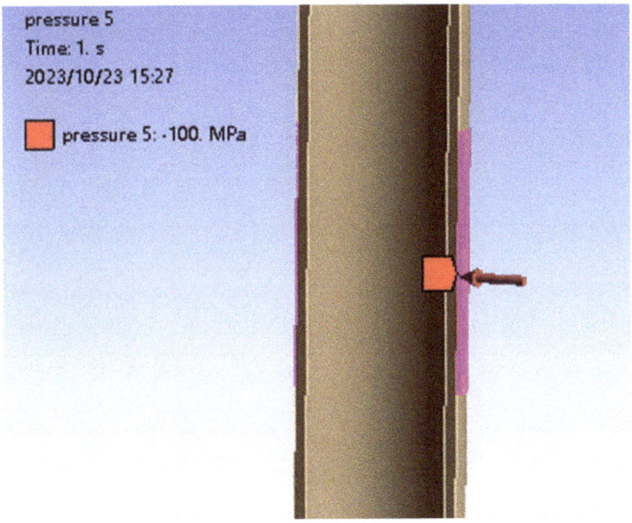

Figure 8. The location of localized load application.

The stress distribution in the casing along the 4 m section of the packer is fitted with a function, as shown in Figure 9.

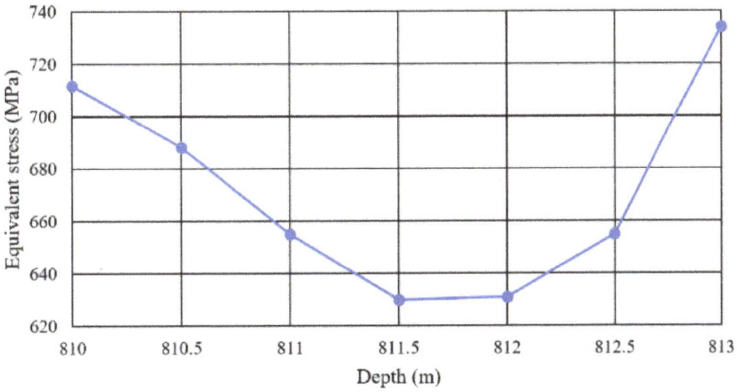

Figure 9. The change in equivalent stress in the casing within the packer section.

At both ends of the packer, there are abrupt changes in stress due to the external load conditions, resulting in stress concentration on both sides and an increase in local stress. As the deformation gradually increases on both sides of the packer, stress concentration occurs, leading to a sharp increase in deformation, making it prone to failure, as shown in Figure 10.

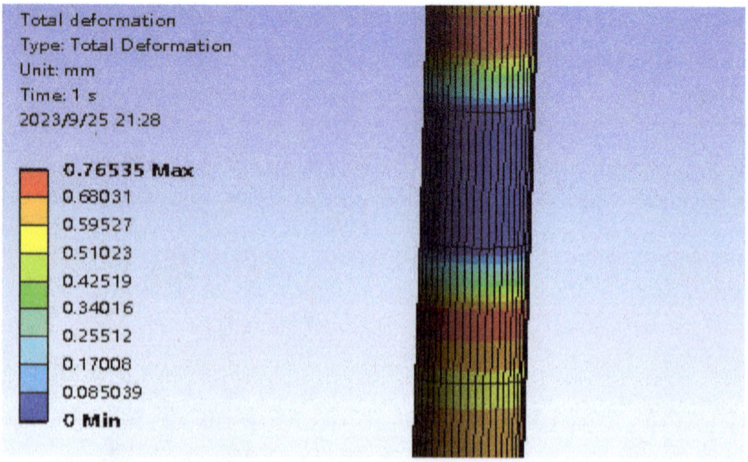

Figure 10. The deformation diagram of the packer section.

Due to the hydration of shale, the cohesion and internal friction angle decrease, resulting in a significant reduction in the shear strength of the rock; therefore, as the water content of the shale increases, the rock's shear strength decreases. In the sand–shale interface, the casing experiences uneven stress, leading to shear failure in this area. Additionally, after shale hydration in the shale section, the shale's supporting capacity decreases, and stress is transferred to the cement sheath and casing. When the stress on the casing exceeds the yield strength of the casing, plastic deformation occurs in the casing. At a depth of 800 m in the wellbore, a simplified approach is used to apply localized external loads to a 2 m segment based on the shale layer thickness and the length of the packer. The external load values are 50, 75, 100, 125, and 150 MPa. The local load is applied by simulating the circumferential pressure to model the expansion and squeezing of the casing and packer at the casing wall after shale hydration. Different results are calculated for each of the different external load values, as shown in Table 6.

Table 6. Different external loads.

Local Load (MPa)	Equivalent Stress (MPa)	Deformation (mm)	Plastic Strain
50	729	0.69	0
75	756	0.7	0.0002
100	760	0.703	0.0008
125	772	0.708	0.0009
150	793	0.72	0.0017

When the external load remains within 50 MPa, there is no yielding in the casing wall. Yielding begins at approximately 70 MPa. As the external load gradually increases beyond 125 MPa, plastic deformation accumulates noticeably. As the localized external load increases, the maximum principal stress, maximum deformation, and plastic strain in the casing gradually increase. However, there are differences in the behavior of the inner and outer walls of the casing. The stress on the inner and outer walls at different locations under external loads is shown in Table 7.

Table 7. Stress on inner and outer walls at different external load locations.

Local External Load (MPa)	Inner Wall Stress (MPa)	Outer Wall Stress (MPa)
50	729	720
75	758	704
100	760	687
125	772	665
150	793	634

5.3. Impact of Different Injection and Production Parameters on Casing Damage Patterns

Controlling the injection and production parameters in steam injection wells is the most easily controllable influencing factor. The study on the impact of steam injection temperature on casing stress in thermal recovery wells primarily focuses on controlling the steam injection temperature, steam injection pressure, and steam injection rate to regulate the wellbore temperature and pressure field. By employing a method of controlling variables, all other well parameters remain constant except for changes in the temperature, pressure, and rate of steam injection. This approach allows for a clear understanding of how steam temperature affects casing stress. Based on the actual field conditions, various steam injection parameters were selected for simulation and solutions to obtain different temperature fields and stress fields. The chosen steam injection parameters include:

(1) Steam injection temperature: 350 °C, 320 °C, 300 °C, 280 °C, and 250 °C.
(2) Steam injection pressure: 8 MPa, 10 MPa, 12 MPa, 15 MPa, and 18 MPa.
(3) Injection rate: 4 t/h, 6 t/h, 8 t/h, 10 t/h, and 12 t/h.

In order to investigate the influence of the steam injection pressure while controlling the steam injection temperature at 350 °C, the simulated Mises stress on the inner casing wall for different steam injection pressures of 8, 10, 12, 15, and 18 MPa are as shown in Table 8.

Table 8. Simulation results under different steam injection pressure conditions.

Steam Injection Pressure (MPa)	Casing Stress (MPa)
8	766
10	768
12	768.6
15	769
18	769.6

By analyzing the Mises stress values under different steam injection pressure conditions, it is observed that changing the steam injection pressure by 10 MPa results in a mere 1% increase in the Mises stress on the inner casing wall. This implies that changing the steam injection pressure has a minimal impact on the Mises stress in the casing. Consequently, it can be concluded that the steam injection pressure has a relatively small influence on the Mises stress on the inner casing wall and, by extension, on casing damage.

Figure 11 illustrates the influence pattern of the steam injection temperature on the casing stress. It can be observed that when the steam temperature is increased from 200 °C to 350 °C, the Mises stress level on the inner casing wall increases by 48%. This indicates that the stress level in the casing is significantly affected by temperature, and changes in the steam temperature have a significant impact on casing stress. Furthermore, at a temperature of 350 °C, the Mises stress on the inner casing wall reached 696 MPa.

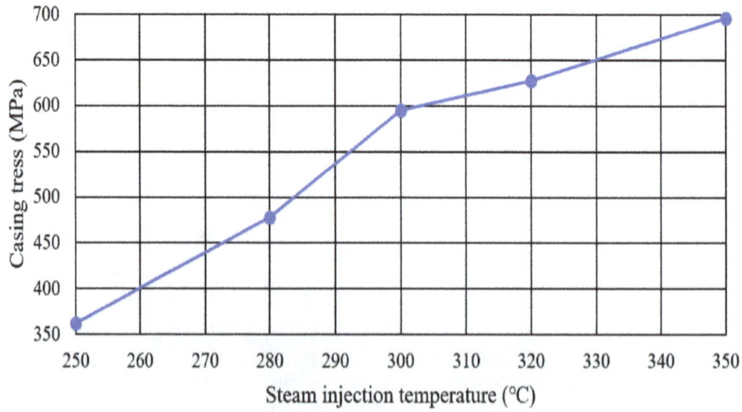

Figure 11. The influence pattern of steam injection temperature on the casing stress.

Figure 12 illustrates the influence pattern of the steam injection rate on the casing stress. It can be observed that when the steam injection rate is increased from 4 t/h to 8 t/h, the Mises stress level on the inner casing wall increases by 26%. This indicates that the stress level in the casing is significantly affected by the steam injection rate, and changes in the injection rate have a certain impact on casing stress. Additionally, at an injection rate of 10 t/h, the Mises stress on the inner casing wall reached its maximum value of 696 MPa. Increasing the injection rate further will not result in a continuous increase in casing stress. The injection rate has a noticeable impact on the overall injection volume. Excessive injection rates do not lead to a significantly more pronounced heating effect.

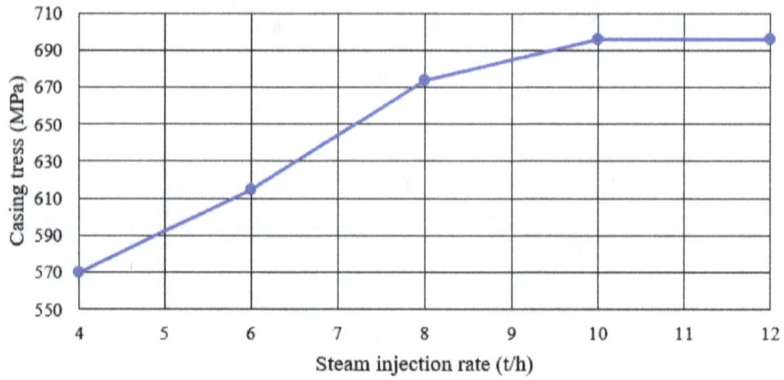

Figure 12. The influence pattern of steam injection rate on the casing stress.

5.4. Impact of Different Insulation Methods and Insulation Effect on Casing Protection and Damage Patterns

During the injection of high-temperature steam into the wellbore, high-temperature thermal stresses often occur. The fundamental reason for this is the differential expansion and constraint of various materials within the wellbore. Materials such as steel casing, cement sheath, and geological formations expand when the temperature rises. Since these materials have different coefficients of thermal expansion and are mutually constrained, they generate thermal stress. Materials with a larger coefficient of thermal expansion will produce greater thermal stress under the same constraints. This method avoids further heat loss along the path. In numerical calculations, a reference temperature of 52 °C for the geological formation and a steam injection temperature of 350 °C are used. To study the influence of insulation methods on the temperature distribution near the casing and, consequently, to determine the casing stress under different insulation methods, this study considers three commonly used insulation measures in engineering. The thermal conductivity coefficients for the casing–cement sheath system under each measure are provided in Table 9.

Table 9. Thermal conductivity of three heat insulation measures.

Insulation Measure	Thermal Conductivity Coefficients (W/m·°C)
Vacuum Insulation Tube + Packer	0.06
Vacuum Insulation Tube + Nitrogen Gas	0.01
Vacuum Insulation Tube + Nitrogen Gas	0.12
Vacuum Insulation Tube + Nitrogen Gas	0.20

The simulation of the annular temperature field under different thermal insulation conditions is shown in Figure 13.

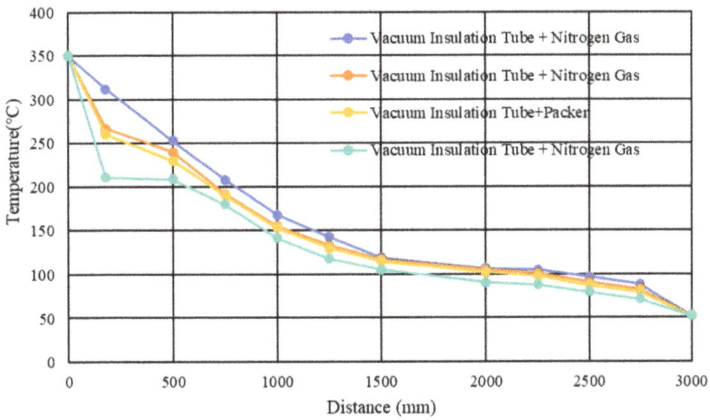

Figure 13. Temperature field distribution under different thermal conductivities.

When the thermal conductivity of the wellbore system changes, the heat transfer effect also changes, which shows that the temperature of the inner and outer walls of the casing is affected by the heat transfer coefficient. The use of low thermal conductivity insulation can significantly reduce the wellbore internal temperature. When the apparent thermal conductivity of the heat insulation tube is smaller, the temperature of the casing wall is lower, and the heat insulation effect is more obvious. Taking the temperature field as the input load, the variation rule of the casing thermal stress state is obtained, as shown in Figure 14.

The thermal insulation coefficient has an obvious influence on the temperature field in the wellbore. The temperature in the wellbore is positively correlated with the equivalent thermal stress. The lower the thermal conductivity coefficient is, the better the thermal insulation effect is. The effect of reducing thermal conductivity is more obvious below 0.1 W/m·°C. The thermal insulation coefficient related to thermal insulation can directly affect the casing temperature field during steam injection. The lower the thermal conductivity coefficient, the lower the casing wall temperature, the better the thermal insulation effect. The maximum thermal stress of the casing also decreases with the decrease of the heat insulation coefficient, showing an obvious positive correlation. The effect of reducing the thermal conductivity is more obvious when the thermal insulation coefficient is below 0.1 W/m·°C. With the process of multiple steam huff-and-huffs, the performance of the thermal insulation pipe will decrease, its apparent thermal conductivity will increase greatly, and the stress of the casing will also increase greatly, which may place the casing above the yield strength and cause casing damage in the overlying strata section.

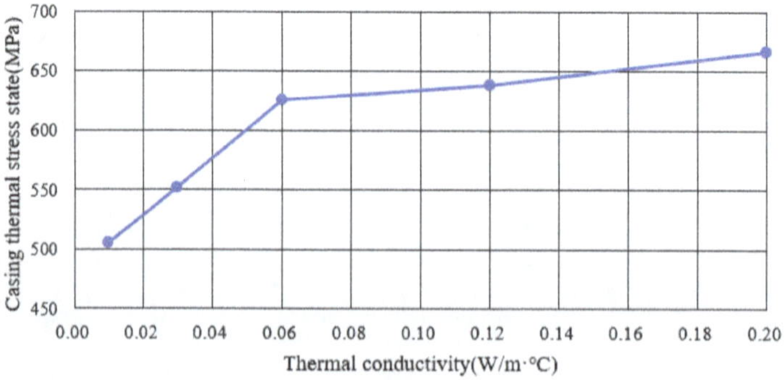

Figure 14. Diagram of variation in casing thermal stress and thermal conductivity.

5.5. *Effect of Cementing Strength on Casing Damage*

Wellbore integrity is a necessary condition to ensure the safe production of oil and gas wells. Compared with the cement ring, casing, and formation body, the interface between the three is a weak link. In production practice, it is found that some oil and gas wells show good cementing quality through logging after cementing; however, there are oil, gas, and water phenomena at the wellhead during actual production. The analysis shows that the cementation failure of the first and second interfaces is caused by the change in temperature and pressure during the operation, which leads to the flow of bottomhole fluid to the surface through the micro-annulus, and the wellbore integrity is damaged. Due to the influence of high temperatures, the thermal deformation properties of the casing and cement ring are more different than those of the cement ring and formation. Therefore, the casing–cement ring interface is more prone to failure than the cement ring–formation interface, and the failure of one interface will cause casing axial elongation. Therefore, the thermal recovery well interface is chosen as the main research object. The device is shown in Figure 15.

Figure 15. Test equipment for cement ring bond strength.

G-grade oil well cement was selected as the test cement. The casing section was first inserted into the cement curing tank to ensure a good fit between the simulated casing and the outer tank. The prepared cement slurry was poured into the wall of the casing pipe and then placed in a water bath for curing for 7 days at a curing temperature of 25 °C. The gland is covered on the top of the casing, and a screw is placed on the upper part of the gland, as shown in Figure 16.

Figure 16. Simulation device maintenance completed.

The press load is transferred to the gland through the pressure plate to ensure that the axis of the simulated casing coincides with the axis of the press. Start the press, adopt the constant displacement loading mode, and the loading speed is 0.05 mm/min. When the pressure increases to a certain value, the interface between the simulated casing and cement will be damaged, and the simulated casing will be extruded from the cement. The maximum load during the loading process will be recorded, and then the above steps will be repeated to record the experimental data. The test data are shown in Figure 17.

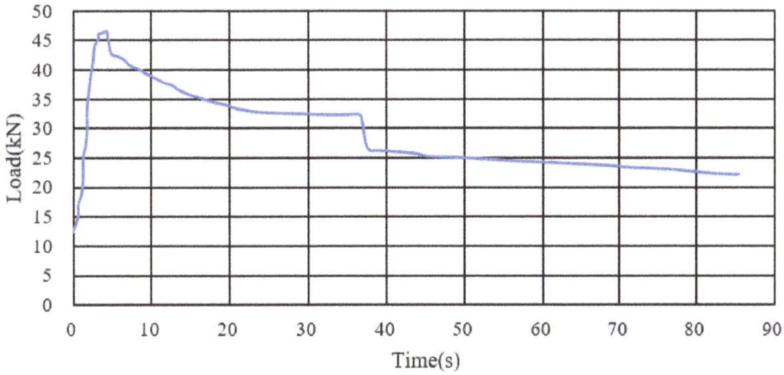

Figure 17. Load–time variation test data.

The maximum pressure recorded during the loading process is converted into bond strength using the following formula:

$$S = F/\pi DH \quad (9)$$

where S is the bond strength at the interface, in MPa; F is the maximum pressure at interface failure, in N; D is the diameter of the simulated casing, in mm; and H is the length of the simulated casing in contact with the cement, in mm. The test result data are shown in the following Table 10.

Table 10. Test results.

Casing Diameter (mm)	Contact Length (mm)	Failure Load (kN)	Bond Strength (MPa)
140	11.7	160	3.11

A simulation of the connection between the cement sheath and the casing was conducted to study casing damage in thermal recovery wells under different junction conditions. Under various junction strength conditions, different degrees of relative slippage occur between the casing and the cement sheath, resulting in casing elongation and wellhead lifting phenomena. Partial loss of the cement sheath can also lead to casing damage in the form of deformation, narrowing, or other damage due to uneven external loads.

As shown in Figures 18 and 19, at different junction strengths, the maximum stress on the wellbore remains at around 682.4 MPa, with a relatively minor impact from the changes in cement sheath bond strength. Under different junction strengths, the maximum displacement at the wellbore increases as the junction strength components increase, with a more significant effect from the variations in cement sheath bond strength—complete detachment results in a substantial elongation. In the absence of detachment or fracture, the bond strength of the cement sheath does not directly affect the thermal stress on the casing wall. However, once a fracture occurs between the cement sheath and casing, it significantly increases the elongation, leading to wellhead uplift.

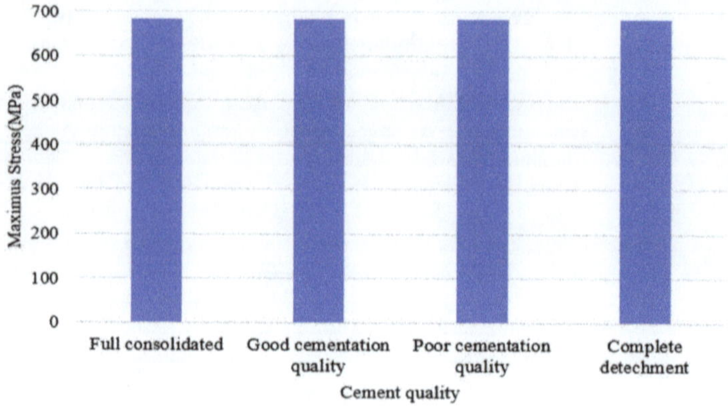

Figure 18. Maximum stress in the casing under different bond qualities.

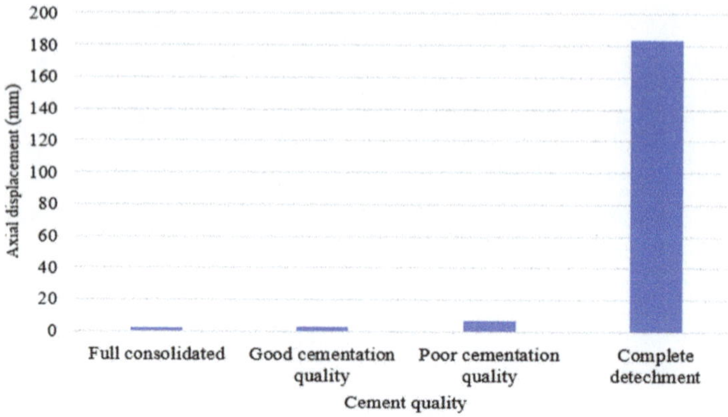

Figure 19. Displacement change with varying consolidation qualities.

5.6. Thermal Stress and Yield Strength of Different Steel Grades of Casing

The N80, TP100H, BG110H, and TP130TT casings were selected as the steel grades of the simulated casing in the thermal production well. According to the experimental results that the property parameters of the relevant casing properties change with the temperature at normal temperature, the material properties at a high temperature can be obtained according to the empirical formula of the strength attenuation of the casing under high-temperature conditions of steam injection, as seen in Table 11.

Table 11. Casing steel strength properties.

Casing Steel Grade	Outer Diameter (mm)	Yield Strength (MPa)	Elastic Modulus (GPa)
TP130TT	177.8	985.1	165
N80	177.8	582.5	146
TP100H	177.8	812.2	176
BG110H	177.8	837.7	180

When using N80 casing, the thermal stress experienced by the casing exceeds the casing's yield limit. However, when using TP100H, TP130TT, and BG110H casings, the entire casing's thermal stress remains within the casing's yield limit. N80 casing exhibits a much larger plastic deformation range compared to other casings, as shown in the Figure 20. The difference among different casings in total elongation is not significant. Simulated responses of different steel grades of casings under thermal loads are shown in the following Figures 21 and 22.

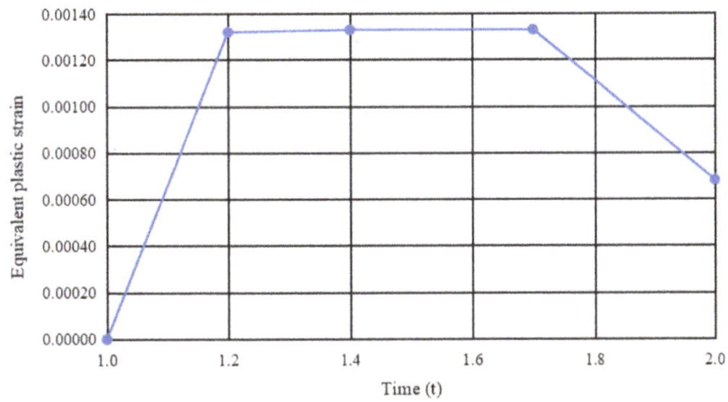

Figure 20. Time-dependent variations in the equivalent plastic strain of N80 casings.

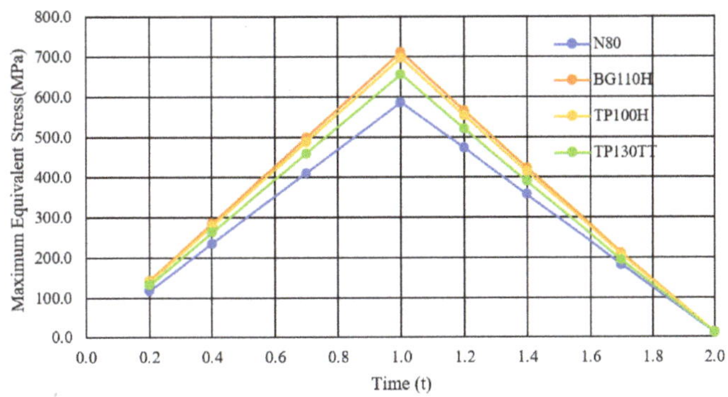

Figure 21. Time-dependent variations in the maximum equivalent stress of casings of different steel grades.

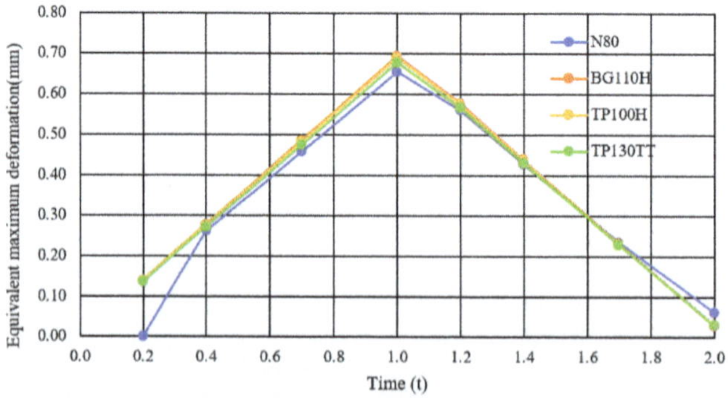

Figure 22. Time-dependent variations in the equivalent maximum deformation of casings of different steel grades.

6. Optimal Design of Steam Injection-Production Well

According to the simulation results obtained in the previous paper, the optimal design of the injection-production parameters can be obtained by fitting the influence law of the injection-production parameters on casing damage as follows: The common steam injection rates of 4, 6, 8, and 10t/h in steam injection-production wells are selected as reference steam injection rates. The effects of different injection and production parameters on casing stress under 350 °C, 320 °C, 300 °C, 280 °C, and 250 °C conditions are obtained, as shown in the following Figures 23–26.

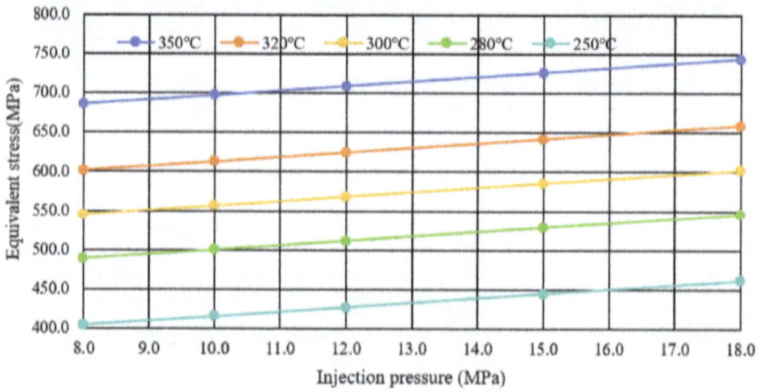

Figure 23. The influence of casing stress with injection temperature at a 4 t/h injection rate.

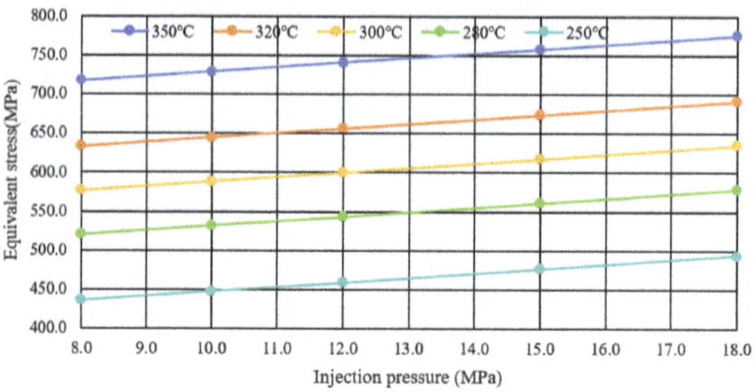

Figure 24. The influence of casing stress with injection temperature at a 6 t/h injection rate.

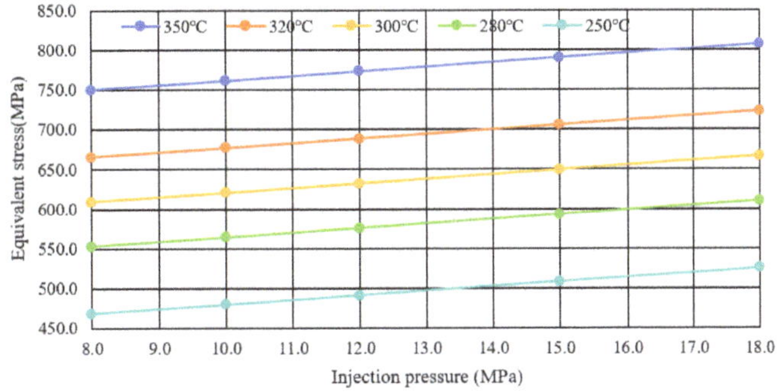

Figure 25. The influence of casing stress with injection temperature at an 8 t/h injection rate.

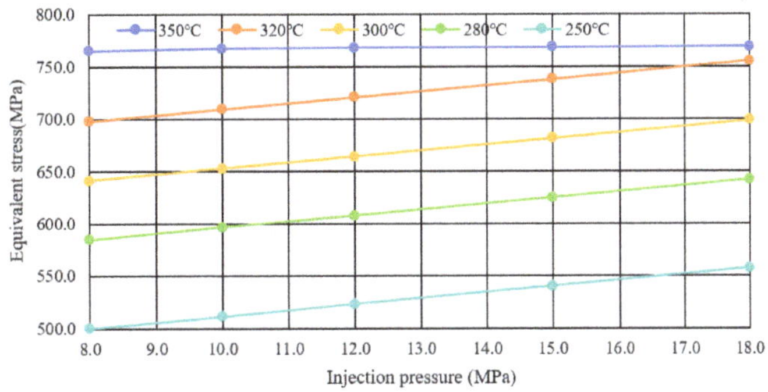

Figure 26. The influence of casing stress with injection temperature at a 10 t/h injection rate.

For a steam injection well with a steam injection rate greater than 10 t/h, the maximum thermal stress is consistent with 10t/h. The steam injection parameters selected in the chart are based on the range of steam injection parameters of the casing-damaged wells. Three casing-damaged wells in this block were selected for the optimal design of injection and production parameters, as shown in the following Table 12.

Table 12. Optimization design of injection and production parameters of casing damage well.

Well No.	Development Method	Wellbore Type	Production Days	Casing Damage Cycles	Casing Steel Grade	Injection Pressure (MPa)	Injection Temperature (°C)	Steam Volume (m^3)
1	Steam Injection	Vertical Well	1329	11	TP100H	14.7	350	33,503
2	Steam Injection	Vertical Well	3253	5	TP100H	8.2	350	35,436
3	Steam Injection	Vertical Well	2099	10	TP100H	17.3	350	52,739

In comparison to the previous optimized design, the optimization strategies for the injection and production parameters for the three wells are as follows: For well 1, reduce the steam injection rate (as it was too high). For well 2, decrease the steam injection temperature and implement better insulation measures to ensure production. For well 3, decrease the steam injection rate (as it was too high). As a result of the optimization of the injection and production parameters, it is expected that the casing damage cycles for these wells will be postponed by 2–3 cycles.

7. Conclusions

(1) When using N80 casing, the entire casing experiences a thermal stress exceeding its yield limit. For the TP100H, BG110H, and TP130TT casings, thermal stress remains within the yield limit. Different casings under steam thermal stress show little difference in cross-sectional deformation and total elongation. It is advisable to avoid using N80-grade steel casing whenever possible. Instead, consider using higher-strength steel casing materials for the production zone and casing designed for thermal recovery, and use casing materials with greater resistance to external loads, such as TP130TT, for the sections exposed to mudrock hydration and other external loading conditions.

(2) Different casings exhibit relatively similar cross-sectional deformation and total elongation under steam thermal stress loads. While an increase in the elastic modulus of the cement sheath may not lead to casing yield, it could induce plastic deformation in the cement sheath-formation structure, potentially affecting the bonding interface of the cement sheath and the casing.

(3) An increase in local external loads leads to an increase in maximum casing principal stress, maximum deformation, and plastic strain. Stress at the casing's inner surface increases with external load, while stress at both ends decreases. The opposite trend is observed for stress at the casing's outer surface. Stress concentration can occur at both ends of the applied pre-stress, potentially leading to casing deformation or bending under non-uniform stress conditions.

(4) Injection parameters, including pressure, have a limited impact on casing stress, with higher injection pressure leading to slightly higher inner wall pressure. The injection temperature significantly affects the casing's temperature field and thermal stress; a lower injection temperature leads to reduced stress on the casing. The injection rate has a significant influence on the overall injection volume, with excessively high injection rates not improving the heating effect. The application of pre-stress can reduce the casing wall thermal stress but may increase stress at the wellhead. Adequate pre-stress can delay the casing damage cycle. Pre-stress exceeding 500 kN has an unclear effect on reducing thermal stress and can increase wellhead tensile stress. Excessive pre-stress may exceed the casing's tensile strength.

(5) For the existing shale oil steam injection wells, optimizing the injection parameters is essential. Controlling the appropriate injection parameters helps in safeguarding the casing and prolonging its lifespan. The application of suitable pre-stress to shallow wells can alleviate casing wall thermal stress and extend the casing's operational life. Additionally, employing better insulation methods, such as vacuum insulation pipes and nitrogen insulation, aids in enhancing steam injection efficiency, reducing heat loss, lowering casing wall temperatures, and effectively controlling thermal stress.

(6) Implementing more effective insulation methods, such as vacuum insulation tubes and nitrogen insulation, can help enhance steam injection efficiency, reduce heat loss, lower the temperature on the casing wall, and effectively control thermal stress. Improving the quality of cementing for the wellbore is essential. In sections with poor cementing quality, it is crucial to prevent significant relative slippage at the interface. In areas prone to slippage between the casing and the external cement sheath, appropriate roughening of the casing's outer surface can be applied to prevent complete detachment of the cement sheath from the casing.

(7) For the existing shale oil steam injection wells, it is advisable to optimize the steam injection parameters, controlling them within a reasonable range to protect the casing and extend the casing's lifespan. Applying appropriate pre-stress to shallow wells can alleviate thermal stress on the casing wall and delay the onset of casing damage.

Author Contributions: Conceptualization, C.K.; Methodology, X.Y.; Software, X.C. (Xiupeng Chen) and Z.H.; Investigation, X.C. (Xueqi Cen) and X.C. (Xinyuan Chen); Resources, S.T.; Data curation, Y.H.; Writing—original draft, Y.Y. All authors have read and agreed to the published version of the manuscript.

Funding: This research was funded by the National Science Foundation of China (Grant No. 52004259).

Acknowledgments: The authors gratefully acknowledge the financial support of the National Science Foundation of China (Grant No. 52004259).

Conflicts of Interest: The authors declare no conflict of interest.

References

1. Zou, C.; Pan, S.; Zhao, Q. On the connotation, challenge and significance of China's "energy independence" strategy. *Pet. Explor. Dev.* **2020**, *47*, 416–426.
2. Du Jinhu, H.S.; Zhenglian, P.; Senhu, L.; Lianhua, H.; Rukai, Z. The types, potentials and prospects of continental shale oil in China. *China Pet. Explor.* **2019**, *24*, 560–568.
3. Liu, G.; Jin, Z.; Zeng, L.; Huang, L.; Ostadhassan, M.; Du, X.; Lu, G.; Zhang, Y. Natural fractures in deep continental shale oil reservoirs: A case study from the Permian Lucaogou formation in the Eastern Junggar Basin, Northwest China. *J. Struct. Geol.* **2023**, *174*, 104913.
4. Qun, L.E.I.; Dingwei, W.E.N.G.; Baoshan, G.U.A.N.; Junfeng, S.H.I.; Bo, C.A.I.; Chunming, H.E.; Qiang, S.U.N.; Huang, R. Shale oil and gas exploitation in China: Technical comparison with US and development suggestions. *Pet. Explor. Dev. Online* **2023**, *50*, 944–954.
5. Ding, J.; Xi, Y.; Jiang, J.; Wang, H.; Li, X.; Li, H. Wellbore Integrity of High Temperature Production of Immature Shale Oil. *Sci. Technol. Eng.* **2021**, *21*, 1752–1757.
6. Mou, Y.; Zhao, H.; Cui, J.; Wang, Z.; Wei, F.; Han, L. Failure Analysis of Casing in Shale Oil Wells under Multistage Fracturing Conditions. *Processes* **2023**, *11*, 2250.
7. Hu, Z.; Yang, J.; Li, W.; Li, S.; Feng, P.; Xin, Y. Research and development of compressible foam for pressure management in casing annulus of deepwater wells. *J. Pet. Sci. Eng.* **2018**, *166*, 546–560.
8. Xiaocong, Y.; Jin, Y.; Changbin, K.; Changpeng, H.; Xudong, W.; Fabin, X.; Chao, Z.; Lei, L.; Liang, H. Unidirectional control mechanism of annulus trapped pressure in deepwater shaft. *Acta Pet. Sin.* **2018**, *39*, 335–340+364.
9. Liu, A.; Yan, T.; Wei, L.; Zhu, Q.; Wang, W.; Liu, S. Influence of Casing Deformation in Shale Oil Horizontal Wells in Jiyang Depression. *China Pet. Mach.* **2023**, *51*, 93–99.
10. Li, J.; Wu, J.; Xie, S.; Chen, L.; Lian, W. Characteristics and Control Methods of Wellbore Integrity Failure for Jimsar Shale Oil. *Xinjiang Oil Gas* **2021**, *17*, 37–43.
11. Zhang, Y. Casing Strength Theoretical and Experimental Study of Thermal Recovery Steam Injection Wells. Ph.D. Thesis, Yanshan University, Qinhuangdao, China, 2008.
12. Li, J.; Lin, C.Y.; Yang, S.C.; Zhi, Y.M.; Chen, S.W. Theoretical solution of thermal stress for casing-cement-formation coupling system. *J. China Univ. Pet.* **2009**, *33*, 63–69.
13. Tian, Z.; Lin, S.; Lei, Q. Problems in the wellbore integrity of a shale gas horizontal well and corresponding countermeasures. *Nat. Gas Ind. B* **2015**, *2*, 522–529. [CrossRef]
14. Zhou, S.; Liu, R.; Zeng, H.; Zeng, Y.; Zhang, L.; Zhang, J.; Li, X. Mechanical characteristics of well cement under cyclic loading and its influence on the integrity of shale gas wellbores. *Fuel* **2019**, *250*, 132–143. [CrossRef]
15. Liu, Y.; Feng, M.; Sun, G.; Liu, Z.; Ma, C.; Zhong, X.; Wang, L.; Liao, Y. Casing Failure Characteristics, Prevention and Control Strate gies for Mature Oilfields. In *IOP Conference Series: Earth and Environmental Science*; IOP Publishing: Bristol, UK, 2019; Volume 267.
16. Geng, L.; Liu, D.; Han, L.; Feng, S. Numerical Analysis of Casing Damage in Heavy Oil Thermal Recovery Wells. In *IOP Conference Series: Earth and Environmental Science*; IOP Publishing: Bristol, UK, 2021; Volume 726.
17. Ji, Y.; Xu, H.; Zhang, J.; Guo, L. Analytical modeling on the geo-stress and casing damage prevention with the thermo-hydro-mechanical (THM coupling of multi-field physics. *J. Therm. Anal. Calorim.* **2020**, *139*, 2719–2737. [CrossRef]
18. Zhai, X.P.; Lou, Y.S.; He, B.S.; Ji, H. Mechanical Analyze of Casing Failure with Sand Production in Loose Sandstone Reservoir. *Appl. Mech. Mater.* **2012**, *2024*, 217–219. [CrossRef]
19. Zhu, J.Q.; Hu, K.S.; Chen, H.Y. Finite Element Analysis of Crack Propagation and Casing Failure Process under Thermal Mechanical Coupling. *Adv. Mater. Res.* **2011**, *1167*, 197–198. [CrossRef]
20. Mohammed, A.I.; Oyeneyin, B.; Atchison, B.; Njuguna, J. Casing structural integrity and failure modes in a range of well types—A review. *J. Nat. Gas Sci. Eng.* **2019**, *68*, 102898. [CrossRef]
21. Kuanhai, D.; Yue, Y.; Yi, H.; Zhonghui, L.; Yuanhua, L. Experimental study on the integrity of casing-cement sheath in shale gas wells under pressure and temperature cycle loading. *J. Pet. Sci. Eng.* **2020**, *195*, 107548. [CrossRef]
22. Ouyang, Y.; Gao, Y.; Wang, X.; Duan, Z. Stress Analysis of the Integrity of Casing-Cement Ring-Structural Consolidation Body by Volume Fracturing. *Adv. Pet. Explor. Dev.* **2018**, *15*, 64–71.

23. Feng, P.Z. Thermal stresses analysis of casing string used in enhanced geothermal systems wells. *J. Groundw. Sci. Eng.* **2016**, *4*, 293–300.
24. Shipping, L. Study on the cause of casing damage and the feasibility of comprehensive prevention. In *E3S Web of Conferences*; EDP Sciences: Les Ulis, France, 2023; Volume 375.
25. Li, Q.; Han, Y.; Liu, X.; Ansari, U.; Cheng, Y.; Yan, C. Hydrate as a by-product in CO_2 leakage during the long-term sub-seabed sequestration and its role in preventing further leakage. *Environ. Sci. Pollut. Res.* **2022**, *29*, 77737–77754. [CrossRef]
26. Li, Q.; Wang, F.; Wang, Y.; Forson, K.; Cao, L.; Zhang, C.; Zhou, C.; Zhao, B.; Chen, J. Experimental investigation on the high-pressure sand suspension and adsorption capacity of guar gum fracturing fluid in low-permeability shale reservoirs: Factor analysis and mechanism disclosure. *Environ. Sci. Pollut. Res.* **2022**, *29*, 53050–53062. [CrossRef] [PubMed]
27. Yang, Y.; Wang, K.; Zhang, L.; Sun, H.; Zhang, K.; Ma, J. Pore-scale simulation of shale oil flow based on pore network model. *Fuel* **2019**, *251*, 683–692. [CrossRef]
28. Li, Z.; Lei, Z.; Shen, W.; Martyushev, D.A.; Hu, X. A Comprehensive Review of the Oil Flow Mechanism and Numerical Simulations in Shale Oil Reservoirs. *Energies* **2023**, *16*, 3516. [CrossRef]
29. Kaleem, W.; Tewari, S.; Fogat, M.; Martyushev, D.A. A Hybrid Machine Learning Approach Based Study of Production Forecasting and Factors Influencing the Multiphase Flow through Surface Chokes. *Petroleum* **2023**. [CrossRef]
30. Xu, J.; Chen, Z.; Zhou, D.; Nie, W.; Li, R. Review on the characteristics of pyrolysis during in-situ conversion of oil shale. *J. Southwest Pet. Univ. (Sci. Technol. Ed.)* **2021**, *43*, 220.

processes

MDPI

Article

A New Method for Measuring Water Holdup of Oil-Water Two-Phase Flow in Horizontal Wells

Yanjun Wang [1,2,*], Jianlong Han [3], Zhiqiang Hao [4], Lijian Zhou [1], Xinjie Wang [1] and Mingwei Shao [1]

[1] School of Information and Control Engineering, Qingdao University of Technology, Qingdao 266525, China; zhoulijian@qut.edu.cn (L.Z.); wangxinjie@qut.edu.cn (X.W.); shaomingwei@qut.edu.cn (M.S.)
[2] Reservoir Performance Monitoring Center, Shengli Oilfield Company, China Petroleum & Chemical Corporation, Dongying 257000, China
[3] Logging and Testing Services Company, Daqing Oilfield Company Limited, Daqing 163453, China; dlts_hanjl@petrochina.com.cn
[4] China National Logging Corporation, Xi'an 710077, China; haozhq@cnpc.com.cn
* Correspondence: wangyanjun@qut.edu.cn; Tel.: +86-532-6805-2205

Abstract: The production profile logging technology of Oil-water two-phase flow is an important support technology for efficient development of horizontal oil wells. Moreover, the water holdup of Oil-water two-phase flow is one of the important dynamic measurement parameters for production profile logging technology. Flow patterns in horizontal oil wells are more complex than those in vertical oil wells, making water holdup measurements more difficult. In China, horizontal oil wells are characterized by low liquid production, high water cut and layered development; hence, the concentrating flow logging scheme is adopted to reduce the effects of flow patterns on measurement. In this paper, we proposed a new method of measuring water holdup of Oil-water two-phase flow for the production profile logging in horizontal oil wells. Firstly, the structure of water holdup sensors and the measuring principle of water holdup were introduced. Secondly, the dynamic test research was carried out on the horizontal well simulation facility by using the test prototype of the water holdup measurement method of Oil-water two-phase flow. Additionally, the response characteristics of water holdup sensors at five inclined angles were emphatically discussed. Finally, the measurement error of water holdup was calculated and evaluated in detail. Experimental results show that the absolute values of measurement error of water holdup are less than 5% while the total flow rate ranged from 10 m^3/d to 200 m^3/d, and the water cut ranged from 30% to 90%, respectively.

Keywords: production profile logging; horizontal well; oil-water two-phase flow; water holdup; conductance sensor

Citation: Wang, Y.; Han, J.; Hao, Z.; Zhou, L.; Wang, X.; Shao, M. A New Method for Measuring Water Holdup of Oil-Water Two-Phase Flow in Horizontal Wells. *Processes* **2022**, *10*, 848. https://doi.org/10.3390/pr10050848

Academic Editor: Yidong Cai

Received: 18 February 2022
Accepted: 21 April 2022
Published: 25 April 2022

1. Introduction

Production profile logging technology of Oil-water two-phase flow is an important support technology for efficient development of horizontal wells. It provides important key reference data for horizontal well stimulation implementation and effect evaluation, such as optimizing injection-production scheme, evaluating fracturing, water plugging, and profile modification effects. Moreover, the water holdup of Oil-water two-phase flow is one of the important dynamic measurement parameters for production profile logging technology. Flow patterns in horizontal oil wells are more complex than those in vertical oil wells, making water holdup measurements more difficult. At present, reliable production profile logging technology of Oil-water two-phase flow in the horizontal oil wells is still undeveloped in China.

Review the literatures published in the past thirty years, it is known that various conductance sensors are widely used for measuring water holdup and identifying flow pattern in Oil-water two-phase flow and gas-Oil-water three-phase flow. The double sensor conductivity probe method was used to measure local parameters in bubbly two-phase flow in horizontal pipe, the local void fraction, the local interfacial area concentration, and bubble interfacial velocity were measured and obtained [1–3]. Ishill and Kim [4] proposed the micro four-sensor conductivity probe for measuring local two-phase flow parameters in vertical upward annulus bubbles flow, and the experiments showed that the error of absolute measurement is less than 10%. Lucas and Mishra [5] described the construction of a local four-sensor conductance probe and the associated electronic measurement circuit, for measuring the local axial, radial and azimuthal velocity components of the gas in bubbly gas-liquid vertical flows with and without swirl. Angeli and Hewitt [6] investigated the flow pattern identification of Oil-water two-phase flow in horizontal pipes with 25.4 mm inner diameter; namely high-speed video recording and determination of the local phase fractions with a high frequency impedance probe, while the continuous phase in dispersed flows was recognized with a conductivity needle probe. Dong et al. [7], Sun and Yang [8], Fang et al. [9] tried to measure separate phase fraction and distinguish flow pattern of Oil-water flows, gas-water flows, gas-Oil-water flows by using the single modality electrical capacitance tomography (ECT), electrical resistance tomography (ERT), or the dual-modality ECT/ERT, the accuracy of separate phase fraction measurement is highly dependent on the quality of image reconstruction.

For the purpose of developing reliable production profile logging instruments in the petroleum industry, Hu et al. [10] firstly invented an impedance watercut meter which consists of four annular conductance sensors to measure water holdup in vertical oil wells, laboratory and field experiments proved that the water holdup measurement range and accuracy are 50~100% and 3%, respectively. Furthermore, Wang et al. [11] optimized the measuring circuit of the meter that overcome artificial replacement of excitation source resistance caused by the salinity of downhole fluid varies widely. Xu et al. [12] proposed a novel conductance sensor array for water holdup measurement in horizontal oil wells of low liquid production of oil and water, namely which consists of 24 needle-like electrodes that are mounted on 12 supporting arms arising from the central shaft. Han et al. [13] used a radial dual-sensors conductance probe array to investigate flow patterns and water holdup in low velocity Oil-water two-phase flow through a vertical upward pipe with 20 mm inner diameter. Vu-Hoang et al. [14] introduced a new logging tool to concretely obtain records of the holdups and velocity profiles along vertical direction of borehole, three types of array sensors consisted of five rotor flow sensors, six conductance probe sensors and six fiber probe sensors, are used to measure separate phase flow rate and holdup of gas, oil, and water. Liu et al. [15] invented an annular array conductance sensor consisted of six annular electrodes for measuring the flow rate and water holdup of Oil-water two-phase flow in vertical production well. Han et al. [16] presented using sixteen mini conductance probe sensors to measure the local oil holdup of oil-in-water emulsion flows of high water cut (80–96%) in vertical upward pipe. Zong et.al [17] proposed using eight annular conductance sensors and five mini conductance probe sensors to study the inclined Oil-water two-phase flow characteristics. Xu et al. [18] designed a bicircular conductance probe sensor array for measuring water holdup in simulation horizontal well for Oil-water two-phase flow.

It is very difficult to accurately measure water holdup due to the complex and variable flow patterns of Oil-water two-phase flow in horizontal oil wells affected by borehole inclination. In early literatures, it is known that the water holdup measurement of Oil-water two-phase flow is affected by flow pattern, which in turn is affected by pipe inclined angle. Mukherjee et al. [19] researched pipe inclination angle effect to water holdup and Oil-water slippage in 38.1 mm inner diameter pipe for inclination angel ranged from $\pm 30^\circ$ to $\pm 90^\circ$. Hill and Oolman [20] studied simulation well inclination angel effect to production profile logging tool response in 152 mm inner diameter pipe. Zavareh et al. [21], Davarzani et al. [22], Tabeling et al. [23], Oddie et al. [24], Lum et al. [25] studied the pipe inclination angel effect to flow patterns of Oil-water two-phase flow. As for water holdup measurement method of Oil-water two-phase flow, annular conductance sensor is suitable for more flow patterns than conductance probe sensor. However, since there is a serious problem in the field application of production profile logging of horizontal oil wells, i.e., the measurement error of water holdup increases due to the failure of the annular array conductance sensor to correct the influence of temperature and mineralization degree of groundwater in real time [26]. Therefore, we proposed a novel method of measuring water holdup of Oil-water two-phase flow in horizontal oil wells. Then, we conducted the dynamic test by using the test prototype of the water holdup measurement method of Oil-water two-phase flow on the horizontal well simulation experimental device at five inclination angles. Finally, we analyzed the experimental phenomena and calculated the measurement error of the water holdup. The research results indicate that the novel method of measuring the water holdup of Oil-water two-phase flow in horizontal oil wells is a simple and accurate measurement method.

2. Measurement Principle of Water Holdup

In the production profile logging area, "water cut" and "water holdup" are commonly used terms, which are defined in reference [27]. The geometry structure of water holdup measurement sensors of Oil-water two-phase flow in horizontal oil well is shown in Figure 1, including annular array conductance sensor A and circumferential array conductance probe sensor B. The inner diameter of the sensor measuring channel is 20 mm. The annular array conductance sensor A consists of four annular conductance electrodes, thereinto, electrodes 1 and 4 are excitation electrodes, electrodes 2 and 3 are measurement electrodes. The inner diameter of each annular conductance electrode is 20 mm, and its thickness and width are 3 mm and 5 mm, respectively. The distance between electrodes 1 and 2 and between 3 and 4 is 80 mm for both. The distance between electrodes 2 and 3 is 50 mm. The distance between sensor A and sensor B is 60 mm in flow direction. The circumferential array conductance probe sensor B consists of four cylindrical conductance probes, for which the outer diameter of each cylindrical conductance probe is 3 mm, and its height is 5 mm. Each electrode is arranged at center angle 90° to its neighbor electrode.

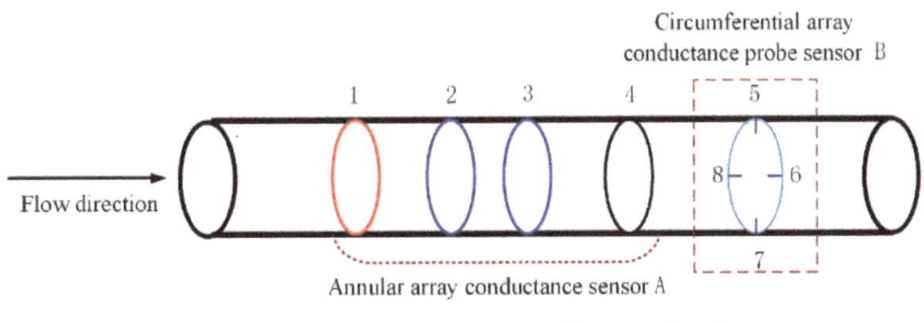

Figure 1. Schematic diagram of water holdup sensors of horizontal well.

The schematic diagram of water holdup test prototype of horizontal oil well in casing pipe is shown in Figure 2, including petal type concentrating flow diverter, water holdup sensors, centralizer and measuring circuit. The measurement principle of water holdup is described as follows: The inner diameter of casing pipe is 125 mm, and the Oil-water two-phase flow flows in the casing pipe. Firstly, the measuring circuit controls petal type concentrating flow diverter to open, the annular space between the test prototype and the casing pipe is enclosed. The Oil-water two-phase fluid flows into the measurement channel of the water holdup sensors by the inlet of the petal type concentrating flow diverter. The mixed-phase conductivity (σ_m) of Oil-water two-phase flow is measured by the annular array conductance sensor A. The specific measurement process is described as follows: The annular conductance sensor 1 and 4 are set as exciting electrodes; and the annular conductance sensor 2 and 3 are set as detecting electrodes, and the corresponding output voltage of detecting electrodes is V_m. Additionally, then the measuring circuit controls petal type concentrating flow diverter to close. The Oil-water two-phase fluid stratifies within the measurement channel of the water holdup sensors due to gravity. The water phase conductivity (σ_w) is measured by the circumferential array conductance probe sensor B. The specific measurement process is described as follows: firstly, the instrument enclosure is set as ground electrode; then, four pairs of exciting-detecting electrodes are formed by four conductance probe sensors and the instrument enclosure; once the measuring circuit determines that one of the four exciting-detecting electrodes is flooded with water, therefore it can be used to measure the water-phase conductivity; additionally, the corresponding output voltage of exciting-detecting electrodes is V_w. The water holdup (Y_w) of Oil-water two-phase flow can be obtained by the following Equation (1):

$$Y_w = \frac{\sigma_m}{\sigma_w} = \frac{V_w}{V_m} \quad (1)$$

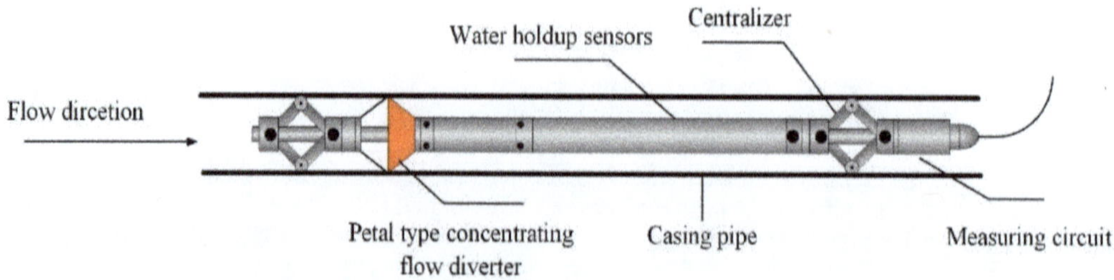

Figure 2. Schematic diagram of water holdup test prototype of horizontal well.

In order to obtain the water cut of Oil-water two-phase flow, it is necessary to calibrate the test prototype under the condition of Oil-water two-phase flow on the multiphase flow experimental device of the horizontal simulation well, and obtain the water cut–water holdup calibration chart; furthermore, water holdup can be converted to water cut through the calibration chart.

3. Experiments and Discussion

On the innovative horizontal well simulation facility, the Oil-water two-phase flow experiment was carried out by using the test prototype. Firstly, the horizontal well simulation facility was introduced. Secondly, the experimental method was described. Then, the experimental results and phenomena were discussed in detail. Finally, the measurement error of water holdup was calculated and evaluated.

3.1. Experimental Facility

The horizontal well simulation facility of Oil-water two-phase flow is located at the Daqing Production Well Logging Institute (DPWLI) in Heilongjiang Province of China. The schematic diagram of the flow loop is shown in Figure 3. As shown in Figure 3, the 125 mm inner diameter transparent Plexiglass pipes are 12 m long, which can be inclined from vertical direction to horizontal direction. In order to simulate the industrial application, the transparent Plexiglass pipes has the same diameter as the casing pipe. The flow rates of oil and water phase can be accurately controlled and metered using adjustment devices and measuring instruments. In addition, the experiments of Oil-water two-phase flow at any inclined angle can be carried out on the horizontal well simulation facility at normal pressure and temperature. More details of the experimental facility are provided in references [27,28]. Figure 4 shows the photo of the horizontal well simulation facility of Oil-water two-phase flow.

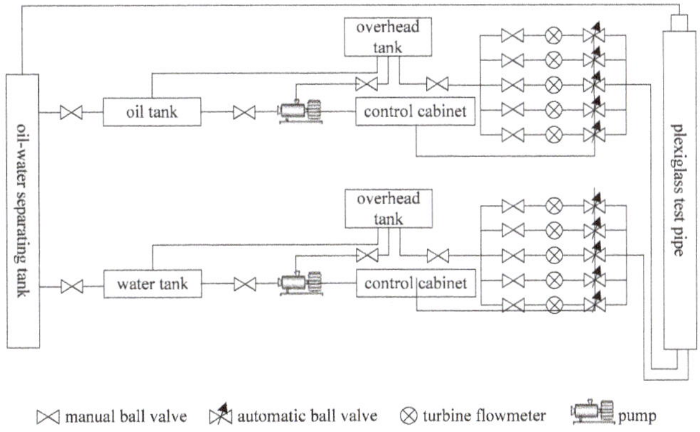

Figure 3. Schematic diagram of horizontal simulation well of Oil-water two-phase flow loop.

Figure 4. Photo of horizontal simulation well of Oil-water two-phase flow.

3.2. Experimental Design

The dynamic experiments of water holdup measurement of Oil-water two-phase flow were carried out by using the test prototype on the horizontal well simulation facility at the DPWLI, in Heilongjiang Province of China. The experimental mediums are tap water and diesel. The diesel density, kinematic viscosity, and interfacial tension are 820.0 kg/m^3, 7.0 mm^2/s and 28.6 mN/m, respectively, at 20 °C. The total flow rate of Oil-water two-phase flow is set as 3 m^3/d, 5 m^3/d, 10 m^3/d, 30 m^3/d, 40 m^3/d, 60 m^3/d, 100 m^3/d, 120 m^3/d, 150 m^3/d and 200 m^3/d, respectively. For each of the total flow rate, the water cut ranged from 30% to 90%, and the water cut adjust step is 10%. The inclination angle of horizontal simulation well is rotated clockwise, and the vertical direction and the horizontal direction are defined as 0° and 90°, respectively. The inclination angle is set as 80°, 83°, 85°, 88° (fluid flows upward) and 90° (fluid flows horizontally), respectively.

When the total flow rate is greater than 30 m^3/d, after adjusting the water cut, the flow stable for 10 min before open the petal type concentrating flow diverter. Instead, when the total flow rate is less than 30 m^3/d, after adjusting the water cut, the flow stable for 30 min before open the petal type concentrating flow diverter. After the petal type concentrating flow diverter is fully opened, continue the flow stable for 3 min, then start to measure the conductivity of Oil-water two-phase flow. The measurement time is set as 3 min, and two data are collected every second. Hence, a total of 360 instantaneous conductivity measurement data of the Oil-water two-phase flow are collected. Furthermore, starting to measure the conductivity of the water phase while the petal type concentrating flow diverter is fully closed. Similarly, the measurement time is set as 3 min, and two data are collected every second. Hence, a total of 360 instantaneous conductivity measurement data of the water phase are collected.

3.3. Discussions of Experimental Results

Figures 5–9 shows the water holdup measurement charts with different flow rates and water cuts under the conditions of the horizontal simulation well inclination angle of 80°, 83°, 85°, 88° and 90°, respectively. The horizontal and vertical coordinates are the set standard flow rate and water cut, respectively, and curves in different colors represent measured water holdup. From the five figures, it indicated that the measurement result of water holdup is not only dependent on the water cut, but also related to the flow rate. The shape of water holdup curves measured at five inclined angles has good consistency. Further analysis of water holdup measurement charts at different inclined angles shows that the relationship between measured water holdup and set standard water cut has obvious regularity. In the low flow rates range, that is, when the total flow rates of oil and water two-phase flow are less than 30 m^3/d, the measured water holdups are not only dependent on the water cuts, but also related to the flow rates. When the set standard water cut remains unchanged, the measured water holdup moves to the direction of high water cut with the decrease in the flow rate. This phenomenon is caused by the Oil-water two-phase flow characteristics. Due to the density difference between the oil and water phase, the oil phase flows at a higher speed than the water phase. Therefore, under the condition of maintaining a certain set standard water cut, the measured water holdup at low flow rate is higher than that at high flow rate. With the decrease in the average flow rate of Oil-water two-phase flow, the difference of oil and water slippage velocity increases compared with the average velocity, especially when the total flow rate is less than 10 m^3/d, the slippage phenomenon is more significant, resulting in the difference between water holdup and water cut becoming larger. At present, other types of instruments of measuring horizontal oil well water holdup in use have difficulty reflecting the slippage phenomenon of Oil-water two-phase flow at low flow rates. For example, for the capacitive water holdup instruments, it is difficult to obtain accurate water holdup and water cut measurement curves because the capacitive electrode is easily affected by oil phase contamination. However, the water holdup measurement method by the combination of annular array conductance sensor and circumferential array conductance probe sensor proposed in this paper can better reflect the Oil-water slippage phenomenon at low flow rates. It indicates that the water holdup measurement method has obvious advantages at low flow rates of Oil-water two-phase flow.

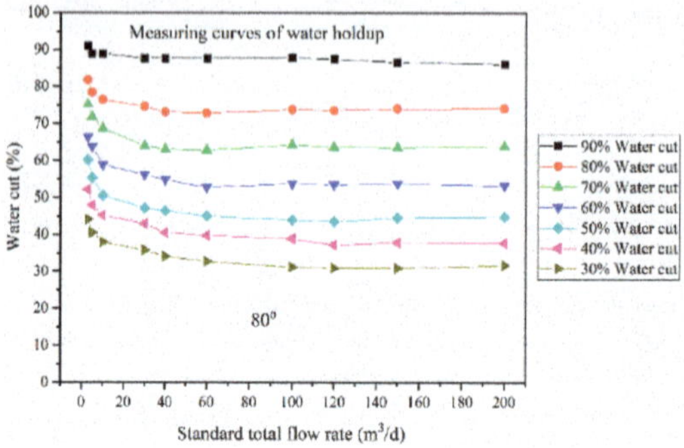

Figure 5. Water holdup measurement chart (inclination angle 80°).

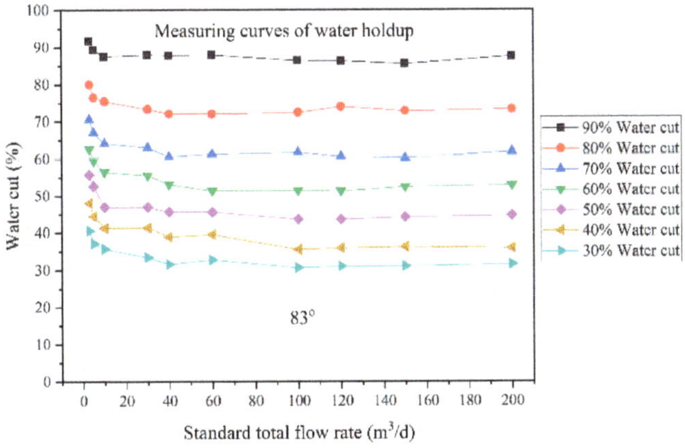

Figure 6. Water holdup measurement chart (inclination angle 83°).

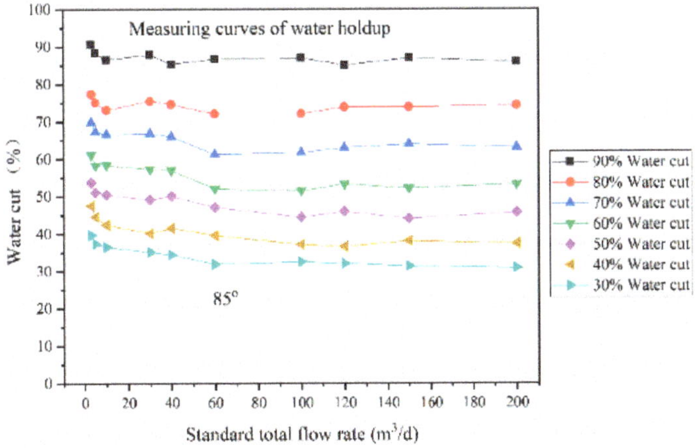

Figure 7. Water holdup measurement chart (inclination angle 85°).

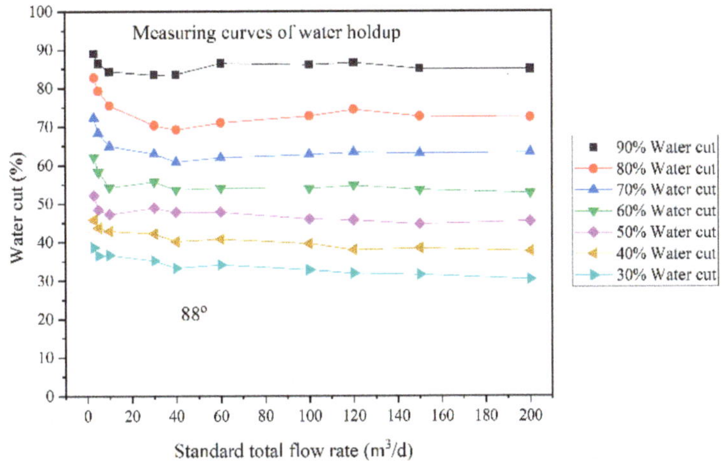

Figure 8. Water holdup measurement chart (inclination angle 88°).

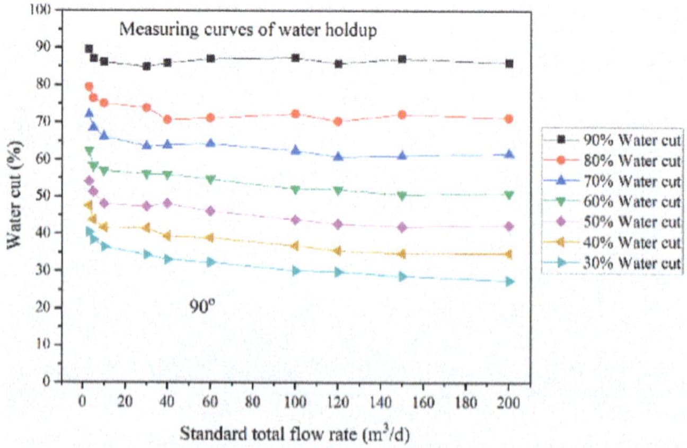

Figure 9. Water holdup measurement chart (inclination angle 90°).

In order to further investigate the influence of horizontal simulation well inclination angles on water holdup measurement at high flow rates, the water holdup measurement chart of five inclination angles is given in Figure 10, while the set standard flow rate ranged from 60 m^3/d to 200 m^3/d, and the set standard water cut ranged from 30% to 90%. It can be seen that the measured water holdup and the standard water cuts are approximately flat curves when the total flow rate is greater than 60 m^3/d and 100 m^3/d, and the standard water cut is greater than 60% and 30%, respectively. Because the oil and water slippage velocity rather small compared with the total average velocity. Therefore, the variation of flow rate has little influence on the water holdup measurement. At this time, the water holdup is very close to the water cut in the test prototype flow channel. The inclination angles have little influence on the water holdup measurement data of the test prototype and can be ignored. However, when the total flow rate of Oil-water two-phase flow is less than 60 m^3/d, the water holdup measurement data of the test prototype is greatly affected by the inclination angles and cannot be ignored. Therefore, the wellbore inclination angle measurement module should be included in the development of horizontal oil well production profile logging combination instrument, and the well logging interpretation model of different inclination angles must be established.

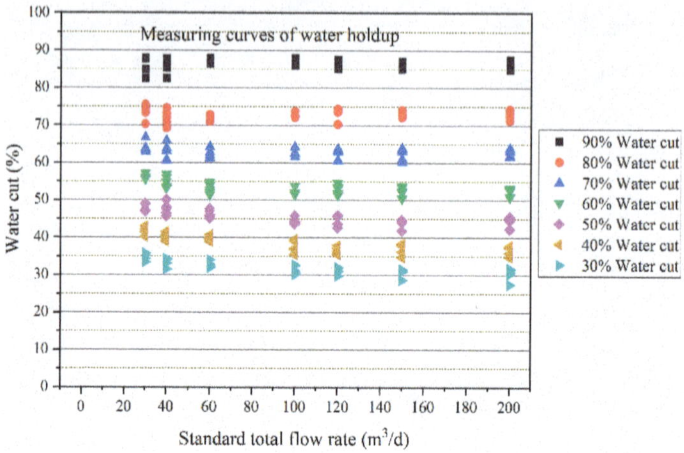

Figure 10. Water holdup measurement chart at five inclination angles.

3.4. Error Calculation of Water Holdup

Furthermore, the measurement error of water holdup was evaluated. Due to the existence of random error, the measured values in the measurement column of equal accuracy are generally different, and they are dispersed around the arithmetic mean value of the measurement column to a certain extent. The degree of dispersion indicates the unreliability of the single measured value, and the standard deviation is used as the evaluation standard for evaluating unreliability. The specific process of error calculation of water holdup is described as follows:

(1) For a set standard flow rate, the average values of Oil-water mixed-phase conductivity and water phase conductivity measured by the test prototype are calculated, while the set standard water cut is 30%, 40%, 50%, 60%, 70%, 80% and 90%:

$$\overline{F}_m = \frac{1}{360}\sum_{i=1}^{360} F_{mi} \tag{2}$$

where F_{mi} is the instantaneous measurement value of mixed-phase conductivity, and $\overline{F}_m$ is the average measurement value of mixed-phase conductivity.

$$\overline{F}_w = \frac{1}{360}\sum_{i=1}^{360} F_{wi} \tag{3}$$

where F_{wi} is the instantaneous measurement value of water phase conductivity, and $\overline{F}_w$ is the average measurement value of water phase conductivity.

(2) The water holdup measured by the test prototype is calculated:

$$Y = \frac{\overline{F}_w}{\overline{F}_m} \times 100\% \tag{4}$$

where Y is the measurement value of water holdup.

(3) The standard deviation of mixed-phase conductivity and water phase conductivity are calculated:

$$E_{\overline{F}_m} = \left[\frac{1}{n-1}\sum_{i=1}^{n=360}\left(F_{mi} - \overline{F}_m\right)^2\right]^{\frac{1}{2}} \tag{5}$$

where $E_{\overline{F}_m}$ is the standard deviation of mixed-phase conductivity.

$$E_{\overline{F}_w} = \left[\frac{1}{n-1}\sum_{i=1}^{n=360}\left(F_{wi} - \overline{F}_w\right)^2\right]^{\frac{1}{2}} \tag{6}$$

where $E_{\overline{F}_w}$ is the standard deviation of water phase conductivity.

(4) The absolute value of measurement error of water holdup is calculated:

$$R_{Y_E} = \left|\frac{E_{\overline{F}_w}}{E_{\overline{F}_m}} \times 100\% - Y\right| \tag{7}$$

where R_{Y_E} is the absolute value of water holdup measurement error.

The calculation results show that the absolute values of water holdup measurement error are less than 5% under the conditions of five inclination angles, while the total flow rate ranged from 10 m^3/d to 200 m^3/d, and the water cut ranged from 30% to 90%, respectively. However, the absolute values of water holdup measurement error are greater than 5% and less than 12% under the conditions of five inclination angles, while the total flow rate ranged from 3 m^3/d to 10 m^3/d, and the water cut ranged from 30% to 90%, respectively.

4. Conclusions

In this paper, a novel method of measuring the water holdup of Oil-water two-phase flow in horizontal oil wells was proposed. An innovative horizontal well simulation facility was used to simulate Oil-water two-phase flow and the proposed method was verified. Hence, the new method indicated important engineering application value because it solved the serious problem of measuring water holdup by using annular conductivity sensors alone, i.e., the measurement error of water holdup increases due to the failure of the annular array conductance sensor to correct the influence of temperature and mineralization degree of groundwater in real time. The experiment was correctly designed, the results were good and competently discussed. The dynamic experiments of water holdup measurement of Oil-water two-phase flow were carried out by using the self-developed test prototype on the horizontal well simulation facility. The measurement data of water holdup of five inclination angles were analyzed in detail, and the analysis results show that the measurement data of water holdup are affected by the wellbore inclination angle. The error calculation results show that the absolute values of water holdup measurement error are less than 5% under the conditions of five inclination angles, while the total flow rate ranged from 10 m^3/d to 200 m^3/d, and the water cut ranged from 30% to 90%, respectively. The research results prove that the novel method of measuring the water holdup of Oil-water two-phase flow in horizontal oil wells is a reliable and accurate measurement method. Future work will continue, the measurement method of water holdup proposed in this paper will be applied in engineering, the wellbore inclination angle measurement module should be included in the development of logging instrument, and the well logging interpretation model of different inclination angle must be established.

Author Contributions: Conceptualization, Y.W.; methodology, Y.W. and J.H.; validation, Y.W., J.H. and Z.H.; investigation, Y.W. and Z.H.; writing—review and editing, L.Z., X.W. and M.S.; project administration, Y.W.; funding acquisition, Y.W. and X.W. All authors have read and agreed to the published version of the manuscript.

Funding: This research was funded by China Postdoctoral Science Foundation (grant number 2020M682229), and Shandong Provincial Natural Science Foundation (grant number ZR2020MF001).

Institutional Review Board Statement: Not applicable.

Informed Consent Statement: Not applicable.

Data Availability Statement: Not applicable.

Conflicts of Interest: The authors declare no conflict of interest.

References

1. Hogsett, S.; Ishii, M. Local two-phase flow measurements using sensor techniques. *Nucl. Eng. Des.* **1997**, *175*, 15–24. [CrossRef]
2. Wu, Q.; Ishii, M. Sensitivity study on double-sensor conductivity probe for the measurement of interfacial area concentration in bubbly flow. *Int. J. Multiph. Flow* **1999**, *25*, 155–173. [CrossRef]
3. Hibiki, T.; Situ, R.; Mi, Y.; Ishii, M. Local flow measurement of vertical upward bubbly flow in an annulus. *Int. J. Heat Mass Transf.* **2003**, *46*, 1479–1496. [CrossRef]
4. Ishill, M.; Kim, S. Micro four-sensor probe measurement of interfacial area transport for bubbly flow in round pipes. *Nucl. Eng. Des.* **2001**, *205*, 123–131.
5. Lucas, G.P.; Mishra, R. Measurement of bubble velocity components in a swirling gas-liquid pipe flow using a local four-sensor conductance probe. *Meas. Sci. Technol.* **2005**, *16*, 749–758. [CrossRef]
6. Angeli, P.; Hewitt, G.F. Flow structure in horizontal oil-water flow. *Int. J. Multiph. Flow* **2000**, *26*, 1117–1140. [CrossRef]
7. Dong, F.; Liu, X.P.; Deng, X.; Xu, L.J.; Xu, L.A. Identification of two-phase flow regimes in horizontal, inclined and vertical pipes. *Meas. Sci. Technol.* **2001**, *12*, 1069–1075. [CrossRef]
8. Sun, J.T.; Yang, W.Q. A dual-modality electrical tomography sensor for measurement of gas-oil-water stratified flows. *Measurement* **2015**, *66*, 150–160. [CrossRef]
9. Fang, L.D.; Wang, P.P.; Zeng, Q.Q.; Li, M.M.; Li, X.T.; Wang, M.; Faraj, Y.; Wang, Q. Measurement of interphase forces based on dual-modality ERT/DP sensor in horizontal two-phase flow gas-water. *Measurement* **2019**, *136*, 703–717. [CrossRef]
10. Hu, J.H.; Liu, X.B.; Zhang, Y.H.; Zhou, J.Q.; Yuan, Z.H.; Qiao, Z.E. Impedance watercut meter and its applications. *Well Log. Technol.* **1999**, *23*, 511–514.

11. Wang, Y.J.; Hu, J.H.; Huang, C.H.; Li, L.; Liu, X.B. Adaptive modules design of impedance watercut meter and its applications. *Well Log. Technol.* **2010**, *34*, 469–472.
12. Xu, W.F.; Xu, L.J.; Cao, Z.; Chen, J.J.; Liu, X.B.; Hu, J.H. Normalized least-square method for water hold-up measurement in stratified Oil-water flow. *Flow. Meas. Instrum.* **2012**, *27*, 71–80. [CrossRef]
13. Han, Y.F.; Jin, N.D.; Zhai, L.S.; Zhang, H.X.; Ren, Y.Y. Flow pattern and holdup phenomena of low velocity oil-water flows in a vertical upward small diameter pipe. *J. Pet. Sci. Eng.* **2017**, *159*, 387–408. [CrossRef]
14. Vu-Hoang, D.; Faur, M.; Marcus, R.; Cadenhead, J.; Besse, F.; Haus, J.; Wadjiri, A. A novel acquisition system for production logging in multiphase horizontal wells. In Proceedings of the AAPG International Conference, Cancun, Mexico, 24–27 October 2004; pp. 24–27.
15. Liu, X.B.; Hu, J.H.; Shan, F.J.; Cai, B.; Sun, X.Y.; Chen, Q.B. Conductance sensor for measurement of the fluid water cut and flow rate in production wells. *Chem. Eng. Comm.* **2010**, *197*, 232–238. [CrossRef]
16. Han, Y.F.; Jin, N.D.; Ren, Y.Y.; Han, T.H.; Huang, Z.H. Measurement of local oil holdup for oil-in-water emulsion flows using multiple mini-conductance probes. *Measurement* **2018**, *121*, 6–18. [CrossRef]
17. Zong, Y.B.; Jin, N.D.; Wang, Z.Y.; Gao, Z.K.; Wang, C. Nonlinear dynamic analysis of large diameter inclined oil-water two phase flow pattern. *Int. J. Multiph. Flow* **2010**, *36*, 166–183. [CrossRef]
18. Xu, L.J.; Zhang, W.; Zhao, J.Y.; Cao, Z.; Xie, R.H.; Liu, X.B.; Hu, J.H. Support-vector-regression-based prediction of water holdup in horizontal oil-water flow by using a bicircular conductance probe array. *Flow Meas. Instrum.* **2017**, *57*, 64–72. [CrossRef]
19. Mukherjee, H.; Brill, J.P.; Beggs, H.D. Experimental study of oil-water flow in inclined pipes. *J. Energy Resour. Technol.* **1981**, *103*, 56–66. [CrossRef]
20. Hill, A.D.; Oolman, T. Production logging tool behavior in two-phase inclined flow. *J. Petrol. Technol.* **1982**, *34*, 2432–2440. [CrossRef]
21. Zavareh, F.; Hill, A.D.; Podio, A.L. Flow regimes in vertical and inclined oil/water flow in pipes. In Proceedings of the SPE Annual Technical Conference and Exhibition, Houston, TX, USA, 2–5 October 1988. [CrossRef]
22. Davarzani, J.; Sloan, L.; Roesner, R.E. Research on simultaneous production logging instruments in multiphase flow loop. In Proceedings of the SPE Annual Technical Conference and Exhibition, Las Vegas, NV, USA, 22–26 September1985. [CrossRef]
23. Tabeling, P.; Pouliquen, O.; Theron, B.; Catala, G. Oil-water flows in deviated pipes: Experimental study and modelling. In Proceedings of the 5th International Conference on Multiphase Flow Production, Cannes, France, 19–21 June 1991; pp. 294–306.
24. Oddie, G.; Shi, H.; Durlofsky, L.J.; Aziz, K.; Pfeffer, B.; Holmes, J.A. Experimental study of two and three phase flows in large diameter inclined pipes. *Int. J. Multiph. Flow* **2003**, *29*, 527–558. [CrossRef]
25. Lum, J.Y.L.; Al-Wahaibi, T.; Angeli, P. Upward and downward inclination oil-water flows. *Int. J. Multiph. Flow* **2006**, *32*, 413–435. [CrossRef]
26. Liu, W.X.; Jin, N.D.; Han, Y.F.; Zhai, L.S.; Chen, X.; Zhang, H.X. Effects of flow patterns and salinity on water holdup measurement of oil-water two-phase flow using a conductance method. *Measurement* **2016**, *93*, 503–514. [CrossRef]
27. Wang, Y.J.; Li, H.Y.; Liu, X.B.; Li, L.; Huang, C.H.; Hu, J.H. A new method of measuring the oil-air-water three-phase flow rate. *Chem. Eng. Comm.* **2020**, *207*, 1–16. [CrossRef]
28. Wang, Y.J.; Li, H.Y.; Liu, X.B.; Zhang, Y.H.; Xie, R.H.; Huang, C.H.; Hu, J.H.; Deng, G. Novel downhole electromagnetic flowmeter for oil-water two-phase flow in high-water-cut oil-producing wells. *Sensors* **2016**, *16*, 1703. [CrossRef] [PubMed]

processes

Article

Optimizing the Gamma Ray-Based Detection System to Measure the Scale Thickness in Three-Phase Flow through Oil and Petrochemical Pipelines in View of Stratified Regime

Abdulilah Mohammad Mayet [1], Tzu-Chia Chen [2,3,*], Seyed Mehdi Alizadeh [4], Ali Awadh Al-Qahtani [1], Abdullah K. Alanazi [5], Nivin A. Ghamry [6], Hala H. Alhashim [7] and Ehsan Eftekhari-Zadeh [8,*]

1 Electrical Engineering Department, King Khalid University, Abha 61411, Saudi Arabia
2 College of Management and Design, Ming Chi University of Technology, New Taipei City 243303, Taiwan
3 International College, Krirk University, 3 Ram Inthra Rd, Khwaeng Anusawari, Khet Bang Khen, Krung Thep Maha Nakhon, Bangkok 10220, Thailand
4 Petroleum Engineering Department, Australian University, West Mishref 13015, Kuwait
5 Department of Chemistry, Faculty of Science, Taif University, P.O. Box 11099, Taif 21944, Saudi Arabia
6 Faculty of Computers and Artificial Intelligence, Cairo University, Giza P.O. Box 12613, Egypt
7 Department of Physics, College of Science, Imam Abdulrahman Bin Faisal University, P.O. Box 1982, City Dammam 31441, Saudi Arabia
8 Institute of Optics and Quantum Electronics, Friedrich Schiller University Jena, Max-Wien-Platz 1, 07743 Jena, Germany
* Correspondence: tzuchiachen1688@gmail.com (T.-C.C.); e.eftekharizadeh@uni-jena.de (E.E.-Z.)

Citation: Mayet, A.M.; Chen, T.-C.; Alizadeh, S.M.; Al-Qahtani, A.A.; Alanazi, A.K.; Ghamry, N.A.; Alhashim, H.H.; Eftekhari-Zadeh, E. Optimizing the Gamma Ray-Based Detection System to Measure the Scale Thickness in Three-Phase Flow through Oil and Petrochemical Pipelines in View of Stratified Regime. *Processes* **2022**, *10*, 1866. https://doi.org/10.3390/pr10091866

Academic Editors: Tianshou Ma and Yuqiang Xu

Received: 15 August 2022
Accepted: 10 September 2022
Published: 15 September 2022

Abstract: As the oil and petrochemical products pass through the oil pipeline, the sediment scale settles, which can cause many problems in the oil fields. Timely detection of the scale inside the pipes and taking action to solve it prevents problems such as a decrease in the efficiency of oil equipment, the wastage of energy, and the increase in repair costs. In this research, an accurate detection system of the scale thickness has been introduced, which its performance is based on the attenuation of gamma rays. The detection system consists of a dual-energy gamma source (^{241}Am and ^{133}Ba radioisotopes) and a sodium iodide detector. This detection system is placed on both sides of a test pipe, which is used to simulate a three-phase flow in the stratified regime. The three-phase flow includes water, gas, and oil, which have been investigated in different volume percentages. An asymmetrical scale inside the pipe, made of barium sulfate, is simulated in different thicknesses. After irradiating the gamma-ray to the test pipe and receiving the intensity of the photons by the detector, time characteristics with the names of sample SSR, sample mean, sample skewness, and sample kurtosis were extracted from the received signal, and they were introduced as the inputs of a GMDH neural network. The neural network was able to predict the scale thickness value with an RMSE of less than 0.2, which is a very low error compared to previous research. In addition, the feature extraction technique made it possible to predict the scale value with high accuracy using only one detector.

Keywords: GMDH neural network; data mining; scale thickness; stratified flow regime; oil products

1. Introduction

The presence of scale inside oil pipes causes problems such as reducing the cross-section of oil pipes, disrupting the performance of oil equipment, increasing the cost and time of repairs, and even emergency shutdown of the oil field. Using an accurate and non-invasive diagnostic system to diagnose this problem and take action to fix it can prevent the above-mentioned problems. In recent years, systems based on gamma-ray attenuation to determine the parameters of the type of flow regime and volume percentages in three-phase [1–3] and two-phase [4–6] fluids have received much attention from researchers. In these researches, different gamma sources have been investigated. Still, the big problem of these researches is the use of two or more detectors, which has increased the cost and complexity of the detection structure. In study [7], researchers proposed a system based on gamma rays to determine volume percentages and type of flow regimes, whose detection structure consisted of a cesium source and two sodium iodide detectors. They simulated a two-phase flow in different volume percentages and three regimes of annular, stratified, and homogeneous using Monte Carlo N Particle (MCNP) code. In order to increase the accuracy, they examined several time characteristics and introduced the best time characteristic with an innovative method. The use of two detectors is the most fundamental gap of this research, which, in addition to imposing a high cost on system design, has increased the complexity of the detection system. In [8], researchers investigated the performance of the GMDH neural network, but the lack of characteristic extraction and the use of raw signals as inputs of the neural network prevented access to high accuracy. In the following research [9], the researchers put the temporal characteristics as the inputs of the GMDH neural network. This neural network has the ability to self-organize and can recognize the network structure and appropriate inputs automatically. They predicted the type of flow regimes and volume percentages with high accuracy. Alamoudi et al. [10] proposed an in-pipe scale value detection system. The proposed detection structure consisted of a dual-energy gamma source and two detectors placed opposite each other on either side of a test pipe. In the test pipe, a two-phase flow of oil and gas in different volume percentages was simulated. They did not check the characteristics of the received signals, which made not only the accuracy of their proposed system not high, but also the structure of the introduced system was complex and expensive. The use of radioisotope devices has problems such as the need to use protective clothing, difficult transportation, inability to turn off, etc.; for this reason, in recent years, the use of X-ray tubes to determine various parameters of Multiphase flows has been studied by researchers. In [11], to determine the type of flow regime and volume percentages in two-phase flows, a system consisting of an X-ray tube and a detector was introduced. They extracted temporal characteristics from the signals received by the detector and considered them as inputs of the MLP neural network. Two neural networks were designed that were responsible for determining the type of flow regimes and the percentage of void fraction separately. In [12], a three-phase flow was simulated in three different regimes and different volume percentages. The simulated structure consisted of an X-ray tube and two sodium iodide detectors. The received signals were processed in the frequency domain, and four frequency characteristics were considered as the inputs of the RBF neural network. Three RBF neural networks were trained, two of which had the task of determining volume percentages, and the other one had the task of determining the type of flow regimes. The X-ray tube has also been used to determine the type and amount of product passing through oil pipelines [13,14]. In [13], four petroleum products were simulated in a two-by-two combination in a test pipe. An X-ray tube was placed on one side of the pipe, and a sodium iodide detector was placed on the other side of the pipe directly in front of the source. The signals received from each simulation without any processing were simultaneously considered as the input of three MLP neural networks, and the volume ratio of each product was predicted with a mean absolute error of less than 2.72. In the next research [14], in order to increase the accuracy of the proposed system in research [14], the received signals were processed using wavelet transform, and the characteristics of approximate and detailed signals were extracted. They increased the accuracy of the detection system by about two times.

Inspired by previous researches, in current research, an attempt has been made to implement a system for detecting the amount of scale inside an oil pipe while a three-phase regime is passing through it. This research has tried to improve two important parameters, such as increasing accuracy and simplifying the detected system, using time characteristic extraction techniques. These are the contributions made by this study:

1. Extracting signal features by the use of statistical formulas.
2. Utilizing a single detector reduces expenses and the complexity of the detection system's structure.

3. Improving scale thickness determination accuracy by the extraction of valuable properties from received signals
4. Determining scale thickness using the GMDH neural network as a self-organizing network.

2. MCNP Simulation Setup

MCNP code has been used to simulate the detection structure. In this simulation, a dual-energy gamma source, a steel pipe, with an inner diameter of 10 cm and a thickness of 0.5 cm, and a sodium iodide detector are used. The simulated structure is shown in Figure 1. The dual-energy source consists of ^{241} Am and ^{133} Ba, are capable of emitting photons with energies of 59 and 356 KeV, respectively. In the test pipe, a three-phase flow consisting of water, oil, and gas was simulated in a stratified flow regime. Different volume percentages in the range of 10% to 80% were investigated for each phase. The considered scale was made of barium sulfate ($BaSO_4$) with a density of 4.5 $g \cdot cm^{-3}$. Seven different scale thicknesses, including 0, 0.05, 1, 1.5, 2, 2.5, and 3 cm, were simulated inside the pipe. The sodium iodide detector with dimensions of 2.54 cm $\times$ 2.54 cm was placed at a distance of 30 cm and directly in front of the gamma source. 252 different simulations were performed, and all the collected data were labeled for use in the next step. The graph of the spectrum received by the detector for two scale thicknesses of 0 cm and 1.5 cm is shown in Figure 2.

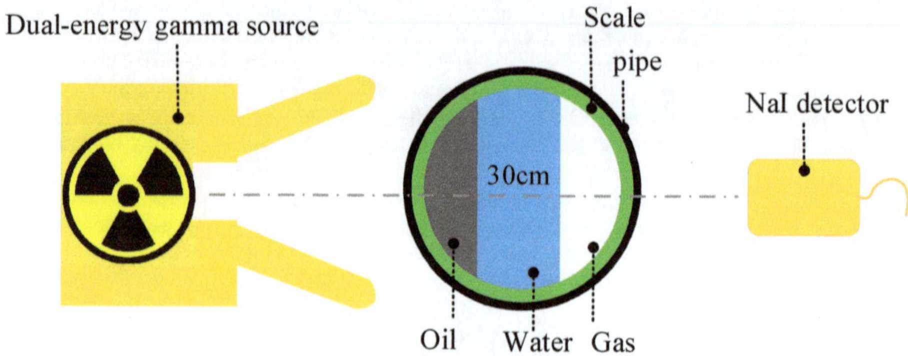

Figure 1. The structure of the simulated detection system.

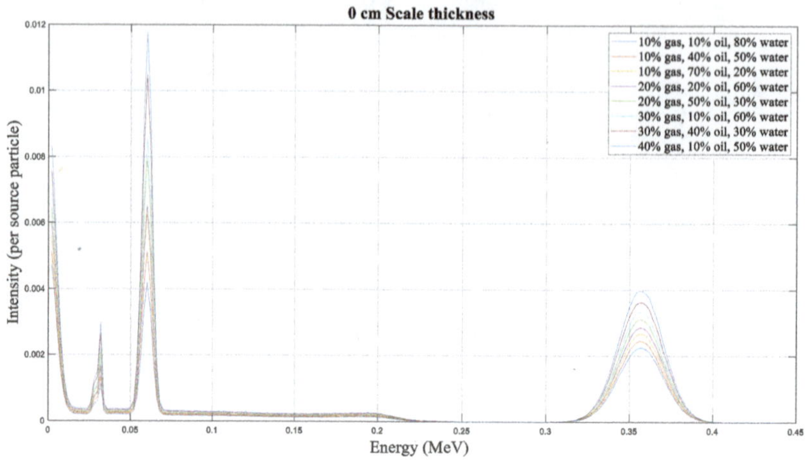

(a)

Figure 2. *Cont.*

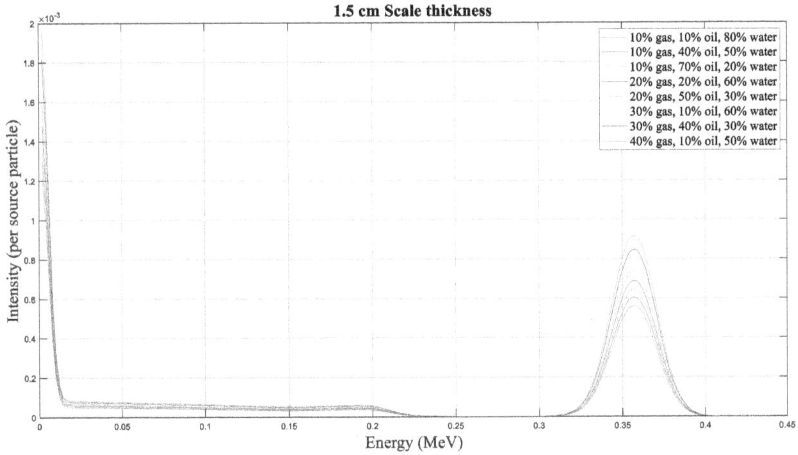

(**b**)

Figure 2. Spectrum received by sodium iodide detector for scale thickness of (**a**) 0 cm and (**b**) 1.5 cm.

LamberteBeer's law is used to determine the attenuation rate when a narrow gamma ray strikes an object.

$$I = I_0 e^{-\mu\rho x} \tag{1}$$

I, I_0, μ and ρ denote the intensity of un-collided, the original photons, the mass attenuation coefficient, and the material density of the absorber, respectively. X denotes how far the beam travels through the absorber. Equation (1) states that different objects respond to gamma radiation in different ways. This variation in behavior determines the kind and amount of a substance in the environment. A 2.54 cm × 2.54 cm NaI detector was used in this research to capture the photons that were transmitted. The detector was used to record photon energy spectra using pulse height tally (Tally F8). Previous analyses have supported the study's reproductive outcomes [6]. Several research laboratory structures were used in this review, and the results from the MCNP code were compared with them. Both were normalized to units to examine exploratory and reenactment data because the Tally output in the MCNP procedure is per source particle. The disparity between the simulation results and the lab setup represented the largest relative error at 2.2%.

3. Time-Domain Feature Extraction

The signals collected in the previous section had many dimensions, and their interpretation was a very complicated and time-consuming task. For this purpose, to simplify and separate the available data, an attempt was made to extract time characteristics. Four time-characteristics with the names of sample mean, sample of summation of square root (SSR), sample skewness, and sample kurtosis were extracted from the signals recorded by the NaI detector. These features have been used to improve the accuracy and structure of the scale thickness detection system since they were first introduced as helpful characteristics in earlier study [7,9,11]. The equations of these features are given below:

- Sample mean:

$$m = \frac{1}{N}\sum_{n=1}^{N} x(n) \tag{2}$$

- Sample of summation of square root (SSR):

$$SSR = \sum_{n=1}^{N} (x(n))^{0.5} \tag{3}$$

- Sample skewness:

$$Skewness = \frac{m_3}{\sigma^3}, \ m_3 = \frac{1}{N}\sum_{n=1}^{N}[x(n) - m]^3 \tag{4}$$

- Sample kurtosis:

$$Kurtosis = \frac{m_4}{\sigma^4}, \ m_4 = \frac{1}{N}\sum_{n=1}^{N}[x(n) - m]^4 \tag{5}$$

Here n is the dataset's values, N denotes the total data number, and x(n) represents the principal signal in the time domain.

The diagram of these four characteristics in terms of volume percentage of gas and oil is shown in Figure 3. As it is clear from this figure, the amount of scale thickness can be separated into different volume percentages. These characteristics have been used to train the GMDH neural network. As mentioned in the previous section, 252 different simulations have been performed, and four time characteristics have been extracted from the signal received from each simulation. Therefore, the available matrix contains 4 rows and 252 columns. The output of the neural network is also the value of the thickness of the scale inside the pipe in centimeters.

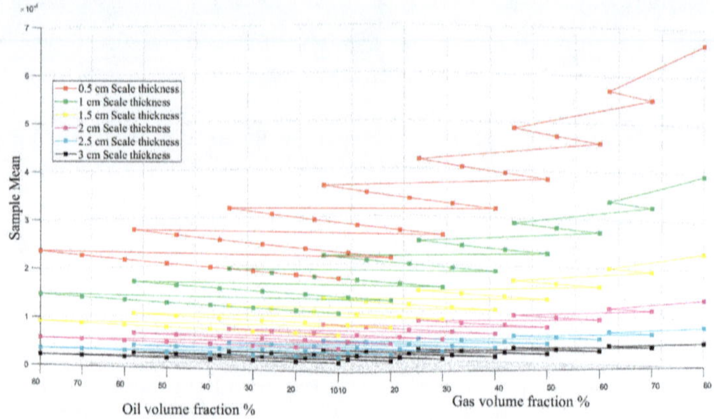

(a)

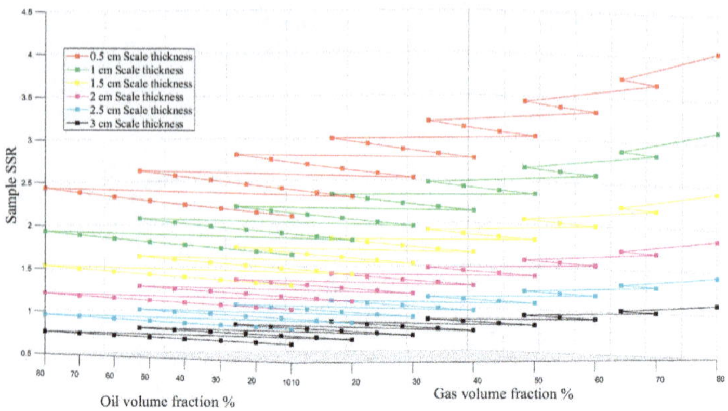

(b)

Figure 3. *Cont.*

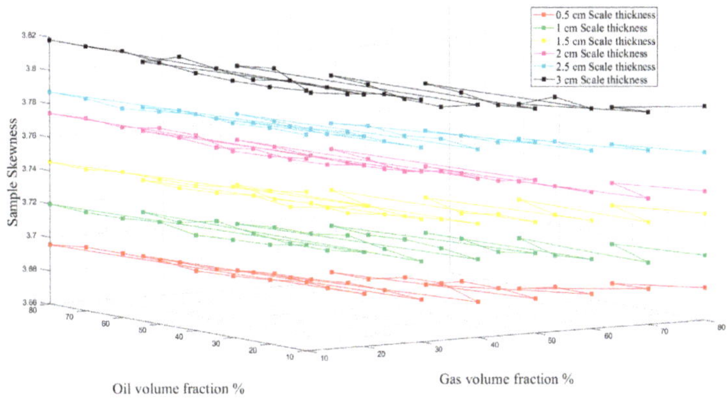

(c)

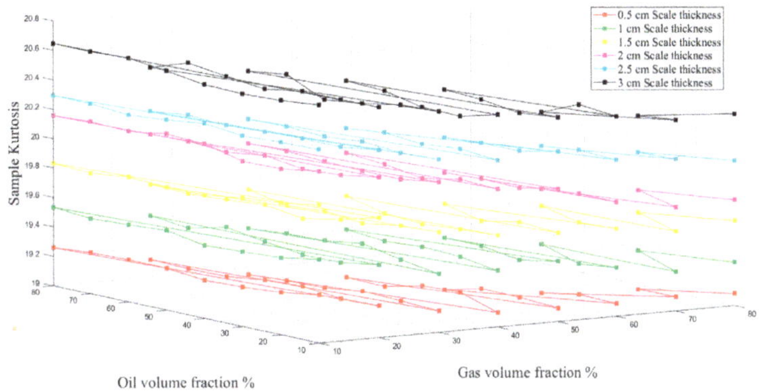

(d)

Figure 3. Extracted temporal characteristics in terms of volume percentage of gas and oil (**a**) sample mean, (**b**) sample SSR, (**c**) sample skewness, and (**d**) sample kurtosis.

4. GMDH Neural Network

In 1968, a Ukrainian mathematician named M.G. Ivakhnenko presented a mathematical model to solve prediction and classification problems and called it Group Method of Data Handling (GMDH) [15]. The model proposed by Ivakhnenko has self-organization capability so that the network structure, effective inputs, number of hidden layers, and number of neurons of hidden layers are automatically selected. In this neural network, the relationship between input and output is described by the Kolmogorov-Gabor polynomial, which is described below.

$$y = a_0 + \sum_{i=1}^{m} a_i x_i + \sum_{i=1}^{m}\sum_{j=1}^{m} a_{ij} x_i x_j + \sum_{i=1}^{m}\sum_{j=1}^{m}\sum_{k=1}^{m} a_{ijk} x_i x_j x_k + \ldots \quad (6)$$

where a $(a_1, a_2, \ldots, a_m)$ are weights or coefficients of vector, X $(x_1, x_2, \ldots, x_m)$ are also vector inputs or the same extracted features, and y is the output of the network. GMDH neural network is implemented in 5 steps, which are as follows.

1. All neural network inputs (extracted characteristics) two at the time and for each $\binom{m}{2}$ admixture are fitted to the quadratic polynomial given in Equation (7). The purpose of this step is to calculate the C coefficients that are obtained with the least

squares algorithm. The output of each quadratic polynomial predicts the desired output value. The task of calculating these polynomials is assigned to the neurons of the neural network.

$$Z = c_1 + c_2x_i + c_3x_j + c_4x_i^2 + c_5x_j^2 + c_6x_ix_j \quad (7)$$

2. The neurons with the most error in predicting the desired output are removed.
3. The neurons selected in the previous step are considered quadratic polynomial inputs described in step one. In this step, polynomials are produced from polynomials, producing a polynomial with a higher order.
4. The second step is repeated, and neurons with high errors are removed. This repetition of the steps and generation of polynomials from polynomials are repeated until the desired error value is obtained.
5. Checking network performance with test data. In the design of neural networks, the major of the data (about 70%) are used for training the neural network, and the rest of the data are used for the final test of the network. The correct performance of the neural network against these data sets ensures that the designed network can show acceptable performance in operational conditions. In order to identify various characteristics in many scientific domains, several studies have employed intelligent computer systems [16–37].

5. Results

A GMDH neural network with four inputs and one output was trained. The inputs of this network were the extracted temporal characteristics, and the network's output was the value of the scale thickness inside the pipe in centimeters. 252 samples were available for the implementation of this network, of which 176 samples were assigned to the training data, and the remaining samples were used for the final test of the neural network. The selection of these samples was done randomly so that the neural network could be trained with all the range of data. According to the self-organizing capability of the GMDH neural network, the structure that had the best accuracy is shown in Figure 4. This network has 3 hidden layers, and the number of selected neurons in these layers was 4, 4, and 2, respectively. A regression diagram and error diagram have been used to show the performance of the designed neural network. In the regression diagram, the yellow line shows the desired output, and the green circles represent the output of the neural network. The closer the yellow line and the green circle are to each other, the more accurate the designed neural network is. The error diagram shows the difference between the network output and the desired output for each sample. The regression and error graphs for the training and testing datasets are shown in Figure 5. In order to obtain the error value, two widely used criteria named Mean Square Error (*MSE*) and Root Mean Square Error (*RMSE*) were calculated with the following equations:

$$MSE = \frac{\sum_{j=1}^{N} (X_j(Exp) - X_j(Pred))^2}{N} \quad (8)$$

$$RMSE = \left[\frac{\sum_{j=1}^{N} (X_j(Exp) - X_j(Pred))^2}{N} \right]^{0.5} \quad (9)$$

In which N indicates data number, '*X* (*Exp*)', and '*X* (*Pred*)' illustrate the experimental and predicted (ANN) values, respectively.

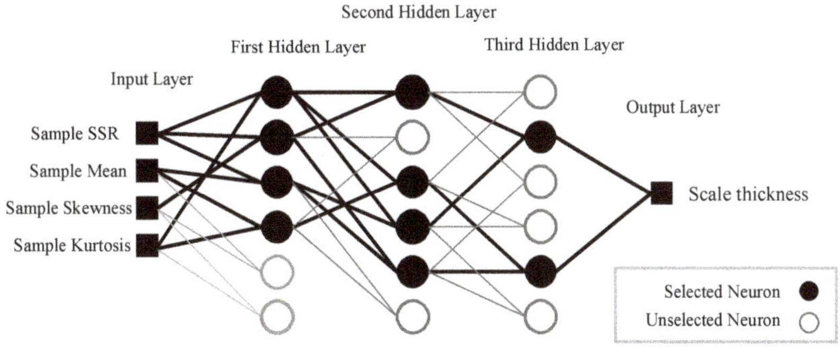

Figure 4. The structure of the trained GMDH neural network.

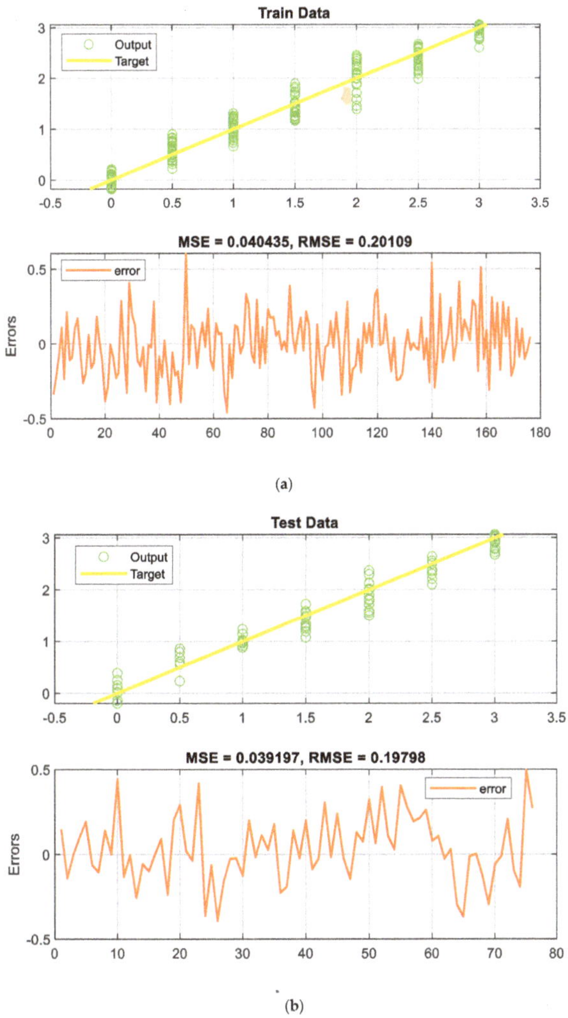

Figure 5. Performance of designed GMDH neural network against (**a**) training and (**b**) testing data.

The comparison table of the output of the neural network with the desired output is given in Table 1. To show the accuracy of the implemented system, Table 2 shows a comparison between the accuracy of current research and previous research. Extracting the appropriate features from the received signals has not only improved the accuracy of the presented system, but also reduced the number of detectors. Reducing the number of detectors, along with reducing the complexity of the system, will significantly reduce the implementation cost of the detection system.

Table 1. Comparison of target values with neural network outputs.

Data Number	*Train Targets*	*Train Outputs*	*Test Targets*	*Test Outputs*
1	0.5000	0.8364	1.5000	1.3493
2	1.0000	1.2139	2.5000	2.6387
3	1.0000	1.0764	1.0000	0.9995
4	3.0000	2.8872	2.0000	1.8995
5	1.5000	1.7345	0	−0.1932
6	1.0000	0.7840	0.5000	0.5605
7	0	0.1079	0.5000	0.6024
8	0	0.0893	2.5000	2.3590
9	2.5000	2.3907	2.0000	1.9970
10	1.0000	0.8268	2.0000	1.5573
11	0	0.0608	2.0000	2.1318
12	1.0000	1.2595	2.5000	2.5019
13	1.0000	1.1938	0	0.2556
14	2.5000	2.4362	2.5000	2.5553
15	0.5000	0.6588	0	0.0966
16	2.5000	2.6227	1.5000	1.4977
17	3.0000	2.8169	3.0000	2.9088
18	3.0000	2.9739	1.0000	1.2369
19	2.5000	2.6138	2.0000	1.7970
20	2.0000	2.3813	2.5000	2.2073
21	2.0000	2.2872	3.0000	2.9807
22	0	0.0054	3.0000	3.0354
23	0.5000	0.5838	1.5000	1.0813
24	1.5000	1.7266	0.5000	0.8614
25	0	0.1918	3.0000	3.0631
26	2.0000	1.7102	0	0.3922
27	0	0.0498	0	0.1574
28	1.5000	1.8249	1.0000	1.0251
29	2.5000	2.0919	0	0.0212
30	2.5000	2.3027	2.0000	2.1242
31	0	−0.1304	2.5000	2.2996
32	2.5000	2.6107	3.0000	3.0133
33	0.5000	0.6473	1.0000	0.8878
34	3.0000	3.0209	1.0000	0.9709
35	0.5000	0.8172	2.0000	1.8204
36	0	0.0083	1.5000	1.7240
37	0.5000	0.5213	0.5000	0.6903
38	1.5000	1.2158	2.5000	2.3593
39	1.5000	1.8909	0	0.0223
40	2.0000	2.0860	1.5000	1.3002
41	2.0000	2.2205	1.5000	1.5859
42	1.0000	0.9735	3.0000	3.0234
43	0	0.1812	1.5000	1.1938

Table 1. *Cont.*

Data Number	*Train Targets*	*Train Outputs*	*Test Targets*	*Test Outputs*
44	0.5000	0.9019	2.0000	2.0133
45	2.0000	2.0582	1.5000	1.2607
46	0	0.2109	3.0000	3.0214
47	1.5000	1.6819	1.0000	1.1423
48	2.0000	2.3901	0	−0.1317
49	0	0.0807	1.0000	0.9227
50	2.0000	1.3950	3.0000	2.6769
51	2.5000	2.6330	3.0000	2.9334
52	2.5000	2.3719	2.5000	2.1041
53	1.5000	1.3983	2.0000	1.8912
54	1.0000	1.1618	1.5000	1.4685
55	0.5000	0.4591	2.0000	1.5945
56	2.5000	2.3568	2.0000	1.7184
57	2.5000	2.5199	3.0000	2.8049
58	3.0000	2.7633	3.0000	2.7848
59	1.0000	1.1078	0.5000	0.2380
60	0	0.1705	3.0000	2.9187
61	1.0000	0.8614	2.5000	2.3903
62	1.0000	0.9309	3.0000	3.0254
63	1.5000	1.4250	1.0000	0.9684
64	0.5000	0.8160	2.0000	2.2965
65	2.0000	2.4583	2.0000	2.3671
66	0.5000	0.5069	0	0.0096
67	0.5000	0.7269	1.0000	0.9977
68	2.5000	2.3730	2.0000	2.1213
69	1.5000	1.3851	0.5000	0.7948
70	3.0000	3.0613	1.5000	1.5571
71	1.0000	1.0067	2.0000	2.0093
72	1.0000	0.6657	1.5000	1.2935
73	1.0000	0.7362	0	0.0936
74	1.0000	1.0683	0.5000	0.6893
75	0	0.1252	2.0000	1.5044
76	1.5000	1.2048	3.0000	2.7268
77	2.0000	2.1610	-	-
78	0.5000	0.3873	-	-
79	0.5000	0.6288	-	-
80	1.5000	1.2658	-	-
81	1.5000	1.3239	-	-
82	0.5000	0.3186	-	-
83	1.0000	0.9420	-	-
84	1.5000	1.4139	-	-
85	3.0000	3.0394	-	-
86	3.0000	2.9834	-	-
87	0	0.0517	-	-
88	3.0000	2.6107	-	-
89	1.0000	0.9446	-	-
90	0.5000	0.5613	-	-
91	0	−0.0763	-	-
92	0	−0.1754	-	-
93	3.0000	3.0456	-	-
94	3.0000	2.7840	-	-
95	2.5000	2.3713	-	-

Table 1. *Cont.*

Data Number	*Train Targets*	*Train Outputs*	*Test Targets*	*Test Outputs*
96	0.5000	0.7758	-	-
97	2.0000	2.4267	-	-
98	2.0000	1.8703	-	-
99	1.0000	1.1189	-	-
100	1.0000	1.2409	-	-
101	1.0000	1.0213	-	-
102	2.0000	2.0038	-	-
103	0	−0.1516	-	-
104	1.5000	1.5227	-	-
105	3.0000	2.7891	-	-
106	1.5000	1.4973	-	-
107	2.0000	2.3414	-	-
108	3.0000	2.9899	-	-
109	2.0000	1.7173	-	-
110	0.5000	0.8231	-	-
111	1.5000	1.6692	-	-
112	0.5000	0.6473	-	-
113	0	−0.0318	-	-
114	1.0000	1.1917	-	-
115	0.5000	0.3629	-	-
116	0.5000	0.4943	-	-
117	0	−0.1362	-	-
118	3.0000	3.0185	-	-
119	2.5000	2.1735	-	-
120	2.5000	2.1380	-	-
121	0	0.0072	-	-
122	2.0000	1.9919	-	-
123	2.5000	2.3027	-	-
124	0	0.0226	-	-
125	1.5000	1.6830	-	-
126	0	−0.0423	-	-
127	0.5000	0.7433	-	-
128	0.5000	0.7383	-	-
129	0	0.1972	-	-
130	3.0000	3.0329	-	-
131	2.5000	2.4056	-	-
132	3.0000	3.0404	-	-
133	2.5000	2.4958	-	-
134	3.0000	3.0254	-	-
135	0	0.0398	-	-
136	3.0000	2.8249	-	-
137	2.5000	2.6042	-	-
138	2.5000	2.4641	-	-
139	0.5000	0.7573	-	-
140	2.0000	1.4576	-	-
141	1.5000	1.7930	-	-
142	2.5000	2.5810	-	-
143	1.5000	1.1687	-	-
144	1.0000	1.1215	-	-
145	3.0000	3.0252	-	-
146	0.5000	0.4367	-	-
147	2.5000	2.2581	-	-
148	0.5000	0.6105	-	-

Table 1. *Cont.*

Data Number	*Train Targets*	*Train Outputs*	*Test Targets*	*Test Outputs*
149	2.5000	2.5026	-	-
150	2.0000	1.5837	-	-
151	3.0000	2.9770	-	-
152	1.5000	1.3390	-	-
153	2.0000	1.9029	-	-
154	1.0000	0.9743	-	-
155	1.5000	1.2099	-	-
156	1.5000	1.2457	-	-
157	2.5000	2.6815	-	-
158	2.5000	1.9872	-	-
159	2.0000	2.1030	-	-
160	0	−0.0921	-	-
161	1.0000	1.3085	-	-
162	1.5000	1.1887	-	-
163	1.5000	1.5302	-	-
164	0.5000	0.2263	-	-
165	0.5000	0.6759	-	-
166	2.5000	2.2218	-	-
167	1.5000	1.4478	-	-
168	2.5000	2.2561	-	-
169	1.0000	1.2136	-	-
170	0.5000	0.6492	-	-
171	0	−0.1685	-	-
172	1.0000	1.0860	-	-
173	1.0000	0.8985	-	-
174	0.5000	0.6025	-	-
175	3.0000	3.0462	-	-
176	3.0000	2.9525	-	-

Table 2. A comparison of the accuracy of the proposed detection system and previous studies.

Ref	*Number of Detectors*	*Source Type*	*Type of Neural Network*	*Maximum MSE*	*Maximum RMSE*
[9]	1	^{137}Cs	GMDH	1.24	1.11
[7]	2	^{137}Cs	MLP	0.21	0.46
[8]	1	^{60}Co	GMDH	7.34	2.71
[38]	2	^{137}Cs	MLP	0.67	0.82
[39]	1	X-Ray tube	MLP	17.05	4.13
[40]	1	^{137}Cs	MLP	2.56	1.6
[41]	1	^{60}Co	RBF	37.45	6.12
[42]	2	^{137}Cs	MLP	1.08	1.04
[current study]	1	**Dual-energy gamma source**	**GMDH**	**0.04**	**0.2**

6. Conclusions

The presence of scale inside the oil pipes will cause significant problems such as reducing the effective diameter of the oil pipes, increasing the energy consumed by the oil pumps, reducing the efficiency, and increasing the repair costs. Therefore, timely detection of the amount of scale inside the oil pipes helps to reduce the mentioned damages. In this research, a non-invasive system based on gamma-ray attenuation was introduced. The introduced system consisted of a dual-energy source of ^{241}Am and ^{133}Ba, a NaI detector, and a steel pipe. A three-phase flow was simulated in a stratified regime consisting of oil, water, and gas in different volume percentages. In all these simulations, the value of different scales in the range of 0 to 3 cm were examined. All these structures and flows passing through the test pipe were simulated by MCNP code. After collecting the received signals from the detector, the feature extraction operation started in the time domain, and four time features with the names of sample mean, sample SSR, sample skewness, and sample kurtosis were extracted from the signals and introduced as the inputs of the GMDH neural network. The trained neural network predicted the thickness of the scale inside the pipe with an RMSE of less than 0.2, which is a very low error compared to previous studies. In addition, the use of feature extraction techniques made it unnecessary to have multiple detectors, and only one detector is enough to determine the thickness of the scale, which has significantly saved the implementation costs. One of the major limitations of this research is that working with radioisotope devices requires the use of protective clothing, and the transportation of these devices is challenging. The use of feature extraction techniques in the frequency domain, time-frequency domain, and the investigation of the performance of other neural networks such as MLP, RBF, and even deep neural networks are highly recommended to the researchers in this field.

Author Contributions: Conceptualization, A.M.M. and T.-C.C.; methodology, S.M.A.; software, T.-C.C. and S.M.A.; validation, A.A.A.-Q., A.K.A. and N.A.G.; formal analysis, A.A.A.-Q.; investigation, N.A.G.; resources, A.K.A.; data curation, H.H.A.; writing—original draft preparation, H.H.A.; writing—review and editing, T.-C.C. and E.E.-Z.; visualization, E.E.-Z.; supervision, A.M.M., T.-C.C. and E.E.-Z.; project administration, A.M.M.; funding acquisition, A.M.M. and E.E.-Z. All authors have read and agreed to the published version of the manuscript.

Funding: This work was supported by the Deanship of Scientific Research at King Khalid University (Grant numbers RGP.1/243/42). The authors acknowledge support from the German Research Foundation and the Open Access Publication Fund of the Thueringer Universitaets-und Landesbibliothek Jena Projekt-Nr. 433052568; the BMBF-Projekt 05P21SJFA2 Verbundprojekt 05P2021 (ErUM-FSP T05).

Institutional Review Board Statement: Not applicable.

Informed Consent Statement: Not applicable.

Data Availability Statement: Not applicable.

Conflicts of Interest: The authors declare no conflict of interest.

References

1. Roshani, G.H.; Karami, A.; Nazemi, E.; Shama, F. Volume fraction determination of the annular three-phase flow of gas-oil-water using adaptive neuro-fuzzy inference system. *Comput. Appl. Math.* **2018**, *37*, 4321–4341. [CrossRef]
2. Roshani, M.; Phan, G.; Roshani, G.H.; Hanus, R.; Nazemi, B.; Corniani, E.; Nazemi, E. Combination of X-ray tube and GMDH neural network as a nondestructive and potential technique for measuring characteristics of gas-oil–water three phase flows. *Measurement* **2021**, *168*, 108427. [CrossRef]
3. Roshani, G.H.; Karami, A.; Nazemi, E. An intelligent integrated approach of Jaya optimization algorithm and neuro-fuzzy network to model the stratified three-phase flow of gas–oil–water. *Comput. Appl. Math.* **2019**, *38*, 1–26. [CrossRef]
4. Hosseini, S.; Taylan, O.; Abusurrah, M.; Akilan, T.; Nazemi, E.; Eftekhari-Zadeh, E.; Roshani, G.H. Application of Wavelet Feature Extraction and Artificial Neural Networks for Improving the Performance of Gas–Liquid Two-Phase Flow Meters Used in Oil and Petrochemical Industries. *Polymers* **2021**, *13*, 3647. [CrossRef]
5. Sattari, M.A.; Korani, N.; Hanus, R.; Roshani, G.H.; Nazemi, E. Improving the performance of gamma radiation based two phase flow meters using optimal time characteristics of the detector output signal extraction. *J. Nucl. Sci. Technol.* **2020**, *41*, 42–54.
6. Nazemi, E.; Roshani, G.H.; Feghhi, S.A.H.; Setayeshi, S.; Zadeh, E.E.; Fatehi, A. Optimization of a method for identifying the flow regime and measuring void fraction in a broad beam gamma-ray attenuation technique. *Int. J. Hydrog. Energy* **2016**, *41*, 7438–7444. [CrossRef]
7. Sattari, M.A.; Roshani, G.H.; Hanus, R.; Nazemi, E. Applicability of time-domain feature extraction methods and artificial intelligence in two-phase flow meters based on gamma-ray absorption technique. *Measurement* **2021**, *168*, 108474. [CrossRef]
8. Roshani, M.; Sattari, M.A.; Ali, P.J.M.; Roshani, G.H.; Nazemi, B.; Corniani, E.; Nazemi, E. Application of GMDH neural network technique to improve measuring precision of a simplified photon attenuation based two-phase flowmeter. *Flow Meas. Instrum.* **2020**, *75*, 101804. [CrossRef]
9. Sattari, M.A.; Roshani, G.H.; Hanus, R. Improving the structure of two-phase flow meter using feature extraction and GMDH neural network. *Radiat. Phys. Chem.* **2020**, *171*, 108725. [CrossRef]

10. Alamoudi, M.; Sattari, M.A.; Balubaid, M.; Eftekhari-Zadeh, E.; Nazemi, E.; Taylan, O.; Kalmoun, E.M. Application of Gamma Attenuation Technique and Artificial Intelligence to Detect Scale Thickness in Pipelines in Which Two-Phase Flows with Different Flow Regimes and Void Fractions Exist. *Symmetry* **2021**, *13*, 1198. [CrossRef]
11. Basahel, A.; Sattari, M.A.; Taylan, O.; Nazemi, E. Application of Feature Extraction and Artificial Intelligence Techniques for Increasing the Accuracy of X-ray Radiation Based Two Phase Flow Meter. *Mathematics* **2021**, *9*, 1227. [CrossRef]
12. Taylan, O.; Sattari, M.A.; Essoussi, I.E.; Nazemi, E. Frequency Domain Feature Extraction Investigation to Increase the Accuracy of an Intelligent Nondestructive System for Volume Fraction and Regime Determination of Gas–Water–Oil Three-Phase Flows. *Mathematics* **2021**, *9*, 2091. [CrossRef]
13. Roshani, G.H.; Ali, P.J.M.; Mohammed, S.; Hanus, R.; Abdulkareem, L.; Alanezi, A.A.; Kalmoun, E.M. Simulation Study of Utilizing X-ray Tube in Monitoring Systems of Liquid Petroleum Products. *Processes* **2021**, *9*, 828. [CrossRef]
14. Balubaid, M.; Sattari, M.A.; Taylan, O.; Bakhsh, A.A.; Nazemi, E. Applications of discrete wavelet transform for feature extraction to increase the accuracy of monitoring systems of liquid petroleum products. *Mathematics* **2021**, *9*, 3215. [CrossRef]
15. Ivakhnenko, A.G. Polynomial theory of complex systems. *IEEE Trans. Syst. Man Cybern.* **1971**, *SMC-1*, 364–378. [CrossRef]
16. Lalbakhsh, A.; Mohamadpour, G.; Roshani, S.; Ami, M.; Roshani, S.; Sayem, A.S.; Alibakhshikenari, M.; Koziel, S. Design of a compact planar transmission line for miniaturized rat-race coupler with harmonics suppression. *IEEE Access* **2021**, *9*, 129207–129217. [CrossRef]
17. Hookari, M.; Roshani, S.; Roshani, S. High-efficiency balanced power amplifier using miniaturized harmonics suppressed cou-pler. *Int. J. RF Microw. Comput. Aided Eng.* **2020**, *30*, e22252. [CrossRef]
18. Lotfi, S.; Roshani, S.; Roshani, S.; Gilan, M.S. Wilkinson power divider with band-pass filtering response and harmonics suppres-sion using open and short stubs. *Frequenz* **2020**, *74*, 169–176. [CrossRef]
19. Jamshidi, M.; Siahkamari, H.; Roshani, S.; Roshani, S. A compact Gysel power divider design using U-shaped and T-shaped res-onators with harmonics suppression. *Electromagnetics* **2019**, *39*, 491–504. [CrossRef]
20. Roshani, S.; Jamshidi, M.B.; Mohebi, F.; Roshani, S. Design and modeling of a compact power divider with squared resonators using artificial intelligence. *Wirel. Pers. Commun.* **2021**, *117*, 2085–2096. [CrossRef]
21. Roshani, S.; Azizian, J.; Roshani, S.; Jamshidi, M.B.; Parandin, F. Design of a miniaturized branch line microstrip coupler with a simple structure using artificial neural network. *Frequenz* **2022**, *76*. [CrossRef]
22. Khaleghi, M.; Salimi, J.; Farhangi, V.; Moradi, M.J.; Karakouzian, M. Application of Artificial Neural Network to Predict Load Bearing Capacity and Stiffness of Perforated Masonry Walls. *CivilEng* **2021**, *2*, 48–67. [CrossRef]
23. Dabiri, H.; Farhangi, V.; Moradi, M.J.; Zadehmohamad, M.; Karakouzian, M. Applications of Decision Tree and Random Forest as Tree-Based Machine Learning Techniques for Analyzing the Ultimate Strain of Spliced and Non-Spliced Reinforce-ment Bars. *Appl. Sci.* **2022**, *12*, 4851. [CrossRef]
24. Zych, M.; Petryka, L.; Kępński, J.; Hanus, R.; Bujak, T.; Puskarczyk, E. Radioisotope investigations of compound two-phase flows in an ope channel. *Flow Meas. Instrum.* **2014**, *35*, 11–15. [CrossRef]
25. Zych, M.; Hanus, R.; Wilk, B.; Petryka, L.; Świsulski, D. Comparison of noise reduction methods in radiometric correlation measurements of two-phase liquid-gas flows. *Measurement* **2018**, *129*, 288–295. [CrossRef]
26. Golijanek-Jędrzejczyk, A.; Mrowiec, A.; Hanus, R.; Zych, M.; Heronimczak, M.; Świsulski, D. Uncertainty of mass flow meas-urement using centric and eccentric orifice for Reynolds number in the range $10{,}000 \leq Re \leq 20{,}000$. *Measurement* **2020**, *160*, 107851. [CrossRef]
27. Mayet, A.; Hussain, M. Amorphous WNx Metal For Accelerometers and Gyroscope. In Proceedings of the MRS Fall Meeting, Boston, MA, USA, 30 November–5 December 2014.
28. Mayet, A.; Hussain, A.; Hussain, M. Three-terminal nanoelectromechanical switch based on tungsten nitride—An amor-phous metallic material. *Nanotechnology* **2016**, *27*, 035202. [CrossRef]
29. Shukla, N.K.; Mayet, A.M.; Vats, A.; Aggarwal, M.; Raja, R.K.; Verma, R.; Muqeet, M.A. High speed integrated RF–VLC data communication system: Performance constraints and capacity considerations. *Phys. Commun.* **2022**, *50*, 101492. [CrossRef]
30. Mayet, A.; Smith, C.E.; Hussain, M.M. Energy Reversible Switching from Amorphous Metal Based Nanoelectromechanical Switch. In Proceedings of the 13th IEEE International onference on Nanotechnology (IEEE-NANO 2013), Beijing, China, 5–8 August 2013; pp. 366–369.
31. Khaibullina, K. Technology to Remove Asphaltene, Resin and Paraffin Deposits in Wells Using Organic Solvents. In Proceedings of the SPE Annual Technical Conference and Exhibition 2016, Dubai, United Arab Emirates, 26–28 September 2016.
32. Tikhomirova, E.A.; Sagirova, L.R.; Khaibullina, K.S. A review on methods of oil saturation modelling using IRAP RMS. *IOP Conf. Ser. Earth Environ. Sci.* **2019**, *378*, 012075. [CrossRef]
33. Khaibullina, K.S.; Korobov, G.Y.; Lekomtsev, A.V. Development of an asphalt-resin-paraffin deposits inhibitor and substan-tiation of the technological parameters of its injection into the bottom-hole formation zone. Period. *Tche Quim.* **2020**, *17*, 769–781.
34. Khaibullina, K.S.; Sagirova, L.R.; Sandyga, M.S. Substantiation and selection of an inhibitor for preventing the formation of asphalt-resin-paraffin deposits. Period. *Tche Quim.* **2020**, *17*, 541–551.
35. Mayet, A.M.; Alizadeh, S.M.; Kakarash, Z.A.; Al-Qahtani, A.A.; Alanazi, A.K.; Alhashimi, H.H.; Eftekhari-Zadeh, E.; Nazemi, E. Introducing a Precise System for Determining Volume Percentages Independent of Scale Thickness and Type of Flow Regime. *Mathematics* **2022**, *10*, 1770. [CrossRef]

36. Mayet, A.M.; Alizadeh, S.M.; Nurgalieva, K.S.; Hanus, R.; Nazemi, E.; Narozhnyy, I.M. Extraction of Time-Domain Charac-teristics and Selection of Effective Features Using Correlation Analysis to Increase the Accuracy of Petroleum Fluid Monitor-ing Systems. *Energies* **2022**, *15*, 1986. [CrossRef]
37. Hanus, R.; Zych, M.; Golijanek-Jędrzejczyk, A. Investigation of Liquid–Gas Flow in a Horizontal Pipeline Using Gamma-Ray Technique and Modified Cross-Correlation. *Energies* **2022**, *15*, 5848. [CrossRef]
38. Hosseini, S.; Roshani, G.H.; Setayeshi, S. Precise gamma based two-phase flow meter using frequency feature extraction and only one detector. *Flow Meas. Instrum.* **2020**, *72*, 101693. [CrossRef]
39. Roshani, M.; Ali, P.J.; Roshani, G.H.; Nazemi, B.; Corniani, E.; Phan, N.H.; Tran, H.N.; Nazemi, E. X-ray tube with artificial neural network model as a promising alternative for radioisotope source in radiation based two phase flowmeters. *Appl. Radiat. Isot.* **2020**, *164*, 109255. [CrossRef]
40. Peyvandi, R.G.; Rad, S.I. Application of artificial neural networks for the prediction of volume fraction using spectra of gamma rays backscattered by three-phase flows. *Eur. Phys. J. Plus* **2017**, *132*, 511. [CrossRef]
41. Roshani Gholam, H.; Ehsan, N.; Farzin, S.; Mohammad, A.I.; Salar, M. Designing a simple radiometric system to predict void fraction percentage independent of flow pattern using radial basis function. *Metrol. Meas. Syst.* **2018**, *25*, 347–358.
42. Roshani, G.H.; Nazemi, E.; Feghhi, S.A.H.; Setayeshi, S. Flow regime identification and void fraction prediction in two-phase flows based on gamma ray attenuation. *Measurement* **2015**, *62*, 25–32. [CrossRef]

processes

Article

Study on the Erosion of Choke Valves in High-Pressure, High-Temperature Gas Wells

Ling Guo [1], Yayong Wang [2], Xiaohui Xu [1], Han Gao [1], Hong Yang [2] and Guoqing Han [2,*]

1 Research Institute of Engineering Technology, Xinjiang Oilfield Company, PetroChina, Karamay 834000, China
2 State Key Laboratory of Petroleum Resources and Prospecting, China University of Petroleum (Beijing), Beijing 102249, China
* Correspondence: cup_706ac@163.com

Abstract: During the process of gas production in high-pressure, high-temperature (HPHT) gas wells, the choke valve, as the most vital component of the surface control equipment, plays a significant role in regulating the output and reducing the fluid pressure to ensure the safety of surface gathering and transportation equipment. High-pressure, high-velocity fluid flow and solid-phase particles cause deterioration of the choke valve. With the enhancement of intelligent and digital oilfields, conventional choke valves have been progressively replaced by electric choke valves. Due to the complex structure of the throttle valve, the flow path and the velocity state of the fluid in the throttle valve, and the distribution law of the erosion fraction are quite distinctive from those in the ordinary throttle valve, meriting further research. In this paper, a simulation of computational fluid dynamics (CFD) was conducted to determine the effects of the pressure distribution, fluid state, divergent particle sizes, and sand volume on the erosion rate of the choke valve. Under various valve openings, the fluid state and the location of high-risk points can be ascertained. The large particle size (diameter greater than 6 mm) of sand and gravel is convenient for causing concentrated erosion in the position of the valve hole, which induces the channel diameter to expand. Fine silt sand (diameter from 0.1 mm to 1 mm) gives rise to relatively uniform abrasion to the choke's current-facing surface. This study can optimize the layout of the choke valve and reduce the cost and number of switching wells, thereby decreasing the frequency of maintenance and the pressure fluctuation's effect on the formation.

Keywords: high-pressure gas wells; CFD simulations; adjustable choke valve; erosion

Citation: Guo, L.; Wang, Y.; Xu, X.; Gao, H.; Yang, H.; Han, G. Study on the Erosion of Choke Valves in High-Pressure, High-Temperature Gas Wells. *Processes* **2022**, *10*, 2139. https://doi.org/10.3390/pr10102139

Academic Editors: Tianshou Ma and Yuqiang Xu

Received: 16 September 2022
Accepted: 17 October 2022
Published: 20 October 2022

1. Introduction

The HT1 well is an HPHT gas well with a tectonic belt on the southern edge of the Dzungar Basin, with a depth of over 7000 m and a closing pressure of more than 120 MPa [1]. As the most essential component of the choke manifold, the choke valve takes part in regulating the output and reducing the fluid pressure to ensure the safety of ground gathering and of transportation equipment and personnel [2]. Choke valve erosion is unavoidable as a result of the high-pressure, high-velocity airflow and the erosion engendered by solid-phase particles. The disclosure of the choke valve's erosion law significantly reduces erosion and optimizes the choke valve's structure [3]. A number of researchers have examined this issue.

Paggiaro et al. [4] experimentally validated a three-dimensional computational fluid dynamics (CFD) throttling flow simulation model. CFD flow and sand grain tracking simulations demonstrate that the predicted flow parameters are consistent with the experimental data, and the simulations can accurately predict the location of erosion hotspots. Chang Z. et al. [5] took advantage of the RNG k-ε turbulence model to numerically calculate the dual disc check valve's three-dimensional transient flow based on dynamic mesh technology. Tuladhar U. et al. [6] investigated how the nozzle exit diameter and the slope of the cut front affect gas flow behavior using the Reynolds averaged Navier–Stokes

(RANS)-based k–ω turbulence model. Andres Pinilla et al. [7] presented a numerical study that describes the working principle of RCP valves in accordance with a computational fluid dynamics (CFD) analysis. However, this study did not address the erosion of choke valves with variable morphology in high-pressure, high-speed fluids.

To address the above deficiencies, this paper focuses on the erosion of the adjustable choke valves at different opening in HPHT gas wells. The flow in the choke valve at various opening degrees was simulated utilizing CFD software, and the influence of the particle size and the sand volume of solid-phase particles on erosion were simulated. It is crucial to maximize the choke valve's structure, as this would reduce field expenses.

2. Numerical Simulation Model of Erosion

2.1. Flow Equation of Liquid Phase

The production of sand is one of the primary challenges associated with the development of HPHT gas wells. The choke valve is more susceptible to sand erosion. Moreover, the flow in the choke valve can be simplified to a gas–solid two-phase flow, where the gas phase is methane and the solid phase is the sand particles of the formation. Since the volume fraction of the solid phase in gas production is typically less than 5%, when utilizing the Fluent simulation, the gas phase can be regarded as a continuous phase and the solid phase can be regarded as a discrete phase.

The flow equations of the gas phase include the equation of continuity, the equation of momentum, and the equation of turbulence model. Specific forms of the equations are as follows [8–10]:

Equation of continuity:

$$\frac{\partial \rho}{\partial t}+\frac{\partial(\rho u_i)}{\partial x_i}=0 \tag{1}$$

where ρ is the density of the gas (continuous phase) and u_i is the velocity component parallel to the axis x_i.

Equation of momentum [11]:

$$\frac{\partial(\rho u_i)}{\partial t}+\frac{\partial(\rho u_i u_j)}{\partial x_j}=-\frac{\partial p}{\partial x_i}+\frac{\partial \tau_{ij}}{\partial x_j}+\rho g+F_i \tag{2}$$

where p is the static pressure, τ_{ij} is the viscous stress tensor, g is the acceleration of gravity, and F_i is the generalized volumetric force.

RNG k-ε turbulence model calculation has an extensive range of applications due to its high accuracy. The model is suitable for the numerical calculation of gas well throttling, and its equations are as follows [8]:

$$\frac{\partial(\rho k)}{\partial t}+\frac{\partial(\rho k u_i)}{\partial x_i}=\frac{\partial\left[\left(\mu+\frac{\mu_t}{\sigma_k}\right)\frac{\partial k}{\partial x_j}\right]}{\partial x_j}+G_k+G_b-Y_m-\rho\varepsilon \tag{3}$$

$$\frac{\partial(\rho \varepsilon)}{\partial t}+\frac{\partial(\rho \varepsilon u_i)}{\partial x_i}=\frac{\partial\left[\left(\mu+\frac{\mu_t}{\sigma_\varepsilon}\right)\frac{\partial \varepsilon}{\partial x_j}\right]}{\partial x_j}+C_{1\varepsilon}\frac{\varepsilon}{k}C_k-C_{2\varepsilon}\rho\frac{\varepsilon^2}{k}-C_{1\varepsilon}C_{3\varepsilon}\frac{\varepsilon}{k}G_b \tag{4}$$

where k is the turbulent flow energy, μ_t is the turbulence viscosity, μ is the fluid dynamic viscosity, G_k is the turbulent flow energy generated by the average speed gradient, G_b is the turbulent flow energy generated by the buoyancy, Y_m is the effect of the expansion of compressible turbulence fluctuation on the overall dissipation rate, ε is the turbulent flow energy consumption dissipation power, σ_k is the turbulent kinetic energy k of the turbulent flow Plant number, and σ_ε is the turbulent flow Plant number with an ε dissipation rate [12].

The default values in Fluent are $C_{1\varepsilon}$ = 1.42, $C_{2\varepsilon}$ = 1.68, and $C_{3\varepsilon}$ = 1.83 [13–15].

2.2. Movement Equation of Particle Phase

Due to the small volume fraction of solid particles in the choke valve, particle collisions can be disregarded, and the DPM (Deformable Part Model) is adopted to simulate the erosion of solid particles.

The kinetic equation for particles in discrete phase models is as follows [16]:

$$\frac{du_P}{\mathrm{dt}} = F_D(u - u_P) + \frac{g(\rho_P - \rho)}{\rho_P} + F_P \tag{5}$$

where u_P is the particle velocity component, ρ_P is the particle density, F_D is the flow resistance of the particle, and F_P is other forces affected by the particle.

Other forces affected by the particles are as follows [17]:

$$F_D = \frac{18_\mu}{\rho_P d_P^2}\frac{C_D Re}{24} \tag{6}$$

$$\mathrm{Re} = \frac{\rho d_P |u - u_P|}{\mu} \tag{7}$$

where d_P is the particle diameter, Re is the relative Reynolds number, and C_D is the resistance coefficient.

2.3. Equation of Erosion

In the evaluation of the erosion of the choke valve, the erosion rate is selected for analysis and comparison; the DPM model is utilized for calculation; the flow equation is solved based on the Euler–Lagrange method, which considers gas as the continuous phase and solid particles as the discrete phase and the change in the particle's state of motion is accounted for by a differential equation to obtain the particle motion trajectory and energy transfer change.

The erosion rate calculation model is as follows [18]:

$$R_{\mathrm{erosion}} = \sum_{P=1}^{N} \frac{m_P C(d_P) f(\alpha) v^{b(v)}}{A_{\mathrm{face}}} \tag{8}$$

where m_P is the average mass flow of the particles, which is a fixed value defined in the inlet face prior to the calculation; N is the number of particles when colliding with the wall of the structure; $C(d_P)$ is a particle diameter function, usually related to the physical properties of the material being eroded, and the value of C is 1.8×10^{-9} [19,20] because the size of the particles in this paper is mono; α is the impact angle between the particle motion path and the wall surface of the structure; $f(\alpha)$ is the angle of the impact function of the particles; v is the relative velocity of the particles; $b(v)$ is a function of the relative velocity of the particles, and the solid particles in this article are quartz sand, which take the value of 2.6 [21–23]; A_{face} is the wall area of the wall surface of the particle collision wall, which is defined during the process of simulation based on the effect of the angle and the velocity of the particles; and R_{erosion} is the erosion quality of the particles on the wall surface of the structure per unit area per unit time. The choke impact angle function $f(a)$ was proposed by Huser and Kvemvold.

When particles strike the valve, the speed of the particles decreases. To reflect the situation, the concept of recovery coefficient is put forward to describe the loss. The speed ratio of the vertical wall before and after the collision is a normal recovery coefficient ε_n, and the ratio of the tangent direction of the wall before and after the collision is the tangential recovery coefficient ε_t.

The equations proposed by A. Forder [24] are used to calculate the recovery coefficients. The expressions are as follows:

$$\varepsilon_n = 0.993 - 0.0307\mathrm{a} + 4.75 \times 10^{-4} a^2 - 2.61 \times 10^{-6} a^3 \tag{9}$$

$$\varepsilon_t = 0.998 - 0.029a + 6.43 \times 10^{-4}a^2 - 3.56 \times 10^{-6}a^3 \quad (10)$$

3. Model Creation with Parameter Setting

3.1. Choke Valve Geometry and Meshing

The choke valve modifies the opening degree by adjusting the tour valve's position relative to the spool. Choke geometries are built by SOLIDWORKS, and the choke valve model consists of three parts: the spool, the base, and the valve. This is shown in Figure 1. The distinctive openings of the choke valve can be simulated by changing the position of the valve and the base. The combined structure is shown in Figure 2. Three rows of four holes with diameters of 3.2 mm, 6.5 mm, and 10.2 mm are staggered across the opening of the spool.

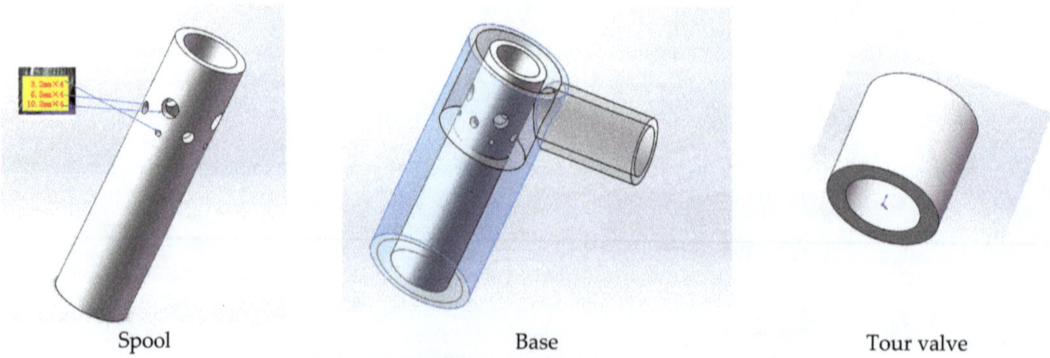

Figure 1. Each member of the choke valve.

Figure 2. Schematic diagram of the structure of the choke valve.

The structure of the combined choke valve is demonstrated in Figure 2.

The extraction of fluids is performed using the ANSYS geometry tools. The fluid model is based on a cylinder structure, and triangular meshing helps ensure the mesh's quality.

3.2. Initial Boundary Condition Setting

The inflow surface, the outflow surface, and the contact surface of the inner wall of the fluid and the choke valve are characterized in the model, as revealed in Figure 3. Formulated using field data, the inlet pressure is set to 80 MPa, the temperature is set to 300 K, the outlet is set to outflow, and the particle state is set to escaping. Predicated on the geometric dimensions of the choke valve and the gas flow rate, the Reynolds number is determined to be 215,463, which is larger than 4000, indicating that the fluid's flow state is turbulent.

The grit is chosen to be calcium sulfate, with a density of 2400 kg/m^3. The continuous phase in a high-pressure gas well is mainly methane, which is less dense, and the effect of gravity can be ignored. Furthermore, the energy equation states are open, avoiding standard wall functions, no-slip conditions, and particles bouncing off after collision avoidance. The amount of momentum reduction is determined by the collision model. Thus, the solution method adopts the SIMPLE algorithm to solve the pressure–velocity coupling, the others adopt the second-order windward format, and the mean residual convergence standard is set to the recommended value of 10^{-5}. High temperature and high pressure frequently occur together, and a high pressure generally indicates a high rate of gas well production. The change in pressure and flow directly affects the quantity of the fluid and particles impacting the nozzle wall. In comparison to a low flow rate, erosion at a high flow rate is more noticeable and causes bigger priority at the production sites. The simulation and conclusions in this paper are more applicable to high-pressure and high-production gas wells.

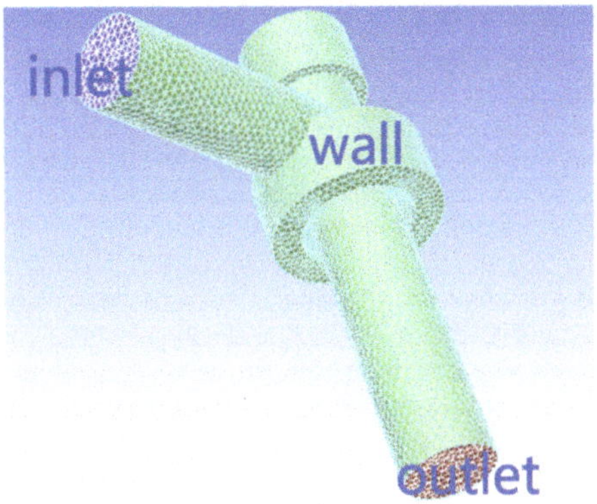

Figure 3. Fluid geometry and meshing.

3.3. Mesh Independence Verification

The mesh independence verification of the fluid domain model in the choke valve is carried out. A too-large grid will lead to large errors or non-convergence of the calculation results, while too dense of a grid will increase the burden of calculation. Under the premise of ensuring the accuracy of the simulation results, the selection of an appropriate mesh size can therefore reduce the cost of calculation. Using the Ansys Fluent software, the erosion wear rate is the statistical average, and the grid independence verification is conducted with the maximum erosion rate as an indicator. The consequence is shown in Figure 4. Under the condition that the number of grids reaches 56,734, the calculation results tend to be stable, and hence, the number of grids is applied in subsequent numerical simulation calculations.

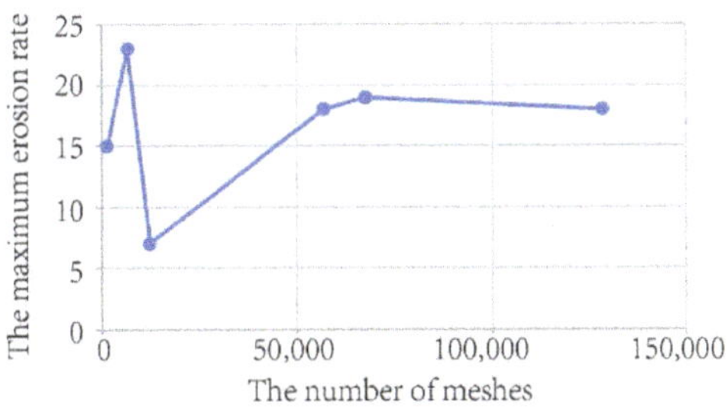

Figure 4. The maximum erosion rate is related to the number of meshes.

4. Choke Valve Erosion Numerical Simulation

4.1. Pure Gas Phase Erosion

The fluid is set to be pure methane without the addition of solid-phase particles, simulating the situation inside a choke valve with a pure gas phase.

4.1.1. Wall Pressure Distribution

The distribution of the pressure in the choke valve is simulated as shown in Figure 5.

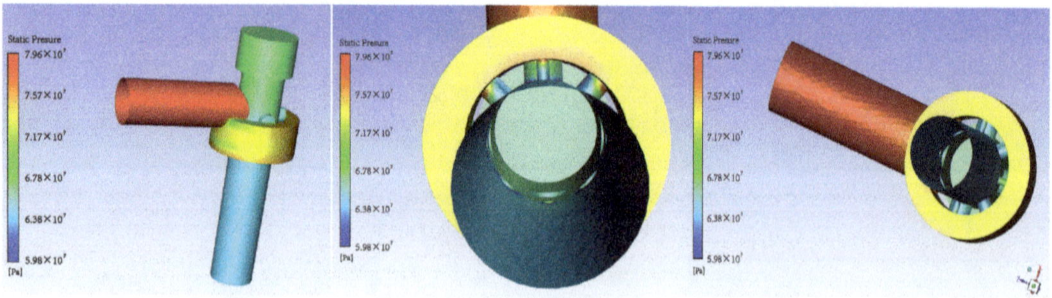

Overall stress profile | High-pressure points | Pressure on the inner wall

Figure 5. Pressure distribution of the choke valve.

By observing the pressure of the wall, the pressure distribution of the choke valve's wall can be obtained. It does seem that the high pressure is located on the outer wall of the spool facing the fluid. A high pressure indicates stronger gas erosion, and the presence of sand can exacerbate erosion at this location, which is a high-risk location that deserves additional attention. Broadly speaking, the pressure on the inner wall of the spool is low, and its erosion is considerably weaker than the outer wall erosion.

4.1.2. Velocity Vectors at Different Openings

In comparison to the traditional choke valve, the adjustability of the cage and sleeve choke valve's opening is its most prominent characteristic. This signifies that in divergent production systems, the structure of the choke valve is divergent, the flow of the fluid is dissimilar, and the inner wall pressure and the highest point of the fluid speed differ significantly. By simulating the distribution of the inner wall pressure of the choke valve under different opening degrees, it is possible to determine the distribution of high-risk sites under various production systems. Velocity vector distribution at different openings are shown in Figure 6.

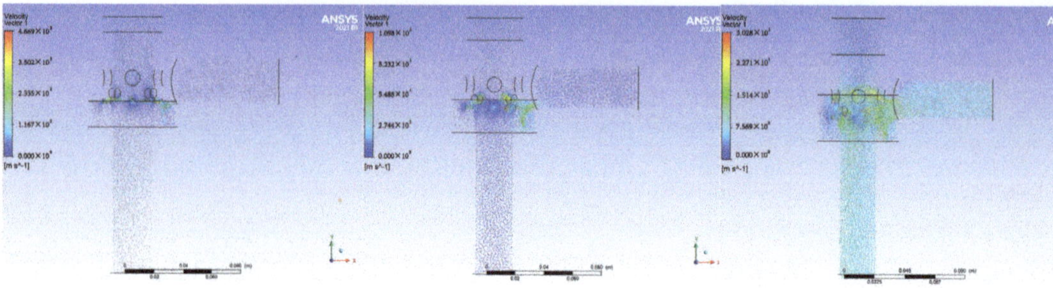

Small opening | Medium opening | Large opening

Figure 6. Velocity vector distribution at different openings.

Small opening: The erosion position is concentrated in the small hole position in the direction of the incoming flow, and while the other three holes also have an erosion phenomenon, the degree of erosion is lighter than the small hole in the incoming flow position.

Medium opening: The erosion position is concentrated along the flow direction of the small and middle holes, while the other three small holes do not show erosion. It should be noted that there is a phenomenon of fluid acceleration in the presence of the narrow slits in the middle hole, which can result in more serious erosion.

Large opening: The erosion position is concentrated in the flow direction of the large hole and the middle hole, and the small hole does not show erosion.

By simulating the distribution of the velocity vector at different openings, the shear stress of the choke valve wall surface can be reflected. It is also possible to recognize the erosion of the choke valve from a distinctive angle. Since the sand is wrapped in gas and moves, the velocity vector of the fluid can show the movement of the gravel to a certain extent. Through the simulation described above, the high-risk locations corresponding to each of the small, medium, and large openings are determined, allowing for the use of harder materials in the manufacture of choke valves.

4.2. Simulation of Sand Erosion

4.2.1. Effect of Gravel Diameter on Wear

In the case of a certain amount of grit, the erosion of the wall surface of the choke valve under dissimilar gravel diameters is simulated, and the consequences are shown in Figure 7.

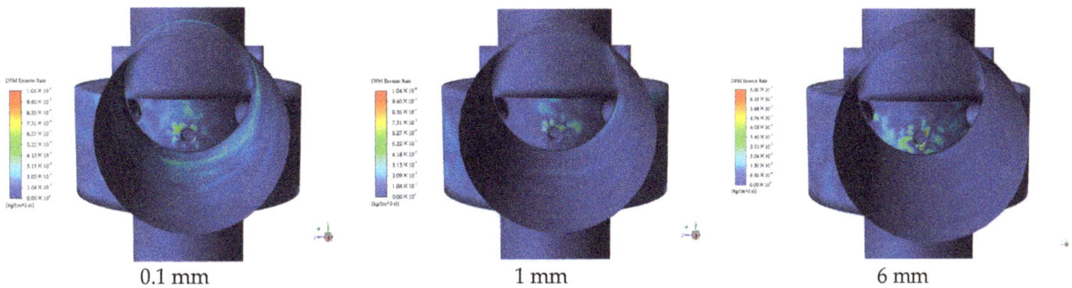

Figure 7. Erosion rate distribution under different gravel diameters.

In fine silt sand with a gravel diameter of 0.1 mm, the erosion area on the incoming flow surface of the valve's spool part is more uniformly distributed, and the phenomenon of erosion is in a larger area. As the particle size of the gravel increases, the erosion position shifts towards the edge of the hole and the erosion area shrinks, converging around the small hole. Consistent with the actual performance in the field, when there are large particles of gravel, the erosion at the small hole is very severe and concentrated at the small hole runner; eventually, the small hole continuously expands, resulting in a change in the throttling effect, and the yield and pressure after throttling have a gradual increase. Due to the fact that the high-pressure gas well generally adopts a secondary throttling, the first-level throttling will block off the large particles of sand and gravel, and only fine silty sand exists at the secondary choke. The erosion of the first-level choke valve is significantly larger than that of the secondary choke valve, indicating that the grit of the large particles has a greater impact on the erosion process.

4.2.2. The Influence of Sand Volume on Wear

The diameter of the gravel is set at 0.1 mm, which can simulate fine silt sand. A simulation of the outcomes is depicted in Figure 8 using varying sand quantities.

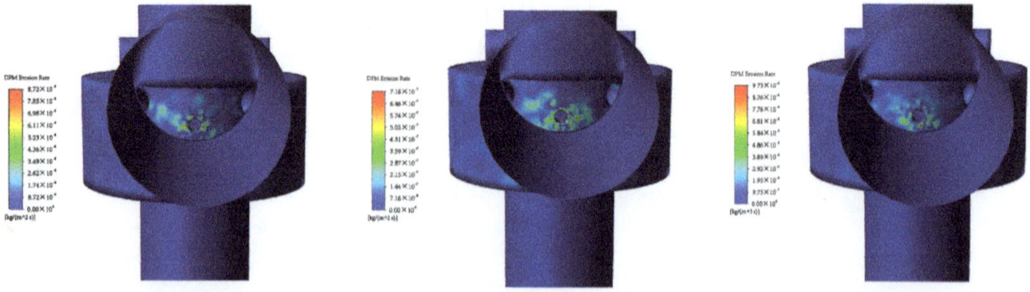

0.002 kg/s 0.02 kg/s 0.2 kg/s

Figure 8. Erosion at different amounts of sand injection.

The maximum erosion rates of 0.002 kg/s, 0.02 kg/s, and 0.2 kg/s are 8.72×10^{-8}, 7.18×10^{-7}, and 9.73×10^{-6}, respectively. The erosion velocity of the choke valve spool surface is within an order of magnitude, and the maximum erosion speed point is randomly distributed around the pore size. The erosion is uniform and correlates positively with the amount of sand produced. The results show that, in the case of fine silt sand, the erosion is positively correlated with the amount of sand, the distribution of the erosion locations does not change, and all of them are more commonly abraded.

4.3. Comparison of Simulation Results with On-Site Results

The simulation results obtained were compared to those of the erosion of a choke valve, which had been in use for more than six months before being disassembled. Comparison of the simulation results and the actual results is shown in Figure 9.

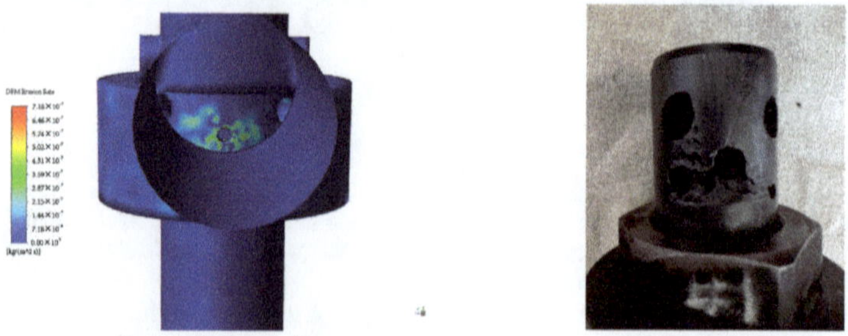

Simulation results Actual results

Figure 9. Comparison of the simulation results and the actual results.

The results of the simulation are consistent with reality. The choke spool's surface that is facing the current is subjected to the most severe deterioration. The small hole continues to expand as the erosion progresses, until it reaches a certain point where the choke valve loses its choke effect. Hence, in a follow-up study to predict the maintenance cycle for the choke valve, it is critical to assess the erosion.

5. Conclusions

(1) The position of the high-risk point of the choke valve under different opening conditions is obtained through simulation, which can help choke manufacturers optimize the shape and material of choke valves.

(2) In the case of a certain amount of sand, the increase in grit particle size concentrates the erosion position at the edge of the hole, which manifests as the concentrated erosion of the edge of the small hole. Consistent with the erosion pit at the edge of

the actual small hole, the grit's large particle size seems to be more significant in the erosion process.

(3) In the case of a certain grain size of sand and gravel (0.1 mm for fine silt sand), and under the condition that the amount of sand changes, the erosion rate of the surface facing the current is positively correlated with the amount of sand. Additionally, the maximum erosion speed point is randomly distributed around the pore size, resulting in a uniform erosion.

Author Contributions: Conceptualization, L.G.; methodology, Y.W.; software, H.Y.; validation, X.X.; formal analysis, L.G.; investigation, Y.W.; data curation, L.G.; writing—original draft preparation, X.X.; writing—review and editing, H.G.; visualization, Y.W.; supervision, G.H.; funding acquisition, G.H. All authors have read and agreed to the published version of the manuscript.

Funding: This research was funded by the National Natural Science Foundation of China grant number No. 52204059.

Institutional Review Board Statement: Not applicable.

Informed Consent Statement: Not applicable.

Data Availability Statement: Not applicable.

Conflicts of Interest: The authors declare no conflict of interest.

Abbreviations

Symbol	Definition
ρ	density (kg/m^3)
u_i	velocity component parallel to the axis x_i ($m \cdot s^{-1}$)
p	pressure (Pa)
τ_{ij}	viscous stress tensor ($N \cdot m^{-2}$)
g	acceleration of gravity ($m \cdot s^{-2}$)
F_i	generalized volumetric force (N)
k	turbulent kinetic energy ($J \cdot kg^{-1}$)
μ_t	turbulence viscosity (Pa·s)
μ	fluid dynamic viscosity (Pa·s)
G_k	turbulent flow energy generated by the average speed gradient
G_b	turbulent flow energy generated by the buoyancy
Y_m	effect of the compressible turbulence fluctuation
ε	turbulent flow energy consumption dissipation power
σ_k	turbulent kinetic energy ($m^2 \cdot s^{-1}$)
σ_ε	turbulent flow Plant number ($m^2 \cdot s^{-1}$)
u_P	particle velocity component
ρ_P	particle density (kg/m^3)
F_D	flow resistance of the particle (N)
F_P	other forces affected by the particle (N)
d_P	particle diameter (mm)
Re	relative Reynolds number
C_D	resistance coefficient
m_P	average mass flow of the particles ($kg \cdot s^{-1}$)
N	number of particles
$C(d_P)$	particle diameter function
$f(a)$	The function of the relative velocity of particles
A_{face}	wall area of the wall surface of the particle collision tube (mm^2)
$R_{erosion}$	erosion quality of particles on the wall surface ($kg \cdot s^{-1} \cdot mm^2$)
ε_n	normal recovery coefficient
ε_t	tangential recovery coefficient

References

1. Wang, Q.; Wang, B.; Yan, L.; Wang, Y.; Luo, J.; Du, G. Design and application of wellhead hydraulic safety control system in Hutan 1 Well. *Xinjiang Oil Gas* **2022**, *43*, 587–591.
2. Kim, N.-S.; Jeong, Y.-H. An investigation of pressure build-up effects due to check valve's closing characteristics using dynamic mesh techniques of CFD. *Ann. Nucl. Energy* **2021**, *152*, 107996. [CrossRef]
3. Yohana, E.; Utomo, T.S.; Sumardi, V.; Laksono, D.A.; Rozi, K.; Choi, K. Simulation sediment transport in development location of a diesel power plant using computational fluid dynamic (CFD) methods. *IOP Conf. Ser. Earth Environ. Sci.* **2021**, *623*, 012006. [CrossRef]
4. Gharaibah, E.; Zhang, Y.; Paggiaro, R.; Friedemann, J.; Oil, G. Prediction of sand erosion in choke valves-CFD model development and validation against experiments. In Proceedings of the Annual Offshore Technology Conference, Rio de Janeiro, Brazil, 29–31 October 2013.
5. Chang, Z.; Jiang, J. Study on transient flow and dynamic characteristics of dual disc check valve mounted in pipeline system during opening and closing. *Processes* **2022**, *10*, 1892. [CrossRef]
6. Tuladhar, U.; Ahn, S.-H.; Cho, D.-W.; Kim, D.-H.; Ahn, S.; Kim, S.; Bae, S.-H.; Park, T.-K. Analysis of gas flow dynamics in thermal cut kerf using a numerical and experimental approach for nozzle selection. *Processes* **2022**, *10*, 1951. [CrossRef]
7. Pinilla, A.; Stanko, M.; Asuaje, M.; Ratkovich, N. In-depth understanding of ICD completion technology working principle. *Processes* **2022**, *10*, 1493. [CrossRef]
8. Tu, Y.; Xv, X.; Yin, H.; Du, J.; Chen, F.; Qiu, J. Study on erosion and wear law of high-pressure pipe sink. *Pet. Mach.* **2018**, *46*, 84–88. [CrossRef]
9. Lai, Z.; Karney, B.; Yang, S.; Wu, D.; Zhang, F. Transient performance of a dual disc check valve during the opening period. *Ann. Nucl. Energy* **2017**, *101*, 15–22. [CrossRef]
10. Yang, Z.; Zhou, L.; Dou, H.; Lu, C.; Luan, X. Water hammer analysis when switching of parallel pumps based on contra-motion check valve. *Ann. Nucl. Energy* **2020**, *139*, 107275. [CrossRef]
11. Deng, F.; Yin, B.; Xiao, Y.; Li, G.; Yan, C. Research on erosion wear of slotted screen based on high production gas field. *Processes* **2022**, *10*, 1640. [CrossRef]
12. Cuamatzi-Meléndez, R.; Rojo, M.H.; Vázquez-Hernández, A.O.; Silva-González, F.L. Predicting erosion in wet gas pipelines/elbows by mathematical formulations and computational fluid dynamics modeling. *J. Eng. Tribol.* **2018**, *232*, 1240–1260. [CrossRef]
13. Alghurabi, A.; Mohyaldinn, M.; Jufar, S.; Younis, O.; Abduljabbar, A.; Azuwan, M. CFD numerical simulation of standalone sand screen erosion due to gas-sand flow. *J. Nat. Gas Sci. Eng.* **2020**, *85*, 103706. [CrossRef]
14. Xiong, J.; Cheng, R.; Lu, C.; Chai, X.; Liu, X.; Cheng, X. CFD simulation of swirling flow induced by twist vanes in a rod bundle. *Nucl. Eng. Des.* **2018**, *338*, 52–62. [CrossRef]
15. Kang, S.K.; Hassan, Y.A. Computational fluid dynamics (CFD) round robin benchmark for a pressurized water reactor (PWR) rod bundle. *Nucl. Eng. Des.* **2016**, *301*, 204–231. [CrossRef]
16. Yang, D.; Zhu, H. Erosion wear analysis of natural gas with sand gas and solid two-phase flow at the elbow pipe. *Pet. Mach.* **2019**, *47*, 125–132. [CrossRef]
17. Karaman, U.; Kocar, C.; Rau, A.; Kim, S. Numerical investigation of flow characteristics through simple support grids in a 1×3 rod bundle. *Nucl. Eng. Technol.* **2019**, *51*, 1905–1915. [CrossRef]
18. Wang, J.; Sun, Q.; Yan, C.; Feng, D.; Tu, Y. Study on the influence of erosion wear of tested surface process elbows. *Pet. Mach.* **2021**, *49*, 88–94. [CrossRef]
19. Bieder, U.; Falk, F.; Fauchet, G. CFD analysis of the flow in the near wake of a generic PWR mixing grid. *Ann. Nucl. Energy* **2015**, *82*, 169–178. [CrossRef]
20. Lospa, A.M.; Dudu, C.; Ripeanu, R.; Dinita, A. CFD evaluation of sand erosion wear rate in pipe bends used in technological installations. *IOP Conf. Ser. Mater. Sci. Eng.* **2019**, *514*, 012009. [CrossRef]
21. Shinde, S.; Kawadekar, D.M.; Patil, P.; Bhojwani, V. Analysis of micro and nano particle erosion by the numerical method at different pipe bends and radius of curvature. *Int. J. Ambient Energy* **2019**, *43*, 2645–2652. [CrossRef]
22. Mouketou, F.N.; Kolesnikov, A. Modelling and simulation of multiphase flow applicable to processes in oil and gas industry. *Chem. Prod. Process Model.* **2018**, *14*, 20170066. [CrossRef]
23. Lisowski, E.; Filo, G.; Rajda, J. Analysis of energy loss on a tunable check valve through the numerical simulation. *Energies* **2022**, *15*, 5740. [CrossRef]
24. Forder, A.; Thew, M.; Harrison, D. A numerical investigation of solid particle erosion experienced within oilfield control valves. *Wear* **1998**, *216*, 184–193. [CrossRef]

processes

Article

Movement Law of Methane Drained by Large-Diameter Borehole Drilling Machine in the Goaf

Yun Lei [1,2]

[1] Shenyang Research Institute, China Coal Technology & Engineering Group Corp, Fushun 113122, China; leiyun_symky@163.com
[2] State Key Laboratory of Coal Mine Safety Technology, Fushun 113122, China

Abstract: To study the movement law of methane in the goaf drained by a large-diameter borehole drilling machine under "U"-shaped ventilation, a simulation on a coal mine was conducted on Fluent to find the optimal spacing between large-diameter boreholes and the most appropriate distance between the borehole and the upper corner. The variation of borehole drilling and the methane concentration in the upper corner were obtained through a field test. Results show that the method of drilling large-diameter boreholes greatly reduces the methane concentration in the goaf and the upper corner, with the optimal borehole spacing being 30 m and the most appropriate distance between the borehole and the upper corner being 15 m. When the large-diameter borehole is drilled 25 m deep down into the goaf, it penetrates into the stress impact area, and the methane concentration increases rapidly, with the maximum being 3.7%. When the borehole is drilled 35 m down into the goaf, the methane concentration slightly decreases as a result of the drainage superposition effect. The methane concentration in the upper corner increases as the borehole is drilled deeper and is farther away from the upper corner. As a result of the drainage superposition effect, the methane concentration in the upper corner varies from 0.32% to 0.51% in a cyclical way.

Keywords: "U" type ventilation; large-diameter borehole; fluent simulation; upper corner; flow field in the goaf

Citation: Lei, Y. Movement Law of Methane Drained by Large-Diameter Borehole Drilling Machine in the Goaf. *Processes* **2022**, *10*, 1669. https://doi.org/10.3390/pr10091669

Academic Editors: Tianshou Ma and Yuqiang Xu

Received: 26 July 2022
Accepted: 18 August 2022
Published: 23 August 2022

1. Introduction

The management of methane in the upper corner has always been a technical issue for high-outburst mines. Traditional ways include direct drainage, such as burying pipes in the goaf and inserting tubes in the upper corner, and indirect drainage [1–6], such as drilling in the roof or draining methane in the upper part of the goaf along the high-level roadway. The combination of one way or another can somewhat reduce methane leakage in the upper corner, but the management effect proves inadequate for single-sided modern mines with large output [7–12]. The ventilation air methane (VAM) method that drains methane along the tail roadway, which used to effectively keep the methane concentration in the goaf and the upper corner at a certain level, has been phased out for safety reasons. In recent years, the method of drilling large-diameter boreholes from the outward staggered roadway has been widely used [13–20]. This alternative not only lowers the risk of methane accumulation, but also reduces the number of crosscuts as well as the cost of construction and maintenance. Domestic scholars have done a lot of research on the use of large-diameter boreholes. Gao Hong et al. [21] found that the drainage effect was relatively ideal when the spacing between large-diameter boreholes was about 15 m or 20 m, and the methane concentration increased by 1.33 times and the pure volume by 54.9% compared with the old way of pipe burying in the goaf. Jia Jinzhang et al. [22] studied the methane drainage technology of drilling large-diameter boreholes from the perspective of construction parameters, borehole protection parameters, and borehole sealing technology, and found the correlation between the methane concentration and the depth down the goaf.

They came to the conclusion that the optimal borehole spacing was 20 m and the opening depth was 1.2 m. Shao Guoan et al. [23] divided the spontaneous combustion "three zones", using large-diameter borehole samples, and found that the lowest safe advancing speed of the working face was 0.46 m/d. The large-diameter boreholes can help keep the methane concentration in the upper corner at a certain level and be used for fire prevention, which means it serves "multi-purpose". Domestic scholars have also studied the movement law of methane in the goaf under the condition of using the traditional pipe-burying method. Yang Qianyi et al. [24] used COMSOL to simulate the variation of methane concentration in the upper corner under different pipe burying drainage parameters. Liu Yanqing et al. [25] used Fluent to simulate the effect of the air supply volume of the working face on the effective depth and area of air leakage in the goaf and found the optimal number of buried pipes in closed crosscuts in the goaf.

Scholars also looked into how methane moves when large-diameter boreholes are drilled compared to the pipe-burying method. However, few studies tried to understand the movement law of methane drained by drilling large-diameter boreholes. For this reason, in this paper, a simulation on a coal mine was made on Fluent to find how the spacing between large-diameter boreholes and the distance between the first borehole and the upper corner affect the movement of methane in the goaf [26]. The results were verified through field test, and the optimal borehole spacing was obtained, which was a reference to real practices.

2. Overview of the Working Face

The working face, which adopted the "U"-shaped ventilation, is located in the No. 1 mining area of 8 # coal seam. The strike length is 1100 m, the incline length is 180 m, and the mean coal thickness is 1.08 m. The coal is mined using one-pass comprehensive mechanized coal mining and roof caving method. In order to prevent the methane from accumulation in the goaf and the upper corner, a drainage roadway is built at 20 m from the air return crossheading, and the large diameter boreholes are drilled in the coal pillar between the return airway and the drainage roadway. The layout of the working face is shown in Figure 1.

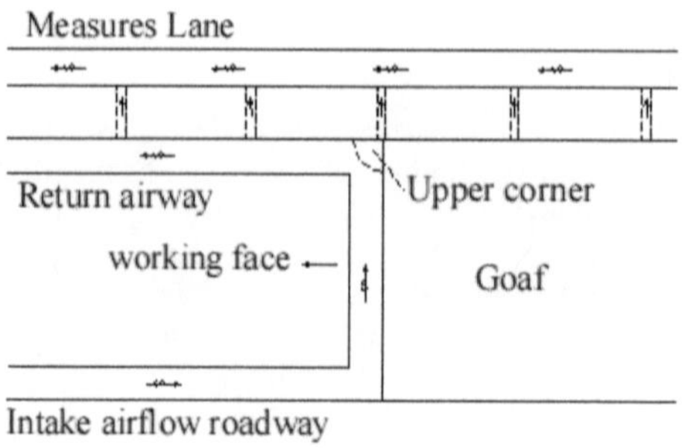

Figure 1. Layout of the working face.

3. Numerical Computing Model

3.1. Numerical Computing Model for Methane Flow

Since the working face and the goaf have complex distribution, it is difficult to model according to the actual size [27–29]. Therefore, the impact of the equipment on the working face was not considered in the modeling, and the model size was simplified as follows: the working face was 180 m in width, the mining spacing was 7 m along the strike direction, the intake airflow roadway was 5 m in width and 3 m in height, the intake airflow roadway was 20 m along the strike direction, and the air volume of the intake airflow roadway was 1800 m^3/min. The goaf was divided into three areas based on the compaction degree of the goaf strata: the natural accumulation area, the stress impact area, and the compaction area [18]. In the simulation, the range of the natural accumulation area, the stress impact area, and the compaction area was 0–20 m, 20–120 m, and 120–270 m, respectively. ICEM CFD was used for the modeling. The model was divided using an unstructured grid (Figure 2) and simulated on Fluent.

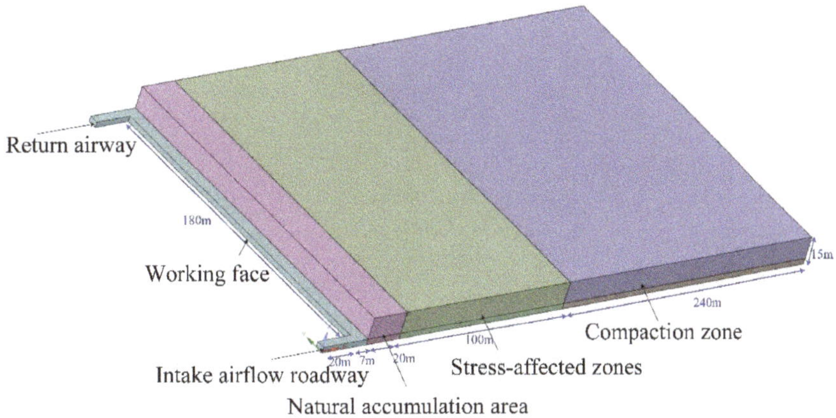

Figure 2. Numerical computing model for methane flow.

To simplify the calculation, the leakage of methane in three areas was regarded as continuous and uniform. Given that each area in the goaf has different compaction degree, the methane leakage source of each area can be determined based on the porosity:

$$Q_i = Q_s n_i V_i / \sum(n_i V_i) \tag{1}$$

where: Q_s is the methane mass source term; n_i is the porosity of each area in the goaf; V_i is the volume of each mass source.

According to the actual parameters of the working face, the parameters of each area in the goaf are shown in Table 1.

Table 1. Parameters of each area.

Area	Working Face	Natural Accumulation Area	Stress Impact Area	Compaction Area
Porosity	-	0.36%	0.24%	0.09%
Source item	2.78×10^{-5} kg/m^3·s	6.66×10^{-75} kg/m^3·s	5.78×10^{-75} kg/m^3·s	4.88×10^{-75} kg/m^3·s

3.2. Simulation Scheme

To analyze the effect of the distribution of large-diameter boreholes and drainage parameters on the methane concentration in the goaf and the upper corner in a quantitative way, and to obtain the optimal borehole spacing and methane drainage parameters, this paper mainly studied the correlation between the three groups of influencing factors and the movement law of methane in the upper corner: (1) With or without large-diameter boreholes; (2) Large-diameter borehole spacing; (3) The distance between the first borehole and the upper corner.

4. Results and Analysis

4.1. With or without Large-Diameter Boreholes

In order to study the effect of large-diameter boreholes on the methane flow in the goaf, this paper simulated the distribution of methane concentration in the goaf and the working face with or without large-diameter boreholes. In the simulation, the large-diameter borehole spacing was 30 m, the first borehole was 15 m down the working face, and the methane drainage volume was 130 m^3/min. The distribution of methane concentration was obtained through simulation (Figure 3).

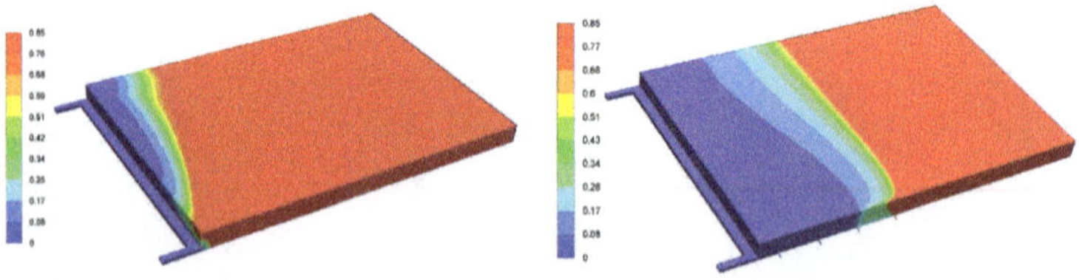

(a) (b)

Figure 3. The distribution of methane concentration. (**a**) Without large-diameter boreholes. (**b**) With large-diameter boreholes.

As can be seen from Figure 3, the large-diameter boreholes have a great effect on the methane concentration in the upper corner. Figure 3a shows that under "U"-shaped ventilation, if no drainage methods are adopted, it can easily cause excessive methane concentration in the upper corner, which will lead to accidents. Figure 3b shows that when the methane drainage flow rate is 130 m^3/min, the methane concentration in both the upper corner and the goaf is significantly reduced, and the methane concentration is reducing in the upper corner, the return airway, and somewhere near the third borehole. The simulation results show that the large-diameter boreholes in the goaf have a good effect on keeping the methane concentration of the working face and the goaf at a certain level and preventing the methane from outburst in the upper corner.

4.2. Large-Diameter Borehole Spacing

In order to study the effect of the large-diameter borehole spacing on the methane drainage effect, four large-diameter boreholes with a spacing L of 10 m, 20 m, 30 m, and 40 m were made, with the first borehole being 15 m down the upper corner and the drainage amount of each borehole being 130 m^3/min. The distribution of methane concentration under different large-diameter borehole spacing is obtained through simulation (Figure 4). The methane concentration along the center line of the return airway is monitored (Figure 5), the monitoring origin of which is the entrance of the return airway.

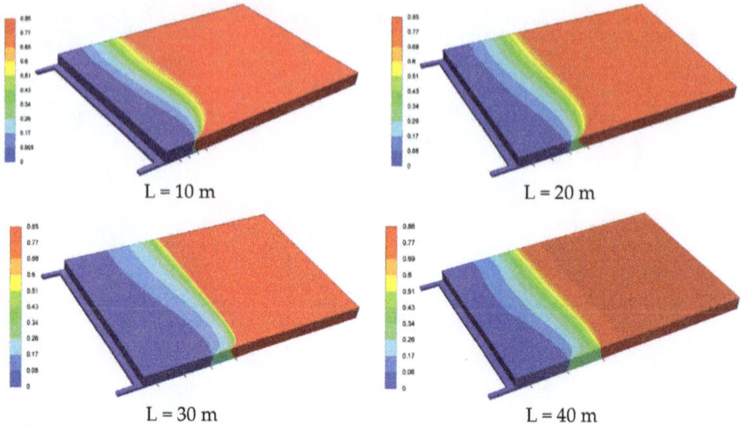

Figure 4. The distribution of methane concentration under different borehole spacing.

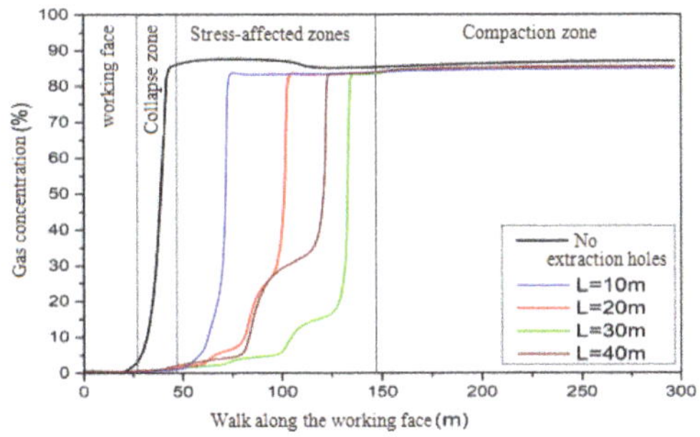

Figure 5. Methane concentration distribution along the center line of the return airway at different borehole spacing.

From Figure 4, we can see that areas with high methane concentration gradually move to the working face as the borehole spacing has decreased. This shows that with the same number of boreholes, increasing the borehole spacing appropriately can improve the methane drainage effect in the goaf. Further observation shows that when the borehole spacing is 10 m, the methane concentration in the goaf rises sharply between the second and the third boreholes; when the borehole spacing is 20 m, the methane concentration in the goaf rise sharply between the third and the fourth boreholes; when the borehole spacing is 30 m, the methane concentration in the goaf rises sharply after crossing the third borehole; when the borehole spacing is 40 m, it is difficult for the air flow of the intake airflow roadway to reach the last borehole located in the compaction area, so the last borehole has little effect of draining methane. This shows that with the same number of boreholes, the smaller the borehole spacing, and the deeper the boreholes. The drainage, therefore, has limited effect, resulting in higher methane concentration. The larger the borehole spacing, the wider the distribution of the boreholes, so the borehole goes deep down the goaf and is far away from the working face, which has little effect on lowering the methane concentration. Therefore, the borehole spacing can be neither too large nor too small. According to the numerical computing results, the optimal borehole spacing under the parameters of this study is 30 m.

Figure 5 shows that when the borehole spacing is 10 m, the methane concentration in the direction of the measuring line rises rapidly at 75 m away from the origin; when the borehole spacing is 20 m, the methane concentration rises rapidly at about 100 m away from the origin; when the spacing is 30 m and 40 m, the methane concentration rises rapidly at about 125 m away from the origin; when the borehole spacing is 30 m, the methane concentration in the goaf is significantly lower than that when the spacing is 10 m, 20 m, and 40 m. This shows that the drainage achieves the best effect at the spacing of 30 m.

4.3. Distance between the Borehole and the Upper Corner

To study how the distance between the first large-diameter borehole and the upper corner affects the drainage effect, four large-diameter boreholes with a spacing of 30 m and a drainage amount of 130 m^3/min were made, and the distance between the first borehole and the upper corner was kept at 5 m, 10 m, 15 m, and 20 m.

As can be seen from Figure 6, when the distance between the first borehole and the upper corner is d = 10 m or d = 20 m, the methane concentration between the third and the fourth boreholes is significantly higher than the condition when the distance is d = 5 m or d = 15 m. As the distance between the borehole and the upper corner increases, the area with low methane concentration is expanding. This shows that increasing the distance between the borehole and the upper corner appropriately helps reduce the methane concentration in the goaf. However, the methane concentration between the third and the fourth boreholes when the distance is d = 20 m is significantly higher than in the other conditions. From the cloud map of the methane concentration in the goaf, we can obtain the optimal distance between the first borehole and the upper corner, which is 15 m.

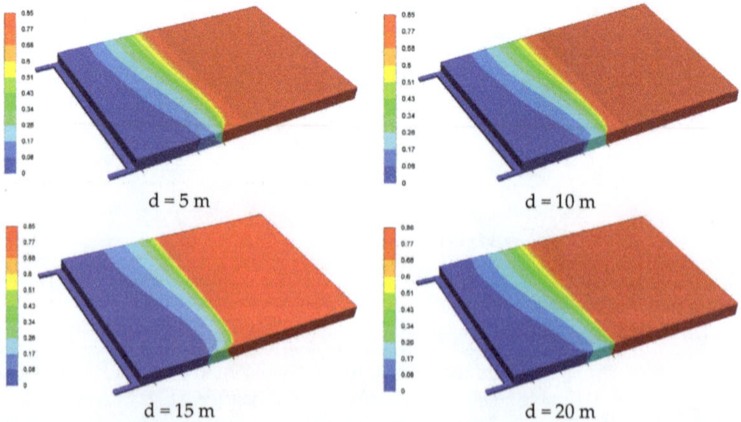

Figure 6. Distribution of methane concentration under different distance between the first borehole and the upper corner.

As can be seen from Figure 7, as the distance between the first borehole and the working face increases, the area with a sharp rise of methane concentration gradually shifts to the compaction area, but the advancing distance does not change significantly. According to the methane concentration on the measuring line and at 75 m away from the origin, the methane drainage effect at different distances between the borehole and the upper corner is almost the same, indicating that changing the distance between the borehole and the upper corner has little influence on the methane concentration in this area. At 100–125 m away from the origin, the drainage effect at d = 15 m and d = 20 m is better than that at d = 5 m and d = 10 m. However, in actual production, the farther the drainage entrance of the goaf is from the working face, the bigger the air leakage radius. Thus, it is easy to result in spontaneous combustion of residual coal in mines with such risks. For this reason, the drainage depth of the goaf with spontaneous combustion risks usually does not exceed 40 m. According to the simulation results, the distance between the borehole and the upper corner has little effect on the regional drainage when the first borehole is 75 m away from the origin (return airway entrance), which equals to 48 m from the upper corner. To ensure safe production, in most mines the draining begins when the distance between the borehole and the upper corner is 5 m, and the drainage achieves the best effect when the distance between the borehole and the upper corner is 15 m.

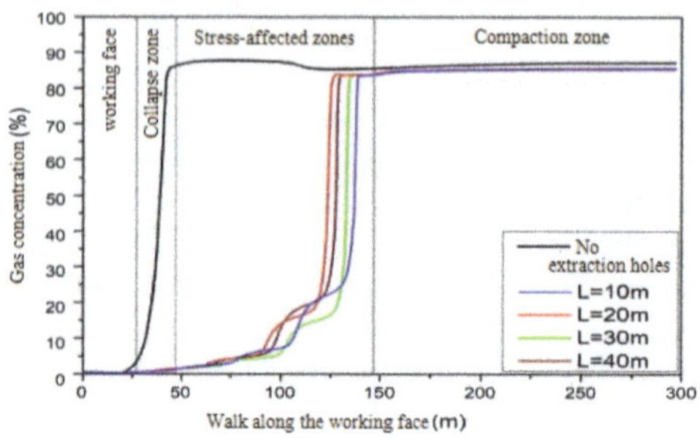

Figure 7. Distribution of methane concentration along the center line of the return airway at different distances between the borehole and the upper corner.

5. Field Test

Scheme Design

As shown in Figure 1, the boreholes with a diameter of 550 mm are drilled in the working face using ZDJ10000L crawler mine drilling machine to reduce the methane concentration in the upper corner. The borehole is 1.5 m deep down the floor, with a dip angle of 5° and borehole spacing of 30 m. In order to prevent the deformation of the borehole, the retaining pipe with a diameter of 426 mm is installed after the boreholes are drilled. The valve is set for each borehole and opened for networking drainage when the borehole is drilled 5 m deep down into the goaf. The blending volume of the single-borehole drainage is about 130 m^3/min. When it is time for backstopping and as the first borehole goes deep down into the goaf, work begins on drilling the second one. When the second borehole is drilled 10 m deep into the goaf, the drilling of the first one stops. Such operation is applied to the rest of the boreholes. When the working face advances 100 m, the drilling of the third borehole suspends. The methane concentration in three boreholes and the upper corner is collected and the variation is shown in Figures 8 and 9.

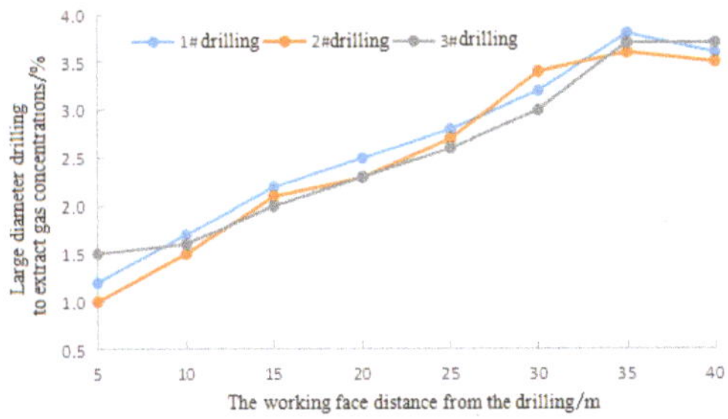

Figure 8. Variation of the methane concentration in three boreholes.

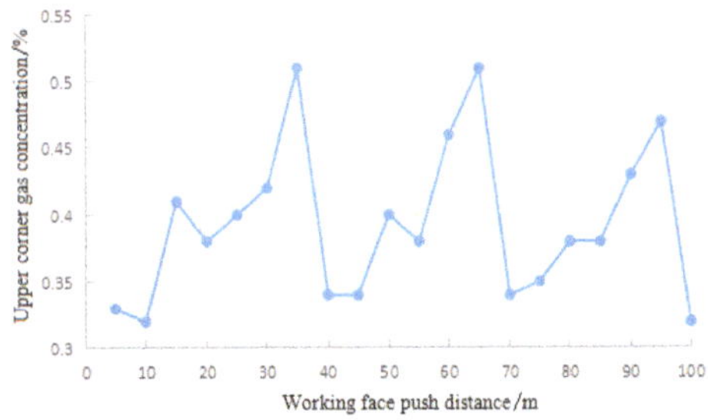

Figure 9. Variation of the methane concentration in the upper corner.

As can be seen from Figure 8, with the advancement of the working face, the borehole gradually goes deep down into the goaf, and the methane concentration in three boreholes increases as the distance between the borehole and the upper corner increases. When the distance between the borehole and the upper corner is about 25 m, the methane concentration in three boreholes increases rapidly. This shows that the borehole has penetrated into the stress impact area, whose medium porosity is smaller than that in the natural accumulation area and is close to the boundary range of the air leakage flow field in the goaf. When the distance between the borehole and the upper corner exceeds 35 m, the methane concentration in three boreholes remains basically stable. Among them, the methane concentration in the first and the second boreholes decreases slightly. This is because the third borehole near the working face begins to work out, resulting in some methane diversion in the goaf.

As can be seen from Figure 9, the methane concentration in the upper corner fluctuates between 0.32% and 0.51%, which is a result of extraction drilling replacement. As the borehole gradually penetrates into the goaf, the ability to control the methane concentration in the upper corner is gradually undermined. When the borehole is drilled about 25 m deep into the goaf, the methane concentration in the upper corner rises rapidly with the maximum reaching 0.51%. When the borehole is drilled 35 m deep into the goaf and the next borehole begins to work out, the methane concentration in the upper corner drops rapidly, and the overall concentration remains within a reasonable range. The influence of negative pressure in large-diameter drilling leads to the change of gas concentration in the upper corner. Drilling large-diameter boreholes in the adjacent drainage roadway can effectively control methane leakage in the upper corner, thus ensuring the safe production of the mine.

6. Conclusions

(1) A reasonable spacing between large-diameter boreholes can effectively change the flow field of the goaf, and the low-pressure area formed at the borehole can manage the methane leakage in the air leakage flow field. Based on the actual geological conditions, physical scale, gas distribution, and other real parameters of the mine, the simulation results show that the optimal spacing of large-diameter boreholes in the goaf is 30 m, and when the boreholes are 15m away from the upper corner, the extraction effect is the best.

(2) The field test shows that when the large-diameter borehole penetrates about 25 m deep into the goaf to reach stress impact area, the methane concentration of the boreholes increases rapidly with the decrease of the porosity (less air leakage), with the maximum being 3.7%. As a result of the drainage superposition effect, when the borehole is drilled 35 m deep into the goaf, the methane concentration of the boreholes slightly decreases.

(3) The methane concentration in the upper corner increases as the distance between the borehole and the upper corner increases. As a result of the drainage superposition effect, the methane concentration in the upper corner changes from 0.32% to 0.51% in a cyclic way.

Funding: This research received no external funding.

Data Availability Statement: The data used to support the findings of this study are available from the corresponding author upon request.

Conflicts of Interest: The authors declare no conflict of interest.

References

1. Si, L.L.; Xi, Y.J.; Wei, J.P.; Li, B.; Wang, H.; Yao, B.; Liu, Y. Dissolution characteristics of gas in mine water and its application on gas pressure measurement of water-intrusion coal seam. *Fuel* **2022**, *313*, 123004. [CrossRef]
2. Özgen Karacan, C.; Martín-Fernández, J.A.; Ruppert, L.F.; Olea, R.A. Insights on the characteristics and sources of gas from an underground coal mine using compositional data analysis. *Int. J. Coal Geol.* **2021**, *241*, 103767. [CrossRef]
3. Liu, P.; Fan, J.Y.; Jiang, D.Y.; Li, J. Evaluation of underground coal gas drainage performance: Mine site measurements and parametric sensitivity analysis. *Process Saf. Environ. Prot.* **2021**, *148*, 711–723. [CrossRef]
4. Chen, X.J.; Zhao, S.; Li, L.Y.; Li, X.; Kang, N. Effect of ambient pressure on gas adsorption characteristics of residual coal in abandoned underground coal mines. *J. Nat. Gas Sci. Eng.* **2021**, *90*, 103900. [CrossRef]
5. Song, R.; Liu, J.J.; Yang, C.; Sun, S. Study on the multiphase heat and mass transfer mechanism in the dissociation of methane hydrate in reconstructed real-shape porous sediments. *Energy* **2022**, *254*, 124421. [CrossRef]
6. Gao, K.; Li, S.N.; Liu, Y.J.; Jia, J.; Wang, X. Effect of flexible obstacles on gas explosion characteristic in underground coal mine. *Process Saf. Environ. Prot.* **2021**, *149*, 362–369. [CrossRef]
7. Yu, D.; Ma, Z.; Wang, R. Efficient smart grid load balancing via fog and cloud computing. *Math. Probl. Eng.* **2022**, *2022*, 3151249. [CrossRef]
8. Zhu, Z.; Wu, Y.; Liang, Z. Mining-Induced stress and ground pressure behavior characteristics in mining a thick coal seam with hard roofs. *Front. Earth Sci.* **2022**, *10*, 843191. [CrossRef]

9. Zhang, L.; Huang, M.; Li, M.; Lu, S.; Yuan, X.; Li, J. Experimental study on evolution of fracture network and permeability characteristics of bituminous coal under repeated mining effect. *Nat. Resour. Res.* **2021**, *31*, 463–486. [CrossRef]
10. Liu, Y.; Zhang, Z.; Liu, X.; Wang, L.; Xia, X. Ore image classification based on small deep learning model: Evaluation and optimization of model depth, model structure and data size. *Miner. Eng.* **2021**, *172*, 107020. [CrossRef]
11. Zhang, G.; Zhang, Z.; Sun, M.; Yu, Y.; Wang, J.; Cai, S. The influence of the temperature on the dynamic behaviors of magnetorheological gel. *Adv. Eng. Mater.* **2022**, 2101680. [CrossRef]
12. Dong, J.; Deng, R.; Quanying, Z.; Cai, J.; Ding, Y.; Li, M. Research on recognition of gas saturation in sandstone reservoir based on capture mode. *Appl. Radiat. Isot.* **2021**, *178*, 109939. [CrossRef]
13. Lei, B.W.; He, B.B.; Zhao, Z.D.; Xu, G.; Wu, B. A method for identifying the fire status through ventilation systems using tracer gas for improved rescue effectiveness in roadway drivage of coal mines. *Process Saf. Environ. Prot.* **2021**, *151*, 151–157. [CrossRef]
14. Wang, K.; Zhang, X.; Wang, L.; Li, L.; Zhang, M.; Zhou, A. Experimental study on propagation law of shock wave and airflow induced by coal and gas outburst in mine ventilation network. *Process Saf. Environ. Prot.* **2021**, *151*, 299–310. [CrossRef]
15. Song, R.; Sun, S.Y.; Liu, J.J.; Feng, X. Numerical modeling on hydrate formation and evaluating the influencing factors of its heterogeneity in core-scale sandy sediment. *J. Nat. Gas Sci. Eng.* **2021**, *90*, 103945. [CrossRef]
16. Yu, D.; Wu, J.; Wang, W.; Gu, B. Optimal performance of hybrid energy system in the presence of electrical and heat storage systems under uncertainties using stochastic p-robust optimization technique. *Sustain. Cities Soc.* **2022**, *83*, 103935. [CrossRef]
17. Liu, Y.; Zhang, Z.; Liu, X.; Wang, L.; Xia, X. Efficient image segmentation based on deep learning for mineral image classification. *Adv. Powder Technol.* **2021**, *32*, 3885–3903. [CrossRef]
18. Zhang, L.; Huang, M.; Xue, J.; Li, M.; Li, J. Repetitive mining stress and pore pressure effects on permeability and pore pressure sensitivity of bituminous coal. *Nat. Resour. Res.* **2021**, *30*, 4457–4476. [CrossRef]
19. Fan, C.; Li, H.; Qin, Q.; He, S.; Zhong, C. Geological conditions and exploration potential of shale gas reservoir in Wufeng and Longmaxi Formation of southeastern Sichuan Basin, China. *J. Pet. Sci. Eng.* **2020**, *191*, 107138. [CrossRef]
20. Hu, S.; Wu, H.; Liang, X.; Xiao, C.; Zhao, Q.; Cao, Y.; Han, X. A preliminary study on the eco-environmental geological issue of in-situ oil shale mining by a physical model. *Chemosphere* **2020**, *287*, 131987. [CrossRef]
21. Gao, H.; Yang, H.W. Super large diameter borehole gob gas extraction technology. *Coal Sci. Technol.* **2019**, *47*, 77–81.
22. Jia, J.Z.; Guo, J. Application of large diameter extraction drilling in goaf gas drainage. *J. Liaoning Technol. Univ. Nat. Sci.* **2020**, *39*, 1–5.
23. Shao, G.A.; Zou, Y.M. Prevention technology of residual coal spontaneous combustion in goaf un-der the condition of large diameter borehole extraction. *Saf. Coal Mines* **2020**, *51*, 74–77.
24. Yang, Q.Y.; Luo, H.G.; Shi, B.M.; Zhang, L.L.; Zhang, H.Z. Numerical study on drainage parameters optimization of buried pipe in goaf based on COMSOL. *J. Saf. Sci. Technol.* **2019**, *15*, 90–95.
25. Shi, G.Q.; Wang, G.Q.; Ding, P.X.; Wang, Y.M. Model and simulation analysis of fire development and gas flowing influenced by fire zone sealing in coal mine. *Process Saf. Environ. Prot.* **2021**, *149*, 631–642. [CrossRef]
26. Zhang, L.; Li, J.; Xue, J.; Zhang, C.; Fang, X. Experimental studies on the changing characteristics of the gas flow capacity on bituminous coal in CO_2-ECBM and N_2-ECBM. *Fuel* **2021**, *291*, 120115. [CrossRef]
27. Shen, X.; Hong, Y.; Zhang, K.; Hao, Z. Refining a distributed linear reservoir routing method to improve performance of the crest model. *J. Hydrol. Eng.* **2017**, *22*, 4016061. [CrossRef]
28. Li, J.B.; Cheng, F.; Lin, G.; Wu, C.L. Improved hybrid method for the generation of ground motions compatible with the multi-damping design spectra. *J. Earthq. Eng.* **2022**, 1–27. [CrossRef]
29. Zhang, G.; Chen, J.; Zhang, Z.; Sun, M.; Yu, Y.; Wang, J.; Cai, S. Analysis of magnetorheological clutch with double cup-shaped gap excited by Halbach array based on finite element method and experiment. *Smart Mater. Struct.* **2022**, *31*, 075008. [CrossRef]

processes

MDPI

Article

Finite-Element Analysis on Energy Dissipation and Sealability of Premium Connections under Dynamic Loads

Yang Yu [1], Yinping Cao [2,*], Zhan Qu [1,3,*], Yihua Dou [2] and Zidi Wang [2]

1 School of Aeronautics, Northwestern Polytechnical University, Xi'an 710072, China; yuyang029xian@163.com
2 Mechanical Engineering College, Xi'an Shiyou University, Xi'an 710065, China; yhdou@vip.sina.com (Y.D.); wangzidi555@163.com (Z.W.)
3 College of Petroleum Engineering, Xi'an Shiyou University, Xi'an 710065, China
* Correspondence: yuyxsy@163.com or caoyinping029@163.com (Y.C.); zhqu@xsyu.edu.cn (Z.Q.)

Abstract: In the process of high flow rate fracture and high gas production, the sealing performance of the premium connection decreases due to the dynamic load and vibration of downhole tubing strings, which may cause accidents. Existing static analysis methods cannot effectively explain this phenomenon. The main objective of this paper is to propose a novel analytical method for evaluating the sealing performance of a premium connection. In this paper, a dynamic model of sealing surfaces of the premium connection is established based on the vibration equation of elastic rod, and the hysteresis characteristics and energy dissipation mechanism of sealing surfaces are analyzed. Considering the influence of spherical radius, internal pressure, axial cyclic load amplitude, and modal vibration, a spherical-conical premium connection finite element model is established to analyze the influence laws of the connection's energy dissipation and sealing performance. The results show that the sealing performance of the premium connection under dynamic load can be effectively analyzed by using energy dissipation theory compared with traditional static contact analysis. Compared with the vibration of the tubing string, the dynamic loads caused by the change of fluid pressure and flow rate in the tubing string have a significant influence on the connection's sealing performance. When the internal pressure and axial cyclic loads are 80 MPa, 400 kN, or 60 MPa and 500 kN respectively, serious plastic deformation occurs in the thread and sealing surfaces, and the energy dissipation of the sealing surfaces increases significantly, which could lead to sealing failure.

Keywords: premium connection; energy dissipation; dynamic load; modal vibration; finite element model

Citation: Yu, Y.; Cao, Y.; Qu, Z.; Dou, Y.; Wang, Z. Finite-Element Analysis on Energy Dissipation and Sealability of Premium Connections under Dynamic Loads. *Processes* **2023**, *11*, 1927. https://doi.org/10.3390/pr11071927

Academic Editors: Tianshou Ma and Yuqiang Xu

Received: 18 May 2023
Revised: 23 June 2023
Accepted: 24 June 2023
Published: 26 June 2023

1. Introduction

The premium connection serves as the connection structure for downhole tubing, connecting tubing to thousands of meters of the tubing string. As shown in Figure 1, the premium connection consists mainly of sealing surfaces, torque shoulders, and threads. The sealing surface is the main sealing structure, the torque shoulder has the function of resisting torque overload and auxiliary sealing, and the threads are used to connect the tubing. Connections are the most dangerous part of the tubing string. Due to complex downhole loads, connections leak frequently and even lead to accidents [1,2]. According to field statistics, connection failure accounts for 85–95% [3,4] of all tubing string failures. Earlier research on the premium connection was mainly based on static load, and the sealing performance was studied by analyzing the equivalent stress, contact pressure, and contact length of the connection. Li et al. [5] used a combination of experimental and theoretical research to study the friction characteristics of bolted connection interfaces. Xu et al. [6–8] used a combination of finite element simulation and theoretical models to calculate the contact pressure and contact area of the sealing surfaces, taking into account the micro profile of the sealing surfaces. The relationship between the average contact pressure, circumferential leakage width, and sealing surfaces was obtained. Nasraoui et al. [9] constructed a simplified numerical model of bolt bonding surfaces and established a contact stiffness matrix. Xu et al. [10] established contact pressure distribution and sealing surfaces length of premium connection under different buckling torques by analytical method and further evaluated the sealing condition of the connection by a comprehensive consideration of various factors. Zhiqian et al. [11] analyzed the sealing performance of the connection by combining the elastic-plastic mechanical contact theory and calculated geometric morphology parameters of the rough surfaces. Chen et al. [12,13] analyzed the effect of preload on the threaded seal and the influence of contact stress at threads, sealing surfaces and shoulders on the sealing performance.

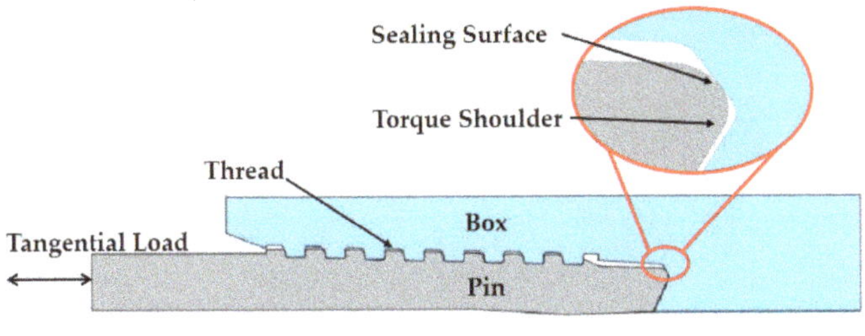

Figure 1. Schematic diagram of the spherical-conical premium connection.

Affected by loads such as the weight of tubing string, internal and external fluid pressure, viscous damping when fluid flows in tubing string and Coulomb friction force, premium connections are subjected to loads such as axial force, internal and external pressure, bending moment, etc. During high flow rate fracturing and high production gas operation, the dynamic load change of tubing caused by the change of fluid pressure and flow rate in tubing causes vibration of tubing [14–16]. Under dynamic load, the sealing surfaces will slip cyclically, resulting in connection loosening or poor sealing performance. However, the traditional static contact analysis method cannot effectively explain this problem. As shown in Figure 1, the sealing surfaces can be considered as two contact surfaces, which are mainly subjected to normal and tangential loads. Normal loads are generated by the interference fit of sealing surfaces and internal and external pressure, which affect contact morphology and pressure of sealing surfaces. Tangential loads are generated by the axial force of the tubing strings and will cause the horizontal slip of the sealing surfaces. Under cyclic loading, the sealing surfaces of premium connections will undergo microslip or gross slip, which will lead to energy dissipation in the structural system [17,18]. As shown in Figure 2, under the action of external incremental load, microslip (stick and slip coexist) occurs on the sealing surfaces, and the slope of the force-displacement curve decreases gradually, resulting in stiffness degradation. As the load increases further, gross slip occurs on the sealing surfaces. At this time, the force–displacement curve approaches the level and the force remains constant, and the gross slip increases gradually [19]. As shown in Figure 3, under cyclic loading, the microslip and gross slip are converted into each other during loading and unloading, resulting in force–displacement hysteresis [17,20,21]. The area of the force–displacement hysteresis curve is the energy dissipation generated during a cycle of loading [22]. Loosening of the connection can be explained by the phenomenon of force–displacement hysteresis of contact surfaces. In the early stage, Iwan [23] built an energy dissipation model of bolted connections based on spring-damped vibration theories. Wang et al. [24] optimized the Iwan model to apply it to the study of energy dissipation in more complex situations. Zhang et al. [25,26] studied the normal stiffness, normal damping, and energy dissipation of the bolt bonding surfaces under different external loads.

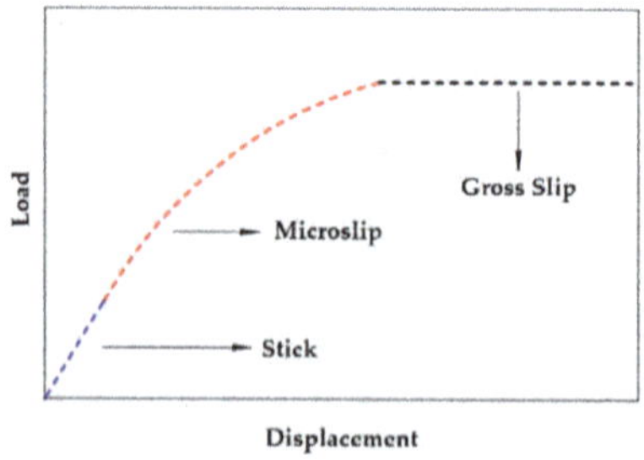

Figure 2. Force–displacement curve of sealing surfaces under single increased load.

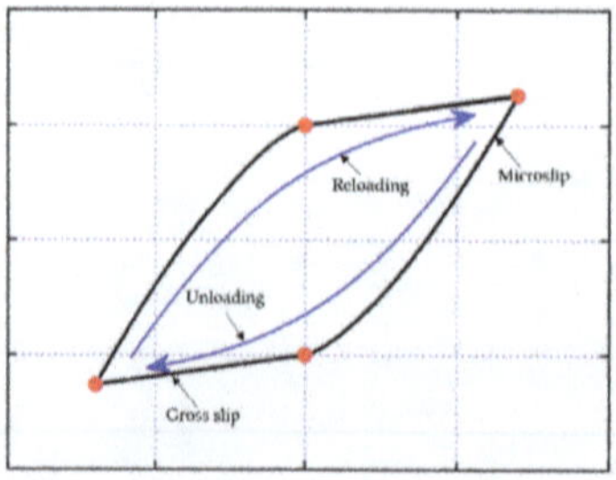

Figure 3. Sealing surfaces force–displacement hysteresis curve under cyclic load.

The premium connection has more complex geometry and load conditions than the bolted joint. At present, there are fewer reports about the sealing property of premium connections under dynamic load. Yang et al. [10] established a contact stress model of sealing surfaces and analyzed the energy dissipation law under different sealing conicities. Yu et al. [27] established a shear layer microslip model of the sealing surfaces at an early stage and analyzed the influence of the shear layer coefficient on the microslip state and energy dissipation of the sealing surfaces. Yu et al. [28]. studied the sealing performance of special threaded joints from a micro level, and based on fractal contact theory, analyzed the effects of fractal dimension and amplitude coefficient on the contact behavior of rough surfaces. The above researches provide some new ideas for the sealing analysis of premium connection, but the results are limited by simplifying models and assuming conditions, such as ignoring the threads and torque shoulders.

In summary, to address the limitations of existing sealing performance evaluation methods, this paper considers the influence of threads and torque shoulders on the contact pressure of the sealing surface of premium connections. On the basis of the traditional evaluation of sealing performance, energy dissipation theory is added to evaluate the sealing performance of premium connections under dynamic loads. We analyze the impact of different parameters on sealing performance under different working conditions based on energy dissipation values. Firstly, based on the vibration equation of the elastic rod, a dynamic model of the connection sealing surfaces is established to analyze the energy dissipation mechanism. Then, based on friction contact analysis and vibration mode analysis, a finite element model (FEM) of spherical-conical premium connection is established by ABQUS numerical simulation software. Finally, the influence of different spherical radii, internal pressure, axial cyclic load amplitudes, and modal shapes on energy dissipation and sealing performance of premium connection under dynamic load is analyzed. The sealing performance evaluation method established based on the energy dissipation theory can provide technical guidance for the design of spherical-conical premium connections and serve as a reference basis for selecting suitable premium connections on oilfield construction sites.

2. Materials and Methods

The energy dissipation of the sealing surfaces is reflected by the force–displacement hysteresis curve. In order to obtain the force–displacement hysteresis curve, a non-linear vibration model of the sealing surfaces must be established first. As shown in Figure 4, the two sealing surfaces of the Pin and the Box are simplified to an elastic rod and a rigid body, respectively, and the contact slip behavior between them is represented by a shear layer without thickness.

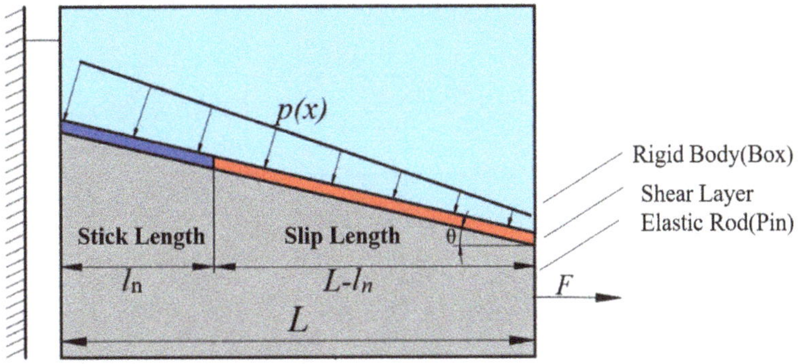

Figure 4. Contact slip model of sealing surfaces.

2.1. Dynamic Model Establishment

Based on the vibration balance equation of the rod, the governing balance equations for the stick and slip regions can be obtained as [18,27]:

Stick regions:

$$EAu''(x) - ku(x)\cos\theta = 0,\ 0 \leqslant x \leqslant l_n \tag{1}$$

Microslip regions:

$$EAu''(x) - \mu p(x)\cos\theta = 0,\ l_{\text{n}} \leqslant x \leqslant L \tag{2}$$

Boundary conditions:

$$EAu'(0) = 0 \quad EA'(L) = F \tag{3}$$

The continuity conditions satisfied at the stick and slip boundary positions, $x = l_n$, is:

$$u(l_n)^- = u(l_n)^+ \quad u\prime(l_n)^- = u'(l_n)^+ \tag{4}$$

In equation:

E—is the Elastic modulus of elastic rod, GPa;
A—is the section area of the elastic rod, mm^2;
θ—is the taper of the sealing surface, °;
k—is the shear layer stiffness in the stick region, MPa;
L—is the horizontal length of the shear layer, mm;
l_n—is the length of the stick zone, mm;
u—is the displacement of a point on the shear layer, mm;
$p(x)$—is the non-uniform normal pressure distribution, N/mm;
F—is the tangential load on the elastic rod, kN;
μ—is the coefficient of friction

The normal pressure decreases linearly in the direction of the elastic rod as follows:

$$p(x) = p_0 - k_p x \tag{5}$$

In formula: P_0—is the maximum normal pressure, N/mm; k_P—is the slope of normal pressure.

From Equations (1) and (2), the boundary condition Equation (3), and the continuity condition (4), the displacements in the stick and slipping regions of the elastic rod are obtained as follows [29]:

$$u(x) = \begin{cases} \frac{\{[\mu k_p(L^2 - l_n{}^2) + 2\mu P_0(l_n - L)]\cos\theta + 2F\}\coth(\eta l_n)}{2\eta EA} & 0 \le x \le l_n \\ a_1 x^3 + a_2 x^2 + a_3 x + a_4 & l_n \le x \le L \end{cases} \tag{6}$$

and:

$$a_1 = \frac{-\mu k_p \cos\theta}{6EA}, a_2 = \frac{\mu P_0 \cos\theta}{6EA}, a_3 = \frac{\mu k_p L^2 \cos\theta - 2\mu P_0 L \cos\theta + 2F}{2EA}$$

$$a_4 = \frac{\{[\mu k_p(L^2 - l_n{}^2) + 2\mu P_0(l_n - L)]\cos\theta + 2F\}\coth(\eta l_n)}{2\eta EA} - \frac{-\mu k_p l_n{}^3 + 3\mu p_0 l_n{}^3 + 6 l_n F - 6\mu p_0 L l_n + 3\mu k_p L^2 l_n}{6EA}$$

$\eta = \sqrt{\frac{k}{EA}}$, it is a parameter related to the stiffness of the interface shear layer.

From Equations (1) and (2), the stress continuity equation at the critical point of stick and slip $x = l_n$ can be obtained [28]:

$$ku(l_n)^- \cos\theta = \mu p(l_n)^+ \cos\theta \tag{7}$$

F that can be obtained by simultaneous Equations (6) and (7):

$$F = \frac{\mu(p_0 - k_p l_n)\tanh(\eta l_n)}{\eta} + \mu p_0(L - l_n)\cos\theta - \frac{\mu k_p(L^2 - l_n{}^2)\cos\theta}{2} \tag{8}$$

In Equation (8), when $l_n = L$, the minimum load at which the microslip occurs can be obtained:

$$F_{\min} = \frac{\mu(p_0 - k_p L)\tanh(\eta L)}{\eta} \tag{9}$$

When $l_n = 0$, the maximum tangential force for gross slip can be obtained:

$$F_{\max} = \mu p_0 L \cos\theta - \frac{\mu k_p L^2 \cos\theta}{2} \tag{10}$$

Based on Masing's hypothesis of steady-state cyclic hysteresis response, the force–displacement curves during unloading and reloading can be obtained [30]. Masing's hypothesis holds that the unloading and reloading of the system's steady-state hysteresis response are geometrically similar to the initial force curve, but only amplified twice.

Elastic rod unloading process [31]:

$$\frac{F_A - F_0}{2} = -k_c\left(\frac{u(L)_0 - u(L)_A}{2} + D_c\right) \tag{11}$$

In the formula, F_0 and F_A are the starting force at unloading and the force during unloading, N; $u(L)_0$ and $u(L)_A$ are the initial displacement at unloading and the displacement during unloading, mm.

Reloading process [31]:

$$\frac{F_A + F_0}{2} = -k_c\left(\frac{u(L)_0 + u(L)_A}{2} + D_c\right) \tag{12}$$

Meanwhile

$$k_c = \frac{\eta EA}{\eta(L-l_n)+\coth(\eta l_n)}$$

$$D_c = \left[2\mu k_p L^3 - 3\mu p_0 L^2 + \frac{3\mu k_p\left(L^2-{l_n}^2\right)-6\mu p_0(L-l_n)}{\eta\tanh(\eta l_n)} + \right.$$

$$\left.\mu k_p {l_n}^3 - 3\mu p_0 {l_n}^2 - 3\mu k_p L^2 l_n + 6u p_0 L l_n\right] / \left[6(L-l_n) + \frac{6}{\eta\tanh(\eta l_n)}\right]$$

From Equations (7) and (8), the force–displacement hysteresis curve of the sealing surfaces under cyclic load can be obtained. The area included in the hysteresis curve is the energy dissipation generated by each cycle of cyclic loading [18].

$$\Delta E_h = 4\int_{l_n}^{L} \mu p(x)[u(x) - u_n(x)]dx \tag{13}$$

where, ΔE_h is the microslip energy dissipation of per cycle, J; $u_n(x)$ is the critical displacement when the stick region is converted into slip region, mm.

2.2. Verification of Dynamic Model

To verify the above model, referring to the sealing surfaces' structure and constraints shown in Figure 4, a FEM of friction contact of the sealing surfaces as shown in Figure 5 is established by ABAQUS. The FEM selects the C3D8I element suitable for the contact analysis. Mesh refinement is applied to particular areas, such as sealing surfaces and shoulder regions, with a mesh size of 0.10. Areas not deemed critical are assigned a relatively rough mesh in order to reduce computational time. Abaqus/Standard Solver is used to analyze finite element models. The interaction attribute is set using the Coulomb friction model to set the contact attribute, the tangential behavior friction formula uses a penalty function, and the normal behavior is set to hard contact [10,32]. Set the Box (rigid body) with higher hardness as the main contact surface and set the Pin (elastic body) as the secondary contact surface. Select a penalty friction formula to simulate a small amount of relative movement during the sticking of the contact surfaces. The left end of the rigid body is set to be fully constrained and a cyclic tangential load is applied at the right end of the elastic body, and a normal pressure is applied at the contact surface. According to the dimensions and material parameters of premium connections and literature [18], the parameters in Table 1 are respectively substituted into the Formulas (11) and (12) and the FEM and the force–displacement hysteresis curves of the two models are calculated (as shown in Figure 6). From the figure, it can be seen that the force–displacement hysteresis curves of both models approximate olive-shaped curves, and the degree of overlap between the two curves is relatively high. Under monotonic loading, the force–displacement curves consist of linear and nonlinear segments. As the load increases, initially, the displacement of both models shows linear increments with nearly the same magnitude. Then, with further increases in load, the displacement increases nonlinearly, with the FEM exhibiting slightly greater increments than the shear layer model, indicating that the FEM dissipates more energy than the shear layer model. The energy dissipation values of the two models are 0.0399 J and 0.0414 J, respectively, with a difference of 3.5%, thus validating the effectiveness of the FEM.

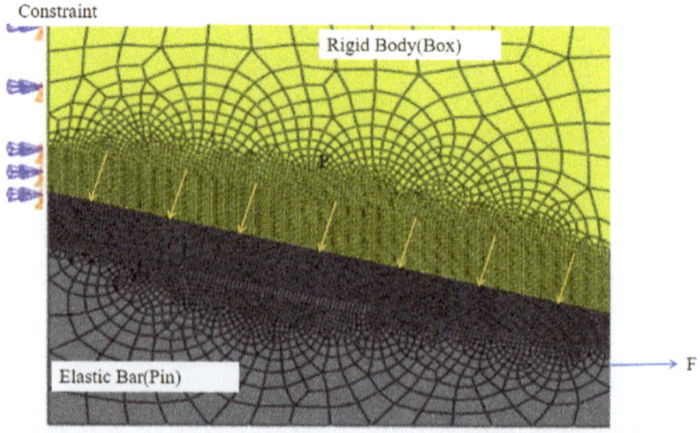

Figure 5. Finite model of friction contact on sealing surfaces.

Table 1. Shear layer model parameters.

Elastic Modulus (GPa)	Cross-Sectional Area of Elastic Rod A (mm^2)	Shear Layer Stiffness k (GPa)	Sealing Surface Cone Angle θ (°)	Contact Length of Sealing Surface (mm)	Friction Coefficient (μ)	Normal Pressure (N/mm)
210	200	10	13.6	2.67	0.1	3.1×10^4

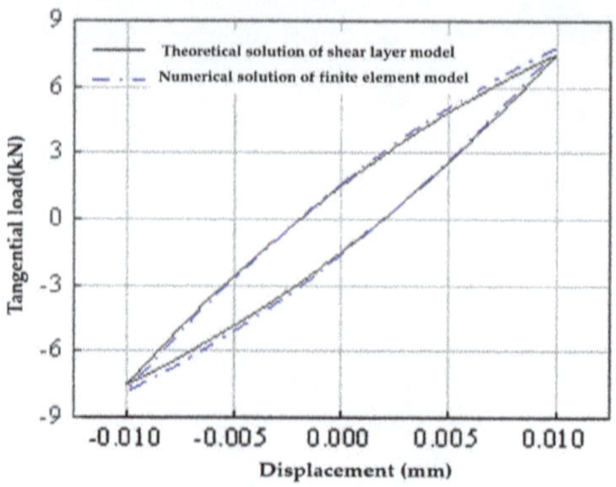

Figure 6. Hysteresis curves of two models.

3. FEM of Premium Connection

From the analysis in the previous section, it is known that the energy dissipation of the sealing surfaces is characterized by a force–displacement hysteresis curve, and the effectiveness of using numerical simulation methods to analyze the energy dissipation of the sealing surfaces is verified. In order to study the influence of sealing structure and downhole loads on energy dissipation and sealing performance of premium connections, a friction contact FEM is established in Section 2—a φ 88.9 × 6.45 P110 full-size spherical-conical premium connection. As shown in Figure 7, the sealing surface of the coupling part is a 1:16 conical surface, the sealing surface of the tubing end is a spherical surface with a radius of 10 mm, the torque shoulder is 25°, the connection thread bearing angle is −3°, the guide surface is 10° and the thread taper is 1:16.

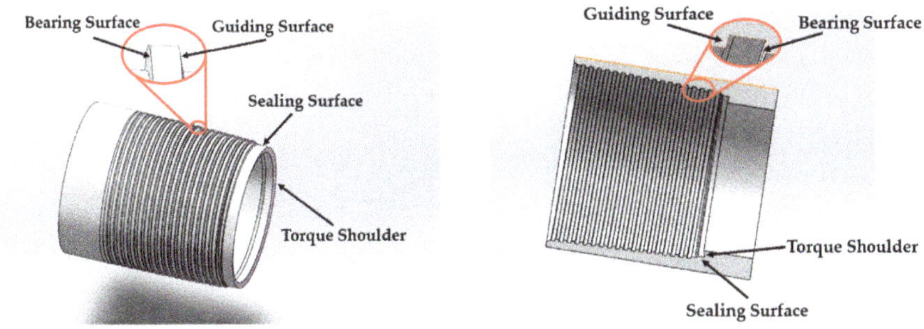

Figure 7. Solid modal of premium connection.

3.1. *FEM for Friction Contact*

Due to the elastic and plastic deformation of the sealing surfaces and thread under the action of make-up and external loads, it is necessary to enable the geometric nonlinearity and material nonlinearity settings in ABAQUS software. The relevant material parameters for premium connections are shown in Table 2.

Table 2. Material parameters of premium connection.

Elastic Modulus	Poisson's Ratio	FrictionCoefficient	Maximum Yield Strength	Strength Limit
210 GPa	0.3	0.1	828 MPa	835 MPa

The C3D8I element is selected for premium connection, and mesh refinement is performed on the threads, sealing surfaces, and torque shoulders, as shown in Figure 8. To accelerate the solution speed of nonlinear analysis and ensure the accuracy of calculation, ABAQUS/Standard solver is selected for solution analysis [33,34].

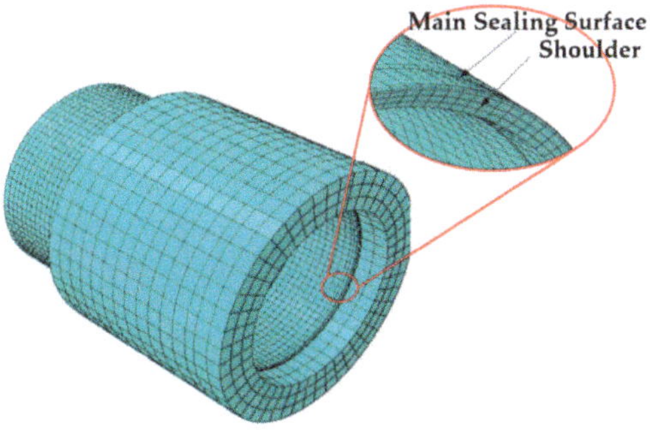

Figure 8. Meshed FEM for premium connection.

After make-up, the premium connection's contact includes the threads, sealing surfaces, and torque shoulder. In nonlinear contact analysis, surface-to-surface contact elements are selected and contact pairs are established according to the method in Section 3. As shown in Figure 9, the coupling's face is set with fixed full constraints, a reference point is arranged at the axis position of the tubing and coupled with the face of the tubing to apply an axial cyclic load to the reference point, and a constant internal pressure is applied to the inner wall of the tubing and the coupling.

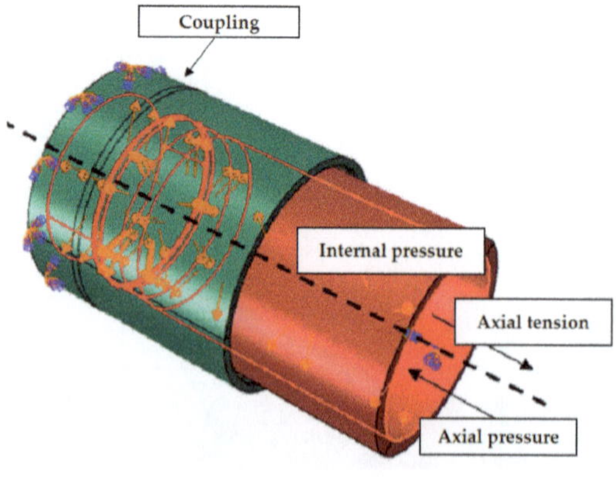

Figure 9. Constraint settings for premium connection.

3.2. *FEM for Vibration*

A FEM of tubing with external threads (each tubing is 9.8 m in length) is established and fully constrained at both ends of the tubing to form a vibration model of the tubing string containing the connection. The material parameters, meshing, and contact settings of the FEM refer to Section 2.1. The finite element meshed model is shown in Figure 10.

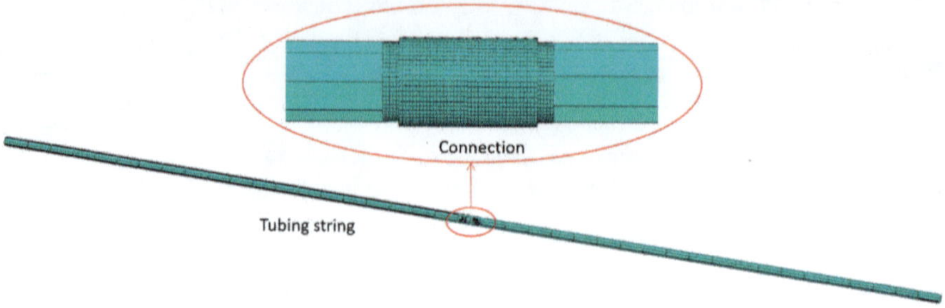

Figure 10. FEM of tubing string with connections.

3.3. *Mesh Independent Verification*

In order to minimize the errors in the calculation results caused by the quality and quantity of the mesh during the analysis process, the paper adopts a mesh-independent verification method. At the sealing surface of the premium connection, this article uses C3D8I elements for mesh division, with mesh sizes of 0.05, 0.1, 0.15, 0.2, and 0.25 mm, respectively [35], the maximum Von Mises stress distribution at the sealing surface of the premium connections under the optimal tightening torque is shown in Figure 11. When the mesh size decreases from 0.25 mm to 0.15 mm, the magnitude of the maximum stress increases monotonically from 0 to 2.12%. When the mesh size is between 0.05 and 0.15 mm, the maximum stress change amplitude at the sealing surface is between 4.0 and 4.4%. The results show that when the mesh size is less than 0.15 mm, the change in mesh size has a small impact on the maximum contact stress value at the sealing surface. As in finite element simulation, the smaller the mesh size, the more accurate the convergence of the obtained results. Taking into account the calculation time and efficiency, this paper adopts a mesh size of 0.10 for analysis.

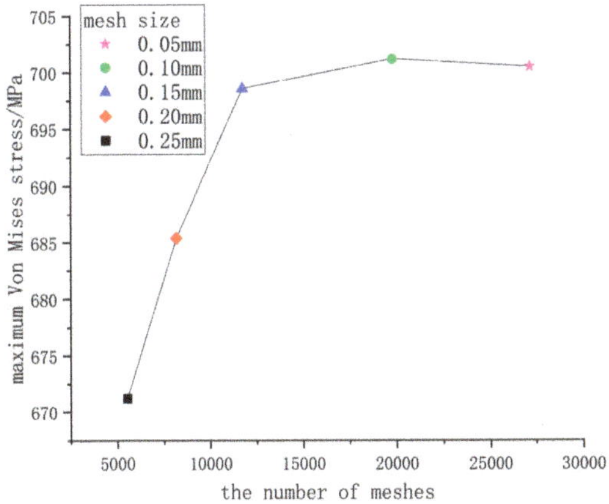

Figure 11. The distribution of maximum von Mises stress under different mesh sizes.

4. Result

Structural parameters and load conditions are the main factors affecting the sealing performance of premium connections. For spherical-conical premium connections, the radius determines the initial contact conditions such as the contact area, length, and contact pressure of the sealing surfaces. In addition, the dynamic load and vibration of the downhole string will change the contact state of the sealing surfaces, threatening the sealing performance of the connection. This section considers the influence of the radius of the sealing surfaces, the pressure inside the tubing, the axial force amplitude on the tubing string (As shown in Table 3), and the modal vibration mode under the axial cyclic load as shown in Figure 12. The finite element method is used to analyze the von Mises stress, contact pressure, and force–displacement hysteresis curve, obtaining the energy dissipation and sealing performance of the premium connection under various influencing factors.

Table 3. Impact parameters.

	Spherical Radius (mm)	Internal Pressure (MPa)	Cyclic Load Amplitude (kN)
I	10, 15, 20	60	400
II	15	60, 70, 80	400
III	15	60	400, 500, 600

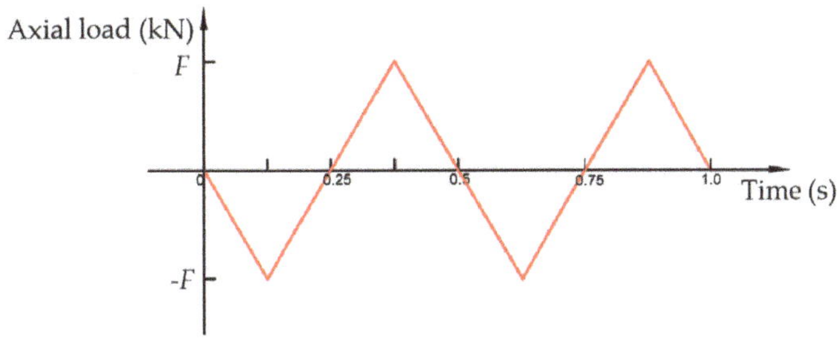

Figure 12. Axial cyclic load.

4.1. Effect of Spherical Radius on Energy Dissipation

Figure 13 shows the von Mises stress nephogram of premium connection with different spherical radii under 60 MPa internal pressure and 400 kN axial cyclic load. As can be seen from Figure 13, under axial compression load, there is significant stress concentration at the thread portion of the connection. When the spherical radius is larger than 15 mm, the maximum von Mises stress of the connection is 834.8 MPa, which is near the strength limit of the material (835 MPa). However, under axial tension load, the overall von Mises stress distribution of the connection is uniform and does not exceed the maximum yield limit of the material (828 MPa). This is because there is a preload at the thread in the axial direction during the make-up of the premium connection. Due to the same axial compression load and preload direction, the von Mises stress in the connection increases, and the stress concentration at the thread is significant.

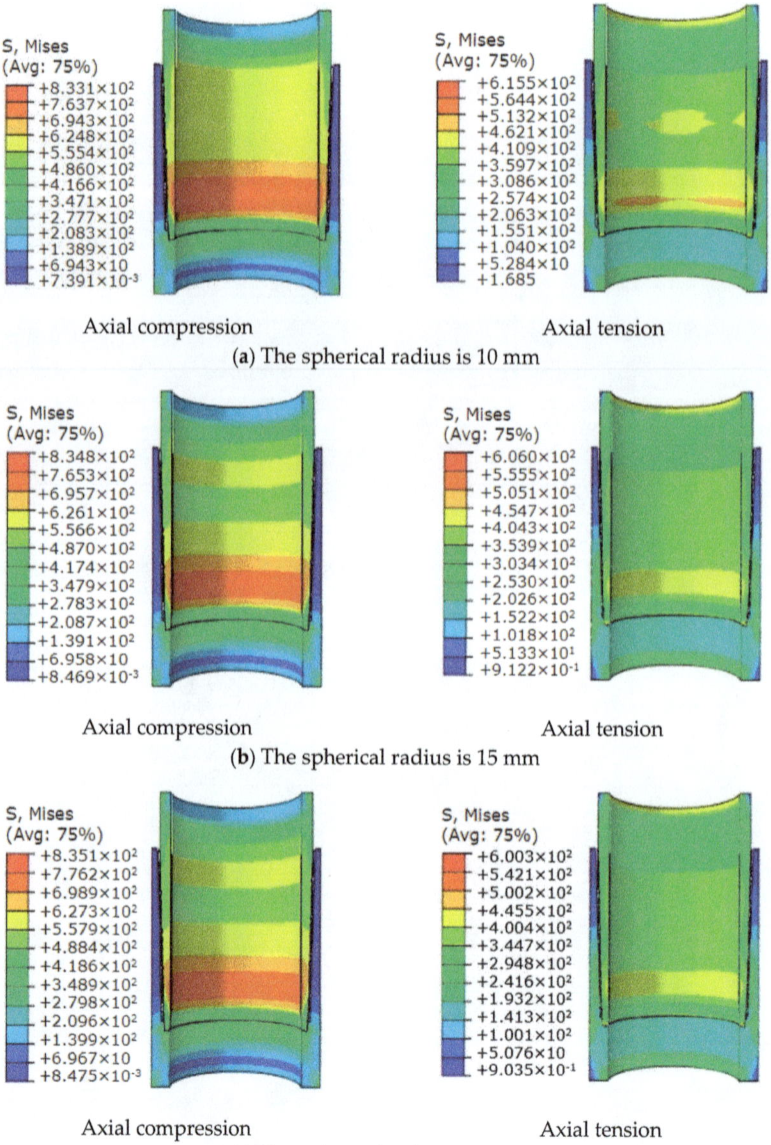

(a) The spherical radius is 10 mm

(b) The spherical radius is 15 mm

(c) The spherical radius is 20 mm

Figure 13. The von Mises stress nephogram for connection with different spherical radii.

Figure 14 shows the contact pressure curves of sealing surfaces with different spherical radii during axial cyclic loading. In the figure, 0–1 s is the internal pressure loading process, and 1 s–2 s is the axial cyclic load loading process. The wave peak is the maximum axial compression load, and the wave trough is the maximum axial tension load. As can be seen from Figure 14, during initial internal pressure loading, the larger the spherical radius, the lower the contact pressure on the sealing surfaces. During the entire axial load cycle, the contact pressure hardly changes with the increase of the spherical radius, and only decreases by 1.8% during axial tension.

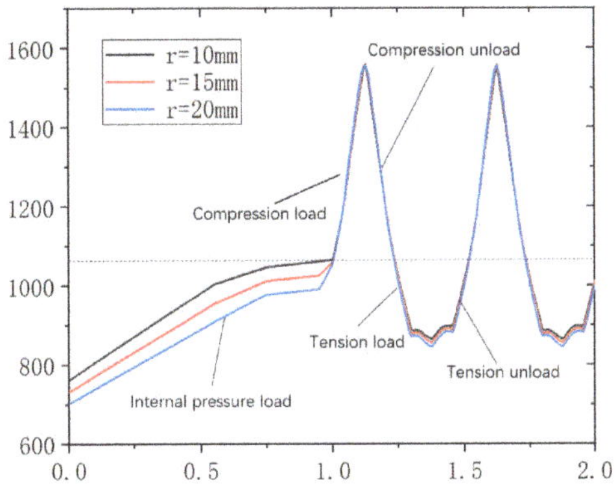

Figure 14. Contact pressure curves at sealing surfaces with different spherical radii.

Extracting the friction force and node displacement at each analysis step on the sealing surfaces within a cycle, the force–displacement hysteresis curves with different spherical radius is obtained as shown in Figure 15, and the gross slip and energy dissipation curves as shown in Figure 16. From Figure 15, it can be seen that the force–displacement hysteresis curve of the spherical-conical seal structure approximates a parallelogram, and its gross slip is much greater than the microslip, indicating that the energy dissipation of the seal structure is mainly caused by gross slip [36].

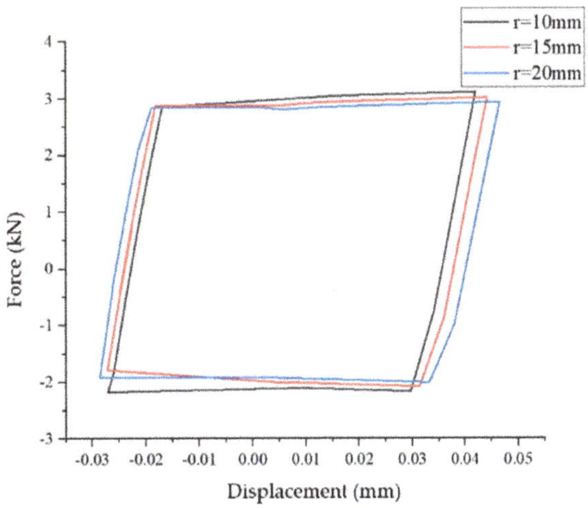

Figure 15. Force–displacement hysteresis curves for different spherical radii.

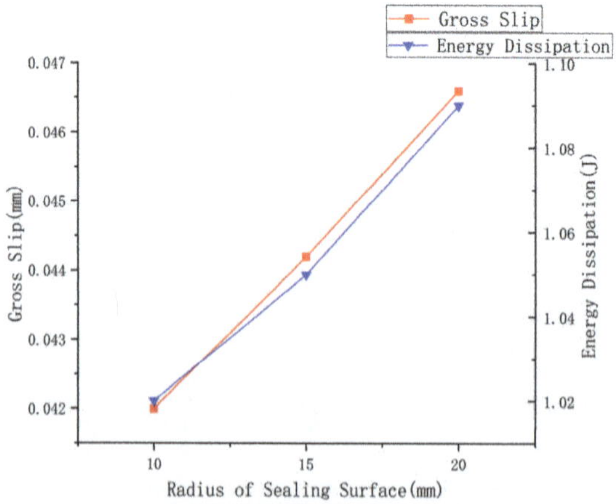

Figure 16. Gross slip and energy dissipation for different spherical radii.

As can be seen from Figure 16, energy dissipation increases as the spherical radius increases. When the spherical radius is increased from 10 mm to 20 mm, the gross slip increases by 5.2% and 5.4%, and the energy dissipation increases by 2.9% and 3.8%, respectively (as shown in Table 4). This is because the larger the spherical radius of the surface, the lower the contact pressure, resulting in a decrease in the friction required for gross slip of the sealing surfaces, and an increase in the amount of gross slip at the connection sealing surfaces. At the same time, since gross slip is the main factor causing energy dissipation, the larger the spherical radius, the greater the energy dissipation value. This will lead to increased wear on the sealing surfaces, which will reduce the sealing performance of the connection.

Table 4. Effect of spherical radius on energy dissipation and gross slip.

Spherical Radius (mm)	Gross Slip		Energy Dissipation	
	Value (mm)	Growth Rate (%)	Value (J)	Growth Rate (%)
10	0.0420		1.02	
15	0.0442	5.2	1.05	2.9
20	0.0466	5.4	1.09	3.8

From the above analysis, it can be seen that under the same axial cyclic load, it is difficult to analyze the impact of the spherical radius on the sealing performance of the connection by the contact pressure of the sealing surfaces; however, the energy dissipation theory can effectively solve this problem.

4.2. *Effect of Internal Pressure on Energy Dissipation*

Figure 17 shows the interaction of 400 kN axial cyclic load with 60 MPa, 70 MPa, and 80 MPa internal pressure. As can be seen from Figure 17, with the increase of internal pressure, the von Mises stress in the connection increases. When the internal pressure is 80 MPa, the maximum von Mises stress of the connection under axial compression load is 844.4 MPa, which exceeds the strength limit of the material. The stress concentration of the thread is a potential danger point.

Figure 18 shows the contact pressure curve of sealing surfaces under the combined action of axial cyclic load and different internal pressures. Table 5 shows the maximum contact pressure of sealing surfaces under different internal pressures. When the internal pressure increases from 60 MPa to 80 MPa, the contact pressure decreases by 0.5% and 4.6%. During the axial compression stage, the contact pressure decreased by 1.5% and 12% respectively (as shown in Table 5). Existing research has shown that the greater the internal pressure [37], the greater the contact pressure on the sealing surfaces. This is because previous studies were mostly based on statics analysis, ignoring the impact of dynamic loads [38]. From Figure 16, it can be seen that serious plastic deformation has occurred between the thread and the sealing surfaces when the internal pressure is 80 MPa. It can be predicted that under the long-term axial cyclic load, the connection will become loose due to the gradual accumulation of plastic deformation [39], which will lead to a decrease in the contact pressure of the sealing surfaces.

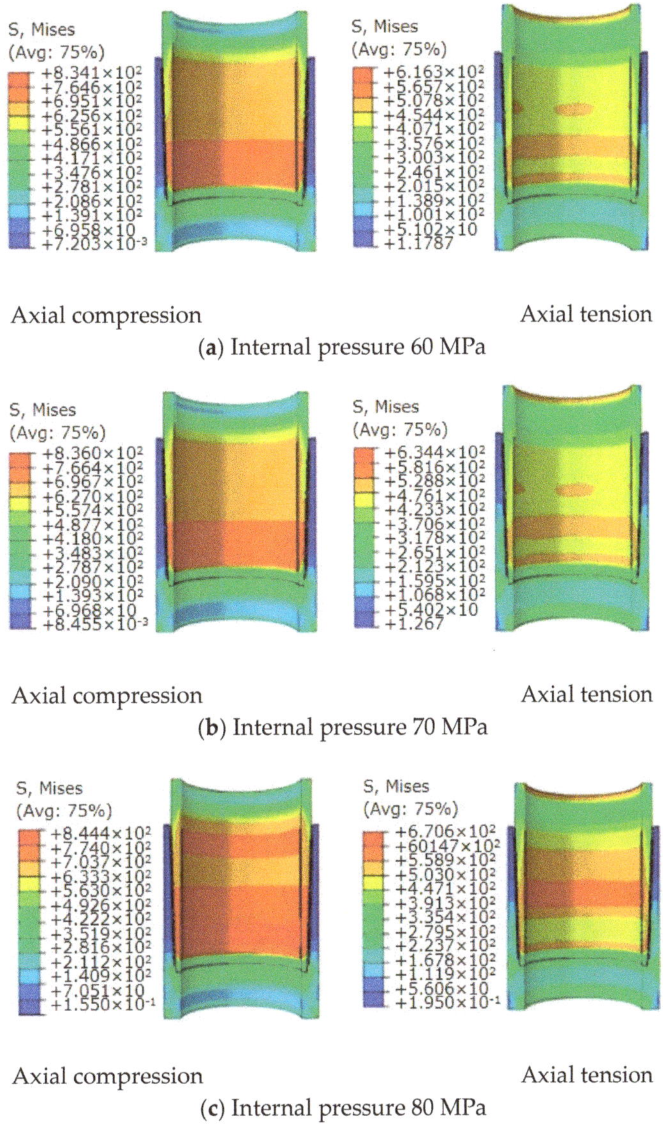

Figure 17. The von Mises stress nephogram of connection under different internal pressures.

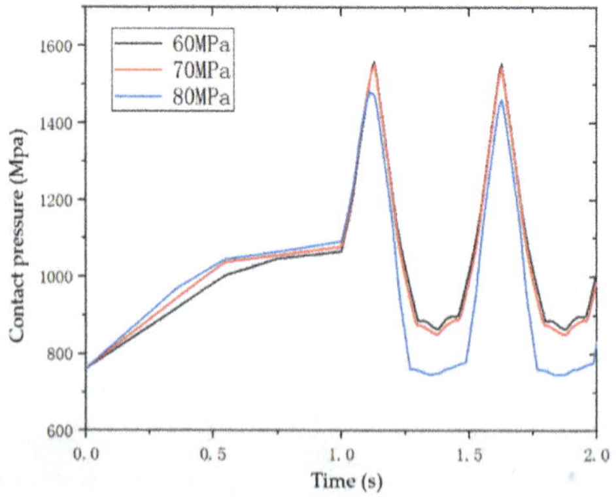

Figure 18. Contact pressure curve for sealing surfaces under different internal pressures.

Table 5. Maximum contact pressure values of sealing surfaces under different internal pressures.

Internal Pressure (MPa)	Maximum Contact Pressure during Axial Compression Stage		Maximum Contact Pressure during Axial Tension Stage	
	Value (MPa)	Growth Rate (%)	Value (MPa)	Growth Rate (%)
60	1559		864.6	
70	1551	−0.5	851.6	−1.5
80	1480.3	−4.6	749.5	−12

Figure 19 is the force–displacement hysteresis curve of the sealing surfaces under the combined action of 400 kN axial cyclic load and different internal pressures. Figure 20 is the corresponding gross slip and energy dissipation curve. It can be seen from Figure 18 that with the increase of internal pressure, the gross slip of sealing surfaces increases, and the energy dissipation increases. The internal pressure increases from 60 MPa to 80 MPa, the gross slip increases by 2.9% and 13.9%, and the energy dissipation increases by 7.8% and 18.2%, respectively (as shown in Table 6). Therefore, when 400 kN axial cyclic load is combined with 80 MPa internal pressure, the energy dissipation of the connection increases significantly, which easily leads to connection loosening and the risk of seal failure exists. In addition, the contact pressure decreases with the loosening of the connection, which explains the decrease of the contact pressure with the increase of internal pressure in the previous section.

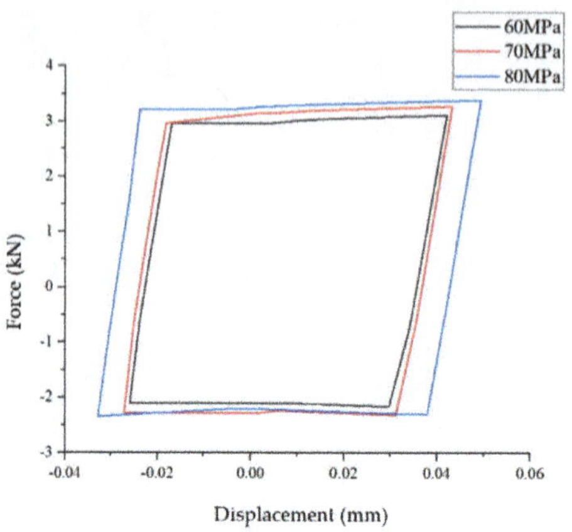

Figure 19. Force–displacement hysteresis curve under different internal pressures.

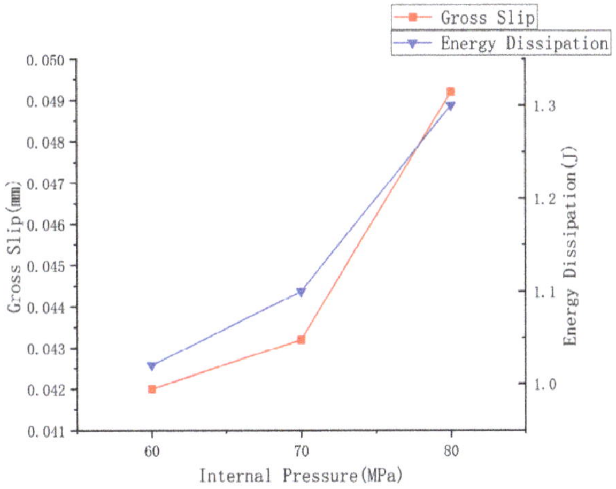

Figure 20. Gross slip and energy dissipation curves under different internal pressures.

Table 6. Effect of internal pressure on gross slip and energy dissipation.

Internal Pressure (MPa)	Gross Slip		Energy Dissipation	
	Value (mm)	Growth Rate (%)	Value (J)	Growth Rate (%)
60	0.0420		1.02	
70	0.0432	2.9	1.10	7.8
80	0.0492	13.9	1.30	18.2

4.3. Effect of Axial Cyclic Load Amplitude on Energy Dissipation

Figure 21 shows the von Mises stress nephogram of the premium connection under 60 MPa internal pressure combined with 400 kN, 500 kN, and 600 kN axial cyclic load, respectively. It can be seen from Figure 18 that the von Mises stress of connection increases with the increase of axial compressive load amplitude. However, the magnitude of the axial tension load has little effect on the von Mises stress of the connection. When the axial cyclic load is 500 kN, the maximum von Mises stresses of the connection under the axial compressive load are 864.1 MPa, which exceeds the strength limit of the material. Therefore, when the axial cyclic load is greater than 500 kN, the thread stress concentration is a potential risk point.

Figure 22 shows the contact pressure curve of the sealing surfaces under the combined action of internal pressure of 60 MPa and different axial cyclic loads. Table 6 shows the maximum contact pressure of the sealing surfaces under different axial cyclic loads. It can be seen from Figure 22 that the contact pressure on the sealing surfaces decreases with the increase of the axial tension load amplitude. Compression load has little effect on the contact pressure of sealing surfaces. Table 6 shows that the amplitude of the axial tension load increases from 400 kN to 600 kN and the contact pressure decreases by 6% and 6.5%, respectively (As shown in Table 7). This is because, under the action of axial tension, the threaded part of the connection deforms, which causes the displacement of the sealing surfaces and the reduction of interference between the spherical and conical surfaces, finally resulting in the reduction of contact pressure.

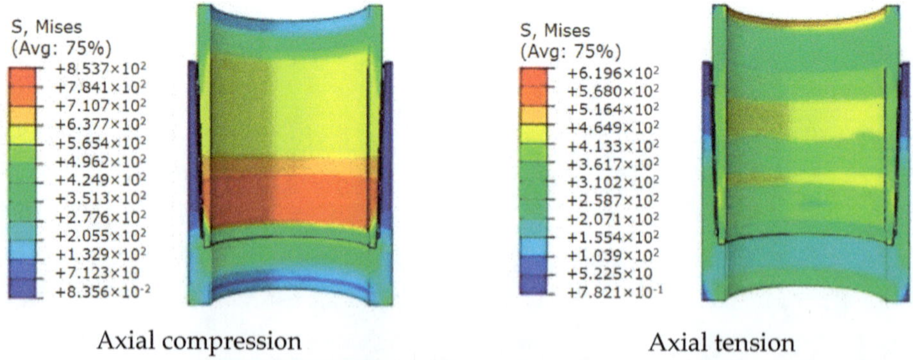

(**a**) Axial cyclic load 400 kN

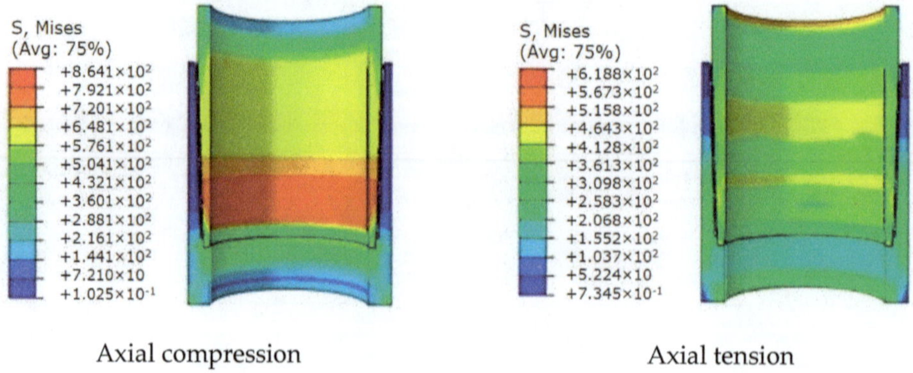

(**b**) Axial cyclic load 500 kN

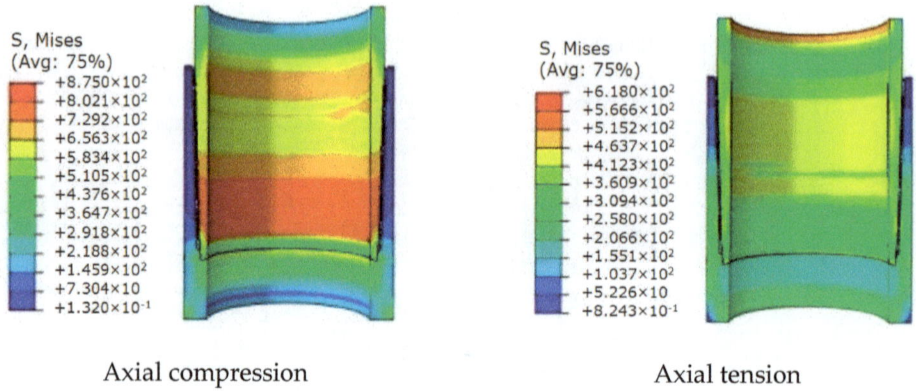

(**c**) Axial cyclic load 600 kN

Figure 21. The von Mises stress nephogram of connection under different cyclic loads.

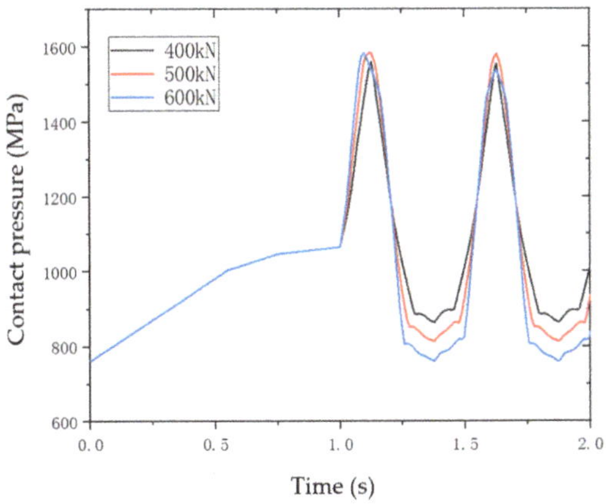

Figure 22. Contact pressure curve at the sealing surfaces under different axial cyclic loads.

Table 7. Maximum contact pressure of sealing surfaces under different axial cyclic loads.

Axial Cyclic Load (kN)	Maximum Contact Pressure during Axial Compression Stage		Maximum Contact Pressure during Axial Tension Stage	
	Value (MPa)	Growth Rate (%)	Value (MPa)	Growth Rate (%)
400	1559.1		866.5	
500	1583.5	−1.5	814.7	−6
600	1584.3	−0.06	761.4	−6.5

Figure 23 shows the force–displacement hysteresis curve of the sealing surfaces under the combined action of internal pressure of 60 MPa and different axial cyclic loads, and Figure 24 shows the corresponding gross slip and energy dissipation curves. It can be seen from Figure 23 that with the increase of the amplitude of the axial cyclic load, the gross slip of the sealing surfaces increases, and the energy dissipation value increases. The axial tension load increases from 400 kN to 600 kN, the gross slip increases by 39.5% and 27.6%, and the energy dissipation increases by 50% and 25%, respectively (as shown in Table 8). Therefore, when the axial cyclic load is 500 kN, the energy dissipation of the connection increases significantly, which is easy to cause connection sealing failure.

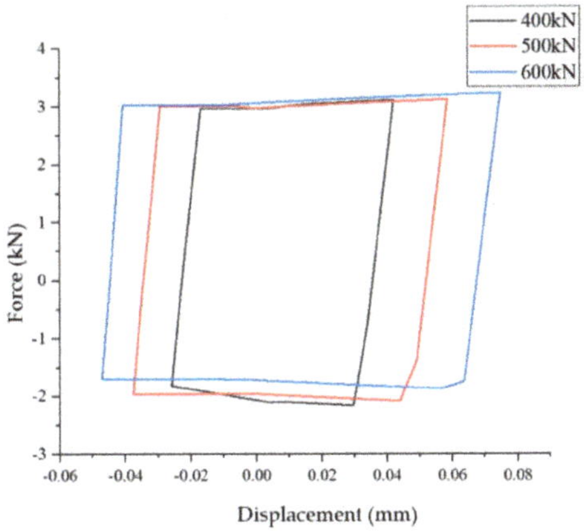

Figure 23. Force–displacement hysteresis curves under different axial cyclic loads.

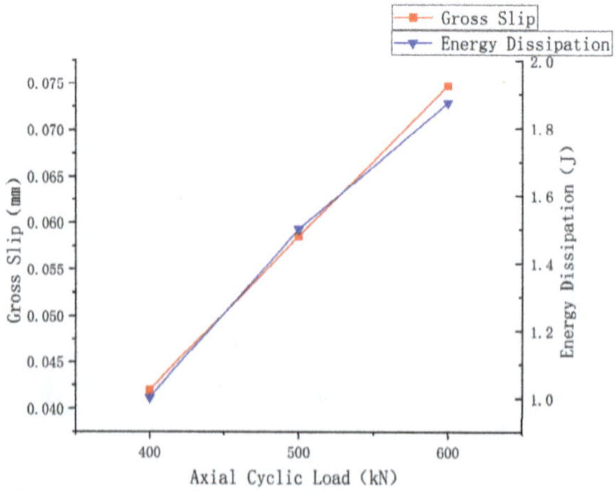

Figure 24. Gross slip and energy dissipation under different axial cyclic loads.

Table 8. Effect of axial cyclic load on gross slip and energy dissipation.

Axial Cyclic Load (kN)	Gross Slip		Energy Dissipation	
	Value (mm)	Growth Rate (%)	Value (J)	Growth Rate (%)
400	0.0420		1.002	
500	0.0586	39.5	1.501	50
600	0.0748	27.6	1.875	25

Literature [34] assessed the load of sealing failure by using contact pressure. The result shows that there is a risk of sealing failure (up to 95% of the sealing limit) when the load is 60 MPa internal pressure combined with a 600 kN tension load. Compared with the results in this section, the axial load causing sealing failure is larger. This is because the influence of gross slip of sealing surfaces is not considered in the literature [40].

4.4. The Influence of Modal Shapes on Energy Dissipation

First, the modal analysis of the tubing string with premium connection is carried out, and then the modal analysis results are applied as displacement fields to the FEM of the premium connection to obtain the contact pressure and force–displacement hysteresis curves of the sealing surfaces in two vibration cycles. Figure 25 shows the first-order and second-order modes of the premium connection. The modal data of the premium connection are extracted and loaded on the connection in the form of cyclic displacement by editing the ABAQUS input file. Cyclic displacement has two cycles and is divided into 100 incremental steps.

Figure 26 shows the contact pressure nephogram at the sealing surfaces of the connection during two displacement cycles, in which every 0.5 s is a cycle, between 0 s and 0.25 s is the axial compression, and between 0.25 s and 0.5 s is the axial tension. As can be seen from Figure 26, the maximum contact pressure under the first-order mode decreased by 8.7 MPa. The maximum contact pressure under the second-order mode decreased by 8.4 MPa. Due to the fact that the second-order modal shape is smaller than the first-order modal shape, the contact pressure at the sealing surfaces decreases with the increase of the vibration mode order.

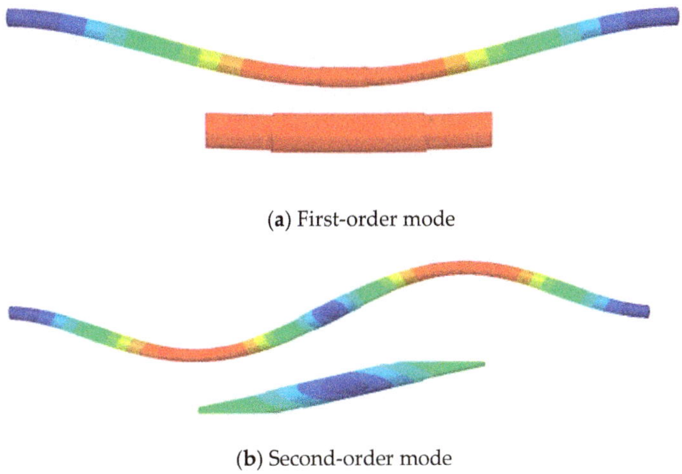

(a) First-order mode

(b) Second-order mode

Figure 25. Mode of premium connection.

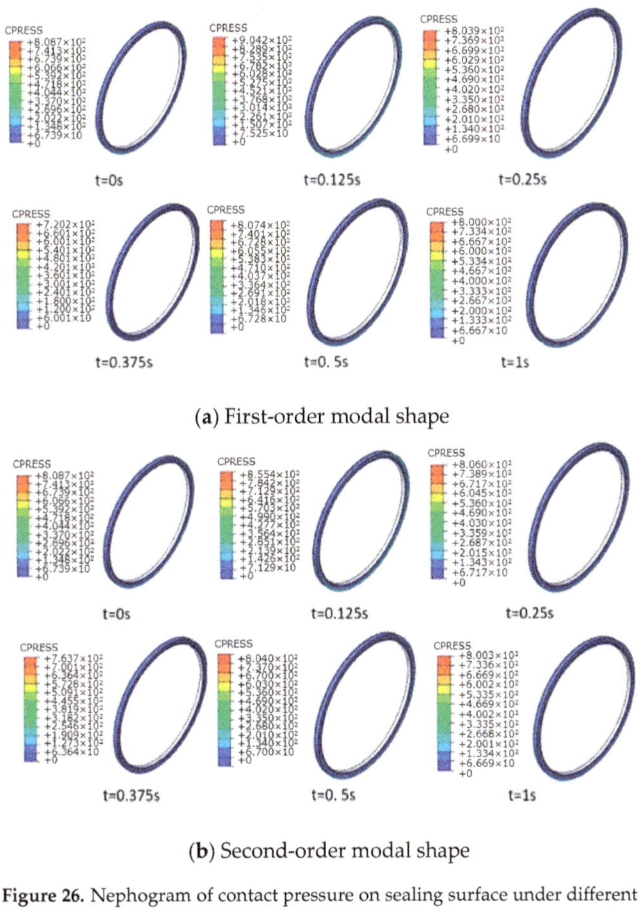

(a) First-order modal shape

(b) Second-order modal shape

Figure 26. Nephogram of contact pressure on sealing surface under different modal shapes.

Figure 27 shows the average contact pressure curve of the sealing surfaces. It can be seen from the figure that the average contact pressure gradually decreases with the increase of the cycle, and the variation amplitude of the average contact pressure under the first-order mode is greater than that under the second-order mode. This phenomenon is also caused by the fact that the second-order modal shape is smaller than the first-order modal shape.

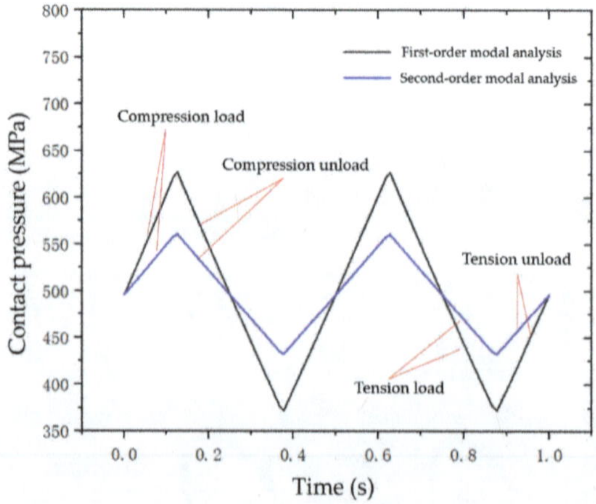

Figure 27. Average contact pressure of sealing surface under different modal shapes.

Extracting the friction force and node displacement of each analysis step within a cycle to obtain the force–displacement hysteresis curve shown in Figure 28. As can be seen from Figure 29, under the vibration modal shape, the force-displacement curve at the sealing surfaces of the spherical-conical premium connection presents a needle-like shape, and there is almost no gross slip of the sealing surfaces. According to Mindlin's [30] theory, the contact edge of the sealing surfaces undergoes microslip and the center of the sealing surfaces is still in the stick state. Due to the fact that the amplitude of the contact pressure change on the sealing surfaces under the first-order mode is greater than that under the second-order mode (Figure 27), and according to Formula (13), it is known that the normal pressure $p(x)$ is positively correlated with the energy dissipation, resulting in the energy dissipation under the first-order mode being greater than that under the second-order mode.

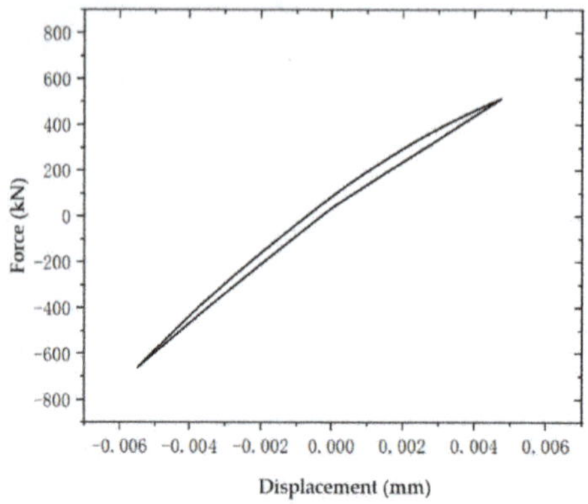

Figure 28. Force–displacement hysteresis curve under first-order modal shape.

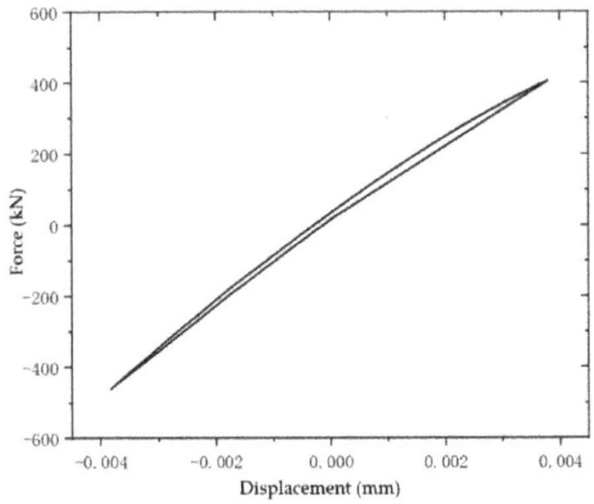

Figure 29. Force–displacement hysteresis curve under second-order modal shape.

Compared with the force–displacement curve of the connection sealing surfaces under axial cyclic load (Figure 15, Figure 19, and Figure 23), only stick-slip occurs on the sealing surfaces under the first-order and the second-order modes and the energy dissipation value is minimal (the difference between these is 10^2 orders of magnitude). This indicates that the dynamic load caused by changes in fluid pressure and flow rate in the tubing has a greater impact on the sealing performance of the connection than the impact of string vibration.

5. Discussion

Through the aforementioned research, it can be found that the method of analyzing the contact pressure of the sealing surface of premium connections under dynamic loading does not effectively explain the decline of connection sealing performance. Firstly, the change in spherical radius under dynamic loading has an insignificant impact on the contact pressure of the sealing surface. Secondly, although internal pressure and contact pressure are positively correlated, the conclusion drawn from traditional analysis methods is that higher internal pressure leads to better sealing performance, contrary to practical situations. It is obvious that all materials have limited strength, exceeding which failure occurs and leads to seal degradation under excessive loading. The method of analyzing the sealing performance of connections using energy dissipation in this article can explain this issue. For spherical-conical premium connections, the gross slip of the sealing surfaces under dynamic load and the microslip of the sealing surfaces under modal vibration mode are analyzed. Both types of slip mentioned above will generate energy dissipation, leading to a decrease in sealing performance.

However, there are still numerous unresolved issues regarding the energy dissipation and sealing performance of premium connections, in particular, the lack of experimental verification. The sealing surfaces of premium connections are typically located in a completely enclosed structure, making it difficult to directly measure the contact pressure and sliding length of the sealing surfaces. This makes it difficult to obtain the force–displacement hysteresis curve of the sealing surfaces. When the tubing string is working downhole, there is liquid or gas in the tubing string, and there is repeated momentum and energy exchange at the interface between the fluid and the tubing string, resulting in fluid–structure interaction excitation, which increases the difficulty of the modal analysis of connections. In addition, the relative slip between the sealing surfaces not only generates energy dissipation but also causes wear, further impacting the sealing performance of the connections. In future studies, the accuracy of FEM analysis will be further validated through force–displacement experiments on connections.

The main purpose of this paper is to propose a novel analytical method for evaluating the sealing performance of a premium connection. This method takes into account the dynamic loads and modal shapes and utilizes finite element analysis to obtain the force-displacement hysteresis curve of the sealing surface. Through analyzing the slip status and energy dissipation of the connection, the impact of dynamic loads on the sealing performance is revealed. The accuracy of the FEM is preliminarily validated by the vibration equation of the rod. The research results can guide the design, manufacturing, inspection, quality control, and production parameter optimization of the premium connections to ensure their sealing performance. In addition, the results can be applied to the production of oil with a high wax content and production processes in difficult conditions [41]. In future research, finite element simulation will be combined with physical experiments to conduct a more in-depth analysis.

6. Conclusions

Based on the dynamic model of sealing surfaces, a spherical-conical premium connection FEM is established to analyze the influence laws of connection energy dissipation and sealing performance under different spherical radii, internal pressure, axial cyclic load amplitude, and mode vibration. The results show that the energy dissipation theory can effectively explain the sealing failure of premium connections under dynamic load. The main conclusions are as follows:

(1) The force–displacement curve of the spherical-conical premium connection under dynamic load is similar to a parallelogram, and its energy dissipation is mainly caused by gross slip. The force–displacement curve under the modal vibration mode assumes a needle-like shape, and its energy dissipation is mainly caused by microslip. Compared with dynamic load, the energy dissipation value of the sealing surfaces under the first-order, the second-order modes is minimal. Therefore, the dynamic load impact on the sealing performance of the connection is greater than string vibration.

(2) During the cyclic load, the spherical radius is positively correlated with energy dissipation but has little impact on the contact pressure of the sealing surfaces. Under the same axial cyclic load, it is impossible to analyze the influence of spherical radius on the sealing performance of connections by contact pressure, but the energy dissipation theory can effectively solve this problem.

(3) Under different cyclic loads, the change amplitude of contact pressure is much smaller than that of energy dissipation. When analyzing the influence of cyclic load amplitude on the sealing performance of connections, the results tend to be conservative in terms of the contact pressure of sealing surfaces.

(4) Under the combined action of internal pressure and axial cyclic load, it will cause plastic deformation accumulation on the thread and sealing surfaces. When the internal pressure and axial cyclic loads are 80 MPa, 400 kN, or 60 MPa, 500 kN, respectively, the thread with stress concentration is a potential danger point. At the same time, the energy dissipation significantly increases, leading to a decline in the sealing performance of the connection which has a risk of sealing failure.

Author Contributions: Conceptualization, Y.Y. and Z.Q.; methodology, Y.Y. and Z.Q.; software, Y.Y. and Y.C.; validation, Y.C. and Y.D.; formal analysis, Y.Y.; investigation, Z.W.; resources, Y.Y.; data curation, Z.W.; writing—original draft preparation, Y.Y.; writing—review and editing, Y.Y., Z.Q., Y.C. and Y.D.; visualization, Y.Y.; supervision, Z.Q. and Y.D.; project administration, Y.Y.; funding acquisition, Z.Q. and Y.D. All authors have read and agreed to the published version of the manuscript.

Funding: This research was funded by the National Natural Science Foundation of China (NSFC) (Nos. 52274006 and 51974246), and the Natural Science Basic Research Program of Shaanxi (No. 2023-JC-YB-337).

Institutional Review Board Statement: Not applicable.

Informed Consent Statement: Informed consent was obtained from all subjects involved in the study.

Data Availability Statement: The data presented in this study are available on request from the corresponding author.

Conflicts of Interest: The authors declare no conflict of interest.

Nomenclature

E	Young's modulus, Gpa
A	constant cross-section area of the elastic bar, mm^2
θ	the taper of the sealing surface, °
k	the shear layer stiffness in the stick region, MPa
L	the horizontal length of the shear layer, mm
l_n	the length of the stick zone, mm
u	the displacement of a point on the shear layer, mm
u'	the velocity of a point on the shear layer, mm/s
u''	the acceleration of a point on the shear layer, mm/s^2
$p(x)$	the non-uniform normal pressure distribution, N/mm
F	the tangential load on the elastic rod, kN
μ	the coefficient of friction
P_0	the maximum normal pressure, N/mm
k_p	the slope of normal pressure
F_0	the starting force at unloading, N
F_A	the force during unloading, N
$u(L)_0$	the initial displacement at unloading, mm
$u(L)_A$	the displacement during unloading, mm
ΔE_h	the microslip energy dissipation per cycle, J
$u_n(x)$	the critical displacement when the stick region is converted into slip region, mm

References

1. Matthews, C. Assessing tubular connection leakage integrity. *World Oil* **2002**, *223*, 82–84.
2. Dou, Y.H.; Cao, Y.P.; Zhang, F.X.; Yang, X.T. Analysis of Influence to the Connect and Seal Ability of Tubing Connection of Inner Pressures. In Proceedings of the Advanced Materials Research, Xi'an, China, 7–8 January 2012; pp. 790–793.
3. Bradley, A.; Nagasaku, S.; Verger, E. Premium connection design, testing, and installation for HPHT sour wells. In Proceedings of the SPE High Pressure/High Temperature Sour Well Design Applied Technology Workshop, The Woodlands, TX, USA, 17–19 May 2005.
4. Xie, J. Numerical evaluation of tubular connections for HPHT applications. In Proceedings of the Baosteel Conference, Shanghai, China, 4–6 June 2013.
5. Li, D.; Botto, D.; Li, R.; Xu, C.; Zhang, W. Experimental and theoretical studies on friction contact of bolted joint interfaces. *Int. J. Mech. Sci.* **2022**, *236*, 107773. [CrossRef]
6. Xu, H.; Yang, B.; Zhang, Z.; Shi, T. Special considerations to calculate joint strength of premium connections. *J. Pet. Sci. Eng.* **2019**, *182*, 106295. [CrossRef]
7. Xu, H.; Zhang, Z.; Xiang, S.; Yang, B.; Shi, T. Leakage Model of Tubing and Casing Premium Connection Based on Sinusoidal Contact Simulation between Rough Surfaces. *Processes* **2023**, *11*, 570. [CrossRef]
8. Yang, B.; Xu, H.; Xiang, S.; Zhang, Z.; Su, K.; Yang, Y. Effects of Make-Up Torque on the Sealability of Sphere-Type Premium Connection for Tubing and Casing Strings. *Processes* **2023**, *11*, 256. [CrossRef]
9. Nasraoui, M.T.; Chakhari, J.; Khalfi, B.; Nasri, M. Modeling and analysis of a bolted joint under tension and shear loads. *Trans. Can. Soc. Mech. Eng.* **2019**, *43*, 376–386. [CrossRef]
10. Xu, H.; Yang, B. A Quantitative Model to Calculate Gas Sealing Capacity and Design Sealing Parameters for Premium Connection. *Math. Probl. Eng.* **2020**, *2020*, 9074381. [CrossRef]
11. Zhiqian, X.; Xiangzhen, Y.; Xiujuan, Y.; Xiaokang, Y.; Mingda, W.; Xiaoyun, Z. Aplication of micro-leakage mechanism for evaluating the sealing performance of non-API casing connections. *Acta Pet. Sin.* **2014**, *35*, 963.
12. Chen, W.; Di, Q.; Zhang, H.; Chen, F.; Wang, W. The sealing mechanism of tubing and casing premium threaded connections under complex loads. *J. Pet. Sci. Eng.* **2018**, *171*, 724–730. [CrossRef]
13. Chen, S.-J.; An, Q.; Zhang, Y.; Gao, L.-X.; Li, Q. Loading analysis on the thread teeth in cylindrical pipe thread connection. *J. Press. Vessel Technol.* **2010**, *132*, 031202. [CrossRef]
14. Wang, Y.; Fan, H.; Zhang, L.; Wei, F.; Feng, G.; Niu, M. Lateral Vibration and Stability of High Temperature and High Pressure Gas Well Test String. *Pet. Mach.* **2011**, *39*, 36–38.
15. Liu, E.; Wang, X.; Zhao, W.; Su, Z.; Chen, Q. Analysis and research on pipeline vibration of a natural gas compressor station and vibration reduction measures. *Energy Fuels* **2020**, *35*, 479–492. [CrossRef]
16. Luo, J.; Zhang, K.; Liu, J.; Mu, L.; Wang, F. The Study of Tubing Vibration Mechanism in High Pressure Gas Well. *World J. Eng. Technol.* **2021**, *9*, 128. [CrossRef]
17. Segalman, D.J. A Four-Parameter Iwan Model for Lap-Type Joints. *J. Appl. Mech.* **2005**, *72*, 752–760. [CrossRef]

18. Sun, D.; Liao, R. Investigation Into the Effect of Interface Characteristics on the Energy Dissipation of the Bolted Joints. *J. Comput. Theor. Nanosci.* **2015**, *12*, 2821–2829. [CrossRef]
19. Wang, D.; Zhang, Z. A four-parameter model for nonlinear stiffness of a bolted joint with non-Gaussian surfaces. *Acta Mech.* **2020**, *231*, 1963–1976. [CrossRef]
20. Segalman, D.J. Modelling joint friction in structural dynamics. *Struct. Control Health Monit.* **2006**, *13*, 430–453. [CrossRef]
21. Tan, L.; Wang, C.; Liu, Y.; Sun, W.; Zhang, W. Study on hysteresis and threaded fitting behavior of bolted joint with non-parallel bearing surface. *Mech. Syst. Signal Process.* **2022**, *168*, 108655. [CrossRef]
22. Oldfield, M.; Ouyang, H.; Mottershead, J.E. Simplified models of bolted joints under harmonic loading. *Comput. Struct.* **2005**, *84*, 25–33. [CrossRef]
23. Iwan, W.D. On a Class of Models for the Yielding Behavior of Continuous and Composite Systems. *J. Appl. Mech.* **1967**, *34*, 612–617. [CrossRef]
24. Wang, S.-A.; Zhu, M.; Xie, X.; Li, B.; Liang, T.-X.; Shao, Z.-Q.; Liu, Y.-L. Finite Element Analysis of Elastoplastic Elements in the Iwan Model of Bolted Joints. *Materials* **2022**, *15*, 5817. [CrossRef] [PubMed]
25. Xueliang, Z.; Yumei, H.; Weiping, F.; Xiaoyue, W.; Yongchao, L. Research on the Normal Damping of Machine Joints. In Proceedings of the International Design Engineering Technical Conferences and Computers and Information in Engineering Conference, Atlanta, GA, USA, 13–16 September 1998.
26. Fu, W.; Huang, Y.; Zhang, X.; Guo, Q. Experimental investigation of dynamic normal characteristics of machined joint surfaces. *J. Vib. Acoust.* **2000**, *122*, 393–398. [CrossRef]
27. Yang, Y.; Zhan, Q.; Yihua, D.; Yinping, C. Analysis of Energy Dissipation on the Sealing Surface of Premium Connection Based on a Microslip Shear Layer Model. *Energies* **2022**, *15*, 8400.
28. Yu, Y.; Qu, Z.; Cao, Y.; Dou, Y.; Li, J. Sealability Analyses of Premium Connections Characterized by a Surface Fractal Function. *Appl. Sci.* **2023**, *13*, 6467. [CrossRef]
29. Menq, C.-H.; Griffin, J.; Bielak, J. The influence of microslip on vibratory response, part II: A comparison with experimental results. *J. Sound Vib.* **1986**, *107*, 295–307. [CrossRef]
30. Menq, C.H.; Bielak, J.; Griffin, J.H. The influence of microslip on vibratory response, part I: A new microslip model. *J. Sound Vib.* **1986**, *107*, 279–293. [CrossRef]
31. Zhang, X.; Huang, Y.; Wen, S. Fractal model of contact stiffness of joint surfaces. *Trans. Chin. Soc. Agric. Mach.* **2000**, *31*, 89–91.
32. Izumi, S.; Yokoyama, T.; Iwasaki, A.; Sakai, S. Three-dimensional Finite Element Analysis on Tightening and Loosening Mechanism of Bolted Joint. *Trans. Jpn. Soc. Mech. Eng. Ser. A* **2005**, *71*, 204–212. [CrossRef]
33. Guangjie, Y.; Zhenqiang, Y.; Qinghua, W.; Zhentong, T. Numerical and experimental distribution of temperature and stress fields in API round threaded connection. *Eng. Fail. Anal.* **2006**, *13*, 1275–1284. [CrossRef]
34. Shen, X.P.; Shen, G.Y. Numerical Analysis on the Expanding Process of Expandable Casing and Contact Evaluation of Its Threaded Connection. In Proceedings of the Applied Mechanics and Materials, Hong Kong, China, 17–18 August 2013; pp. 632–638.
35. Nagata, S.; Fujita, S.; Sawa, T. FEM Stress Analysis and Leakage Behavior of Pipe-Socket Threaded Joints Subjected to Bending Moment and Internal Pressure. In Proceedings of the ASME 2020 Pressure Vessels & Piping Conference, Virtual, Online, 3 August 2020.
36. Song, Y.; McFarland, D.M.; Bergman, L.A.; Vakakis, A.F. Effect of pressure distribution on energy dissipation in a mechanical lap joint. *AIAA J.* **2005**, *43*, 420–425. [CrossRef]
37. Murtagian, G.; Fanelli, V.; Villasante, J.; Johnson, D.; Ernst, H. Sealability of stationary metal-to-metal seals. *J. Trib.* **2004**, *126*, 591–596. [CrossRef]
38. Ferjaoui, A.; Yue, T.; Wahab, M.A.; Hojjati-Talemi, R. Prediction of fretting fatigue crack initiation in double lap bolted joint using Continuum Damage Mechanics. *Int. J. Fatigue* **2015**, *73*, 66–76. [CrossRef]
39. Dou, Y.; Li, Y.; Cao, Y.; Yu, Y.; Zhang, J.; Zhang, L. FE simulation of sealing ability for premium connection based on ISO 13679 CAL IV tests. *Int. J. Struct. Integr.* **2020**, *12*, 138–148. [CrossRef]
40. Ilyushin, Y.V.; Novozhilov, I. Temperature field control of a metal oil-well tubing for producing of high-paraffin oil. In Proceedings of the 2020 XXIII International Conference on Soft Computing and Measurements (SCM), St. Petersburg, Russia, 27–29 May 2020; pp. 149–152.
41. Martirosyan, A.V.; Kukharova, T.V.; Fedorov, M.S. Research of the Hydrogeological Objects' Connection Peculiarities. In Proceedings of the 2021 IV International Conference on Control in Technical Systems (CTS), Saint Petersburg, Russia, 21–23 September 2021; pp. 34–38.

processes

Article

A New Experimental Method for Acid Pretreatment in Perforated Horizontal Wells: A Case Study of Mahu Conglomerate Reservoir

Wenting Jia [1], Jianye Mou [1,*], Guifu Wang [2], Xiaowei Li [1], Xinliang Wang [1] and Xinfang Ma [1]

1 College of Petroleum Engineering, China University of Petroleum (Beijing), Beijing 102249, China; 2020310156@student.cup.edu.cn (W.J.); 2021215162@student.cup.edu.cn (X.L.); 2022215162@student.cup.edu.cn (X.W.); maxinfang@cup.edu.cn (X.M.)
2 CNPC Engineering Technology R&D Company Limited, Beijing 102206, China; wanggfdr@cnpc.com.cn
* Correspondence: moujianye@cup.edu.cn

Abstract: The Mahu Oilfield in Xinjiang, China, is the world's largest conglomerate oilfield, with a daily output of more than 8000 tons. In the fracturing of perforated horizontal wells, the breakdown pressure is high, resulting in difficulties in their treatment. Acid pretreatment has been applied to reduce the breakdown pressure in the field, but with poor performance. Few studies have been conducted on how acid pretreatment affects the breakdown pressure under perforation conditions in the Mahu conglomerate reservoir. Also, existing fracturing or acid pretreatment experiments cannot simulate perforation well. Therefore, a new method was developed to make perforations by hydro jetting the fracturing specimens (300 mm × 300 mm × 300 mm) first. It can achieve specified perforation parameters, including the perforation angle, position, and length. Subsequently, true triaxial hydraulic fracturing experiments were conducted on the conglomerate specimens obtained from the Mahu area. The effects of the acid pretreatment on the fracture initiation and breakdown pressure were investigated by injecting the perforation section of the conglomerate using various acid systems. The study results showed that the perforation made by the new apparatus was extremely close to the perforation on-site. The acid pretreatment effectively dissolved minerals and could decrease the breakdown pressure down to 7.7 MPa. A combination of 6%HF + 10%HCl was recommended for the acid pretreatment in the Mahu conglomerate reservoir. A total of 60 min acid–rock contact time is necessary for sufficient rock dissolution. The mechanism of the acid pretreatment to decrease the breakdown pressure is that the rock dissolution by the acid reduces the rock tensile strength and increases the permeability. The raised permeability increases the fluid pressure of the reservoir near the wellbore so as to reduce the breakdown pressure of the formation. The research results provide technical support for reducing construction difficulty and optimizing parameters in the Mahu Oilfield.

Keywords: acid pretreatment; new perforation method; triaxial fracturing; lowering breakdown pressure; conglomerate

Citation: Jia, W.; Mou, J.; Wang, G.; Li, X.; Wang, X.; Ma, X. A New Experimental Method for Acid Pretreatment in Perforated Horizontal Wells: A Case Study of Mahu Conglomerate Reservoir. *Processes* **2023**, *11*, 3353. https://doi.org/10.3390/pr11123353

Academic Editors: Dicho Stratiev, Tianshou Ma and Yuqiang Xu

Received: 31 October 2023
Revised: 22 November 2023
Accepted: 30 November 2023
Published: 2 December 2023

1. Introduction

The Mahu conglomerate reservoir is unconventional with poor physical properties and a low effective permeability [1]. There are some problems for the Mahu conglomerate reservoir development, such as a high breakdown pressure and difficulty in the fracture initiation of horizontal wells. The breakdown pressure can be reduced by changing the stress state near the well or the rock properties [2]. Acidizing, abrasive perforating, high-energy gas fracturing, and perforation parameter optimization are essential methods to disrupt the formation stress for pretreatment [3]. In implementing acid pretreatment technology, acid with an aggressive chemical dissolution effect on reservoir rocks should be selected [4]. The commonly used acid systems consist of hydrochloric acid and mud acid (a mixture of hydrofluoric and hydrochloric acid). Hydrochloric acid in mud acid can dissolve carbonates and iron–aluminum compounds, while hydrofluoric acid can dissolve clay and silicates in the formation. It is utilized to treat reservoirs with low carbonate content and high clay content. Generally, according to the reservoir mineral components to optimize the acid system [5,6]. A high acid concentration will raise the corrosion risk of pipes and loosen bonding near the wellbore [7]. It is therefore crucial to select an acid system reasonably. This paper assesses two types of acid to pretreat rock samples. The acid pretreatment effectiveness was clarified by analyzing the pressure curves and mineral composition variation before and after acidizing.

The triaxial experiment is frequently conducted to study the fracture propagation law. However, most triaxial experiments are carried out under open hole conditions but rarely under perforation conditions. Chen et al. [8] conducted triaxial simulation experiments on outcrop and artificial rock samples with open hole completions. They discussed the effects of in situ stress, natural fractures, and other factors on hydraulic fracture propagation. Yao et al. [9] utilized a large-scale true triaxial fracturing experimental system to study the hydraulic fracture propagation in naturally fractured formations. Concrete specimens were used in the experiments under open hole conditions. Hou et al. [10] carried out a triaxial hydraulic fracturing experiment using limestone rock samples under open hole conditions. They investigated the influence of the in situ stress difference, variable injection rate, and acid treatment on the fracture propagation law in horizontal well fracturing. Due to the influencing factors of the rock's mechanical properties, Tan et al. [2] selected shale rock samples to implement large-scale triaxial experiments. Hydrochloric acid was used to pretreat the open hole interval of the samples. The effect of the shale–acid reaction on the hydraulic fracture initiation and propagation was investigated. Wang et al. [11] performed triaxial acid fracturing experiments on fractured and fracture-cavity type carbonate rocks under open hole conditions. So far, few approaches have simulated perforation well in experiments. Popular methods for perforation simulations are prefabricated holes and slotting. The prefabricated holes or slots on the well pipes could be plugged during cementing. The slot cut after cementing is very different to perforation. To simulate perforation, Shan et al. [12], Feng et al. [13], and Zhang et al. [14] drilled holes in the steel wellbore and plugged it with paper during cementing. The paper can be dissolved, so the holes are retained. Wu et al. [15] and Yu et al. [16] drilled holes with a specific diameter on the steel wellbore, inserted a certain length of a straight cylinder into the hole, and mixed cement and quartz sand to form a concrete sample. The effects of the perforation phase angle, perforation density, and perforation cluster spacing on the fracture morphology were investigated. Hou et al. [17], Liu et al. [18], Liang et al. [19], and Zhang et al. [20] cut slots inside the fracturing specimens to simulate perforation in triaxial fracturing experiments. Fu et al. [21] established a perforation method using a perforation gun in large rock samples (762 mm × 762 mm × 914 mm) and revealed the fracture initiation law under perforating conditions. However, the perforating gun still cannot be applied to rock samples in laboratory triaxial experiments (300 mm × 300 mm × 300 mm).

In this study, we conducted a systematic experimental study on how acid pretreatment lowers the breakdown pressure in perforated horizontal wells in the Mahu conglomerate reservoir. The rock samples were cut from the outcrop of the formation. A new method was first established to generate perforation in the rock samples by hydraulic jetting. Subsequently, acid pretreatment was performed on the perforations. Afterward, the triaxial fracturing experiments were carried out on the rock samples. By analyzing the experimental results, the effectiveness of the acid pretreatment in lowering the breakdown pressure in the Mahu sag was validated. In addition, the acid system, as well as treatment parameters, were optimized.

2. New Perforation Method in Rock Samples

Generally, laboratory hydraulic fracturing experiments are conducted with open hole completion [22–24]. Perforation is hard to implement in experimental sample dimensions

(300 mm × 300 mm × 300 mm). Although slots or pre-drilled holes can be made on the steel tube, they may be plugged during cementing, so there is no perforation outside the tube. To overcome the shortcomings of conventional triaxial fracturing experiments, we developed a new method for perforation with hydraulic jetting.

2.1. Perforation Simulation Apparatus

Figure 1 shows the physical drawing and schematic diagram of the hydraulic jetting perforation apparatus. This perforation system comprises a pressurized tank, water jet assembly, nozzle, protective cover, high-pressure pipeline, and low-pressure water supply pipeline. The operating pressure of the apparatus is 25 MPa with a maximum injection rate of high-pressure water of 15 L/min. An inlet and an outlet pressure gauge are installed to monitor the pressure status. Since the internal diameter of the steel casing used in fracturing experiments is generally between 22 mm and 28 mm, and the outer diameter of the hydraulic jetting gun is 23 mm, it applies to all fracturing samples with a inner casing diameter greater than 25 mm. The fracturing sample can be perforated using a 80-mesh garnet sand abrasive mixed in high-pressure water. Variable perforation parameters, including perforation angle, length, and depth can be achieved. A specific perforation angle can be achieved by rotating the hydraulic nozzle and different perforation lengths are obtained by controlling the jetting time.

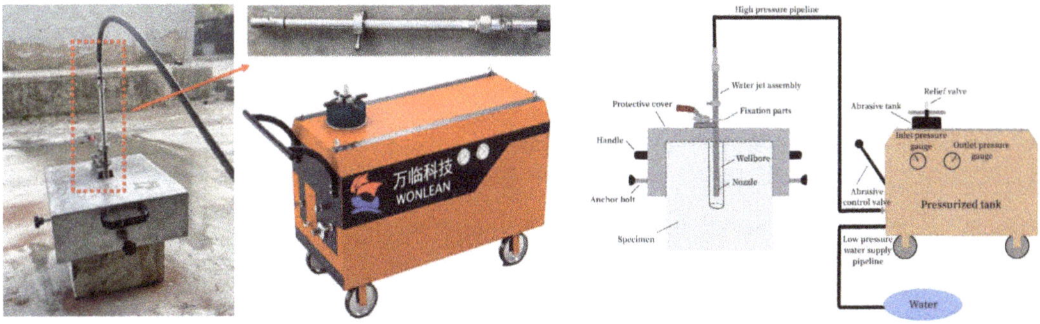

(a) (b)

Figure 1. Hydraulic jetting perforation apparatus. (**a**) Hydraulic jetting perforation apparatus; (**b**) schematic diagram of hydraulic jetting perforation apparatus.

2.2. Reservoir Information

The Mahu conglomerate reservoir depth ranges from 2500 to 3900 m. The Triassic Baikouquan formation (T_1b) and Permian Urho formation (P_3w) are the primary development horizons [25–28]. The distribution of the Baikouquan formation (T_1b) is generally stable. The average thickness of the Baikouquan formation is 138 m, and the average thickness of T_1b_1, T_1b_2, and T_1b_3 are 46.0 m, 47.0 m, and 45.0 m, respectively. The rock mineral composition of the T_1b_2 section of the Baikouquan formation is mainly composed of quartz (35.5%), followed by feldspar (27.5%) and clay (23.2%), and a small amount of calcite (10.3%). The average porosity is 9.78%, and the average permeability is 2.4×10^{-3} μm^2. This paper conducts research based on the Ma 18 well area data.

2.3. Specimen Preparation

Five outcrops (Figure 2a) were excavated in Block Ma18. The brief specimen preparation procedures are as follows: Firstly, the outcrops were cut into cubic specimens of 300 mm × 300 mm × 300 mm (Figure 2b). Secondly, a borehole with a 34 mm diameter and a 250 mm depth was drilled in the center of each specimen. Thirdly, a steel pipe (Figure 3a) with an outer diameter of 30 mm, an internal diameter of 25 mm, and a length of 250 mm was placed inside the borehole, and the annulus between the borehole and the steel pipe was bonded by high-strength epoxy resin to simulate the cementing process, as shown in Figure 3b.

(a) (b)

Figure 2. Rock samples. (**a**) Natural outcrops excavated on site; (**b**) description of what is contained in the second panel. Figures should be placed in the main text near to the first time they are cited.

(a) (b)

Figure 3. Wellbore and cementing. (**a**) Simulated wellbore made of steel; (**b**) rock samples after cementing.

2.4. Perforating Process

After 24 h cementation, the perforating operation is performed; the brief perforation processes are as follows: (1) determine the in situ stress direction and mark the perforation orientation according to the experimental schedule (Figure 4a); (2) locate and orient the nozzle with specified parameters in the borehole using the fixation parts and the protective cover (Figure 4b); (3) perform hydraulic jetting for a specified time (Figure 4c). It is worth noting that the relationship between hydraulic jetting time and perforation length varies for different wellbore materials and rocks, which should be obtained in advance through tests. Table 1 shows the relationship between perforation length and hydraulic jetting time for the specimen in this study.

(a) (b) (c)

Figure 4. Perforating procedure. (**a**) Mark the perforation orientation on the surface; (**b**) fix the protective cover with specific parameters; (**c**) work in process.

Table 1. The relationship between jetting time and perforation length for the Mahu conglomerate rock.

Number	Perforation Length	Jetting Time
1	1.8 cm	4 min
2	2 cm	7 min
3	2.5 cm	7 min 30 s
4	3 cm	10 min
5	4 cm	18 min 14 s
6	5 cm	20 min

Figure 5 displays the perforation hole morphology comparison of on-site and laboratory settings. Clearly, the holes made from hydraulic jetting can simulate the perforation in the reservoir well in terms of perforation length and phase angle. The new laboratory perforating method restored the in situ perforation sufficiently. To further illustrate the improvement of the new perforation method, we compared slotting and hydraulic jetting for the perforation simulation in the experiments in terms of the mechanics (taking horizontal well as an example).

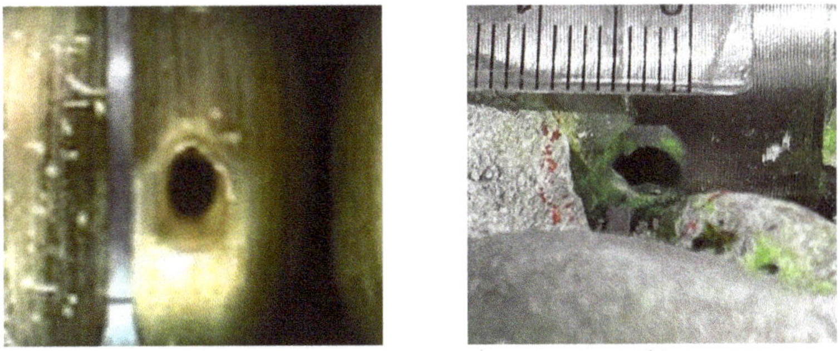

(a) (b)

Figure 5. Perforation hole morphology. (**a**) The perforation hole took place underground in the field, showing an irregular circular hole; (**b**) the perforation hole made by the new water jetting apparatus, showing a similar morphology with the hole on site.

Figure 6 illustrates the difference between the two approaches in terms of the force condition. Figure 6a reveals the perforation made by the slotting. σ_{h1} represents the minimum horizontal principal stress, σ_{h2} represents the maximum horizontal principal stress, σ_z represents the vertical stress, and S_t represents the tensile strength. The circular fractures formed by slotting are highly affected by vertical stress, and the mechanical failure mode is tensile failure. Figure 6b displays the perforation holes made by the new perforating apparatus. The perforation wall is subjected to circumferential stress (also known as the 'hoop stress') along the perforation tunnel [29], and this mode of stress aligns with that of the field, which is closer to the actual perforation state.

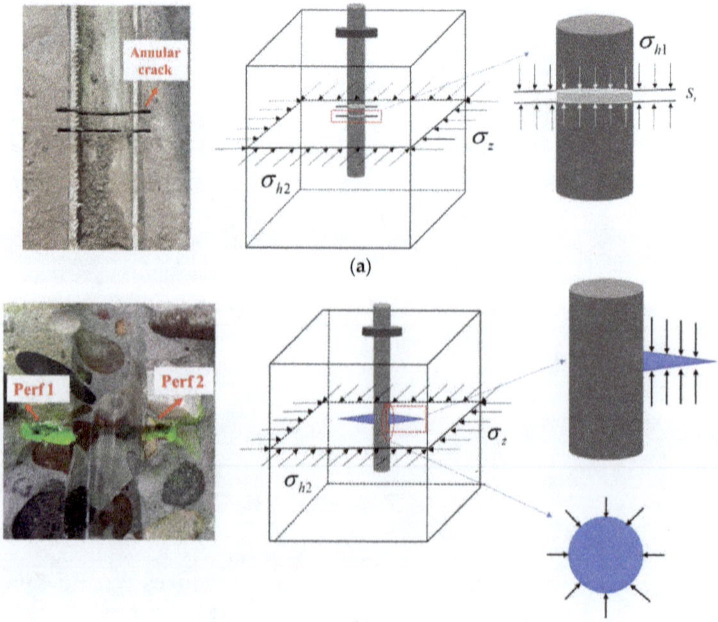

Figure 6. Comparison of two kinds of perforations. (**a**) The perforations by slotting; (**b**) the perforation by hydraulic jetting.

In this study, the perforation diameter was 5 to 6 mm. The perforation length was 4 to 5 cm. The perforation phase angle is defined as the acute angle between the axis of the perforation and the horizontal principal stress direction, comprising 0°, 30°, 45°, and 60°. The perforation depth is 12 cm from the wellhead, as shown in Figure 7.

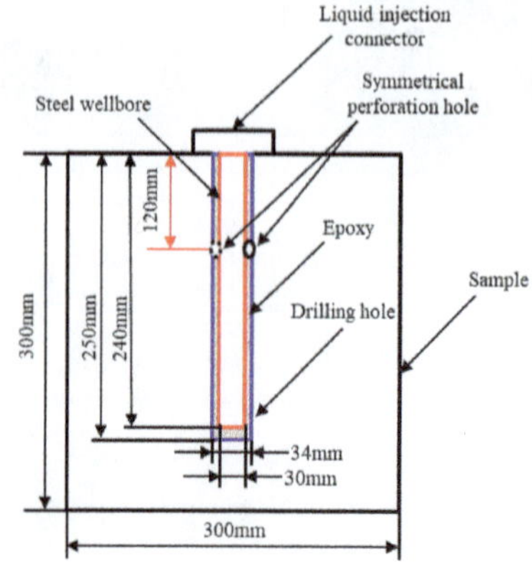

Figure 7. Schematic of perforation.

3. Acid Pretreatment for Fracturing

Before fracturing, the specimens were pretreated by different acid types (Figure 8). A 3% corrosion inhibitor was added to the acid solution to protect the steel pipe. The 15 min, 30 min, 60 min, 90 min, and 120 min injection times were tested to study the influence of acidizing time. The other four specimens without acidizing were used as the control groups.

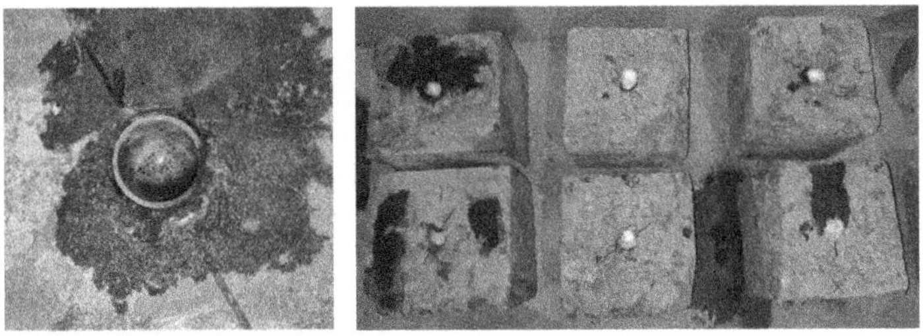

Figure 8. Acid pretreatment.

4. Triaxial Hydraulic Fracturing Experiment

4.1. Mineral Composition and Rock Mechanical Properties of Rock Samples

We conducted a series of property experiments on the Mahu conglomerate outcrops. The results of the X-ray diffraction tests indicated that the mineral composition is mainly quartz (35.29%), followed by feldspar (25.12%) and clay (21.22%). The carbonate content is relatively low (9.52%). Mud acid can effectively dissolve feldspar and clay, which is recommended for acid pretreatment in the Mahu area. Additionally, the average permeability is 2.2×10^{-3} μm^2, the average porosity is 10.85%, the average Young's modulus is 25.19 GPa, the average Poisson's ratio is 0.27, and the tensile strength of the rock is 5.54 MPa. The mineral composition and physical property parameters of the outcrops are generally consistent with the geological data of the T_1b_2 section of the Baikouquan formation, which validates that the outcrop obtained in the field has similar properties to the formation and the experimental results are reliable.

4.2. Triaxial Fracturing Experimental Procedures

4.2.1. Experimental Apparatus

A large true triaxial hydraulic fracturing system can perform the fracturing experiment on a 300 mm cubic rock sample, as shown in Figure 9. The triaxial pressure loading system can achieve three principal stresses on the cube rock samples. Considering the in situ stress (σ_{hmin} = 63 MPa σ_{Hmax} = 75 MPa, and σ_v = 102 MPa) and the performance of the equipment, σ_{hmin} was set as 5 MPa, σ_{Hmax}, 17 MPa, and σ_v, 25 MPa in this study. The fracturing fluid viscosity, μ (15 mPa·s), is taken from the field data. The injection rate was set based on the geometrical similarity principle shown in Equation (1) [30–34].

$$\frac{Q_F}{A_F \times H_F} = \frac{Q_M}{A_M \times H_M}, \tag{1}$$

where Q is the displacement, m^3/min; A is the cross-sectional area of the wellbore, m^2; H is the fracture length, m; the subscript M denotes the modeling parameters; the subscript F denotes the field parameters. Equation (1) illustrates that the dimension ratio should be equaled for the laboratory and the field. According to data in the field, the injection rate Q is approximately 12–14 m^3/min, and the injection rate in the experiment was set at 130 mL/min. The detailed experimental scheme is in Table 2.

Figure 9. Triaxial hydraulic fracturing apparatus.

Table 2. Breakdown pressure results under different experimental conditions.

Number	Perforation Angle	Acid System	Acidizing Time (min)	Breakdown Pressure (MPa)
1	60°	-	-	15.65
2	45°	-	-	15.4
3	30°	-	-	13.42
4	0°	-	-	13.77
5	0°	10%HCl	30	9.8
6	0°	3%HF + 10%HCl	30	9.23
7	0°	6%HF + 10%HCl	30	5.45
8	0°	8%HF + 10%HCl	30	4.55
9	0°	12%HCl	30	9.1
10	0°	3%HF + 12%HCl	30	8.2
11	0°	6%HF + 12%HCl	30	6
12	0°	8%HF + 12%HCl	30	5.08
13	0°	15%HCl	30	9.67
14	0°	3%HF + 15%HCl	30	9.51
15	0°	6%HF + 15%HCl	30	6.05
16	0°	8%HF + 15%HCl	30	4.97
17	0°	6%HF + 10%HCl	15	12.85
18	0°	6%HF + 10%HCl	60	3.16
19	0°	6%HF + 10%HCl	90	3.3
20	0°	6%HF + 10%HCl	120	3

4.2.2. Experimental Procedure

The procedure of the triaxial fracturing experiment is summarized as follows: (1) Load the cube rock sample into the chamber and connect the pipelines. (2) Apply three principal stresses from the x, y, and z axes through a hydraulic pressure system. (3) Start the injection pump at a constant rate and record the injection pressure until the specified time. (4) Open the fracturing specimen to inspect the fracture pattern and analyze the results.

4.3. Experimental Results and Analysis

Twenty sets of experiments were conducted. The results are displayed in Table 2. The average breakdown pressure for the specimens without acidizing ($1^{\#}$~$4^{\#}$) was 14.56 MPa, while the average breakdown pressure of the acidified specimens ($5^{\#}$~$20^{\#}$) was 6.87 MPa, with a decline of 7.7 MPa. The acid pretreatment strongly influenced the breakdown pressure with a positive effect. All the fracturing specimens were opened along with the hydraulic fractures to observe the fracture initiation morphology.

4.3.1. Effect of Acidizing

It is essential to select an optimal ratio of HF to HCl and concentrations for a particular formation, which directly determines the effectiveness of the acid pretreatment. Two kinds of acid solutions, hydrochloric acid and mud acid, were tested in the experiment. Both acids can lower the breakdown pressure. However, the breakdown pressure of the three groups of specimens treated with hydrochloric acid is still relatively high (average 9.52 MPa), and the treatment results are unsatisfactory. Specimens $5^{\#}$ and $9^{\#}$ developed natural fractures near the wellbore, showing lower breakdown pressure values compared to specimen $13^{\#}$ (Figure 10). The breakdown pressures of specimens $5^{\#}$ and $9^{\#}$ should be higher than the current values (9.8 MPa and 9.1 MPa), without natural fractures developing.

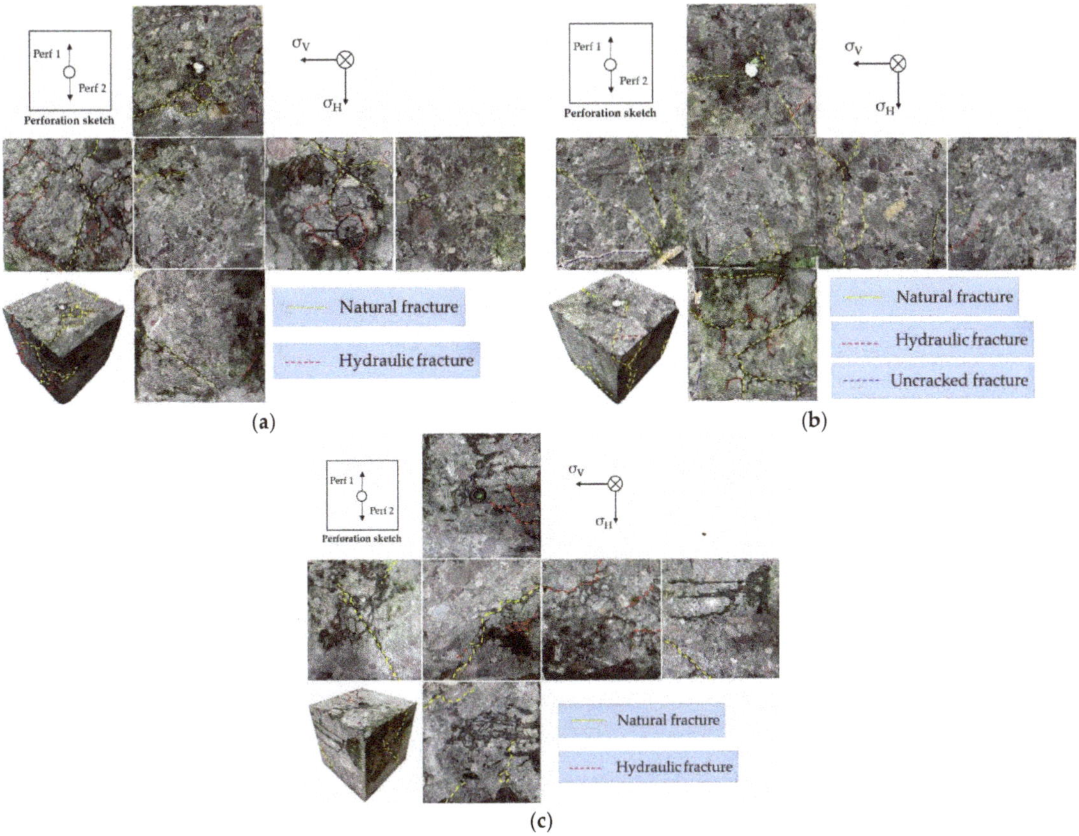

Figure 10. Fracture geometry of specimens. (**a**) Multiple natural fractures (the yellow dotted line) developed near the wellbore observed in specimen $5^{\#}$; (**b**) several natural fractures (the yellow dotted line) developed near the wellbore, and one uncracked fracture after experiment was observed in specimen $9^{\#}$; (**c**) no natural fractures developed near the wellbore in specimen $13^{\#}$, several natural fractures observed on the surface away from wellbore only affect the hydraulic fractures propagation, not initiation.

Mud acid is more effective in the Mahu depression. The reason for this is that the carbonate content in this area is low, but the feldspar content is relatively high. HCl dissolves the carbonates only, and mud acid dissolves both carbonate and silica-aluminates. Therefore, more minerals were dissolved by the mud acid. Multiple comparative experiments were conducted with different concentrations of acid solution to further optimize the acid system suitable for this area. The acidizing effectiveness remained stable with a weak regularity when the concentration of the hydrochloric acid changed, since the carbonate content is low and natural outcrops have a strong heterogeneity and different physical properties. Nevertheless, hydrochloric acid is essential, and is combined with hydrofluoric acid to form mud acid, which maintains a low pH value. Moreover, hydrochloric acid can promote the dissolution of hydrofluoric acid. Therefore, considering the economic cost of development, a lower concentration of hydrochloric acid is recommended. When the concentration of hydrochloric acid is fixed, the concentration of hydrofluoric acid increases from 0% to 8%, and the breakdown pressure declines rapidly (Figure 11). When the concentration of the hydrofluoric acid varies from 6% to 8%, the breakdown pressure rarely changed.

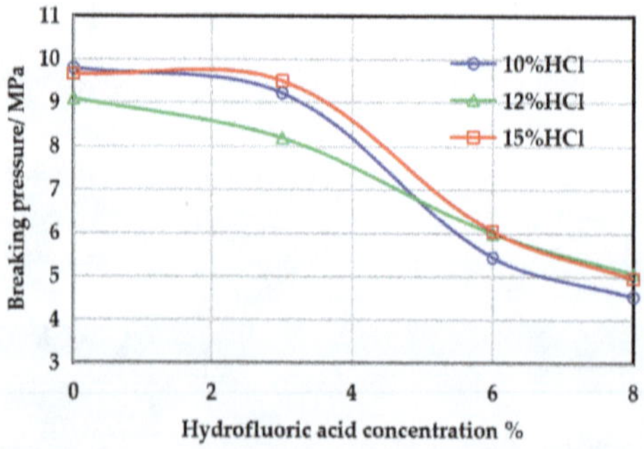

Figure 11. Effect of acid composition on breakdown pressure.

Xue et al. [35] conducted experiments on the dissolution rates of HCl, HF, and mixed solutions of HCl and HF at different acid concentrations and treatment times. The results showed that the dissolution rate was highest at the HF concentration of 9%. High concentrations of HF could lead to a decline in the dissolution rate, since the reaction between HF and clay and feldspar also involves second-order and third-order reactions. Furthermore, they usually produce precipitation (mainly silica gel precipitation). Due to the variation between the HF concentration of 6% and 8% in this experimental result (Figure 11), a combination of 10%HCl + 6%HF mud acid is preferred.

The acid–rock contact time is a crucial factor in the acid pretreatment design as well. Based on the mineral composition of the Mahu sag and the above experimental results, 6%HF + 10%HCl mud acid was selected to investigate the acid treatment effect. Under the same experimental conditions, the breakdown pressure decreases with the rise in the acidizing time, especially for a shorter contact time (for instance, less than 30 min). Increasing the contact time from 15 min to 30 min diminished the breakdown pressure by 64.6%. Above 60 min, the dissolution rate decreased, and the breakdown pressure almost no longer continued to decline and maintained a steady value. The reason for this is that the amount of soluble minerals is certain. After a particular time, soluble minerals are completely dissolved, leaving insoluble minerals. At this time, even if the acid pretreatment time continues to increase, there will be no more obvious dissolution, resulting in no further decrease in the breakdown pressure. It is worth noting that the breakdown pressure of specimen $19^{\#}$ is slightly higher than that of specimen $18^{\#}$. There are no natural fractures developed near the wellbore of both samples. However, the gravel distribution of specimen $19^{\#}$ is more complex, and more different sizes of gravel were found near the wellbore. The strong heterogeneity of the conglomerate gives rise to some randomness in the experimental results. Still, after 60 min of acidizing, the overall performance was outstanding, and the breakdown pressure remained almost constant. This relationship curve is illustrated in Figure 12. Compared with the case without acidizing, the breakdown pressure of the specimen with 30 min effective contact time was lowered by about 10 MPa. Furthermore, the breakdown pressure witnessed a stable trend after 60 min. Therefore, an adequate acid–rock contact time of 60 min is recommended.

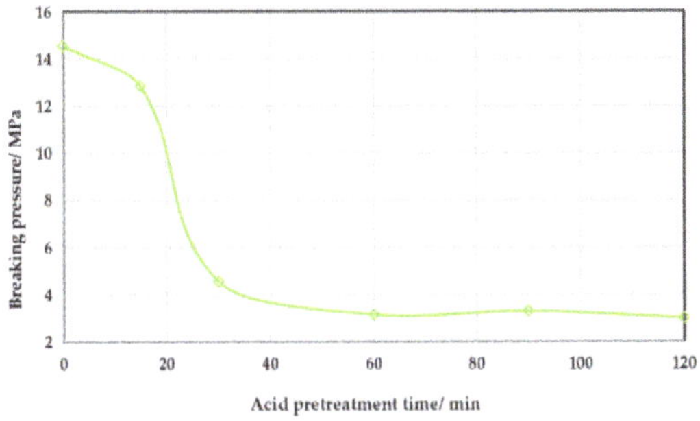

Figure 12. Effect of acid treatment time on breakdown pressure.

To further explore the mechanism of the acid pretreatment lowering the breakdown pressure, XRD mineral composition tests were conducted on rock debris scraped near the perforations (within a radial 1 cm range) of fracturing specimens $4^{\#}$, $5^{\#}$, $8^{\#}$, $17^{\#}$, $18^{\#}$, $19^{\#}$, and $20^{\#}$, respectively. The test results are shown in Table 3. After the hydrochloric acid treatment, the content of carbonate rock decreases. In contrast, the content of feldspar remains stable (specimen $5^{\#}$), as carbonate can be quickly dissolved by hydrochloric acid, but feldspar cannot. The reaction between HCl and carbonates is

$$C_aCO_3 + 2HCl \rightarrow CaCl_2 + CO_2 + H_2O, \tag{2}$$

$$C_aMg(CO_3)_2 + 4HCl \rightarrow CaCl_2 + MgCl_2 + 2CO_2 + 2H_2O \tag{3}$$

Table 3. Mineral composition tests results for different acidizing times.

Number	Acid System	Acidizing Time (min)	Quartz (%)	Clay (%)	Feldspar (%)	Calcite (%)	Other (%)
4	-	-	35.29	21.22	25.12	9.52	8.85
5	10%HCl	30	34.55	18.91	26.89	8.05	11.6
17	6%HF + 10%HCl	15	32.5	18.13	25.75	9.18	14.44
8	6%HF + 10%HCl	30	37.5	16.33	22.09	8.17	17.91
18	6%HF + 10%HCl	60	37.8	15.19	21.86	7.85	17.3
19	6%HF + 10%HCl	90	38.48	14.67	21.36	7.67	17.82
20	6%HF + 10%HCl	120	38.26	14.55	21.15	7.55	18.49

Therefore, the acidizing performance could be better if only hydrochloric acid was applied. In specimens $17^{\#}$, $8^{\#}$, $18^{\#}$, $19^{\#}$, and $20^{\#}$, the feldspar and clay react with the HF due to the fluoride ion (F-), showing a great performance. Clay minerals mainly include kaolinite, montmorillonite, illite, and chlorite [36]. The reaction between kaolinite and H^+ i:

$$AlSi_4O_{10}(OH)_8 + 24HF + 4H^+ = 4AlF_2{}^+ + 4SiF_4 + 18H_2O, \tag{4}$$

The reaction between montmorillonite and H^+ is

$$AlSi_8O_{20}(OH)_4 + 40HF + 4H^+ = 4AlF_2{}^+ + 8SiF_4 + 24H_2O, \tag{5}$$

Feldspar is unstable, with many types, including potassium feldspar, albite, and anorthite. The reaction between potassium feldspar and hydrofluoric acid is

$$KAlSi_3O_8 + 22HF = KF + AlF_3 \downarrow + 3H_2SiF_6 + 8H_2O, \tag{6}$$

The reaction between albite and hydrofluoric acid is

$$NaAlSi_3O_8 + 22HF = NaF + AlF_3 \downarrow +3H_2SiF_6 + 8H_2O, \quad (7)$$

The reaction between anorthite and hydrofluoric acid is

$$CaAl_2Si_2O_8 + 22HF = CaF_2 \downarrow +2AlF_3 \downarrow +2H_2SiF_6 + 8H_2O, \quad (8)$$

The acid treatment time was increased from 15 to 120 min, and the proportion of the carbonate, feldspar, and clay minerals decreased significantly. Wherever the acid solution reached, the soluble minerals would be rapidly dissolved, and there were still areas where the acid was not covered in the experiment, so the proportion of these minerals stayed relatively high. The action distance of the acid is limited. Accordingly, it is crucial to accurately select a targeted acid composition in the acid pretreatment design.

In addition to the in situ stress, the tensile strength affects significantly the breakdown pressure. Therefore, lowering the rock tensile strength is vital to reducing the breakdown pressure. We also cut ten cores with a diameter of 25.4 mm and a thickness of 13 mm using the outcrops, and compared the tensile strength variation before and after acidizing to clarify the mechanism of the acid pretreatment lowering the breakdown pressure. Table 4 provides the tensile strength test results. Group number (a) represents the samples without acidizing, group number (b) represents the samples after acidizing. The average tensile strength of the unacidified and acidized cores are 5.54 MPa and 3.8 Mpa, with a decline ratio of 31.4%. In essence, the acid pretreatment can destroy the strength of the rock and cement to decrease the rock tensile strength, while creating more seepage channels, increasing the porosity and permeability, thereby effectively lowering the breakdown pressure.

Table 4. Experimental parameters and results.

Number	Tensile Strength (Before) (MPa)	Number	Acid System	Contact Time (min)	Tensile Strength (After) (MPa)
(1)a	4.72	(1)b	6%HF + 10%HCl	10	4.53
(2)a	4.33	(2)b	6%HF + 10%HCl	20	3.59
(3)a	6.83	(3)b	6%HF + 10%HCl	30	4.64
(4)a	5.39	(4)b	6%HF + 10%HCl	60	3.13
(5)a	6.42	(5)b	6%HF + 10%HCl	120	3.08

4.3.2. Effect of Perforation Angle

To avoid interference with other parameters, we kept the other parameters the same when analyzing the effect of the perforation orientation angle. A perforation orientation angle of 0°, 30°, 45°, and 60° were used to investigate the impacts of the perforation angle on the fracture initiation and breakdown pressure. Figure 13 displays the pressure curves of fracturing specimens $1^{\#}$ to $4^{\#}$. There is an apparent peak on each pressure curve, which indicates the breaking of the fracturing specimen and gives the breakdown pressure. The breakdown pressure declined by 14.2% when the perforation rotated from 60 degrees to 0 degrees. The overall trend of the breakdown pressure is consistent with the results of numerical simulation studies [37–40]. The ideal perforation orientation angle is zero, corresponding to the minimum breakdown pressure [41]. Therefore, the larger the perforation angle, i.e., the greater the angle between the perforation axis and the maximum horizontal principal stress, the more energy needs to be consumed to counteract the increasing stress and the more difficult it is to crack. To remove the noise of multiple perforations, the perforation just shoots from one angle, different to the field, in which the perforation shoots from several directions around the wellbore. To decrease the breakdown pressure, ensure that there are perforations with an angle of 0° in the field treatments.

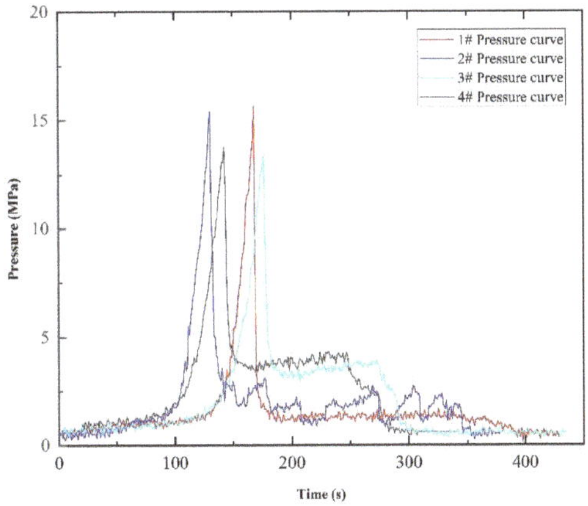

Figure 13. Pressure curves of unacidified samples with apparent peak.

Additionally, the content, distribution, particle size, and hardness of gravel in conglomerate reservoirs can strongly influence fracture initiation. The experimental results showed that the breakdown pressure declined with the decrease in the perforation orientation angle, except in specimens $3^{\#}$ and $4^{\#}$. The perforation angle varied from 30 degrees to 0 degrees, and the breakdown pressure increased by 0.25 MPa. Figure 14a illustrates the fracture geometry in specimen $4^{\#}$. The fracture initiates at the perforation base and wall. There is a large piece of gravel closing on the perforation wall, causing an obstruction, and the difficulty of fracture initiation rises. According to the tracer distribution, the fractures are obstructed by gravel and reorient during the initiation process, resulting in a breakdown pressure increase. Comparatively, there is no gravel block around the perforation in specimen $3^{\#}$. The tracer can be observed around the perforation, indicating that the fractures initiate and propagate at the perforation base in each direction (Figure 14b). Therefore, the breakdown pressure of specimen $4^{\#}$ is supposed to be lower than that of specimen $3^{\#}$ if there were no gravel blocks. Consequently, for heterogeneous conglomerates, when the gravel distribution near the wellbore is similar, the breakdown pressure can be changed by turning the perforation angle without considering the natural fracture. In a conglomerate characterized by strong heterogeneity, when the gravel distribution near the well is quite different, the perforation angle has little effect on the breakdown pressure. Affected by the larger gravel strength, the fracture initiation process may be blocked to generate different paths. In this special case, the gravel distribution is the dominant factor affecting the breakdown pressure.

(a) (b)

Figure 14. A picture of fracture initiation inside the samples. (**a**) Gravels near the wellbore observed after opening specimen $4^{\#}$; (**b**) there are no major gravels observed in specimen $3^{\#}$.

5. Field Application of Acid Pretreatment

The lithology of the Baikouquan formation is mainly gray conglomerate, gravelly coarse sandstone, sandy conglomerate, and medium coarse conglomerate [42,43]. Based on the geological data, the medium conglomerate is dominant, with a volume fraction of 61.36%, followed by small conglomerate and fine conglomerate, with a volume fraction of 10.07% and 7.29%, respectively. Sandstone is less, with a volume fraction of 8.22% [44]. The minimum horizontal principal stress of the reservoir is 63 MPa, Young's modulus is 26,038 MPa, Poisson's ratio is 0.28, and the tensile strength is 5.1 MPa. The acid pretreatment was applied for over 50 horizontal wells in the Ma 18 block.

Figure 15 displays the operation curves of the unacidified and acidized horizontal well. The red line represents the injection pressure, and the green line represents the pump rate. Figure 15a indicates the curve of the horizontal well M1 without the acid pretreatment. The pressure curve has an apparent peak, giving a breakdown pressure of 73.19 MPa. Figure 15b is for well M2 with acid pretreatment. After acidizing, the pressure curve has a tiny peak, significantly decreasing the fracturing difficulty. The breakdown pressure is 65.5 MPa, a proportion of 10.5% lower than the unacidified well M1.

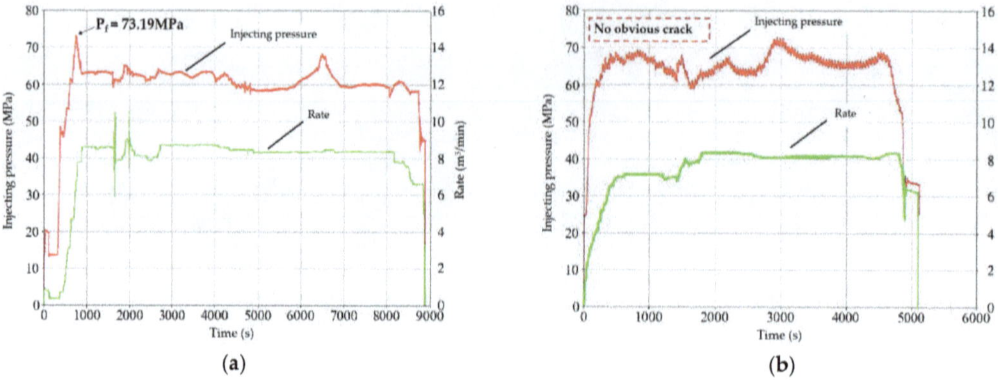

Figure 15. Pressure curves of horizontal well treatments. (**a**) Injection pressure curve indicates a peak point with a high breakdown pressure value; (**b**) there is no obvious crack showing on the injection curve with a low breakdown pressure value.

6. Conclusions

This paper developed a new perforation method and conducted a systematic study on how acid pretreatment lowers the breakdown pressure in the Mahu conglomerate reservoirs. By analyzing the experimental results, the following conclusions were reached:

1. The novel perforation method by water jetting in fracturing specimens can better simulate perforation conditions such as the perforation length and angle for the acid pretreatment and triaxial fracturing experiments.
2. The perforation angle has a significant effect on the breakdown pressure. A 0° perforation angle results in the minimum breakdown pressure. It is necessary to ensure there are perforation holes with a 0° perforation angle relative to the maximum horizontal stress for a low breakdown pressure.
3. Acid pretreatment can effectively decrease the breakdown pressure down to 7.7 MPa under experimental conditions. The mechanism of the acid pretreatment to decrease the breakdown pressure is that rock dissolution by the acid reduces the rock tensile strength and increases its permeability. The raised permeability increases the fluid pressure of the reservoir near the wellbore so as to reduce the breakdown pressure of the formation.

4. For the Mahu conglomerate reservoir with a low carbonate content, an acid system of 6%HF + 10%HCl with 60 min plus acid contact time is recommended for the acid pretreatment.
5. The field application of the acid pretreatment to horizontal well fracturing in the Mahu conglomerate reservoirs showed that the acid pretreatment was successful and the breakdown pressure was lowered by 10.5%.

Author Contributions: Conceptualization, W.J. and J.M.; methodology, J.M.; software, X.L.; validation, J.M., G.W. and X.M.; formal analysis, W.J.; investigation, W.J.; resources, X.W.; data curation, X.W.; writing—original draft preparation, W.J.; writing—review and editing, W.J. and J.M.; visualization, G.W.; supervision, X.M.; project administration, J.M.; funding acquisition, J.M. and X.M. All authors have read and agreed to the published version of the manuscript.

Funding: This research received no external funding.

Data Availability Statement: The raw data required to reproduce these findings cannot be shared at this time as the data also form part of an ongoing study.

Conflicts of Interest: Author Guifu Wang was employed by the CNPC Engineering Technology R&D Company Limited. The remaining authors declare that the research was conducted in the absence of any commercial or financial relationships that could be construed as a potential conflict of interest.

References

1. Zhu, J.; Dong, Y.; Zhu, Y.; Zhang, D.; Hui, J.; Wang, W. Sedimentary reservoir evolution characteristics of lower Triassic Baikouquan formation in Ma18 well area of Mahu sag. *J. Hebei Geo Univ.* **2022**, *45*, 32–41.
2. Yun, J.; Qin, G.J.; Xu, F.Y.; Li, X.; Zhong, N.; Wu, W. Development and utilization prospects of unconventional natural gas in China from a low-carbon perspective. *Acta Pet. Sin.* **2012**, *33*, 526–532.
3. Tan, P.; Jin, Y.; Han, L.; Shan, Q.; Zhang, Y.; Chen, G.; Zhou, Y. Influencing mechanism of acidification pretreatment on hydraulic fracture for deep fractured shale reservoirs. *Chin. J. Geotech. Eng.* **2018**, *40*, 384–390.
4. Wang, Q.; Wen, J.Q.; Yang, Z.C. Application effect analysis of acid fracturing technology in low permeability oilfield development. *Oil Gas Prod.* **2022**, *48*, 32–34+51.
5. Zhang, J.F. The Research on Potential Damage and Acid System of Ultra-Low Permeability Sandstone Reservoir in Shanshan Oilfield. Master's Thesis, Yangtze University, Hubei, China, 2023.
6. Zhang, J.; Li, T.; Wu, J.; Guan, Y.; Xu, M.; Dan, Z.; Zhou, M. Sensitivity evaluation of ultra-low permeability sandstone reservoir and development of acidizing stimulation fluid. *Spec. Oil Gas Reserv.* **2022**, *29*, 166–174.
7. Xiong, T. Geological Characteristics of Carbonated Sandstone Reservoir and Acidification Optimization Technology. Master's Thesis, China University of Petroleum (Beijing), Beijing, China, 2018.
8. Chen, M.; Pang, F.; Jin, Y. Experiments and analysis on hydraulic fracturing by a large-size triaxial simulator. *Chin. J. Rock Mech. Eng.* **2000**, *19*, 868–872.
9. Yao, F.; Chen, M.; Wu, X.; Zhang, G. Physical simulation of hydraulic fracture propagation in naturally fractured formations. *Oil Drill. Prod. Technol.* **2008**, *3*, 83–86.
10. Hou, B.; Tan, P.; Chen, M.; Yuan, L.; Xiong, Z.; Xu, C. Experimental investigation on propagation geometry of hydraulic fracture in compact limestone reservoirs. *Chin. J. Geotech. Eng.* **2016**, *38*, 219–225.
11. Wang, Y.Z.; Hou, B.; Zhang, K.P.; Zhou, C.L.; Liu, F. Laboratory true triaxial acid fracturing experiments for carbonate reservoirs. *Pet. Sci. Bull.* **2020**, *3*, 412–419.
12. Shan, Q.L.; Jin, Y.; Han, L.; Zhang, R.X. Influence of spiral perforation parameters on fracture geometry near horizontal wellbores. *Pet. Sci. Bull.* **2017**, *1*, 44–52.
13. Feng, F.; Yang, L.S.; Liu, Z.H.; Shen, K. Experimental study on initiation and propagation law of sandstone with axial prefabricated cracks. *Saf. Coal Mines* **2018**, *49*, 18–21.
14. Zhang, R.X.; Hou, B.; Shan, Q.L.; Tan, P.; Wu, Y.; Guo, X.F. Parameter optimization of spiral perforations in horizontal well with tight sandstone reservoir. *Chin. J. Geotech. Eng.* **2018**, *40*, 2143–2147.
15. Wu, Y.; Hou, B.; Han, H.F.; Zhou, X. Study on the optimization of helical perforation parameters for horizontal wells in the condition of high horizontal stress difference. *Chin. J. Undergr. Space Eng.* **2019**, *15*, 226–231.
16. Yu, R.; Zhang, Y.; Zheng, B.; Yang, W.; Tian, Y.; Liu, B. Experimental study on the effects of perforation phasing on fracturing pressure and fracture propagation of thin interbeds. *Drill. Fluid Complet. Fluid* **2020**, *37*, 110–115.
17. Hou, Z.K.; Yang, C.H.; Wang, L.; Liu, P.J.; Guo, Y.T.; Wei, Y.L.; Li, Z. Hydraulic fracture propagation of shale horizontal well by large-scale true triaxial physical simulation test. *Rock Soil Mech.* **2016**, *37*, 407–414.

18. Liu, N.Z.; Zhang, Z.P.; Zou, Y.S.; Ma, X.F.; Zhang, Y.N. Propagation law of hydraulic fractures during multi-staged horizontal well fracturing in a tight reservoir. *Pet. Explor. Dev.* **2018**, *45*, 1059–1068. [CrossRef]
19. Liang, J.; Liu, J.; Xu, Q.Y.; Zhang, L. Study on rock fracture initiation pressure under different hydraulic fracturing parameters. *Coal Technol.* **2020**, *39*, 87–89.
20. Zhang, S.C.; Li, S.H.; Zou, Y.S.; Li, J.M.; Ma, X.F.; Zhang, X.H.; Wang, Z.F.; Wu, S. Experimental study on fracture height propagation during multi-stage fracturing of horizontal wells in shale oil reservoirs. *J. China Univ. Pet. (Ed. Nat. Sci.)* **2021**, *45*, 77–86.
21. Fu, H.F.; Huang, L.K.; Zhang, F.S.; Xu, Y.; Cai, B.; Liang, T.C.; Wang, X. Effect of perforation technologies on the initiation and propagation of hydraulic fracture. *Chin. J. Rock Mech. Eng.* **2021**, *40*, 3163–3173.
22. Wang, Y.H.; Fu, H.F.; Liang, T.C.; Wang, X.; Liu, Y.Z.; Peng, Y.; Yang, L.F.; Tian, Z.H. Large-scale physical simulation experiment research for hydraulic fracturing in shale. In Proceedings of the SPE Middle East Oil and Gas Show and Conference, Manama, Bahrain, 8 March 2015.
23. Bai, J.; Martysevich, V.; Walters, H.; Dusterhoft, R.; Matzar, L.; Sansil, M. Laboratory-scale hydraulic fracturing: Experiment and numerical modeling. In Proceedings of the U.S. Symposium on Rock Mechanics, Houston, TX, USA, 26 June 2016.
24. Fu, H.F.; Zhang, Y.M.; Wang, X.; Yan, Y.Z.; Guan, B.S.; Liu, Y.Z.; Liang, T.C.; Weng, D.W. New stimulation technology research based on impulse fracturing reservoir. *Chin. J. Rock Mech. Eng.* **2017**, *36*, 4008–4017.
25. Lei, D.W.; Chen, G.Q.; Liu, H.L.; Liu, X.; Abrimiti, T.K.; Cao, J. Study on the forming conditions and exploration fields of the Mahu giant oil (gas) province, Junggar Basin. *Acta Geol. Sin.* **2017**, *91*, 1604–1619.
26. Kang, X.; Hu, W.; Cao, J.; Wu, H. Controls on reservoir quality in fan-deltaic conglomerate: Insight from the Lower Triassic Baikouquan Formation, Junggar Basin, China. *Mar. Pet. Geol.* **2019**, *103*, 55–75. [CrossRef]
27. Sang, L.X.; Liu, J.; Wang, G.W.; Wang, S.T.; Li, Q.; Zhang, Y.L. High-quality reservoirs prediction of fan delta in the Triassic Baikouquan Formation in west slope of Mahu sag, Junggar Basin. *J. Palaeogeogr. (Chin. Ed.)* **2020**, *22*, 1053–1064.
28. Jiang, Q.P.; Kong, C.X.; Li, W.F.; Qiu, Z.G.; Lu, Z.Y.; Chang, T.Q.; Liu, K.; Chen, D.L.; Li, S.; Yuan, X.G. Sedimentary characteristics and evolution law of a lacustrine large-scale fan delta: A case study from the Triassic Baikouquan formation on the west slope of Mahu sag. *Acta Sedimentol. Sin.* **2020**, *38*, 923–932.
29. Waters, G.; Weng, X. The Impact of Geomechanics and Perforations on Hydraulic Fracture Initiation and Complexity in Horizontal Well Completions. In Proceedings of the Spe Technical Conference & Exhibition, Dubai, United Arab Emirates, 26–28 September 2016.
30. Detournay, E. Mechanics of hydraulic fractures. *Annu. Rev. Fluid Mech.* **2016**, *48*, 311–339. [CrossRef]
31. Dontsov, E.V. An approximate solution for a plane strain hydraulic fracture that accounts for fracture toughness, fluid viscosity, and leak-off. *Int. J. Fract.* **2017**, *205*, 221–237. [CrossRef]
32. Madyarov, A.; Prioul, R.; Zutshi, A.; Seprodi, N.; Groves, D.; Pei, J.; Wong, S.W. Understanding the impact of completion designs on multi-stage fracturing via block test experiments. In Proceedings of the 55th U.S. Rock Mechanics/Geomechanics Symposium, Houston, TX, USA, 20–23 June 2021.
33. Bunger, A.P.; Jeffrey, R.G.; Detournay, E. Application of scaling laws to laboratory-scale hydraulic fractures. In Proceedings of the 40th U.S. Symposium on Rock Mechanics (USRMS), Anchorage Alaska, Alaska, 25–29 June 2005.
34. Zou, Y.S.; Shi, S.Z.; Zhang, S.C.; Li, J.M.; Wang, F.; Wang, J.C.; Zhang, X.H. Hydraulic fracture geometry and proppant distribution in thin interbedded shale oil reservoirs. *Pet. Explor. Dev.* **2022**, *49*, 1025–1032. [CrossRef]
35. Xue, L.; Zhu, J.G.; Guo, C.F.; Ni, X.M.; Li, Y.; Cao, Y.X. Experimental optimization of concentration ratio of buffering acid used for removing plugging of coal seam. *Saf. Coal Mines* **2021**, *52*, 72–77.
36. Deng, Y.; Xue, R.J.; Guo, J.C. The mechanism of high-pressure high-temperature and low permeability acid pretreatment to reduce fracturing pressure. *J. Southwest Pet. Univ. (Sci. Technol. Ed.)* **2011**, *33*, 125–129+199.
37. Wang, S.L.; Dong, K.X.; Dong, H.Y. Effect analysis of perforating parameters upon initiation pressure in low permeability reservoir. *Oil Drill. Prod. Technol.* **2009**, *31*, 85–89.
38. Yu, J.Y.; Shen, F.; Gu, Q.H.; Ren, S.S. Influence of perforation parameters on hydraulic fracturing of fracture pressure in horizontal well. *Pet. Geol. Recovery Effic.* **2011**, *18*, 105–107.
39. Xue, S.F.; Sun, C.H.; Yu, H.B.; Sun, F. Effect of spiral perforation parameters on formation fracture pressure. *Well Test.* **2015**, *24*, 11–13+73.
40. Yang, Y.M.; Li, X.; Wang, Z.; Ju, Y. Influence of perforation parameters on propagation laws of hydraulic fracture in heterogeneous sandstones. *China Civ. Eng. J.* **2022**, *55*, 1–9.
41. Andreas, B.; Chavez, F.J.; Sergey, N.; Nihat, G.; Olga, A.; Sergey, C.; Dmitry, K.; Vasily, L. Impact of Perforation Tunnel Orientation and Length in Horizontal Wellbores on Fracture Initiation Pressure in Maximum Tensile Stress Criterion Model for Tight Gas Fields in the Sultanate of Oman. In Proceedings of the SPE Middle East Oil and Gas Show and Conference, Manama, Bahrain, 8–11 March 2015.
42. Liu, X.J.; Xiong, J.; Liang, L.X.; You, X. Rock mechanics characteristics and fracture propagation mechanism of glutenite reservoir in Baikouquan Formation of Mahu Sag. *Xinjiang Pet. Geol.* **2018**, *39*, 83–91.

43. Tang, Y.; Xu, Y.; Li, Y.Z.; Wang, L.B. Sedimentation model and exploration significance of large-scaled shallow retrogradation fan delta in Mahu Sag. *Xinjiang Pet. Geol.* **2018**, *39*, 16–22.
44. Qin, J.H.; Wang, J.G.; Li, S.Y.; Li, S.; Dou, Z.; Peng, S.M. Characteristics and formation mechanism of hydraulic fractures in tight conglomerate reservoirs of Triassic Baikouquan formation in Mahu Sag. *Lithol. Reserv.* **2023**, *35*, 29–36.

processes

MDPI

Article

Experimental Tests on In Situ Combustion Using Dynamic Ignition Simulation System in High-Temperature and High-Pressure Conditions

Yuchuan Yin [1], Xinyuan Chen [2,*], Xiaocong Yu [2], Deji Liu [1], Chao Chen [1], Xiaosong Zhou [1], Xiaohui Li [1], Lidong Zhang [1] and Changbin Kan [2,*]

1 Research Institute of Engineering Technology, Tuha Oilfield Branch Company, PetroChina, Beijing 838200, China; yinyuchuan@sohu.com (Y.Y.)
2 School of Earth Resources, China University of Geosciences, Wuhan 430074, China
* Correspondence: chenxinyuan@cug.edu.cn (X.C.); kanchangbin@cug.edu.cn (C.K.)

Abstract: The study of crude oil oxidation characteristics is fundamental to the design of ignition in situ combustion. Experimentation is the most crucial method for studying the oxidation characteristics of crude oil. Aiming to address the challenges posed by high temperature, high pressure, and rapid temperature changes during the combustion of crude oil, a dynamic simulation system for high-temperature and high-pressure ignition is designed. In order to study the oxidation characteristics of the crude oil ignition process, we conducted experiments using a high-temperature and high-pressure dynamic ignition simulation device. The experiments focused on determining the ignition point of crude oil under different pressure conditions, oil–water ratios, heating rates, gas injection rates, and other relevant characteristics. The kinetic model for the oxidation process of crude oil ignition was established. The kinetic parameters were calculated for different ignition conditions and the apparent activation energy for each oxidation stage was determined. Additionally, the stability of in situ combustion was evaluated under various ignition parameters. The results show that the Arrhenius curves for crude oil exhibit noticeable differences in the HTO (high-temperature oxidation) and LTO (low-temperature oxidation) regions. The curves demonstrate good linearity in the HTO region, with correlation coefficients exceeding 0.9. Moreover, the apparent activation energies in the HTO region range from 8.01 to 26.7 kJ/mol. The apparent activation energies and finger front factors were calculated for the HTO stage under different pressure conditions. The results showed that, as the pressure increased, the autoignition point, inflection point temperature, and apparent activation energy of the crude oil decreased. This suggests that increasing the pressure can enhance the HTO of the crude oil. The spontaneous ignition point of the crude oil exhibited an upward trend as the heating rate increased. Additionally, the maximum temperature during the combustion process generally increased with the heating rate, reaching a maximum temperature of 453.1 °C. The tests demonstrated that the simulation system is capable of real-time monitoring and recording of oxidation parameters during the combustion process of crude oil. This system can provide essential data for project implementation and numerical simulation.

Keywords: crude oil; dynamic ignition; in-situ combustion; HTHP; experimental Tests; HTO; LTO; numerical simulation

Citation: Yin, Y.; Chen, X.; Yu, X.; Liu, D.; Chen, C.; Zhou, X.; Li, X.; Zhang, L.; Kan, C. Experimental Tests on In Situ Combustion Using Dynamic Ignition Simulation System in High-Temperature and High-Pressure Conditions. *Processes* **2024**, *12*, 52. https://doi.org/10.3390/pr12010052

Academic Editors: Albert Ratner and Dicho Stratiev

Received: 13 November 2023
Revised: 11 December 2023
Accepted: 18 December 2023
Published: 25 December 2023

1. Introduction

In situ combustion is an oil extraction technology that can effectively enhance the recovery rate of oil reservoirs [1,2]. Heavy oil has poor fluidity in formation and is difficult to extract, so thermal extraction methods such as steam huff and puff, and steam flooding are commonly employed. After many rounds of steam injection, the water content in the heavy oil reservoir increases, leading to a decrease in production. In situ combustion, as a replacement development method to improve recovery after steam huff and puff, has the advantages of a high recovery rate and low heat loss. It is worth noting that, for low-permeability light oil reservoirs, in situ combustion can heat the reservoir skeleton and fluids through heat conduction and heat convection. This process prompts the fluids in the pores to expand and discharge on their own, thereby achieving enhanced recovery. The initiation of combustion is the most critical step in the in situ combustion process. The time it takes to ignite and establish a combustion front in the formation can vary, ranging from a few days to several weeks. This duration depends on the chemical composition and physical properties of the crude oil. Crude oils vary in terms of their chemical composition and physical properties, and, as a result, their oxidizing characteristics also differ. In situ combustion is different from conventional combustion. It is the process of replacing crude oil with combustion in a high-pressure porous medium. This process is extremely complex, involving both chemical and physical factors [3–5]. To study the combustion characteristics of in situ combustion, it is necessary to obtain the dynamics of the temperature field and other essential data. However, the process of in situ combustion takes place underground, making it impossible for researchers to obtain such dynamic information directly. Therefore, the aforementioned information must be obtained through indoor experiments.

In-house experiments are essential for studying in situ combustion, and allow for important qualitative and quantitative conclusions to be drawn. Thermal analysis experimental methods are commonly used in the laboratory. Thermal analysis experimental methods, such as TGA (Thermogravimetric Analysis) and DSC (Differential Scanning Calorimetry), are commonly employed in laboratory settings to investigate the oxidation mechanism of crude oil. In addition, in situ combustion indoor simulation experiments are also utilized for this purpose. In terms of studying crude oil oxidation using thermal analysis methods, Kok et al. utilized TGA to examine the oxidation characteristics of various fractions of crude oil at different heating rates [6]. They also analyzed the kinetic parameters using the iso-conversional method. Li et al. conducted TGA/DSC experiments to investigate the oxidation behaviors of three types of crude oils, and observed variations in oxidation behaviors between heavy oils and dilute oils [7]. Additionally, Zhao et al. explored the oxidation behavior of crude oil through oxidizing tube experiments [8]. The oxidation behavior of crude oil during the LTO (low-temperature oxidation) stage was studied, and the results indicated that various factors, including pressure, oxidation time, water saturation, and clay type, significantly influence the oxidation process of crude oil. Hu et al. conducted research on the oxidation characteristics of light oil, rock chips, and oil-bearing rock chips using TG/DTG and DTA techniques. They also analyzed the catalytic role of clay minerals in the oxidation of crude oil. Gundogar et al. conducted TGA/DSC experiments to analyze the oxidation behavior of six crude oils [9]. They also examined the oxidation characteristics of different crude oils in the LTO and HTO regions and calculated their kinetic parameters. Shokrlu et al. studied the impact of nickel ions on the combustion of crude oils using TGA and infrared spectroscopy [10]. The results indicated that nickel ions had a significant catalytic effect on the LTO of crude oils. It was found that the presence of nickel ions reduced the reaction activation energy from 16.9 kJ/mol to 10.9 kJ/mol. Ni et al. analyzed the kinetic properties of light oil oxidation in the presence or absence of rock chips using TGA [11]. The findings revealed that the clay medium exhibited a catalytic effect on the oxidation of heavy oil, resulting in a reduction in the apparent activation energy of the crude oil. Khelkhal et al. investigated the influence of an iron catalyst on the oxidation of heavy oil, using TGA, DSC, and SEM (scanning electron microscope) [12]. The results showed that the catalyst promoted both LTO and HTO (high-temperature oxidation). Yang et al. investigated the oxidation behavior of crude oil containing Ottawa sand, iron oxide, and pyrite using accelerated rate calorimetry (ARC). Temperature profiles and detailed kinetic data from the ARC tests were compared to analyze the oxidation characteristics of crude oil containing various mineral particles in the ARC vessel. The objective was to investigate the catalytic effect of iron minerals on the oxidation of crude oil. Thermal analysis experiments, which monitor changes in a single physical quantity to reflect crude oil oxidation, are easy to perform. However, they cannot be generalized to represent the overall oxidation behavior of crude oil. Additionally, it is challenging to maintain consistency between the experimental process and the actual conditions of the formation [13–15].

Unlike thermal analysis experimental devices, RTO (ramped temperature oxidation kinetic cell), CT (combustion tube), PMTEC (porous medium thermo-effect cell), and other devices can simulate the oxidation behavior of crude oil in a porous medium under the action of dynamic airflow [16–18]. The RTO test can obtain some basic parameters in ISC. Zhao et al. used RTO to measure the oxidation kinetic characteristics of different boiling fractions of Karamay crude oil in a porous medium [19]. They then compared it with the whole crude oil and used the iso-conversional method to analyze the results of RTO for each boiling fraction, obtaining the activation energy characteristics of each boiling fraction. Pu et al. used CT to study the low-temperature oxidation (LTO) of heavy crude oil in a porous medium [20]. They also conducted a comprehensive assessment of the in situ combustion (ISC) performance of the heavy oil and the alteration of oil properties before and after ISC. CT scanning enables the acquisition of crucial parameters such as the temperature distribution at the combustion front, variations in off-gas components, and the recovery rate during the ISC process [21]. However, CT experiments are often large in scale. For instance, the CT developed by the University of Calgary has a combustion tube with a diameter of 0.1 m and a length of 1.825 m. Each group of repulsion experiments takes more than 24 h, making the operation process very complicated [22]. This complexity requires a significant amount of manpower and material resources. Yuan et al. developed PMTEC, which allows for the study of crude oil combustion behavior in a porous medium by directly monitoring the temperature signal, similarly to TGA. PMTEC has the advantages of saving time and materials [17,23]. However, it cannot simulate the high-pressure environment in the formation. The ISC process is accompanied by high temperature and high pressure, so the simulation device needs to have high-temperature and high-pressure resistance, as well as high sealing.

2. System Components and Design

The high-temperature and high-pressure dynamic ignition simulation system is used to simulate the ignition state of crude oil under formation conditions. It needs to fulfill the following functions: studying the impact of reservoir parameters (such as porosity, permeability, crude oil saturation, reservoir temperature, etc.) and injection parameters (such as gas injection rate, gas injection pressure, etc.) on the underground combustion of crude oil; examining the influence of additives (solid, liquid) on the ignition of crude oil; and collecting dynamic information during the process of crude oil combustion (such as temperature, pressure, component data, etc.) to analyze the kinetic parameters of crude oil oxidation and provide data for numerical simulation of ISC.

The simulation system can accurately control the heating-up program, collect temperature field pressure field data, and collect real-time exhaust gas component data. It consists of six systems, including ignition simulation system, safety protection system, discharge system, online exhaust gas monitoring system, exhaust gas treatment and discharge system, and data acquisition and control system. The principle of the high-temperature and high-pressure dynamic ignition simulation system is shown in Figure 1.

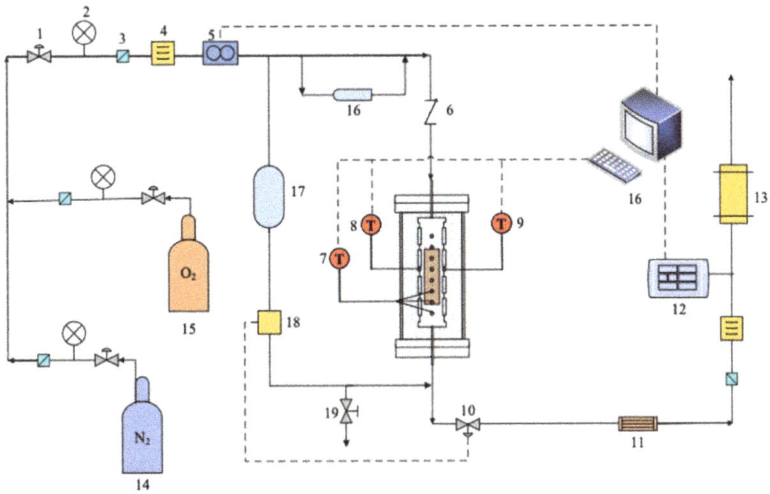

Figure 1. Schematic diagram of high-temperature and high-pressure dynamic ignition simulation device. Note: 1—Pressure regulating valve; 2—Pressure gauge; 3—Filter; 4—Dryer; 5—Mass flow controller; 6—Check valve; 7—Internal thermocouple in core; 8—External thermocouple; 9—Temperature-control thermocouple; 10—Back-pressure control valve; 11—Gas–liquid separator; 12—Online exhaust monitoring system; 13—Wash gas cylinder; 14—Nitrogen cylinder; 15—Oxygen cylinder; 16—Liquid injector; 17—Intermediate vessel; 18—Back-pressure controller; 19—Pressure relief valve.

2.1. *Ignition Simulation System*

2.1.1. Combustion Tube Structure Design

The combustion tube is 37 cm long with an inner diameter of 2.5 cm and is characterized by a high length-to-diameter ratio (L/D). Due to the end effect, the temperature at the two ends is generally lower than that in the center. However, the axial temperature of the combustion tube with a large L/D ratio is more uniform. There are 11 temperature probes inside the core at 1.5 cm intervals, as illustrated in Figure 2. Thermocouples are also installed in the middle of the outer wall of the combustion tube to monitor its temperature. Additionally, thermocouples are placed on the opposite side of the outer wall of the combustion tube. These thermocouples are connected to the temperature control system, allowing for temperature regulation through the collection of temperature data. Both ends of the combustion tube are sealed with a double-layer high-temperature graphite ring sealing structure. The combustion tube is made of GH4169 alloy, which exhibits excellent high-temperature resistance and thermal fatigue resistance. There is a heat-insulating layer on the outside of the combustion tube and this layer is in close proximity to the outer wall of the combustion tube. The heating layer is connected to the control module through a specialized cable. The control module determines the output power of the heating layer in order to accurately regulate the heating of the core. The heating layer can reach a maximum temperature of 500 °C with a sampling resolution of 0.1 °C.

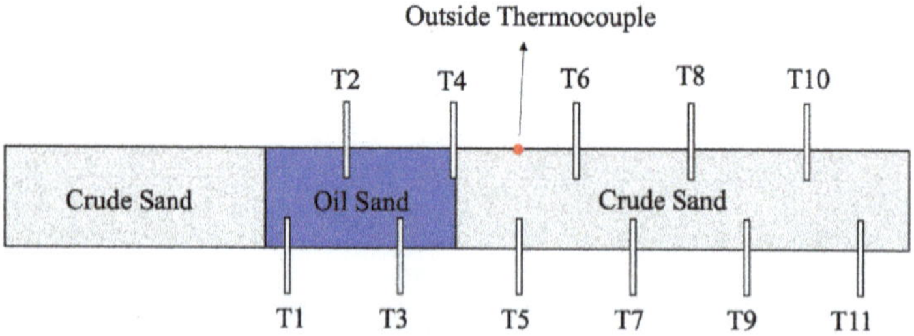

Figure 2. Combustion tube structure.

2.1.2. Temperature Control Design

The heating power is precisely controlled through the thermocouple feedback loop, which can maximize the simulation of the stratum heat storage process. The temperature control module includes the control module, processor module, and communication serial port module. As shown in Figure 3, the control module adopts the combination of fuzzy theory and PID control, taking the ignition simulation system as the controlled object, the J-type thermocouple located in the middle of the outer wall of the combustion pipe as the sensor, and the heating layer as the actuator. The temperature information collected by the thermocouple is fed back to the controller, which adjusts the PID parameters online after self-tuning by adding fuzzy control and outputs the optimal control parameters, thus adjusting the output power of the heating layer to keep the output temperature consistent with the target temperature, and the temperature control accuracy is ±0.1 °C.

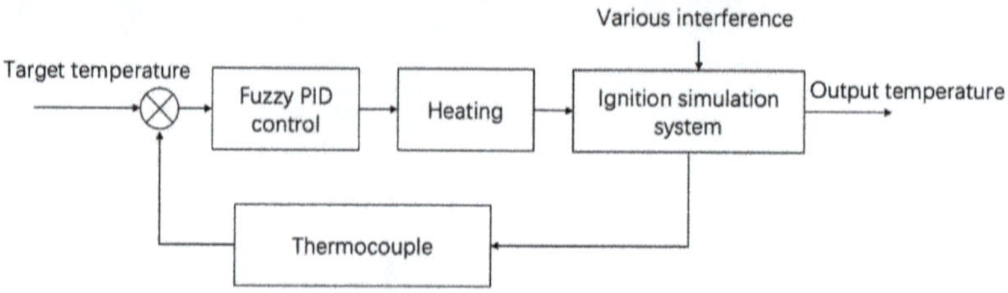

Figure 3. Block diagram of temperature control design.

2.1.3. Combustion Tube Sealing and Stability Design

The two ends of the combustion tube are sealed with double-layer graphite rings that are resistant to high temperatures. The structure of the sealing device is divided into five parts: mold sleeve, compression block, mandrel, wedge-shaped graphite ring, gasket, and nut. The die sleeve has two different diameters of internal threads to connect the combustion pipe body and the tightening block. The mandrel's role is to compact the core. The lower part of the ring limit can make contact with the wedge-shaped graphite ring, while the upper part has external threads for connecting the nut. Additionally, the interior of the mandrel has a through hole that serves as a gas channel. The compacting block has an external thread that connects it to the mold sleeve through the connection. The mandrel, wedge-shaped graphite ring, ring gasket, and compression block can be sequentially assembled into a combination of parts. The mold sleeve is used to secure the assembled parts and, at this point, the core is compressed. However, due to the loose pressing of the wedge-shaped graphite ring, the sealing is not effective. To improve the sealing, it is necessary to tighten the nut, which will expand the wedge-shaped graphite ring and ensure close contact with the inner wall of the combustion tube for high-pressure sealing. According to the test results, the ignition simulation system can withstand pressures exceeding 10 MPa.

2.2. Online Exhaust Gas Monitoring System

An online exhaust gas monitoring system can monitor the exhaust gas in real time and has an alarm function. The online exhaust gas monitoring system includes the main instrument box, gas monitoring controller, and external serial port, among other components. The main instrument box is responsible for collecting data on exhaust gas components. The main instrument box is responsible for collecting data on exhaust gas components. It is equipped with built-in sensors for CH_4, O_2, H_2S, CO, CO_2, H_2, and N_2 gases. These sensors are connected in series through a hose. The parameters of the gas sensors are shown in Table 1. When the concentration of a specific gas exceeds or falls below the predetermined alarm threshold, it will activate either the high-level alarm or the low-level alarm, thereby ensuring the safety of the experimental procedure.

Table 1. Gas sensor measurement parameters.

Gases	Measuring Range	Resolution
CH_4	(0–100)% LEL	1% LEL
O_2	(0–30.0)% LEL	0.1% LEL
H_2S	(0–100) PPM	1 PPM
CO	(0–1000) PPM	1 PPM
CO_2	(0–30.0)% LEL	0.1% LEL
H_2	(0–100)% LEL	1% LEL
N_2	(0–100)% LEL	1%LEL

2.3. Injection and Discharge System

The function of the injection system is to simulate the gas circulation state under stratum conditions and it has four roles: first, supplying oxygen and gas to the system; second, controlling the flow; third, controlling back pressure; and fourth, injecting liquid. The flow control section consists of a gas mass flow meter and a flow controller. The gas mass flow meter is connected to the flow indicator using a specialized cable. The back-pressure control section consists of a back-pressure control valve, a back-pressure controller, and an intermediate container. The back-pressure controller collects pressure information and transmits it to a PC through a specialized cable. The PC then outputs control parameters based on the pressure information, and the back-pressure controller adjusts the opening and closing of the valve.

The tail gas generated in the in situ combustion ignition experiment contains water vapor, core particles, and oil, among other substances. In order to prevent pipeline clogging and damage to downstream equipment, it is necessary to treat the tail gas. The treatment process includes depressurization, filtration, dehydration, drying, and other steps.

3. Experimental Section

3.1. Materials Preparation

Experimental samples: Tuha oilfield light crude oil with viscosity of 0.74 mPa-s was used in the ignition experiment. The crude oil used for all the experiments was the same light crude oil and the properties of the crude oil are shown in Table 2.

Table 2. Oil properties.

Oil Properties		Value
API gravity (°)		41.1
Viscosity (mPa·s) 50 °C		0.74
SARA fractions (%)	Saturates	84.35
	Aromatics	11.31
	Resins	4.09
	Asphaltenes	0.25

Pre-experiment preparation: 400-mesh quartz sand was mixed with crude oil and brine in the specified ratio. The mixture was then filled into the middle section of the combustion tube, ensuring it was compacted. The two ends of the core were filled with 70-mesh quartz sand. Fine mesh screens were placed near the air inlet and outlet to prevent sand particles from entering the pipeline and causing blockages.

3.2. Experimental Procedure

1. Close the outlet valve and inject nitrogen to bring the back pressure to 4 MPa.
2. Once the pressure has stabilized, open the outlet valve and adjust the mass flow controller to maintain a flow rate of 300 mL/min. Keep the flow at this rate for 5 min.
3. If it is necessary to add liquid additives, place the additives in the liquid cylinder of the liquid injector. Inject nitrogen at a flow rate of 500 mL/min and continue injecting for 5 min. Then, decrease the flow rate to 300 mL/min and switch to air once stabilization is achieved. If there is no need to add liquid additives, switch to using air directly.
4. Once the flow rate has stabilized, begin heating and set the heating rate on the PC. Start recording the dynamic parameters of the combustion process.
5. When all temperature measurements at the measurement points remain stable for 30 min, close the cylinder valve to stop gas injection. Slowly adjust the pressure relief valve to ensure that the system pressure drops to atmospheric pressure.
6. Wait for the device to cool down and then disassemble the core, which is divided into sections at 3 cm intervals, weighed and sealed for later analysis and processing.

3.3. Experimental Test

In order to study the characteristics of crude oil oxidation and the factors that influence it during the in situ combustion process, ignition experiments were conducted to evaluate the factors that influence the spontaneous combustion of light crude oil under dynamic gas flow. The experimental parameters were designed as shown in Table 3.

Table 3. Experimental parameter table.

Serial Number	Experimental Sample	Oil–Water Ratio	Sample Quality g	Heating Rates °C/min	Gas Injection Rate mL/min	Gas Injection Pressure MPa
1 2 3	light crude oil	7:3	5.8	4	300	1 2 4
4 5	light crude oil	7:3	5.8	4	200 400	4
6 7	light crude oil	6:4 5:5	5 4.1	4	300	4
8 9 10	light crude oil	7:3	5.8	1 2 6	300	4

3.3.1. Blank Heating Experiment

In the ignition simulation experiment, in order to simulate the actual formation conditions, the temperature of the combustion tube should be consistent with the reservoir temperature. Therefore, a blank experiment was designed to verify the thermal tracking compensation effect of the experimental device. The temperature difference between the core wall temperature and the warming rate is illustrated in Figure 4. As the warming rate increases, the temperature difference also increases. This is due to the airflow carrying away the heat from the core and the porous medium requiring time for heat transfer. Therefore, a temperature difference exists. At the heating rate of 0.5 °C/min, the temperature difference is only 6.3 °C.

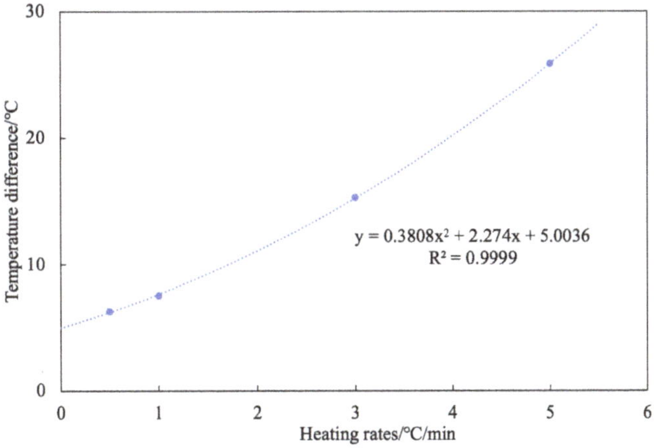

Figure 4. Relationship curve between temperature difference and heating rate.

3.3.2. Ignition Test

The purpose of the ignition simulation experiment is to obtain various key parameters in the oxidation process of crude oil for subsequent numerical simulation and combustion analysis. The data from the ignition simulation experiment were organized to generate curves depicting the core temperature and oxygen consumption rate over time, as illustrated in Figure 5.

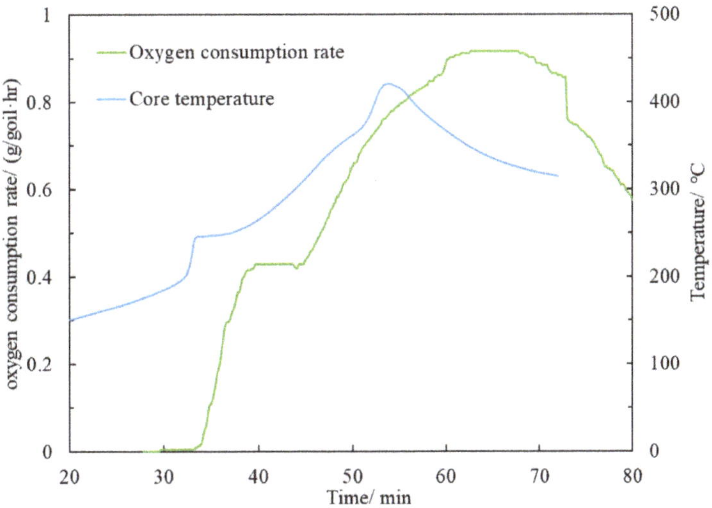

Figure 5. Curve of core temperature and oxygen consumption rate in ignition experiment.

According to the theory of thermal spontaneous combustion, combustion occurs when the exothermic factors of a reaction interact with the heat dissipation factors of the system [24]. When the rate of exothermic reaction exceeds the rate of heat dissipation in the system, the reaction rate is further accelerated, indicating that the reactants have entered the combustion state. The definition of spontaneous combustion point is when the temperature at the measurement point exceeds and continues to exceed the temperature of the wall. With the continuous exothermic LTO, the heat storage of the reaction system prompts the crude oil to undergo spontaneous oxidation. At 102 min, the core temperature reaches 189.1 °C, causing the experimental oil samples to reach the ignition temperature. As a result, the crude oil oxygen consumption rate and temperature characteristics undergo significant changes. The O_2 in the tail gas is completely consumed and a large amount of O_2 is involved in the reaction. This leads to a rapid rise in temperature at the front end of the oil sands section and the movement of the thermal leading edge can be observed in the temperature curves [25].

The experimental results were processed to obtain the spontaneous ignition point and the maximum temperature under different heating rates, pressures, gas injection rates, and oil–water ratios, as shown in Figures 6–9.

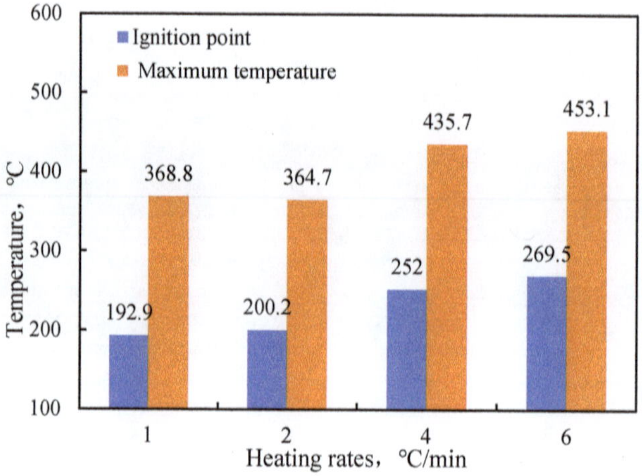

Figure 6. Spontaneous ignition point and maximum temperature of crude oil at different heating rates.

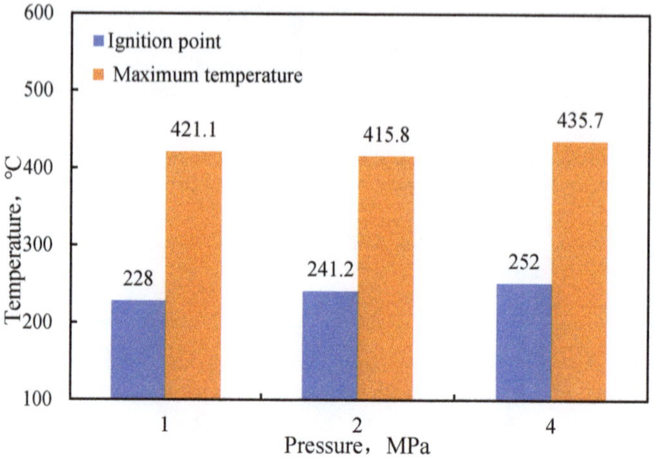

Figure 7. Spontaneous ignition point and maximum temperature of crude oil at different pressures.

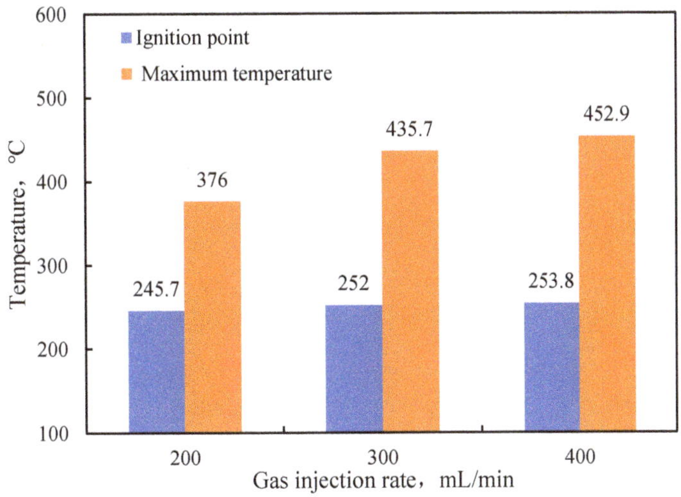

Figure 8. Spontaneous ignition point and maximum temperature of crude oil at different gas injection rates.

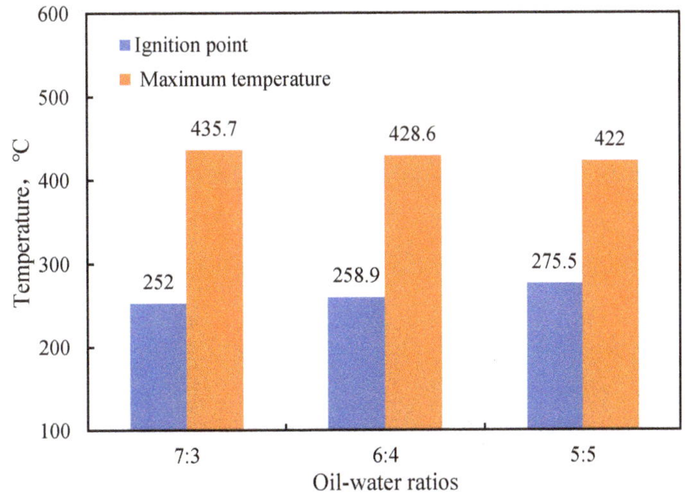

Figure 9. Spontaneous ignition point and maximum temperature of crude oil at different oil–water ratios.

As shown in Figure 6, the auto-ignition point of crude oil exhibited an upward trend as the heating rate increased. Additionally, the maximum temperature during combustion generally increased with the heating rate, reaching a maximum temperature of 453.1 °C. This is because the slower rate of temperature rise prolongs the low-temperature oxidation stage of crude oil. Although some of the light crude oil undergoes phase change and volatilization, a larger portion of the crude oil undergoes oxidation. This increases the viscosity and fuel content of the oil, promoting the movement of the crude oil to LTO and subsequently lowering the auto-ignition point. As shown in Figure 7, the spontaneous ignition point of crude oil does not change significantly with an increase in pressure. However, the maximum temperature is slightly higher compared to the maximum temperature observed in the lower pressure experiment at 4 Mpa pressure. This is due to the fact that the oil sample is lighter, making it easier to volatilize. The increase in pressure inhibits the volatilization of the lighter components. At high pressure, the oil may undergo a violent oxidation reaction at the gas-phase or gas-phase and liquid-phase interface, resulting in an increased exothermic rate. As shown in Figure 8, the spontaneous ignition point of crude oil remains relatively unchanged with an increase in gas injection rate. However, the maximum temperature does increase as the gas injection rate increases. As shown in Figure 9, the spontaneous combustion point of crude oil decreases with an increase in oil saturation, while the maximum temperature exhibits an increasing trend.

3.3.3. Apparent Activation Energy Analysis

Crude oil oxidation kinetic parameters can be further calculated using a high-temperature and high-pressure ignition simulator [16]. The rate of change of the reactant concentration, obtained from the derivation, is used to characterize the crude oil oxidation rate according to the law of mass action [26]:

$$-\frac{dC}{dt} = kC^n \quad (1)$$

where C is the concentration of the reactant, mol/L; t is the time, s; k is the rate function of Arrhenius equation; and n is the number of reaction stages, which is taken as 1 here.

The rate function k can be expressed by the Arrhenius equation:

$$k = Ae^{-\frac{E}{RT}} \quad (2)$$

where A is the prefinger factor, L/(s·kPa); E is the activation energy, kJ/mol; R is the gas constant, 8.314 J/(mol-K); and T is the temperature, K. In order to ensure accurate calculations, the temperature of the core section is determined by taking the weighted average of the temperatures measured by the three thermocouples located in the core section.

Since oxygen serves as the oxidizing agent in the process of crude oil oxidation, the calculations of the kinetic parameters were conducted by using the oxygen concentration as an indicator of reactant generation. This involved incorporating Equation (2) into Equation (1):

$$-\frac{dC}{dt} = C^n Ae^{-\frac{E}{RT}} \quad (3)$$

The inner diameter of the combustion tube is $\phi = 2.5 \times 10^{-2}$ m, the cross-sectional area is $S = 4.91 \times 10^{-4}$ m^2, and the length of the core is L = 5 cm. Assuming that the injected gas is an ideal gas, the equation for the rate of oxygen consumption of a unit volume of an oil-bearing core located at X at a temperature of T is known from the theory of crude oil oxidation:

$$\frac{dC_{O_2}}{dt} = -\frac{dC_{O_2}^x}{dx} \cdot \frac{Q}{S} \quad (4)$$

where $dC_{O_2}^x$ is the consumption of oxygen in dx section, mol/m^3 and Q is the gas flow rate, m^3/s.

Substitute Equation (4) into Equation (3) and take the natural logarithm to obtain the following:

$$\ln\left(ln\left(\frac{C_o}{C_L}\right)\right) = ln\left(\frac{SLA}{Q}\right) - \frac{E}{R} \cdot \frac{1}{T} \quad (5)$$

where C_o is the molar concentration of oxygen at the inlet end, mol/m^3 and C_L is the molar concentration of oxygen at the outlet end, mol/m^3.

The temperature and exhaust gas component data from experiments 1 to 3 were processed to obtain the curve of $\ln\left(ln\left(\frac{C_o}{C_L}\right)\right)$ versus T^{-1} under different pressure conditions, as shown in Figure 10.

As depicted in Figure 10, the relationship between $\ln\left(ln\left(\frac{C_0}{C_L}\right)\right)$ and T^{-1} is not entirely linear as the system temperature increases. Instead, it exhibits a distinct stage characterized by an inflection point. The curves only demonstrate better linearity when the temperature surpasses a specific inflection point temperature. This is because the reaction mechanism of crude oil in different oxidation stages varies. In the LTO oxidation stage, the oxidation reaction is not the sole factor, as it is accompanied by physical changes such as phase change and volatilization. Therefore, it becomes challenging to demonstrate the linear relationship depicted in the Arrhenius equation. When the temperature exceeds the inflection point, the oxidation reaction is rapidly enhanced and enters the high-temperature oxidation (HTO) stage. As the temperature rises, the water and light components in the core become volatile. This leads to a change in temperature and concentration of the output material, which is primarily caused by HTO [6]. The temperature range of the HTO stage is about 230~400 °C, the linearity of the curve is improved, and all correlation coefficients are above 0.9.

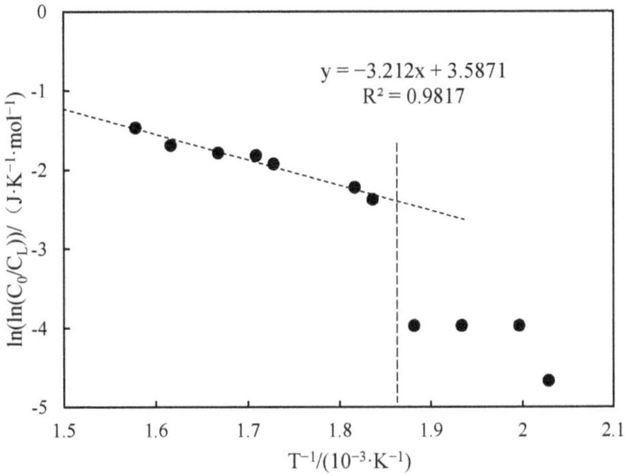

(a)

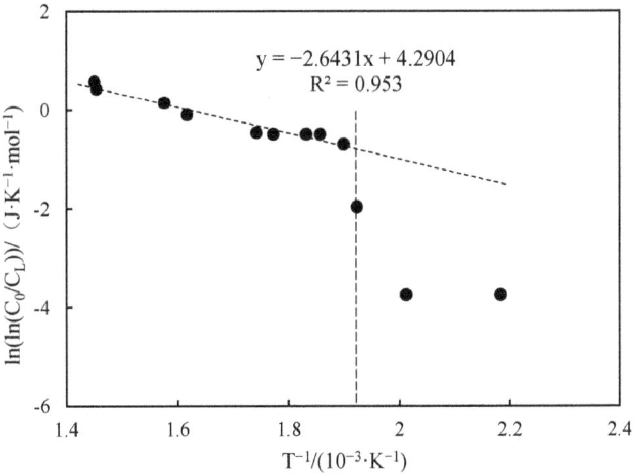

(b)

Figure 10. *Cont.*

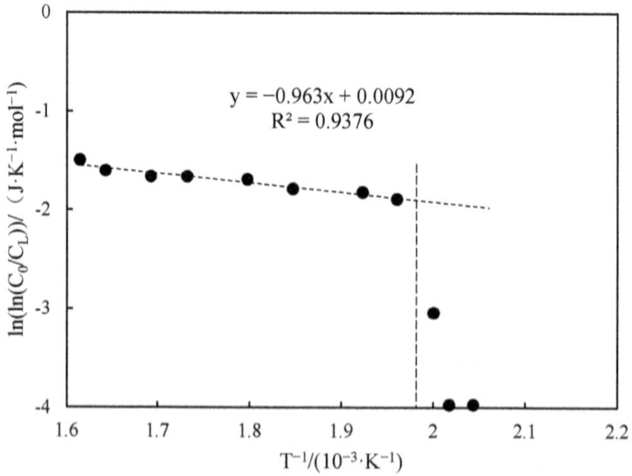

(c)

Figure 10. Variation curve of $\ln\left(ln\left(\frac{C_0}{C_L}\right)\right)$ with T^{-1} at different pressures. (**a**) 1 MPa. (**b**) 2 MPa. (**c**) 4 MPa.

The curve of the HTO stage was linearly fitted, and the apparent activation energy, finger front factor, and inflection point temperature were determined based on the fitted curve. The calculation results are shown in Table 4. From the data in the table, it can be observed that, as the pressure increases, the inflection point temperature decreases and the apparent activation energy decreases. This suggests that, during the combustion initiation stage, increasing the pressure raises the partial pressure of oxygen and enhances the heat transfer efficiency of the medium, resulting in a lower apparent activation energy and facilitating the oxidation of the crude oil [27].

Table 4. Calculated oxidation kinetic parameters at various pressures.

Number	Fitted Linear Slope	Fitted Linear Intercept	Correlation Coefficient	Inflection Temperature (°C)	Apparent Activation Energy (kJ·mol^{-1})	Prefactor (s^{-1})	Pressures (MPa)
1	−3.212	3.587	0.981	267	26.70	8.41	1
2	−2.643	4.290	0.953	253	21.97	16.99	2
3	−0.963	0.009	0.937	234	8.01	0.23	4

The temperature and exhaust gas component data from experiments 4 to 5 were processed to obtain the curves of $\ln\left(ln\left(\frac{C_0}{C_L}\right)\right)$ versus T^{-1} under the conditions of different injection velocities, which are shown in Figure 11. The curves were processed to obtain the oxidation kinetic parameters, as shown in Table 5.

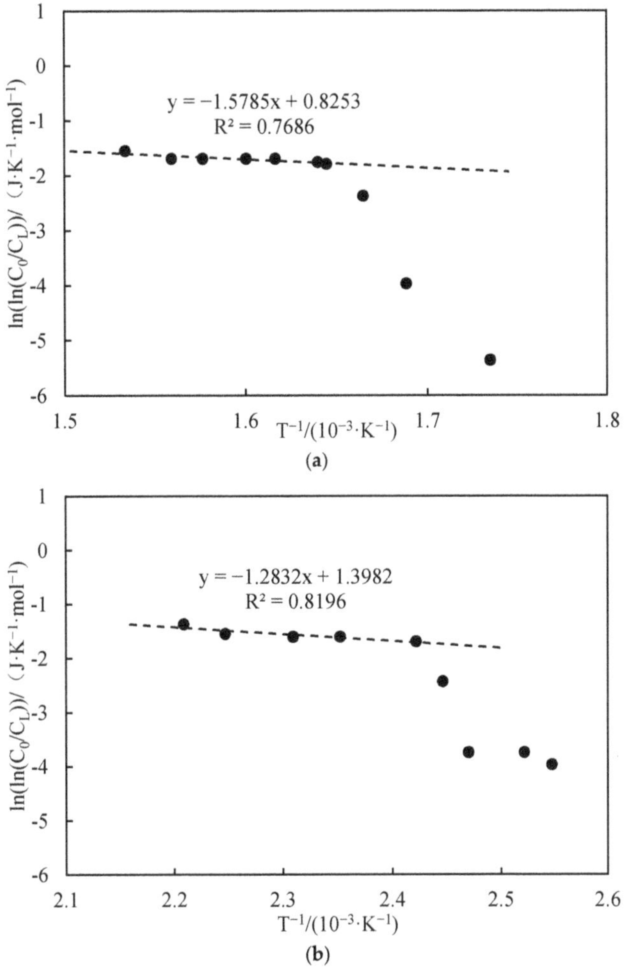

Figure 11. Variation curve of $\ln\left(ln\left(\frac{C_o}{C_L}\right)\right)$ with T^{-1} at different gas injection rates. (**a**) 200 mL/min. (**b**) 400 mL/min.

Table 5. Calculated results of oxidation kinetic parameters at various gas injection rates.

Number	Fitted Linear Slope	Fitted Linear Intercept	Correlation Coefficient	Inflection Temperature (°C)	Apparent Activation Energy (kJ·mol^{-1})	Prefactor (s^{-1})	Gas Injection Rate (mL/min)
4	−1.578	0.825	0.768	333	13.12	0.53	200
3	−0.963	0.009	0.937	234	8.01	0.23	300
5	−1.283	1.398	0.819	140	10.67	0.94	400

When the gas injection rate is 200 mL/min, the correlation coefficient of its high-temperature section is only 0.7686, indicating a poor fit. Additionally, the apparent activation energy is higher compared to the experiment with a high gas injection rate. The inflection point temperature is also as high as 333 °C, suggesting insufficient oxygen supply at this time. As a result, the accumulation of heat generated by HTO is slower, causing the inflection point to be reached only at a relatively higher temperature. Consequently, the apparent activation energy is also higher. When the gas injection rate reaches 400 mL/min, the inflection point temperature is only 140 °C. At this point, the heat generated by LTO is greater than the reservoir warming and heat loss [28]. Therefore, in the in situ combustion ignition process, attention should be paid to the design of the air injection speed during ignition. The injection air speed must be controlled within a certain range. If it is too low, it cannot ensure the maintenance of low-temperature oxidation. On the other hand, if it is too high, it is likely to sweep the formation, causing cooling. The cold air carries LTO heat into the depths of the formation and the accumulation of heat cannot meet the ignition requirements. Consequently, the success of the ignition is bound to decline.

The temperature and exhaust gas component data from experiments 6 and 7 were processed to obtain the curve of $\ln\left(ln\left(\frac{C_o}{C_L}\right)\right)$ versus T^{-1} under different oil–water ratios, as shown in Figure 12. The curves were processed to obtain the oxidation kinetic parameters, as shown in Table 6.

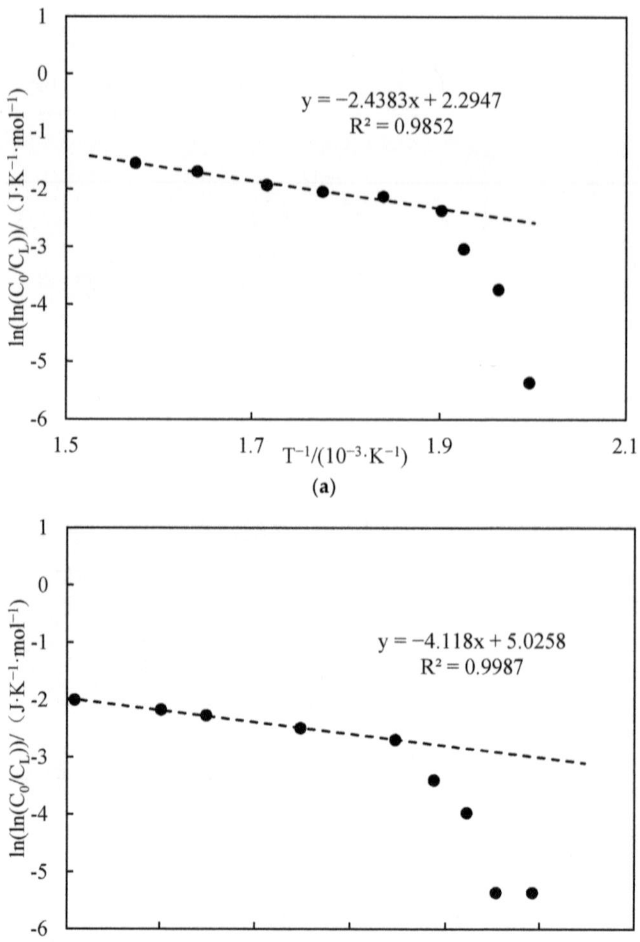

Figure 12. Variation of $\ln\left(ln\left(\frac{C_o}{C_L}\right)\right)$ with T^{-1} at different oil–water ratios. (**a**) oil-water ratios 6:4. (**b**) oil-water ratios 7:3.

Table 6. Calculated results of oxidation kinetic parameters at various oil–water ratios.

Number	Fitted Linear Slope	Fitted Linear Intercept	Correlation Coefficient	Inflection Temperature (°C)	Apparent Activation Energy (kJ·mol^{-1})	Prefactor (s^{-1})	Oil–Water Ratio
6	−2.438	2.294	0.985	253	20.27	2.31	6:4
7	−4.118	5.025	0.998	260	34.24	35.47	5:5
3	−0.963	0.009	0.937	234	8.01	0.23	7:3

From Figure 12, it can be seen that the crude oil can still reach the combustion state even under high water saturation conditions (50%). With the increase in oil saturation, both the inflection point temperature and apparent activation energy also increase.

The temperature and exhaust gas component data from experiments 8 to 10 were processed to obtain the curves of $\ln\left(ln\left(\frac{C_o}{C_L}\right)\right)$ versus T^{-1} under the conditions of different heating rates, as shown in Figure 13. The curves were processed to obtain the oxidation kinetic parameters, as shown in Table 7.

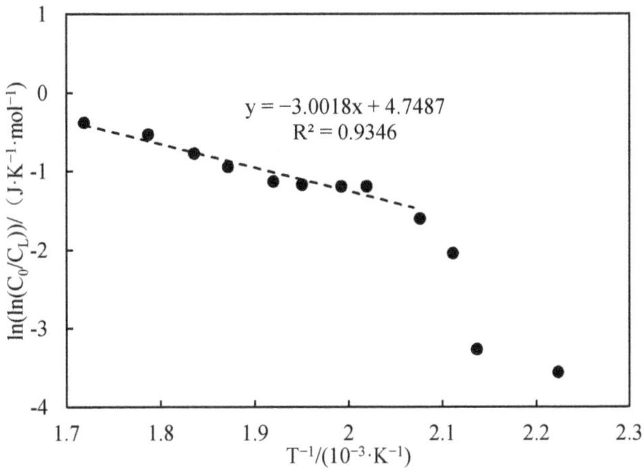

(a)

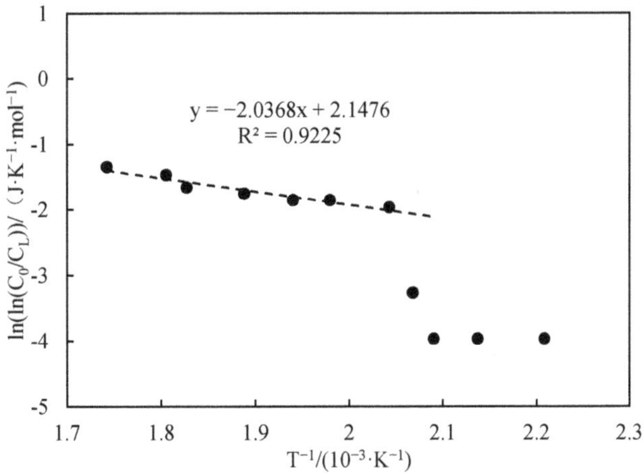

(b)

Figure 13. *Cont.*

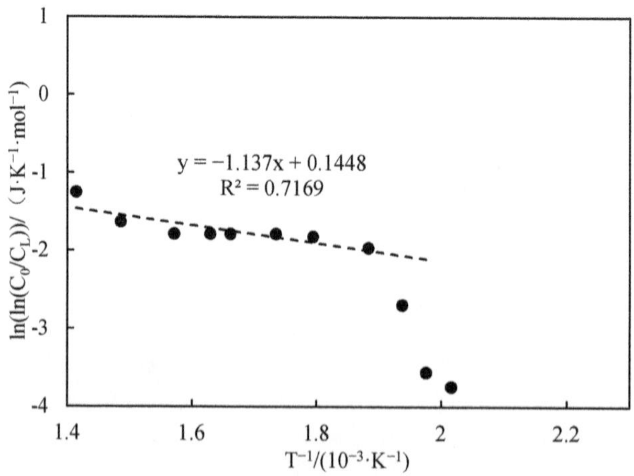

(c)

Figure 13. Variation curves of $\ln\left(ln\left(\frac{C_o}{C_L}\right)\right)$ with T^{-1} at different heating rates. (**a**) heating rates 1 °C/min. (**b**) heating rates 2 °C/min. (**c**) heating rates 6 °C/min.

Table 7. Calculated results of oxidation kinetic parameters at various heating rates.

Number	Fitted Linear Slope	Fitted Linear Intercept	Correlation Coefficient	Inflection Temperature (°C)	Apparent Activation Energy (kJ·mol^{-1})	Prefactor (s^{-1})	Heating Rates
8	−3.001	4.748	0.934	209	24.96	26.88	1
9	−2.036	2.147	0.922	217	16.93	1.99	2
10	−1.137	0.144	0.716	258	9.45	0.39	6
3	−0.963	0.009	0.937	234	8.01	0.23	4

As can be seen from Figure 13, the inflection point temperature rises with the increase in heating rate and the apparent activation energy generally shows a decreasing trend. With the increase in heating rate, the onset temperature of crude oil HTO gradually shifts backward and the thermal hysteresis phenomenon becomes more pronounced. This is because an increase in the heating rate results in a shorter duration of the fuel deposition stage. When the heating rate is low, the crude oil and oxygen can be well contacted and react in the LTO oxidation reaction stage. However, when the heating rate is increased, the speed of the crude oil oxidation reaction will accelerate, resulting in a shorter oxidation time for the different oxidation stages. This phenomenon is known as thermal hysteresis [29]. When the heating rate is 6°C/min, the correlation coefficient of the HTO curve is only 0.7169, which indicates a poor fit. This suggests that, as the heating rate increases, the reaction time for each reaction stage of the crude oil becomes shorter and shorter. Consequently, the reactions become less sufficient, leading to the overlapping of oxidation stages.

4. Conclusions

7. The experimental system can achieve real-time monitoring of crude oil combustion in situ. The combination of fuzzy theory and PID control enables thermal tracking compensation, achieving a temperature control accuracy of ±0.1 °C. This device is designed to meet the experimental requirements of high-pressure, rapid temperature changes and complex gas production components during crude oil combustion. It has a pressure resistance of 10 MPa, a sampling resolution of 0.1 °C, and the capability to monitor and record the content of seven tail gas components in real time.

8. Functional experiments showed that the variation of $\ln\left(ln\left(\frac{C_0}{C_L}\right)\right)$ with T^{-1} during the experimental process exhibited a distinct stage bounded by an inflection point. The linearity of the curves improved only when the temperature exceeded a specific inflection point temperature. The apparent activation energies and finger front factors were calculated for the HTO stage under different pressure conditions. The results showed that, as the pressure increased, the auto-ignition point, inflection point temperature, and apparent activation energy of the crude oil decreased. This indicates that increasing the pressure can promote the HTO of the crude oil. The spontaneous ignition point of the crude oil exhibited an upward trend as the heating rate increased. Additionally, the maximum temperature during the combustion process generally rose with the increase in heating rate, reaching a maximum temperature of 453.1 °C.
9. Using the experimental data obtained from the high-temperature and high-pressure dynamic ignition simulation system, it is possible to calculate the kinetic parameters of crude oil oxidation. This calculation provides essential data support for the implementation of in situ combustion and numerical simulation.

Author Contributions: Conceptualization, Y.Y. and X.Y.; writing—review and editing, X.C.; methodology, C.K. and D.L.; data curation, X.Z.; writing—original draft preparation, Y.Y. and C.C.; investigation, X.L. and L.Z. All authors have read and agreed to the published version of the manuscript.

Funding: The study received funding from the National Science Foundation of China (Grant No. 52004259).

Data Availability Statement: All data used to support the findings of this study are available from the corresponding author on request.

Conflicts of Interest: Authors Yuchuan Yin, Deji Liu, Chao Chen, Xiaosong Zhou, Xiaohui Li, and Lidong Zhang were employed by the Tuha Oilfield Branch Company, PetroChina. The remaining authors declare that the research was conducted in the absence of any commercial or financial relationships that could be construed as a potential conflict of interest.

References

1. Shah, A.; Fishwick, R.; Wood, J.; Leeke, G.; Rigby, S.; Greaves, M. A review of novel techniques for heavy oil and bitumen extraction and upgrading. *Energy Environ. Sci.* **2010**, *3*, 700–714. [CrossRef]
2. Changfeng, X.; Wenlong, G.; Jihong, H. Research and Application of Fire-Flooding Technologies in Post-Steam Injected Heavy Oil Reservoir. In Proceedings of the International Petroleum Technology Conference, Beijing, China, 26–28 March 2013.
3. Hamdy, M.; Mohamed, A.A.; Curran, H. Combination Effect of in-Situ Combustion and Exhaust Gases Recirculation on 1D Combustion Tube: Numerical Approach. *Combust. Sci. Technol.* **2023**, *195*, 3296–3309. [CrossRef]
4. Farahani, M.D.; Abdrabou, M.; Zhang, H.; Zhu, J.; Wang, F.; Lee, K.; Zheng, Y. In situ burning of crude oils using iron oxide nanoparticles as additives. *Fuel* **2022**, *330*, 125568. [CrossRef]
5. Yang, L.F.; Sheng, J.J. Pyrite effects on the oxidation of in situ crude oil. *J. Pet. Sci. Eng.* **2022**, *209*, 109812. [CrossRef]
6. Kok, M.V.; Iscan, A.G. Catalytic effects of metallic additives on the combustion properties of crude oils by thermal analysis techniques. *J. Therm. Anal. Calorim.* **2001**, *64*, 1311–1318. [CrossRef]
7. Li, J.; Mehta, S.A.; Moore, R.G.; Ursenbach, M.G. New Insights Into Oxidation Behaviours of Crude Oils. *J. Can. Pet. Technol.* **2009**, *48*, 12–15. [CrossRef]
8. Zhao, J.-Z.; Jia, H.; Pu, W.-F.; Wang, L.-L.; Peng, H. Sensitivity Studies on the Oxidation Behavior of Crude Oil in Porous Media. *Energy Fuels* **2012**, *26*, 6815–6823. [CrossRef]
9. Gundogar, A.S.; Kok, M.V. Thermal characterization, combustion and kinetics of different origin crude oils. *Fuel* **2014**, *123*, 59–65. [CrossRef]
10. Hamedi Shokrlu, Y.; Maham, Y.; Tan, X.; Babadagli, T.; Gray, M. Enhancement of the efficiency of in situ combustion technique for heavy-oil recovery by application of nickel ions. *Fuel* **2013**, *105*, 397–407. [CrossRef]
11. Ni, J.; Jia, H.; Pu, W.; Jiang, H.; Yang, J.; Ren, Q. Thermal kinetics study of light oil oxidation using TG/DTG techniques. *J. Therm. Anal. Calorim.* **2014**, *117*, 1349–1355. [CrossRef]
12. Khelkhal, M.A.; Eskin, A.A.; Sitnov, S.A.; Vakhin, A.V. Impact of Iron Tallate on the Kinetic Behavior of the Oxidation Process of Heavy Oils. *Energy Fuels* **2019**, *33*, 7678–7683. [CrossRef]
13. Jia, H.; Zhao, J.-Z.; Pu, W.-F.; Liao, R.; Wang, L.-L. The Influence of Clay Minerals Types on the Oxidation Thermokinetics of Crude Oil. *Energy Sources Part A Recovery Util. Environ. Eff.* **2012**, *34*, 877–886. [CrossRef]

14. Li, J.; Mehta, S.A.; Moore, R.G.; Ursenbach, M.G.; Zalewski, E.; Ferguson, H.; Okazawa, N.E. Oxidation and ignition behaviour of saturated hydrocarbon samples with crude oils using TG/DTG and DTA thermal analysis techniques. *J. Can. Pet. Technol.* **2004**, *43*, 45–51. [CrossRef]
15. Li, J.; Mehta, S.A.; Moore, R.G.; Zalewski, E.; Ursenbach, M.G.; Van Fraassen, K. Investigation of the oxidation behaviour of pure hydrocarbon components and crude oils utilizing PDSC thermal technique. *J. Can. Pet. Technol.* **2006**, *45*, 48–53. [CrossRef]
16. Jiao, A.; Huang, P.; Qian, X.; Feng, C.; Liu, Z. A chemically assisted ignition mathematical model based on the theory of thermal ignition. *Energy Rep.* **2022**, *8*, 1358–1369. [CrossRef]
17. Yuan, C.D.; Sadikov, K.; Varfolomeev, M.; Khaliullin, R.; Pu, W.; Al-Muntaser, A.; Mehrabi-Kalajahi, S.S. Low-temperature combustion behavior of crude oils in porous media under air flow condition for in-situ combustion (ISC) process. *Fuel* **2020**, *259*, 116293. [CrossRef]
18. Ren, S.R.; Greaves, M.; Rathbone, R.R. Air injection LTO process: An IOR technique for light-oil reservoirs. *SPE J.* **2002**, *7*, 90–99. [CrossRef]
19. Zhao, R.; Chen, Y.X.; Huang, R.P.; Castanier, L.M.; Kovscek, A.R. An experimental investigation of the in-situ combustion behavior of Karamay crude oil. *J. Pet. Sci. Eng.* **2015**, *127*, 82–92. [CrossRef]
20. Pu, W.F.; Wang, L.-L.; Peng, X.-Q.; Li, N.; Zhao, S. Effects of aromatics, resins, and asphaltenes on oxidation behavior and kinetics of heavy crude oil. *Pet. Sci. Technol.* **2020**, *38*, 815–822. [CrossRef]
21. Yang, M.; Liu, Y.S.; Lu, N.; Chai, M.J.; Wang, S.; Feng, Q.H.; Chen, Z.X. Integration of ramped temperature oxidation and combustion tube tests for kinetic modeling of heavy oil in-Situ combustion. *Energy* **2023**, *274*, 127435. [CrossRef]
22. Abu, I.I.; Moore, R.G.; Mehta, S.A.; Ursenbach, M.G.; Mallory, D.G.; Almao, P.P.; Ortega, L.C. Upgrading of Athabasca Bitumen Using Supported Catalyst in Conjunction With In-Situ Combustion. *J. Can. Pet. Technol.* **2015**, *54*, 220–232. [CrossRef]
23. Zhao, S.; Pu, W.; Varfolomeev, M.A.; Yuan, C.; Mehrabi-Kalajahi, S.S.; Saifullin, E.R.; Sadikov, K.; Talipov, S. Low-temperature combustion characteristics of heavy oils by a self-designed porous medium thermo-effect cell. *J. Pet. Sci. Eng.* **2020**, *195*, 107863. [CrossRef]
24. Shouman, A.R. A review of one aspect of the thermal-explosion theory. *J. Eng. Math.* **2006**, *56*, 179–184. [CrossRef]
25. Niu, B.L.; Ren, S.; Liu, Y.; Wang, D.; Tang, L.; Chen, B. Low-Temperature Oxidation of Oil Components in an Air Injection Process for Improved Oil Recovery. *Energy Fuels* **2011**, *25*, 4299–4304. [CrossRef]
26. Ambalae, A.; Mahinpey, N.; Freitag, N. Thermogravimetric studies on pyrolysis and combustion behavior of a heavy oil and its asphaltenes. *Energy Fuels* **2006**, *20*, 560–565. [CrossRef]
27. Fadaei, H.; Castanier, L.; Camp, A.M.M.; Debenest, G.; Quintard, M.; Renard, G. Experimental and Numerical Analysis of In-Situ Combustion in a Fractured Core. *SPE J.* **2011**, *16*, 358–373. [CrossRef]
28. Kuppe, G.J.M.; Mehta, S.A.; Moore, R.G.; Ursenbach, M.G.; Zalewski, E. Heats of combustion of selected crude oils and their SARA fractions. *J. Can. Pet. Technol.* **2008**, *47*, 38–42. [CrossRef]
29. Clara, C.; Durandeau, M.; Quenault, G.; Nguyen, T.-H. Laboratory studies for light-oil air injection projects: Potential application in Handil field. *SPE Reserv. Eval. Eng.* **2000**, *3*, 239–248. [CrossRef]

processes

MDPI

Article

Application of a Deep Learning Network for Joint Prediction of Associated Fluid Production in Unconventional Hydrocarbon Development

Derek Vikara [1] and Vikas Khanna [1,2,*]

[1] Department of Civil and Environmental Engineering, University of Pittsburgh, Pittsburgh, PA 15261, USA; dmv42@pitt.edu
[2] Department of Chemical and Petroleum Engineering, University of Pittsburgh, Pittsburgh, PA 15261, USA
* Correspondence: khannav@pitt.edu; Tel.: +1-(412)-624-9867

Abstract: Machine learning (ML) approaches have risen in popularity for use in many oil and gas (O&G) applications. Time series-based predictive forecasting of hydrocarbon production using deep learning ML strategies that can generalize temporal or sequence-based information within data is fast gaining traction. The recent emphasis on hydrocarbon production provides opportunities to explore the use of deep learning ML to other facets of O&G development where dynamic, temporal dependencies exist and that also hold implications to production forecasting. This study proposes a combination of supervised and unsupervised ML approaches as part of a framework for the joint prediction of produced water and natural gas volumes associated with oil production from unconventional reservoirs in a time series fashion. The study focuses on the pay zones within the Spraberry and Wolfcamp Formations of the Midland Basin in the U.S. The joint prediction model is based on a deep neural network architecture leveraging long short-term memory (LSTM) layers. Our model has the capability to both reproduce and forecast produced water and natural gas volumes for wells at monthly resolution and has demonstrated 91 percent joint prediction accuracy to held out testing data with little disparity noted in prediction performance between the training and test datasets. Additionally, model predictions replicate water and gas production profiles to wells in the test dataset, even for circumstances that include irregularities in production trends. We apply the model in tandem with an Arps decline model to generate cumulative first and five-year estimates for oil, gas, and water production outlooks at the well and basin-levels. Production outlook totals are influenced by well completion, decline curve, and spatial and reservoir attributes. These types of model-derived outlooks can aid operators in formulating management or remedial solutions for the volumes of fluids expected from unconventional O&G development.

Keywords: long short-term memory; Midland Basin; *k*-means clustering; associated gas; water production; oil and gas

Citation: Vikara, D.; Khanna, V. Application of a Deep Learning Network for Joint Prediction of Associated Fluid Production in Unconventional Hydrocarbon Development. *Processes* **2022**, *10*, 740. https://doi.org/10.3390/pr10040740

Academic Editors: Yidong Cai and Tianshou Ma

Received: 6 March 2022
Accepted: 8 April 2022
Published: 11 April 2022

1. Introduction

The continued pursuit for reliable, affordable, and secure supplies of energy accentuates the necessity for continued research into ways to economically and efficiently access the vast amount of unconventional natural gas and oil resources that exist. Over the last decade and a half, the application of horizontal drilling techniques coupled with advanced, multi-stage hydraulic fracturing technologies has facilitated the widespread development of unconventional oil and gas (O&G) reservoirs (such as shale and tight oil reserves) [1], resulting in a revolution in the energy landscape [2–4], particularly in the United States (U.S.).

Hydraulic fracturing methods make use of injected liquids under high pressure to generate breakages in subsurface formations and are usually implemented where low permeability conditions exist. The fracturing fluid is composed of a base fluid, typically water, constituting >98 percent of the total fluid volume [5] with the remaining contribution coming from proppant and chemical additives. The goal of the hydraulic fracturing process is to promote the generation of new fractures in the tight hydrocarbon-bearing rock formations inherently low in both permeability and porosity while simultaneously augmenting the size, magnitude, and connectivity of existing fractures to stimulate oil and/or gas flow to wells [6–8]. Once the hydraulic fracturing process is completed, the high in situ pressures within the reservoir as compared to the lower bottomhole pressure in the wellbore (which can be managed via artificial lifting) prompts fluids to migrate towards the well and be produced at the surface. The fluid that returns to the surface may contain a combination of hydrocarbons (oil and/or gas) and water, in addition to injected chemical additives from the hydraulic fracturing process, as well as naturally occurring materials such as brines, metals, and radioactive materials [9]. Each constituent requires some form of management, depending heavily on the intended endues of each, which may include sale to market as a commodity, reuse as part of site operations, or treatment and disposal.

Horizontal wells drilled and completed in shale gas and tight oil formations make up the preponderance of hydrocarbon production in the United States. Specifically, crude oil production from tight formations alone reached 6.5 million barrels per day in the U.S. through 2018, accounting for 61 percent of the total oil produced in the U.S. The U.S. Energy Information Administration (EIA) indicates that use of horizontal wells accounted for 96 percent of the overall U.S. crude oil production from tight formations by the end of 2018 [10]. A recent surge in the development of tight oil reserves located in the Permian Basin in western Texas and eastern New Mexico (41 percent of total tight oil production in the U.S. in 2018) has led to considerable growth in overall U.S. crude oil production [11].

While unconventional oil (and gas) resources remain critically important in the pursuit towards energy security, challenges persist in effectively forecasting their production potential. For instance, productivity in unconventional reservoirs is known to be responsive to the nature and effectiveness of the interactions between wellbore design, completion and stimulation processes, and the inherent irregularities in reservoir conditions. As a result, fluid production responses can be highly disparate across: (1) An entire O&G play [12]; (2) wells on a given pad targeting the same formation; or even (3) the different perforation stages of single well's lateral component [13]. Production forecasts hold implications on the strategic decisions made by the O&G sector. For instance, resulting production outlooks, depending on the long-term trajectories of fluid volumes produced, can prompt macro-scale consequences such as potential fluctuations in oil and/or gas market prices and associated impacts on the environment [14]. Additionally, forecasts can influence micro-scale outcomes that ultimately shape a wide range of operating and maintenance scenarios for field operators or even affect company profit margins. Reservoir modeling and simulation are commonly used to inform decision makers regarding the potential production response and long-term performance of hydraulically fractured horizontal wells in unconventional reservoirs. These approaches can be costly in terms of the time and computational resources needed to execute effectively [15,16]. Furthermore, difficulties exist in attaining sufficient levels of geological data at the well level [17] to sufficiently reflect the diversity in reservoir conditions needed to model fluid flow. This challenge intensifies when the interest spans to multi-well performance evaluation at the field-scale or larger.

Given the computational resources that are typically widely available and the emergence of O&G digital datasets that include features associated with well completion, stimulation, and production, many have taken to machine learning (ML) and data analytics as a compliment to existing approaches for O&G production analysis [18–20]. ML-based tactics can provide additional analytical functionality to traditional reservoir simulation methods. They have proven effective in accurately and reliably modeling circumstances involving highly complex systems where variable conditions are known to be prominent, not uncommon to wellbore/reservoir relationship interactions in unconventional O&G development. Additionally, they offer expeditious predictive capability, allowing practitioners to quickly generate multiple realizations thereby enabling greater insight into the systems modeled [21].

A number of potential use cases exist where ML has been applied as part evaluating the effects of hydraulic fracturing designs on hydrocarbon production in unconventional reservoirs. As an example, several studies utilize static productivity indicators that reflects cumulative production under a fixed time duration (i.e., six months or one year) as response variables [22–26] to evaluate potential well response to various hydrofracking completion designs. The use of static response variables enabled straightforward evaluation of input feature impact rating and ranking, as well as sensitivity evaluation. The findings from these studies have proven insightful in identifying key production drivers representative to the study areas evaluated, as well as effective in approximating well productivity potential given the associated completion design and placement choices. However, the findings are not directly translatable to applications in the oil and gas space requiring more dynamic, temporal-dependent considerations. Well history matching, hydrocarbon production forecasting, and facilitating data-driven production outlook scenarios are examples that come to mind [27–30].

Many studies are taking focus on using ML for dynamic reservoir analysis by evaluating time series-based topics, such as oil or gas production over the life of producing wells. These studies are leveraging empirical data that includes daily or monthly cumulative hydrocarbon production values over all or a portion of each well's productive life. Many of the relevant studies apply deep learning ML strategies in order to capture and generalize the intrinsic temporal or time sequence-based properties within the data. Findings from recent studies indicate that the deep learning approaches applied have been exceedingly effective at predicting dynamic production trends accurately on holdout data. The results of which suggests that these approaches hold substantial implications and potential viability in production forecasting.

To gain further comprehension on O&G-related time series analysis using ML, we provide a short review of relevant studies works that have focused on this topic. A study by Jie et al. developed two deep learning models to predict daily gas production from a single well completed in the Sichuan basin in China [31]. The researchers developed artificial neural network- (ANN) based models using: (1) A fully-connected multilayer perceptron (MLP)-based ANN with a single hidden layer and (2) a long-short term memory- (LSTM) based ANN with stacked LSTM layer architecture. Empirical data for daily gas production over a three-year period was used for analysis. The first 900 dataset observations were used for model training and the last 100 observations were used for holdout model performance testing. Input data included the data features (assumed at daily resolution) of oil pressure, casing pressure, daily water production, cumulative gas production, cumulative water production, and water-gas ratio. Results indicated prediction error of 1.56 percent for the LSTM-based model and upwards of 9.66 percent for the MLP-ANN. Sagheer and Kotb implemented deep LSTM architectures to estimate monthly oil production for two oil fields; one was the Tarapur Block of Cambay Basin to the west of Cambay Gas Field in India and the other in the Huabei oilfield in China [32]. They demonstrated the predictive effectiveness in stacking LSTM layers as part of network architecture when long interval temporal dependencies may exist as compared to model performance when shallow neural network architectures are used. Additionally, the researchers noted that their LSTM-based model outperformed counterpart formulations explored that were based on deep recurrent neural networks (RNN) and Deep Gated Recurrent Unit models. The work performed by Liu et al. included the development of an ensemble empirical mode decomposition (EEMD) based LSTM learning network capable of time series forecasting of oil production. Case studies were performed using empirical field data from the SL and JD oilfields, China [33]. Their proposed EEMD-LSTM configuration outperformed other model types developed under ensembles between EEMD and MLP-based artificial neural networks and EEMD with support vector machine.

Collectively, these studies demonstrate the utility and capability of deep learning-based ML (with noted effectiveness of LSTM) for time series hydrocarbon production prediction. The knowledge gained through these works provides both a foundation as well as an opportunity to extend these approaches to other aspects critical to O&G development where: (1) Dynamic, temporal dependencies exist; (2) said aspects possess significant connotations to production forecasting; and (3) that have not been extensively explored in previous research. An obvious need that meets these criteria would be to possess the ability for assessing the potential volumes of the associated water and natural gas produced in tandem with crude oil. Many operators targeting oil-rich unconventional reservoirs are faced with the challenge of managing large volumes of water and natural gas that are often co-produced. Limited natural gas processing and pipeline takeaway capacity can force operators to resort to venting or flaring produced natural gas.

Venting is the direct release of natural gas produced from O&G operations to the atmosphere. Flaring involves the controlled combustion of produced natural gas at the wellhead, converting methane to carbon dioxide and water vapor. From an environmental standpoint, flaring is less detrimental than venting given that carbon dioxide is 25 to 28 times less potent as a greenhouse gas than methane over a 100-year period [34,35]. According to the EIA, the quantities of natural gas vented or flared from O&G wells in the U.S. reached record levels in 2019 averaging 1.48 billion cubic feet per day (Bcf/day) (1.3 percent of the total natural gas volume produced) [36]. Texas and North Dakota contributed nearly 85% (1.3 Bcf/day) of all reported flaring and/or venting (only Texas contributed to gas venting) of produced natural gas. Produced water is often managed via disposal through deep well underground injection.

The injection of large volumes of waste water from O&G operations has been strongly correlated to the increased frequency of occurrence of induced seismic events including magnitude 2+ earthquakes, particularly in Oklahoma, Ohio, Arkansas, West Virginia, and Texas [37]. Literature suggests that many are working to generate solutions and reuse options for associated gas and water production [38–42]—but a need exists to be able to effectively quantify and forecast produced volumes of both natural gas and water to best inform the development of management or remedial solutions as well as grasp the potential environmental implications for planned O&G development [43].

We propose a combination of supervised and unsupervised ML approaches as part of a framework that can reliably estimate both produced water volumes and natural gas associated with oil production in a time series fashion. This type of predictive modeling capability is expected to be useful towards (1) informing well operators as part of developing strategies to ensure the effective management, treatment, or potential reuse based on the volumes and quantities of produced fluids, and (2) supplementing hydrocarbon production outlooks with additional fluid volumes in time series fashion. Additionally, this work offers a novel complement to other noted O&G machine learning-based predictive models from literature; largely achieved through its joint prediction functionality, capability to either reproduce or forecast cumulative volumes of natural gas and water produced alongside oil at the well level, and its applicability centered towards a major oil and gas producing play. In addition, the ensemble of the supervised and unsupervised elements of this work enables a means to rapidly forecast oil, water, and natural gas production at the well level as influenced by operational development considerations.

The focus of this study is on the Permian Basin region of the U.S. The region holds enormous consequence regarding domestic oil and gas production. According to a report by the Texas Independent Producers & Royalty Owners Association, yearly crude oil production in the Permian Basin has grown by 1.2 billion barrels since 2009, resulting in a 371% increase in oil output over the last ten years [44]. This overall growth has enabled the Permian to become the world's top-producing oil field [45]. While the region itself is a major producer of both oil and gas, the basin currently faces several challenges. These include: (1) Steeper well decline rates and lower initial production (IP) values as development is moving to non-core regions; (2) associated natural gas production has outpaced pipeline takeaway capacity, which has led to an increase in flaring and venting practices; and (3) produced water volumes and associated management costs are both on the rise [43,46,47]. Combined, these impacts threaten to potentially lower the Permian's overall production potential while consequently increasing the environmental burden associated with O&G operations. Therefore, an opportunity exists to propose research targeted towards these specific challenges and would provide beneficial outcomes to both potentially improving recovery and estimation of the types and volumes of fluids produced at the well level—each of which require specific management strategies and bear potential environmental implications.

2. Data and Methods

The focus of this study is to generate a ML-based prediction model capable of time series joint prediction of associated natural gas and water that are produced alongside oil as part of unconventional hydrocarbon development (three-stream production example presented in Figure 1). Secondarily, this study aims to demonstrate the utility of such a model as a compliment to existing O&G operational management strategies.

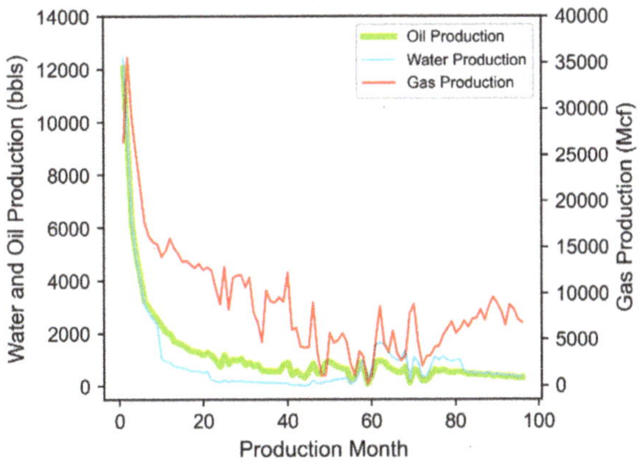

Figure 1. Example of oil, water, and natural gas production data for a horizontal well in northern Reagan County, Texas producing from Wolfcamp A and placed at a total vertical depth of 7713 feet below ground surface. Timeseries fluid data was acquired from vendor Drilling/Enverus [48].

The model is based on a deep neural network architecture leveraging LSTM layers in order to accommodate time-dependent conditions in the data and be proficient towards multi-output prediction. The model development workflow, described throughout the following subsections, is interconnected with several data preprocessing steps that includes data sub-division, engineering of new features, outlier removal, data standardization, and feature selection. The model would have the functionality to not only replicate well production history (the primary focus of many existing time series O&G analyses), but also enable forward-based fluid production forecasts for existing wells throughout their remaining productive lives, as well as be used to predict fluid volumes in time series fashion at new (i.e., theoretical) well sites where no historic production data exists. Additionally, the ML-based model proposed here is intended to be applicable across multiple producing reservoirs, focusing on the "Wolfberry" pay zones (highlighted in Upper Spraberry through Cisco/Cline [Wolfcamp D] reservoirs in Table 1). Such a model will help provide a data-driven approach for a more holistic evaluation towards field development where multiple producing reservoir options are co-located. The volatility that exists in oil and gas market prices and supply and demand encourages operators to remain informed to the best extent possible of potential risks and opportunities they may face over both the short and long term [49]. The inherent challenges facing the Permian suggests that field development decision making is complex. Overall, this study proposes a modeling tool that works towards helping inform complex field decision choices by scaling up model outputs via a single predictive model.

2.1. Study Area

The study area for this work focuses on the Midland Basin, one of the major sub-basins of the larger Permian Basin. The Permian Basin (Permian) is an extensive sedimentary basin and major O&G-producing region geographically located in West Texas and the neighboring areas of southeastern New Mexico. The Permian spans roughly 75,000 square miles and comprises greater than 7000 fields in West Texas alone [50]. The Permian has been important in the U.S. energy economy for nearly a century. According to the EIA, the Permian has produced hydrocarbons for approximately 100 years and has supplied more than 35.6 billion barrels of oil and roughly 125 trillion cubic feet of natural gas (data as of January 2020). The Permian accounted for approximately 35 percent of the total U.S. crude oil production and over 13% of the total U.S. natural gas production in 2019 [51]. It is expected to remain one of the largest hydrocarbon-producing regions in the world with remaining reserves on the order of 46 trillion cubic feet of natural gas and over 11 billion barrels of oil [52]. The Permian contains several sub-basins and platforms that include the westernmost Delaware Basin, Central Basin Platform, and the easternmost Midland Basin [53]. The extent of the Central Platform and Midland sub-basins as well as the eastern edge of the Delaware Basin is shown in Figure 2.

The Midland Basin is the eastern subbasin of the larger Permian Basin and is bordered by carbonate platforms such as the Central Basin Platform, Eastern shelf, and Northern shelf. The basin is at its deepest on its western edge and shallows to the east. Towards its southernmost portion, basin's formations start to thin towards the Ozona Arch—an extension of the Central Basin Platform [53]. The stratigraphy within the Midland Basin is characterized as containing several stacked geologic sequences that offer hydrocarbon producing potential. Two stratigraphic sections within portions of the Leonardian and Wolfcampian epochs that have been a focus of substantial O&G development are the Spraberry (along with the Dean) and the Wolfcamp formations; collectively referred as the "Wolfberry" [54] (Table 1). The stratigraphic groupings that make up the Wolfberry series of reservoirs serve as the primary producing pay zones of interest evaluated in this study.

Table 1. Stratigraphic description for a subset of the Midland Basin, Texas. The producing reservoirs of interest to this study are highlighted (Spraberry in pink and Wolfcamp in yellow). This figure was generated from collective content compiled from lithostratigraphic interpretations of the Permian Basin from several literature sources [51,53,55–60].

<table>
<tr><th>Era</th><th>Period</th><th>Epoch</th><th>Local Series</th><th>Stratigraphic/Formation Name</th><th>Reservoir Operational Name</th></tr>
<tr><td rowspan="14">Paleozoic</td><td rowspan="9">Permian</td><td rowspan="2">Guadalupian</td><td rowspan="2">Ward</td><td>San Andreas</td><td>San Andreas</td></tr>
<tr><td>San Angelo/Glorieta</td><td>San Angelo/Glorieta</td></tr>
<tr><td rowspan="5">Leonardian</td><td colspan="2">Clearfork</td><td>Upper Leonard</td></tr>
<tr><td rowspan="3">Wichita</td><td>Upper Spraberry</td><td rowspan="3">Spraberry</td></tr>
<tr><td>Lower Spraberry</td></tr>
<tr><td>Dean</td></tr>
<tr><td>Lower Leonard</td><td>Wolfcamp</td><td>Wolfcamp A</td></tr>
<tr><td rowspan="2">Wolfcampian</td><td colspan="2" rowspan="2">Wolfcamp</td><td>Wolfcamp B</td></tr>
<tr><td>Wolfcamp C</td></tr>
<tr><td rowspan="4">Pennsylvanian</td><td>Virgilian</td><td colspan="2">Cisco/Cline</td><td>Wolfcamp D</td></tr>
<tr><td>Missourian</td><td colspan="2">Canyon</td><td>Canyon</td></tr>
<tr><td>Des Moinesian</td><td colspan="2">Strawn</td><td>Strawn</td></tr>
<tr><td>Atokan</td><td colspan="2">Atoka/Bend</td><td>Atoka/Bend</td></tr>
</table>

The Lower Permian aged (Leonardian epoch) Spraberry and Dean formations are made up of interbedded turbidite sands, laminated siltstone, carbonate, and organic-rich shales [57]. The Spraberry consists of upper- and lower-unit intervals [61,62] (certain interpretations include a middle Sprayberry and Jo Mill as well [63,64])—the Dean formation is located stratigraphically beneath the Lower Spraberry. Each stratigraphic unit is distinguished by its lithologic composition. For instance, each of the three formations consists of thick sequences of fine-grained sandstones and siltstones that lie on top of an equally thick lower unit made up of black shales and dark carbonates [65]. The formations are known to be generally under-pressured (averaging 800–900 psi [5.4–6.1 MPa]) with matrix porosity ranging from 6 to 15 percent, matrix permeability below 10 md, and are highly naturally fractured [54,66,67]. The average true vertical depth to the top of the Upper Spraberry unit is roughly 6800 feet across the Midland Basin. The complete section from the top of the Upper Spraberry to the base of the Dean ranges in thickness between 1200 and 1870 feet [54]. Similar to other unconventional hydrocarbon plays, productivity in the Spraberry fluctuates across the basin [68].

The early Permian aged (Wolfcampian-Leonardian epoch) Wolfcamp is described as a mixed siliciclastic-carbonate succession with stacked stratigraphic units comprising of cyclic gravity flow deposits—each separated by mudstone and siltstone [51]. The Wolfcamp is described by Sutton [54] as a dual-lithology system consisting of organic-rich shale with interbedded limestone. Lower reservoir quality portions of the Wolfcamp are associated with the presence of grainy carbonate facies, whereas higher reservoir quality portions have been tied to the occurrence of siliceous mudstones [69]. The entire section of the Wolfcamp ranges in porosity between 2 and 12 percent with average permeability near 10 millidarcies (mD) [51]. The formation varies substantially across the Midland Basin in terms of depth, thickness, and lithologic composition. The Wolfcamp is at its deepest near the center of the Midland Basin, measuring approximately 12,000 feet deep. It shallows substantially towards the edges of the basin, varying in depth from 4000 to 7000 feet [54]. The thickness of the entire section of the Wolfcamp averages around 1800 feet. The Wolfcamp is extensive throughout the Permian Basin and is considered one of the most abundant unconventional O&G plays worldwide. The Wolfcamp formation has been appealing to O&G operators given its stacked configuration, in which multiple thick hydrocarbon-producing zones exist in sequence [70]. The stacked intervals of the Wolfcamp formation are called benches—from shallow to deep they are referred to as A, B, C, and D. Each bench has shown to be different in terms of its overall lithology, fossil content, total organic carbon content, and thermal maturity [71]. Saller et al. (1994), Blomquist (2016), and Peng et al. (2020) provide detail on the geologic composition of the Wolfcamp and various benches within and therefore the differentiation is not described at length here [72–74]. Recent development efforts in the Midland Basin are preferentially targeting the more oil-rich Wolfcamp A and B (roughly 95 percent of total Wolfcamp production) opposed to the more gas-rich Wolfcamp benches C and D [71,75].

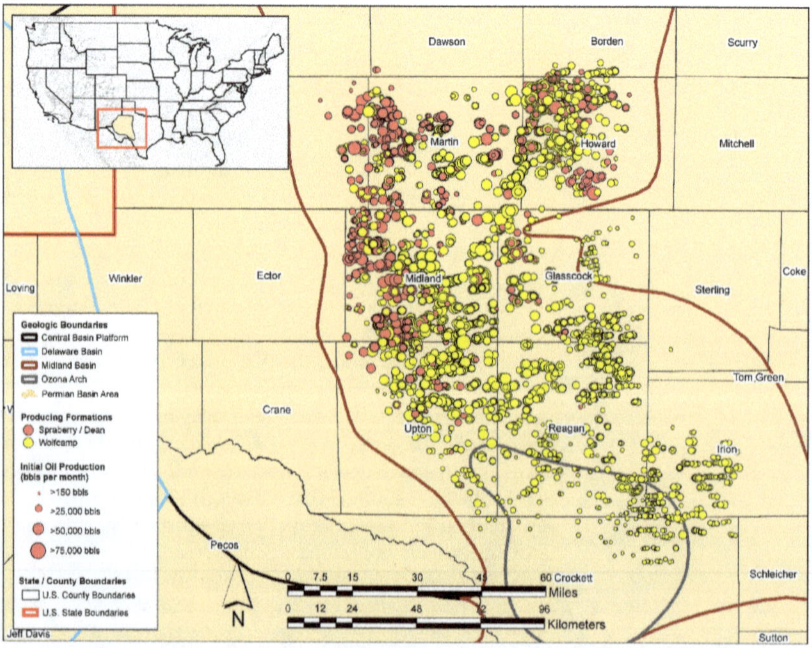

Figure 2. Map of the study area in the Midland Basin, Texas. Well data used for the study was acquired from DrillingInfo/Enverus [76]. The geographic information system (GIS) layers applied to support the generation of this figure were acquired from the University of Texas at Austin [77] and United States Geological Survey [78].

The Permian region and associated sub-basins have been known to produce large volumes of natural gas and water that are co-produced with oil. A study by Kondash et al. has noted that Permian Basin wells have increased the water used per well as part of hydraulic fracturing operations from 30,800 barrels per well in 2011 up to 267,325 barrels per well in 2016—a 770 percent increase [79]. The flowback and produced water volumes during that same timespan had increased over 400 percent; averaging 56,610 barrels per well in 2011 to over 232,700 barrels per well in 2016. Specifically, in the Midland Basin, waste water disposal volumes derived from O&G operations have steadily increased since 2011, reaching approximately 4.5 billion barrels per day in 2017 [80].

In 2017, flaring and venting of natural gas in the Permian basin in Texas and New Mexico was estimated at nearly 300 million cubic feet per day (MMcfd), roughly 4.4 percent of the total gas produced that year. In that same year, the Midland Basin produced approximately 1019 billion cubic feet (Bcf) of natural gas, and flared 24 Bcf of that total (2.35 percent of all gas produced) [81]. In 2019, flaring and venting of natural gas in the Permian reached an all-time record high based on the year's third quarter estimates, averaging 752 MMcfd (275 Bcf total) [82]. The Midland Basin portion of 2019 flaring ranged from approximately 150 to 290 MMcfd [83].

Well data leveraged for this study (described further in Section 2.2) are grouped based on the associated targeted producing reservoirs listed in Table 1. Wells are tabbed as either "Spraberry/Dean" or "Wolfcamp" dependent upon their associated Stratigraphic/Formation Name. The wells used as part of this study are plotted in Figure 2; they are colored based on their associated producing formation and sized based on each well's initial oil production (in barrels [bbls]/month).

2.2. Study Data Overview and Data Processing

Much of the well completion and production-related data used for this study is acquired from the O&G data vendor DrillingInfo/Enverus [76]. Other features were derived through feature engineering to further supplement the available feature dataset. The dataset contains features related to well production performance attributes, Arps decline curve attributes [84], well completion attributes, and spatial and reservoir attributes—all specific to horizontal production wells spanning the Spraberry/Dean and Wolfcamp producing intervals (highlighted in Table 1) in the Midland Basin with drilling initiation dates within the 1 January 2010 to 30 June 2020 timeframe. The dataset includes a combination of static (well data that does not change over the well's productive lifetime) and dynamic features (well data with temporal dependencies—mostly three-stream production data) for the wells meeting these screening criteria. This database query yields data for approximately 6480 wells in total in which each well has data reported for all features of interest (both static and dynamic features) and duplicate entries are omitted. No attempts at data interpolation with respect to missing values occurs in this study.

The distributions of the static study features of interest are evaluated to screen and remove potential outlying well data and refine the overall dataset. Their distributions are presented in Figure 3. Data outside of +/− 3 standard deviations from a given feature's mean value (grey margins within subplots in Figure 3) are considered outlying and possibly highly influential on ML model response [85,86]; even if distributions are not explicitly gaussian. All outlying data is removed from the static and dynamic contributions to the dataset (approximately 270 wells had features meeting outlying criteria). The resulting dataset consists of 6210 wells in total extending across 12 Texas counties, the extent of which is plotted in Figure 2 and the descriptive statistics for features from these wells are summarized in Table 2.

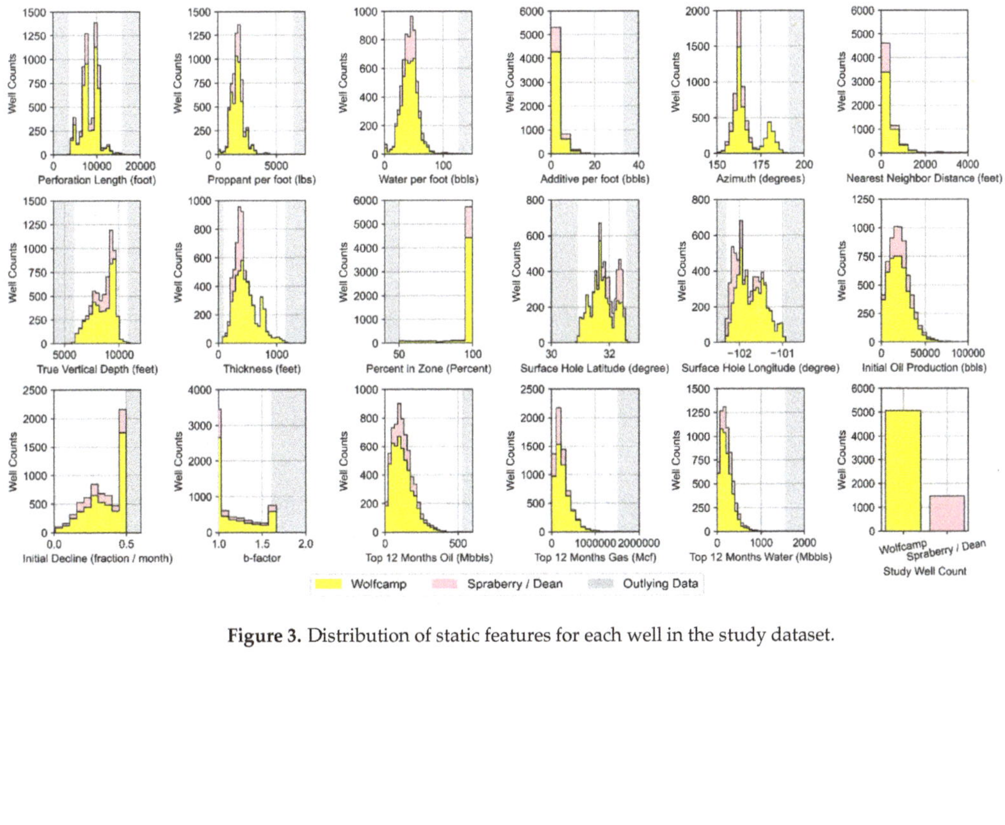

Figure 3. Distribution of static features for each well in the study dataset.

Table 2. Statistical summary of the study dataset features evaluated.

Dataset Features	Data Group	Static	Dynamic	Mean	Median	Standard Deviation
Monthly Oil (bbls)	Well Performance Attributes		X	4863	2429	6448
Monthly Gas (Mcf) [1]			X	12,500	7906	13,846
Monthly Water (bbls)			X	8510	3572	13,496
Top 12 Months Gas (Mcf)		X		251,286	207,532	182,648
Top 12 Months Oil (bbls)		X		124,320	114,314	70,210
Top 12 Months Water (bbls)		X		226,856	197,664	157,721
EUR Gas (MMcf)		X		1,732,470	1,171,682	1,722,215
EUR Oil (bbls)		X		449,302	380,333	326,663
Initial Oil Production (bbls) [2]	Decline Curve Attributes	X		20,807	19,675	11,593
Initial Decline (fraction/month)		X		0.35	0.36	0.13
b-factor		X		1.2	1.0	0.2
Timestep Cumulative (months)			X	25.3	21	18.8
Perforation Length (foot)	Well Completion Attributes	X		8480	8302	1959
Proppant per foot (lbs)		X		1732	1718	548
Water per foot (bbls)		X		43	44	14
Additive per foot (bbls)		X		2.9	2.4	2.4
Azimuth (degrees) [3]		X		166	163	8
Nearest Well Distance (feet)		X		438	231	838
Percent in Zone (percent)		X		97	100	10
True Vertical Depth (feet)	Spatial and Reservoir Attributes	X		8571	8828	993
Thickness (feet)		X		460	415	188
Surface Hole Latitude (degrees)		X		31.8253	31.7971	0.4093
Surface Hole Longitude (degrees)		X		−101.7740	−101.8346	0.3204

[1] Mcf = thousand cubic feet. [2] DrillingInfo/Enverus quantifies initial oil production as the cumulative production volume observed during a given well's first full month of production [48]. [3] All wellbore azimuth trajectories based on true north = 0 degrees.

The features within each data group from Table 2 have a specific role as part of the hydraulic fracturing and oil/gas production process. The breadth of data features available within the study dataset affords the opportunity to explore a multitude of aspects related to unconventional oil and gas production in the Midland Basin. Data groupings and their associated features are briefly described next.

- ***Well Performance Attributes:*** These features relate to fluid production for wells in the study dataset. The dynamic features within the data group represent summation of the three-stream (oil, gas, and water) empirically-derived monthly values at the well level provided by DrillingInfo/Enverus. Data for these dynamic features is available for each month in a given well's productive lifetime. Therefore, the volume of this data varies across wells depending on when they began production and how long wells are kept online. The "Top 12-months" static features for oil, gas, and water were derived via summation of the 12 largest observed values for each well based on monthly dynamic feature data. This approach has been implemented in our prior work [12,23] and has proven to effectively represent productivity potential for unconventional wells that may or may not have been subject to disruptions to their production time series profiles. Both the Top 12-months Oil and Gas features correlate strongly to well level estimated ultimate recover (EUR) as indicated in Figure 4. The static EUR features represent an estimation of the technically recoverable reserves at the well level. They are calculated by DrillingInfo/Enverus [87] using a combination of historic production data and a combination of Arps decline curve models [84].
- ***Decline Curve Attributes:*** These features are inherent to decline curve analyses based on the Arps decline curve model [84]. The Arps model can be used to evaluate oil and/or gas declining production rates over time. Time-dependent reduction in hydrocarbon production can be attributed to reduced reservoir pressure as well as the relative change in the volumes of the produced fluids. The approach can also be

used to forecast hydrocarbon production into the future. The Arps approach is based on fitting a mathematical decline model (either exponential, hyperbolic, or harmonic) to empirical observations of an asset's (i.e., well) performance history [88]. Well features related to initial (oil) production, the initial decline, and degree of curvature (b-factor) are the parameters related to the Arps model. Values for these features for each well in the study dataset have been determined by Drillinginfo/Enverus [87]. The DrillingInfo/Enverus approach solves for the most appropriate Arps model parameters that minimize the sum of squared errors based on empirical production values for a given well [87]. DrillingInfo/Enverus restricts b-factors between 0 and 2. The b-factor is typically greater than 1 in unconventional shale plays given the inherent low permeability rock matrix and resulting extended duration of transient flow [89]; potentially a derivative of the bulk of empirical observations with shorter producing timeframes [90].

- ***Well Completion Attributes:*** These features pertain to each well's design and completion attributes as it relates to well placement, orientation, and hydraulic fracturing design. The major hydraulic fracturing design features include the length of the perforated interval contacting the reservoir and the volume of proppant, water, and additive used for hydraulic fracturing normalized to a per foot of perforated interval basis. Proppant includes solids that may vary in size, shape or material type. They typically consist of sand or engineered materials (i.e., resin-coated sand or high-strength ceramic materials such as sintered bauxite) and are used to keep reservoir fractures open and conductive following hydraulic fracturing [91]. Additives may serve a variety of functions, with examples including the assurance of effective transport of water and proppant downhole and throughout the reservoir, as well as to ensure sustained hydrocarbon recovery after hydraulic fracturing. Specific components can tend to vary from one well to another and from operator to operator. However, example constituents include acids, friction reducers, biocides, pH adjusters, scale inhibitors, iron stabilizers, corrosion reducers, gelling agents, and cross-linking agents [92,93]. Other important well design characteristics captured in the dataset relate to the wellbore lateral orientation, spacing distance to nearby wells, and the portion of the horizontal perforated length within the targeted producing reservoir zone of interest. The directional alignment (reflected by azimuth) is often a design choice by field operators; one that is driven by the natural orientation of in situ stresses in targeted reservoir producing zones. Horizontal segments of wells that are drilled along the minimum horizontal stress often produce transverse fractures following horizontal fracturing. This form of fracturing may improve drainage efficiency. As a result, well laterals oriented properly on azimuth given natural in situ stress regimes may experience higher productivity [5,92]. Well azimuth was approximated based on the geographic orientation between each well's surface hole latitude and longitude and lateral toe latitude and longitude. Well spacing may provide insight into the field operator's anticipated drainage area based on the applied water and proppant intensity. Additionally, spacing-related data can be helpful in determining if closely-spaced wells suffer from possible interference from hydraulic fracturing operations (i.e., frack hits) or effects from parent/child well interactions [94,95] from nearby wells. We approximated the nearest well distance for each well in the dataset using the haversine formula and bottom hole latitude and longitude coordinates to its closest well neighbor prior to any dataset reduction. Percentage in zone is a metric which provides an indication of the wellbore geo-steering efficiency of the horizontal lateral component. DrillingInfo/Enverus provides this data readily for each well. Wells with a high portion of their perforated segment in the targeted producing zone are more likely to be better producers than those wells expected to deviate substantially off target. Each feature in this data group is treated as static. In actuality, many of these features, such as proppant, water, and additive per foot, could essentially vary over the life of any given well due to refracturing campaigns.

- ***Spatial and Reservoir Attributes:*** The features included attempt to best approximate the variability that may exist in the geologic conditions which influence hydrocarbon prominence and producibility that span the reservoirs of interest across the study domain. True vertical depth and thickness (i.e., reservoir thickness) are provided from DillingInfo/Enverus for each well. However, other relevant geologic characteristics that are known to influence hydrocarbon production, such as total organic carbon, porosity, hydrocarbon and/or water saturation, thermal maturity, reservoir pressure, existence of fracture networks, and capacity of the reservoir(s) to be hydraulically fractured [96–99], are not directly or readily available in bulk. Additionally, many of these features are dynamic in nature and change over the duration of hydrocarbon production (such as fluid saturation and pressure in the reservoir), while others essentially remain static (such as porosity and thermal maturity) [100]. Each well's locational data (surface latitude and longitude) is used as a contingency means to approximate geologic conditional variability known to vary spatially across the study area—an approach widely used in other ML-based model development efforts occurring over large spatial horizons [22,26,27,101].

A correlation matrix using Pearson's Product-Moment Correlation is presented in Figure 4 which provides quantitative indication of the linear relationship between each of the various static features of interest. The analysis represented in Figure 4 is informative specifically due to the fact that: (1) it suggests how attributes correspond to other attributes, as well as with potential model outputs; and (2) it serves as a diagnostic check on data quality to ensure data features are related in a fashion that is intuitive and confirmatory based on heuristic understanding of the Midland Basin.

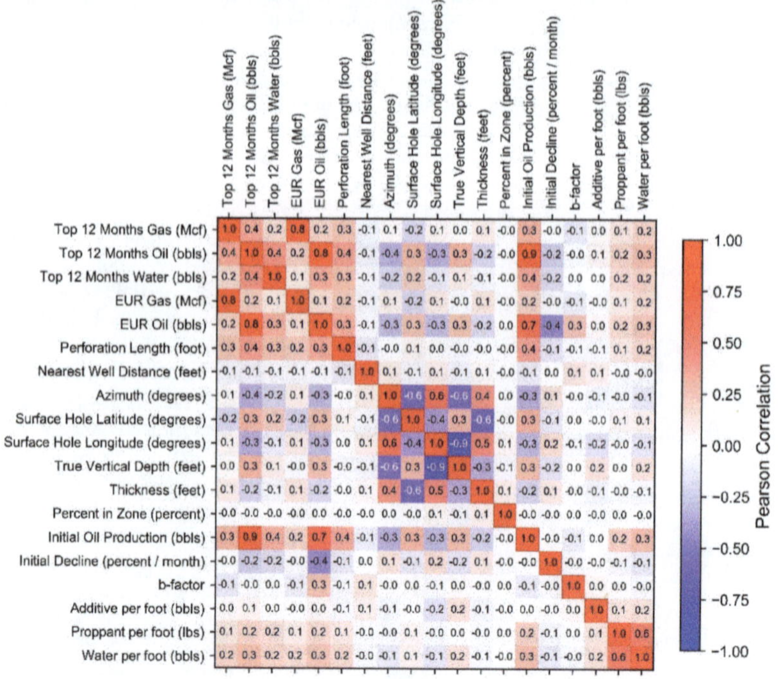

Figure 4. Pearson correlation matrix for the static dataset features evaluated.

The Pearson correlations alone highlight a number of noteworthy trends. For instance, Figure 4 shows several positive relationships between many of the well performance attributes representing fluid production with well completion attributes specific to hydraulic fracturing design. The attributes of top 12-month oil, water, and gas, as well as the estimated EUR per well for both oil and gas are all positively correlated with increasing values of perforation length, proppant, and water per foot. These relationships suggest greater production results from well completion and hydraulic fracturing design upscaling; a concept noted by others [22–24]. Additionally, the decline curve attributes show correlation to both the well performance and well completion attribute features. Initial oil production is mostly positively correlated to these attributes, while initial decline (for oil), as expected is negatively correlated. The b-factor component is mostly uncorrelated to all features in the dataset with the exception of a positive correlation to oil EUR, and therefore holds influence over a well's longer-term productive profile. Finally, worth noting are the correlations associated with reservoir thickness and true vertical depth based on well location in the basin. Moving west to east in the basin (based on surface hole latitude), Figure 4 suggests the reservoirs become both shallower and thinner. In contrast, reservoirs trend thicker and deeper when moving south to north (based on surface hole longitude). These correlations are as expected based on interpretations of Midland Basin reservoir depth and thickness isopaches and interpretations generated by the EIA [53,102], Hamlin and Baumgardner [61], and Blomquist [74]. Based on this analysis, the dataset following outliers removed appears representative and suitable for use in ML model development.

2.3. Data Preprocessing Prior to Model Training and Testing

An important data preprocessing step is applied that scales attribute data to consistent ranges in order to (1) afford equal consideration to all attributes, (2) improve training efficiency and, (3) increase numerical stability of the resulting models [103]. The data scaling approach was implemented to both the static and time series parameters prior to use in the following feature selection and ML model development steps (described in Sections 2.4 and 2.5). For the feature selection and clustering, input and response features were standardized to Z-values (Z) per Equation (1). For model training regarding the time series joint associated fluid production model, all features were scaled (i.e., normalized in this case) between 0 and 1 using linear mapping via Equation (2):

$$Z = \frac{x - \mu}{\sigma} \tag{1}$$

$$x_{normalized} = \frac{x - min_x}{max_x - min_x} \tag{2}$$

where x represents feature values, μ is the feature mean value, σ is the feature standard deviation, and min_x and max_x represent the respective minimum and maximum values for each dataset feature. The Z-score standardization step in Equation (1) rescales data for each parameter to a standard normal distribution with a mean of 0 and a standard deviation of 1. The data transformation from Equation (2) is used as a variant to the zero mean, unit variance standardization from Equation (1). The authors have gleaned from recent experience the effectiveness of 0 to 1 scaling in deep learning ML applications [104–107] and are therefore applied it here. Predictions using finalized ML models are rescaled to their normal unit ranges.

Following data standardization and/or normalization, project dataset features were apportioned and merged into distinct dataset aggregates for use dependent upon the machine learning model workflow they would be applied against. The workflows include feature selection (Section 2.4), clustering (Section 2.5.1), or joint time series prediction (Section 2.5.2). Each workflow utilized a distinct aggregate of the full project dataset. However, the data features that were carried forward to each framework were largely dependent on the results from the feature selection, described in Section 2.4.

2.4. Feature Selection Approach

Features (i.e., variables) that are strongly correlated are therefore linearly dependent and may have almost correspondingly similar (if positively correlated) or opposing (if negatively correlated) effects on dependent variables of interest. The Pearson correlation metric (presented in Figure 4) is limited to assessing linear relationships concerning two features. However, important functional relationships between two or more features may exist which may not be linear in nature. This can be true even if Pearson correlation coefficients are close or equal to 0 [108].

Feature selection involves a systematic process to down-select a subset of the most relevant features within the study dataset that strongly contribute to the ML model prediction response. Utilizing fewer features (and eliminating redundant or non-informative features) enables ML algorithms to train faster and more efficiently as well as decreases the likelihood of ML algorithms overfitting to irrelevant input features [109]. This study utilized recursive feature elimination with cross validation (RFECV) as a feature selection approach. The objective was to establish a final set of input features from the variety available per Table 2 that would be commonly applied as part of both the clustering evaluation and the development of the time series joint associated fluid production model.

The feature elimination component of the RFECV process searches for a subset of features by starting with all features in the training dataset and fitting a ML algorithm which is used as the estimator [109,110]. The estimator is trained on the original set of features considered. A total of 14 input features (i.e., x data) are included in this study which comprise variables associated with the "Well Completion Attributes", the "Spatial and Reservoir Attributes", and the Top 12-months Oil listed in Table 2, as well as two categorical variables that label the production wells evaluated based on their producing reservoir group—either the Wolfcamp or Spraberry/Dean formations. Two features were used as responses (i.e., y data) which comprise of the Top 12-month Water and Top 12-Month Gas. Static data (e.g., Top 12-month Water or Gas) was used exclusively as part of the RFECV instead of dynamic time series data (e.g., Monthly Water) in order to enable more efficient training of the estimator model. The importance of each feature is acquired following model training. The feature(s) with the lowest importance are then pruned from original set of features [111,112]. The procedure is recursively repeated on the pruned set and resulting model accuracy is calculated for each iteration—the process continues until a single feature remains. The desired number of features can then be established [112,113]; typically set at the number of features that maximizes model performance, or where the inclusion of additional features does not substantially improve model performance.

Random forest (RF) was used as the estimator in the RFECV process for this study. RF-based models are considered advantageous in RFECV [114], most notably since they possess the ability to measure the importance of each feature [115] based on mean decrease impurity (described effectively by Hur et al. [116]). Prior to use in RFECV, the RF estimator's hyperparameters were tuned via k-fold cross-validation using five folds. In this process, four folds of the training dataset are amassed to train models, and the remaining fifth fold was used to test (i.e., validate) the performance of resulting prediction models. The step was repeated so that each fold was ultimately used once for model validation while the other $k-1$ folds constitute the training set [117]. An exhaustive grid search occurs as part of the cross-validation loop to tune hyperparameters. The RF estimator formulated on all 14 input data features is built on four folds training data for distinctive hyperparameter combinations evaluated [118] as part of the grid search. Trained models were then used to make predictions against held out fifth fold validation data. The process is repeated for each combination of hyperparameters evaluated. The RF-specific hyperparameters tuned as part of cross-validation includes (1) the number of trees in each forest ensemble and (2) the minimum number of samples needed to split an internal node. The maximum depth corresponding to each tree (i.e., limits the number of nodes in each tree) was unbounded. The RF hyperparameter combination that provided for the best prediction accuracy while avoiding over or underfitting was used for RFECV.

The RFECV process also involved k-fold cross-validation using five folds. For each of the five RFECV fold iterations, 14 RF models were generated with the feature subset size decreasing from 14 to 1. Resulting prediction model performance was evaluated by explained variance per Equation (3) which can effectively evaluate the multi-output response nature of the RF estimator.

$$\text{explained\_variance}(y, \hat{y}) = 1 - \frac{Var\{y - \hat{y}\}}{Var\{y\}} \tag{3}$$

where $\hat{y}$ is the predicted value, y is the observed value, and *Var* is the variance (or square of the standard deviation). The selected feature set from this process was then utilized as the input features for performing the clustering analysis as well as for the time series-based joint associated fluid production model. The results from RFECV is then used to inform the feature sets used for both the clustering and time series machine learning steps of this study (described in Sections 2.5.1 and 2.5.2 respectively).

2.5. Machine Learning Model Development and Evaluation

This section describes the various ML approaches implemented as part of this study, the contribution of each towards the study objectives, and how their performance accuracy was quantified. The ML approaches utilized included both supervised and unsupervised methods, as well as the use of deep learning. Static data features that remained following RFECV step were incorporated in ML-based workflows. Python (version 3) and packages within the scikit-learn library [119] and Keras [120] were leveraged as part of the ML workflow implementation.

2.5.1. Clustering Evaluation

The majority of the static features within the study dataset underwent evaluation via k-means clustering [121], an unsupervised ML approach, prior to the development of the joint associated fluid production model. This step was intended to identify congregations of closely related wells based on their well completion, decline, well performance, and spatial and reservoir attributes (Table 2). The goal of this step was to be able to harvest Arps Decline properties (b-factor, initial production, and initial decline discussed previously) and well completion attributes representative of given clusters; from which oil production forecasts can be generated at the well level.

The k-means clustering process determines an optimal number of clusters based on the input dataset features incorporated. Assuming dataset A of V-dimensional entities $a_i \in A$, for $i = 1, 2, \ldots, N$, with N being the number of data entities in the dataset, k-means creates K number non-empty separate clusters $S = \{S_1, S_2, \ldots, S_K\}$ proximal to centroids $C = \{c_1, c_2, \ldots, c_K\}$, by iteratively minimizing the sum of the within-cluster sum of squared distances (W_K, show in Equation (4)) between each centroid and the data entities associated [122].

$$W_K = W(S, C) = \sum_{k=1}^{K} \sum_{i \in S_k} d(a_i, c_k) \tag{4}$$

The term $d(a_i, c_k)$ in (4) is the distance between data entity a_i and the associated centroid location c_k. In this study, k-means analysis was performed over a wide arbitrary range of values set to $K = 1$ through 30 to ensure sufficient volumes of clusters are evaluated to determine an optimal.

Two heuristic algorithms were applied to determine the optimal number of clusters—the Elbow method [123] and Hartigan's Rule [124]. The Elbow method can be used to visually evaluate W_k as a function of the number of clusters. The optimal number of clusters occurs at the point in which adding another cluster does not result in a substantial improvement to W_k. However, determining the optimal number of clusters through a visual determination approach such as the Elbow Method can be highly subjective to the evaluator's judgement. Hartigan's Rule provides an alternative cluster determination approach and is based on comparing the resulting Hartigan's Index, which is a ratio between the Euclidean within-cluster sum of squared error based on k number of clusters (i.e., W_k) to that based on $k + 1$ clusters (W_{k+1}). The rule utilizes the notion that when clusters are effectively separated, Hartigan's Index ($H(K)$) becomes ≤ 10 and is taken as k to be the optimal number of clusters.

The optimal number of clusters was determined based on the resulting $H(K)$ for each $K = 1$ through 30 evaluated. The Elbow Method was applied in tandem to provide a visual heuristic complement to the resulting optimal K derived from Hartigan's Rule.

2.5.2. Time Series Joint Associated Fluid Production Model

For forecasting under time series circumstances, a deep learning neural network based on Long Short-Term Memory was developed for the joint prediction of associated water and natural gas production as part of oil production operations (referred to as the joint associated fluid production model [model]). The model objective is to provide the capability to reproduce as well as forecast water and natural gas volumes produced at a given well at monthly resolution based on the well's: (1) Monthly oil production volume; (2) explicit spatial and reservoir attributes (limited to the Spraberry/Dean and Wolfcamp Formations) in the Midland Basin; (3) specific well completion attributes; (4) producing month number (i.e., Timestep Cumulative data per Table 2), and (5) prior three-stream (oil, gas, and water) production volumes relative to current time (t) = month$_{t-1}$, month$_{t-2}$, month$_{t-3}$, and month$_{t-4}$.

LSTM are variants of Recurrent Neural Networks (RNN) which include memory functions that enable networks to learn long-term dependencies. The conceptual basis behind RNN is to utilize information where sequential dependencies exist so that output response is influenced by prior, yet relevant elements in sequence. The inherent RNN "memory" feedback component provides differentiation from "feedforward" neural networks (e.g., multilayer perceptron) where input data are independent from one another and strictly flow from input to output [125]. As a result, RNNs are effective in evaluating sequences of data, but are subject to gradient vanishing and struggle to handle longer-term sequential dependencies [126]. LSTM is a choice RNN-based architecture for dealing with noted shortcomings under circumstances where temporal dependencies exist that span over several time steps. Additionally, LSTMs have been shown to outperform and be advantageous to traditional-based algorithms for time series forecasting such as autoregressive integrated moving average (ARIMA) models [127,128].

The LSTM concept was first introduced by Hochreiter and Schmidhuber in 1997 [129] and subsequently expanded and adapted by other since. LSTMs utilize a memory cell structure (Figure 5) to handle both shorter and long-term dependencies in time series datasets [130]. Short-term memory is captured as input from previous timestep cell output (h_{t-1}). The long-term memory component is reflected in the cell state (C_{t-1}). LSTM memory cells have the ability to add or omit information to the cell state (i.e., $C_{t-1} \rightarrow C_t$), but only does so through carefully regulated structures called gates. Network gates consist of either sigmoid or hyperbolic tangent (tanh) activation coupled with pointwise multiplication operations.

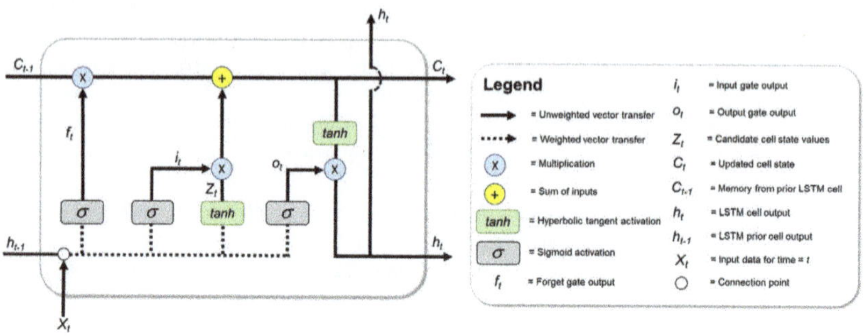

Figure 5. Example schematic of an LSTM cell. Figure concept compiled from concepts presented in Kwak & Hui [131], Olah [132], and Poornima & Pushpalatha [133].

Given the input data vector at time step t (X_t) and the previous time step LSTM cell output (h_{t-1}) instituted, the hidden state output for current LSTM cell (h_t) is calculated per the sequence discussed in the following bullets [129,134]:

- First, the forget game (f_t) is utilized to determine information that becomes omitted away from the cell state. New information introduced to the LSTM memory cell via h_{t-1} and X_t undergoes sigmoid transformation, the result of which is output between 0 (becomes fully omitted) and 1 (becomes fully included) for each number in the cell state C_{t-1} per Equation (5).

$$f_t = \sigma\left(U_f X_t + W_f h_{t-1} + b_f\right) \quad (5)$$

- The second step involves determining new information to be stored in the cell state; this step occurs through two separate parts. The input gate (i_t) applies sigmoid activation to h_{t-1} and X_t and is used to inform values that will be updated in the cell state per Equation (6). Additionally, tanh activation generates a vector of new candidate values (Z_t), which could be included in the cell state per Equation (7).

$$i_t = \sigma(U_i X_t + W_i h_{t-1} + b_i) \quad (6)$$

$$Z_t = tanh(U_z X_t + W_z h_{t-1} + b_z) \quad (7)$$

- The prior cell state C_{t-1} is updated with new information to a new cell state C_t, via Equation (8):

$$C_t = f_t C_{t-1} + i_t Z_t \quad (8)$$

- The final step generates output (h_t) that leverages memory from the cell. The output is a function of the new cell state C_t that undergoes some filtering via tanh activation as well as from output from the output gate (o_t). The mathematical expressions for these steps are presented in Equations (9) and (10).

$$o_t = \sigma(U_o X_t + W_o h_{t-1} + b_o) \quad (9)$$

$$h_t = o_t \times \tanh\left(C_t\right) \quad (10)$$

The equation variables pertaining to U and W include, respectively, the weights to the input data (X_t) and recurrent (h_{t-1}) vectors. The b term is the bias for each gate.

Model architecture (Table 3) and hyperparameter settings were ultimately determined via trial and error opposed to a more systematic approach such as cross-validation (CV) with grid-search. The deep learning-based model requires a fairly extensive training duration (trained on a personal computer requiring approximately five seconds to train per epoch), therefore a holistic grid-search approach with CV to refine hyperparameter settings was not considered practical. Ultimately, the model network consists of four hidden units comprised of two stacked LSTM layers in a recurrent network fashion and two dense layers. All hidden layers utilize sigmoid activation. The stacked LSTM architecture was used given the noted successes demonstrated from comparable studies such as Sagheer and Kotb, Utgoff and Stracuzzi, and Jie et al. that found improved modeling generalization with deep, stacked structures over shallower architectures [31,32,135]. The hidden layer sizes were set to vary as a function of the input size (input shape = 24 features) by 2×, 4×, 4×, and 2× accordingly. A masking layer was used as the network input layer. The masking layer facilitates the omission of timesteps as part of sequence processing where input data are noted as missing. Prior to model training, well-level input data time series sequences are encoded via zero padding (post) into consistent sequence lengths [136]. Setting consistent sequence lengths for each well enables contiguous batch sizes as part of model training and resetting of LSTM cell states following each batch. The masking layer informs the network to skip timesteps where all input data = 0. The output layer enables regression-based prediction and is a dense layer with linear activation consisting of two neurons; one handling the predicted response for natural gas production and the other handling the predicted response for water production. All neurons are fully connected (no dropout applied) between model layers.

Table 3. Summary of network architecture for the joint associated fluid production model.

Layer Type	Activation	Output Shape	Trainable Parameters
Masking	Not Applicable	(None, 1, 24)	0
LSTM	Sigmoid	(None, 1, 48)	14,016
LSTM	Sigmoid	(None, 1, 96)	55,680
Dense	Sigmoid	(None, 1, 96)	9312
Dense	Sigmoid	(None, 1, 48)	4656
Dense	Linear	(None, 1, 2)	194

The inclusion of the dynamic well performance attributes of monthly oil, gas, and water production results in a dataset size with 561,661 observations (224,421 of which are not subject to zero padding) spanning 5561 wells at monthly resolution. Wells with less than 12 months of production data were omitted from model training. The portion of the project dataset used as part of the joint associated fluid production model development was randomly segmented into training, validation, and testing datasets through an 80/10/10 percentage-based split. This approach implements a training, validation, and testing split that maintains the temporal order of observations from the project dataset by keeping the entire productive timeframe for a given well intact. For instance, 10 percent of the dataset wells (based on American Petroleum Institute well ID number) were selected at random to isolate a test dataset. All associated static and dynamic data was appropriately cross-referenced to each well for use in model development. The same process was conducted on the remainder of the dataset to isolate an additional 10 percent to serve as a validation dataset. The data from the remaining 80 percent of the wells was used for training as part of model training.

Early stopping was applied as an additional regularization step to combat overfitting. This approach monitors the predictive performance of the model for every epoch during training against predictions on the held-out validation set (56,156 observations; 21,732 of which are not subject to zero padding) as a proxy for generalizing error. Model training was discontinued when validation error was minimized conditional to the use of a patience tolerance of 25 epochs. Model weight optimization was determined under mini-batch gradient descent using the "Adam" adaptive learning rate optimization algorithm [137], a batch size = 101 which is equal to the sequence length for each well with zero padding applied, and epochs = 1000. The learning rate was set at 0.0001. Keras default settings for first and second-momentum estimate decay rates as well as epsilon were used as part of Adam implementation. Once trained, model performance accuracy was evaluated on the 10 percent subset holdout test data (56,156 observations; 23,044 of which are not subject to zero masking). This step also provided additional confirmation that models were not over or underfit. The performance metrics used as part of model training, early stopping, and testing evaluation are discussed in Section 2.5.3.

The model is easily employed to replicate a given well's historic water and gas production with the use of required input data for the given month of interest. To generate prediction forecasts for future time instances, we employed a recursive prediction approach as explained by Ji et al. [138]. This strategy involves implementing the model in a $t + 1$ one step ahead prediction functionality under multiple iterations through the desired prediction horizon ($t + h$); where the prediction for the prior month (t) is used as an input for making a prediction for the following month ($t + 1$). Assuming well completion attributes do not change over time, these input features can be simply carried forward for all timesteps predicted. However, oil production is a dynamic, time-dependent input and required for forecasting water and gas volumes. Therefore, oil production forecasts that serve as inputs to the model must be derived from another means; potentially reservoir simulation output, a separate ML oil production predictive model, or even though analytical methods proposed by the likes of Fetkovich [88] and Arps [84].

2.5.3. Model Performance Evaluation

Our model performance was evaluated for the supervised learning-based joint associated fluid production model in two specific instances; (1) during model training against both the training and validation data sets and (2) through analysis of goodness-of-fit for simulated predictions against the test dataset. During model training, mean squared error (*MSE*) is used as the loss function. Performance of the model is quantified by *MSE* at each epoch against both the training and validation datasets; the latter provides an overall generalization error estimate as well as an indication to potential overfitting if training and validation *MSEs* begin to diverge substantially [139]. *MSE* is mathematically represented in Equation (11):

$$MSE = N^{-1}\sum_{i=1}^{N}(y_i - \hat{y}_i)^2 \tag{11}$$

where N represents the length of the dataset, y_i is the observed value, and $\hat{y}_i$ is the simulated or predicted response value.

The finalized joint associated fluid production model prediction performance was evaluated by making predictions against the test dataset. A combination of *MSE*, root mean squared error (*RMSE*), and R^2 are used to evaluate model performance accuracy. *RMSE* corresponds to the mean error between predicted and observed values and reflects the variance of errors independent of sample size. As with *MSE*, smaller *RMSE* values are associated with reduced mean error between predicted and ground-truth data compared to model predictions where higher *RMSE* values occur [115]. *RMSE* provides a compliment to *MSE* and R^2, one expressed in the units of the response variable(s) of interest. The R^2 metric signifies the degree of correlation between simulated and observed values and is defined as the regression sum of squares ($SS_{Regression}$) divided by the total sum of squares (SS_{Total}). R^2 values are proportional to the data being evaluated and range between 0 and 1—higher values represent smaller variations between the ground truth data and predicted values and lower values may suggest little to no correlation exists. *RMSE* and R^2 are described mathematically in Equation (12) and Equation (13) respectively:

$$RMSE = \sqrt{N^{-1}\sum_{i=1}^{N}(y_i - \hat{y}_i)^2} \tag{12}$$

$$R^2 = \frac{SS_{Regression}}{SS_{Total}} = 1 - \frac{\sum_{i=0}^{N-1}(y_i - \hat{y}_i)^2}{\sum_{i=0}^{N-1}(y_i - \overline{y})^2} \tag{13}$$

The overbar above variables per Equation (13) indicates the mean value for the complete dataset of ground truth observations considered.

2.6. Oil Forecasting

Monthly oil production estimates are needed in order to predict the associated gas and water production for wells in the study area using the LSTM-based deep learning time series joint associated fluid production model. We utilized the Arps decline curve model [84] to enable oil forecasts, either for (1) new (theoretical) wells where no historic production exists or (2) to extend historical production for existing wells. The Arps hyperbolic decline model, common for lower permeability shale production [140], is applied per Equation (14) to forecast oil production at the well level:

$$q = \frac{q_i}{(1 + bD_it)^{\frac{1}{b}}} \tag{14}$$

where q is the monthly oil production (bbls/month), q_i is the initial oil flow rate (bbls/month), b is the decline component which is dimensionless, D_i is the initial decline constant (fraction/month), and t is the production month (month). The Arps models have shown to provide for reliable hydrocarbon history matches (even in cases with $b > 1$) and affords simplicity in their use [141]. However, the hyperbolic model can tend to over approximate reserves when extrapolated without constraints to long-term transient flow considerations [140,142]. Therefore, in this study, Equation (14) is only applied to forecast oil in short durations (limited to 60 months).

3. Results and Discussion

The following subsections outline key results as part of model development, evaluation, and application associated with the various machine learning workflows applied throughout the study to enable joint associated fluid production time series prediction capability.

3.1. Feature Selection Results

The feature selection step using RFECV and feature importance evaluation helps establish final sets of input features that can be applied as part of both the clustering evaluation and the development of the time series joint associated fluid production model. Results from this analytical step are described here, but can be found in detail in Appendix A. Informed from the findings from RFECV and importance evaluation, two distinct dataset aggregates (in addition to the set used for feature selection) are created; one for clustering and another for the time series-based joint associated fluid production model training and testing (Table 4).

Table 4. Summary of feature inclusion for the various dataset aggregates. Each feature is demarcated for inclusion into the associated dataset aggregates as an input feature (x) or a response feature (y).

Dataset Features	Data Group	Feature Selection	Clustering	Joint Time Series Prediction
Monthly Oil (bbls) (t through $t-4$)	Well Performance Attributes			x
Monthly Gas (Mcf) (t through $t-4$)				y
Monthly Water (bbls) (t through $t-4$)				y
Top 12 Months Gas (Mcf)		y	x	
Top 12 Months Oil (bbls)		x	x	
Top 12 Months Water (bbls)		y	x	
EUR Gas (MMcf)				
EUR Oil (bbls)				
Initial Oil Production (bbls)	Decline Curve Attributes		x	
Initial Decline (fraction/month)			x	
b-factor			x	
Timestep Cumulative (months)				x
Perforation Length (foot)	Well Completion Attributes	x	x	x
Proppant per foot (lbs)		x	x	x
Water per foot (bbls)		x	x	x
Additive per foot (bbls)		x	x	x
Azimuth (degrees)		x	x	x
Nearest Well Distance (feet)		x	x	x
Percent in Zone (percent)		x		
True Vertical Depth (feet)	Spatial and Reservoir Attributes	x	x	x
Thickness (feet)		x	x	x
Surface Hole Latitude (degrees)		x	x	x
Surface Hole Longitude (degrees)		x	x	x
Wolfcamp (yes/no)		x		
Spraberry/Dean (yes/no)		x		

Table 4 highlights the specific dataset features that make up each dataset aggregate. Based on findings from RFECV, 11 static features were selected and three omitted from the feature selection dataset for consideration in analysis moving forward. The down-selection includes omission of the features with the three lowest values of feature importance; which include percent in zone and the two categorical variables demarcating wells completed in either the "Spraberry/Dean" or "Wolfcamp" formations. The remaining data features are used for each of the following associated subsequent project tasks described in Section 3.2 (clustering) and Section 3.3 (the joint time series associated fluid production model).

3.2. Cluster Analysis

The results from the k-means clustering analysis are exhibited in Figure 6. Clustering results are presented in the context of both the Elbow method and Hartigan's rule; both of which are used in tandem to select a representative number of well clusters from the study dataset where adding another cluster does not result in any substantial improvement to within-cluster sum of squared error. The visual heuristic results for the Elbow method suggest an appropriate cluster count falls somewhere between roughly 17 and 21 clusters (Figure 6A). The Hartigan Solution in Figure 6B explicitly identifies 18 clusters as optimal, and that adding the 19th cluster (where 19 is the $k + 1$ cluster where the Hartigan Index ratio between k and $k + 1$ is ≤ 10) results in negligible reductions to within-cluster sum of squared error.

Wells within the study dataset were mapped to their corresponding cluster and then plotted to inspect clustering distribution across the study area (Figure 7). An initial observation is that the resulting distribution of well clusters appears influenced by more so than just three-dimensional placement characteristics. For instance, clusters five and 14 (dark green and dark purple respectively) span a large area and occur over a variety of burial depths. Although the specific reasoning for cluster assignment is not analyzed in detail as part of this study, it is likely that non-spatial features related to well completion design, well performance, and reservoir thickness were influential for the commonalities of wells in these clusters. However, in certain cases, wells within certain clusters are in close spatial proximity. This seems true for cluster eight (red) in the southern portion of the basin as well as cluster 15 (light yellow) in the northeast portion of the basin. Table A2, presented in Appendix C in this study, provides a summary of descriptive statistics for wells making up each cluster.

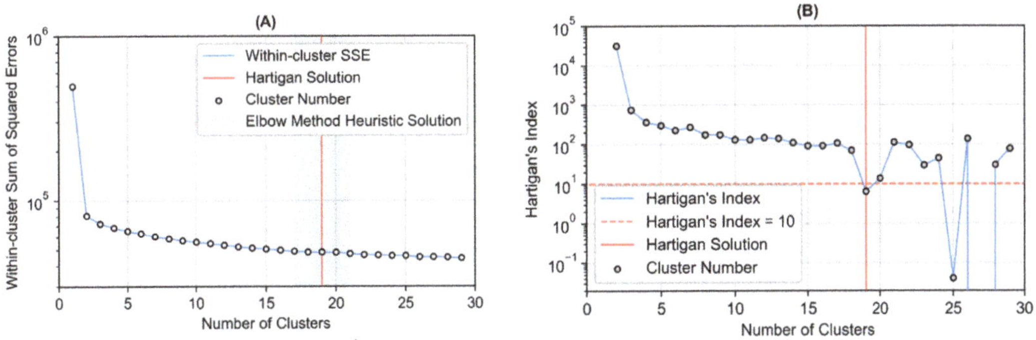

Figure 6. Elbow diagrams from k-means clustering analysis. The top figure (**A**) represents the total within-cluster sum of squared errors based on the number of clusters evaluated. The lower figure (**B**) shows the resulting Hartigan's Index as a function of the numbers of clusters evaluated.

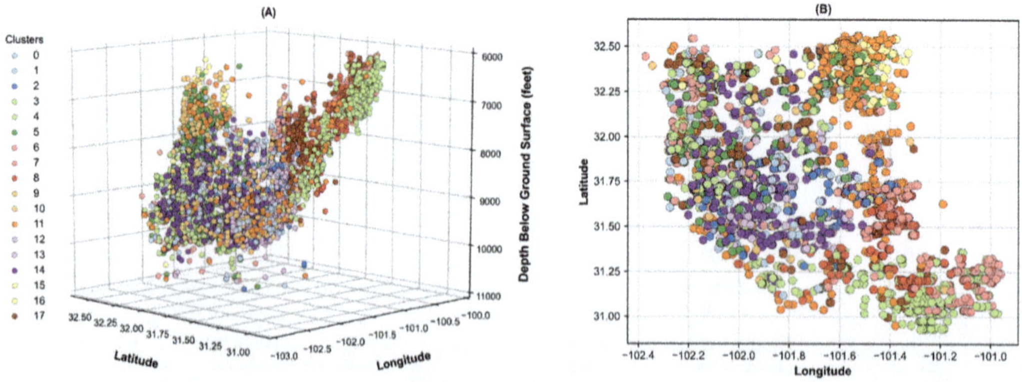

Figure 7. Well data demarcated by color corresponding to one of the 18 clusters (labeled 0–17 based on Python's zero-based indexing). The top (**A**) is a three-dimensional representation of well data location which features placement along burial depth. The bottom (**B**) is a top-down depiction featuring well location by latitude and longitude coordinates only.

Arps decline properties can be extracted that are representative of the wells common to each cluster. These properties can then be used to forecast oil production at the well level using the Arps model per Equation (14). Figure 8 shows the distribution of the Arps decline properties for each cluster. Based on the distribution of these properties across clusters, oil production trends, and therefore associated gas and water, are expected to vary across clusters as well.

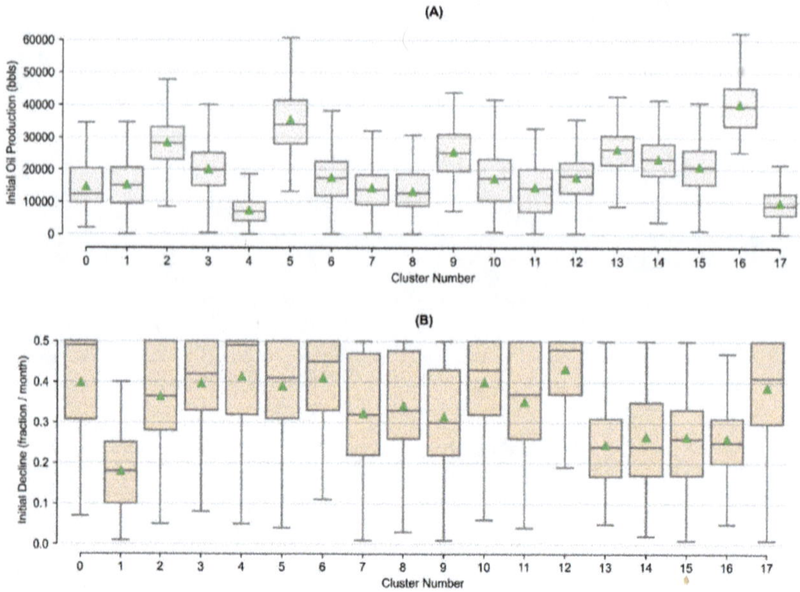

Figure 8. *Cont.*

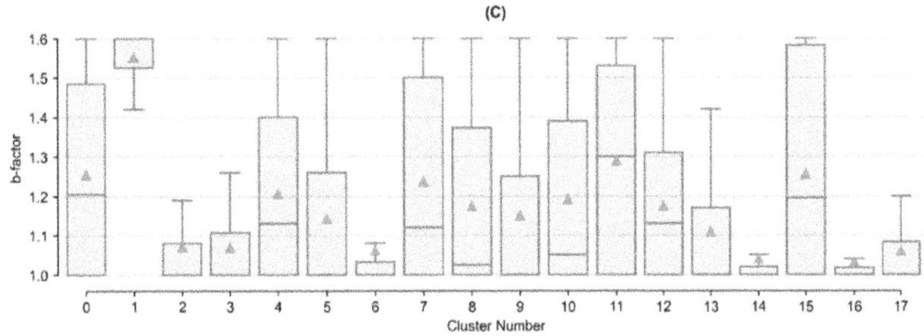

Figure 8. Box-and-whisker plots of Arps decline curve attributes calculated for wells within each cluster; including (**A**) initial oil production, (**B**) initial decline, and (**C**) b-factor. Boxes extends from the 25th to 75th quantile values of the data. A line occurs at the median (50th quantile). Green triangles occur at the mean value. Whiskers extend to the minimum and maximum values of the data absent outliers.

Multiple one-way Analysis of Variance (ANOVA) were conducted to evaluate the similarity or disparity of the Arps decline properties within and across each cluster as a way to statistically infer and differentiate variability in oil production trends across clusters. ANOVA is a parametric statistical technique used to compare different datasets—specifically equality associated with their means and the relative variance between them [143–145]. In this case, the independent variable evaluated was the cluster number, which included 18 levels [0 through 17]. The dependent variables included initial oil production, initial decline, and b-factor. Null hypotheses are rejected at a significance level of $\alpha = 0.05$. ANOVA can provide insights into the overall significance of the well clusters and corresponding Arps decline properties, but the test cannot inform exactly where differences lie. Following ANOVA, Tukey's Test [143,146] are used post-hoc to compare pairs of means for Arps decline attributes for which null hypotheses are rejected across each of 18 well clusters. The overall significance level is assumed $\alpha = 0.05$ for testing pairwise mean comparisons. ANOVA results yielded significant variation for all Arps attributes among well cluster as a condition, $p < 0.05$. No Arps attribute was determined to be insignificant based on well cluster groupings. Therefore, a Tukey's test was performed for each of the three Arps attributes across the 18 well clusters. The post hoc Tukey's test (Table A1—shown in Appenix B) highlights which clusters, and therefore Arps decline attributes, differed significantly from cluster to cluster at $\alpha = 0.05$. Clusters in Table A1 (shown in Appendix B) that do not share a Tukey's Group are considered significantly different from each other. The Tukey's Group lettering [A through L] are order based on the cluster with the highest mean value for the given attribute of interest relative to the other Tukey's Groups. Tukey's test results indicate that out of 18 different clusters, there are 12 statistically different cluster given initial oil production groupings (A–L), only eight statistically different cluster exist regarding initial decline (A–H), and 10 statistically different clusters in regard to b-factor (A–J). From an Arps model perspective, higher oil productivity is tied to larger values of initial oil production and b-factor and smaller values of initial decline. The analysis of variance and Tukey's pairwise comparison tests are performed using Minitab 18 Statistical Software.

3.3. Joint Associated Fluid Production Model Training and Performance

The predictive performance of the model as a function of training epoch is presented in Figure 9. The figure depicts the associated model loss (as *MSE* where model predictions, training data, and validation data values are in normalized form between 0 and 1) following the update of network weights prompted by new estimates of the error gradient following each training epoch. Given the consistency of the trends in validation and training loss, the model appears to demonstrate a suitable fit to the training data with no suggestion of over or underfitting, indicating the model's overall effectiveness at generalizing associated fluid production. The application of early stopping ended model training after 918 epochs, resulting in a minimal generalization gap between training (1.16×10^{-4} MSE) and validation (1.15×10^{-4} MSE) performance.

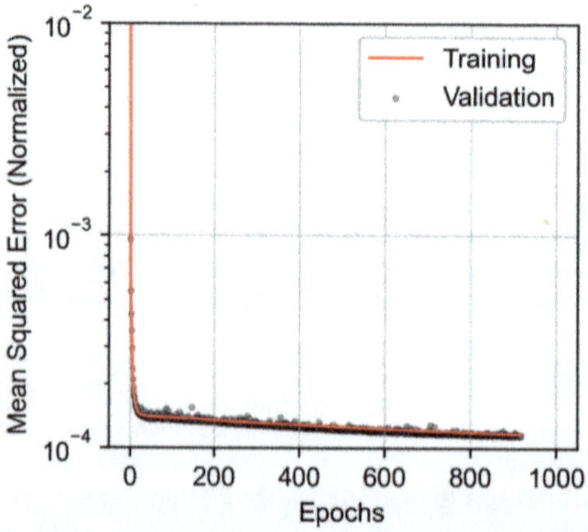

Figure 9. Learning curves for the joint associated fluid production model over training epochs.

The model's predictive performance summary against both the training and test dataset set is compared in Table 5. Performance metrics presented in Table 5 are based on the response data transformed from normalized states per Equation (2) back into original units (Mcf and bbls) relative to each fluid stream. Overall, there is little disparity for model performance between the training and held-out test data, as well as marginal difference in the model's ability to predict either water or gas.

Table 5. Model results for prediction on the training and test dataset.

Predicted Value	Training Data			Test Data		
	R^2	*MSE*	*RMSE*	R^2	*MSE*	*RMSE*
Monthly Gas (Mcf)	0.930	7.63×10^6	2762	0.931	7.54×10^6	2746
Monthly Water (bbls)	0.914	6.72×10^6	2593	0.899	7.35×10^6	2710
Joint Prediction (Monthly Water and Gas)	0.922	7.17×10^6	2679	0.915	7.44×10^6	2728

The prediction performance is visually compared with observed data from the test dataset in Figure 10. The parity plots (Figure 10A,C) provide a visual depiction of the model's prediction to actual observed water or gas production on a monthly basis. The R^2 metric (listed in Table 5) is used to quantify the correlation of actual to predicted monthly production data as part of the comparison in Figure 10. Model performance that would perfectly generalize production trends would have an R^2 of one, and all data would fall exactly along the black dotted lines (i.e., 1-to-1 match) provided for reference. The model's joint predictive capability is fairly strong overall; however, the model is slightly more accurate at predicting monthly gas on holdout data compared to water. Data is color coded by producing formation and sized by the production month to provide visual indicators for potential glaring trends in residual patterns. Fortunately, none seem to exist given that no irregularities in model residuals for either formation occur based upon visual inspection of the Figure 10 parity plots.

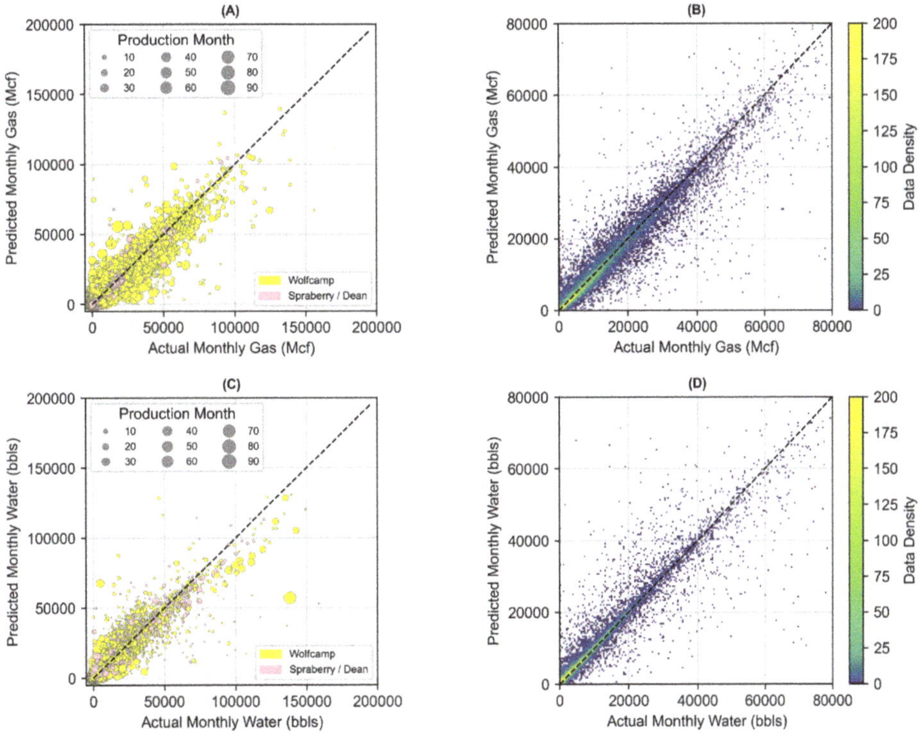

Figure 10. Parity plots of model performance comparing predicted values for monthly gas (**A**) or water (**C**) against actual values (i.e., observations) for wells in the test dataset. Additionally, the density of data within plot area pixels is provided for gas (**B**) and water (**D**). Density plots zoom to focus on the 0 to 80,000 bbls or Mcf fluid volume range where the majority of test data occurs.

Figure 10B,D also features visual depictions of the density of data within each pixel of the x and y plotting area. Pixel coloration is based on the amount of data at a given x and y pixel. Viewed in isolation, the parity plots can be a bit challenging to assess the distribution of data around the 1-to-1 line given the large volume of data presented within and the spread throughout the plotting space. The density plots emphasize where higher aggregations of data fall and where model residual (variation from the 1-to-1 line) are most prominent. The majority of monthly gas and water predictions compared to test data actuals fall along the 1-to-1 line and residuals appear evenly distributed at all fluid production volumes. Density plots are zoomed in to focus on the 0 to 80,000 bbls or Mcf fluid volume range where the majority of test data occurs.

Figure 11 shows replication of the production history for water and gas for four different randomly selected wells within the test dataset. Predictions using the joint associated fluid production model stop when known production observations end. Solid lines in Figure 11 depict actual production data for oil (green), water (blue), and gas (red) from each of the four wells. Red and blue dots indicate prediction responses for LSTM-based joint associated fluid production model. For reference, a brief review of each well evaluated in Figure 11 is provided in the bullets below:

- **Well 1:** Located in central Martin County producing from the Lower Spraberry with an 8409-foot perforated length, and placed at a total vertical depth of 9334 feet below ground surface.

- **Well 2:** Located in northern central Midland County producing from the Wolfcamp B with a 7142-foot perforated length, and placed at a total vertical depth of 9673 feet below ground surface.
- **Well 3:** Located in southeastern Midland County producing from the Wolfcamp B with a 6722-foot perforated length, and placed at a total vertical depth of 9383 feet below ground surface.
- **Well 4:** Located in western Martin County producing from the Wolfcamp C with a 4855-foot perforated length, and placed at a total vertical depth of 10,031 feet below ground surface.

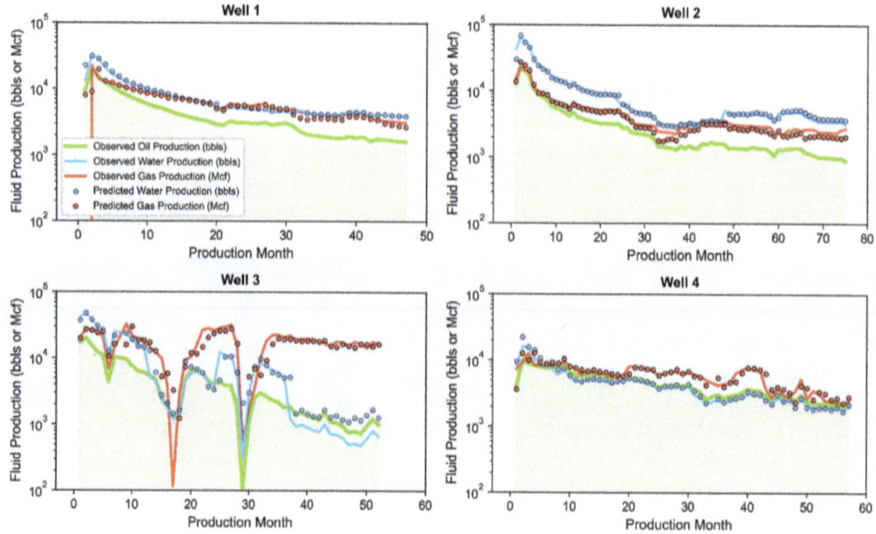

Figure 11. Replication of production history using the joint associated fluid production model for four test dataset wells.

Prediction results in Figure 11 are encouraging given the favorable replications of water and gas production profiles, even under circumstances that include irregular production trends. Worth noting is that the actual production trends for oil, water, and gas for each of the four wells evaluated are dissimilar in nature, yet the model is effective in replicating production profiles. Noted discrepancies in predictions to actual monthly flows seem to most commonly occur when highly transient (i.e., spikes or rapid falloffs) events transpire. However, given that the model input features are heavily dependent on prior timestep flows for oil, water, and gas, the model appears to adjust to transient events in making next timestep predictions.

Results to this point have been based on comparison of model prediction to replicate known production flows from wells within the test dataset. However, one of the functionalities of a time series-based model lies in the ability to forecast into the future where no observations exist. We implement the model under a recursive multi-step forecasting strategy as a way to predict gas and water production trends past existing wells' known producing timeframes, as well as for generating production outlooks for new, theoretical well sites. Under this strategy, the model is used to make a prediction at time t, then the predicted values are appended to the input dataset to serve as prior month flow input data for predicting at time $t + 1$. Oil predictions via the Arps model are incorporated as part of the input dataset to enable prediction at time t, $t + 1$, through $t + h$ where h = the total producing months prediction horizon. This process is repeated in a recursive manner until the $t + h$ is reached. A simple exponential forecast smoothing function [147] is applied at $t > 24$ months where the $t + 1$ prediction is a sum of model's $t + 1$ estimate plus the prior *t* value in a weighted 90/10 percentage contribution. Past the $t > 24$ months producing timeframe, observed monthly water and gas values for wells in the study dataset are frequently at the scale (or lower) of the model prediction error (roughly 2500 to 2600 *RMSE* in Mcf or bbls per month as per Table 5). The smoothing approach ensures stability in the forward predictions as part of the recursive implementation of forecasts.

Since the joint associated fluid production model is a purely data driven model, it may be limited at making sound predictions for: (1) circumstances where low quantities of data to train models exists; and (2) timeframes that extend far beyond the extent of the production durations for wells in the training data. Over 80 percent of the wells in the study dataset have well production timeframes less than 60 months (Figure 12). After 60 months, the volume of well data becomes sparse, especially for Spraberry/Dean wells. Additionally, as discussed in Section 2.6, the application of the Arps model over the long-term with high b-factors using the hyperbolic model may overestimate hydrocarbon production. Plus, the recursive prediction strategy can suffer from error accumulation and propagation, particularly when the forecasting horizons increase [148,149]. These potential limitations serve as the basis for setting our constraint to limit forecasts to shorter-term predictions.

Results in Figure 13 show forecasted production for oil, water, and gas for four different wells; three of which (Wells A, B, and C) are existing wells selected from the test dataset and the fourth (Well D) is a theoretical well based on the dataset mean values for input features common to Cluster 13 (see Table A2 in Appendix C). Cluster 13 was selected as an example for analysis here since it contains a realitvely large mean initial oil production and encompases a substantial portion of the well count from the study dataset; the majority of which are Wolfcamp wells. Forecasts using the joint associated fluid production model intentionally stop at 50 months under all cases regardless of well production history. Solid lines in Figure 13 depict actual production data for oil (green), water (blue), and gas (red). Red, green, and blue dots represent the montly forecasts for the Arps (oil) and joint associated fluid production model (water and gas).

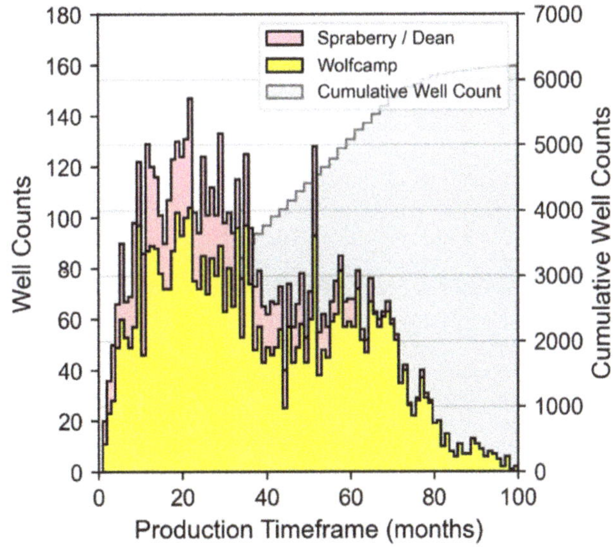

Figure 12. Stacked (left *y*-axis) and cumulative (right *y*-axis) histograms of well counts within the study dataset based on the production timeframe for each well.

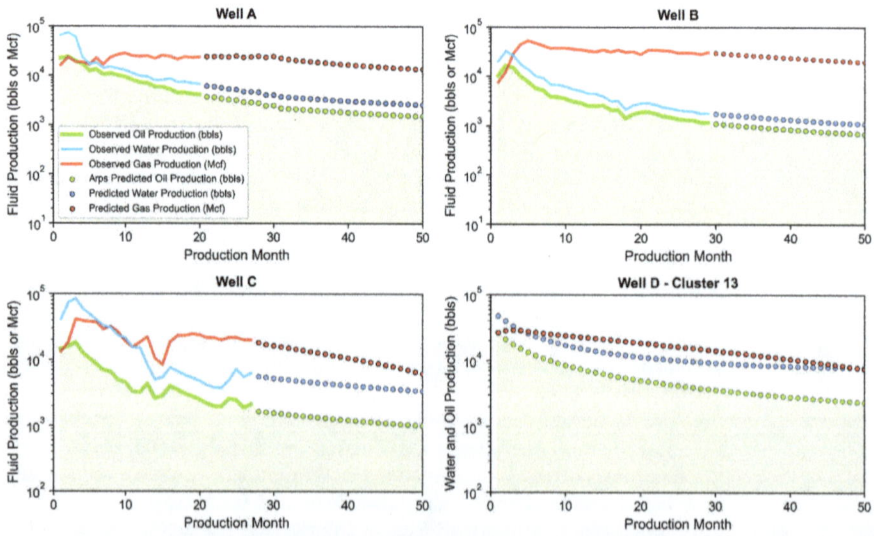

Figure 13. Gas and water prediction forecast using the joint associated fluid production model leveraging oil forecast outlooks generated from the Arps model.

For reference, a brief review of each well evaluated in Figure 13 is provided in the bullets below:

- **Well A:** Located in northern Upton County producing from the Wolfcamp A with a 7745-foot perforated length, and placed at a total vertical depth of 9476 feet below ground surface.
- **Well B:** Located in western Irion County producing from the Wolfcamp B with a 10,114-foot perforated length, and placed at a total vertical depth of 6709 feet below ground surface.
- **Well C:** Located in southern Glasscock County producing from the Wolfcamp A with a 10,261-foot perforated length, and placed at a total vertical depth of 7976 feet below ground surface.
- **Well D:** Theoretical well representative of Cluster 13 (see Table A2 in Appendix C for specifics) based on a 9870-foot perforated length, an initial monthly oil production of 26,324 bbls, and placed at a total vertical depth of 9128 feet below ground surface.

4. Oil, Gas, and Water Production Outlook

The joint associated fluid production model has been applied in combination with the Arps model to generate oil, gas, and water production outlooks for each of the 18 clusters identified in Section 3.2 (Table 6). The outlooks were generated at the well level for a single hypothetical well representing each cluster. The hypothetical wells that represent each cluster are attributed well completion, decline curve, and spatial and reservoir attributes set to the cluster's mean value for each. Outlooks include the cumulative first and five-year estimates for production totals (Table 7). The suite of data presented in Table A2 in the Appendix C is a digest of attribute statistics (most notably mean, standard deviation, and interquartile range [IQR]) as well as cumulative production outlook estimates from the combination of the Arps and joint associated fluid production models for each cluster. Additionally, this collection of data is intended to serve as a guiding resource for assessing the potential volumes of produced fluids associated with oil production in the Midland Basin based on well completion design considerations and placement within the basin.

Table 6. Inventory of first year and cumulative five-year production estimates for a hypothetical representative well within each Midland Basin Well Cluster.

Response Feature	Outlook Year	Midland Basin Well Cluster Number: 0 through 8								
		0	1	2	3	4	5	6	7	8
Cumulative Oil (Mbbls)	1st year	77	111	147	100	38	181	86	82	73
	5-years	154	282	275	183	74	346	156	169	145
Cumulative Gas (Bcf)	1st year	0.16	0.20	0.25	0.15	0.12	0.27	0.25	0.13	0.22
	5-years	0.29	0.58	0.62	0.23	0.23	0.60	0.76	0.21	0.79
Cumulative Water (Mbbls)	1st year	154	230	268	181	102	304	200	162	182
	5-years	289	587	545	347	175	659	358	324	328
Response Feature	**Outlook Year**	**Midland Basin Well Cluster Number: 9 through 17**								
		9	10	11	12	13	14	15	16	17
Cumulative Oil (Mbbls)	1st year	141	89	80	87	160	135	129	237	50
	5-years	281	173	168	167	328	265	279	465	91
Cumulative Gas (Bcf)	1st year	0.26	0.14	0.12	0.15	0.31	0.22	0.22	0.34	0.12
	5-years	0.57	0.19	0.16	0.27	0.91	0.50	0.45	0.85	0.27
Cumulative Water (Mbbls)	1st year	271	185	170	171	306	249	265	373	111
	5-years	574	364	287	332	684	515	621	879	178

Table 7. Summary of the highest and lowest predicted production totals and associated cluster groups.

Metric	Oil Production		Natural Gas Production		Water Production		Gas-to-Oil		Water-to-Oil	
	Mbbls	Cluster	Bcf	Cluster	Mbbls	Cluster	Bcf/Mbbl	Cluster	Mbbl/Mbbl	Cluster
Highest 1st year	237	16	0.34	16	377	16	0.0014	16	1.59	16
Highest 5 years	465	16	0.91	13	879	16	0.0020	11	1.89	11
Lowest 1st year	38	4	0.12	4 and 11	102	4	0.0032	4	2.68	4
Lowest 5 years	74	4	0.16	11	175	4	0.0022	8	2.36	4

The predictions for each cluster appear aligned to typical volumes of in-field production trends for wells in the Midland Basin. For instance, our predicted production totals in Table 7 when compared in the context of water-to-oil and gas-to-oil ratios appear in range with those reported in literature [49,150–152]. For instance, the ratios from estimated production throughout the first producing year from Table 7 values range from approximately 1.57 to 2.68 bbls/bbls for water-to-oil across clusters (with a mean of 2.03]) and 1.43 to 3.15 thousand cubic feet (Mcf)/bbl for gas-to-oil across clusters (with a mean of 1.94). Cumulative produced water and gas (to-oil) estimates after 5-years or production are in the ranges reported by Rassenfross [49] and Kondash et al. [79] respectively. Additionally, the predictions capture increasing gas-to-oil ratio trends as wells becomes older [153]; not uncommon to unconventional plays, particularly when production causes reservoir pressures to fall below the bubble-point [154].

Table 7 highlights several major takeaways from the digest presented in Table 7; particularly the cluster groups estimated to have the highest or lowest totals for (1) oil, gas, and water production per well, as well as (2) associated fluids normalized to a barrel of oil produced. The results indicate that Cluster 16 is the best oil producer for the first producing year and through five years of production. Cluster 16 also is noted to be comparatively efficient versus other clusters based on the ratio of associated fluids volumes produced with oil; particularly for the first year of production. Cluster 4 is the lowest oil producing cluster and highly inefficient regarding the associated fluids volumes produced with oil. Cluster 1 produces some of the largest volumes of associated water and gas, but only produces oil near the average for all clusters. As a result, Cluster 1 is one of the most inefficient clusters in terms of oil to gas and oil to water production ratios in addition to Cluster 4. Clusters 3, 5, 11, and 16 are noted as relatively more "efficient" clusters than others based on their higher oil to gas and oil to water producing ratios for both the first producing year and through 5 years. Overall, clusters 1, 4, 6, 8, and 17 appear to be the least efficient regarding associated fluid production normalized to oil.

We performed one last analytical case study using the data in Table 7 to generate production volume outlooks in regards to associated fluid production in the Midland Basin. Specifically, first and five-year production outlooks are generated at the basin-level under three development scenarios that comprise of a new fleet of wells built on different contributions of wells common to certain cluster groups. The scenarios include:

- **Scenario 1:** high efficiency development—25 percent contribution of wells from clusters 3, 5, 11, and 16
- **Scenario 2:** low efficiency development—20 percent contribution of wells from clusters 1, 4, 6, 8, and 17
- **Scenario 3:** diversified development—contribution of wells from each cluster randomly assigned under equal probability per cluster

An average of 1842 wells have been spud per year in Spraberry/Dean and Wolfcamp formations in the Midland Basin from 2017 to 2019 based on the study dataset. The generated outlooks under each of the three scenarios evaluated are therefore based on a theoretical new well fleet of 1842 wells in each scenario. Production outlooks for oil, water, and gas volumes produced from the new well fleet in the first year and through five years of production are shown in Figure 14.

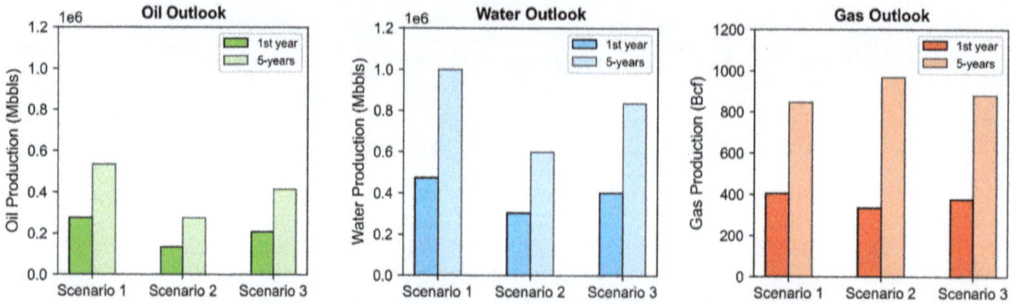

Figure 14. Oil, water, and gas production volumes under three different development scenarios for the Midland Basin. Each scenario assumes 1842 new wells drilled and completed.

First year production volumes range from approximately 132,000 to 275,000 Mbbls oil, 304,000 to 473,000 Mbbls water, and 335 to 405 Bcf of gas across the three scenarios constructed. Production volumes through five years extend from 276,000 to 535,000 Mbbls oil, 599,000 to 1,000,000 Mbbls of water, and 847 to 969 Bcf of gas. Results emphasize the notion that development choices regarding well design and placement (varied here by clusters implemented) have considerable implications on resulting fluid production outlooks. Worth noting is that under Scenario 1, where well deployment is limited to the clusters with the highest oil to associated fluid efficiencies, the largest volumes of associated water are produced compared to other scenarios. Associated gas, however, is the lowest out of all three scenarios. On the other hand, well development under Scenario 2 results in the lowest comparative volume of oil produced, but also generates the highest 5-year volumes of associated gas compared to other scenarios. Additionally, produced fluid volumes are likely to scale accordingly based on the number of wells that come online. Additional deployment scenario analyses could be explored using data in Table A2 to evaluate the influence of coupled well design, placement, and volume on produced fluid outlooks. Based on the approximate percentage of gas flared to gas produced in the Midland Basin per Leyden (2.35 percent of total), roughly 20 to 23 Bcf of gas would be flared over the five-years of production based on the results presented in Figure 14.

While this is a relatively straightforward example, it is nonetheless effective for quantifying produced volumes of both natural gas and water based on potential O&G development considerations. The outlooks can aid operators when formulating management or remedial solutions for the volumes of fluids expected. However, this analysis only includes production outlooks for the new wells considered and does not incorporate legacy production from wells producing prior to the installation of the new well fleet or those wells that come online afterwards. Production outlooks for natural gas or oil are highly dependent on a multitude of factors, including the typical production profiles of individual wells over time, the cost of drilling and operating those wells, the prospective economic return generated by those wells, the prevailing economic conditions related to O&G supply and demand, the intensity in which new wells are drilled, completed, and turned online, and the available prospective area remaining for a given play [29,155–157]. Forecasting associated water and gas would also be subject to similar factors. Therefore, alternative scenario formulations could be used to reflect different basin development outlooks than the one's analyzed here.

5. Conclusions

In this paper, we have introduced a data-driven modeling framework that combines supervised and unsupervised ML approaches. The findings from this study suggest that the approach and combination of machine learning strategies provides for a capable time series predictive model that can be used to either reproduce or forecast cumulative volumes of natural gas and water produced alongside oil at the well level. The intent of the supervised learning component was to produce a deep learning-based model with the capability to generate reliable estimates of produced water and natural gas in a time series manner based on well completion and placement decisions. The unsupervised learning aspect established groupings of related wells, enabling a straightforward method to deduce Arps Decline, well completion, and reservoir and spatial attributes characteristic of each cluster group. The ensemble of the supervised and unsupervised elements of this work facilitates a means to forecast oil, water, and natural gas production at the well level as influenced by specific development considerations. Well level three-stream production volumes can be used to scale up outlooks at the pad, field, or basin-level (as demonstrated in Section 4). The framework has been applied to the producing extent of the "Wolfberry" within the Midland Basin. However, since the overall analytical approach is based on readily available datasets common to public sources, it could be easily modified for use in other mature unconventional O&G producing regions.

Major environmental concerns regarding shale O&G development are associated to water usage, induced seismicity via wastewater disposal, and flaring (and possible venting) of produced natural gas. The framework presented in this study can be leveraged to help support the formulation of management and/or remedial strategies based on the volumes of fluids expected from unconventional O&G development operational conditions. Study results have highlighted the variability in noted water and gas volumes produced depending on wellbore design and placement considerations—a finding which suggests that forecasting is a nontrivial task. Table 6, Table 7, and Table A2 provide quantitative insight that can reduce the burden in estimating associated fluid production for future wells. Data compiled in Table A2 summarizes the potential volumes of produced fluids associated with oil production across the study area given well completion design considerations and placement within the basin. These data can be used to build out three-stream fluid production outlooks for the Midland Basin. Forward-looking production outlooks for oil, water, and natural gas as highlighted in Figure 14 are highly dependent on the nature of well design and placement considerations of the subsequent fleet of wells (as well as legacy production from existing wells). However, many of these design choices that would determine the composition of the out-year fleet of wells can be strongly influenced by external economic or market-driven factors.

Potential follow-on work could be beneficial in addressing possible limitations and imposed constraints in the research presented here; as well as build off of the opportunities this study creates. For instance, the within-cluster variation in decline curve, well completion, and spatial and reservoir attribute data noted in Table A2 affords the opportunity towards a more stochastic analytic approach as a complement to the deterministic strategy using mean values presented in this study. A potential area for improvement to the study in regards to the model development pertains to limited access to geologic data which could be used as inputs. Readily available geologic data at the well level in large volumes is uncommon. Nevertheless, the inclusion of additional geologic characteristics that are known controlling factors to unconventional oil and gas recovery [158] may provide added utility in data-driven ML modeling. Additionally, our study was without access to key time series data pertaining to how wells were operated (i.e., choke, bottom-hole pressure, lift type), the result of which presents a challenge in integrating the human element as part of the forecasting component. In regards to forecasting oil production, gradual or abrupt changes in the producing rate of a well due to reservoir depletion, fluctuation in bottom-hole producing pressure, and changes in conditions in or immediately adjacent to the wellbore are not directly considered when using the Arps models alone. Lastly, potential model performance improvement gains might be realized thorough the development of separate models for predicting monthly water and gas individually instead of in joint fashion.

Author Contributions: Conceptualization, D.V.; methodology, D.V. and V.K.; validation, D.V. and V.K.; formal analysis, D.V.; investigation, D.V. and V.K.; resources, D.V. and V.K.; data curation, D.V.; writing—original draft preparation, D.V.; writing—review and editing, V.K.; visualization, D.V.; supervision, V.K. All authors have read and agreed to the published version of the manuscript.

Funding: This research received no external funding.

Institutional Review Board Statement: Not applicable.

Informed Consent Statement: Not applicable.

Data Availability Statement: The study well data utilized for this study was obtained from the O&G data vendor DrillingInfo/Enverus through a subscription service. Therefore, the study dataset cannot be made publicly available.

Acknowledgments: The authors would like to thank Enverus and DrillingInfo for their generosity in providing access to the data used as part of this study. Additionally, the authors would like to thank William Harbert of the University of Pittsburgh's Department of Geology and Environmental Science and Carla Ng of the University of Pittsburgh's Civil & Environmental Engineering Department for the constructive reviews which greatly improved the quality of manuscript.

Conflicts of Interest: The authors declare no conflict of interest associated with this research.

Unit Conversions: The units used in this manuscript are commonly used industry standards for the oil and gas sector in the United States. Conversion factors to the international system of units are as follows: 1 barrel = 0.159 cubic meters; 1 foot = 0.3048 m; 1 cubic foot = 0.0283 cubic meters; 1 square mile = 2.589 square kilometers.

Appendix A. Feature Selection Results Overview

The hyperparameter combination selected via grid search cross-validation for the RF estimator used as part of RFECV included a formulation of 5050 trees and a minimum of two samples to split an internal node.

Figure A1 depicts the predictive performance of the RF estimator based on the number of features employed as part of training and cross-validation testing. For this study, explained variance for each model iteration across the range of features selected are normalized relative to the number of features included resulting in the highest explained variance.

As a result, the number of features resulting in the estimator with the highest explained variance has value equal to 1, and all others less than 1. Once the number of features is reduced below six, the estimator's predictive performance begins to diminish as more features are omitted as part of estimator training. In contrast, estimator performance gains are marginal at best when the number of features included in training are greater than six; with an optimal range between six and eleven features.

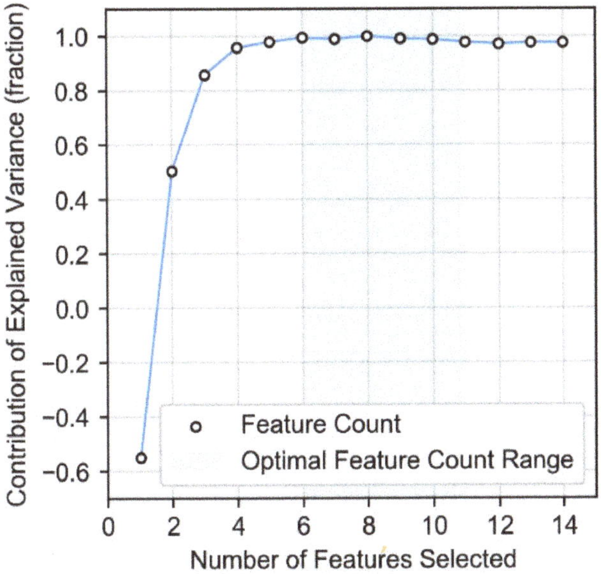

Figure A1. Effect of feature inclusion relative to the highest feature count score.

The ranking importance of each feature based on the estimator formulation with all 14 features included as part of training is presented in Figure A2. The ranking is based on the "relative" importance of each feature to that of the feature with the highest importance. The values for importance for each feature are normalized relative to the most import feature then scaled by 100. As a result, the feature with the highest importance has a value equal to 100, and all others less than 100. Examination of the feature importance ranking and magnitude indicates that oil production (reflected as Top 12 Months Oil) is the most important estimator feature for joint prediction of Top 12 Months Water and Top 12 Months Gas (static proxies for Monthly Gas and Monthly Water dynamic data features). The Top 12 Months Oil static data feature serves as a proxy for Monthly Oil, which is a dynamic feature that changes with time. The following three features (latitude, longitude, and true vertical depth) specify the three-dimensional coordinates for well horizontal placement within the basin. This finding suggests that the associated geological characteristics of the producing reservoirs which vary spatially and with burial depth are important contributors to the associated fluid response. Feature ranks five through seven (perforation length, water per foot, and proppant per foot) are notable well completion design attributes.

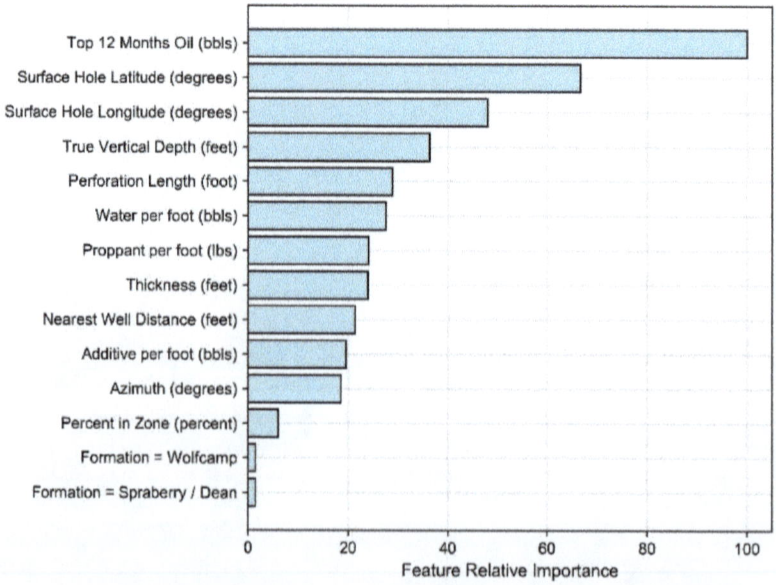

Figure A2. Summary of feature importance for the RF estimator used as part of RFECV.

The feature importance values in Figure A2 are used in concert with RFECV results from Figure A1 to inform the feature selection process. As a result, 11 static features are selected and three omitted from feature selection dataset for consideration in the clustering and time series analysis. This down-selection includes omission of the features with the three lowest values of importance; which include percent in zone and the two categorical variables demarcating wells completed in either the "Spraberry/Dean" or "Wolfcamp" formations. The removal of three features and inclusion of the remaining 11 coincide with the RFECV upper bound feature range count presented in Figure A1 where explained variance remains high.

As mentioned in Section 3.1, the feature selection step helps to systematically finalize sets of input features that can be applied as part of both the clustering evaluation (Section 3.2) and the development of the time series joint associated fluid production model (Section 3.3).

Appendix B. Tukey's Test Results on Arps Attributes by Cluster

The results from the Tukey's test performed one each of the three Arps attributes across the 18 well clusters is presented in Table A1. The post hoc Tukey's test highlights which clusters, and therefore Arps decline attributes, differed significantly from cluster to cluster at $\alpha = 0.05$.

Table A1. Descriptive statistics and results from Tukey's test on decline curve attributes across well clusters.

Cluster Number	Initial Oil Production (bbls)				Initial Decline (Fraction/Month)			b-Factor		
	Count	Mean	Stdev.	Tukey's Group	Mean	Stdev.	Tukey's Group	Mean	Stdev.	Tukey's Group
0	84	14,816	7459	H, I, J, K	0.40	0.12	A, B, C, D	1.25	0.24	B, C, D, E
1	259	15,364	7653	J, K	0.18	0.09	H	1.55	0.08	A
2	246	28,382	7826	C	0.36	0.12	D, E	1.07	0.14	I, J
3	594	20,148	7935	G	0.40	0.11	B	1.07	0.13	I, J
4	460	7481	4835	M	0.41	0.12	A, B	1.21	0.22	C, D, E, F
5	574	35,577	10,694	B	0.39	0.11	B, C, D	1.14	0.20	G, H
6	328	17,625	7588	H, I	0.41	0.10	A, B	1.06	0.13	I, J
7	609	14,442	7392	K	0.32	0.14	F	1.24	0.25	B, C
8	230	13,408	7594	K	0.34	0.12	E, F	1.17	0.22	D, E, F, G
9	173	25,506	8124	D, E	0.32	0.13	F	1.15	0.23	F, G, H
10	515	17,353	8606	H, I, J	0.40	0.11	B	1.19	0.23	E, F
11	101	14,630	8813	I, J, K	0.35	0.13	C, D, E, F	1.29	0.24	B
12	485	17,666	6449	H	0.43	0.08	A	1.18	0.18	F, G
13	304	26,324	6777	C, D	0.25	0.11	G	1.11	0.18	H, I
14	554	23,346	7579	E, F	0.27	0.13	G	1.04	0.09	J
15	160	20,971	8156	F, G	0.26	0.12	G	1.26	0.25	B, C
16	346	40,342	8293	A	0.26	0.10	G	1.03	0.08	J
17	188	9959	5386	L	0.39	0.11	B, C, D	1.06	0.11	I, J

Appendix C. Well Cluster Statics and Production Outlooks

Table A2. Inventory of descriptive statics, 1st year, and cumulative 5-year production estimates for a hypothetical representative well within each Midland Basin Well Cluster.

Data Group	Dataset Feature	Statistic	Midland Basin Well Cluster Number: 0 through 8								
			0	1	2	3	4	5	6	7	8
Well Completion Attributes	Perforation Length (foot)	Mean	6782	8791	9593	7928	7719	10,061	9177	7139	9663
		Stdev.	1990	1770	1319	1665	1563	1502	1856	1565	1763
		IQR	2985	2704	1132	2378	1065	756	2581	1545	1756
	Proppant per foot (lbs)	Mean	1818	1698	1764	1659	1303	1845	1938	1477	2283
		Stdev.	570	391	331	349	413	389	428	395	394
		IQR	487	468	319	430	495	367	459	517	546
	Water per foot (bbls)	Mean	46.8	45.6	50.7	40.6	28.2	47.8	49.6	37.2	51.4
		Stdev.	15.3	10.3	9.0	12.2	8.3	10.1	10.9	10.6	9.3
		IQR	9.8	12.4	11.6	15.5	9.1	13.0	13.1	13.2	8.7
	Additive per foot (bbls)	Mean	12.1	2.8	2.9	4.1	1.8	2.6	3.5	3.2	2.1
		Stdev.	3.9	1.8	1.5	3.1	1.3	1.6	3.3	1.8	1.2
		IQR	3.9	2.5	2.0	4.9	2.1	2.3	2.5	2.6	1.3
	Azimuth (degrees)	Mean	165.1	162.4	162.6	162.6	180.3	162.6	178.9	162.6	180.8
		Stdev.	7.0	3.7	3.7	3.6	3.4	3.3	5.9	3.3	3.1
		IQR	3.2	4.2	3.4	4.1	4.3	4.0	5.1	2.3	4.1
	Nearest Well Distance (feet)	Mean	844	288	254	261	550	259	523	382	388
		Stdev.	1013	384	307	324	473	269	441	556	428
		IQR	1185	307	277	267	519	295	413	390	453

Table A2. *Cont.*

Data Group	Dataset Feature	Statistic	Midland Basin Well Cluster Number: 0 through 8								
			0	1	2	3	4	5	6	7	8
Decline Curve Attributes	Initial Oil Production (bbls)	Mean	14,816	15,364	28,382	20,148	7481	35,577	17,625	14,442	13,408
		Stdev.	7459	7653	7826	7935	4835	10,694	7588	7392	7594
		IQR	10,635	11,155	10,173	10,197	5895	13,455	10,748	9233	9916
	Initial Decline (fraction/month)	Mean	0.40	0.18	0.36	0.40	0.41	0.39	0.41	0.32	0.34
		Stdev.	0.12	0.09	0.12	0.11	0.12	0.11	0.10	0.14	0.12
		IQR	0.20	0.15	0.22	0.17	0.18	0.19	0.17	0.26	0.22
	b-factor	Mean	1.25	1.55	1.07	1.07	1.21	1.14	1.06	1.24	1.17
		Stdev.	0.24	0.08	0.14	0.13	0.22	0.20	0.13	0.25	0.22
		IQR	0.50	0.08	0.08	0.11	0.40	0.26	0.04	0.50	0.39
Spatial and Reservoir Attributes	True Vertical Depth (feet)	Mean	8924	8964	8947	9310	7112	8811	7150	9020	7460
		Stdev.	752	630	626	470	741	785	620	771	577
		IQR	798	848	884	563	963	1296	1018	1174	686
	Thickness (feet)	Mean	443	398	471	320	774	375	633	374	553
		Stdev.	157	137	115	96	146	108	207	134	191
		IQR	209	148	137	148	59	136	338	185	361
	Surface Hole Latitude (degrees)	Mean	31.64	31.92	31.70	32.08	31.15	32.08	31.38	31.98	31.32
		Stdev.	0.32	0.28	0.17	0.26	0.12	0.28	0.23	0.29	0.14
		IQR	0.44	0.45	0.19	0.47	0.19	0.47	0.38	0.56	0.19
	Surface Hole Longitude (degrees)	Mean	−101.93	−101.93	−101.81	−102.08	−101.33	−101.87	−101.26	−101.90	−101.58
		Stdev.	0.28	0.19	0.22	0.14	0.23	0.24	0.18	0.29	0.16
		IQR	0.32	0.29	0.32	0.21	0.26	0.42	0.30	0.53	0.15
	Wolfcamp	Count	68	168	223	315	456	419	326	445	230
	S.berry/Dean	Count	16	91	23	280	4	155	2	164	0
Production Forecast per Well	Cumulative Oil (Mbbls)	1st year	77	111	147	100	38	181	86	82	73
		5-years	154	282	275	183	74	346	156	169	145
	Cumulative Gas (Bcf)	1st year	0.16	0.20	0.25	0.15	0.12	0.27	0.25	0.13	0.22
		5-years	0.29	0.58	0.62	0.23	0.23	0.60	0.76	0.21	0.79
	Cumulative Water (Mbbls)	1st year	154	230	268	181	102	304	200	162	182
		5-years	289	587	545	347	175	659	358	324	328
Data Group	**Dataset Feature**	**Statistic**	**Midland Basin Well Cluster Number: 9 through 17**								
			9	**10**	**11**	**12**	**13**	**14**	**15**	**16**	**17**
Well Completion Attributes	Perforation Length (foot)	Mean	8307	7677	7253	7225	9870	8762	9448	9972	7417
		Stdev.	1814	1970	2103	1794	1155	1469	1892	1333	1605
		IQR	2711	2933	4212	2361	563	2313	2612	740	1549
	Proppant per foot (lbs)	Mean	3281	1728	1609	1441	1752	1812	1787	1828	1342
		Stdev.	775	464	535	547	336	359	412	490	394
		IQR	872	677	759	687	305	413	507	407	532
	Water per foot (bbls)	Mean	71.4	39.4	40.6	36.6	49.4	48.8	44.9	48.0	32.8
		Stdev.	19.7	11.2	14.9	14.0	8.0	10.7	12.5	10.2	8.6
		IQR	23.2	13.8	17.3	18.9	8.7	9.6	15.5	10.6	7.8
	Additive per foot (bbls)	Mean	4.9	2.1	4.2	2.1	2.2	2.3	2.1	2.9	1.9
		Stdev.	2.9	1.5	3.4	1.3	1.4	1.5	1.5	1.7	1.2
		IQR	3.0	1.8	3.8	1.6	2.0	2.3	2.0	1.9	2.0

Table A2. *Cont.*

Data Group	Dataset Feature	Statistic	Midland Basin Well Cluster Number: 0 through 8								
			0	1	2	3	4	5	6	7	8
	Azimuth (degrees)	Mean	163.2	162.6	168.0	162.2	162.9	162.4	163.3	162.8	180.8
		Stdev.	5.5	2.9	8.8	3.3	3.7	3.4	3.3	3.6	2.1
		IQR	4.6	2.5	17.6	3.6	3.5	3.8	3.2	4.4	2.2
	Nearest Well Distance (feet)	Mean	395	392	5658	303	328	243	486	278	343
		Stdev.	542	473	1942	354	308	306	764	270	438
		IQR	419	404	2770	285	386	259	511	316	438
Decline Curve Attributes	Initial Oil Production (bbls)	Mean	25,506	17,353	14,630	17,666	26,324	23,346	20,971	40,342	9959
		Stdev.	8124	8606	8813	6449	6777	7579	8156	8293	5386
		IQR	11,762	12,799	13,555	9324	9246	9785	10,744	11,795	6533
	Initial Decline (fraction/month)	Mean	0.32	0.40	0.35	0.43	0.25	0.27	0.26	0.26	0.39
		Stdev.	0.13	0.11	0.13	0.08	0.11	0.13	0.12	0.10	0.11
		IQR	0.21	0.18	0.25	0.13	0.15	0.18	0.17	0.11	0.21
	b-factor	Mean	1.15	1.19	1.29	1.18	1.11	1.04	1.26	1.03	1.06
		Stdev.	0.23	0.23	0.24	0.18	0.18	0.09	0.25	0.08	0.11
		IQR	0.26	0.39	0.54	0.31	0.17	0.02	0.59	0.02	0.09
Spatial and Reservoir Attributes	True Vertical Depth (feet)	Mean	9078	7883	8340	9238	9128	9123	7751	8963	7523
		Stdev.	609	568	1101	465	424	511	727	587	723
		IQR	784	527	1950	555	540	673	956	962	961
	Thickness (feet)	Mean	503	384	463	541	653	380	369	356	477
		Stdev.	209	103	183	145	176	103	111	86	151
		IQR	244	132	224	168	289	119	110	115	254
	Surface Hole Latitude (degrees)	Mean	31.83	32.23	31.80	31.68	31.60	31.91	32.33	32.09	31.39
		Stdev.	0.33	0.30	0.48	0.16	0.16	0.27	0.24	0.24	0.18
		IQR	0.56	0.47	0.87	0.24	0.26	0.43	0.21	0.39	0.27
	Surface Hole Longitude (degrees)	Mean	−101.90	−101.58	−101.70	−101.89	−101.82	−102.01	−101.62	−101.94	−101.38
		Stdev.	0.20	0.14	0.32	0.15	0.12	0.16	0.22	0.19	0.17
		IQR	0.25	0.12	0.56	0.18	0.15	0.19	0.27	0.37	0.17
	Wolfcamp	Count	137	356	89	459	301	321	88	227	188
	S.berry/Dean	Count	36	159	12	26	3	223	72	119	0
Production Forecast per Well	Cumulative Oil (Mbbls)	1st year	141	89	80	87	160	135	129	237	50
		5-years	281	173	168	167	328	265	279	465	91
	Cumulative Gas (Bcf)	1st year	0.26	0.14	0.12	0.15	0.31	0.22	0.22	0.34	0.12
		5-years	0.57	0.19	0.16	0.27	0.91	0.50	0.45	0.85	0.27
	Cumulative Water (Mbbls)	1st year	271	185	170	171	306	249	265	373	111
		5-years	574	364	287	332	684	515	621	879	178

References

1. U.S. Department of Energy. *Ethane Storage and Distribution Hub in the United States*; U.S. DOE: Washington, DC, USA, 2018.
2. Pirog, R.; Ratner, M. *Natural Gas in the U.S. Economy: Opportunities for Growth*; Congressional Research Service: Washington, DC, USA, 2012.
3. U.S. Energy Information Administration. Today in Energy: Both Natural Gas Supply and Demand Have Increased from Year-Ago Levels. U.S. Department of Energy. 4 October 2018. Available online: https://www.eia.gov/todayinenergy/detail.php?id=37193 (accessed on 31 March 2019).
4. Clemente, J.U.S. Natural Gas Demand for Electricity Can Only Grow. Forbes. 15 January 2019. Available online: https://www.forbes.com/sites/judeclemente/2019/01/15/u-s-natural-gas-demand-for-electricity-can-only-grow/#27b0ba844c74 (accessed on 31 March 2019).

5. Aadnøy, B.; Looyeh, R. *Petroleum Rock Mechanics*, 2nd ed.; Gulf Professional Publishing: Oxford, UK, 2019.
6. United States Geological Survey, U.S. Department of the Interior. What Is Hydraulic Fracturing? Available online: https://www.usgs.gov/faqs/what-hydraulic-fracturing?qt-news_science_products=0#qt-news_science_products (accessed on 21 November 2020).
7. Hyman, J.; Jiménez-Martínez, J.; Viswanathan, H.; Carey, J.P.M.; Rougier, E.; Karra, S.; Kang, Q.; Frash, L.; Chen, L.; Lei, Z.; et al. Understanding hydraulic fracturing: Amulti-scale problem. *Philos. Trans. Ser. A Math. Phys. Eng. Sci.* **2016**, *374*, 20150426. [CrossRef] [PubMed]
8. Aminzadeh, F. Hydraulic Fracturing, An Overview. *J. Sustain. Energy Eng.* **2019**, *6*, 204–228. [CrossRef]
9. U.S. Environmental Protection Agency. The Process of Unconventional Natural Gas Production. 22 January 2020. Available online: https://www.epa.gov/uog/process-unconventional-natural-gas-production (accessed on 21 November 2020).
10. Perrin, J. Horizontally Drilled Wells Dominate U.S. Tight Formation Production. U.S. Energy Information Administration. 6 June 2019. Available online: https://www.eia.gov/todayinenergy/detail.php?id=39752 (accessed on 12 December 2020).
11. van Wagener, D.; Aloulou, F. Tight Oil Development Will Continue to Drive Future U.S. Crude Oil Production. U.S. Energy Information Administration, 28 March 2019. Available online: https://www.eia.gov/todayinenergy/detail.php?id=38852 (accessed on 12 December 2020).
12. Vikara, D.; Remson, D.; Khanna, V. Machine learning-informed ensemble framework for evaluating shale gas production potential: Case study in the Marcellus Shale. *J. Nat. Gas Sci. Eng.* **2020**, *84*, 103679. [CrossRef]
13. U.S. Department of Energy. *Quadrennial Technology Review 2015—Chapter 7: Advancing Systems and Technologies to Produce Clearner Fuels*; U.S. Department of Energy: Washington, DC, USA, 2015.
14. Mehrotra, R.; Gopalan, R. Factors Influencing Strategic Decision-Making Process for the Oil/Gas Industris of UAE—A study. *Int. J. Mark. Financ. Manag.* **2017**, *5*, 62–69.
15. Mo, S.; Zhu, Y.; Zabaras, N.; Shi, X.; Wu, J. Deep convolutional encoder-decoder networks for uncertainty quantification of dynamic multiphase flow in heterogeneous media. *Water Resour. Res.* **2019**, *55*, 703–728. [CrossRef]
16. Esmaili, S.; Mohaghegh, S. Full field reservoir modeling of shale assets using advanced data-driven analytics. *Geosci. Front.* **2016**, *7*, 11–20. [CrossRef]
17. McGlade, C.; Speirs, J.; Sorrell, S. Methods of estimating shale gas resources—Comparison, evaluation and implications. *Energy* **2013**, *59*, 116–125. [CrossRef]
18. Baaziz, A.; Quoniam, L. How to use Big Data technologies to optimize operations in Upstream Petroleum Industry. In Proceedings of the 21st World Petroleum Congress, Moscow, Russia, 15–19 June 2014.
19. Bettin, G.; Bromhal, G.; Brudzinski, M.; Cohen, A.; Guthrie, G.; Johnson, P.; Matthew, L.; Mishra, S.; Vikara, D. *Real-Time Decision Making for the Subsurface Report*; Carnegie Mellon University Wilson E. Scott Institute for Energy Innovation: Pittsburgh, PA, USA, 2019.
20. Mishra, S.; Lin, L. Application of Data Analytics for Production Optimization in Unconventional Reservoirs: A Critical Review. In Proceedings of the Unconventional Resources Technology Conference, Austin, TX, USA, 24–26 July 2017.
21. Abubakar, A. *Potential and Challenges of Applying Artificial Intelligence and Machine-Learning Methods for Geoscience*; Society of Exploration Geophysicists: Houston, TX, USA, 2020.
22. Wang, S.; Chen, S. Insights to fracture stimulation design in unconventional reservoirs based on machine learning modeling. *J. Pet. Sci. Eng.* **2019**, *174*, 682–695. [CrossRef]
23. Vikara, D.; Remson, D.; Khanna, V. Gaining Perspective on Unconventional Well Design Choices through Play-level Application of Machine Learning Modeling. *Upstream Oil Gas Technol.* **2020**, *4*, 100007. [CrossRef]
24. Shih, C.; Vikara, D.; Venkatesh, A.; Wendt, A.; Lin, S.; Remson, D. *Evaluation of Shale Gas Production Drivers by Predictive Modeling on Well Completion, Production, and Geologic Data*; National Energy Technology Laboratory: Pittsburgh, PA, USA, 2018.
25. Wang, S.; Chen, Z.; Chen, S. Applicability of deep neural networks on production forecasting in Bakken shale reservoirs. *J. Pet. Sci. Eng.* **2019**, *179*, 112–125. [CrossRef]
26. LaFollette, R.; Izadi, G.; Zhong, M. Application of Multivariate Analysis and Geographic Information Systems Pattern-Recognition Analysis to Produce Results in the Bakken Light Oil Play. In Proceedings of the SPE Hydraulic Fracturing Technology Conference, The Woodlands, TX, USA, 2 February 2013.
27. Montgomery, J.; O'Sullivan, F. Spatial variability of tight oil well productivity and the impact of technology. *Appl. Energy* **2017**, *195*, 334–355. [CrossRef]
28. Browning, J.; Tinker, S.; Ikonnikova, S.; Gulen, G.; Potter, E.; Fu, Q.; Horvath, S.; Patzek, T.; Male, F.; Fisher, W.; et al. Barnett study determines full-field reserves, production forecast. *Oil Gas J.* **2013**, *111*, 88–95.
29. Ikonnikova, S.; Browning, J.; Gulen, G.; Smye, K.; Tinker, S. Factors influencing shale gas production forecasting: Empirical studies of Barnett, Fayetteville, Haynesville, and Marcellus Shale plays. *Econ. Energy Environ. Policy* **2015**, *4*, 19–35. [CrossRef]
30. U.S. Energy Information Administration. *Annual Energy Outlook 2020*; U.S. Department of Energy: Washington, DC, USA, 2020.
31. Jie, L.; Junxing, C.; Jiachun, Y. Prediction on daily gas production of single well based on LSTM. In Proceedings of the SEG 2019 Workshop: Mathematical Geophysics: Traditional vs. Learning, Beijing, China, 5–7 November 2019.
32. Sagheer, A.; Kotb, M. Time series forecasting of petroleum production using deep LSTM recurrent networks. *Neurocomputing* **2019**, *323*, 203–213. [CrossRef]

33. Liu, W.; Liu, W.; Gu, J. Forecasting oil production using ensemble empirical model decomposition based Long Short-Term Memory neural network. *J. Pet. Sci. Eng.* **2020**, *189*, 107013. [CrossRef]
34. U.S. Department of Energy. *Natural Gas Flaring and Venting: State and Federal Regulatory Overview, Trends, and Impacts*; Office of Fossil Energy—Office of Oil and Natural Gas: Washington, DC, USA, 2019.
35. Myhre, G.; Shindell, D.; Bréon, F.; Collins, W.F.J.; Huang, J.; Koch, D.; Lamarque, J.; Lee, D.; Mendoza, B. Anthropogenic and Natural Radiative Forcing. In *Climate Change 2013: The Physical Science Basis. Contribution of Working Group I to the Fifth Assessment Report of the Intergovernmental Panel on Climate Change*; Cambridge University Press: Cambridge, UK; New York, NY, USA, 2013.
36. U.S. Energy Information Administration. Natural Gas Annual. U.S. Department of Energy, 30 September 2020. Available online: https://www.eia.gov/naturalgas/annual/ (accessed on 10 December 2020).
37. United States Geological Survey, U.S. Department of the Interior. ANSS Comprehensive Earthquake Catalog (ComCat) Documentation. Available online: https://earthquake.usgs.gov/data/comcat/ (accessed on 11 December 2020).
38. Scanlon, B.; Reedy, R.; Xu, P.; Engle, M.; Nicot, J.; Yoxtheimer, D.; Yang, Q.; Ikonnikova, S. Can we beneficially reuse produced water from oil and gas extraction in the U.S.? *Sci. Total Environ.* **2020**, *717*, 137085. [CrossRef]
39. Kah, M. Columbia Global Energy Dialogue: Natural Gas Flaring Workshop Summary. Columbia Center on Global Energy Policy, 30 April 2020. Available online: https://www.energypolicy.columbia.edu/research/global-energy-dialogue/columbia-global-energy-dialogue-natural-gas-flaring-workshop-summary (accessed on 12 December 2020).
40. van Bedolla, L.; Cai, W.; Martin, Z.; Yu, F. *Technology and Policy Solutions to Reduce Harmful Natural Gas Flaring*; Columbia University School of International and Public Affairs: New York, NY, USA, 2020.
41. Tavakkoli, S.; Lokare, O.; Vidic, R.; Khanna, V. Shale gas produced water management using membrane distillation: An optimization-based approach. *Resour. Conserv. Recycl.* **2020**, *158*, 104803. [CrossRef]
42. Shamlou, E.; Vidic, R.; Khanna, V. Optimization-based modeling and economic comparison of membrane distillation configurations for application in shale gas produced water treatment. *Desalination* **2022**, *526*, 115513. [CrossRef]
43. Oil & Gas Journal. Permian Gas Flaring, Venting Reaches Record High. 4 June 2019. Available online: https://www.ogj.com/general-interest/hse/article/17279037/permian-gas-flaring-venting-reaches-record-high (accessed on 31 July 2020).
44. *Texas Independent Producers and Royalty Owners Association*; A Decade of the Permian Basin; Texas Independent Producers & Royalty Owners Association: Austin, TX, USA, 2020.
45. The American Oil & Gas Reporter. Importance of Permian Basin Is Delineated in TIPRO Report. February 2020. Available online: https://www.aogr.com/magazine/markets-analytics/importance-of-permian-basin-is-delineated-in-tipro-report (accessed on 26 July 2020).
46. McEwen, M. Wood Mackenzie Analysts: Permian Faces Multiple Challenges. MRT.com, 28 July 2019. Available online: https://www.mrt.com/business/oil/article/Wood-Mackenzie-analysts-Permian-faces-multiple-14149600.php#photo-17926034 (accessed on 31 July 2020).
47. Vaucher, D. No Free Lunch—The Water Challenges Facing Operating Companies in the Permian Basin. IHS Markit, 4 November 2019. Available online: https://ihsmarkit.com/research-analysis/no-free-lunch-the-water-challenges-facing-companies-permian.html (accessed on 31 July 2020).
48. DrillingInfo/Enverus. DI Research Products Glossary. Enverus. Available online: http://help.drillinginfo.com/robohelp/robohelp/server/general/projects/DI%20Desktop%20Online%20Manual/DI_Analytics/Other_Resources/DI_Research_Products_Glossary.htm (accessed on 15 November 2020).
49. Rassenfoss, S. Rising Tide of Produced Water Could Pinch Permian Growth. *J. Pet. Technol.* **2018**. Available online: https://pubs.spe.org/en/jpt/jpt-article-detail/?art=4273 (accessed on 29 November 2020).
50. Railroad Commission of Texas. Permian Basin Information. 11 November 2020. Available online: https://www.rrc.state.tx.us/oil-gas/major-oil-and-gas-formations/permian-basin-information/ (accessed on 25 November 2020).
51. U.S. Energy Information Administration. *Permian Basin Part 2: Wolfcamp Shale Play of the Midland Basin—Geology Review*; U.S. Department of Energy: Washington, DC, USA, 2020.
52. U.S. Energy Information Administration. U.S. Crude Oil and Natural Gas Proved Reserves, Year End-2018. U.S. Department of Energy: Washington, DC, USA, 13 December 2019. Available online: https://www.eia.gov/naturalgas/crudeoilreserves/ (accessed on 25 November 2020).
53. U.S. Energy Information Administration. *Permian Basin Wolfcamp Shale Play: Geology Review*; U.S. Department of Energy: Washington, DC, USA, 2018.
54. Sutton, L. Permian Basin Geology: The Midland Basin vs. the Delaware Basin Part 2. Enverus, 23 December 2014. Available online: https://www.enverus.com/blog/permian-basin-geology-midland-vs-delaware-basins/ (accessed on 11 November 2020).
55. Yang, K.; Dorobeck, S. The Permian Basin of West Texas and New Mexico: Tectonic History of a "Composite" Foreland Basin and its Effects on Stratigraphic Development. In *Stratigraphic Evolution of Foreland Basins*; SEPM Society for Sedimentary Geology: Tulsa, OK, USA, 1995; Volume 52.
56. Roberts, J. *GDS Geological Column: Geological Data Service*; Geological Data Service: Dallas, TX, USA, 1989.
57. University of Texas at Austin. *Wolfberry and Spraberry Play of the Midland Basin*; Bureau of Economic Geology: Austin, TX, USA. Available online: http://www.beg.utexas.edu/research/programs/starr/unconventional-resources/wolfberry-spraberry (accessed on 2 September 2020).

58. Wilson, G. Midland Basin Wolfcamp Horizontal Development. In Proceedings of the AAPG DPA Forum Midland Playmaker, Midland, TX, USA, 14 January 2015.
59. R. King & Co. Permian Basin Strategraphic Charts & Province MaUndated. Available online: https://rkingco.com/wp-content/uploads/2014/12/PermianBasinStratChart.jpg (accessed on 2 September 2020).
60. Vikara, D.; Khanna, V. Machine learning classification approach for formation delineation at the basin-scale. *Pet. Res.* **2021**. [CrossRef]
61. Hamlin, H.; Baumgardner, R. *Wolfberry (Wolfcampian-Leonardian) Deep-Water Depositional Systems in the Midland Basin: Stratigraphy, Lithofacies, Reservoirs, and Source Rocks*; Part Number RI0277; University of Texas Bureau of Economic Geology: Austin, TX, USA, 2012.
62. Schmitt, G. Genesis and Depositional History of Spraberry Formation, Midland Basin, Texas. *AAPG Bull.* **1954**, *38*, 1957–1978.
63. Hunter, G.; Šegvić, B.; Zanoni, G.; Omodeo-Salé, S.; Adatte, T. Evaluation of Shale Source Rocks and Clay Mineral Diagenesis in the Permian Basin, USA: Inferences on Basin Thermal Maturity and Source Rock Potential. *Geosciences* **2020**, *10*, 381.
64. James, A. Evaluating and Hy-Grading Wolfcamp Shale Opportunities in the Midland Basin. AAPG Search and Discovery Article #110213. In Proceedings of the AAPG DPA Forum Midland Playmaker, Midland, TX, USA, 14 January 2015.
65. Handford, C. Sedimentology and Genetic Stratigraphy of Dean and Spraberry Formations (Permian), Midland Basin, Texas. *AAPG Bull.* **1981**, *65*, 1602–1616.
66. Lorenz, J.; Sterling, J.; Schechter, D.; Whigham, C.; Jensen, J. Natural fractures in the Spraberry Formation, Midland basin, Texas: The effects of mechanical stratigraphy on fracture variability and reservoir behavior. *AAPG Bull.* **2002**, *86*, 505–524.
67. Marshall, J. Spraberry Reservoir of West Texas1: GEOLOGICAL NOTES. *AAPG Bull.* **1952**, *36*, 2189–2191.
68. Shattuck, B. Spraberry Fields Forever. Forbes, 8 September 2017. Available online: https://www.forbes.com/sites/woodmackenzie/2017/09/08/spraberry-fields-forever/?sh=245b4309655a (accessed on 26 November 2020).
69. Murphy, R. Depositional Systems Interpretation of Early Permian mixed Siliciclastics and Carbonates, Midland Basin, Texas. Master's Thesis, University of Indiana, Bloomington, Indiana, 2015.
70. Gaswirth, S. Assessment of Undiscovered Continuous Oil and Gas Resources in the Wolfcamp Shale of the Midland Basin, West Texas. In Proceedings of the AAPG Annual Convention and Exhibition, Houston, TX, USA, 2–5 April 2017.
71. U.S. Energy Information Administration. *EIA Updates Geological Maps of Midland Basin's Wolfcamp Formation*; U.S. Department of Energy: Washington, DC, USA, 24 November 2020. Available online: https://www.eia.gov/todayinenergy/detail.php?id=46016 (accessed on 25 November 2020).
72. Saller, A.; Dickson, A.; Boyd, S. Cycle Stratigraphy and Porosity in Pennsylvanian and Lower Permian Shelf Limestones, Easten Central Basin Platform, Texas. *AAPG Bull.* **1994**, *78*, 1820–1842.
73. Peng, J.; Milliken, K.; Fu, Q.; Janson, X.; Hamlin, S. Grain assemblages and diagenesis in organic-rich mudrocks, Upper Pennsylvanian Cline shale (Wolfcamp D), Midland Basin, Texas. *AAPG Bull.* **2020**, *104*, 1593–1624. [CrossRef]
74. Blomquist, P. *Wolfcamp Horizontal Play Midland Basin, West Texas*; IHS Markit, IHS Geoscience Webinar Series; HIS: London, UK, 2016.
75. U.S. Energy Information Administration. *The Wolfcamp Play Has Been Key to Permian Basin Oil and Natural Gas Production Growth*; U.S. Department of Energy: Washington, DC, USA, 16 November 2018. Available online: https://www.eia.gov/todayinenergy/detail.php?id=37532 (accessed on 25 November 2020).
76. Enverus. DrillingInfo Web A2020. Available online: https://www.enverus.com/products/di-web-app/ (accessed on 1 November 2020).
77. University of Texas at Austin—Bureau of Economic Geology. Integrated Synthesis of the Permian Basin: Data and Models for Recovering Existing and Undiscovered Oil Resources from the Largest OIl-Bearing Basin in the U.S. Jackson School of Geosciences. 2008. Available online: http://www.beg.utexas.edu/resprog/permianbasin/gis.htm (accessed on 9 September 2020).
78. United States Geological Survery. How to Use the National Map Services—Large Scale Base Map Dynamic Services. Available online: https://viewer.nationalmap.gov/help/HowTo.htm (accessed on 2 September 2020).
79. Kondash, A.; Lauer, N.; Vengosh, A. The intensification of the water footprint of hydraulic fracturing. *Sci. Adv.* **2018**, *4*, eaar5982. [CrossRef]
80. Bruant, R. Permian Water Outlook. B3 Insight. 26 February 2019. Available online: http://www.gwpc.org/sites/default/files/event-sessions/Produced%20Water%20-%20Rob%20Bruant_0.pdf (accessed on 12 December 2020).
81. Leyden, C. Satellite Data Confirms Permian Gas Flaring Is Double What Companies Report. Environmental Defense Fund, 24 January 2019. Available online: http://blogs.edf.org/energyexchange/2019/01/24/satellite-data-confirms-permian-gas-flaring-is-double-what-companies-report/ (accessed on 13 December 2020).
82. Abramov, A.; Bertelsen, M. Permian Gas Flaring Reaches yet Another High. Rystad Energy, 5 November 2019. Available online: https://www.rystadenergy.com/newsevents/news/press-releases/permian-gas-flaring-reaches-yet-another-high/ (accessed on 24 December 2020).
83. Agerton, M.; Gilbert, B.; Upton, G. *The Economics of Natural Gas Flaring in U.S. Shale: An Agenda for Research and Policy*; Rice University's Baker Institute for Public Policy: Houston, TX, USA, 2020.
84. Arps, J. Analysis of Decline Curves. *Trans. AIME* **1945**, *160*, 228–247. [CrossRef]
85. Miller, J. Short Report: Reaction Time Analysis with Outlier Exclusion: Bias Varies with Sample Size. *Q. J. Exp. Psychol. Sect. A* **1991**, *43*, 907–912. [CrossRef]

86. Ilyas, I.; Chu, X. *Data Cleaning*; Association for Computing Machinery: New York, NY, USA, 2019.
87. DrillingInfo. Pre-Calculated, Proprietary EUR Database from DrillingInfo—White Paper. May 2016. Available online: https://www.enverus.com/wp-content/uploads/2017/11/WP_EUR_Customer-print.pdf (accessed on 22 November 2020).
88. Fetkovich, M.; Fetkovich, E.; Fetkovich, M. Useful Concepts for Decline Curve Forecasting, Reserve Estimation, and Analysis. *SPE Reserv. Eng.* **1996**, *11*, 13–22. [CrossRef]
89. Martin, E. Behaviour of Arps Equation in Shale Plays. LinkedIn, 29 March 2015. Available online: https://www.linkedin.com/pulse/behavior-arps-Equation-shale-plays-emanuel-mart%C3%ADn/ (accessed on 22 November 2020).
90. Jimenez, R. Using Decline Curve Analysis, Volumetric Analysis, and Baysian Methodology to Quantify Uncertainty in Shale Gas Reserves Estimates. Master's Thesis, Texas A&M University, College Station, TX, USA, 2012.
91. U.S. Environmental Protection Agency. *Analysis of Hydraulic Fracturing Fluid Data from the FracFocus Chemical Disclosure Registry 1.0*; U.S. EPA Office of Research and Development: Washington, DC, USA, 2015.
92. Arthur, J.; Bohm, B.; Coughlin, B.; Layne, M. Evaluating Implications of Hydraulic Fracturing in Shale Gas Reservoirs. In Proceedings of the 2009 SPE Americas E&P Environmental & Safety Conference, San Antonio, TX, USA, 23–25 March 2009.
93. Saba, T.; Mohsen, F.; Garry, M.; Murphy, B.; Hilbert, B. *White Paper Methanol Use in Hydraulic Fracturing*; Exponent: Maynard, MA, USA, 2012.
94. Manchanda, R.; Bhardwaj, P.; Hwang, J.; Sharma, M. Parent-Child Fracture Interference: Explanation and Mitigation of Child Well Underperformance. In Proceedings of the Society of Petroleum Engineering Hydraulic Fracturing Technology Conference and Exhibition, The Woodlands, TX, USA, 23–25 January 2018.
95. Kumar, A.; Shrivastava, K.; Elliott, B.; Sharma, M. Effect of Parent Well Production on Child Well Stimulation and Productivity. In Proceedings of the Society of Petroleum Engineers Hydraulic Fracturing Technology Conference and Exhibition, The Woodlands, TX, USA, 27–29 October 2020.
96. Wang, H. What Factors Control Shale-Gas Production and Production-Decline Trend in Fractured Systems: A Comprehensive Analysis and Investigation (SPE-179967-PA). *SPE J.* **2017**, *22*, 562–581. [CrossRef]
97. Kurison, C.; Kuleli, H.S.; Mubarak, M. Unlocking well productivity drivers in Eagle Ford and Utica unconventional resources through data analytics. *J. Nat. Gas Sci. Eng.* **2019**, *71*, 102976. [CrossRef]
98. Zobak, M.; Arent, D. Shale Gas: Development Opportunities. *Bridge Emerg. Issues Earth Resour. Eng.* **2014**, *44*, 16–23.
99. Liu, W.; Zhang, G.; Cao, J.; Zhang, J.; Yu, G. Combined petrophysics and 3D seismic attributes to predict shale reservoirs favorable areas. *J. Geophys. Eng.* **2019**, *16*, 974–991. [CrossRef]
100. Chakra, N.C.; Song, K.; Gupta, M.; Saraf, D. An innovative neural forecast of cumulative oil production from a petroleum reservoir employing higher-order neural networks (HONNs). *J. Pet. Sci. Eng.* **2013**, *106*, 18–33. [CrossRef]
101. Schuetter, J.; Mishra, S.; Zhong, M.; LaFollette, R. Data Analytics for Production Optimization in Unconventional Reservoirs. In Proceedings of the Unconventional Resources Technology Conference, San Antonio, TX, USA, 20–22 July 2015.
102. U.S. Energy Information Administration. *Maps: Oil and Gas Exploration, Resources, and Production*; U.S. Department of Energy: Washington, DC, USA, 23 April 2020. Available online: https://www.eia.gov/maps/maps.htm#permian (accessed on 25 November 2020).
103. Shanker, M.; Hu, M.; Hung, M. Effect of data standardization on neural network training. *Omega* **1996**, *24*, 385–397. [CrossRef]
104. Kumar, Y.; Bello, K.; Sharma, S.; Vikara, D.; Remson, D.; Morgan, D.; Cunha, L. Neural Network-Based Surrogate Models for Joint Prediction of Reservoir Pressure and CO_2 Saturation. In Proceedings of the 2020 SMART Annual Review Meeting—Virtual Poster Sessions, Pittsburgh, PA, USA, 27–31 March 2020.
105. Bacon, D. Fast Forward Model Development Using Image-to-Image Translation. In Proceedings of the 2020 SMART Annual Review Meeting—Virtual Poster Sessions, Pittsburgh, PA, USA, 27–31 March 2020.
106. Cao, X.H.; Stojkovic, I.; Obradovic, Z. A robust data scaling algorithm to improve classification accuracies in biomedical data. *BCM Bioinform.* **2016**, *17*, 359. [CrossRef]
107. Liu, J. Potential for Evaluation of Interwell Connectivity under the Effect of Intraformational Bed in Reservoirs Utilizing Machine Learning Methods. *Geofluids* **2020**, *2020*, 1651549. [CrossRef]
108. Aggarwal, R. Ranganathan, Common pitfalls in statistical analysis: The use of correlation techniques. *Perspect Clin. Res.* **2016**, *7*, 187–190.
109. Brownlee, J. Recursive Feature Elimination (RFE) for Feature Selection in Python. Machine Learning Mastery, 25 May 2020. Available online: https://machinelearningmastery.com/rfe-feature-selection-in-python/ (accessed on 9 October 2020).
110. Darst, B.; Malecki, K.; Engelman, C. Using recursive feature elimination in random forest to account for correlated variables in high dimensional data. *BMC Genet.* **2018**, *19*, 65. [CrossRef] [PubMed]
111. Guyon, I.; Weston, J.; Barnhill, S.; Vapnik, V. Gene Selection for Cancer Classification Using Support Vector Machines. *Mach. Learn.* **2002**, *46*, 389–422. [CrossRef]
112. Kuhn, M.; Johnson, K. *Feature Engineering and Selection: A Practical Approach for Predictive Models*; CRC Press, Taylor & Francis Group: Boca Raton, FL, USA, 2020.
113. Scikit Learn. sklearn.feature_selection_RFE. Available online: https://scikit-learn.org/stable/modules/generated/sklearn.feature_selection.RFE.html (accessed on 9 October 2020).
114. Svetnik, V.; Liaw, A.; Tong, C.; Culberson, C.; Sheridan, R.; Feuston, B. Random Forest: A Classification and Regression Tool for Compound Classification and QSAR Modeling. *J. Chem. Inf. Comput. Sci.* **2003**, *43*, 1947–1958. [CrossRef] [PubMed]

115. Breiman, L. Random Forests. *Mach. Learn.* **2001**, *45*, 5–32. [CrossRef]
116. Hur, J.; Ihm, S.; Park, Y. A Variable Impacts Measurement in Random Forest for Mobile Cloud Computing. *Wirel. Commun. Mob. Comput.* **2017**, *2017*, 6817627. [CrossRef]
117. Refaeilzadeh, P.; Tang, L.; Liu, H. Cross-Validation. In *Encyclopedia of Database Systems*; Liu, L., Özsu, M.T., Eds.; Springer: Boston, MA, USA, 2009.
118. Hutter, F.; Hoos, H.; Leyton-Brown, K. An Efficient Approach for Assessing Hyperparameter Importance. In Proceedings of the 31st International Conference on Machine Learning, Beijing, China, 21–26 June 2014.
119. Pedregosa, F.; Varoquaux, G.; Gramfort, A.; Michel, V.; Thirion, B.; Grisel, O.; Blondel, M.; Prettenhofer, P.; Weiss, R.; Dubourg, V.; et al. Scikit-learn: Machine Learning in Python. *J. Mach. Learn. Res.* **2011**, *12*, 2825–2830.
120. Chollet, F.; Keras. 2015. Available online: https://github.com//fchollet/keras (accessed on 15 January 2021).
121. MacQueen, J. Some Methods for Classification and Analysis of Multivariate Observations. In Proceedings of the Fifth Berkeley Symposium on Mathematical Statistics and Probability, Volume 1: Statistics, no., Berkeley, CA, USA, 1 January 1967; University of California Press: Berkeley, CA, USA, 1967; pp. 281–297.
122. de Amorim, R.; Henning, C. Recovering the number of clusters in data sets with noise features using feature rescaling. *Inf. Sci.* **2015**, *324*, 126–145. [CrossRef]
123. Bholowalia, P.; Kumar, A. EBK-Means: A Clustering Technique based on Elbow Method and K-Means in WSN. *Int. J. Comput. Appl.* **2014**, *105*, 17–24.
124. Hartigan, J. *Clustering Algorithms*; J. Wiley & Sons: New York, NY, USA, 1975.
125. Dematos, G.; Boyd, M.; Kermanshahi, B.; Kohzadi, N.; Kaastra, I. Feedforward versus recurrent neural networks for forecasting monthly japanese yen exchange rates. *Financ. Eng. Jpn. Mark.* **1996**, *3*, 59–75. [CrossRef]
126. Hochreiter, S. The Vanishing Gradient Problem during Learning Recurrent Neural Nets and Problem Solutions. *Int. J. Uncertain. Fuzziness Knowl. Based Syst.* **1998**, *6*, 107–116. [CrossRef]
127. Siami-Namini, S.; Tavakoli, N.; Namin, A. A Comparison of ARIMA and LSTM in Forecasting Time Series. In Proceedings of the 2018 17th IEEE International Conference on Machine Learning and Applications (ICMLA), Orlando, FL, USA, 17–20 December 2018; pp. 1394–1401.
128. Elsaraiti, M.; Merabet, A. A Comparative Analysis of the ARIMA and LSTM Predictive Models and Their Effectiveness for Predicting Wind Speed. *Energies* **2021**, *14*, 6782. [CrossRef]
129. Hochreiter, S.; Schmidhuber, J. Long Short-Term Memory. *Neural Comput.* **1997**, *9*, 1735–1780. [CrossRef] [PubMed]
130. Greff, K.; Srivastava, R.; Koutnik, J.; Steunebrink, B.; Schmidhuber, J. LSTM: A Search Space Odyssey. *IEEE Trans. Neural Netw. Learn. Syst.* **2017**, *28*, 2222–2232. [CrossRef] [PubMed]
131. Kwak, H.; Hui, P. Deep Health: Deep Learning for Heath Informatics reviews, challenges, and opportunities on medical imaging, electronic health records, genomics, sensing, and online communication health. *arXiv* **2019**, arXiv:1909.00384.
132. Olah, C. Understanding LSTM Networks. Colah's Blog. 27 August 2015. Available online: http://colah.github.io/posts/2015-08-Understanding-LSTMs/ (accessed on 6 December 2020).
133. Poornima, S.; Pushpalatha, M. Prediction of Rainfall Using Intensified LSTM Based Recurrent Neural Network with Weighted Linear Units. *Atmosphere* **2019**, *10*, 668. [CrossRef]
134. Gers, F.; Schmidhuber, J.; Cummins, F. Learning to Forget: Continual Prediction with LSTM. *Neural Comput.* **1999**, *12*, 2451–2471. [CrossRef]
135. Utgoff, P.; Stracuzzi, D. Many-Layered Learning. *Neural Comput.* **2002**, *14*, 2497–2529. [CrossRef]
136. Rio, A.L.; Nonell-Canals, A.; Vidal, D.; Perera-Lluna, A. Evaluation of Cross-Validation Strategies in Sequence-Based Binding Prediction Using Deep Learning. *J. Chem. Inf. Modeling* **2019**, *59*, 1645–1657.
137. Kingma, D.; Ba, J. Adam: A Method for Stochastic Optimization. In Proceedings of the 3rd International Conference for Learning Representations, San Diego, CA, USA, 12 November 2014.
138. Ji, Y.; Hao, J.; Reyhani, N.; Lendasse, A. *Direct and Recursive Prediction of Time Series Using Mutual Information Selection*; IWANN 2005, LNCS 3512; Springer: Berlin/Heidelberg, Germany, 2005; pp. 1010–1017.
139. Carney, J. Cunningham, The Epoch Interpretation of Learning. *IEEE Trans. Neural Netw.* **1998**, *8*, 111–116.
140. Manda, P.; Nkazi, D.B. The Evaluation and Sensitivity of Decline Curve Modeling. *Energies* **2020**, *13*, 2765. [CrossRef]
141. Paryani, M.; Ahmadi, M.; Awoleke, O.; Hanks, L. Decline Curve Analysis: A Comparative Study of Proposed Models Using Improved Residual Functions. *J. Pet. Environ. Biotechnol.* **2018**, *9*, 1–8.
142. Okouma, V.; Symmons, D. Practical Considerations for Decline Curve Analysis in Unconventional Reservoirs—Application of Recently Developed Time-Rate Relations. In Proceedings of the Society of Petroleum Engineers Hydrocarbon, Economics, and Evaluation Symposium, Calgary, AB, Canada, 9–24 September 2012.
143. Montgomery, D. *Design and Analysis of Experiments*, 9th ed.; John Wiley & Sons, Inc.: Hoboken, NJ, USA, 2017.
144. Armstrong, R.; Eperjesi, F.; Gilmartin, B. The application of analysis of variance (ANOVA) to different experimental designs in optometry. *Ophathalmic Physiol. Opt.* **2002**, *22*, 248–256. [CrossRef] [PubMed]
145. Sawyer, S. Analysis of Variance: The Fundamental Concepts. *J. Man. Manip. Ther.* **2009**, *17*, 27E–38E. [CrossRef]
146. Tukey, J. *The Collected Works of John W. Tukey Volume III*; Multiple Compairsons: 1948–1983; Chapman and Hall: New York, NY, USA, 1983.
147. Brown, R. *Exponential Smoothing for Predicting Demand*; Arthur D. Little Inc.: Cambridge, MA, USA, 1956.

148. Taieb, S.B.; Bontempi, G. Recursive Multi-step Time Series Forecasting by Perturbing Data. In Proceedings of the 11th IEEE International Conference on Data Mining, Vancouver, BC, Canada, 11–14 December 2011.
149. Fox, I.; Ang, L.; Jaiswal, M.; Pop-Busui, R.; Wiens, J. Deep Multi-Output Forecasting: Learning to Accurately Predict Blood Glucose Trajectories. In Proceedings of the 24th ACM SIGKDD International Conference on Knowledge Discovery & Data Mining, London, UK, 19–23 August 2018; pp. 1387–1395.
150. Scanlon, B.; Reedy, R.; Male, F.; Walsh, M. Water Issues Related to Transitioning from Conventional to Unconventional Oil Production in the Permian Basin. *Environ. Sci. Technol.* **2017**, *51*, 10903–10912. [CrossRef] [PubMed]
151. Laurentian Research. Understanding GOR in Unconventional Play: Permian and Beyond. Seeking Alpha, 9 August 2017. Available online: https://seekingalpha.com/article/4096835-understanding-gor-in-unconventional-play-permian-and-beyond (accessed on 26 December 2020).
152. Flumerfelt, R. The Wolfcamp Shale: Technical Learnings to Date and Challenges Going Forward. In Proceedings of the 10th Annual Ryder Scott Reserves Conference, Houston, TX, USA, 17 September 2014.
153. Shale Newsletter. Is the Permian Getting Gassier? Not Necessarily in 2020. Rystad Energy: Oslo, Norway, February 2020. Available online: https://www.rystadenergy.com/newsevents/news/newsletters/UsArchive/shale-newsletter-feb-2020/ (accessed on 26 December 2020).
154. Lee, J. Death by Bubble Point: Fact or Fantasy? In Proceedings of the 2018 Ryder Scott Reserves Conference, Calgary, AB, Canada, 1 July 2018.
155. U.S. Energy Information Administration. *Assumptions to AEO2020*; U.S. Department of Energy: Washington, DC, USA, 29 January 2020. Available online: https://www.eia.gov/outlooks/aeo/assumptions/ (accessed on 27 December 2020).
156. Persaud, A.J.; Kumar, U. An eclectic approach in energy forecasting: A case of Natural Resources Canada's (NRCan's) oil and gas outlook. *Energy Policy* **2001**, *29*, 303–313. [CrossRef]
157. Browning, J.; Ikonnikova, S.; Male, F.; Gulen, G.; Smye, K.; Horvath, S.; Grote, C.; Patzek, T.; Potter, E.; Tinker, S. Study forecasts gradual Haynesville production recovery before final decline. *Oil Gas J.* **2015**, *113*, 64–71.
158. Qian, K.; He, Z.; Liu, X.; Chen, Y. Intelligent prediction and integral analysis of shale oil and gas sweet spots. *Pet. Sci.* **2018**, *15*, 744–755. [CrossRef]

processes

Article

Neural Network Model for Permeability Prediction from Reservoir Well Logs

Reda Abdel Azim [1,*] and Abdulrahman Aljehani [2]

1 Petroleum Engineering Department, American University of Kurdistan, Sumel 42003, Iraq
2 Faculty of Earth Sciencs, King Abdulaziz University, Jeddah 21589, Saudi Arabia
* Correspondence: reda.abdulrasoul@auk.edu.krd

Abstract: The estimation of the formation permeability is considered a vital process in assessing reservoir deliverability. The prediction of such a rock property with the use of the minimum number of inputs is mandatory. In general, porosity and permeability are independent rock petrophysical properties. Despite these observations, theoretical relationships have been proposed, such as that by the Kozeny–Carmen theory. This theory, however, treats a highly complex porous medium in a very simple manner. Hence, this study proposes a comprehensive ANN model based on the back propagation learning algorithm using the FORTRAN language to predict the formation permeability from available well logs. The proposed ANN model uses a weight visualization curve technique to optimize the number of hidden neurons and layers. Approximately 500 core data points were collected to generate the model. These data, including gamma ray, sonic travel time, and bulk density, were collected from numerous wells drilled in the Western Desert and Gulf areas of Egypt. The results show that in order to predict the permeability accurately, the data set must be divided into 60% for training, 20% for testing, and 20% for validation with 25 neurons. The results yielded a correlation coefficient (R^2) of 98% for the training and 96.5% for the testing, with an average absolute percent relative error (AAPRE) of 2.4%. To validate the ANN model, two published correlations (i.e., the dual water and Timur's models) for calculating permeability were used to achieve the target. In addition, the results show that the ANN model had the lowest mean square error (MSE) of 0.035 and AAPRE of 0.024, while the dual water model yielded the highest MSE of 0.84 and APPRE of 0.645 compared to the core data. These results indicate that the proposed ANN model is robust and has strong capability of predicting the rock permeability using the minimum number of wireline log data.

Keywords: neural network; permeability; weight curves; dual water; well logging; machine learning

Citation: Abdel Azim, R.; Aljehani, A. Neural Network Model for Permeability Prediction from Reservoir Well Logs. *Processes* **2022**, *10*, 2587. https://doi.org/10.3390/pr10122587

Academic Editors: Qingbang Meng and Albert Ratner

Received: 10 November 2022
Accepted: 2 December 2022
Published: 4 December 2022

1. Introduction

The reservoir characterization process plays an important role in assessing the economic success of reservoir development. Reservoir characterization is a complex process since most reservoirs are heterogeneous due to the depositional environment and nature of rock. Porosity and permeability are key parameters to assess volume and flow behavior in reservoirs. Despite their importance, permeability (in particular) is difficult to estimate from well logs because of its dynamic nature, which led researchers to propose several methods to estimate permeability. Chehrazi et al. (2012) [1] proposed a comprehensive study for permeability prediction using theoretical and soft computing models. In the theoretical model, porosity and initial water saturation are used as inputs. The main drawback of the presented model is the difficulty in obtaining the permeability from laboratory-measured core data.

Well log interpretations are widely used to estimate porosity and permeability values at various depths due to its minimum cost compared to the coring process. In addition, well log data provide a solution to the lack of continuity information from core samples [2,3].

Lucia et al. (2013) [4] showed that petrophysical heterogeneity is commonly found in carbonate reservoirs, and it is demonstrated by the wide variation in porosity–permeability cross plots of core analysis data. Research has shown that basic rock fabrics control petrophysical heterogeneity; within rock-fabric facies, porosity and permeability have little spatial correlation and are widely variable at the scale of inches and feet. Petrophysical rock typing (PRT) and permeability prediction are of great significance for various disciplines of the oil and gas industry. One of the most important usages of rock typing is predicting unknown reservoir properties, specifically permeability in un-cored intervals. Coring from several wells is often unavoidable and an essential task to obtain basic data on the field. However, coring in all wells of large-scale fields or from all zones of interest in a single well poses a substantial financial burden. Permeability can also be calculated using empirical relationship between core-measured porosity and permeability [5,6].

Hasanusi et al. (2012) [7] presented an effective technique for carbonate reservoir characterization using hybrid seismic rock physics, statistics, and an artificial neural network. This methodology integrates various data sets to produce the coherence correlation among input data and their target. The data set consisted of core (i.e., lithology, lithofacies, fracture intensity, fracture width, and porosity), well log (gamma ray, density, water saturation, porosities, sensitivities, etc.), multi-attribute seismic (either pre-stack or post-stack) of different vintages of 2D seismic lines, and seismic rock physics. The whole array of input data was trained together using natural workflow which is also combined with statistic and artificial neural network.

The available numerical equations for permeability estimation are unreliable and strongly dependent on core analyses, which are costly and time consuming. In addition, wire-line-collected information has critical issues, including missing data during the logging process due to excessive temperature, pressure, and operator errors that limit the operation [8,9]. Therefore, the need to seek alternative ways to predict porosity and permeability is highly recommended. This study presents a novel technique that integrates core and well log various data to generate a suitable artificial neural network model that can overcome the abovementioned concerns. The artificial neural network (ANN) technique is one of the latest techniques available to the petroleum industry for porosity and permeability prediction [10–13]. The presented literature review shows that numerous models have been developed to estimate rock permeability. These models suffer from numerous shortcomings, including a poor ability to precisely predict the permeability in Egyptian oil fields. Therefore, the purpose of this study is to present a new model by using an ANN with the back propagation algorithm using FORTRAN language to propose a new correlation for accurately estimating rock permeability of different oil fields located in Egypt and, consequently, predicting precisely other reservoir properties in the case of missing core and wireline data.

In order to fulfill this purpose, more than 500 core and well logs points were collected from Egyptian oil fields. The data are used to develop the ANN model in the direction of predicting the formation permeability. The proposed novel correlation incorporates parameters including gamma ray, porosity, and travel sonic time. In addition, a new convergence structure is presented to accelerate the performance of the proposed ANN model. Furthermore, in this study, the weight visualization curves (WV-curves) technique was used in optimizing the network architecture.

2. ANN Application for Porosity–Permeability Prediction

Malki et al. (1996) [14] used a self-organizing algorithm to classify well logs for lithology prediction to predict porosity and grain density. Smith et al. (1999) [15] proposed a neural network algorithm for porosity, permeability, and grain density predictions. The authors used gamma ray, neutron porosity, and sonic travel time as inputs. The predicted petrophysical properties were compared to the collected core data, and the errors were evaluated based on certain tolerance. Osborne (1992) [16] used a back propagation neural net to predict permeability by using porosity and reservoir flow units as input data. The input data was divided using approximately 10% for training and 10% for testing process. The model robustness is not valid as the model is developed from the same training data. Osborne concluded that the predicted permeability from neural net model provided superior values to those from regression model. Zhou and Wu (1993) [17] presented a comprehensive study for porosity prediction from well logs using regression and neural nets techniques. The study concluded that the neural nets produced the best results. Jian et al. (1995) [18] presented a case study for porosity–permeability prediction by comparing genetic and nongenetic approaches. Other studies used various machine learning techniques to predict porosity and permeability values at different depths [19–24].

Khayer et al. (2022) [25] proposed an efficient method for image segmentation using logistic function method for seismic attributes estimation. The ANN model was used in identifying a complex relationship between rock properties and wireline information. Rezaee et al. (2012), Wang et al. (2013), and Anifowose et al. (2013) [26–28] proposed an ANN model based on the back propagation algorithm for porosity prediction using genetic and nongenetic approaches. The models used several inputs in designing a suitable ANN model for their predictions.

It is worth noting that a multiple regression technique (MLR) was performed by Wendt et al. [29] to predict permeability from well logs. Wendt et al. (1986) [29] concluded that using the MLR technique as a predictive model resulted in a poor data distribution and a narrower than the original data set. Rogers et al. (1995) [30] mentioned the same conclusion regarding permeability prediction from the regression technique compared to neural networks. In addition, Rogers et al. (1995) [30] showed that neural networks do not direct the prediction to the mean and the extreme values outside the range of the training data. In addition, the main advantage of neural network techniques over multiple linear regression (MLR) is that they reproduce a minor nonlinearity embedded in the common log to porosity and permeability transforms.

Another algorithm used in the prediction of various well logs is the convolutional neural network (CNN) [31]. The main disadvantage of a CNN is the large number of training data needed for the CNN to be effective. In addition, CNNs tend to be a much slower than ANN models. Overfitting, exploding gradient, and class imbalance are the major challenges while training the model using CNN technique. Zhang et al. (2018) [32] proposed a cascaded long short-term memory (C-LSTM) based on the LSTM technique. The study by Zhang et al. (2018) [32] concluded that, although LSTM-based models can generate well logs, the technique has a poor prediction accuracy, as the LSTM technique does not perform well on small training data sets. Chen et al. (2019) [33] proposed an ensemble neural network (ENN) to address this issue, which offers advantages in small data problems, but it is not suitable for handling sequential data.

The main drawback of the backpropagation algorithm used in this study is the convergence or the local minima problem and a slow performance [34,35]. Nevertheless, this study proposed a new convergence technique to speed up the performance of the network by adding an acceleration factor (see Section 3.1).

3. Artificial Neural Network

One machine learning algorithm is the artificial neural network (ANN), which mimics the human central nervous system [36]. An ANN consists of organized layers containing single units and artificial neurons that are connected through weight functions [37,38]. There are different types of neural networks, and they can be differentiated depending on the neurons transfer functions, learning rules, and connected formula.

A complex computational framework is performed in ANNs to predict the output responses. Furthermore, an ANN uses massive parallel connections between a nonlinear parameterized and bounded function, which are referred to as neurons [39,40]. The neurons are designed in a way that defines the network architecture using multilayer perception (MLP), where neurons are assembled in continuous layers. Using MLP, neurons in each layer share the same inputs without intersecting with each other.

Although the number of hidden layers and neurons in each layer is arbitrary [41], an increasing number of neurons can cause overfitting. On the contrary, decreasing number of neurons may result in a poor network performance. Perhaps the main advantage of using an ANN over other methods is that it can process a larger number of data sets [42]. Figure 1 shows a typical ANN structure consisting of an input layer, hidden layer, bias unit, and output layer.

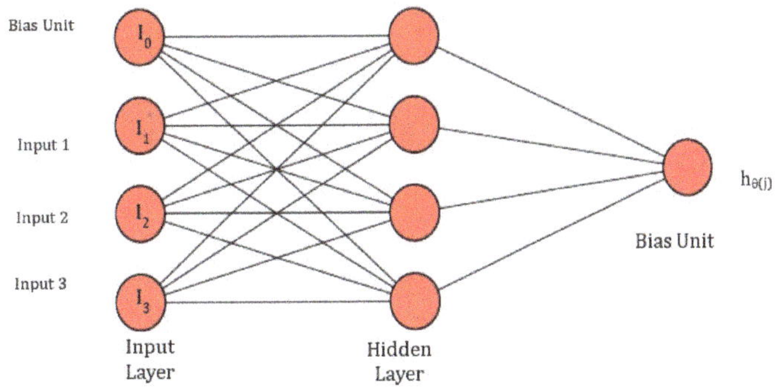

Figure 1. Schematic showing the ANN structure used in this study for one hidden layer.

The output function, $h_{(x)}$, in Figure 1 is calculated as:

$$h_{\theta(x)} = a_1^3 = g\left(\theta_{10}^{(2)} a_0^{(2)} + \theta_{11}^{(2)} a_1^{(2)} + \theta_{12}^{(2)} a_2^{(2)} + \theta_{13}^{(2)} a_3^{(2)}\right) \quad (1)$$

where (g) is the sigmoid function and can be calculated as:

$$g(z) = \frac{1}{(1 + e^{-z})} \quad (2)$$

The activation function for each neuron is vectorized in a matrix of Z as:

$$\begin{Bmatrix} Z_1^{(2)} = \theta_{10}^{(1)} I_0 + \theta_{11}^{(1)} I_1 + \theta_{12}^{(1)} I_2 + \theta_{13}^{(1)} I_3 \\ Z_2^{(2)} = \theta_{20}^{(1)} I_0 + \theta_{21}^{(1)} I_1 + \theta_{22}^{(1)} I_2 + \theta_{23}^{(1)} I_3 \\ Z_3^{(2)} = \theta_{30}^{(1)} I_0 + \theta_{31}^{(1)} I_1 + \theta_{32}^{(1)} I_2 + \theta_{33}^{(1)} I_3 \end{Bmatrix} \quad (3)$$

The size of the neural network (number of hidden layers and number of neurons) determines the degree of complexity in the ANN. However, Soroush et al. (2015) [43] argue that an ANN should be designed with a sufficient level of complexity to avoid data being over fitted.

The neuron network is trained using an algorithm to minimize the error between the network output values and the target values. This was achieved by an iterative process to find the optimum values of the weights and biases. There are many algorithms presented in literature to train the network, and the most well-known training algorithm is the Levenberg–Marquardt (LM).

3.1. Back Propagation Algorithm

This study used a back propagation algorithm (BP) for the developed ANN model and the gradient of the error function. The term back propagation is used to describe the multilayer perception of the ANN architecture [44]. The error function of a specific input pattern set can be defined as:

$$MSE = \sqrt{\frac{\sum_1^{n1} \sum_1^{n2} (x_p - y_p)}{n_1.n_2}} \quad (4)$$

where MSE is the mean square error, n_1 and n_2 are number of training outputs and neurons, respectively; x_p and y_p are the target and estimated outputs, respectively.

The backpropagation algorithm has numerous limitations, including a slow convergence, inability to handle multiple objectives, and a high probability of being trapped in the local minima during a training process [45]. Therefore, this study presents a new convergence technique to speed up the network by adding an acceleration factor, as follows:

$$w(t+1) = w(t) + \beta[\Delta w(t)] + \alpha[w(t-1)] \tag{5}$$

where α is the energy constant; w is the weight; t is the increment by 1 for each epoch; and β is the learning constant. This constant is used to effectively increase step size to reduce abrupt gradient changes. The learning and momentum constants are set in a range of 0 to 1.

4. Methodology

4.1. Collected Data Analysis

Approximately 500 core data points were used and collected from various fields in Egypt to design and develop the ANN model. The data contained three inputs, including sonic travel time (DT), gamma ray (GR), and bulk density (RHOB). The input parameters are used in the training process, while the output is permeability. The data set is normalized in a range of 0 to 1, and the statistical analysis for the collected data is presented in Table 1.

Table 1. Statistical analysis of input and output data.

Parameter	Sonic Travel Time (DT) (μs/ft)	GR	Bulk Density (g/cc)	Permeability (md)	Porosity (v/v)
Minimum	43.302	18.188	2.618	0.1251	0
Maximum	63.641	66.151	2.933	24.491	0.143
Standard deviation	75.10102	223.6315	1.5488	116.582	0.633
Skewness	1.217074	0.376381	−0.087	1.4221	1.620
Mean	48.5665	36.4085	2.801	1.9400	0.0168

4.2. Analyzing the Collected Data

Distribution of the Inputs

The data set was divided into 60% for training and 20% for testing. The BP algorithm was used to minimize the resulting error between the actual and target outputs with the log sigmoid function. The BP learning algorithm provides an exceptional result with an R2 of 0.9806 and MSE of 0.024 compared to other algorithms, including a gradient descent (GD) and a stochastic gradient descent (SGD), which produced an MSE of 0.25 and 0.39, respectively. Stochastic gradient descent (SGD) introduces the momentum for the weight update technique. Figure 2 presents a comprehensive flow chart for the ANN model used in this study. The ANN model parameters including several hidden layers; the neurons and training/testing ratio were optimized to increase the robustness of the developed model. It can be seen from Figure 2 that during the network training process, the overall error was reduced during the training process using the updated connection weight. This weight updating process was performed two ways: epoch updating and stochastic updating. In the epoch updating, all weight changes were added for the input patterns before completing the updating process. The main advantage of the epoch updating process is ensuring the stability and reliability of the learning algorithm. In addition, no problems were encountered during the network convergence. Table 2 summarizes the optimized parameters used in this study.

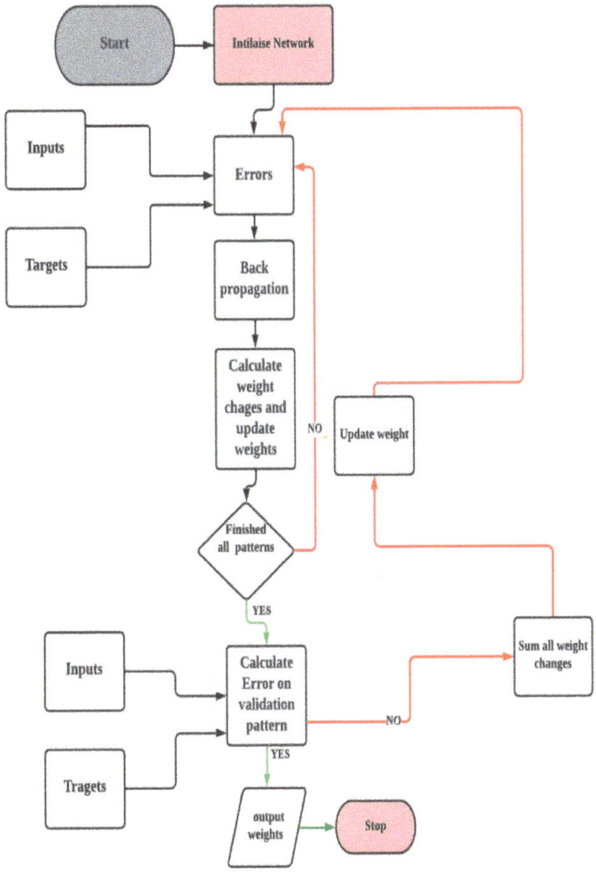

Figure 2. Flow chart showing the ANN model process.

Table 2. ANN optimized parameters (OPs).

Parameter	Tested	OP
Neurons numbers	5–30	25
Hidden layers	1–3	1
Algorithm function	tansig/logsig	logsig
Learning rate	0.001–0.8	0.08

Further, an analysis process was performed to assess the dependency of the outputs (i.e., permeability and porosity) to the inputs of sonic time (DT), bulk density (RHOB), and gamma ray (GR) using a correlation coefficient (CC). Figure 3 shows that the permeability and porosity values were intensely dependent on the DT, GR, and RHOB with CC of 0.262, −0.385, and −0.319 for porosity and CC of 0.806, −0.316, and −0.133 for permeability, respectively. An average correlation coefficient (avr-CC) was calculated using the absolute values of CC for DT, GR, and RHOB. Figure 3 also shows that permeability had a higher avr-CC with the input parameter of 0.4.

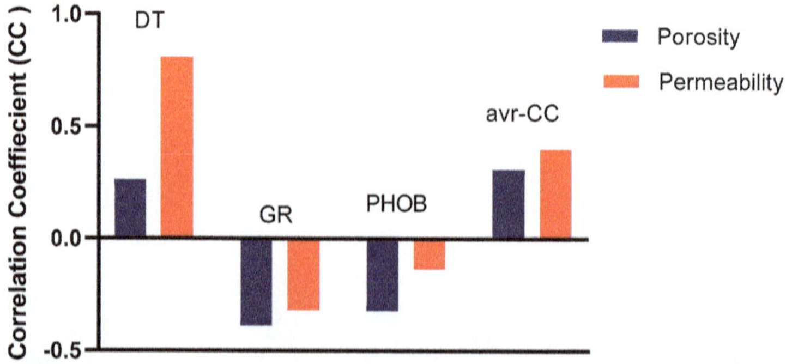

Figure 3. Correlation coefficient for inputs and outputs.

4.3. Optimizing the ANN Model Parameters

One of the main challenges during the learning of a neural net is to find the optimum network parameters, including the choice of input variables, initial weights, and the number of hidden layers. Choosing the number of inputs is a straightforward process, while optimizing the number of hidden layers requires the use of nonlinear inputs to avoid using more than one hidden layer in some cases. The size of the hidden layer and the training epochs require the training of the network to obtain the maximum performance of the unseen data. Therefore, in this study, the weight visualization curve technique was used to select the number of the input and hidden neurons. The weight values were used to calculate the average contribution of a neuron in a layer to a neuron in the next layer [46].

$$Pij = \frac{|W_{ij}|}{\sum|W_{ij}|} \cdot 100\% \tag{6}$$

where P_{ij} is the average contribution of neuron i in a layer to neuron j in the next layer; W is the weight between connections.

The weight visualization curves (WV-curves) technique was used to select a proper input patterns for the training process to save computational time. The use of the weight visualization curves technique has not been discussed before; therefore, in this study, the performance of the WV technique was modified to optimize the network architecture parameters, including the selection of the number of input and hidden neurons. In this study, the WV-technique uses the nonlinear inputs to choose the optimum parakeets based on the learning behavior of different network configurations.

Using the equation below to measure the average contribution of an input variable to the hidden layer as:

$$A = \frac{\sum_{j=1}^{n2}|W_{ij}|}{\sum_{k=0}^{n1}\sum_{j=1}^{n2}|W_{ij}|} \tag{7}$$

where A is the average contribution of the input variable i; n^1 and n^2 are the number of neurons in the input layer and the hidden layer, respectively.

In Figure 4, the WV-curves show that the weights of RHOB and RHOB2 (RHOB × RHOB) and DT and DT2 (DT × DT) to the hidden neurons had almost the same behavior. This conclusion led to using one curve; RHOB and RHOB2 (similarly DT and DT2) can be omitted. Next, it was noted that there are five hidden neurons also contributed closely to the same amount to all the output neurons. Therefore, 5 neurons could be removed from the hidden layer, and a 3-25-1 configuration (i.e., RHOB, DT, and GR) was handled in the network architecture.

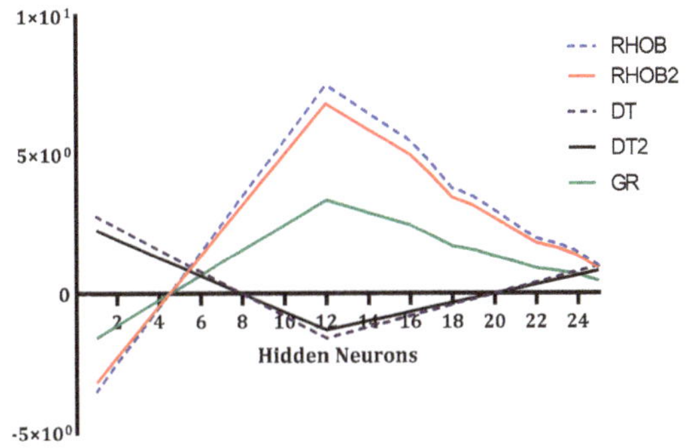

Figure 4. WV-curves for the 1-25-1 configuration.

The data set from the various wells located in Egypt was collected. The data set consist of 500 patterns of core and well logs. The cross plot of the sonic travel time (DT) and bulk density (RHOB) is displayed in Figure 5. It can be seen from this figure (Figure 5) that there was an inverse relationship between the two variables. The data including inputs of sonic time (DT), bulk density (RHOB), and gamma ray (GR), while the output is permeability.

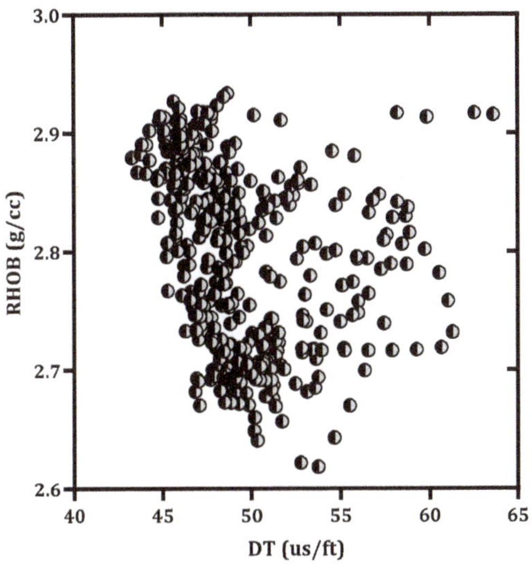

Figure 5. Relationship between the bulk density and sonic travel time of the collected data.

In this study, numerous trials were performed to optimize the network parameters. The first trial used only two independent inputs, RHOB and DT. The results show that 5000 epochs with four hidden layers resulted in an R2 of 92%. In the second trial, three inputs namely DT, RHOB, and GR were used, and the results show that 1500 epochs with 25 hidden neurons were used to obtain the highest network performance with an R2 of 98% and better convergence during the training. Figure 6 shows how the network was optimized using various trials. Each trial was repeated 10 times using different initial random weights.

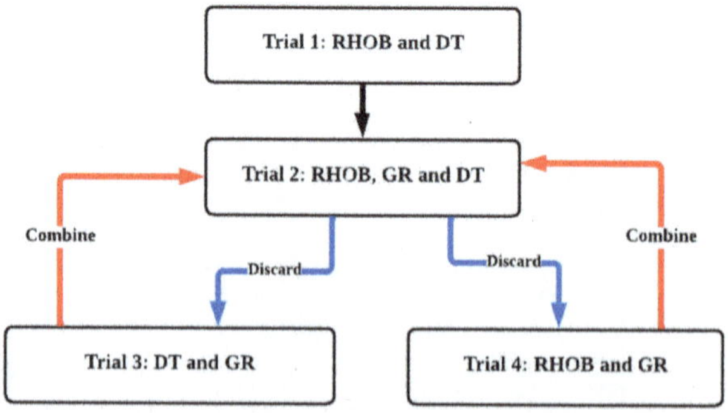

Figure 6. Relationship between the different optimization trials.

Therefore, trial number 2 was used as the final configuration, as it exhibited the least number of epochs and the highest R2 value. Table 3 shows the average contribution of each input to the network. The results show that sonic time (DT), bulk density (RHOB), and gamma ray (GR) were equally important to the network since all variables contributed the same amount.

Table 3. Contribution percentage of each input variables including bias.

Parameter	Contribution (%)
Density	31.2
Sonic time	29.2
GR	30.4
Bias	9.2

5. Results and Discussion

Using the optimized parameters, including 25 neurons and one hidden layer in the training and testing process, the data set was divided as 60% for training, 20% testing, and 20% for validation. Figure 7a shows the relationship between gamma ray (GR) and neutron porosity, while Figure 7b shows the Poro-Perm relationship for the collected data. Figure 8a,b show a cross plot of the predicted permeability versus the core permeability for the training and testing processes. The average absolute percent relative error (AAPRE) for the training was 98%, while the testing AAPRE was 96.5%. These results show the reliability of the proposed ANN model. In addition, Figure 9 shows a comparison between core permeability and permeability values extracted from the ANN model based on the inputs. It can be seen from Figure 8a,b that a good matching was achieved with a minimum square error (MSE) of 0.024.

Using the results of the training and testing processes, a mathematical correlation was created to show the relationship between the permeability and the sonic time (DT), bulk density (RHOB), and gamma ray (GR) to be used in the forecasting of the permeability in the Western Desert and Gulf wells. The weights and biases for the generated equation are provided in Table 4.

The novel correlation generated using the ANN for the permeability estimation is given by:

$$k_n = \left[\sum_{i=1}^{N} w_{2i} \tan(sig)\left(\sum_{j=1}^{J} w_{li,j}x_j + b_{ij}\right)\right] + b_2 \tag{8}$$

$$k_n = \left[\sum_{i=1}^{N} w_{2i}\left(\frac{1}{1 + exp^{(-(GR.w_{1j,1}+DT.w_{1j,2}+RHOB.w_{1j,3})+b_1)}}\right)\right] + b_2 \tag{9}$$

where k_n is the normalized permeability; $(w_{2,\,i})$ is the vector weight between the hidden layer and output layer; $(w_{1,\,j})$ is the vector weight connect the input and the hidden layer; j is the neuron number; b_1 is the biases vector for the input layer; and b_2 is for the output layer, (sig) is the sigmoid function, gamma ray (GR), sonic travel time (DT), and bulk density ($RHOB$).

The extracted k equation can be attained by de-normalizing kn as follows:

$$k = k_n + 0.125 \tag{10}$$

In conclusion, it can be seen from the results that the ANN can predict permeability values at different locations using the proposed ANN model and extracted correlation. The presented correlation in this study proposes a solution for companies in Egypt to precisely predict the permeability values without using ANN software and that can lead to saving extensive computational time.

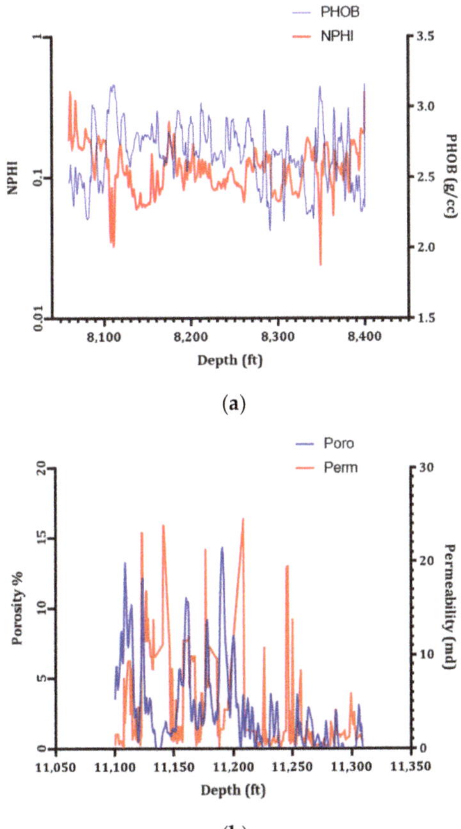

Figure 7. (**a**) Relationship between GR and NPHI; (**b**) Poro-Perm cross plot of the collected data.

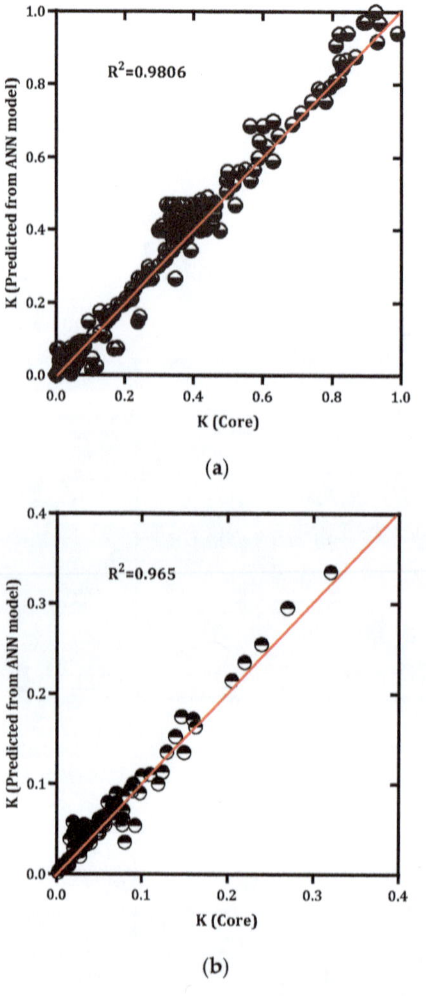

Figure 8. The predicted permeability from the ANN versus core permeability: (**a**) training; (**b**) testing.

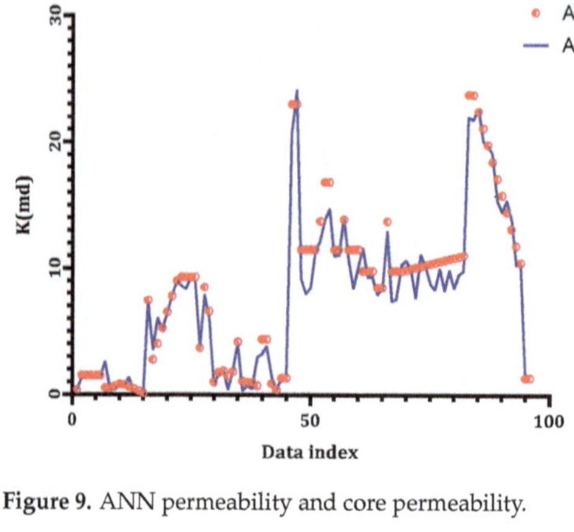

Figure 9. ANN permeability and core permeability.

Table 4. Weights and biases for the generated correlation from Equation (9).

Hidden Neuron	Weights (w_1) GR	DT	RHOB	Weights (w_2)	Hidden Layer Bias (b1)	Output Layer Bias (b2)
1	−0.2625	−0.2925	−0.08	0.0425	−0.12	6.08
2	0	0	0	0	0	
3	0	0	0	0	0	
4	0.015	0.08	0.095	−0.0975	−0.6225	
5	−0.0975	0.0525	0.035	0.025	−0.01	
6	−0.1075	0.04	−0.1225	−0.0875	0.005	
7	−0.11	0.025	−0.015	−0.0725	−0.005	
8	0.11	0.0225	−0.0325	−0.11	0.0025	
9	0.075	0.0475	−0.055	−0.0675	0.0075	
10	−0.035	0.0725	−0.0625	0.0675	−0.0075	
11	0.085	−0.045	0.075	0.0575	0.01	
12	−0.325	−0.685	0.05	0.115	−0.0125	
13	0.0775	0.02	0.06	0.02	−0.0075	
14	−0.1075	−0.0875	0.055	0.02	0.0025	
15	0.1225	0.03	−0.065	−0.08	−0.0125	
16	−0.05	−0.1	0.1175	−0.02	0.005	
17	0.0525	−0.005	−0.1125	−0.12	−0.0075	
18	0.085	−0.0875	−0.08	0.095	0.0075	
19	−0.0575	0.0675	0.0425	0.1125	0.0125	
20	0.01	0.0725	−0.015	0.0575	−0.005	
21	0.2125	0.62	0.02	0.115	0.005	
22	−0.0525	0.08	−0.1125	0.05	−0.0025	
23	−0.025	0.115	0.0125	−0.1025	0.0075	
24	−0.0875	0.1225	−0.0075	−0.095	−0.0125	
25	0.035	−0.0775	0.0425	0.1125	0	

Validation of the ANN Model

In order to validate the developed ANN model, a new data set was used to perform the task. These data were unseen during the training process. The generated correlation was used to predict core permeability using the measured sonic time, gamma ray, and bulk density, and the collected data that used during the validation were for wells located in the Western Desert of Egypt. Numerous published correlations were used in the validation to estimate core permeability as well.

These correlations are most widely used for permeability prediction in the Egyptian oil fields. The correlations were the dual water model [47] and Timur's model [48];

The dual water model is:

$$k = (100 \times \phi_e(1 - S_{wi})/S_{wi})^2 \tag{11}$$

where ϕ_e is the effective porosity; S_{wi} is the initial water saturation. The initial water saturation values at each depth were collected from wireline log data. While the Timur's equation is given as:

$$k = (a \times \phi_e^b) \times S_{wi}^2 \tag{12}$$

Coefficients a and b were determined statistically and had a range of 2–5. The main drawback in Timur's equation is the wide range of coefficients a and b that are used for the permeability estimation. Furthermore, water saturation values must be available.

Figure 10 shows the cross plot for the predicted and actual core permeability using the ANN model developed in this study. It can be seen from Figure 10 that the R2 was 0.94 with an MSE of 0.0325 and an AAPRE of 0.024 (see Table 3). Figure 10b shows that the results of the predicted permeability using the dual water model could not predict core permeability and yielded an MSE of 0.84 with an AAPRE of 0.645 and an R2 of 0.165. In addition, Timur's equation provided a poor result for core permeability values for the same data set, with the highest MSE of 0.95 and AAPRE of 1.35 and the lowest R2 of 0.045 (see Figure 10c). Table 5 summarizes the statistical analysis of the validation process.

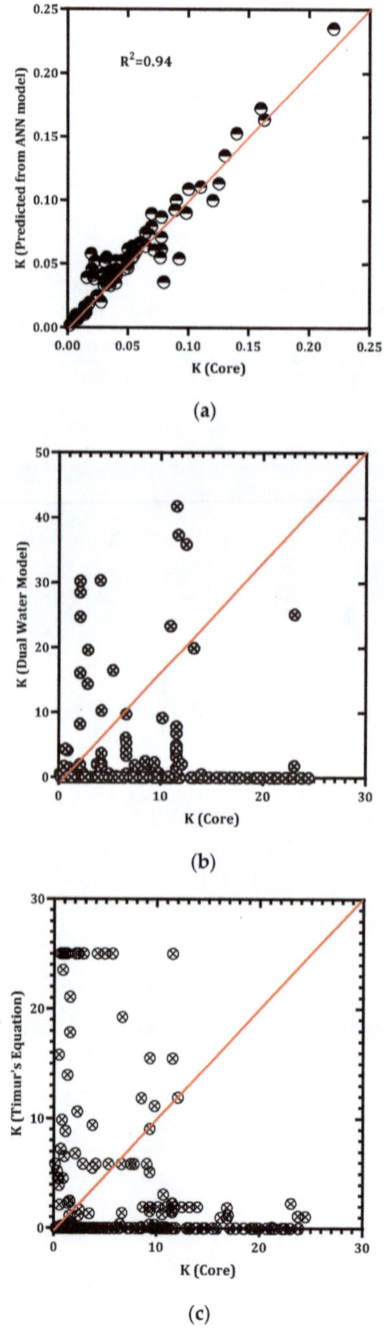

Figure 10. Scatter cross plot showing a comparison of the predicted permeability using the developed ANN model and core permeability: (**a**) ANN model; (**b**) Dual water model; (**c**) Timur's equation.

Table 5 shows that the estimated permeability values by the ANN model had the lowest MSE of 0.035 and AAPRE of 0.024, while the dual water model yielded the highest MSE of 0.84 and APPRE of 0.645 compared to the core data. These results indicate that the proposed ANN model is robust and has strong capability of predicting rock permeability using a minimum number of wireline log data. The reason behind poor permeability values when using the Timur and dual water models was the need for initial water saturation values (S_{wi}). Due to the limitations of these models, ANN techniques have become a more adaptable alternative in this problem domain. Therefore, the presented ANN model and correlations can be used for better forecasting and without the need of S_{wi} values.

Table 5. Statistics analysis for well-known correlations and the ANN model.

Correlation	AAPRE	MSE	Correlation Coefficient (R2)
Dual water model	0.645	0.84	0.165
Timur	0.82	1.35	0.045
This study's ANN	0.024	0.035	0.94

Figure 11 shows a comparison between the actual core permeability and that from the Timur's and dual water models. This clarifies that the published correlations do not have the ability to predict core permeability for different formation lithologies and various hydraulic flow units.

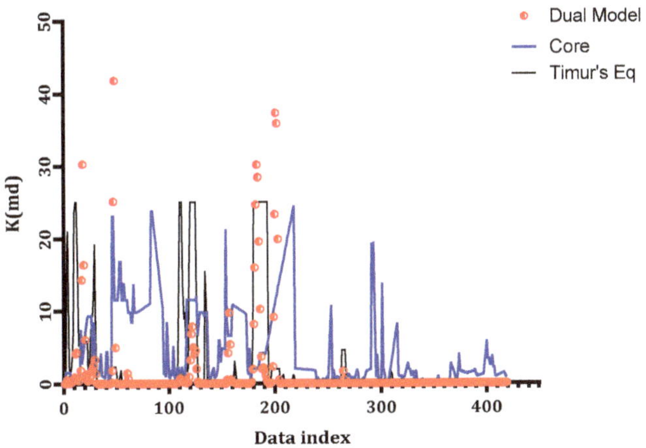

Figure 11. Core permeability vs. the predicted permeability for the numerous correlations used in this study.

6. Conclusions

(1) This study presented a novel correlation for accurately estimating the formation permeability with different lithologies and flow units located in the western part of Egypt using a comprehensive ANN model. The ANN model could forecast the core permeability with a high accuracy of 98%.

(2) The use of weight visualization curves (WV) technique is discussed in the literature; however, this study improved the performance of the WV technique to optimize the network architecture parameters, including the selection of the number of input and hidden neurons.

(3) The ANN model used the weight visualization curve technique to optimize the network parameters in conjunction with the backpropagation algorithm and a learning rate of 0.08.

(4) A comparison study was performed using well-known correlations. The study showed that these correlations had a deficiency in estimating core permeability for various lithologies, and to obtain better forecasting, the data must be divided into flow units.
(5) The proposed ANN and novel correlation may facilitate the issue of permeability prediction that requires using ANN software, and the correlation would be evaluated further using large number of oil fields with various lithologies.
(6) It is highly recommended that future research tests the proposed ANN model and correlation on different wells in the Gulf area of Egypt to show its robustness. In addition, the authors are working on developing another SVR code using FORTRAN language in the comparison process for obtaining more accurate weights and biases of the derived correlations.
(7) In addition, this study anticipated a solution for companies in Egypt to predict the permeability precisely without using ANN software.

Author Contributions: R.A.A.; Conceptualization, Software, Data creation, writing original draft and A.A.; Methodology, Data creation, Writing—review & editing. All authors have read and agreed to the published version of the manuscript.

Funding: This research work was funded by Institutional Fund Projects under grant no. (IFPIP: 415-145-1443).

Acknowledgments: This research work was funded by Institutional Fund Projects under grant no. (IFPIP: 415-145-1443). The authors gratefully acknowledge technical and financial support provided by the Ministry of Education and King Abdulaziz University, DSR, Jeddah, Saudi Arabia.

Conflicts of Interest: The authors declare no conflict of interest.

References

1. Chehrazi, A.; Rezaee, R. A systematic method for permeability prediction, a Petro-Facies approach. *J. Pet. Sci. Eng.* **2012**, *82*, 1–16. [CrossRef]
2. Helle, H.B.; Bhatt, A. Fluid saturation from well logs using committee neural networks. *Pet. Geosci.* **2002**, *8*, 109–118. [CrossRef]
3. Lim, J.S.; Kim, J. Reservoir porosity and permeability estimation from well logs using fuzzy logic and neural networks. In Proceedings of the SPE Asia Pacific Oil and Gas Conference and Exhibition, Perth, Australia, 18 October 2004. [CrossRef]
4. Lu, H.; Tang, H.; Wang, M.; Li, X.; Zhang, L.; Wang, Q.; Zhao, Y.; Zhao, F.; Liao, J. Pore structure characteristics and permeability prediction model in a cretaceous carbonate reservoir, North Persian Gulf Basin. *Geofluids* **2021**, *2021*, 8876679. [CrossRef]
5. Lis-Śledziona, A. Petrophysical rock typing and permeability prediction in tight sandstone reservoir. *Acta Geophys.* **2019**, *67*, 1895–1911. [CrossRef]
6. Fadhil, D.T.; Yonus, W.A.; Theyab, M.A. Reservoir characteristics of the Miocene age formation at the Allas Dome, Hamrin Anticline, Northern Iraq. *MMD J.* **2020**, *14*, 17–23. [CrossRef]
7. Hasanusi, D.; Wijeya, R.; Shahab, W.; Endar, B.; Nurhandoko, B. Fracture and Carbonate Reservoir Characterization Using Sequential Hybrid Seismic Rock Physics and Artificial Neural-Network: A Case Study of Tiaka Field. In Proceedings of the International Conference & Exhibition, Melbourne, Australia, 14 September 2015.
8. Ayoub, M.A.; Raja, A.I.; Almarhoun, M. Evaluation of below bubble point viscosity correlations & construction of a new neural network model. In Proceedings of the Asia Pacific Oil and Gas Conference and Exhibition, Jakarta, Indonesia, 30 October 2007. [CrossRef]
9. Lopes, R.L.; Jorge, A.M. Assessment of predictive learning methods for the completion of gaps in well log data. *J. Pet. Sci. Eng.* **2018**, *162*, 873–886. [CrossRef]
10. Wills, E. AI vs. MACHINE LEARNING: The Devil Is in the Details. *Mach. Des.* **2019**, *91*, 56–60.
11. Jakhar, D.; Kaur, I. Artificial intelligence, machine learning and deep learning: Definitions and differences. *Clin. Exp. Dermatol.* **2020**, *45*, 131–132. [CrossRef]
12. Azim, R.A. Prediction of multiphase flow rate for artificially flowing wells using rigorous artificial neural network technique. *Flow Meas. Instrum.* **2021**, *76*, 101835. [CrossRef]
13. Abdel Azim, R. Application of artificial neural network in optimizing the drilling rate of penetration of western desert Egyptian wells. *SN Appl. Sci.* **2020**, *2*, 1177. [CrossRef]
14. Malki, H.A.; Baldwin, J.L.; Kwari, M.A. Estimating permeability by use of neural networks in thinly bedded shaly gas sands. *SPE Comput. Appl.* **1996**, *8*, 58–62. [CrossRef]
15. Smith, K.A. Neural networks for combinatorial optimization: A review of more than a decade of research. *Inf. J. Comput.* **1999**, *11*, 15–34. [CrossRef]

16. Osborne, D.A. Permeability estimation using a neural network: A case study from the Roberts unit, Wasson field, Yoakum County, Texas. *AAPG Bull. Am. Assoc. Pet. Geol. United States* **1992**, *76*, 4.
17. Zhou, C.D.; Wu, X.L.; Cheng, J.A. Determining reservoir properties in reservoir studies using a fuzzy neural network. In Proceedings of the SPE Annual Technical Conference and Exhibition, Houston, TX, USA, 3–6 October 1993. [CrossRef]
18. Jian, F.X.; Taggart, I.J. A critical comparison of neural networks and discriminant analysis in lithofacies, porosity and permeability predictions. *J. Pet. Geol.* **1995**, *18*, 191–206.
19. Huang, Z.; Shimeld, J.; Williamson, M.; Katsube, J. Permeability prediction with artificial neural network modeling in the Venture gas field, offshore eastern Canada. *Geophysics* **1996**, *61*, 422–436. [CrossRef]
20. Huang, Z.; Williamson, M.A. Determination of porosity and permeability in reservoir intervals by artificial neural network modelling, offshore Eastern Canada. *Pet. Geosci.* **1997**, *3*, 245–258. [CrossRef]
21. Helle, H.B.; Bhatt, A.; Ursin, B. Porosity and permeability prediction from wireline logs using artificial neural networks: A North Sea case study. *Geophys. Prospect.* **2001**, *49*, 431–444. [CrossRef]
22. Rwechungura, R.; Dadashpour, M.; Kleppe, J. Application of particle swarm optimization for parameter estimation integrating production and time lapse seismic data. In Proceedings of the SPE Offshore Europe Oil and Gas Conference and Exhibition, Aberdeen, UK, 6–8 September 2011. [CrossRef]
23. Saputro, O.D.; Maulana, Z.L.; Latief, F.D.E. Porosity log prediction using artificial neural network. *J. Phys. Conf. Ser.* **2016**, *739*, 012092. [CrossRef]
24. Ahmadi, M.A.; Chen, Z. Comparison of machine learning methods for estimating permeability and porosity of oil reservoirs via petro-physical logs. *Petroleum* **2019**, *5*, pp.271–284. [CrossRef]
25. Khayer, K.; Roshandel Kahoo, A.; Soleimani Monfared, M.; Tokhmechi, B.; Kavousi, K. Target-Oriented Fusion of Attributes in Data Level for Salt Dome Geobody Delineation in Seismic Data. *Nat. Resour. Res.* **2022**, *31*, 2461–2481. [CrossRef]
26. Rezaee, R.; Saeedi, A.; Clennell, B. Tight gas sands permeability estimation from mercury injection capillary pressure and nuclear magnetic resonance data. *J. Pet. Sci. Eng.* **2012**, *88*, 92–99. [CrossRef]
27. Wang, Y.; Zhang, K.; Wu, N. Numerical investigation of the storage efficiency factor for CO_2 geological sequestration in saline formations. *Energy Procedia* **2013**, *37*, 5267–5274. [CrossRef]
28. Anifowose, F.; Labadin, J.; Abdulraheem, A. A least-square-driven functional networks type-2 fuzzy logic hybrid model for efficient petroleum reservoir properties prediction. *Neural Comput. Appl.* **2013**, *23*, 179–190. [CrossRef]
29. Wendt, W.A.; Sakurai, S.T.; Nelson, P.H. Permeability prediction from well logs using multiple regression. In *Reservoir Characterization*; Academic Press: Cambridge, MA, USA, 1986; pp. 181–221.
30. Rogers, S.J.; Fang, J.H.; Karr, C.L.; Stanley, D.A. Determination of lithology from well logs using a neural network. *AAPG Bull.* **1992**, *76*, 731–739.
31. Zhong, Z.; Carr, T.R.; Wu, X.; Wang, G. Application of a convolutional neural network in permeability prediction: A case study in the Jackson burg -Springtown oil field, West Virginia, USA. *Geophysics* **2019**, *84*, B363–B373. [CrossRef]
32. Zhang, D.; Yuntian, C.H.E.N.; Jin, M.E.N.G. Synthetic well logs generation via Recurrent Neural Networks. *Pet. Explor. Dev.* **2018**, *45*, 629–639. [CrossRef]
33. Chen, Y.; Chang, H.; Meng, J.; Zhang, D. Ensemble Neural Networks (ENN): A gradient-free stochastic method. *Neural Netw.* **2019**, *110*, 170–185. [CrossRef]
34. Gori, M.; Tesi, A. On the problem of local minima in backpropagation. *IEEE Trans. Pattern Anal. Mach. Intell.* **1992**, *14*, 76–86. [CrossRef]
35. Abdel Azim, R.; Hamada, G. Novel Correlation for Calculating Water Saturation in Shaly Sandstone Reservoirs Using Artificial Intelligence: Case Study from Egyptian Oil Fields. *ACS Omega* **2022**, *7*, 29666–29674. [CrossRef]
36. Bell, J. *Machine Learning: Hands-on for Developers and Technical Professionals*; John Wiley & Sons: Hoboken, NJ, USA, 2020.
37. Agatonovic-Kustrin, S.; Beresford, R. Basic concepts of artificial neural network (ANN) modeling and its application in pharmaceutical research. *J. Pharm. Biomed. Anal.* **2000**, *22*, 717–727. [CrossRef]
38. Giri Nandagopal, M.S.; Selvaraju, N. Prediction of liquid–liquid flow patterns in a Y-Junction circular microchannel using advanced neural network techniques. *Ind. Eng. Chem. Res.* **2016**, *55*, 11346–11362. [CrossRef]
39. Mesbah, M.; Soroush, E.; Kakroudi, M.R. Predicting physical properties (viscosity, density, and refractive index) of ternary systems containing 1-octyl-3-methyl-imidazolium bis (trifluoromethylsulfonyl) imide, esters and alcohols at 298.15 K and atmospheric pressure, using rigorous classification techniques. *J. Mol. Liq.* **2017**, *225*, 778–787.
40. Ghaffari, A.; Abdollahi, H.; Khoshayand, M.R.; Bozchalooi, I.S.; Dadgar, A.; Rafiee-Tehrani, M. Performance comparison of neural network training algorithms in modeling of bimodal drug delivery. *Int. J. Pharm.* **2006**, *327*, 126–138. [CrossRef] [PubMed]
41. Vakili, M.; Yahyaei, M.; Kalhor, K. Thermal conductivity modeling of graphene nanoplatelets/deionized water nanofluid by MLP neural network and theoretical modeling using experimental results. *Int. Commun. Heat Mass Transf.* **2016**, *74*, 11–17.
42. Hippert, H.S.; Pedreira, C.E.; Souza, R.C. Neural networks for short-term load forecasting: A review and evaluation. *IEEE Trans. Power Syst.* **2001**, *16*, 44–55. [CrossRef]
43. Soroush, E.; Mesbah, M.; Shokrollahi, A.; Rozyn, J.; Lee, M.; Kashiwao, T.; Bahadori, A. Evolving a robust modeling tool for prediction of natural gas hydrate formation conditions. *J. Unconv. Oil Gas Resour.* **2015**, *12*, 45–55. [CrossRef]
44. Bishop, C.M. Neural networks and their applications. *Rev. Sci. Instrum.* **1994**, *65*, 1803–1832. [CrossRef]

45. Santos, F.M.; Silva, J.T.; Giardini, A.C.; Rocha, P.A.; Achermann, A.P.; Alves, A.S.; Britto, L.R.; Chacur, M. Neural mobilization reverses behavioral and cellular changes that characterize neuropathic pain in rats. *Mol. Pain* **2012**, *8*, 1744–8069. [CrossRef]
46. Wong, P.M.; Taggart, I.J.; Gedeon, T.D. Use of neural network methods to predict porosity and permeability of a petroleum reservoir. *AI Appl.* **1995**, *9*, 27–37.
47. Ahmed, U.; Crary, S.F.; Coates, G.R. Permeability estimation: The various sources and their interrelationships. *J. Pet. Technol.* **1991**, *43*, 578–587. [CrossRef]
48. Timur, A. An investigation of permeability, porosity, and residual water saturation relationships. In Proceedings of the SPWLA 9th Annual Logging Symposium, New Orleans, LA, USA, 23–26 June 1968.

processes

MDPI

Article

Finite Element and Neural Network Models to Forecast Gas Well Inflow Performance of Shale Reservoirs

Reda Abdel Azim [1,*] and Abdulrahman Aljehani [2]

[1] Petroleum Engineering Department, American University of Kurdistan, Sumel 42003, Iraq
[2] Faculty of Earth Sciences, King Abdulaziz University, Jeddah 21589, Saudi Arabia
* Correspondence: reda.abdulrasoul@auk.edu.krd

Abstract: Shale gas reservoirs are one of the most rapidly growing forms of natural gas worldwide. Gas production from such reservoirs is possible by using extensive and deep well fracturing to contact bulky fractions of the shale formation. In addition, the main mechanisms of the shale gas production process are the gas desorption that takes place by diffusion of gas in the shale matrix and by Darcy's type through the fractures. This study presents a finite element model to simulate the gas flow including desorption and diffusion in shale gas reservoirs. A finite element model is used incorporated with a quadrilateral element mesh for gas pressure solution. In the presented model, the absorbed gas content is described by Langmuir's isotherm equation. The non-linear iterative method is incorporated with the finite element technique to solve for gas property changes and pressure distribution. The model is verified against an analytical solution for methane depletion and the results show the robustness of the developed finite element model in this study. Further application of the model on the Barnett Shale field is performed. The results of this study show that the gas desorption in Barnett Shale field affects the gas flow close to the wellbore. In addition, an artificial neural network model is designed in this study based on the results of the validated finite element model and a back propagation learning algorithm to predict the well gas rates in shale reservoirs. The data created are divided into 70% for training and 30% for the testing process. The results show that the forecasting of gas rates can be achieved with an R^2 of 0.98 and an MSE = 0.028 using gas density, matrix permeability, fracture length, porosity, PL (Langmuir's pressure), VL (maximum amount of the adsorbed gas (Langmuir's volume)) and reservoir pressure as inputs.

Keywords: Langmuir; shale; gas; neural; finite element

Citation: Abdel Azim, R.; Aljehani, A. Finite Element and Neural Network Models to Forecast Gas Well Inflow Performance of Shale Reservoirs. *Processes* **2022**, *10*, 2602. https://doi.org/10.3390/pr10122602

Academic Editors: Tianshou Ma and Yuqiang Xu

Received: 3 November 2022
Accepted: 29 November 2022
Published: 5 December 2022

1. Introduction

Recently, shale gas reservoirs have been considered essential resources, used to supply the world with part of the required energy to compensate for the depletion of conventional reservoirs. Shale gas reservoirs are not similar to conventional reservoirs; these types of formations include conductive natural fractures, narrow thickens and infinite lateral extension. As a result of these properties, only horizontal wells are usually used to increase the gas production rates. In addition, such reservoirs typically have ultra-low permeability; therefore, hydraulic fracturing technology is used to exploit the production zones by creating fracture networks around the wellbore. The gas stored in shale reservoirs is in both free and adsorbed phases; consequently, the production behavior drastically changes through the reservoir's lifetime [1–3].

Based on different field and lab data, 50% of gas in place is stored as adsorbed gas [4,5] and this amount is quantified through Langmuir isotherms [6–11]. The Langmuir isotherms are always used to describe the relationship between methane adsorption on shale surface and gas pressure through ignoring the variation in reservoir temperature.

The gas adsorption in shale systems is controlled by the TOC (Total Organic Carbon), organic matter type, clay minerals and thermal maturity. As the TOC content is raised, the gas capacity increases. The presence of gas in shale in micro-fractures and on surface shale grains during the gas production process leads to gas releasing that provides additional gas in place. Gas production takes place by diffusion in shale matrices while it takes place by Darcy's flow type in the fractures [12–17]. Due to the extra low permeability of shale formations, the continuous effects of pressure transients during well production complicate the gas production forecasting process. Hence, this study presents a simulation model to estimate the shale gas volume in place and to predict surface gas flow rates.

During the production process, free gas is produced at the beginning from natural fractures, then the matrix feeds the fracture network. In turn, the matrix is fed by adsorbed gas existing in the nanopores. In the presented simulation model in this study, organic matter is assumed to be located in the matrix while inorganic matters (free gas) are stored in the micro-fractures. The model also based on the assumption that the gas desorption from organic matter feeds the matrix only, and never contacts the fracture directly, in which the gas flows out of the fractures and not from matrix in a direct way. A vertical well is used in the simulation model for the production process through a horizontal hydraulic fracture.

In this study, the Langmuir's isotherm is used to define the relationship between the gas storage capacity and pressure of the reservoir, and it could be given by:

$$V_E = V_L \frac{P}{P + P_L} \tag{1}$$

where V_E is the gas content, P is the reservoir pressure, P_L is the Langmuir's pressure and V_L is the maximum amount of the adsorbed gas (Langmuir's volume).

The gas content (V_E) in the reservoir rock is affected by several factors including mineral composition and organic matter. The organic matter is considered more important than the mineral composition as it controls the amount of surface area available for adsorption. The organic matter's features include total organic carbon content (TOC) and thermal maturity, and the Langmuir's volume is a function of both. In the shale gas production process, gas desorption cannot be ignored because at low reservoir pressures, most of the gas production comes from desorbed gas. The desorption is the reverse process of adsorption and with a decline in reservoir pressure and production of free gas, the adsorbed gas desorbs from the matrix surface to preserve the reservoir equilibrium and ensure that the reservoir pressure is maintained for a long period. The desorption process starts when the reservoir pressure falls below critical desorption pressure [18,19]. The rate of desorbed gas is controlled by the reservoir permeability, and it has a significant effect on the surface gas production rate.

The desorbed gas volume can be defined by:

$$V_{des} = V_L V_b \rho_R \frac{P}{P + P_L} \tag{2}$$

where V_b is reservoir bulk volume and ρ_R is the shale density at initial reservoir pressure.

The desorbed gas rate into the matrix pore space can be given by:

$$-\frac{\partial V_{des}}{\partial t} = -V_L V_b \rho_R \frac{1}{(P + P_L)^2} \frac{\partial P}{\partial t} \tag{3}$$

This equation gives the volumetric rate in scf/sec and the negative sign indicates that the adsorbed gas decreases as gas desorbs into the matrix.

The advancement in drilling in shale reservoirs for petroleum production using hydraulic fracturing has become a major energy resource worldwide. Such an advancement requires an accurate petroleum production forecast due to the high cost of these technologies. This concern has led scientists around the world to search for an economically feasible project to accurately estimate petroleum production from such reservoirs.

Hydraulic fracturing parameters are the main controlling factor to accurately predict shale gas production and have attracted scientists to develop prediction models. A response surface methodology (RSM) model was proposed by Yu and Sepehrnoori [20] to estimate the net present value (NPV) of a shale gas project. The authors used reservoir parameters and petrophysical properties such as porosity, permeability, reservoir pressure and thickness and fracture properties (spacing, conductivity, half length). The authors did not include a crucial parameter in determining the initial gas in place, namely the initial gas saturation and adsorbed gas content. Later, Nguyen-Le et al. [21] used the parameters investigated by [20], the initial gas saturation and absorbed gas content to develop an economic indicator for the evaluation of shale gas production potential. In their model, they assumed that hydraulic fracturing generates planner fractures. However, in naturally fractured reservoirs, hydraulic fracturing may activate the existing natural fracture system [22–26].

The advancement of artificial neural networks (ANN) for optimizing and predicting petroleum production from shale reservoirs has encouraged scientists to use the technology to accurately predict production. Kim et al. [27] used eleven hydraulic fractures as well as reservoir parameters to predict shale gas production. The authors used matrix-fracture coupling and the diffusion coefficient as input controlling parameters for the prediction model, while these parameters have no direct impact on shale gas production. Furthermore, the authors, similar to Yu and Sepehrnoori [20], ignored the initial gas saturation and adsorbed gas content. Li and Han [28] developed an ANN model to predict the pressure decline parameter using fracture and reservoir properties. The predicted parameters along with the logistic growth decline curve model can be used to reconstruct a production profile. On the other hand, the authors collected the data from vertical oil wells which induce a single stage longitude fracture to predict the model [29]. Multi-stage fracturing techniques are usually carried out in a horizontal well where it is divided into many stages. Moreover, each fracture stage is further divided into more than one fracture cluster. This technique helps in increasing the productivity and contact area between the reservoir and wells. Hence, the predicted model highly depends on the type of wells (i.e., vertical or horizontal).

One model widely used in the literature to predict the petroleum production profile is the decline curve analysis (DCA), due to its simplicity and efficiency. The principle of DCA is to fit the production history data by tuning its decline parameters until the error between the predicted and real data is minimized. Once the minimum fitting error is reached, the DCA model, along with a set of decline parameters, is used to predict future production. One important parameter for accurate prediction of the DCA model is the availability of large amounts of production data (i.e., time). Nelson et al. [30] used the DCA model to predict shale gas production from 48 months of production data. Bashier [31] and Zuo et al. [32] predicted shale gas production using the DCA model with a minimum of 60 months of production history. Odi et al. [33] predicted shale gas production using the DCA model with 36 months of production history. It can be noted from the previous studies that the average production time used for shale gas production predictions ranges from 3–5 years.

Therefore, this study presents an accurate numerical simulation model to estimate the gas production rate and gas volume in place for both adsorbed and free gas. The model is based on a finite element technique using 8-node quadrilateral elements. The gas flow equations are generated by using continuity and Darcy's equations. In addition, an artificial neural network model is proposed to develop a novel correlation for predicting the gas rate and inflow performance using the proposed simulation finite element model. The developed correlation is based on numerous inputs including gas density, matrix permeability, fracture length, porosity, PL (Langmuir's pressure), VL (maximum amount of the adsorbed gas (Langmuir's volume)) and reservoir pressure. This novel correlation will be used in assessing gas shale production avoiding the complexity existing in simulation models.

2. Description of the Flow Simulation Model

A mathematical model is derived in this study to simulate gas flow in shale reservoirs. The gas is considered in a free state in the porous media and adsorbed in the shale matrix.

The model is developed for single-phase fluid flow in a 2-dimensional space. In the developed mathematical model, gas is produced under constant bottomhole pressure. The gas flows from the matrix to the horizontal fracture and directly to the vertical well.

The continuity equation is used as follows:

$$\frac{\partial}{\partial x}(\rho_g u_g) = -\frac{\partial}{\partial t}(\phi\rho_g) \tag{4}$$

where ρ_g is the gas density and u_g is the gas velocity.

The continuity equation can be written in terms of formation volume factor as:

$$\frac{\partial}{\partial x}(\beta_g u_g) - q_g = -\frac{\partial}{\partial t}(\phi\beta_g) \tag{5}$$

where q_g is the source/sink and;

$$u_g = -\frac{ck_x}{\mu_g}\frac{\partial p_g}{\partial x} \tag{6}$$

where c is the conversion factor, k_x is the formation permeability in mD, μ_g is the gas viscosity and p_g is the gas pressure.

By introducing Darcy's velocity in Equation (4), the equation can be written as:

$$\frac{\partial}{\partial x}\left(\frac{ck_x\beta_g}{\mu_g}\frac{\partial p_g}{\partial x}\right) + q_g = \phi\frac{\partial p_g}{\partial t}\frac{\partial \beta_g}{\partial p_g} \tag{7}$$

In case of two-dimensional flows, the equation will be:

$$\frac{\partial}{\partial x}\left(\frac{ck_x\beta_g}{\mu_g}\frac{\partial p_g}{\partial x}\right) + \frac{\partial}{\partial z}\left(\frac{ck_z\beta_g}{\mu_g}\frac{\partial p_g}{\partial z}\right) + q_g = \phi\frac{\partial p_g}{\partial t}\frac{\partial \beta_g}{\partial p_g}. \tag{8}$$

The amount of adsorbed gas can be treated as a sources term (injection well); therefore, by using Equations (3) and (8), it can be written as:

$$\frac{\partial}{\partial x}\left(\frac{ck_x\beta_g}{\mu_g}\frac{\partial p_g}{\partial x}\right) + \frac{\partial}{\partial z}\left(\frac{ck_z\beta_g}{\mu_g}\frac{\partial p_g}{\partial z}\right) + V_L\rho_R\frac{1}{(p+p_L)^2}\frac{\partial p}{\partial t} = \phi\frac{\partial p_g}{\partial t}\frac{\partial \beta_g}{\partial p_g} \tag{9}$$

Introducing a weak formulation, Equation (6) for fluid flow through matrix is described as follows:

$$\int_\Omega w\ \phi\frac{\partial Bg}{\partial t}d\Omega = \int_\Omega w\left(\frac{\partial}{\partial x}\left(\frac{ck_x Bg}{\mu}\frac{\partial p}{\partial x}\right) + \frac{\partial}{\partial z}\left(\frac{ck_z Bg}{\mu}\frac{\partial p}{\partial z}\right) + \right)d\Omega + \int_\Gamma w^T\ q\ d\Gamma \tag{10}$$

where ($w = w\ (x, y, z)$) is a trial function.

Using the finite element method for discretization with respect to time and space results in:

$$\begin{gathered}\int_\Omega (\phi c_t {N_p}^T N_p d\Omega)\left(\vec{P}^i - \vec{P}^{i-1}\right) + \left(V_L\rho_R\frac{1}{(p+p_L)^2}{N_p}^T N_p d\Omega\right)\left(\vec{P}^i - \vec{P}^{i-1}\right) \\ +\Delta t^i\left[\int_\Omega\left(\frac{ck_x Bg}{\mu_g}\frac{\partial {N_p}^T}{\partial x}\frac{\partial N_p}{\partial x} + \right)d\Omega\right]\vec{P}^i + \Delta t^i\left[\int_\Omega\left(\frac{ck_z Bg}{\mu_g}\frac{\partial {N_p}^T}{\partial z}\frac{\partial N_p}{\partial z}\right)d\Omega\right]\vec{P}^i \\ + \int_\Gamma {N_p}^T \mathrm{q}\ d\Gamma = 0.0\end{gathered} \tag{11}$$

where:

$$\vec{P}^T = (p_1\ p_2\ \cdots\ p_n) \tag{12}$$

$$\vec{N}_p{}^T = (N_1\ N_2\ \cdots\ Nn) \tag{13}$$

$$\vec{N_u} = \begin{bmatrix} N_1 \ 0 \ N_2 \ldots \ 0 \\ 0 \ N_1 \ 0 \ldots N_N \end{bmatrix} \tag{14}$$

where N_p and N_u are the shape function for pressure and displacement, respectively, n is the number of nodes, Γ is the domain boundary and P is pressure nodal value.

Galerkin's finite element method is used to discretize the equations as follows:

$$A.(p) + s.(p) + H.(p) = 0 \tag{15}$$

where:

$$A = \int_{ve} [B]^T \left(k \frac{P_L}{(P_L + P)^2} \right) N_p dV \tag{16}$$

$$S = \int_{ve} [N_p]^T \left(\frac{\beta_g V_L P_L}{(P_L + P)^2} \right) N_p dV \tag{17}$$

$$H = \int_{ve} [\nabla N_p]^T \left(\frac{k}{\mu_g} \right) \nabla N_p dV \tag{18}$$

3. Validation of the Numerical Model

Figure 1 shows the model geometry with different boundary conditions. The inner boundary of the reservoir is set to a wellbore pressure of 1000 psi while the outer boundary condition is set as no flow boundary, ignoring the minimum and maximum horizontal stresses. The well is vertical with one hydraulic fracture used to exploit the gas stored. A plain strain is assumed in this model and this assumption is valid when one of the dimensions is very large when compared to other two. In order to validate the developed numerical simulator, a 2-D finite element mesh is used with 5000 elements and eight nodes (see Figure 2). The mesh is generated with a fracture intersecting the wellbore. The fracture length is 100 ft and the parameters of the 2-D model used for the simulation are presented in Table 1. The reservoir pressure is set as 5500 psi with very low permeability of 0.01 md to simulate the actual conditions in shale reservoirs.

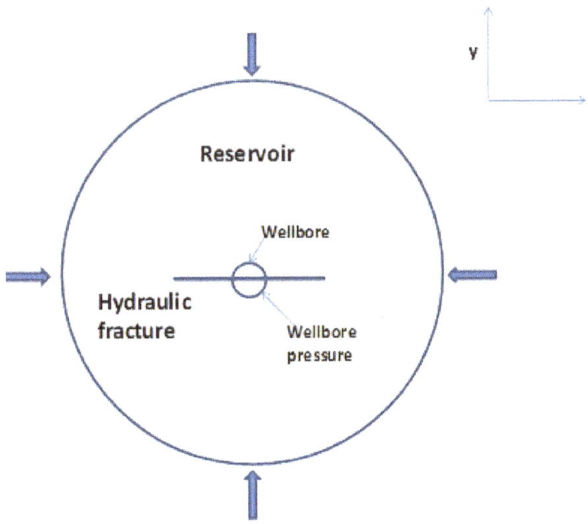

Figure 1. Schematic of the fracture reservoir used for production model.

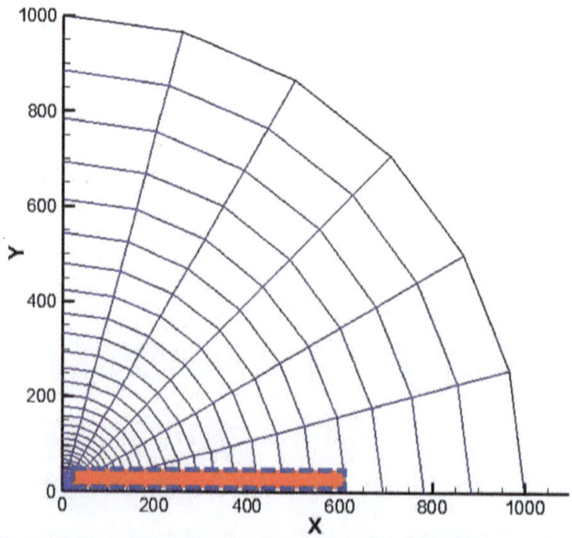

Figure 2. 2-D mesh used for the validation of the numerical model developed in this study.

Table 1. Parameters used in the verification of the developed numerical model.

Parameter	Value
Langmuir's volume	0.08 scf/lb
Langmuir's pressure	5100 psi
Gas viscosity	0.01 cp
Gas density	6 lb/cf
Fracture length	1 m
Fracture permeability	500 md
Initial reservoir pressure	5500 psi
Wellbore pressure	1000 psi
Formation permeability, kx	0.01 md
Formation permeability, ky	0.01 md
Wellbore radius	0.1 m
Reservoir outer radius	1000 m

Only a one-quarter model is selected to take advantage of the reservoir symmetry as shown in Figure 2. Figure 3 shows in detail the gas distribution pressure inside the fracture and the shale matrix. It can be seen from Figure 3 that the gas pressure is initially depleted inside the hydraulic fracture; consequently, pressure depletion occurs on the matrix surface. Figure 4 shows the comparison between the analytical and numerical solution for the gas pressure distribution inside the matrix and the hydraulic fracture. It can be seen from these figures that the gas pressure changes with production time across the wellbore to the reservoir boundary are in a good agreement with the analytical solution. Hence, the developed model in this study can be used to address numerous problems in gas shale reservoirs.

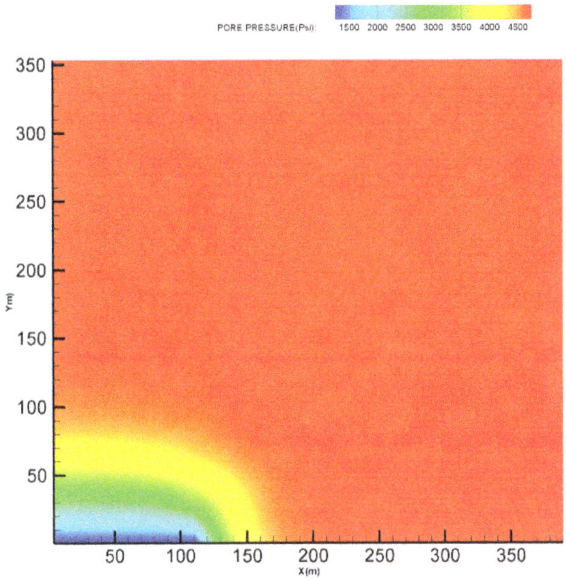

Figure 3. The gas pressure distribution around the wellbore and inside the fracture after 10 days of production for Pi = 5500 psi, Pwf = 1000 psi and k-0.01 md.

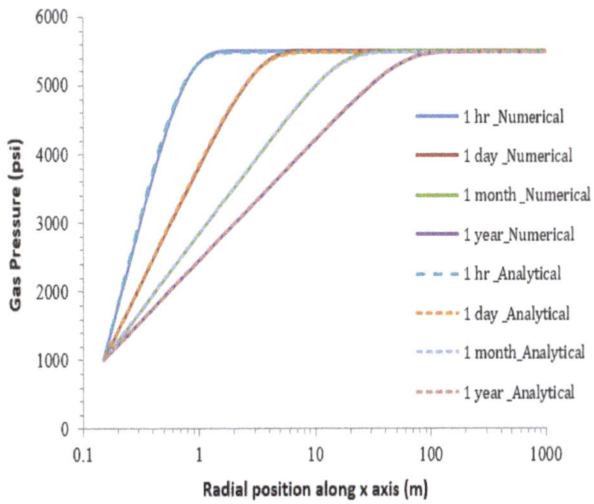

Figure 4. The comparison between analytical and numerical solution of shale gas pressure for Pi = 5500 psi, Pwf = 1000 psi and k-0.01 md.

4. Results and Discussion

A case study is taken from Barnett Shale to estimate the volume of the reservoir adsorbed and free gas. In addition, the developed simulator in this study will evaluate the gas shale flow rates with time. A sensitivity study is performed as well to show the effect of various fluid properties on gas production rates. The collected data are presented in Table 2. The developed simulation model in this study assumes that the gas is stored in natural fractures and pores and it is adsorbed in organic matter.

Table 2. Shale gas reservoir properties used in the verification of the developed numerical model.

Parameter	Value
Gas viscosity	0.019 cp
Gas formation volume factor	1.35 scf/rcf
Langmuir's volume	0.099 scf/lb
Langmuir's pressure	2696 psi
Initial compressibility	5.3×10^{-5} psi^{-1}
Gas density	6.4/lb/cf
Fracture length	20 ft
Fracture permeability	500 md
Initial reservoir pressure	3100 psi
Wellbore pressure	2550 psi
Formation permeability, Kx	0.001 md
Formation permeability, Ky	0.001 md
Porosity	0.05
Reservoir outer radius	1000 m

Figure 5 shows the amount of the adsorbed and free gas versus the change in the reservoir pressure, calculated using the Langmuir's isotherm for Barnett Shale. As it can be seen from this figure, a significant amount of adsorbed gas exists in the 20 ft fracture used in the finite element mesh (see Figures 1 and 2).

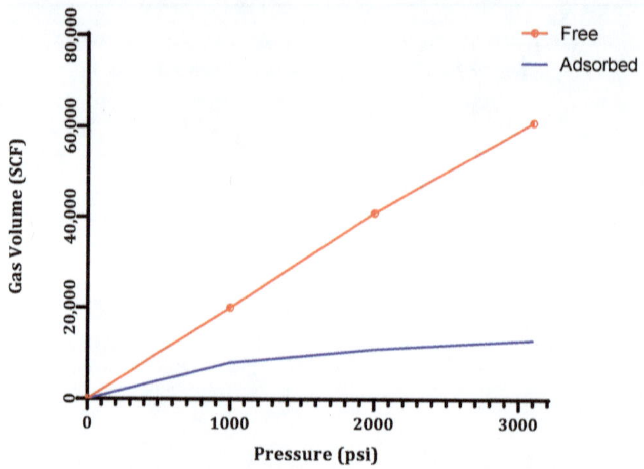

Figure 5. Free gas and adsorbed gas content for Barnett Shale versus reservoir pressure.

Using the permeability of 0.001 md as homogenous reservoir property, the gas production profile is obtained, as shown in Figure 6. The gas production profile shows a long term well performance of 18 years, and this performance is required during the production process in shale gas reservoirs. The gas well is produced at constant bottom hole pressure of 2550 psi at the initial production stage. The free gas exists in the natural fracture and pores. The gas is produced continuously from the fracture until the pressure in the matrix reaches the critical desorption pressure. Then, the adsorbed gas feeds the porous area through the desorption process at a rate depending on the pressure change. Hence, the production rate stabilizes after certain period of production when the total amount of adsorbed gas is desorbed into the fracture near the wellbore (see Figure 6). The production rate is initially started with 180 scf/s and this value is very close to what has been estimated by Wang [34] (211 scf/s). The difference between this study's initial gas rate estimation and that of Wang [34] arises from the different geometry and fracture network used in the study by Wang [34].

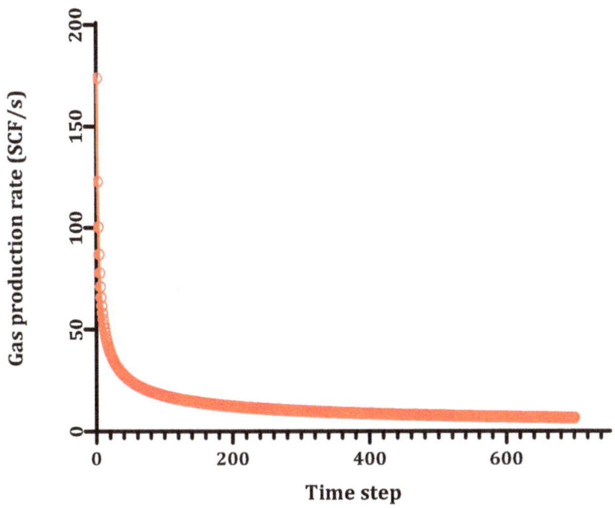

Figure 6. Gas production rate versus reservoir pressure.

Figure 7 shows the cumulative gas production. As it can be seen from Figure 7, the cumulative gas production is initially low due to small the time step size used at the beginning of the simulation run. Then the cumulative gas production volume increases with decreasing the pressure around the hydraulic fracture and the porous matrix. The total volume produced after 12 years of gas production is 80 MMSCF in comparison to the work of Wang [34], which found 88 MMSCF. Estimations of cumulative gas volume using this simulation study and the study of Wang [34] are in a good agreement. Wang [34] uses a subsurface fracture network map which gives more gas volume during the production process.

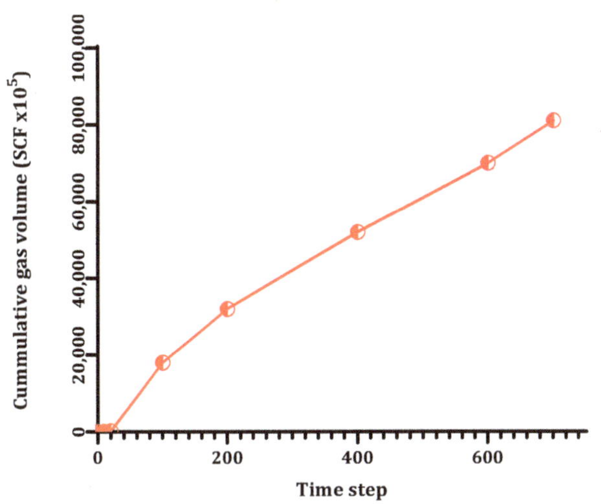

Figure 7. Barnett Shale cumulative gas production volume over 18 years.

Figure 8 shows the variation in gas pressure change due to gas production in two mesh elements, one is close to the wellbore and the other is far from the fracture and wellbore. Figure 8 shows various behaviors of the elements with time. In the early stages of the free gas production, the gas is produced from the cell/element near the wellbore and the closest to the fracture. Hence, the pressure drop decreases faster (see Figure 8). Next, during the period of production, the pressure stabilizes due to it reaching the phase of critical gas desorption. However, the pressure in the cell/element far from the wellbore displays a slow rate of reduction due to low matrix permeability.

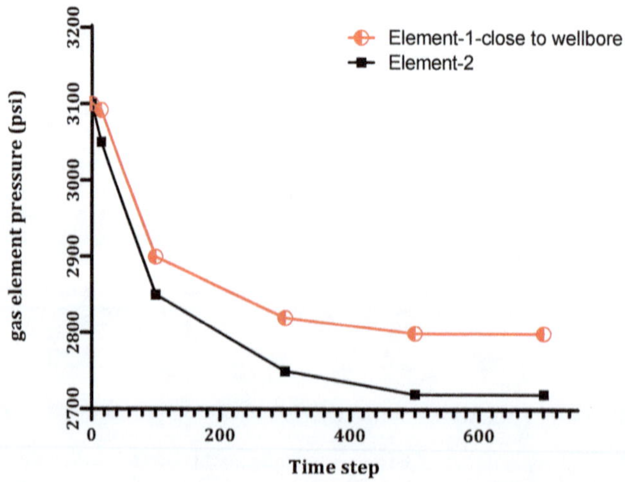

Figure 8. Pressure drop behavior for two elements used in the simulation model.

It can be observed from Figure 9 that the gas pressure diffusion behavior is transient within the hydraulic fracture and matrix at early production stage and reached the boundary after two months of production and the flow regime changed to a pseudo steady-state, dominant for the rest of reservoir production time.

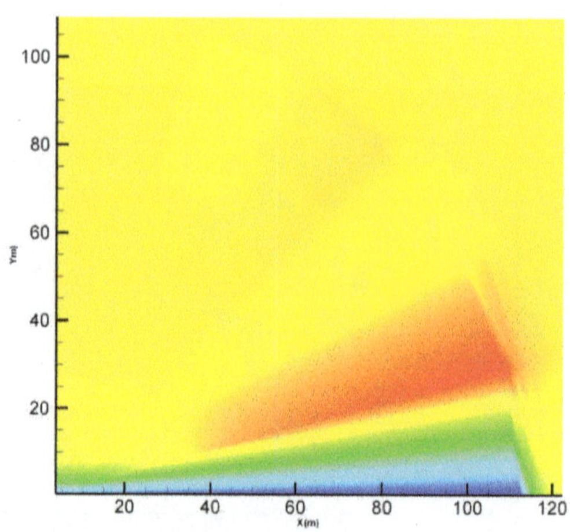

Figure 9. Gas pressure behavior at the early stage of the production process for Pi = 3100 psi, Pwf = 2550 psi, k-0.001 md and fracture length = 20 ft. The blue color is for P = 2550 psi and the red color is for P = 3100 psi.

Figure 10 shows the gas desorption volume for the two cells/elements (with an element/cell close to the fracture and another far from the wellbore). As it can be seen from Figure 10, the gas production starts in the element that is close to the wellbore; hence, its average pressure will be depleted faster. Next, the pressure in the adjacent element will start to deplete to the critical gas desorption pressure. Afterwards, the cell will feed the gas to the pore space and this behavior depends on the reservoir heterogeneity. This means a different behavior might be observed in the desorption process due to a difference in permeability and porosity of the elements.

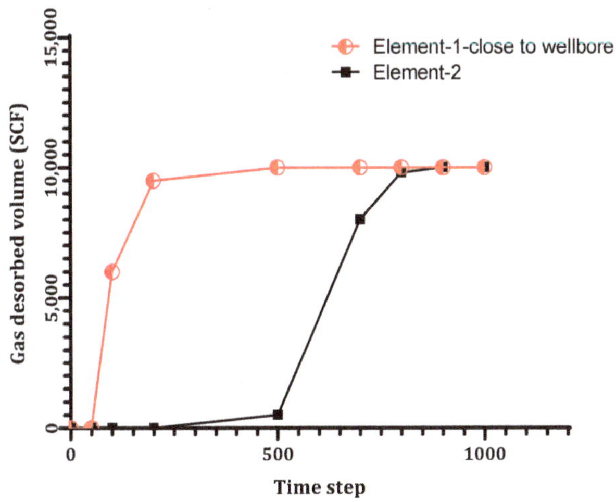

Figure 10. Gas desorbed volume for two elements used in the simulation model.

A sensitivity study was performed to test the effect of the fracture length on the gas production rates. The fracture lengths used were 20 ft and 60 ft. Figure 11 shows the production profile in both cases (with fracture lengths of 20 and 60 ft). It can be seen from Figure 11 that a low production rate profile with fracture length = 20 ft occurred. In addition, the critical desorption pressure is achieved at a later stage for fracture with a length = 60 ft. It can be observed from Figure 11 that the gas production rate is sustained for a long period during the production process with a fracture length = 60 ft. Moreover, it can be observed from Figure 11 that the gas production rate declines 20–35% per month at the beginning of the production process and then the declining rate stabilizes at around 7% after 2 years of production (100-time step) which is a typical scenario for unconventional reservoirs.

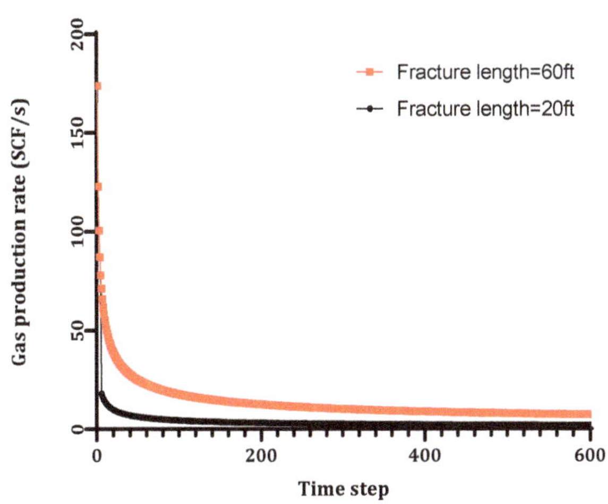

Figure 11. Comparison between production profiles for different fracture lengths.

Since the amount of adsorbed gas depends on the gas density, a simulation run i performed under different gas densities of 5 and 25 lb/cf, respectively. The analysis i Figure 12 shows that the gas production profile is sustained longer in the case of a highe gas density, as the gas in this case is densely packed into the organic matter; thus, a highe amount of adsorbed gas will exist. Therefore, production of more gas is expected wit a high gas density and this parameter will be taken as an essential input parameter in th ANN model to develop a new correlation for gas rate prediction.

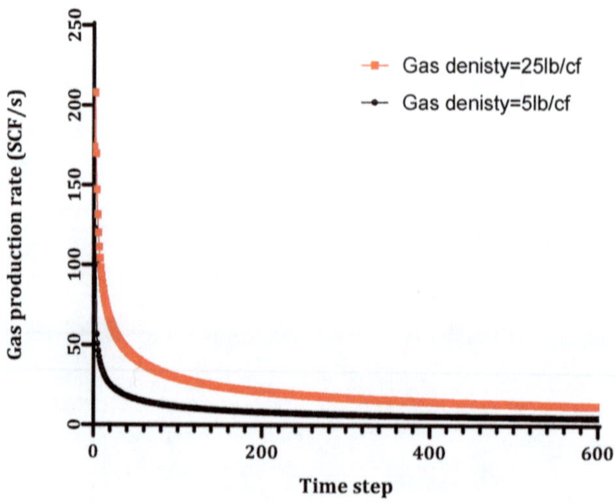

Figure 12. Comparison between production profiles for different gas densities.

5. Neural Network for Gas Rate Prediction

Numerous machine learning algorithms have been used to predict the gas well performance in shale reservoirs. Li et al. [35] used the Neural Based Decision Tree (NDT) learning model for gas production prediction where ANN outperformed NDT. Clarke et al. [36] used a high order neural network for gas well rate forecasting, while Klie [37] used surrogate models for well rate furcating. Fulford et al. [38] combined a supervised learning algorithm with calibrated bias to improve the posterior distribution of forecasts.

Nguyen [39] used the various fracture and reservoir properties to predict the production decline parameters by developing an ANN model. The ANN model was based on data collected from fractured vertical wells which induce a single-stage longitudinal fracture. Syed et al. [40,41] used the unsupervised machine learning algorithm to model and analyze the shale gas production.

Numerous authors have used machine learning algorithms including the neural network (ANN), random forest (RF), function networks (FNs), adaptive neuro-fuzzy inference system (ANFIS) and support vector machine (SVM) to predict the gas shale rates during the production process [42,43].

The back propagation (BP) learning algorithm was used in this study as a supervised algorithm. During the backward propagation, the errors are sent backward though the hidden and input layers. Then, the error values are used to update the weight in the artificial neural network (ANN). The error value at the output neuron can be defined as a quadratic cost function:

$$E_p = \frac{1}{2}\sum_{1}^{n2}(x_p - y_p)^2 \tag{19}$$

where E is the error vector for the training pattern p.

A gradient decent method [44,45] is used in this study in order to modify the weight of the connection and compute the delta weight vector to minimize the cost function. A stopping criterion is set in the ANN for a fixed error threshold, for a fixed number of allowable epochs and for the use of validation data [46–48]. These criteria are very sensitive to the input parameters and if these parameters are not chosen properly, the results will show poor performance due to excessive training. Notably, the target outputs in the training are usually scaled in the interval of (0 and 1) for logistic function.

During the learning phase, it is mandatory to test the performance of the neural net at each single epoch. Therefore, the mean square error *MSE* in Equation (20) is used to show the network performance.

$$MSE = \sqrt{\frac{\sum_1^{n1} \sum_1^{n2} (x_p - y_p)}{n_1 . n_2}} \tag{20}$$

where n_1 and n_2 are number of training output neurons, respectively. x_p and y_p are the target and calculated outputs, respectively.

This section will show how the validated simulator developed in this study is used to construct the ANN model by creating almost 300 data points to be used in the training and testing processes. The inputs include gas density, matrix permeability, fracture length, porosity, PL (Langmuir's pressure), VL (maximum amount of the adsorbed gas (Langmuir's volume)) and reservoir pressure. The target output is the gas production rate (see Figure 13).

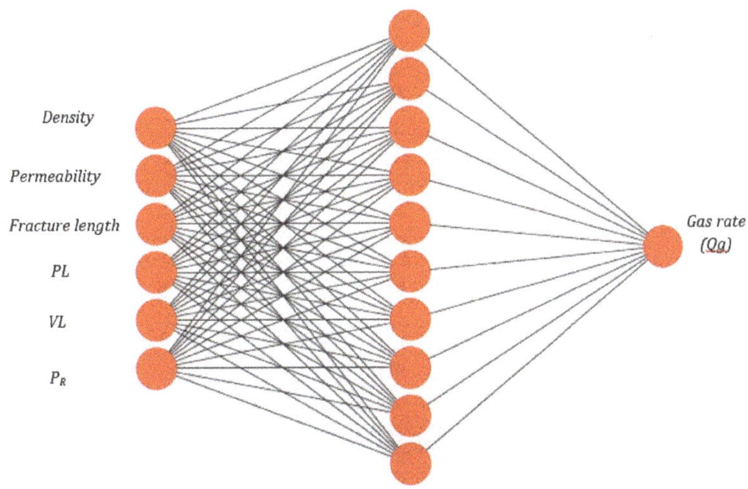

Figure 13. Artificial neural network architecture used in this study.

Date Description

Around 300 points are created using the validated finite element model in this study to design and develop the ANN model. The data contains six inputs including gas density, matrix permeability, fracture length, porosity, PL (Langmuir's pressure), VL (maximum amount of the adsorbed gas (Langmuir's volume)) and reservoir pressure that are used in the training process, while the output is the gas production rate. The statistical analysis of the collected data is presented in Table 3.

The data set is divided into two groups for training and testing processes. A total of 70% is assigned for the training and 30% for the testing. The Back Propagation learning algorithm is used to minimize the results error between actual and target outputs with the log sigmoid function. The BP learning algorithm provides exceptional results with an R^2 of

0.98 and a MSE = 0.018 for training (see Figure 14), and an R^2 = 0.99 for the testing proce (see Figure 15).

Table 3. Statistics analysis of the input data.

Parameter	Gas Density (lb/cf)	Matrix Permeability (md)	Fracture Length (ft)	Langmuir's Volume (VL, psi)	Langmuir's Pressure (PL, psi)	Reservoir Pressure (psi)
Max	18	0.099491	60	0.099662	2795	3996
Min	4	6.08×10^{-5}	10	0.000322	701	1002
Standard Deviation	4.211135	0.028379	14.31894	0.028638	608.8278	866.6465
Skewness	−0.04798	−0.10025	−0.18278	−0.00804	0.010806	0.038504
Mean	11	0.053936	38	0.050758	1760	2446

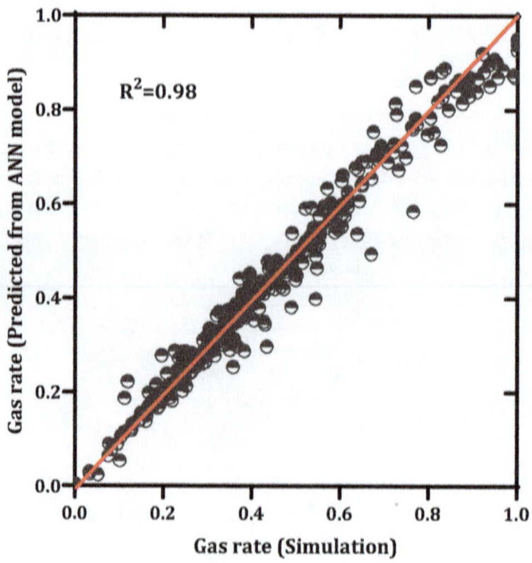

Figure 14. The predicted *Qg* from ANN versus simulation results for training.

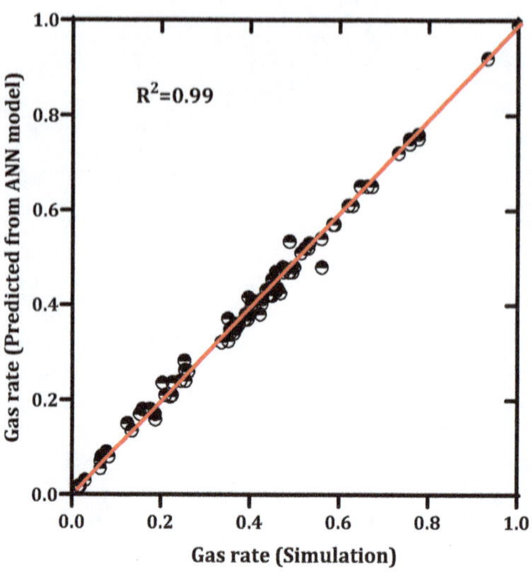

Figure 15. The predicted *Qg* from ANN versus simulation results for testing.

Using the results of training and testing processes, a mathematical correlation is produced to show the relationship between the Q_g and inputs to be used in the forecasting of shale gas rates.

The novel correlation generated using ANN for Q_g estimation is given by:

$$Q_{gn} = \left[\sum_{i=1}^{N} w_{2i} \tan sig\left(\sum_{J=1}^{J} w_{1i,j}x_j + b_{1j}\right)\right] + b_2 \quad (21)$$

$$Q_{gn} = \left[\sum_{i=1}^{N} w_{2i}\left(\frac{1}{1 + exp^{-(Gas\ density \times w_{1,j,1} + Frac/L \times w_{1,j,2} + K_{mat} \times w_{1,j,3} + V_L \times w_{1,j,4} + P_L \times w_{1,j,5} + P_R \times w_{1,j,6}) + b_1}}\right)\right] + b_2 \quad (22)$$

where Q_{gn} is the normalized gas rate, $(w_{2,i})$ is the vector weight between the hidden layer and output layer, $(w_{1,j})$ is the vector weight connecting the input and the hidden layer, j is the neuron number, b_1 is the biases vector for the input layer and b_2 for the output layer. The extracted Q_g equation can be attained by de-normalizing Q_{gn} as follows:

$$Q_g = 160 \times Q_{\mathrm{gn}} + 12 \quad (23)$$

Table 4 shows the weights and bias for the correlation. The proposed correlation in Equation (22) can be used to estimate and predict the gas flow rate for the shale reservoirs using the above-mentioned inputs. Figure 16 shows the comparison between the simulated gas rates and the gas rates values extracted from the ANN model based on inputs. It can be seen from Figure 16 that a good match is achieved with a minimum square error (MSE) of 0.018. The correlation proves that the designed ANN model is reliable and helps in reducing the computational time used by the numerical simulation model which requires extensive mathematics knowledge.

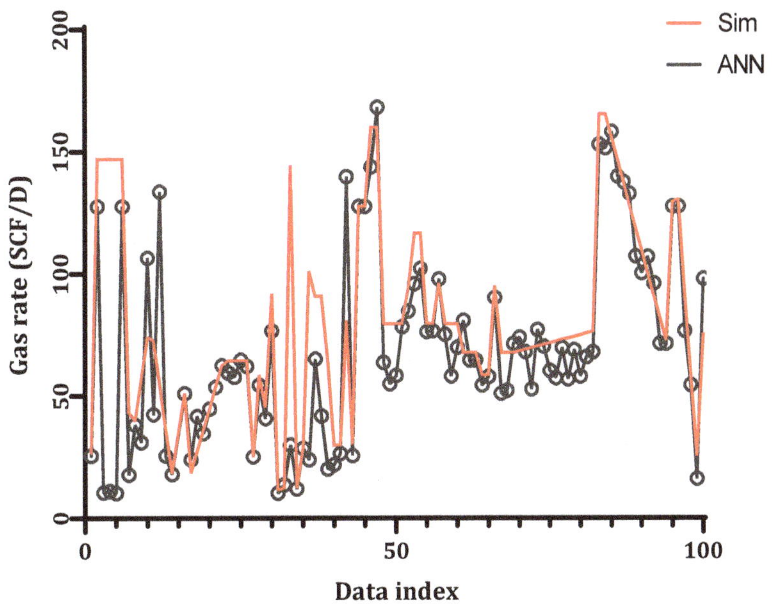

Figure 16. ANN gas flow rate vs. simulated gas flow rate.

Table 4. Weights and biases for the generated correlation Equation (22).

Neuron Number	Input and Hidden Layers Weights (w_1)						Hidden and Output Layers Weights (w_2)	Hidden Layer Bias (b_1)	Output Layer Bias (b_2)
1	$-1.34 \times 10^{+00}$	$-1.53 \times 10^{+00}$	1.04×10^{-01}	4.56×10^{-01}	$1.04 \times 10^{+00}$	$-2.73 \times 10^{+00}$	$-1.99 \times 10^{+00}$	4.93×10^{-02}	
2	8.02×10^{-01}	-3.40×10^{-01}	6.41×10^{-01}	1.31×10^{-01}	-5.66×10^{-01}	2.98×10^{-01}	$-2.53 \times 10^{+00}$	$-1.57 \times 10^{+01}$	
3	3.86×10^{-01}	8.12×10^{-01}	-1.18×10^{-01}	-6.67×10^{-02}	-2.47×10^{-01}	$-1.99 \times 10^{+00}$	9.62×10^{-02}	$-2.92 \times 10^{+00}$	
4	-4.72×10^{-01}	7.89×10^{-02}	-3.63×10^{-02}	1.10×10^{-01}	-7.36×10^{-02}	$1.35 \times 10^{+00}$	4.07×10^{-01}	3.42×10^{-01}	
5	-1.07×10^{-01}	2.10×10^{-01}	1.04×10^{-01}	2.09×10^{-01}	2.59×10^{-01}	$-2.75 \times 10^{+00}$	$1.23 \times 10^{+00}$	9.37×10^{-01}	-0.4621008
6	-1.26×10^{-01}	-2.53×10^{-01}	-1.61×10^{-01}	3.92×10^{-02}	1.20×10^{-01}	$1.55 \times 10^{+00}$	8.74×10^{-01}	$1.16 \times 10^{+00}$	
7	-1.30×10^{-01}	-1.69×10^{-01}	2.73×10^{-01}	-9.72×10^{-02}	4.10×10^{-02}	-9.88×10^{-01}	-4.86×10^{-01}	-4.20×10^{-01}	
8	-1.70×10^{-01}	-2.73×10^{-01}	3.09×10^{-01}	6.93×10^{-02}	-3.07×10^{-01}	$2.19 \times 10^{+00}$	6.41×10^{-01}	6.57×10^{-01}	
9	5.08×10^{-01}	5.04×10^{-01}	2.60×10^{-01}	2.18×10^{-01}	-1.20×10^{-01}	9.73×10^{-02}	4.79×10^{-02}	$-1.80 \times 10^{+00}$	
10	-3.55×10^{-02}	-8.34×10^{-03}	-3.27×10^{-01}	5.66×10^{-02}	3.37×10^{-01}	$-1.22 \times 10^{+01}$	-6.69×10^{-01}	$-1.39 \times 10^{+00}$	

6. Conclusions

This study presents a finite element model to simulate the shale reservoir response in the gas production process for free and adsorbed gas within the shale matrix and the fracture. The mass conservation equation used to develop the gas transport equation includes both free and absorbed gas. The adsorbed gas volume is calculated through Langmuir isotherms. The results show that the long term well performance is achieved during the gas production process and the existing adsorbed gas led to a stabilization in the production after an initial drop when the matrix elements reached critical desorption pressure.

Simulation results of Barnett Shale field show that the evolution of gas pressure strongly depends on the gas desorption process near the wellbore during gas production.

In addition, an ANN is created to forecast the gas production rates for different characteristics of the shale reservoirs. The number of hidden layers and neurons used in the ANN are one and ten, respectively. The results show that the developed ANN in this study predicts the gas rate precisely and these results led to the conclusion that the ANN can be used in some cases for saving computational time.

Author Contributions: Conceptualization, Software, Data creation, writing original draft R.A.A.; Methodology, Data creation, Writing—review & editing A.A. All authors have read and agreed to the published version of the manuscript.

Funding: This research work was funded by Institutional Fund Projects under grant no. (IFPIP: 423-145-1443).

Acknowledgments: This research work was funded by Institutional Fund Projects under grant no. (IFPIP: 423-145-1443). The authors gratefully acknowledge technical and financial support provided by the Ministry of Education and King Abdulaziz University, DSR, Jeddah, Saudi Arabia.

Conflicts of Interest: The authors declare that they have no conflict of interest.

References

1. Baihly, J.; Altman, R.; Malpani, R.; Luo, F. Shale gas production decline trend comparison over time and basins. In Proceedings of the SPE Annual Technical Conference and Exhibition, Florence, Italy, 19–22 September 2010. [CrossRef]
2. Darishchev, A.; de Nancy, E.N.; Lemouzy, P.; Rouvroy, P. On simulation of flow in tight and shale gas reservoirs. In Proceedings of the SPE Unconventional Gas Conference and Exhibition, Muscat, Oman, 28–30 January 2013. [CrossRef]
3. Tolmachev, O.M.; Urunov, A.A.; Muminova, S.N.; Dvoichenkova, G.P.; Davydov, I.A. Review of unconventional hydrocarbon resources: Production technologies and opportunities for development. *Min. Miner. Depos.* **2020**, *14*, 113. [CrossRef]
4. Alexander, T.; Baihly, J.; Boyer, C.; Clark, B.; Waters, G.; Jochen, V.; Le Calvez, J.; Lewis, R.; Miller, C.K.; Thaeler, J.; et al. Shale gas revolution. *Oilfield Rev.* **2011**, *23*, 40–57.
5. Alharthy, N.; Al Kobaisi, M.; Torcuk, M.A.; Kazemi, H.; Graves, R. Physics and modeling of gas flow in shale reservoirs. In Proceedings of the Abu Dhabi International Petroleum Conference and Exhibition, Abu Dhabi, United Arab Emirates, 11–14 November 2012. [CrossRef]
6. Anderson, D.M.; Nobakht, M.; Moghadam, S.; Mattar, L. Analysis of production data from fractured shale gas wells. In Proceedings of the SPE Unconventional Gas Conference, Pittsburgh, PA, USA, 23–25 February 2010. [CrossRef]
7. Andrade, J.; Civan, F.; Devegowda, D.; Sigal, R. Accurate simulation of shale-gas reservoirs. In Proceedings of the SPE Annual Technical Conference and Exhibition, Florence, Italy, 19–22 September 2010. [CrossRef]
8. Arthur, J.D. Hydraulic fracture for natural gas well of the Marcellus Shale. In Proceedings of the Ground Water Production Council Annual Forum, Cincinnati, OH, USA, 21–24 September 2008; Volume 88, pp. 15–20.
9. Asef, M.; Farrokhrouz, M. *Shale Engineering: Mechanics and Mechanisms*; CRC Press: Boca Raton, FL, USA, 2013; Volume 12, pp. 14–18.
10. Bello, R.O.; Wattenbarger, R.A. Modelling and analysis of shale gas production with a skin effect. *J. Can. Pet. Technol.* **2010**, *49*, 37–48. [CrossRef]
11. Soeder, D.J. The successful development of gas and oil resources from shales in North America. *J. Pet. Sci. Eng.* **2018**, *163*, 399–420. [CrossRef]
12. Berawala, D.S. Modelling of Gas Production from Tight Shale Formations: An Innovative Approach. Master's Thesis, University of Stavanger, Stavanger, Norway, 2015.
13. Knudsen, B.R. Production Optimization in Shale Gas Reservoirs. Master's Thesis, Institutt for Teknisk Kybernetikk, Trondheim, Norway, 2010.
14. Lee, S.J.; Kim, T.H.; Lee, K.S. Type curves for pressure transient analysis of horizontal wells in shale gas reservoirs. In Proceedings of the SPE Middle East Oil and Gas Show and Conference, Manama, Bahrain, 19–21 March 2013. [CrossRef]

15. Mengal, S.A. Accounting for Adsorbed Gas and Its Effect on Production Bahavior of Shale Gas Reservoirs. Ph.D. Thesis, Texas A & M University, College Station, TX, USA, 2010.
16. Moghanloo, R.G.; Hosseinipour, S.S. Mechanistic Modeling of Fluid Flow in Shale. In Proceedings of the SPE/AAPG/SEG Unconventional Resources Technology Conference, Denver, CO, USA, 25–27 August 2014. [CrossRef]
17. Olorode, O.M. Numerical Modeling of Fractured Shale-Gas and Tight-Gas Reservoirs Using Unstructured Grids. Ph.D. Thesis, Texas A & M University, College Station, TX, USA, 2012.
18. Olorode, O.; Wang, B.; Rashid, H.U. Three-dimensional projection-based embedded discrete-fracture model for compositional simulation of fractured reservoirs. *SPE J.* **2020**, *25*, 2143–2161. [CrossRef]
19. Azim, R.A. A poroelastic numerical model for simulation of hydraulic fracture propagation: Application to Upper Safa formation-Western Desert-Egypt. *Pet. Res.* **2020**, *5*, 39–51. [CrossRef]
20. Nguyen-Le, V.; Shin, H. Development of reservoir economic indicator for Barnett Shale gas potential evaluation based on the reservoir and hydraulic fracturing parameters. *J. Nat. Gas Sci. Eng.* **2019**, *66*, 159–167. [CrossRef]
21. Sesetty, V.; Ghassemi, A. Hydraulic fracture propagation in naturally fractured anisotropic shale. In Proceedings of the 52nd U.S. Rock Mechanics/Geomechanics Symposium, Seattle, WA, USA, 17–20 June 2018.
22. Ren, L.; Zhan, S.; Zhou, D.; Su, Y.; Wang, W.; Chen, M.; Jing, C.; Sun, J. Propagation simulation and structural characterization of multiple hydraulic fractures in naturally fractured unconventional hydrocarbon reservoirs. *J. Nat. Gas Sci. Eng.* **2020**, *83*, 103557. [CrossRef]
23. Ru, Z.; Hu, J.; Madni, A.S.; An, K. A study on the optimal conditions for formation of complex fracture networks in fractured reservoirs. *J. Struct. Geol.* **2020**, *135*, 104039. [CrossRef]
24. Zhang, H.; Sheng, J.J. Numerical simulation and optimization study of the complex fracture network in naturally fractured reservoirs. *J. Petrol. Sci. Eng.* **2020**, *195*, 107726. [CrossRef]
25. Vishkai, M.; Gates, I.D. Geomechanical characterization of naturally fractured formation. In Proceedings of the 52nd U.S. Rock Mechanics/Geomechanics Symposium, Seattle, WA, USA, 17–20 June 2018.
26. Qin, M.; Yang, D.; Chen, W.; Xia, X. Hydraulic fracturing network modeling based on peridynamics. *Eng. Fract. Mech.* **2021**, *247*, 107676. [CrossRef]
27. Kim, K.; Ju, S.; Ahn, J.; Shin, H.; Shin, C.; Choe, J. Determination of key parameters and hydraulic fracture design for shale gas productions. In Proceedings of the Twenty-fifth International Ocean and Polar Engineering Conference, Kona, HI, USA, 21–26 June 2015.
28. Li, Y.; Han, Y. Decline curve analysis for production forecasting based on machine learning. In Proceedings of the SPE Symposium: Production Enhancement and Cost Optimisation, Kuala Lumpur, Malaysia, 7–8 November 2017. [CrossRef]
29. Zhang, L. Fractured vertical wells in shale gas reservoirs without SRV. *Dev. Pet. Sci.* **2019**, *66*, 73–114. [CrossRef]
30. Nelson, B.; Belyadi, F.; Mashayekhi, A.; Aminian, K.; Ameri, S. Predicting long-term production behavior of the Marcellus shale. In Proceedings of the SPE Western North American and Rocky Mountain Joint Meeting, Denver, CO, USA, 16–18 April 2014. [CrossRef]
31. Bashir, M.O. Decline curve analysis on the woodford shale and other major shale plays. In Proceedings of the SPE Western Regional Meeting, Anchorage, AK, USA, 23–26 May 2016. [CrossRef]
32. Zuo, L.; Yu, W.; Wu, K. A fractional decline curve analysis model for shale gas reservoirs. *Int. J. Coal Geol.* **2016**, *163*, 140e8. [CrossRef]
33. Odi, U.; Bacho, S.; Daal, J. Decline curve analysis in unconventional reservoirs using a variable power law model: A Barnett shale example. In Proceedings of the SPE/AAPG/SEG Unconventional Resources Technology Conference, Denver, CO, USA, 25–27 July 2019. [CrossRef]
34. Wang, H. What factors control shale-gas production and production-decline trend in fractured systems: A comprehensive analysis and investigation. *SPE J.* **2017**, *22*, 562–581. [CrossRef]
35. Li, X.; Chan, C.W.; Nguyen, H.H. Application of the Neural Decision Tree approach for prediction of petroleum production. *J. Pet. Sci. Eng.* **2013**, *104*, 11–16. [CrossRef]
36. Clark, A.J.; Lake, L.W.; Patzek, T.W. Production forecasting with logistic growth models. In Proceedings of the SPE Annual Technical Conference and Exhibition, Denver, CO, USA, 30 October–2 November 2011. [CrossRef]
37. Klie, H. Physics-based and data-driven surrogates for production forecasting. In Proceedings of the SPE Reservoir Simulation Symposium, Houston, TX, USA, 23–25 February 2015.
38. Fulford, D.S.; Bowie, B.; Berry, M.E.; Bowen, B.; Turk, D.W. Machine learning as a reliable technology for evaluating time/rate performance of unconventional wells. *SPE Econ. Manag.* **2016**, *8*, 23–39. [CrossRef]
39. Nguyen-Le, V.; Shin, H. Artificial neural network prediction models for Montney shale gas production profile based on reservoir and fracture network parameters. *Energy* **2022**, *244*, 123150. [CrossRef]
40. Syed, F.I.; AlShamsi, A.; Dahaghi, A.K.; Neghabhan, S. Application of ML & AI to model petrophysical and geo-mechanical properties of shale reservoirs–A systematic literature review. *Petroleum* **2020**, *8*, 158–166.
41. Syed, F.I.; Alshamsi, M.; Dahaghi, A.K.; Neghabhan, S. Artificial lift system optimization using machine learning applications. *Petroleum* **2020**, *8*, 219–226. [CrossRef]
42. Alarifi, S.A.; Miskimins, J. A new approach to estimating ultimate recovery for multistage hydraulically fractured horizontal wells by utilizing completion parameters using machine learning. *SPE Prod. Oper.* **2021**, *36*, 468–483. [CrossRef]

43. Abdelgawad, K.; Elkatatny, S.; Moussa, T.; Mahmoud, M.; Patil, S. Real-time determination of rheological properties of spud drilling fluids using a hybrid artificial intelligence technique. *J. Energy Resour. Technol.* **2019**, *141*, 032908. [CrossRef]
44. Rumelhart, D.E.; Hinton, G.E.; Williams, R.J. Learning representations by back-propagating errors. *Nature* **1986**, *323*, 533–536. [CrossRef]
45. Caudill, M. Neural networks primer, Part III. *AI Expert* **1988**, *3*, 53–59.
46. Hush, D.R.; Horne, B.G. Progress in supervised neural networks. *IEEE Signal Process. Mag.* **1993**, *10*, 8–39. [CrossRef]
47. Guo, C. Flow Mechanisms and Numerical Simulation of Gas Production from Shale Reservoirs. Ph.D. Thesis, Missouri University of Science and Technology, Rolla, MO, USA, 2015.
48. Chang, O.C.Y. Integrated 3-Dimensional Modeling of Proppant Transport through Hydraulic Fracture Network in Shale Gas Reservoir. Ph.D. Thesis, The Pennsylvania State University, State College, PA, USA, 2016.

processes

MDPI

Article

Blank Strip Filling for Logging Electrical Imaging Based on Multiscale Generative Adversarial Network

Qifeng Sun [1,*], Naiyuan Su [1], Faming Gong [1] and Qizhen Du [2]

1 Qingdao Institute of Software, College of Computer Science and Technology, China University of Petroleum (East China), Qingdao 266580, China
2 Key Laboratory of Deep Oil and Gas, China University of Petroleum (East China), Qingdao 266580, China
* Correspondence: sunqf@upc.edu.cn

Abstract: The Fullbore Formation Micro Imager (FMI) represents a proficient method for examining subterranean oil and gas deposits. Despite its effectiveness, due to the inherent configuration of the borehole and the logging apparatus, the micro-resistivity imaging tool cannot achieve complete coverage. This limitation manifests as blank regions on the resulting micro-resistivity logging images, thus posing a challenge to obtaining a comprehensive analysis. In order to ensure the accuracy of subsequent interpretation, it is necessary to fill these blank strips. Traditional inpainting methods can only capture surface features of an image, and can only repair simple structures effectively. However, they often fail to produce satisfactory results when it comes to filling in complex images, such as carbonate formations. In order to address the aforementioned issues, we propose a multiscale generative adversarial network-based image inpainting method using U-Net. Firstly, in order to better fill the local texture details of complex well logging images, two discriminators (global and local) are introduced to ensure the global and local consistency of the image; the local discriminator can better focus on the texture features of the image to provide better texture details. Secondly, in response to the problem of feature loss caused by max pooling in U-Net during down-sampling, the convolution, with a stride of two, is used to reduce dimensionality while also enhancing the descriptive ability of the network. Dilated convolution is also used to replace ordinary convolution, and multiscale contextual information is captured by setting different dilation rates. Finally, we introduce residual blocks on the U-Net network in order to address the degradation problem caused by the increase in network depth, thus improving the quality of the filled logging images. The experiment demonstrates that, in contrast to the majority of existing filling algorithms, the proposed method attains superior outcomes when dealing with the images of intricate lithology.

Keywords: electrical imaging logging; blank strip filling; GAN; U-Net

Citation: Sun, Q.; Su, N.; Gong, F.; Du, Q. Blank Strip Filling for Logging Electrical Imaging Based on Multiscale Generative Adversarial Network. *Processes* **2023**, *11*, 1709. https://doi.org/10.3390/pr11061709

Academic Editor: Yidong Cai

Received: 24 April 2023
Revised: 30 May 2023
Accepted: 31 May 2023
Published: 2 June 2023

1. Introduction

Electrical imaging logging is an effective technical tool for reservoir logging evaluation. The two-dimensional image of the well perimeter obtained by using electrical imaging logging can reflect the structure and characteristics of the wellbore wall more intuitively and clearly, addressing geological challenges that cannot be resolved using conventional logging techniques [1,2]. By correlating electrical imaging data with the geological features they reflect, it is possible to identify rock types, as well as divide and compare stratigraphic layers [3–5]. Similarly, by combining core data with electrical imaging well logging data, it is possible to directly analyze sedimentary structures, identify sedimentary environments, and distinguish main sedimentary units, revealing provenance and paleo water flow direction [6,7]. However, due to the structure of the well body and the structure of the electrical imaging logging instrument, when scanning along the well wall, the well perimeter coverage cannot reach 100%, and a white band is produced on the electrical logging image [8]. The filling of these blank strips is necessary in order to facilitate the subsequent work of geologists.

Currently, the traditional methods for well log image inpainting can generally be classified into two main categories: structure-based methods and multipoint geostatistical methods, such as the Filtersim algorithm [9,10]. The structure-based filling algorithms are mainly interpolation algorithms, such as inverse distance [11] weighting and Lagrangian interpolation [12]. Jing et al. [11] used the resistivity values of adjacent pole plates between the blank regions in order to interpolate the inverse distance of the blank regions to achieve the information filling of the blank regions. Although this method is simple and fast, the continuity of the filled area with the known area is poor [13]. The interpolation algorithm is simple in principle and fast in processing, but there are obvious traces of processing, and the overall visual effect is poor and cannot meet the actual engineering needs [14]. The Filtersim algorithm [15] is a set of multipoint geostatistical methods based on the principles of filtering. It employs the template-matching principle to "paste" the most similar templates into the target inpainting region, resulting in improved computational efficiency [16]. For example, Hurley et al. [17] applied the Filtersim simulation algorithm from multipoint geostatistics to blank strip filling and achieved good results in logging images of simple stratigraphic structures; Sun et al. [13] applied the inverse distance weighting interpolation method and the multipoint geostatistical Filtersim method to fill the blank strip. However, due to the sequential simulation used by the Filtersim algorithm in filling, which has a stochastic nature, the results are not good after filling the multilayered rational regions where structural features dominate.

With the continuous research and development of deep learning theory in recent years, the advantages of convolutional neural networks in image processing have gradually emerged [18,19]. Owing to the powerful ability of neural networks to capture image information, some successful applications have been achieved in image processing [20–23]. In recent years, researchers have gradually applied neural networks to the processing of electrical imaging logs. For example, Wang et al. [24] preliminarily applied neural networks to image completion in electric logging, drawing on the idea of deep neural network structures designed by humans to capture a large amount of prior statistical information about bottom-level images [25]. They constructed an encoder–decoder network model, which extracts features through the encoding layer and restores the image through the decoding layer. However, because of the simplicity of the applied model, the filling effect of electric logging images was not ideal. Zhang et al. [26] also utilized the U-Net network to complete electric logging images, which improved the filling effect to some extent. However, for intricate sand and gravel imagery, the contour information cannot be suitably reconstructed, thereby engendering an inadequate filling outcome. Du et al. [27] extended Wang's [24] approach by incorporating attention mechanisms to constrain the feature extraction process, enabling a focus on important features and resulting in further enhancement of the filling effect.

Drawing upon the literature examined previously, it can be observed that the application of neural network technology in blank interval filling methods is relatively limited, and the targeting is not sufficiently strong. After an extensive literature review, it was found that generative adversarial networks (GANs) [28] have found extensive applications in the domain of image generation. Thanks to their ability to simulate complex functional relationships and their powerful generation capabilities, several papers have used GANs to process image data and have achieved good results [29–32]. Residual Network (ResNet) [33] is another highly influential network following the three classic CNNs: AlexNet, VGG, and GoogleNet. Owing to its simplicity and practical advantages, ResNet has been widely used in fields such as detection, classification, and recognition [34–37]. Dilated Convolution [38,39] has been demonstrated to enhance the modeling capacity of deep learning neural networks more effectively for geometric transformations. The dilated convolution operation has the advantage of being able to capture a larger receptive field compared to traditional convolutions, thereby preserving the spatial features of the image well without sacrificing image detail information [40].

In order to address the problem of filling blank strips in structurally complex electrical imaging, we propose a method for filling the blank strips in electrical imaging images based on multiscale generative adversarial network architecture and apply this method to fill the blank strips in complex electrical imaging images. The major contributions are outlined as follows:

1. The proposed method utilizes generative adversarial network architecture with U-Net as the generator to enhance the feature extraction capability of the network. In addition, two discriminators, i.e., global and local discriminators, are introduced to capture the overall image and local texture information of complex electrical imaging, respectively, which leads to better texture details in the completed image.
2. The introduction of residual networks enhances gradient propagation and addresses the issue of network degradation with increasing depth, thereby improving the quality of the electrical imaging image filling.
3. The use of dilated convolution instead of conventional convolution in neural networks helps to better preserve the spatial features of the image, thus improving the reconstruction of complex electric logging images, especially in terms of contour features.

Based on the experimental results, it can be inferred that the image filled by the network model proposed by us is more consistent with the contextual content, and shows significant improvements in the details and textures of the image compared to traditional filling methods and basic deep generative networks.

2. Construction of the Blank Strip Filling Model

2.1. Improved U-Net-Based Model

The deep generative network model we used for blank strip filling in logging electrical imaging is shown in Figure 1. The model is based on GAN architecture, consisting of two main modules: a generator and a discriminator.

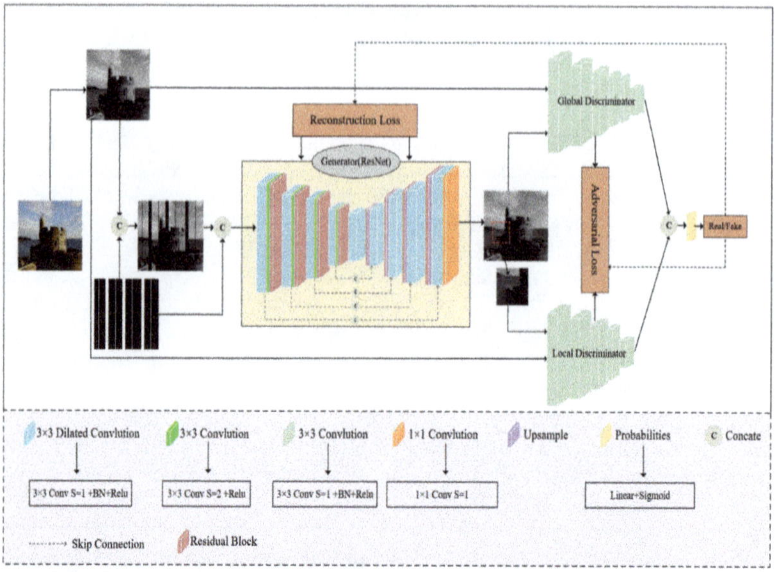

Figure 1. The network architecture diagram of the blank stripe filling model.

The generator is based on encoder–decoder network architecture. The encoder consists of five modules, each of which is composed of a convolutional layer, batch normalization layer, Rectified Linear Unit (ReLU) activation function, and residual network connection layer. The introduction of a residual network simplifies the learning process, enhances gradient propagation, and solves the degradation problem of the network, caused by increasing depth of the network, thereby further improving the quality of image inpainting. The specific structure is shown in Figure 2. The encoder employs a 3 × 3 convolution kernel, and replaces pooling layers by using convolutional layers with a stride of 2 for down-sampling. Batch normalization layers are added in order to alleviate the problem of "vanishing/exploding gradients" in the network. The decoder part consists of four modules, each of which is composed of an up-sampling layer, a skip connection layer, a convolution layer, a normalization layer, an activation function layer, and a residual connection layer. The up-sampling scale of the decoding layer is 2, with a convolution kernel size of 3 × 3 and a stride of 1.

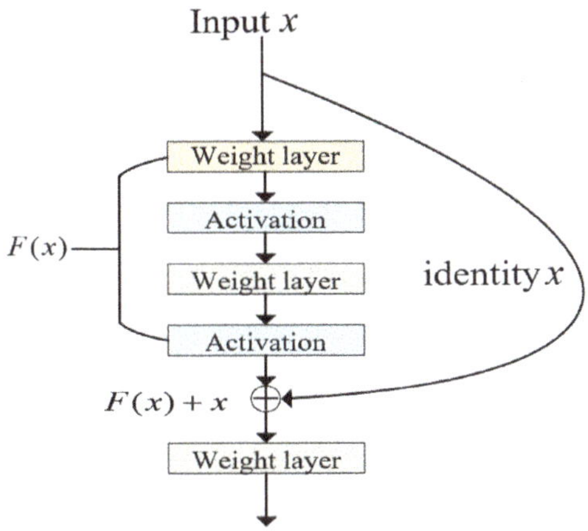

Figure 2. The network architecture of ResNet and the short connection.

The discriminator consists of two modules: a global discriminator and a local discriminator. The global discriminator has 5 modules, where the first 4 modules consist of convolutional layers, batch normalization layers, and ReLU activation functions, and the last module consists of a Flatten layer, fully connected layer, and Sigmoid activation function. The local discriminator has the same architecture as the global discriminator. The global discriminator evaluates the entire image from a global perspective, while the local discriminator focuses more on the details of the image to provide better image details. Alternating training between the generator and discriminator can improve the quality of image inpainting.

In order to achieve better inpainting results, we modified the convolution method in the encoder–decoder section by introducing dilated convolutions with which to replace regular convolution layers. This increases the receptive field to better handle variations in image details, optimizing the model structure and enabling inference of texture feature information for a single image.

2.2. Residual Network Structure

ResNet was proposed by Kaiming He in 2015. At that time, it was widely believed that deeper neural networks would lead to better performance. However, researchers found that increasing network depth actually led to worse performance, known as the problem of network degradation, and gradient vanishing was identified as a key factor. ResNet provided a solution to this problem by introducing residual connections and achieved excellent results in the 2015 ImageNet image recognition challenge, which had a profound impact on the subsequent design of deep neural networks.

As is widely recognized, in the context of convolutional neural networks (CNNs), the matrix representation of an image serves as its most fundamental feature, which is utilized as the input to the CNNs. The CNNs function as information extraction processes, progressively extracting highly abstract features from low-level features. The greater the number of layers in the network, the greater the number of abstract features that can be extracted, which consequently yields more semantic information. In the case of conventional CNNs, augmenting the network depth in a simplistic manner can potentially trigger issues such as vanishing and exploding gradients. The solutions to the problems of vanishing and exploding gradients usually involve normalized initialization and intermediate normalization layers. However, this can lead to another problem—the degradation problem—in which, as the network depth increases, the accuracy of the training set saturates or even decreases. This is dissimilar to overfitting, which generally exhibits superior performance on the training set.

ResNet proposed a solution to address network degradation. In the event that the subsequent layers of a deep network are identity mappings, the model would experience a decline in performance to that of a shallow network. Therefore, the challenge now is to learn the identity mapping function. However, it is challenging to directly fit some layers to a potential identity mapping function $H(x) = x$, which is likely the reason why deep networks are difficult to train. In order to address this issue, ResNet is designed with $H(x) = F(x) + x$, as shown in Figure 2. We can transform it into learning a residual function, $F(x) = H(x) - x$. As long as $F(x) = 0$, it constitutes an identity mapping $H(x) = x$. Here, $F(x)$ is the residual, for which fitting is undoubtedly easier.

ResNet offers two approaches to addressing the degradation problem: identity mapping and residual mapping. Identity mapping refers to the "straight line" portion in the Figure 2, while residual mapping refers to the remaining "non-straight line" portion. $F(x)$ represents the pre-summing network mapping, while $H(x)$ represents the post-summing network mapping of the input x. In order to provide an intuitive example, suppose we map 5 to 5.1. Before introducing the residual, we have $F'(5) = 5.1$. After introducing the residual, we have $H(5) = 5.1$, where $H(5) = F(5) + 5$ and $F(5) = 0.1$. Both F' and F represent network parameter mappings, and the mapping after introducing the residual is more sensitive to changes in the output. For example, if the output s changes from 5.1 to 5.2, the output of the mapping F' increases by 2% (i.e., $1/51$). However, for the residual structure, when the output changes from 5.1 to 5.2, the mapping F changes from 0.1 to 0.2, resulting in a 100% increase. It is evident that the change in output has a greater effect on adjusting the weights for the latter, resulting in better performance. This residual learning structure is achieved through forward neural networks with shortcut connections, where the shortcut connection performs an equivalent mapping without introducing extra parameters or elevating computational complexity. The entire network can still be trained end-to-end through backpropagation.

2.3. Dilated Convolution

The convolutional modules used in conventional convolutional neural networks have a fixed structure, which limits their ability to model geometric transformations [40]. As a result, when applied to images of sandstone with complex structures and textures, the performance is often poor. In order to address this issue, we employ dilated convolution to replace the regular convolutional layers, optimizing the network model and improving its ability to model geometric transformations.

Typical convolutional neural network algorithms commonly leverage pooling and convolution layers to expand the receptive field, while also reducing the resolution of the feature map. Subsequently, up-sampling methods such as deconvolution and unpooling are used to restore the image size. Due to the loss of accuracy caused by the shrinking and enlarging of feature maps during the down-sampling and up-sampling process, there is a need for an operation that can increase the receptive field while maintaining the size of the feature map. This operation can replace the down-sampling and up-sampling operations. In response to this need, the dilated convolution was introduced.

Different from normal convolution, dilated convolution introduces a hyperparameter called the "dilation rate", which defines the spacing between values in the convolution kernel when processing the data. From left to right, Figure 3a–c are independent convolution operations. The large boxes represent the input images (with a default receptive field of 1), the black circles represent 3 × 3 convolution kernels, and the gray areas represent the receptive field after convolution. Figure 3a represents the normal convolution process (with a dilation rate of 1), resulting in a receptive field of 3. Figure 3b represents the dilation rate of 2 for the dilated convolution, resulting in a receptive field of 5. Figure 3c represents the dilation rate of 3 for the dilated convolution, resulting in a receptive field of 7.

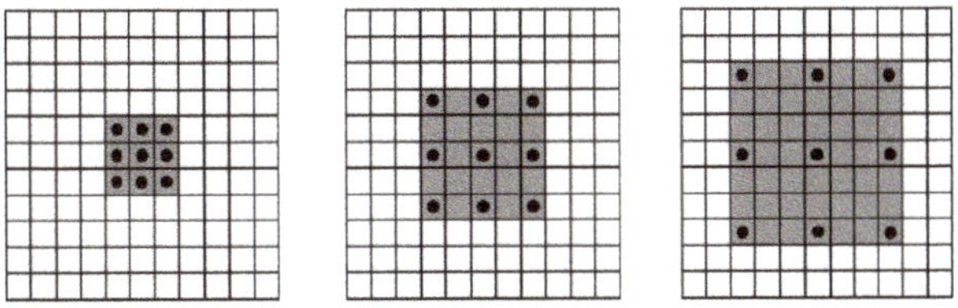

(**a**) (**b**) (**c**)

Figure 3. Illustration of Receptive Fields of Dilated Convolutions with Different Dilation Rates: (**a**) represents the receptive field when the dilation rate is 1; (**b**) represents the receptive field when the dilation rate is 2; (**c**) represents the receptive field when the dilation rate is 3.

From Figure 3, it can be seen that with the same 3 × 3 convolution, the effect of convolutions equivalent to 5 × 5, 7 × 7, etc., can be achieved. Dilated convolutions possess the ability to enlarge the receptive field, whilst avoiding an increase in the number of parameters (number of parameters = convolutional kernel size + bias). Assuming the convolutional kernel size of the dilated convolution is k and the dilation rate is d, then its equivalent convolutional kernel size k' can be calculated using the following formula, for example, for a 3 × 3 convolutional kernel, k = 3:

$$k' = k + (k-1) \times (d-1) \quad (1)$$

$$Out = \frac{In - k' + 2 \times padding}{stride} + 1 \quad (2)$$

The formula involves several parameters: "Out" denotes the dimensions of the output feature map, "In" denotes the size of the input feature map, k' denotes the equivalent kernel size computed in the first step, stride denotes the step size, and padding denotes the size of the padding. Usually, the padding size is designed to be the same as the dilation rate in order to ensure that the output feature map size remains unchanged. However, simply stacking dilated convolutions with the same dilation rate can cause grid effects, as shown in Figure 4.

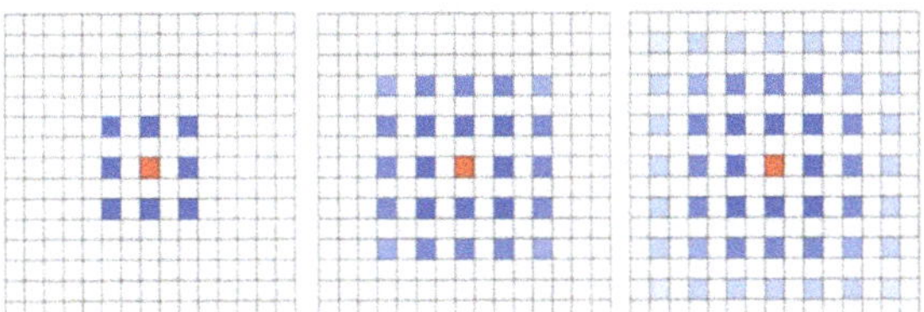

Figure 4. The grid effect of dilated convolution.

As shown in Figure 4, stacking multiple dilated convolutions with a dilation rate of 2 can lead to the problem reflected in the figure. It is evident that the kernel exhibits discontinuity, thereby implying that not all pixels are involved in the computation, and thus leading to a reduction in the continuity of information. In order to address this issue, Panqu Wang proposed the design principle of HDC (Hybrid Dilated Convolution). The solution is to design the dilation rates in a zigzag structure, such as [1,2,5], and satisfy the following formula:

$$M_{i+1} = \max[M_{i+1} - 2r_i,\ M_{i+1} - 2(M_{i+1} - r_i),\ r_i] \tag{3}$$

The variable r_i represents the dilation rate for layer i and M_i represents the maximum dilation rate in layer i, assuming n total layers and $M_n = r_n$ by default. If we use a kernel of size $k \times k$, our goal is to ensure that $M_2 \leq k$, so that all the holes can be covered using dilation rate 1, which is equivalent to standard convolution.

3. The Algorithm of Blank Stripe Filling Based on Multiscale Generative Adversarial Network

3.1. Algorithm Principle

Current deep neural network-based image restoration methods typically require a large amount of training data [41], a requirement which is not suitable for filling the blank strips in well logging electric images. For electrical imaging logging, it is impossible to obtain a complete image of the formation, and obtaining a large amount of image data of the wellbore and surrounding wells is also a significant challenge in engineering. Ulyanov et al. [25] pointed out that neural network architectures themselves can capture low-level statistical distributions, and that the original image can be restored using only the corrupted image. Building upon this idea, this paper minimizes the function $E(f_\theta(z); x_0)$ by optimizing the deep convolutional network model parameters θ, and implements the filling of blank strips through alternating training of the generator and discriminator. The specific algorithmic principle is shown in Figure 5.

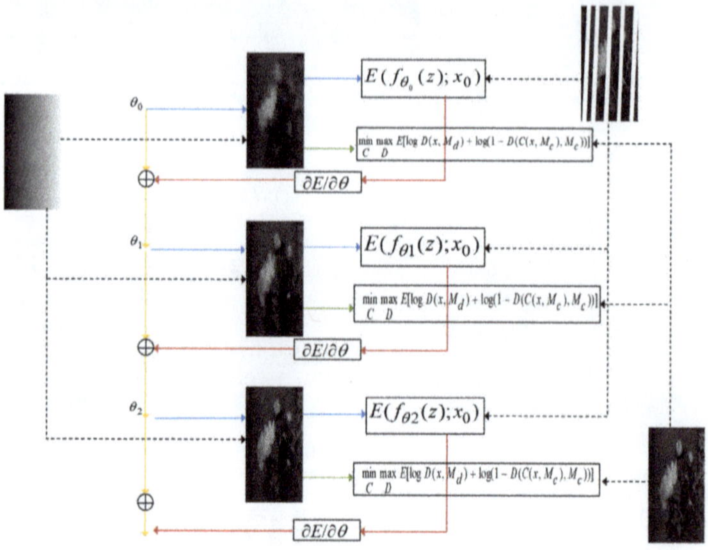

Figure 5. The schematic diagram of the blank stripe filling network algorithm.

In Figure 5, the leftmost image represents the input to the network model, which is initialized with randomly initialized model parameters θ_0. The network output x' is obtained from input z through the function $x' = f_{\theta_0}(z)$. The image x' is used to calculate the loss term $E(x', x_0)$ for the filling task, and is input together with the original image into the discriminative network to calculate the GAN loss. Here, D represents the discriminant function, M_d is a random mask used to randomly select an image patch for the local discriminator, C represents the generative network function, and M_c represents the missing area. We aim for the discriminator to not be able to distinguish between the generated images and the real original images, thus obtaining a complete image with realistic texture details. The loss gradient is computed using the gradient descent algorithm to obtain new weights θ_1 and this process is repeated until the optimal θ^* is found. Finally, the complete image is obtained using the formula $x' = f\theta_*(z)$.

3.2. Algorithm Flow

Step 1: Generating mask images.

In order to simulate the missing data in real well logging images, we generate corresponding masks by scanning the real electric well logging images. As modern electrical imaging logging software typically sets blank areas in the image data as a constant value when converting data from electrode measurements to image data, the pixel values of blank areas are fixed. Therefore, we implement the detection of blank areas in electrical imaging logging images through a point-by-point scanning method using PyCharm software. The RGB pixel values of the scanned blank areas are set to (255, 255, 255), and the pixel values of other non-blank areas are set to (0, 0, 0), thus obtaining the mask image of the logging image. The original logging image and its corresponding mask image are shown in Figure 6.

(a) (b)

Figure 6. Log image and its corresponding mask image: (**a**) log image and (**b**) mask image.

Step 2: Generating the image to be filled.

To ensure that colors outside of the electrical imaging scale bar do not appear after filling, this paper converts the electrical imaging log image to grayscale as the image to be filled, and the output of the network model is also a 1-channel grayscale image. The grayscale image is subtracted from the original grayscale image and multiplied by its corresponding mask to obtain the image to be filled, as shown in Figure 7.

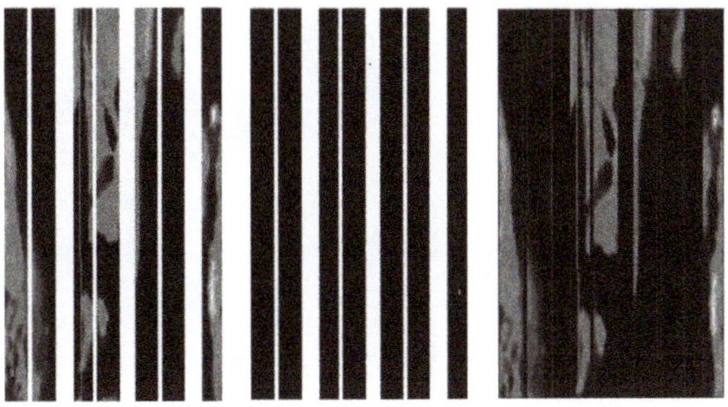

(a) (b) (c)

Figure 7. Grayscale image of the electric logging to be filled in: (**a**) the original grayscale image; (**b**) mask image; and (**c**) image to be filled.

Step 3: Training of the generative network.

For the training of the generative network, we utilized the Places365-Standard dataset [42]. This dataset comprises approximately 1.8 million images, covering 365 distinct scene categories, such as beaches, forests, offices, kitchens, and more. The images in the dataset were collected from the internet, resulting in a wide range of visual diversity and complexity. The image sizes vary from as low as a few tens of kilobytes to as high as several tens of megabytes. This characteristic makes Places365-Standard a challenging dataset suitable for evaluating and training machine learning models in real-world scenarios. To obtain Places365-Standard dataset, it can be downloaded from its official website.

From the dataset, we randomly selected 30,000 images for the training set and 6000 images for the validation set. These images were preprocessed and cropped to a uniform size of 256 × 256 pixels, and the to-be-filled images were produced using the methods from the previous two steps. The output data of the generative network, i.e., the filled grayscale image and the input grayscale image with the mask image, were multiplied, and the mean squared error (MSE) of the pixel values in the missing regions was calculated. The formula for MSE loss is as follows:

$$MSE = \frac{1}{n}\sum_{i=1}^{n}(Y_i' - Y_i)^2 \quad (4)$$

In the formula, n represents the number of pixels, Y_i' represents the output of the generative network, and Y_i represents the original image.

The MSE function has a smooth and continuous curve that is differentiable everywhere, making it suitable for use with gradient descent algorithms, and is a commonly used loss function. Additionally, as the error decreases, the gradient also decreases, which is conducive to convergence. Even with a fixed learning rate, it can converge quickly to the minimum value. The Adam stochastic gradient descent (SGD) optimization algorithm was used to train the generative network and update the parameter θ. Thereafter, we updated the network parameters through the backpropagation of the generation model. The training process was repeated until the loss value reached an acceptable range.

Step 4: Training of the discriminative network.

After the training of the generative network is completed, a randomly cropped section of the generative network output is used as the input for the local discriminator, while the entire generated image is used as the input for the global discriminator. During the training of the discriminator, the binary cross-entropy (BCE) is used as the loss function.

The utilization of the BCE loss facilitates the discriminator in effectively distinguishing between real and generated samples. By minimizing the BCE loss, the discriminator gradually improves its classification ability, thereby enabling the generator to produce more realistic samples, which is formulated as follows:

$$BCE = -\frac{1}{N}\sum_{i=1}^{N} y_i \cdot \log(p(y_i)) + (1 - y_i) \cdot \log(1 - p(y_i)) \quad (5)$$

In the formula, y represents the binary label of 0 or 1, and $p(y)$ represents the probability that the output belongs to the y label. The discriminator is trained by feeding both the original and generated images produced by the generative network into the discriminative network, and the binary cross-entropy (BCE) loss is computed. This process is repeatedly performed until the loss value reaches an acceptable range. We hope that the discriminator cannot distinguish between complete images and real images, thereby obtaining complete images with realistic texture details.

Step 5: Training the generative and discriminative networks jointly.

After the discriminative network is trained, a joint loss function is used to alternately train the generative network and the discriminative network. The joint loss function is shown as follows:

$$\min_{C}\max_{D} E[L(x, M_c) + \alpha \log D(x, M_d) + \alpha \log(1 - D(C(x, M_c), M_c))] \quad (6)$$

In the formula, α is a weighted hyperparameter used to balance the MSE loss and GAN loss. In this step, we first train the discriminative network to correctly distinguish between the completed images and the real images. Thereafter, the generative network is trained to generate results that cannot be distinguished as fake by the discriminative network. We repeat this alternating training process between the discriminative and generative networks until the loss values reach an acceptable range.

Step 6: Generation of final images.

After the entire training process is completed, we only need to use the generative network to input the incomplete image and obtain the completed image. The output image at this time is still a 1-channel grayscale image, which is eventually converted to a color image by comparing the grayscale values on the electric imaging scale, thereby completing the filling task of electric imaging well logging.

In order to facilitate a better understanding of the key steps and logic of the algorithm, we have included a pseudocode representation of the algorithm, as shown in Algorithm 1. This pseudocode serves as a concise and structured description of the algorithm's implementation, enabling readers to grasp its essential components and flow more readily.

Algorithm 1 Training procedure of the ResGAN-Unet

```
while iterations t < T_train do
  Sample a minibatch of images x from training data.
  Input masks M_c for each image x in the minibatch.
  if t < T_c then
    Update the generation C with the weighted MSE
    Loss (Equation (4)) using (x, M_c).
  else
    Generate masks M_d with random hole for each
    image x in the minibatch.
    Update the discriminators D with the binary cross
    entropy loss with both (C(x, M_c),M_c) and (x, M_d).
    if t > T_C + T_D then
      Update the generative network C with the
      joint loss gradients (Equation (6)) using (x, M_c), and D.
    end if
  end if
end while
```

In Algorithm 1, " T_{train}" denotes the total number of iterations for the network, while " T_C" and " T_D" respectively denote the iteration counts for the generative network and the discriminative network.

4. Experimental Results and Analysis

4.1. Experimental Environment

The network model proposed in this paper was implemented in a PyTorch deep learning framework and run on an NVIDIA RTX A5000 GPU server with a virtual environment configured with Python 3.9 and CUDA 11.6. The batch size during training was set to 8, and the Adam optimizer [43] was used with a learning rate of 0.001. The experiment was iterated 5000 times.

4.2. Experiment on Natural Image Inpainting

In order to validate the effectiveness of the proposed algorithm, we first conducted image inpainting experiments on missing natural images.

4.2.1. Introduction to the Dataset

For the natural image experiment, the testing dataset used is still the Places365-Standard dataset. We randomly selected 10 images from the dataset as the original images for this paper's experiment. In order to make the experimental results more convincing, we used strip masks similar to those in logging-while-drilling during the experiments.

One of the natural images utilized for the experiment in this paper is depicted in Figure 8, where Figure 8a is the original image in the dataset. We converted the color image to a grayscale image for processing, as shown in Figure 8b, while Figure 8c represents the mask image used in the filling experiment, where white areas denote missing parts of the image, and black areas denote preserved image information. The original image subtracted from the corresponding pixel-wise multiplication of the original image with the mask image is used as the image to be filled in the experiment, as shown in Figure 8d.

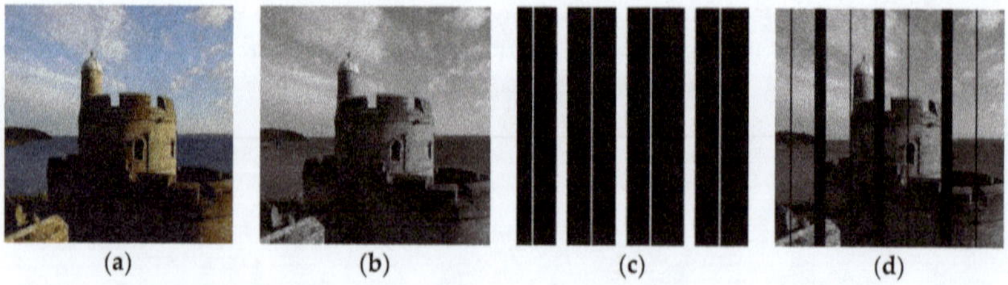

Figure 8. Example of natural image to be inpainted: (**a**) original image; (**b**) grayscale image; (**c**) mask image; and (**d**) inpainting image.

4.2.2. Experimental Results and Analysis

The filling results of the example image in Figure 8, using four different network models, are shown in Figure 9. We can observe from Figure 9 that the conventional encoder–decoder network (Figure 9a) [16] shows obvious filling traces, as indicated by the red dotted lines. Compared to Figure 9a, the use of the U-Net network structure Figure 9b brings some improvement to filling traces, but pixel loss is still severe. With the Res-Unet network and the addition of dilated convolutions (Figure 9c), the pixel loss issue is significantly improved, as indicated by the dotted lines in the figure, and the overall filling quality of the image is improved. After incorporating two discriminators into the base model (Figure 9c) to form a GAN (Figure 9d), the details of image inpainting have been improved, as is evident in the marked regions. The filling results of the strip edges have been significantly enhanced, leading to a notable improvement in overall performance.

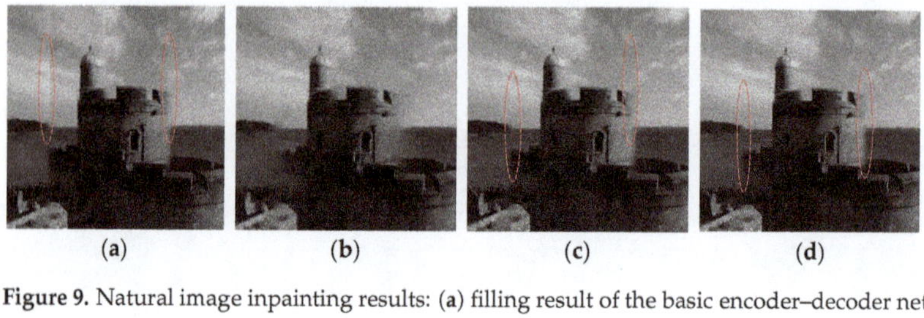

Figure 9. Natural image inpainting results: (**a**) filling result of the basic encoder–decoder network; (**b**) filling result of the U-Net network; (**c**) filling result of the Res-Unet network; and (**d**) filling result of the ResGAN-Unet network.

Further quantitative analysis was conducted on the experimental results by calculating evaluation metrics, such as SSIM (Structural Similarity Index), PSNR (Peak Signal-to-Noise Ratio), MSE (Mean Squared Error), and FID (Fréchet Inception Distance) for both the 10 original images and the corresponding generated images using the proposed models. The average results for each metric across the 10 images are shown in Table 1. From the table, it can be observed that the proposed algorithm consistently outperforms the conventional model methods, thereby validating the advantages of the proposed models.

Table 1. Comparison of Evaluation Metrics for Image Inpainting Results using Different Models.

Models	SSIM (%)	PSNR	MSE	FID
Basic Encoder–Decoder Model	93.78	28.18	110.53	72.42
U-Net Model	94.04	28.48	92.15	61.08
Res-Unet Model	94.58	29.86	70.11	55.23
ResGAN-Unet Model	**94.77**	**29.89**	**66.61**	**50.12**

Here, SSIM is a structured image quality evaluation index that measures the structural similarity between two images. The closer the SSIM value is to 1, the higher the similarity with the original image. PSNR, or peak signal-to-noise ratio, is the most widely used objective image quality evaluation index, where a larger value indicates less distortion. MSE represents the mean square error between the current image X and the reference image Y, with a higher value indicating more severe image distortion. In order to evaluate the metrics of GAN-generated images, a new metric called Fréchet Inception Distance (FID) has been introduced. FID quantifies image quality by comparing the feature distributions of generated images and real images. A smaller FID value indicates that the generated images are closer to the real images in terms of their features and distributions, demonstrating higher realism and fidelity.

As shown in Table 1, the values of SSIM and PSNR increase gradually, while the value of MSE and FID decreases gradually, indicating that the network model's filling effect is becoming better and better.

4.3. Experiment on Electrical Logging Image Inpainting

4.3.1. Introduction to the Dataset

Regarding the well log image dataset, we employed actual recorded electric imaging images from a logging company, specifically from Well C2 in the Bo Block, located approximately 700 m underground. This dataset encompasses a variety of well log images with different types of color calibration, including both dynamic and static color scales.

4.3.2. Experimental Results and Analysis

The filling results of two logging images are shown in Figure 10. The two original logging images are shown in Figure 10a, and the corresponding filling results using the basic encoder–decoder network are shown in Figure 10b. From the filling results, it can be observed that both the upper and lower images in Figure 10b show filling traces, as marked by the blue dashed lines. The continuity of the texture in the images is poor, and there is a clear sense of boundary, indicating an overall poor performance, which is not conducive to subsequent geological work.

In order to improve the filling quality of well logging images, firstly, the conventional neural network structure was replaced with a U-Net network structure in the basic encoder–decoder network. Secondly, in the U-Net network, residual blocks were added and the original convolutional layers were replaced with dilated convolutions. Lastly, two discriminators were further introduced to form a generative adversarial network structure on the previous network. The experimental results are shown in Figure 11.

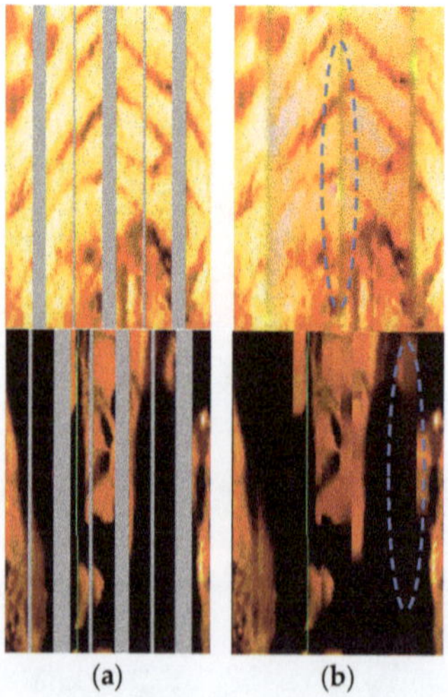

Figure 10. Image filled by basic model: (a) shows the original electrical resistivity well logging image; (b) shows the final image after being filled using the basic encoder–decoder network.

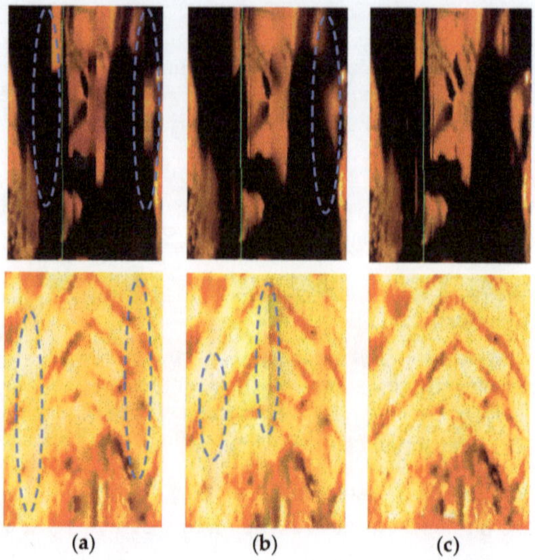

Figure 11. Comparison of filling results of different improved network models: (a) filling result of the U-Net network; (b) filling result of the Res-Unet network; and (c) filling result of the ResGAN-Unet network.

The filling results obtained by replacing the basic encoder–decoder structure with the U-Net network structure are illustrated in Figure 11a. It can be seen that, compared with the original network, the filling effect has been improved to some extent. The continuity of the image texture has been enhanced, and the filling traces at the edge of the overall blank strip have been improved. However, there are still filling traces, as indicated by the blue dotted line in the figure.

The filling results are displayed in Figure 11b after residual networks and dilated convolutions were introduced on U-Net structure. The results show that the introduction of residual networks significantly improves the filling trace of blank strips, enhances the clarity of boundaries between different structures in the image, and makes the filling of the same structure more natural. This indicates that the introduction of residual networks helps to capture the overall structural information of the image, resulting in more accurate filling. However, there are still some shortcomings in the filling results regarding some details, as marked in Figure 11b.

The filling results of the generative adversarial network (GAN), composed of U-Net and residual network structures with both global and local discriminators, are shown in Figure 11c. From the result in Figure 11c, it can be observed that the filling effect of the image has reached its optimal level after training the GAN. The filling traces of the overall blank stripe have become barely distinguishable, which indicates that the multiscale discriminators can provide good texture details of the complete image at different scales. After training with the two discriminators, the filling of the details in the image has been significantly improved, thereby further enhancing the overall visual effect, which is beneficial to the subsequent work of formation division and lithology detection.

The filling results of three additional electrical logging images are shown in Figure 12. The original well logging images before filling are represented by Figure 12a, while the best filling results obtained using other models are displayed in Figure 12b. The corresponding filling results obtained using the proposed model in this paper are shown in Figure 12c. From these images, it can be observed that, for relatively simply structured images, such as the first and third images in Figure 12a, the results obtained by other models are acceptable. However, for complex-structured images, such as the second image, the filling results achieved by the proposed method in this paper are significantly superior to those of other models. They can accurately and realistically fill the missing content in the blank intervals, with almost imperceptible traces at the edges of the intervals. Overall, the visual effect of the filled images has been greatly improved.

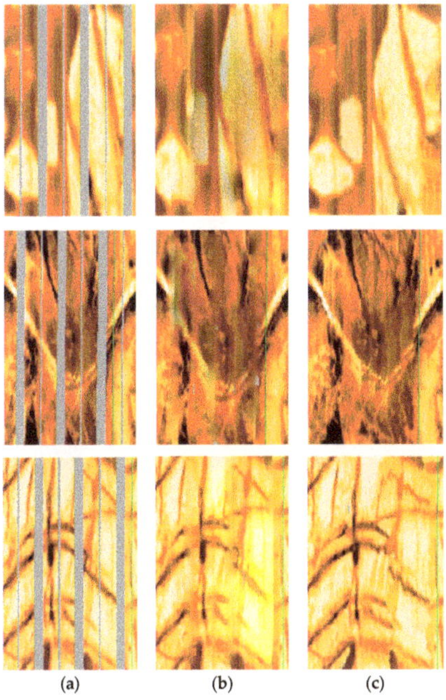

Figure 12. Demonstration of final filling results of other logging images: (**a**) the original image; (**b**) the best results obtained by applying other models for filling; and (**c**) the results obtained by applying the model proposed in this paper for filling.

5. Conclusions

In order to address the issue of filling in blank strips in complex well logging images, we improved the basic encoder–decoder network in the U-Net architecture by simultaneously introducing residual networks and multiscale discriminators. By learning a large number of examples of filling blank bands, and through network model calculation based on the original image and marked area to be filled, the filling of blank bands was achieved. The improved model was used to conduct image inpainting experiments on natural images and complex well logging images. Both experiments demonstrated that the proposed deep learning network can effectively fill in missing data in images, and outperforms conventional encoder–decoder networks in metrics, such as SSIM, PSNR, MSE and FID. This indicates that:

(1) The introduction of residual networks helps to preserve the integrity of information and solve the problem of network degradation. It better captures the overall structure of the image and, thus, more accurately fills in the overall content of the image;
(2) The incorporation of the multiscale discriminator ensures the global and local consistency of the image. The global discriminator is better at filling in the overall image in a global sense, while the introduction of a local discriminator better fills in the texture details of the image.

After conducting experiments on multiple electric imaging well logging images, the results show that the algorithm proposed in this paper can effectively fill the blank strips of complex electric imaging well logging images. The filling traces are barely noticeable, and the image exhibits good continuity and texture, with clear edge contours, which is beneficial for professional geological interpretation in the later stage. The proposed algorithm provides technical support for actual exploration in well logging. For instance, it aids geologists in efficiently identifying distinct geological layers and lithologies, as well as inferring underground geological structures and structural characteristics. Additionally, in tasks such as deep learning-based stratigraphic segmentation, it facilitates the work of geology experts in conveniently and accurately annotating datasets, thereby establishing a solid foundation for model training.

Nevertheless, in the research of filling blank strips in electrical well logging, there are still challenges to be addressed, such as dealing with complex geological structures, improving the stability and robustness of filling results, etc. Future research on filling blank strips in electrical well logging can incorporate more geological information in order to enhance the filling effectiveness. For example, introducing stratigraphic information and rock distribution models can help achieve more realistic geological representation in the filling results. Therefore, it is necessary to explore new algorithms and methods in order to further improve the accuracy and reliability of filling blank intervals in electrical well logging.

Author Contributions: Conceptualization, Q.S. and N.S.; methodology, N.S.; software, N.S. and Q.S.; validation, N.S.; formal analysis, N.S. and F.G.; investigation, Q.S. and N.S.; data curation, N.S.; writing—original draft preparation, N.S.; writing—review and editing, Q.S. and Q.D.; visualization, N.S.; supervision, Q.S. and Q.D.; funding acquisition, Q.S., F.G. and Q.D. All authors have read and agreed to the published version of the manuscript.

Funding: This research was funded by the National Natural Science Foundation of China, grant number 41930429, and the CNPC Major Science and Technology Project (ZD2019-183-006).

Data Availability Statement: Not applicable.

Acknowledgments: We would like to thank the reviewers for their valuable comments to improve the quality of the paper.

Conflicts of Interest: The authors declare no conflict of interest.

References

1. Hassan, S.; Darwish, M.; Tahoun, S.S.; Radwan, A.E. An integrated high-resolution image log, sequence stratigraphy and palynofacies analysis to reconstruct the Albian–Cenomanian basin depositional setting and cyclicity: Insights from the southern Tethys. *Mar. Pet. Geol.* **2022**, *137*, 105502. [CrossRef]
2. Gao, J.; Jiang, L.; Liu, Y.; Chen, Y. Review and analysis on the development and applications of electrical imaging logging in oil-based mud. *J. Appl. Geophys.* **2019**, *171*, 103872. [CrossRef]
3. Zhang, Z.B.; Xu, W.; Liu, Y.H.; Zhou, G.Q.; Wang, D.S.; Sun, B.; Feng, J.S.; Liu, B. Understanding the Complex Channel Sand Reservoir from High-Definition Oil-Base Mud Microresistivity Image Logs: Case Study from Junggar Basin. In Proceedings of the Offshore Technology Conference, Houston, TX, USA, 2–5 May 2022.
4. Zhou, F.; Oraby, M.; Luft, J.; Guevara, M.O.; Keogh, S.; Lai, W. Coal seam gas reservoir characterisation based on high-resolution image logs from vertical and horizontal wells: A case study. *Int. J. Coal Geol.* **2022**, *262*, 104110. [CrossRef]
5. Luo, X.; Pang, X.; Dongxu, S.U.; Hui, L.U.; Zhang, N.; Wang, G. Recognition of Complicated Sandy Conglomerate Reservoir Based on Micro-Resistivity Imaging Logging: A Case Study of Baikouquan Formation in Western Slope of Mahu Sag, Junggar Basin. *Xinjiang Pet. Geol.* **2018**, *39*, 1.
6. El-Gendy, N.H.; Radwan, A.E.; Waziry, M.A.; Dodd, T.J.; Barakat, M.K. An integrated sedimentological, rock typing, image logs, and artificial neural networks analysis for reservoir quality assessment of the heterogeneous fluvial-deltaic Messinian Abu Madi reservoirs, Salma field, onshore East Nile Delta, Egypt. *Mar. Pet. Geol.* **2022**, *145*, 105910. [CrossRef]
7. Zhang, C.; Fan, T.; Meng, M.; Jun, W.U. Geological Interpretation of Ordovician Carbonate Reservoir in Tahe Oilfield: Application of Imaging Logging Technology. *Xinjiang Pet. Geol.* **2018**, *39*, 1.
8. Ni, L.Q.; Xu, H.Q.; Li, Q.W.; Li, G.J.; Xu, C.H. Ultrasonic Logging Image Restoration Based on Texture. *Well Logging Technol.* **2010**, *34*, 4.
9. He, K.; Lu, W.X.; Shen, C.N.; Huang, W.R. Exemplar-Based Inpainting Algorithm with Rotation and Scaling Transformation. *Laser Optoelectron. Prog.* **2018**, *55*, 031006.
10. Ma, P.S.; Li, S.H.; Lu, C.S.; Huang, D.W.; Duan, D.P.; Lu, Y.; Ding, F.; Huang, X. Multi-point geostatistical method based on mode method clustering. *J. Nat. Gas Geosci.* **2020**, *31*, 9.
11. Jing, M.; Wu, J. Fast image interpolation using directional inverse distance weighting for real-time applications. *Opt. Commun.* **2013**, *286*, 111–116. [CrossRef]
12. Occorsio, D.; Ramella, G.; Themistoclakis, W. Lagrange–Chebyshev Interpolation for image resizing. *Math. Comput. Simul.* **2022**, *197*, 105–126. [CrossRef]
13. Sun, J.M.; Zhao, J.P.; Lai, F.Q.; Chen, H. Methods to Fill in the Gaps Between Pads of Electrical Logging Images. *Well Logging Technol.* **2011**, *35*, 532–537.
14. Jini, P.; Rajkumar, K. Image Inpainting Using Image Interpolation—An Analysis. *Rev. Geintec-Gest. Inov. E Tecnol.* **2021**, *11*, 1906–1920.
15. Zhang, T.F. Filter-Based Training Pattern Classification for Spatial Pattern Simulation. Ph.D. Thesis, Stanford University, Stanford, CA, USA, 2006.
16. Zhang, T. Reconstruction Method and Implementation of Porous Media Based on Multi-Point Geostatistics. Ph.D. Thesis, University of Science and Technology of China, Hefei, China, 2006.
17. Hurley, N.F.; Zhang, T. Method to Generate Full-Bore Images Using Borehole Images and Multipoint Statistics. *SPE Reserv. Eval. Eng.* **2011**, *14*, 204–214. [CrossRef]
18. Li, Y.L. Research and application of deep learning in image recognition. In Proceedings of the 2022 IEEE 2nd International Conference on Power, Electronics and Computer Applications, Shenyang, China, 21–23 January 2022.
19. Zhang, J.; Li, C.; Yin, Y.; Zhang, J.; Grzegorzek, M. Applications of artificial neural networks in microorganism image analysis: A comprehensive review from conventional multilayer perceptron to popular convolutional neural network and potential visual transformer. *Artif. Intell. Rev.* **2023**, *56*, 1013–1070. [CrossRef]
20. Gecer, B.; Aksoy, S.; Mercan, E.; Shapiro, L.G.; Weaver, D.L.; Elmore, J.G. Detection and classification of cancer in whole slide breast histopathology images using deep convolutional networks. *Pattern Recognit.* **2018**, *84*, 345–356. [CrossRef]
21. Lin, C.J.; Li, Y.C.; Lin, H.Y. Using Convolutional Neural Networks Based on a Taguchi Method for Face Gender Recognition. *Electronics* **2020**, *9*, 1227. [CrossRef]
22. Xiang, H.; Zou, Q.; Nawaz, M.A.; Huang, X.; Zhang, F.; Yu, H. Deep learning for image inpainting: A survey. *Pattern Recognit.* **2023**, *134*, 109046. [CrossRef]
23. Gong, F.; Li, C.; Gong, W.; Li, X.; Song, T. A Real-Time Fire Detection Method from Video with Multifeature Fusion. *Comput. Intell. Neurosci.* **2019**, *2019*, 1939171. [CrossRef]
24. Wang, Z.F.; Gao, N.; Zeng, R.; Du, X.F.; Du, X.R.; Chen, S.Y. A Gaps Filling Method for Electrical Logging Images Based on a Deep Learning Model. *Well Logging Technol.* **2019**, *43*, 578–582.
25. Ulyanov, D.; Vedaldi, A.; Lempitsky, V. Deep Image Prior. In Proceedings of the 2017 IEEE Conference on Computer Vision and Pattern Recognition, Honolulu, HI, USA, 21–26 July 2017.
26. Zhang, H.; Sima, L.Q.; Wang, L.; Che, G.Q.; Guo, Y.H.; Yang, Q.Q. Blank Strip Filling Method for Resistivity Imaging Image Based on Convolution Neural Network. *Prog. Geophys.* **2021**, *36*, 2136–2142.

27. Du, C.; Xing, Q.; Zhang, J.; Wang, J.; Liu, B.; Wang, Y. Blank strips filling for electrical logging images based on attention-constrained deep generative network. *Prog. Geophys.* **2022**, *37*, 1548–1558.
28. Goodfellow, I.; Pouget-Abadie, J.; Mirza, M.; Xu, B.; Warde-Farley, D.; Ozair, S.; Courville, A.; Bengio, Y. Generative adversarial networks. *Commun. ACM* **2020**, *63*, 139–144. [CrossRef]
29. Liu, M.; Wei, Y.; Wu, X.; Zuo, W.; Zhang, L. Survey on leveraging pre-trained generative adversarial networks for image editing and restoration. *Sci. China Inf. Sci.* **2023**, *66*, 151101. [CrossRef]
30. Wang, Z.; Zhang, Z.; Dong, L.; Xu, G. Jitter Detection and Image Restoration Based on Generative Adversarial Networks in Satellite Images. *Sensors* **2021**, *21*, 4693. [CrossRef]
31. Zhao, M.; Liu, X.; Liu, H.; Wong, K. Super-Resolution of Cardiac Magnetic Resonance Images Using Laplacian Pyramid based on Generative Adversarial Networks. *Comput. Med. Imaging Graph.* **2020**, *80*, 101698. [CrossRef]
32. Belmeskine, R.; Benaichouche, A. Inpainting borehole images using Generative Adversarial Networks. *arXiv* **2023**, arXiv:2301.06152.
33. He, K.; Zhang, X.; Ren, S.; Sun, J. Deep Residual Learning for Image Recognition. In Proceedings of the 2016 IEEE Conference on Computer Vision and Pattern Recognition, Las Vegas, NV, USA, 26 June–1 July 2016.
34. Mo, J.; Zhang, L.; Feng, Y. Exudate-based diabetic macular edema recognition in retinal images using cascaded deep residual networks. *Neurocomputing* **2018**, *290*, 161–171. [CrossRef]
35. Zhou, Z.; Kuang, W.; Wang, Z.; Huang, Z.-L. ResNet-based image inpainting method for enhancing the imaging speed of single molecule localization microscopy. *Opt. Express* **2022**, *30*, 31766–31784. [CrossRef]
36. Zhang, J.; Zheng, Y.; Qi, D.; Li, R.; Yi, X.; Li, T. Predicting citywide crowd flows using deep spatio-temporal residual networks. *Artif. Intell.* **2018**, *259*, 147–166. [CrossRef]
37. Das, B.; Saha, A.; Mukhopadhyay, S. Rain Removal from a Single Image Using Refined Inception ResNet v2. *Circuits Syst. Signal Process.* **2023**, *42*, 3485–3508. [CrossRef]
38. Yu, F.; Koltun, V. Multi-Scale Context Aggregation by Dilated Convolutions. In Proceedings of the 2016 International Conference on Learning Representations, San Juan, Puerto Rico, 2–5 May 2016.
39. Wu, J.; Shi, Y.; Wang, W. Fault imaging of seismic data based on a modified u-net with dilated convolution. *Appl. Sci.* **2022**, *12*, 2451. [CrossRef]
40. Jaderberg, M.; Simonyan, K.; Zisserman, A. Spatial transformer networks. *Adv. Neural Inf. Process. Syst.* **2015**, *28*, 2017–2025.
41. Su, J.; Xu, B.; Yin, H. A survey of deep learning approaches to image restoration. *Neurocomputing* **2022**, *487*, 46–65. [CrossRef]
42. Zhou, B.; Khosla, A.; Lapedriza, A.; Torralba, A.; Oliva, A. Places: An Image Database for Deep Scene Understanding. *J. Vis.* **2016**, *17*, 296. [CrossRef]
43. Kingma, D.; Ba, J. Adam: A Method for Stochastic Optimization. *arXiv* **2014**. [CrossRef]

processes

MDPI

Article

A Novel Pattern Recognition based Kick Detection Method for Offshore Drilling Gas Kick and Overflow Diagnosis

Yang Xu [1,2], Jin Yang [1,*], Zhiqiang Hu [3], Dongsheng Xu [1], Lei Li [1] and Chao Fu [1]

1 College of Safety and Ocean Engineering, China University of Petroleum-Beijing, Beijing 102249, China
2 SINOPEC International Petroleum Exploration & Production Corporation, Beijing 100029, China
3 SINOPEC Research Institute of Petroleum Engineering Co., Ltd., Beijing 102206, China; huzhiq.sripe@sinopec.com
* Correspondence: yjin@cup.edu.cn

Abstract: In offshore drilling, accidents such as gas invasion, overflow, and kicks are unavoidable, and they can escalate into blowouts and other catastrophic events, resulting in casualties and significant economic losses. Therefore, ensuring drilling safety requires precise monitoring of gas invasion and overflow. Currently, most overflow monitoring methods used at drilling sites are based on threshold criteria. However, the monitoring parameters obtained during actual drilling operations often contain noise signals, which makes it challenging for threshold-based methods to achieve a balance between improving accuracy and minimizing false positives. This paper proposes a novel method called Pattern-Recognition-based Kick Detection (PRKD) for diagnosing overflow in offshore drilling. The PRKD method utilizes the overflow evolution process by integrating multiphase flow calculations, data filtering theory, pattern recognition theory, the Bayesian framework, and other theoretical models. By analyzing the shape and wave characteristics of the curves, PRKD effectively detects and monitors gas intrusion and overflow based on single parameters. Through case analysis, it is demonstrated that the proposed method achieves high precision in monitoring drilling overflow while maintaining a low false positive rate. By combining advanced computational techniques with pattern recognition algorithms, PRKD improves the accuracy and reliability of kick detection, enabling proactive responses to potential risks, protecting the environment and human lives, and optimizing drilling operations. The case analysis shows that by integrating the probabilistic information of pre-drilling kicks and various characteristic parameters, when the noise amplitude is less than 8 L/s, the PRKD model exhibits superior detection performance. Moreover, when the noise amplitude is 16 L/s, the PRKD model detects the continuous overflow approximately 200 s after the actual overflow occurs and predicts a 95.8% probability of overflow occurrence at the specified location, meeting the on-site requirements. The gas invasion monitoring method proposed in this paper provides accurate diagnostic results and a low false positive rate, offering valuable guidance for gas invasion monitoring in drilling operations.

Keywords: offshore drilling; overflow; kick; gas invasion; pattern recognition; threshold method

Citation: Xu, Y.; Yang, J.; Hu, Z.; Xu, D.; Li, L.; Fu, C. A Novel Pattern Recognition based Kick Detection Method for Offshore Drilling Gas Kick and Overflow Diagnosis. *Processes* **2023**, *11*, 1997. https://doi.org/10.3390/pr11071997

Academic Editor: Albert Ratner

Received: 20 May 2023
Revised: 20 June 2023
Accepted: 27 June 2023
Published: 3 July 2023

1. Introduction

In the process of offshore drilling, if gas invasion and overflow are not detected in time, a blowout can rapidly occur. Blowouts are often the most dangerous of the numerous drilling accidents. To accomplish safe and efficient drilling, reduce downhole accidents, and lower drilling costs, it is necessary to excel at early monitoring, early detection, and early treatment of overflow [1–5]. In terms of theory, there are currently two types of overflow detection techniques used in deepwater drilling: the threshold method and predictive systems. The threshold method involves establishing a threshold value for the identifying parameters detected during the kick process. When the detection parameter exceeds the specified value, an alarm activates and a kick will be detected. For instance, if the increment of the mud pool is 1 m^3, when the judged kick occurs under a certain working condition, if the increment is less than 1 m^3, the drilling is considered normal, while, if the increment is greater than 1 m^3, overflow is considered to have occurred [6–8]. Maus L.D. et al. [9] pointed out that the incremental monitoring of the mud pit can be used as a parameter for gas invasion monitoring. The mud pit increment is the result of integrating the difference between inflow and outflow rates over time. J.M. Speers and G.F. Gehrig [10] proposed a system for monitoring wellbore influx or losses during drilling using flow rate differentials. B.T. Anfinsen et al. [11] reported that although the sensitivity of mud pit increment monitoring to gas invasion is controversial, it can detect the infiltration of fluids even at very low flow rates. D. Fraser et al. [12] proposed two major parameters related to wellbore influx. The first parameter is the Kick Detection Volume (KDV), which represents the volume of invaded formation fluid before the wellbore influx is identified. The second parameter is the Kick Response Time (KRT), which refers to the time required for well control operations once the wellbore influx is detected. Nayeem A.A. et al. [13] introduced a method for monitoring wellbore influx based on changes in downhole parameters such as mass flow rate, pressure, density, and drilling fluid conductivity. The major advantage of this method is its short detection time and quick response to wellbore influx. It allows for rapid identification of influx events, enabling timely well control actions.

The second method is the simulation prediction method, which simulates the "theoretical value" of the overflow characteristic parameters under normal conditions (no overflow) in real time using computer software. If the "measured value" exceeds a predetermined threshold of the "theoretical value" during the drilling process, it is considered an overflow [14–16]. Moreover, there are additional overflow diagnosis methods, such as the BP neural network method and the probabilistic analysis method, but their applications are limited. Liao Mingyan et al. [17] developed a drilling process state monitoring and fault diagnosis method based on a BP neural network model. This method focused on six parameters: drilling pressure, pump pressure, pump rate, rotary speed, torque, and drilling speed. The neural network model possesses fault tolerance, self-learning, and adaptive capabilities, allowing it to simulate complex nonlinear mappings and process data in parallel with high real-time performance. However, neural networks often require many training samples to achieve optimal performance. Roar et al. [18] combined theoretical flow models with artificial intelligence techniques to identify hidden patterns in time-series data and address limitations in physical models, thereby reducing false alarm rates. David Hargreaves et al. [19] addressed the limitations of the threshold method and proposed a new approach for wellbore influx diagnosis based on pattern recognition theory and Bayesian discrimination. This method utilized statistical techniques to handle noise issues and simulated gas influx and non-influx events to avoid false alarms or missed detections due to ambiguous data. Pournazari et al. [20] developed a pattern recognition system for rapid analysis and real-time diagnosis of drilling events using the SAX (Symbolic Aggregate Approximation) method. This system enables fast and accurate wellbore influx diagnosis.

Although various kick monitoring methods have been proposed, many are unsuitable for offshore deepwater drilling sites, where wellheads are usually placed on the seabed and kick detection is necessary to prevent gas from entering the riser. Common kick monitoring techniques, such as the flow back velocity method, mud pool increment method, etc., rely on the volume expansion of gas in the invading wellbore. Under high-pressure conditions in the wellbore, natural gas has a high solubility and density in the drilling fluid, as well as a small volume after invasion, making it difficult to locate. Simultaneously, deepwater drilling conditions are complex, the fluctuation range of kick monitoring parameters is large, and the noise level is high. The traditional discriminant technique based on the threshold value method has low kick diagnosis accuracy and a high false positive rate, which poses significant difficulties for the early monitoring of the deep kick [21,22].

Currently, the threshold method is widely used for kick monitoring in the drilling field. However, the monitoring parameters collected during actual drilling processes often contain significant noise signals, making it challenging to achieve a balance between accuracy improvement and reducing false positives based solely on threshold judgment. To address this issue, this study proposes a novel and reliable kick monitoring method called PRKD (Pattern-Recognition-based Kick Detection) for offshore drilling. PRKD is established by integrating multiphase flow calculations, data filtering theory, pattern recognition theory, and the Bayesian framework. It focuses on a single-parameter gas intrusion monitoring approach, utilizing the time series of kick monitoring parameters as the research object. By combining the single-parameter process identification and diagnosis mechanism with pattern recognition techniques, PRKD offers an innovative solution for kick detection and has several advantages: (1) Maintaining accuracy in kick diagnosis while reducing false positives: PRKD ensures the accuracy of kick detection while minimizing the occurrence of false alarms. Real-time data matching enables early identification of kicks, allowing proactive measures to be implemented promptly. (2) Minimizing the impact of data noise on judgment results: The PRKD method incorporates data filtering techniques to mitigate the effects of noise signals present in the monitoring parameters. This helps in achieving more reliable and accurate kick detection. (3) Acquiring basic kick patterns from multiple sources: The PRKD method allows for obtaining basic kick patterns from various channels, reducing the dependence on the accuracy of computational software results. These patterns can be continuously updated in real time and can be customized based on the experience and expertise of technical personnel. (4) Utilizing multiple available data sources: By integrating various data sources, PRKD achieves probabilistic output results, enhancing the overall accuracy of kick detection. This comprehensive approach improves the reliability of the monitoring system. (5) Potential for inferring additional kick information: The PRKD method has the capability to infer other relevant kick information beyond basic detection. This further expands the scope of kick monitoring and provides a more comprehensive understanding of the drilling process. By incorporating sophisticated computational techniques and pattern recognition algorithms, PRKD significantly enhances the reliability and accuracy of kick detection. This empowers drilling operations with the ability to proactively implement measures to mitigate potential risks, safeguard the environment, and protect human lives. The integration of PRKD into drilling practices optimizes operational efficiency while ensuring the safety and well-being of all involved parties.

2. The Basic Overflow Pattern

In the PRKD method, the "ruler" used to assess overflow events is a pattern or trend of change rather than a single threshold. In the past, it was believed that an increase in the outlet flow rate indicated a possible overflow event, but, an increase in the outlet flow rate could also be caused by starting the pump. If a threshold is used to make decisions, it is challenging to differentiate between overflow events and pump start events. In fact, both overflow and pump start events can increase the outlet flow rate, but their "trends" of change are different. During overflow, the outlet flow rate increases linearly, whereas it increases abruptly and then maintains a specific value during pump start. Similarly, traditional methods focus on "increasing to a certain specific value" for the basic event of "the outlet flow rate increase", whereas the PRKD method is concerned with "what the trend the increase follows, linear increase or sudden change".

Using change trends as the basic element has the following three advantages:

(1) Reduces the number of false alarms. The use of a threshold value alone cannot distinguish excess events from similar events.

(2) Significantly decreases the dependence on the accuracy of calculation software. To achieve high calculation accuracy, it is necessary to precisely describe the physical processes and calculation parameters, and even a minor deviation can result in significant errors, which is frequently challenging in deepwater drilling environments. For instance, when the friction coefficient is 0.001 and 0.0011, the calculated value of the bottom hole pressure will result in significant errors, but the influence of the changing trend of the calculated value of the bottom hole pressure can be neglected.

(3) Other stratigraphic information can be retrieved. For example, the increment pressure and mud pool increment trend can be utilized to retrieve formation pressure data. The gas invasion process is like the unstable well test process. Therefore, stratigraphic information can be represented based on the change in gas invasion response parameters (standpipe pressure (SPP), outlet flow rate, mud pool increment, etc.), thus realizing stratigraphic information inversion before shut-in, which plays an important role in the density design of kill fluid.

The basic pattern can be obtained through various channels. If the drilling block contains a large amount of geological history data of the stratigraphic profile, the basic overflow pattern can be determined by fitting the statistical data. If historical data are limited, the basic overflow pattern can be simulated for the well. In addition, drilling engineers can establish the basic pattern based on their experience. Furthermore, the basic overflow pattern can be updated in real time based on the well overflow events that have occurred.

Theoretically, most overflow characteristic drilling parameters can be depicted by segmented polynomial functions and sudden change functions. Hargreaves et al. [23] found that after a wellbore surge occurs, the difference between inlet and outlet flow rates increases linearly, and the mud increment changes with an increase to a quadratic polynomial form. Reitsma et al. [24] reported that after a deepwater managed pressure drilling surge occurs, the casing pressure increases linearly and the pore pressure increases linearly before decreasing linearly. Based on the results of multiphase flow simulation and the expertise of experts, the basic patterns of characteristic parameters such as the flow rate differential, mud pit increment, casing pressure, pore pressure change, and mechanical drilling speed during wellbore surges and other accidents can be determined, as shown in Figures 1–4 below.

(1) Flow rate differential

- ① Steady State: the flow rate differential remains within a relatively stable range, indicating a balanced flow of drilling fluid entering the wellbore and returning to the surface.
- ② Kick State: the flow rate differential increases as a larger volume of gas returns to the surface through the wellbore.
- ③ Pump on State: the flow rate differential increases as a large volume of mud is injected into the wellbore.
- ④ Loss State: the flow rate differential decreases as some drilling fluids are absorbed or leaked into the formation.

(2) Mud Pit Increment

- ① Steady State: The mud pit increment is maintained within an appropriate range to replenish the drilling fluid consumed during the drilling process. Without mud replenishment, the mud pit increment will decrease as drilling fluid is consumed.
- ② Kick State: the mud pit increment increases to compensate for the decrease in mud volume due to gas invasion
- ③ Pump on State: the mud pit increment decreases to accommodate the additional mud volume pumped into the wellbore.
- ④ Loss State: the mud pit increment decreases to compensate for the lost drilling fluid.

(3) Casing Pressure (CP) and Standpipe Pressure (SPP) Changes

- ① Steady State: casing pressure and standpipe pressure changes are minimal, indicating no abnormal influx or pressure variations.
- ② Kick State: CP and SPP changes increase due to the added pressure from gas influx. As well control measures are implemented, the SPP is expected to decrease, while the CP is expected to further increase.
- ③ Pump on State: CP and SPP changes may increase instantaneously as the wellbore pressure rises due to the large volume of mud injection.

④ Loss State: CP and SPP changes decrease due to the pressure loss in the wellbore caused by losses.

(4) Rate of Penetration (ROP) Changes

① Steady State: the ROP remains relatively stable, ensuring good drilling progress.

② Kick State: If the density of the drilling fluid is not adjusted promptly when gas kick, it can lead to a decrease in the bottom hole pressure compared to the formation pressure or a reduction in the positive pressure difference between the bottom hole pressure and formation pressure, resulting in an increase in the ROP.

③ Pump on State: When pump on operations begin, the increased flow rate of drilling fluid circulating through the drill string can enhance the hydraulic horsepower at the bit. This can result in an increased bit penetration rate and, therefore, an accelerated ROP.

④ Loss State: When drilling fluid is lost to the formation, it reduces the hydrostatic pressure exerted by the drilling fluid column. As a result, the effective weight on the bit decreases, leading to a decrease in the ROP.

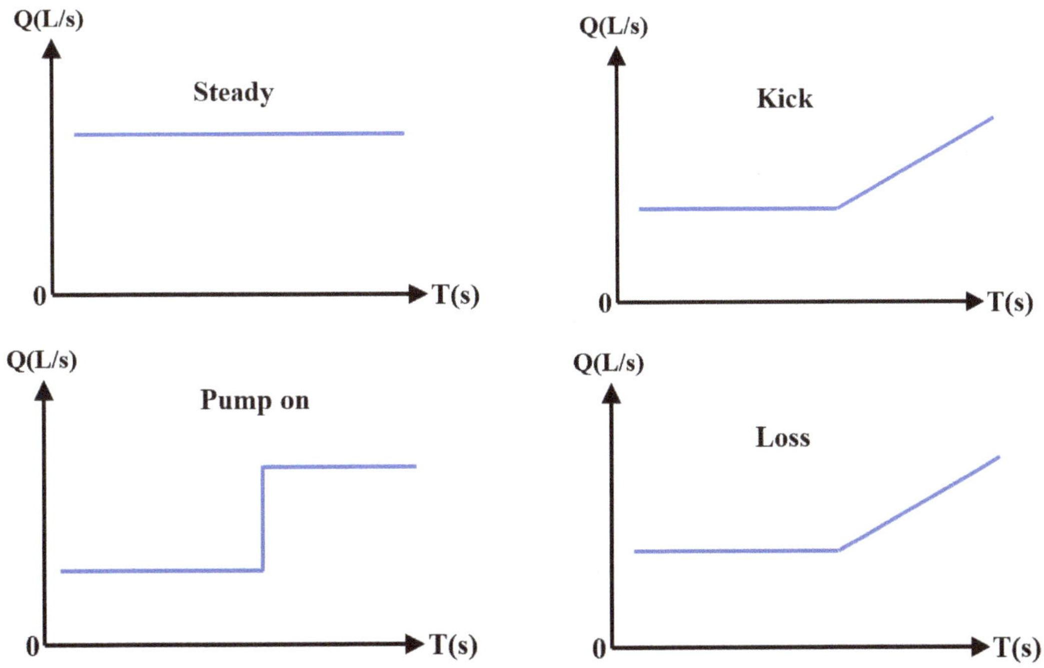

Figure 1. Pattern of flow rate differential in overflow and other events.

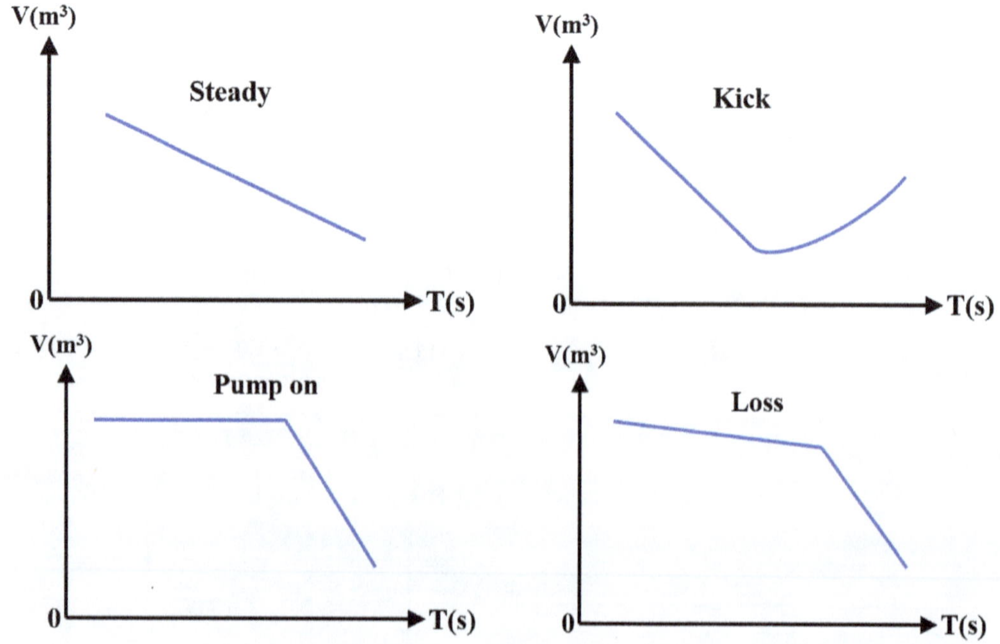

Figure 2. Pattern of incremental change of the mud pit during overflow and other events.

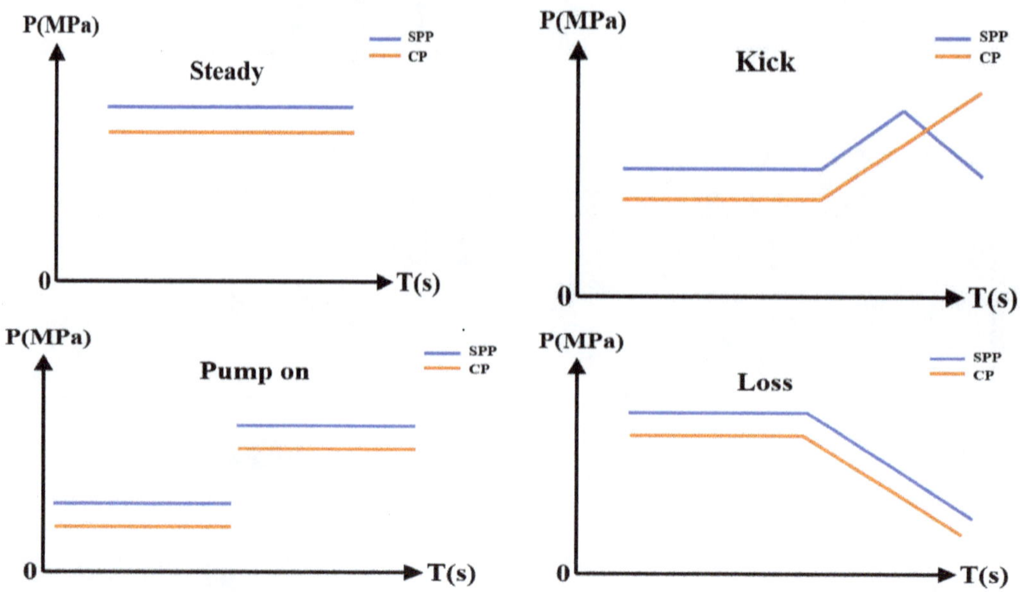

Figure 3. Pattern of CP and SPP changes in overflow and other events.

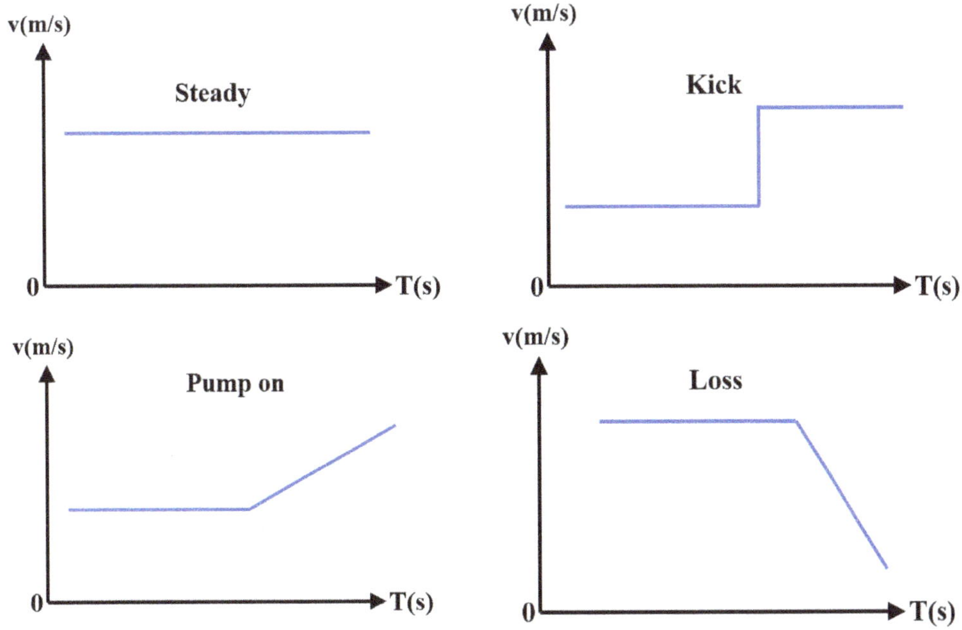

Figure 4. Pattern of ROP changes during overflow and other events.

3. The Principle of Data Processing and Filtering

The objective of data processing and filtration is to eliminate data noise without distorting the overall trend. The processing of overflow characteristic data begins with data normalization to reflect the data trend to the maximum extent, followed by Kalman filtering to determine the data noise.

Kalman filtering is a recursive estimation algorithm introduced by Kalman, which introduces the state space model into filtering theory and derives a set of recursive estimation algorithms to overcome the drawback that previous filtering theories cannot make unlimited use of past data and are not suitable for real-time processing. Kalman filtering seeks a set of recursive estimation algorithms based on the best criterion of minimum mean square error. Additionally, it has been documented in the field of overflow detection. The two main equations are the discrete state equation and the observation equation [25].

$$X(K) = F(k,k-1) \times X(K-1) + T(K,K-1) \times U(K-1) \quad (1)$$

$$Y(K) = H(k) \times X(K) + N(K) \quad (2)$$

where X(K) and Y(K) are the state vector and observation vector, respectively, at time k, F(k, k − 1) is the state transition matrix, U(K) is the dynamic noise at time k, T(K, K − 1) is the system control matrix, H(k) is the observation matrix at time k, and N(K) is the observation noise at time k.

For each surge characteristic parameter, the data processing and filtering process is as follows [26]:

Step 1: Data normalization Y(k)

$$Y(K) = (Y(k) - Y_{\min})/(Y_{\max} - Y_{\min}) \quad (3)$$

Step 2: Calculate the predicted covariance matrix

$$C(k)^{\wedge} = F(k,k-1) \times C(k) \times F(k,k-1)' + T(k,k-1) \times Q(k) \times T(k,k-1)' \quad (4)$$

$$Q(k)^{\wedge} = U(k) \times U(k)' \quad (5)$$

Step 3: Calculate the Kalman gain matrix

$$K(k) = C(k)^{\wedge} \times H(k)^{\wedge} \times \left[H(k) \times C(k)' \times H(k)' + R(k)\right]^{-1} \quad (6)$$

$$R(k) = N(k) \times N(k)^{1} \quad (7)$$

Step 4: Estimate update

$$X(k)^{\sim} = X(k)^{\wedge} + K(k) \times \left[Y(k) - H(k) \times K(k)^{\wedge}\right] \quad (8)$$

Step 5: Calculate the updated estimated covariance matrix

$$C(k)^{\sim} = [I - K(k) \times H(k)] \times C(k)^{\wedge} \times [I - K(k) \times H(k)]' + K(k) \times R(k) \times K(k)' \quad (9)$$

Step 6: Set parameters and repeat steps 2–6.

$$X(k+1) = X(k)^{\sim} \quad (10)$$

$$C(k+1) = C(k)^{\sim} \quad (11)$$

Using flow wave data as an example, the data curve before and after filtering is compared, as shown in Figures 5 and 6.

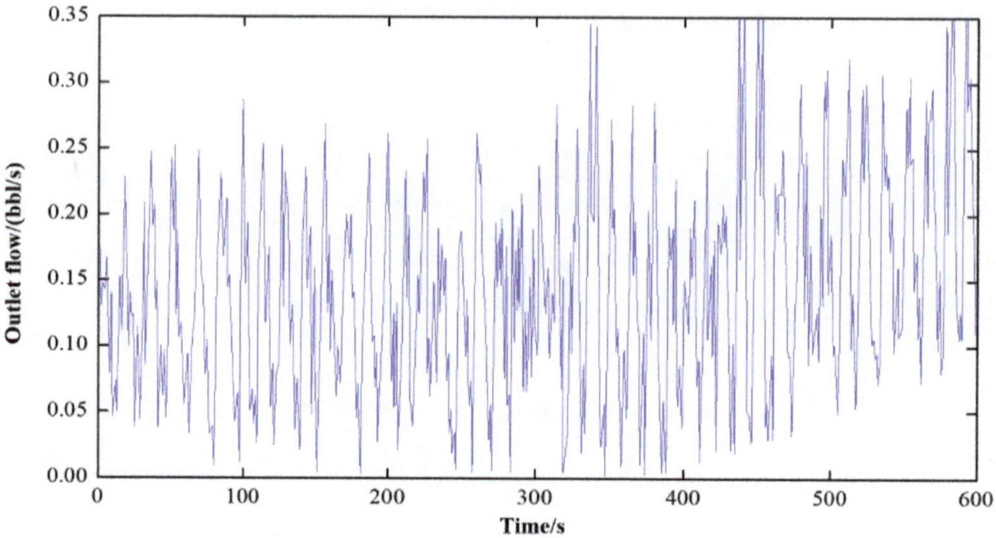

Figure 5. Flow wave data before data processing.

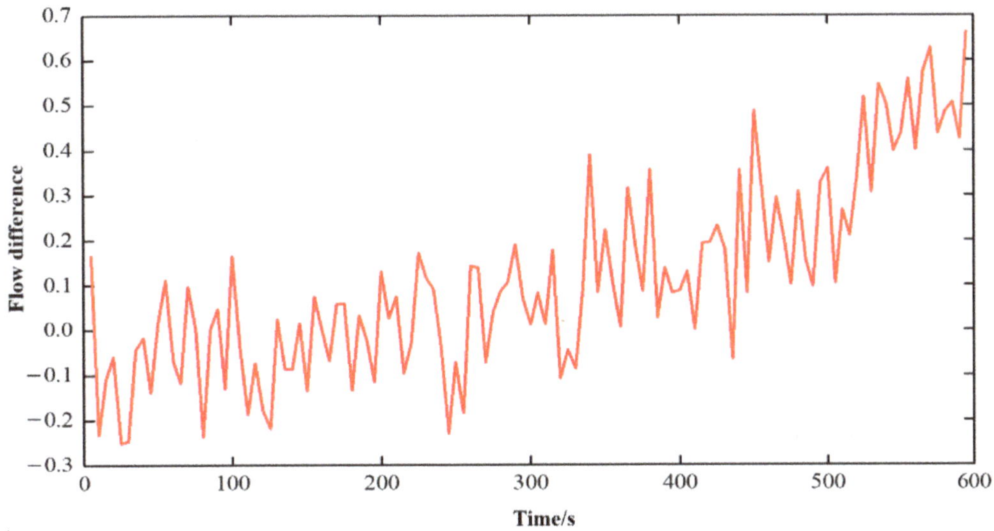

Figure 6. Flow wave data after data processing.

4. Pattern Recognition Representation Methods

Pattern recognition in the PRKD method consists of comparing the measured characteristic parameter vector to the basic overflow pattern and measuring the degree of similarity. In contrast to common function fitting, the challenge here is to avoid "entanglement" in the change in specific values and to select the changing trend and turning point as the characteristic parameters. Figure 7 depicts the measured flow variation of the drilling inlet and outlet. Despite the large difference between the two curves at different times A and B, the matching degree between the two curves and the overflow mode in the PRKD method should be 100%.

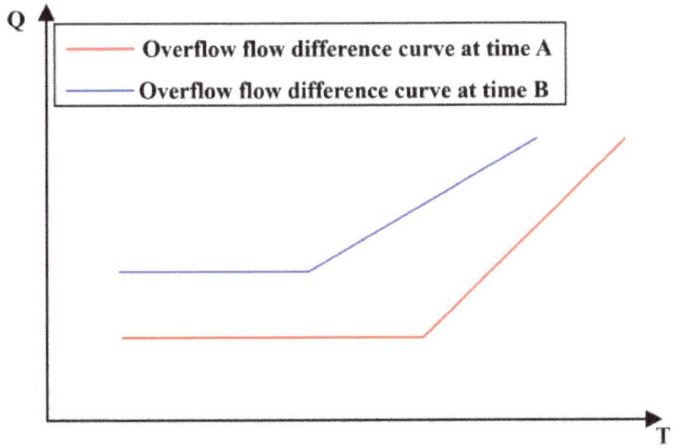

Figure 7. Overflow flow difference curve.

4.1. Pattern Classification

The difficulty in pattern classification lies in the extraction and matching of feature vectors. A feature vector is a group of observed or preliminary calculated characteristic values that serves as the basis for pattern classification. The basic overflow mode in the

PRKD method is primarily composed of a piecework polynomial function or mutation function, and the feature vector of any basic mode is as follows:

$$Vector = (K_{all}, m, x_1, k_1, x_2, k_2, \ldots, x_{m-1}, k_{m-1}, k_m) \quad (12)$$

where K_{all} is the overall change trend of the curve; m is the turning point of the curve; $x_i(i = 1, 2, \ldots, m-1)$ is the position of the segment point in the pattern; and $k_i(i = 1, 2, \ldots, m)$ is the trend change of the curve in paragraph i.

In the basic overflow mode, the "kick" event against the increment of the mud pool is taken, as an example, as m = 1, and the feature vector is:

$$Vector = (K_{all}, 1, x_1, k_1, k_2) \quad (13)$$

K_{all} decreases first and then increases, going from a negative value to a positive value; m represents the number of inflection points in the curve; x_1 approximates the time of occurrence of overflow, which is the position of the curve's inflection points; k_1 represents a linear change trend with a negative slope; and k_2 represents a quadratic polynomial function with a slope that transitions from negative to positive.

Another crucial advantage of the PRKD method is that feature vectors in pattern recognition can invert the kick information. Similarly, taking the "kick" event of the increment of the mud pool as an example, x_1 and k_2 can reflect the beginning of the overflow moment and the overflow velocity. The kick process and formation information can be retrieved by combining other characteristic changes.

4.2. Optimal Matching Algorithm

The PRKD method's pattern recognition bases its success on identifying the optimal solution. The fundamental concept is to arrange all possible combinations into a tree in a specific order and then search along the tree to avoid superfluous calculations, thereby ensuring that the algorithm is efficient, fast, and capable of real-time data processing. The process of building an algorithm involves the following parameters and operations:

(1) The root node is all features (level 0), one feature is discarded on each node, and each leaf node represents a variety of selection combinations. In the PRKD method, the first-level leaf node is the segment number, the second-level leaf node is the curve information of each segment, and so on, as shown in Figure 8.

(2) Record the maximum criterion function value of the currently searched leaf nodes and set the initial value to 0 in order to avoid the same combination of branches and leaves in the entire tree. In the PRKD method, the matching degree of each combination must be recorded.

(3) P is defined as the value of the real-time matching degree. At each level, the feature that is least likely to be discarded is placed on the leftmost side, and the search starts from the right side. In the PRKD method, from left to right, each leaf node is x_i, K, the change in K, and function value.

(4) If the left level of the abandoned feature is not below this node, search for the leaf node, update the value of $P_{max,}$ and then go back to the previous branch.

(5) If $P < P_{max}$ on the node, then do not search down, but instead retrace upwards. Each retrace will put back the abandoned feature (put it back on the list to be discarded). If the process has retraced to the top (root) and cannot search further down, then the leaf node of $P = P_{max}$ is the solution.

Based on steps A to E, the optimal search algorithm is formulated, and the optimal solution is searched based on various basic modes.

4.3. Measures of Pattern Matching Degree

The PRKD method must also address the problem of quantifying the degree of match between the optimal solution and the basic mode. In the similarity measurement based on time series, Euclidean distance is the most basic measurement method. The advantage of Euclidean distance is that it can represent the matching degree of a curve value based on wave amplitude; however, its ability to recognize sequence shape is poor, and it is easily disturbed by noise information.

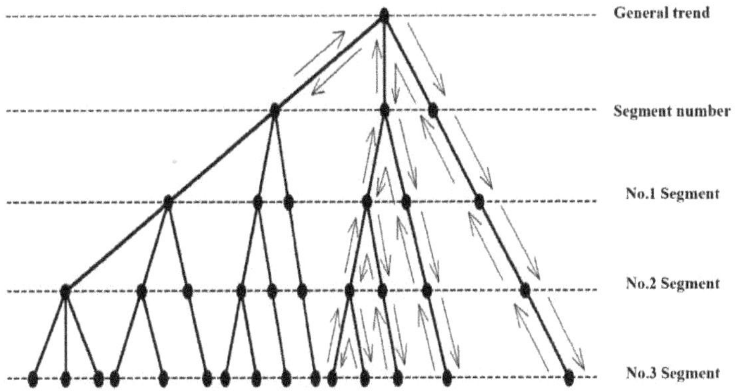

Figure 8. Schematic diagram of the pattern matching method.

Regarding morphological feature matching, several studies based on radian distance, slope distance, morphological features, similarity, and other aspects have been conducted. In this paper, the improved slope distance concept is proposed for measuring the matching degree of the curve shape. The matching degree of the final model should be based on the fluctuation amplitude, change trend, and time span. The matching degree measurement standard is the product of the Euclidean distance and the improved slope distance.

It is assumed that the time series S of wellbore detection characteristic parameters is:

$$S = \{(y_0, y_1, t_1), (y_1, y_2, t_2), \ldots, (y_{i-1}, y_i, t_i), \ldots, (y_{n-1}, y_n, t_n)\} \tag{14}$$

$$S = \{(k_1, t_1), (k_2, t_2), \ldots, (k_i, t_i), \ldots, (k_n, t_n)\} \tag{15}$$

where (y_{i-1}, y_i, t_i) is each segment; y_{i-1} is the starting point of the segment; y_i is the segmented endpoint; t_i is the initial moment of the segment; and k_i is the slope of the segment.

Thus, the Euclidean distance of each parameter value of the time series S and another series S′ is:

$$D_o(S, S') = sqrt\left[\sum_{i=1}^{n} (y_i - y_i')^2\right] \tag{16}$$

Obviously, Euclidean distance can reflect the amplitude of fluctuation between two sequences. Similarly, the slope distance between two sequences is:

$$D_k(S, S') = \left|\sum_{i=1}^{n} \Delta t_i (k_i - k_i') / t_n\right| \tag{17}$$

The slope distance D_k has effective anti-noise properties and can intuitively describe the trend of sequence changes. In fact, in addition to a trend change, the time span should also be considered when determining morphological similarity; therefore, this paper proposes an improved slope distance based on time weighting:

$$D_{KM}(S, S') = \left|\sum_{i=1}^{n} \Delta t_i W_i (k_i - k_i')\right| \tag{18}$$

$$W_i = \frac{(D_i - D_{\min}) \times a}{D_{\max} - D_{\min}} + (1 - a) \tag{19}$$

where $W_i \in [a, 1]$ is the time weighting of paragraph i, $W_i \in [a, 1]$, $a \in [0.1, 1]$ and $D_i = \max(|y_{i-1} - y_{i-1}'|, |y_i - y_i'|)$ is the fluctuation value.

Based on the fluctuation amplitude, change trend, and time span, this paper proposes that the distance between the monitoring data in the PRKD method and the optimal solution is:

$$C_r = D_O \times D_{KM} \tag{20}$$

Based on this, the similarity probability between the optimal solution of an event and the basic mode should be inversely proportional to the distance C_r, and the similar probability vector of the basic mode can be written as:

$$[P_1, P_2, \ldots, P_M] = \left[\frac{1}{c_r^1} / \sum_{i=1}^{M} \left(\frac{1}{c_r^i}\right), \frac{1}{c_r^2} / \sum_{i=1}^{M} \left(\frac{1}{c_r^i}\right), \ldots, \frac{1}{c_r^M} / \sum_{i=1}^{M} \left(\frac{1}{c_r^i}\right)\right] \tag{21}$$

where M represents the type of basic mode. For example, there are four basic events, namely, kick, state, pump on, and loss, aiming at the incremental change of the mud pool, namely, M = 4.

Based on the above pattern matching method, the possible probability of each optimal solution in Figure 9 can be obtained as follows:

$$[P_{kick}, P_{pump\ on}, P_{state}, P_{loss}] = [74.886, 17.456, 7.657, 0]/100 \tag{22}$$

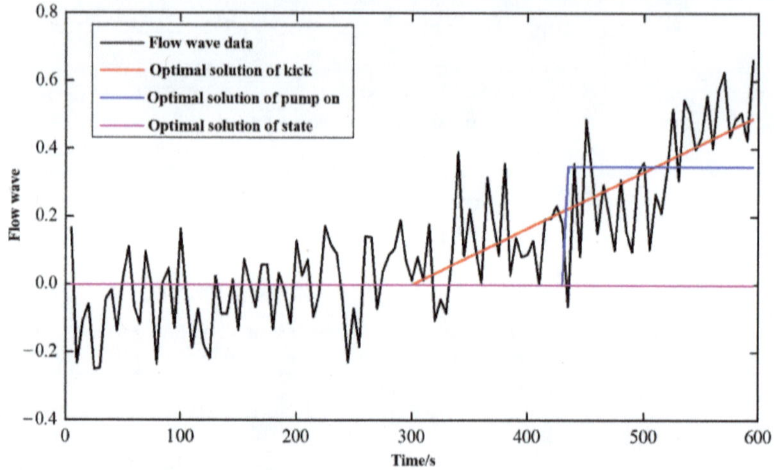

Figure 9. Matching result of the optimal solution of flow wave data.

There is a 71.886% chance of flooding, a 17.456% chance of pumping, a 7.657% chance of drilling properly, and a zero chance of loss.

5. Bayesian Framework

In this paper, the Bayesian framework is adopted to coordinate the prior information and likelihood information and realize the output form of the probability. The expression for the Bayes formula is:

$$P(A_i|B) = \frac{P(B|A_i)P(A_i)}{\sum_{i=1}^{n} P(B|A_i)P(A_i)}, i = 1, 2, \ldots, n \tag{23}$$

where A_i are different kinds of detection parameters; B is the overflow event; $P(A_i)$ is the prior probability; and $P(B|A_i)$ is the likelihood function.

(1) Prior information: Kick prior probability is the kick probability derived from the uncertain profile of formation pressure. Prior probabilities for other events, such as state events and pump on events, can be obtained from statistics. If the prior information of each parameter is missing, its influence is neglected.
(2) Likelihood function: the similarity measure between kick characteristic data and the basic model of overflow, which is obtained by multiplying the PRKD model results with the weight vector of overflow characteristic parameters.
(3) Posteriori probability: posteriori probability is proportional to the product of prior probability and the likelihood function, and can be obtained by normalization on this basis.

6. Case Analysis

The monitoring data that were obtained in a well kick event, including the inlet/outlet flow differential, mud pool increment, Stand Pipe Pressure (SPP), casing pressure (CP), and Rate of Penetration (ROP) data, were applied, as shown in Figure 10. The PRKD model was used to diagnose the overflow without taking the prior information of each parameter into account, and the rule of the calculated results was analyzed.

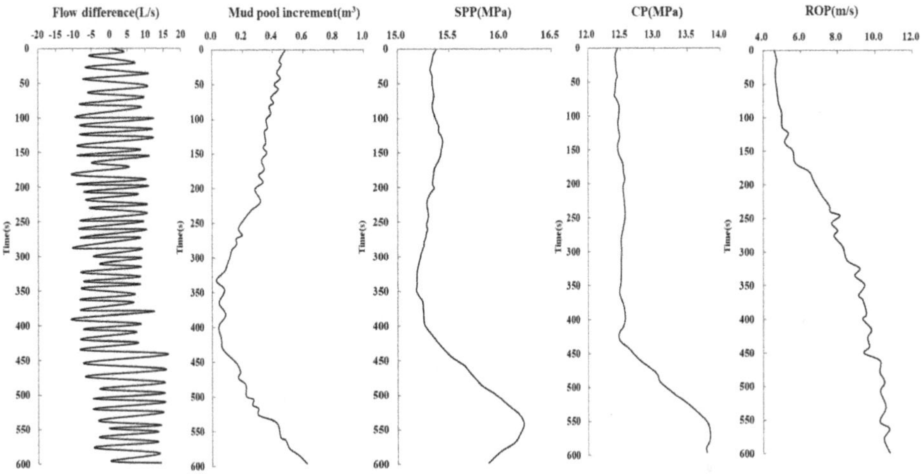

Figure 10. Change in kick characteristic parameters before and after 0–600 s overflow.

The range of real-time data automatically processed by the PRKD model can be set manually, but too much data will affect the computing speed, and too little data will not accurately reflect the overflow process. In consideration of the duration of the general overflow process, the calculation process time range was set to 10 min; thus, the PRKD model automatically collected data from the previous 10 min for real-time diagnosis.

6.1. Basic Pattern Matching Results

Figures 11–14 show the results of data mode matching for outlet flow difference, mud pool increment, SPP, and ROP at different times. As depicted in the figures, the PRKD model can automatically match the changing trend of each detection parameter based on the evolution process of overflow, and can automatically identify overflow and other working conditions with a high degree of accuracy. Using the optimal matching result of mud pool increment as an example, the matching result is the normal drilling mode at 200 s and 400 s. At 600 s, the matching result is overflow mode.

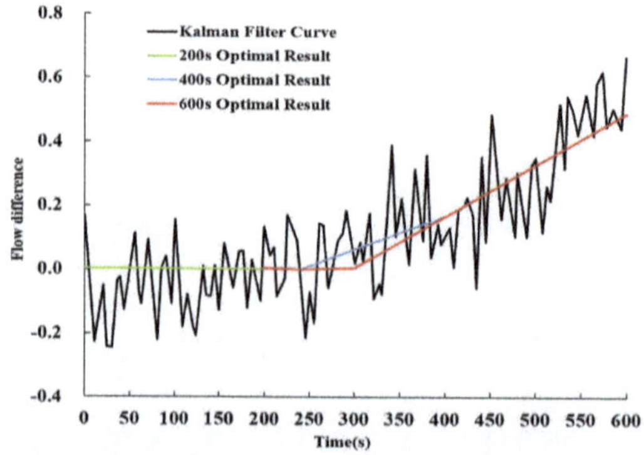

Figure 11. Flow difference optimal matching result.

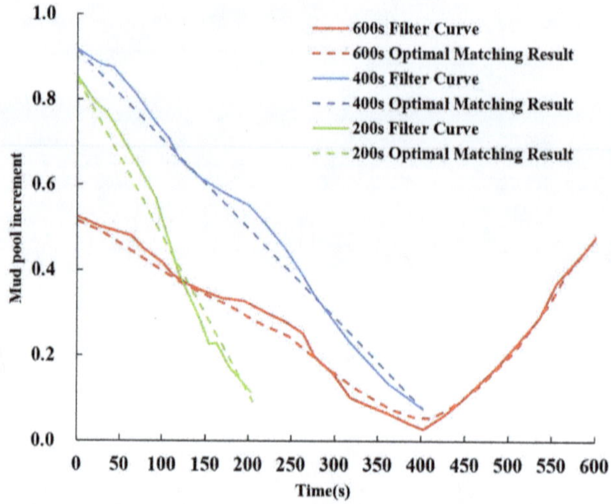

Figure 12. Optimal matching results of mud pool increment.

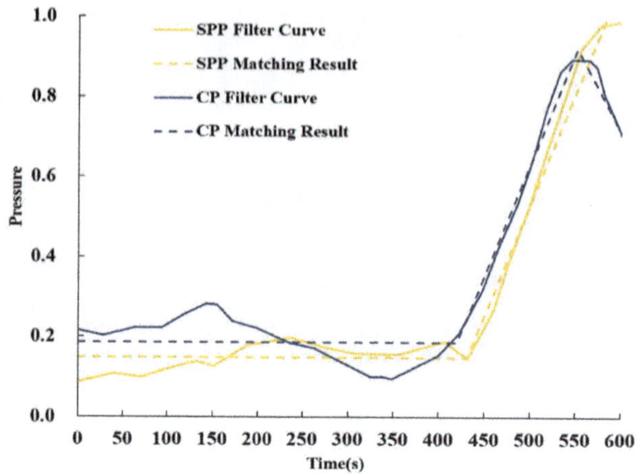

Figure 13. Optimal matching results of SPP and CP.

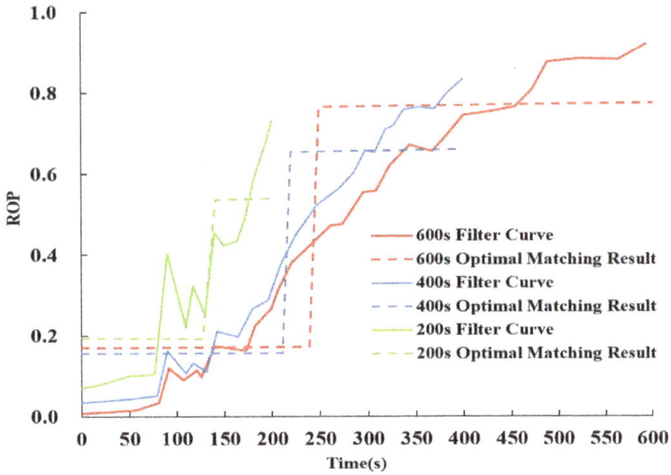

Figure 14. Optimal matching results of ROP.

Different overflow monitoring parameters have different sensitivities to the overflow evolution process, so the pattern recognition results at different times may vary. Taking the flow difference and mud pool increment data in this example as an example, at 400 s, the matching result of flow difference data is overflow, while the matching result of mud pool increment is no overflow. Obviously, the former diagnosis is timelier, which is consistent with standard drilling practices.

6.2. Overflow Probability Analysis

Figure 15 illustrates the analysis curve for overflow probability based on kick characteristic parameters. As shown in the figure, the probability of overflow diagnosis for each parameter increases progressively with increasing time. In this case, the overflow "trend" is diagnosed by ROP, flow differential, pressure, and mud pool increment, in descending order.

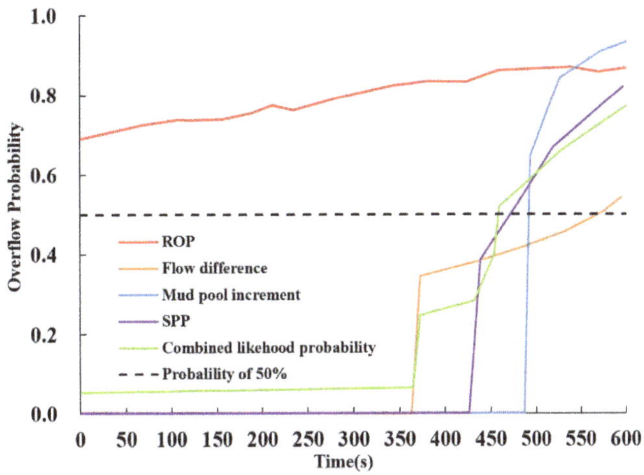

Figure 15. Overflow probability analysis curve based on kick characteristic parameters.

Although the PRKD method can reduce the impact of data noise to the maximum extent possible, the probability result of a single type of detection data will still be disturbed by noise, making it challenging to propose a "threshold probability" applicable to all conditions for a single type of data. In general, the larger the level of data noise, the smaller the probability (lower threshold probability) of successfully judging overflow. The comprehensive probability of various available types of data can reduce the impact of data noise from a single type of detection data. Therefore, "comprehensive diagnosis probability of more than 50%" is adopted as the overflow judgment standard in this paper. This means that by integrating all available data, if the probability of an overflow is greater than the probability of all other modes, it is judged to be an overflow. If only a single type of data is available, 50% is also recommended as the "overflow threshold probability", but should be appropriately modified according to data noise. In the figure, the time when the combined probability is equal to 50%—the time when overflow is detected—is approximately 465 s.

6.3. Comparison with Traditional Methods

(1) Comparison of result accuracy

Inlet and outlet flow difference data is the most used type of overflow monitoring data. Figure 16 shows a comparison between the diagnostic results of the traditional threshold method for traffic difference data and the PRKD model. As shown in the figure, the traditional threshold method has low diagnostic accuracy and a high false positive rate when applied to the example data, whereas the PRKD model yields superior diagnostic results. When the threshold value is 10 L/s and 15 L/s, the false positive rate of the diagnosis results of the threshold method is very high in 0–400 s. When the threshold value is 20 L/s, the diagnostic results of the threshold method are discontinuous and have a low degree of accuracy. The figure depicts overflow at 460 s and non-overflow at 470 s. Compared with the threshold method, the PRKD model found overflow at 465 s, and in the form of probabilities, which can provide additional reference information for engineers.

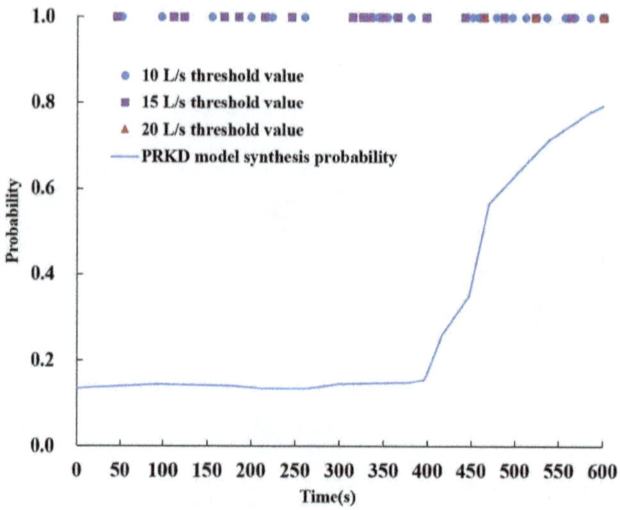

Figure 16. Comparison between the PRKD model and the traditional threshold method for overflow diagnosis.

(2) The influence of data noise

Figure 17 illustrates the results of applying noise with amplitudes of 1, 2, 4, 8, and 16 L/s to the standard overflow mode data in order to investigate the impact of different levels of data noise on the PRKD model. The probability curve of kick diagnosis with the

PRKD method under different noise levels is shown in Figure 18. According to the standard overflow mode data, the overflow occurred at a time of 300 s. With the increase in noise intensity, the time of overflow diagnosis was gradually extended. In this case, the PRKD model with a noise amplitude of less than 8 L/s has superior detection performance. When the noise amplitude is 16 L/s, the PRKD model detects the continuous overflow time approximately 200 s after the real overflow, which satisfies engineering practice requirements.

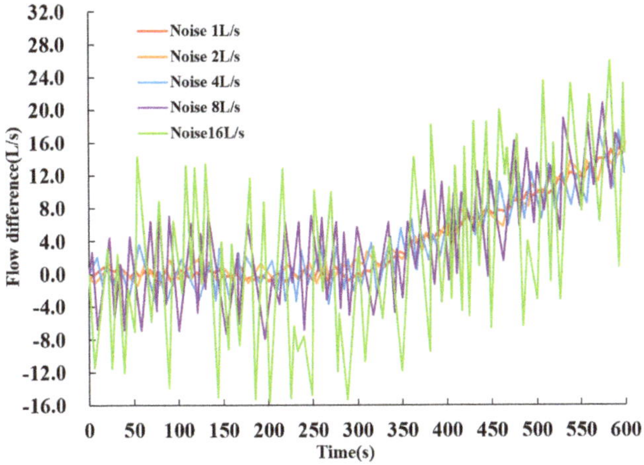

Figure 17. Flow difference curves under different levels of noise.

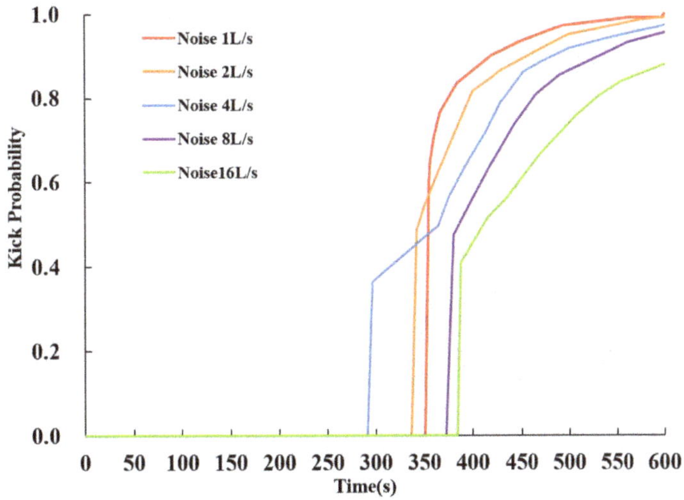

Figure 18. Kick probability distribution curves of the PRKD method under different noise levels.

6.4. Bayesian Probability Analysis

The case data are as follows: The calculated pre-drilling kick probability is 84.56% when the drilling depth is 3500 m. The relative prior probabilities of state, pump on, and loss events are assumed to be 91%, 7%, and 2%, respectively, based on field experience. The available data are flow differential data, mud pool increment data, pressure data, and ROP data.

(1) Prior probability: basic events can be divided into overflow events and non-overflow events, which can be obtained according to pre-drilling kick probability calculation and prior data:

$$\left[P_{kick}, P_{state}, P_{pump\ \mathrm{on}}, P_{loss}\right]_{prior} = [84.560\%, 14.046\%, 1.175\%, 0.219\%] \tag{24}$$

(2) Likelihood function: the PRKD method is used to obtain the likelihood probability of each overflow characteristic parameter:

$$\left[P_{kick}, P_{state}, P_{pump\ \mathrm{on}}, P_{loss}\right]^{deltflow}_{likehood} = [62.506\%, 25.414\%, 12.080\%, 0.000\%] \tag{25}$$

$$\left[P_{kick}, P_{state}, P_{pump\ \mathrm{on}}, P_{loss}\right]^{pit\ \mathrm{gain}}_{likehood} = [97.8\%, 2.200\%, 0.000\%, 0.000\%] \tag{26}$$

$$\left[P_{kick}, P_{state}, P_{pump\ \mathrm{on}}, P_{loss}\right]^{SPP}_{likehood} = [78.900\%, 17.750\%, 3.350\%, 0.000\%] \tag{27}$$

$$\left[P_{kick}, P_{state}, P_{pump\ \mathrm{on}}, P_{loss}\right]^{ADP}_{likehood} = [74.300\%, 21.290\%, 4.410\%, 0.000\%] \tag{28}$$

$$\left[P_{kick}, P_{state}, P_{pump\ \mathrm{on}}, P_{loss}\right]^{ROP}_{likehood} = [82.400\%, 4.900\%, 12.700\%, 0.000\%] \tag{29}$$

The normalized weight vector of flow difference data, mud pool increment data, pressure data, and ROP data is obtained by using the method of middle hierarchical analysis:

$$w = [0.5104, 0.2574, 0.1276, 0.0645, 0.0401] \tag{30}$$

By synthesizing all overflow parameters, the overflow likelihood function vector is:

$$\begin{aligned} \left[P_{kick}, P_{state}, P_{pump\ \mathrm{on}}, P_{loss}\right]_{likehood} &= \sum_{i=1}^{5} w_i \left[P_{kick}, P_{state}, P_{pump\ \mathrm{on}}, P_{loss}\right]^{i}_{likehood} \\ &= [0.7538, 0.1715, 0.0747, 0.0000] \end{aligned} \tag{31}$$

(3) Posterior distribution:

$$\begin{aligned} \left[P_{kick}, P_{state}, P_{pump\ \mathrm{on}}, P_{loss}\right]_{posterior} &= \left[P_{kick}, P_{state}, P_{pump\ \mathrm{on}}, P_{loss}\right]_{prior} \times \left[P_{kick}, P_{state}, P_{pump\ \mathrm{on}}, P_{loss}\right]_{likehood} \\ &= [0.9580, 0.0408, 0.0012, 0.0000] \end{aligned} \tag{32}$$

Therefore, the overflow probability at this location is 95.8% when the pre-drilling kick probability information and the PRKD prediction results of each characteristic parameter are combined.

7. Conclusions

(1) A single-parameter gas intrusion monitoring method for offshore drilling called PRKD based on pattern recognition is established by combining multiphase flow calculations, data filtering theory, pattern recognition theory, and the Bayesian framework. By integrating sophisticated computational techniques with pattern recognition algorithms, PRKD enhances the reliability and precision of kick detection. This enables the implementation of proactive measures to mitigate potential risks, protecting the environment and human lives while optimizing drilling operations.

(2) Although the PRKD method can minimize the impact of data noise to the maximum extent possible, the probability result obtained from a single type of detection data

will still be disturbed by noise. Combining the comprehensive probability of various available types of data and adopting "comprehensive diagnosis probability over 50%" as the overflow judgment standard, can meet the requirements of engineering practice.

(3) The traditional threshold method has low accuracy and a high false positive rate in diagnosing kicks, while the PRKD model shows better diagnostic results in the chosen case study. When the threshold value is set at 10 L/s, the threshold method has a high false positive rate in the range of 0 to 400 s. When the threshold value is set at 10 and 15 L/s, the threshold method lacks continuity and has low accuracy. For example, at 460 s in the graph, it diagnoses an overflow, while at 470 s it diagnoses a non-overflow. Compared to the threshold method, the PRKD model detects the overflow at 465 s and provides the output in the form of probabilities, which can provide engineers with more reference information.

(4) In the case analysis of standard overflow pattern data, the occurrence of overflow is at 300 s. As the intensity of noise increases, the time at which overflow is diagnosed gradually extends. In this case, the PRKD model performs well when the noise amplitude is below 8 L/s. Under severe conditions, with a noise amplitude of 16 L/s, the PRKD model detects continuous overflow approximately 200 s after the actual overflow occurs (at around 500 s), which meets the requirements of engineering practice.

(5) Based on the case analysis, the PRKD method combines the probabilistic information of pre-drilling kicks and various characteristic parameters to predict a 95.8% probability of overflow occurrence at the specified location, which satisfies the requirements in the field. The gas invasion monitoring method proposed in this study delivers accurate diagnostic results with a low false positive rate, thereby providing valuable guidance for gas invasion monitoring in drilling operations.

(6) Through case studies, it has been observed that the proposed kick monitoring method in this paper can accurately and rapidly detect the occurrence of well kicks. Additionally, this method exhibits good noise resistance and outperforms traditional threshold-based methods in terms of lower false positive rates and higher monitoring accuracy. Furthermore, this method is not limited to kick monitoring alone but can also be applied for real-time monitoring of other drilling conditions such as wellbore leakage. However, due to the difficulty in obtaining a large amount of field data, the PRKD kick monitoring model established in this study is still not perfect. Future work will focus on further improving the PRKD kick monitoring model.

Author Contributions: Writing—original draft preparation, Y.X., J.Y., Z.H. and D.X.; Writing—review and editing, Y.X., J.Y., Z.H., D.X., L.L. and C.F.; Supervision, Y.X., J.Y. and Z.H. All authors have read and agreed to the published version of the manuscript.

Funding: Research and Development Program of China (No. 2022YFC2806401).

Data Availability Statement: The authors confirm that the data supporting the findings of this study are available within the article.

Acknowledgments: The authors would like to acknowledge the support of the National Key Research and Development Program of China (Grant No. 2022YFC2806401).

Conflicts of Interest: The authors declare that they have no conflict of interest.

References

1. Liang, H.; Zou, J.; Liang, W. An early intelligent diagnosis model for drilling overflow based on GA–BP algorithm. *Clust. Comput.* **2019**, *22*, 10649–10668. [CrossRef]
2. Yin, Q.; Yang, J.; Tyagi, M.; Zhou, X.; Hou, X.; Wang, N.; Tong, G.; Cao, B. Machine learning for deepwater drilling: Gas-kick-alarm Classification using pi-lot-scale rig data with combined surface-riser-downhole monitoring. *SPE J.* **2021**, *26*, 1773–1799. [CrossRef]
3. Jin, Y.; Chao, F.; Shujie, L.; Weiguo, Z.; Renjun, X.; Yi, W. Key technological innovation and practice of well construction in ultra-deepwater shallow formations. *Acta Pet. Sin.* **2022**, *43*, 1500–1508.
4. Zhang, Z.; Sun, B.; Wang, Z.; Pan, S.; Lou, W.; Sun, D. Intelligent well killing control method driven by coupling multiphase flow simulation and real-time data. *J. Pet. Sci. Eng.* **2022**, *213*, 110337. [CrossRef]

5. Yang, J.; Wu, S.; Tong, G.; Wang, H.; Guo, Y.; Zhang, W.; Zhao, S.; Song, Y.; Yin, Q. Acoustic prediction and risk evaluation of shallow gas in deep-water areas. *J. Ocean. Univ. China* **2022**, *21*, 1147–1153. [CrossRef]
6. Jiang, H.; Liu, G.; Li, J.; Zhang, T.; Wang, C.; Ren, K. Numerical simulation of a new early gas kick detection method using UKF estimation and GLRT. *J. Pet. Sci. Eng.* **2019**, *173*, 415–425. [CrossRef]
7. Yin, Q.; Yang, J.; Borujeni, A.T.; Shi, S.; Sun, T.; Yang, Y.; Geng, Y.; Xia, Q.; Wu, X.; Zhao, X. Intelligent Early Kick Detection in Ultra-Deepwater High-Temperature High-Pressure (HPHT) Wells Based on Big Data Technology. In Proceedings of the 29th International Ocean and Polar Engineering Conference, Honolulu, HI, USA, 16–21 June 2019; OnePetro: Richardson, TX, USA, 2019.
8. Hu, Z.; Yang, J.; Li, W.; Li, S.; Feng, P.; Xin, Y. Research and development of compressible foam for pressure management in casing annulus of deepwater wells. *J. Pet. Sci. Eng.* **2018**, *166*, 546–560. [CrossRef]
9. Maus, L.D.; Tannich, J.D.; Ilfrey, W.T. Instrumentation requirements for kick detection in deep water. *J. Pet. Technol.* **1979**, *31*, 1029–1034. [CrossRef]
10. Speers, J.M.; Gehrig, G.F. Delta flow: An accurate, reliable system for detecting kicks and loss of circulation during drilling. *Spe Drill. Eng.* **1987**, *2*, 359–363. [CrossRef]
11. Anfinsen, B.T.; Rolv, R.; Rogaland, R. Sensitivity of earlykick detection parameters in fullscale gas kick experiments with oil- and water-based drilling muds. In Proceedings of the IADC/SPE Drilling Conference, New Orleans, LA, USA, 18–21 February 1992. SPE23934.
12. Fraser, D.; Lindley, R.; Moore, D.; Vander Staak, M. Early kick detection methods and technologies. In Proceedings of the SPE Annual Technical Conference and Exhibition, Amsterdam, The Netherlands, 27–29 October 2014. SPE170756.
13. Nayeem, A.A.; Venkatesan, R.; Khan, F. Monitoring of down-hole parameters for early kick detection. *J. Loss Prev. Process Ind.* **2016**, *40*, 43–54. [CrossRef]
14. Yin, Q.; Yang, J.; Tyagi, M.; Zhou, X.; Wang, N.; Tong, G.; Xie, R.; Liu, H.; Cao, B. Downhole quantitative evaluation of gas kick during deepwater drilling with deep learning using pi-lot-scale rig data. *J. Pet. Sci. Eng.* **2022**, *208*, 109136. [CrossRef]
15. Yang, H.; Li, J.; Liu, G.; Wang, C.; Jiang, H.; Luo, K.; Wang, B. A new method for early gas kick detection based on the consistencies and differences of bottomhole pressures at two measured points. *J. Pet. Sci. Eng.* **2019**, *176*, 1095–1105. [CrossRef]
16. Hu, Z.; Yang, J.; Wang, L.; Hou, X.; Zhang, Z.; Jiang, M. Intelligent identification and time-efficiency analysis of drilling operation conditions. *Oil Drill. Prod. Technol.* **2022**, *44*, 241–246.
17. Liao, M.Y. Drilling state monitoring and fault diagnosis based on multi-parameter fusion by neural network. *J. China Univ. Pet.* **2007**, *31*, 149–152.
18. Nybo, R.; Bjorkevoll, K.S.; Rommetveit, R. Spotting a false alarm. integrating experience and real-time analysis with artificial intelligence. In Proceedings of the SPE Intelligent Energy International Conference and Exhibition, Amsterdam, The Netherlands, 25–27 February 2008. SPE112212.
19. David, H.; Stuart, J.; Ben, J. Early kick detection for deepwater drilling: New probabilistic methods applied in the field. In Proceedings of the SPE Annual Technical Conference and Exhibition, New Orleans, LA, USA, 30 September–3 October 2001. SPE71369.
20. Pournazari, P.; Ashok, P.; van Oort, E.; Unrau, S.; Lai, S. Enhanced kick detection with low-cost rig sensors through automated pattern recognition and real-time sensor calibration. In Proceedings of the SPE Middle East Intelligent Oil and Gas Conference and Exhibition, Abu Dhabi, United Arab Emirates, 15–16 September 2015. SPE176790.
21. Al-Obaidi, A.R.; Mohammed, A.A. Numerical Investigations of Transient Flow Characteristic in Axial Flow Pump and Pressure Fluctuation Analysis Based on the CFD Technique. *J. Eng. Sci. Technol. Rev.* **2019**, *12*, 70–79. [CrossRef]
22. Muojeke, S.; Venkatesan, R.; Khan, F. Supervised data-driven approach to early kick detection during drilling operation. *J. Pet. Sci. Eng.* **2020**, *192*, 107324. [CrossRef]
23. Hargreaves, D.; Jardine, S.; Jeffryes, B. Early kick detection for deepwater drilling: New probabilistic methods applied in the field. In Proceedings of the SPE Annual Technical Conference and Exhibition, Denver, CO, USA, 30 October–2 November 2011; OnePetro: Richardson, TX, USA, 2001.
24. Reitsma, D. A simplified and highly effective method to identify influx and losses during Managed Pressure Drilling without the use of a Coriolis flow meter. In Proceedings of the SPE/IADC Managed Pressure Drilling and Underbalanced Operations Conference and Exhibition, Kuala Lumpur, Malaysia, 24–25 February 2010; OnePetro: Richardson, TX, USA, 2010.
25. Geekiyanage, S.C.; Ambrus, A.; Sui, D. Feature selection for kick detection with machine learning using laboratory data. In Proceedings of the International Conference on Offshore Mechanics and Arctic Engineering, Glasgow, Scotland, UK, 9–14 June 2019; American Society of Mechanical Engineers: New York, NY, USA, 2019; Volume 58875.
26. Bang, J.; Mjaaland, S.; Solstad, A.; Hendriks, P.; Jensen, L.K. Acoustic gas kick detection with wellhead sonar. In Proceedings of the SPE Annual Technical Conference and Exhibition, New Orleans, LA, USA, 25–28 September 1994; OnePetro: Richardson, TX, USA, 1994.

processes

MDPI

Article

Prediction of Leakage Pressure during a Drilling Process Based on SSA-LSTM

Dong Chen [1], Baolun He [2,*], Yanshu Wang [1], Chao Han [3], Yucong Wang [2] and Yuqiang Xu [2,*]

1 Sinopec Matrix Corporation, Qingdao 266071, China
2 National Key Laboratory of Deep Oil and Gas, China University of Petroleum (East China), Qingdao 266580, China
3 Geosteering & Logging Research Institute, Sinopec Matrix Corporation, Qingdao 266071, China; hansuper713@hotmail.com
* Correspondence: s21020020@s.upc.edu.cn (B.H.); xuyuqiang@upc.edu.cn (Y.X.)

Abstract: Drilling-fluid loss has always been one of the challenging issues in the field of drilling engineering. This article addresses the limitations of a single fluid-loss pressure mechanism model and the challenges in predicting positive drilling-fluid-loss pressure. By categorizing fluid losses of various types encountered during drilling, different geological formations associated with distinct mechanisms are considered. The actual drilling-fluid density in the wellbore at the time of fluid-loss occurrence is taken as a reference value for calculating the positive drilling-fluid-loss pressure of the already drilled well. Building upon this foundation, a combined model utilizing the Sparrow Search Algorithm (SSA) and Long Short-Term Memory (LSTM) neural network is constructed. This model effectively explores the intricate nonlinear relationship between well logging, logging engineering data, and fluid-loss pressure. By utilizing both data from the already drilled wells and upper formation data from ongoing drilling, precise prediction of positive drilling formation fluid-loss pressure can be achieved. Case studies demonstrate that the approach established in this paper, incorporating upper formation data, reduces the average absolute percentage error of fluid-loss pressure prediction to 2.4% and decreases the root mean square error to 0.0405. Through the synergy of mechanistic models and data-driven techniques, not only has the accuracy of predicting positive drilling formation fluid-loss pressure has been enhanced, but also valuable insights have been provided for preventing and mitigating fluid losses during drilling operations.

Keywords: mechanism model; leakage pressure; SSA-LSTM; during the drilling process

Citation: Chen, D.; He, B.; Wang, Y.; Han, C.; Wang, Y.; Xu, Y. Prediction of Leakage Pressure during a Drilling Process Based on SSA-LSTM. *Processes* **2023**, *11*, 2608. https://doi.org/10.3390/pr11092608

Academic Editor: Qingbang Meng

Received: 28 July 2023
Revised: 23 August 2023
Accepted: 29 August 2023
Published: 1 September 2023

1. Introduction

Deep and ultradeep oil and gas resources represent the primary frontier for future energy supply. However, due to the intricate and ever-changing geological conditions in these depths, coupled with inadequate predrilling knowledge of subsurface formations, drilling operations frequently encounter challenges such as leaks, surges, collapses, and sticking, among other downhole complexities. Among these issues, wellbore leakage has emerged as one of the most prevalent complexities in recent years. This not only escalates drilling costs and diminishes drilling efficiency but also poses a significant risk of wellbore collapse, surges, and even blowouts, leading to major safety incidents [1–3]. At the drilling site, the occurrence of fluid loss is typically determined by monitoring changes in drilling-fluid volume and the flow rate at the wellhead. Wang [4] discussed the leakage judgment under different working conditions and found that the leakage law of drilling fluid also varies under different working conditions. Sun et al. [1] pointed out that the leakage channel, leakage pressure, and leakage rate are the three main characteristics of oil-well leakage. Accurately determining the nature of the leakage will strongly support the scientific screening of leak prevention and blockage technologies. However, due to the complexity of underground rock formations and the complexity and variability of pressure systems, accurately predicting the leakage pressure of complex and volatile formations remains a challenge, and there is currently no complete set of leakage pressure prediction methods [5,6].

In previous studies, the rupture pressure as a safe drilling-pre-fluid density window can no longer meet engineering requirements, and it is necessary to use accurate leakage pressure as its upper limit. In the traditional leakage pressure mechanism model, Zhang et al. [7] established a mechanical model for the intersection of cracks and the wellbore based on the stress state of the wellbore-surrounding rock. However, the solution of this model is complex, and its applicability is poor. Subsequently, Fang [8] proposed a method for calculating formation leakage pressure based on logging data, and Zhai et al. [9] conducted research on the prediction and control model of shale-induced fracture leakage pressure. They proposed a dynamic model for leakage pressure based on leakage time, rate, and other factors, and demonstrated the accuracy of the model. However, the occurrence of drilling-fluid leakage is not only related to the formation leakage pressure and wellbore pressure but also to construction parameters and human factors. Therefore, traditional methods based on wellbore pressure balance have certain limitations.

With the continuous indepth integration of machine learning, big data, and other technologies in the field of petroleum engineering, the data-driven method for identifying and early warning of lost circulation risk shows obvious advantages over traditional model methods. Mohammad Sabah et al. [10] conducted research on intelligent prediction models such as decision trees (DT), adaptive neural fuzzy inference systems (ANFIS), artificial neural networks (ANN), and genetic algorithms multilayer perception hybrid artificial neural networks (GA-MLP), and confirmed that machine learning has certain advantages in predicting leakage. Pang et al. [11] selected 16 comprehensive logging parameters with the strongest correlation with drilling loss rate for model training and established a complex relationship between logging parameters and mud loss rate through a subgaussian mixed density network, confirming that the model can evaluate drilling loss risk in real-time. Matinkia et al. [12] validated multiple models using logging data, and the results showed that convolutional neural networks (CNN) models have significant advantages in feature extraction, especially for data with high volatility such as logging data. Song et al. [13] conducted research on the LSTM and back propagation (BP) combined model, and took the formation pore pressure of two wells (the whole well section is considered as the true value) as the training set and one as the verification, proving the feasibility and accuracy of the model. In summary, there are two major difficulties in constructing a leakage pressure profile. First, the model has numerous parameters that are difficult to determine, resulting in poor generalization ability; The second is that the model is simple but lacks accuracy. However, it is obvious that a single model can no longer meet the needs of safe drilling in the project. This paper builds a leakage pressure profile for the whole well section by evaluating the actual leakage of drilled wells and using different mechanism models for different types of leakage. By constructing a drilled leakage pressure profile using mechanism methods, it serves as a machine-learning data sample for predicting leakage pressure while drilling in the drilling formation. The selected LSTM model can perform feature learning well. Considering the impact of model hyperparameters on prediction results, this paper uses the SSA algorithm to optimize hyperparameters, striving to achieve higher accuracy of the model during the drilling process.

2. Methodology

2.1. Applicability Analysis

This study is based on ultradeep wells on land and aims to calculate the leakage pressure of various fractures, karst caves, and high permeability formations using various mechanism models. The actual data of leakage is included in the mechanism model validation, providing more accurate leakage pressure data for the drilled formations. The SSA algorithm optimizes the parameter search by simulating sparrows' foraging behavior and is suitable for various optimization problems. The LSTM model performs well in time-series data analysis; therefore, it also has potential value in other tasks that require time correlation modeling, such as weather forecasting, financial data analysis, and other fields. In addition, this method provides insights for research in other geological engineering fields. This cross-domain migration potential will help expand the applicability of our method. The SSA-LSTM model used provides a good solution for predicting the leakage pressure in forward drilling.

2.2. Background Introduction to Block L

The high and steep structures in this block belong to the mountain front and mountain area [14], with adverse surface terrain conditions and underground structures coexisting. The strong folding and orogeny not only make the terrain height difference change greatly but also make the high and steep structure appear from the surface, making the stratum dip angle very large. Due to long-term erosion of surface water and weathering, fractures and karst caves are developed, and lost circulation is very serious. According to the statistical data of the leakage layer rock core in high and steep structures, the opening size, distribution shape, and filling material of the cracks vary greatly, with significant differences [15], but generally speaking, there are the following rules: from Shaximiao to Xujiahe, the leakage channels are mainly porous leakage channels. When drilling into poorly cemented sandstone and mudstone formations, permeability leakage often occurs; on the main part of the structure, due to long-term surface-water erosion and weathering, the leakage channels are mainly karst caves and large fractures, the width of the fractures is generally more than 10 mm, and most of them are nonfilling inclined fractures with an inclination of more than 30° and extend far to the depth of the stratum. Some of them are connected to the surface. When encountering such lost circulation channels, very serious well losses occur, and handling them is also quite difficult [16]. In addition, the geological environment of the block also has old strata exposed to the surface, weathered mountain gravel, and mostly broken surface of the old strata. Due to differences in the sedimentary age, sedimentation time, and fragmentation of the surface rocks, there are significant differences in the horizontal distribution of strata, resulting in the frequent occurrence of leakage in mid-to-shallow formations. When encountering faults during drilling, there is a phenomenon of venting, which can be confirmed by logging curves to indicate the development of fractures in this section, which can easily lead to malignant leakage.

There are a total of 12 wells in the selected block. Based on the analysis of the actual drilling situation, almost every layer of the wells in Block L, from the Penglai town formation to the Changxing Formation, has experienced well leakage, with a depth ranging from the surface to below 5000 m. However, the severity of well leakage varies greatly among different formations. According to the reasons for the leakage, it can be divided into fracturing leakage, expansion leakage, and differential pressure leakage. After detailed statistics, a total of 259 leaks occurred (multiple leaks at the same depth in continuous operations are recorded as one leak), and, according to different operating conditions, 66% of them were lost during the drilling process, as shown in Figure 1. The main layer is the Qianfo Cliff Formation, with a depth concentration of 1800–3300 m. The leakage type is mainly differential pressure leakage, up to 69%, as shown in Figure 2.

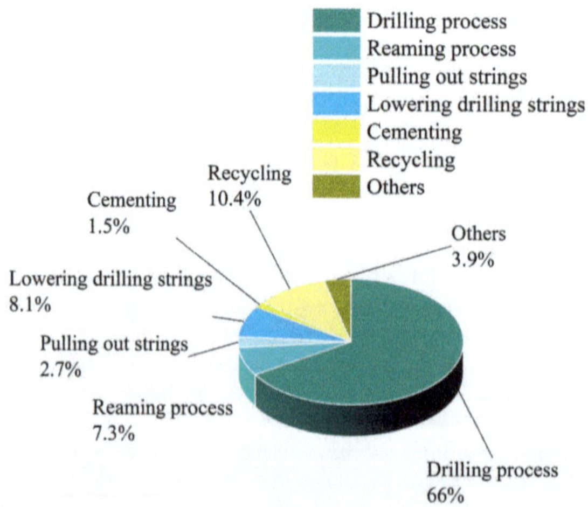

Figure 1. Analysis of Leakage Conditions in Block L.

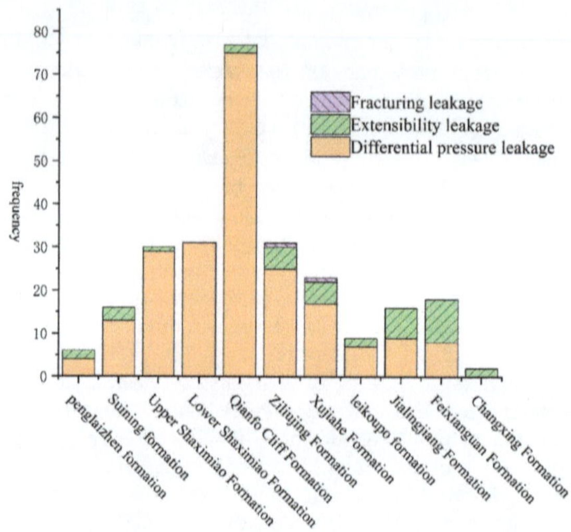

Figure 2. Layered Group Statistics for Different Types of Leakage.

2.3. Drilled Leakage Pressure Model

In response to the above statistical results, it is necessary to conduct indepth research and construct a set of leakage pressure prediction methods during the drilling process. This article takes the cause of leakage as the classification standard and uses different mechanism models for different leakage formations. Since regardless of the type of leakage, the leakage pressure can be regarded as the drilling fluid column pressure in the wellbore when the formation experiences leakage, which means that the drilled leakage pressure profile calculated based on traditional mechanism models needs to be corrected. Research has shown that the values of some model coefficients are also a challenge, and the examples provided in this article have already provided some model parameters in Section 3.1. The specific process is shown in Figure 3.

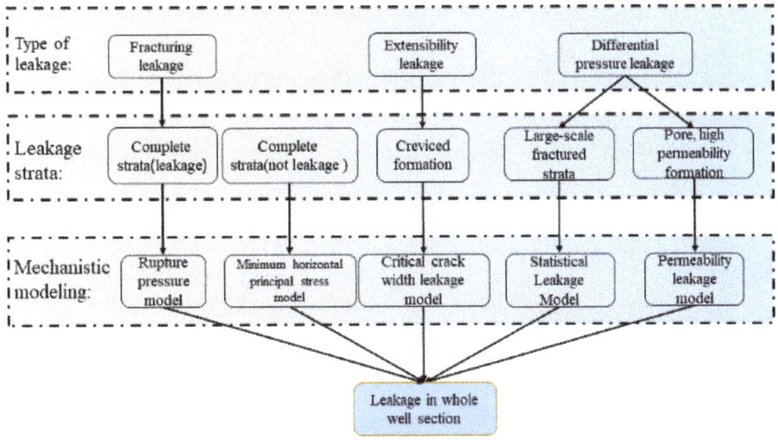

Figure 3. Construction process of the leakage pressure mechanism model for the entire well section.

2.3.1. Critical Crack Width Leakage Model

For extended leakage [17], a critical crack width leakage model is adopted, which assumes the existence of a critical crack width. When the crack width is less than the critical crack width, the drilling fluid will form a sealing layer with a certain pressure-bearing capacity inside the crack, and the drilling fluid will filter normally; when the fracture width exceeds the critical fracture width, it will be converted into fracture leakage. The deformation of cracks follows a power function form, and the flow of drilling fluid in cracks follows a cubic law. The relationship between crack width and effective stress is as follows [17]:

$$\omega = \omega_0 \left\{ A\left[\left(\frac{\sigma}{\sigma_0} \right)^a + 1 \right] \right\}^{-1} \tag{1}$$

where, ω is the dynamic width of the crack, mm; ω_0 is the crack width when the wellbore pressure is equal to the formation pressure, mm; σ is the effective stress on the vertical crack surface, MPa; σ_0 is the effective stress on the vertical fracture surface when the wellbore pressure is equal to the formation pressure, MPa; A and a is an undetermined coefficient; therefore, there is no reason.

Taking a single vertical joint as an example, ignoring the stress concentration around the wellbore, the relationship between the wellbore fluid column pressure and the effective normal stress on the fracture surface is obtained as follows [15]:

$$\sigma = \sigma_h - p'_f \tag{2}$$

where, p'_f is the effective liquid column pressure in the wellbore, MPa; σ_h is the minimum horizontal principal stress, MPa.

Combine Equations (1) and (2) above to obtain the relationship between the dynamic width of fractures and the pressure of the drilling-fluid column:

$$\omega = \omega_0 \left\{ A\left[\left(\frac{\sigma_h - p'_f}{\sigma_h - P_p} \right)^a + 1 \right] \right\}^{-1} \tag{3}$$

From the equation, it can be seen that there is a positive correlation between the dynamic width of fractures and the wellbore fluid column pressure. When the crack width reaches the critical crack width, fractured leakage occurs underground. The leakage pressure calculation model [17] based on the critical crack width ω_c is:

$$P_{L1} = \sigma_h - \left(\frac{\omega_0}{A\omega_c} - 1\right)^{\frac{1}{a}} (\sigma_h - P_p) \quad (4)$$

where, P_{L1} is the crack propagation pressure, MPa; ω_c is the critical crack width, mm; P_p is the formation pore pressure, MPa.

2.3.2. Permeability Leakage Model

For permeability leakage [18], drilling-fluid leakage can be reflected by the Poromechanics formula. Assuming that the borehole is regular, it can be regarded as cylindrical. Since the drilling fluid is generally non-Newtonian fluid, when the Bingham model is used to describe the Rheology Constitutive equation of the drilling fluid, the relationship between the pressure and flow rate of drilling-fluid seepage can be expressed as the following formula [18]:

$$P_{L2} = P_p + \frac{Q_L \times 10^3 \eta_p}{2\pi kh} \ln\frac{r_e}{r_h} + \frac{7}{6000}\tau_0\sqrt{\frac{\phi}{5k}}(r_e - r_h) \quad (5)$$

where, Q_L is the average flow rate of drilling-fluid leakage, L/s; η_p is the plastic viscosity of Bingham plastic, Pa·s; τ_0 is the yield stress of Bingham plastic, Pa·s; ϕ is the porosity of the formation; therefore, there is no reason; P_{L2} is the permeability leakage pressure, MPa; k is the formation permeability, D; r_h is the wellbore radius, m; r_e is the leakage radius of the formation, m; h is the thickness of the leakage layer, m.

2.3.3. Fracture Pressure Model

Usually, due to excessive pressure in the drilling-fluid column or rapid increase in drilling-fluid density in the well, a large amount of pressure is formed, exceeding the maximum pressure-bearing capacity of the weak layer underground, leading to rock fracture and the formation of cracks, or the expansion of closed cracks in the rock, resulting in leakage [19]. The intact formation did not experience fracturing during the drilling process, therefore, its value is approximately equal to the fracturing pressure value, as shown in Equation (7). However, based on the analysis of well history data in this block, when the practical drilling-fluid density is much less than, fracturing leakage still occurs, which is mostly related to insufficient pressure-bearing capacity when drilling to thin and weak layers or lithological interfaces. Therefore, this article modifies Equation (6) by assigning a correction coefficient.

$$P_{L3} = P_f = 3\sigma_h - \sigma_H - \alpha P_p + S_t \quad (6)$$

$$P_{L3} = K_i P_f = K_i(3\sigma_h - \sigma_H - \alpha P_p + S_t) \quad (7)$$

where, P_f is the formation fracture pressure, MPa; S_t is the tensile strength of the rock, MPa; P_{L3} is the fracturing leakage pressure, MPa; α is the effective stress coefficient, dimensionless; K_i Correction coefficient for different depths, with a value range of 0.55~0.92.

2.3.4. Statistical Leakage Model

Differential pressure leakage refers to the presence of large-scale fractures, karst caves, and fracture karst-cave networks connected to the wellbore. Multiple pressure systems coexist in the L block where fractures intersect. In order to obtain the differential pressure of drilling-fluid leakage, the first step is to refer to well history data and classify the leakage based on the cause. Then, the leakage pressure difference and leakage rate are calculated and fitted for this type of differential pressure leakage; the fitting model is as follows [8]:

$$P_{L4} = P_p + \Delta p = P_p + KQ^n \quad (8)$$

where, P_{L4} is the differential leakage pressure, MPa; Δp is the leakage pressure difference, MPa; Q is the leakage rate, m^3/h; n is the fitting coefficient, dimensionless.

2.3.5. Minimum Horizontal Principal Stress Model

In traditional understanding, a complete formation without leakage can only occur when hydraulic fracturing occurs during drilling. However, for the development of fractures in this block, the stability of the rock layer is poor, and the leakage pressure is much lower than the fracturing pressure. Therefore, this article adopts the minimum horizontal principal stress model for the formation without leakage accidents. The minimum horizontal principal stress model believes that, for the formation with bedding, joints, and closed fractures, the fluid pressure that causes the crack to open only needs to overcome the ground stress on the vertical crack surface; that is, the leakage pressure is approximately equal to the minimum horizontal principal stress [20].

$$P_{L5} = \sigma_h = (\frac{\mu}{1-\mu} + \omega)(\sigma_z - \alpha P_p) + \alpha P_p \tag{9}$$

where, P_{L5} the leakage pressure determined for the small horizontal principal stress model, MPa; ω is the stress coefficient of horizontal construction, dimensionless; μ is Poisson's ratio, dimensionless; σ_z is the pressure of the overlying rock layer, MPa.

2.4. Data-Driven Approach

The process of oil drilling is a continuous process of generating data and deepening our understanding of reservoirs. The acquisition of underground data is very valuable, and how to fully explore the connections between data has become an urgent problem to be solved. Incorporating logging and engineering data into leak-pressure prediction provides a new approach. This article uses data from drilled wells as training samples, and it is a feasible method to calculate the leakage pressure of drilling wells. However, in order to achieve ideal prediction accuracy, adding data from the upper strata of the drilling well (previous drilling) to the previous samples will effectively improve model accuracy. At the same time, using an SSA algorithm to optimize LSTM hyperparameters can eliminate the blindness of manually setting parameters and improve timeliness.

2.4.1. SSA-LSTM Model

SSA [21] is an algorithm proposed in 2020, inspired by sparrows' predatory and antipredatory behavior. This algorithm has advantages such as avoiding falling into local optima, fast convergence speed, high convergence accuracy, and strong search ability. The algorithm and its variants have good performance in continuous optimization problems. The modeling process of this algorithm can be summarized as follows: assuming that the position of each sparrow is $x = \{x_1, x_2, \ldots, x_D\}$, its fitness is $f = f[x_1, x_2, \ldots, x_D]$, the population of sparrows is set to m, and n sparrows with the best population position are selected as producers in each generation, while the remaining m-n are selected as scavengers. Compared to other sparrows, individuals with higher fitness will prioritize the discovery of food. In addition, producers will always find abundant food and provide directions for all scavengers to search for food. Therefore, producers will obtain a larger search range. Therefore, the producer's location update is as follows (10). If R_2 is less than ST, then there are no predators and producers begin to conduct more extensive searches; If R_2 is greater than or equal to ST, it indicates that a certain number of sparrows have discovered predators, and an alarm is issued. All sparrows in the population must fly to safer areas to forage.

$$X_{i,j}^{t+1} = \begin{cases} X_{i,j}^{t} \cdot \exp(\frac{-i}{\alpha \cdot iter_{\max}}), R_2 < ST; \\ X_{i,j}^{t} + Q \cdot L, R_2 \geq ST; \end{cases} \tag{10}$$

where, $X_{i,j}$ is the position information of the i th sparrow in the j th dimension; t is the current number of iterations; α is a random number between $(0, 1]$; $iter_{\max}$ is a constant; R_2 is the alarm value, which belongs to the range [0, 1]; ST is the safety threshold, which belongs to the range [0.5, 1]; Q is the value of a simple random distribution; L is a $1 \times d$ matrix with each element of 1.

The position of the scavengers is shown in Equation (11) below. When $i > n/2$, it means that the i th scavenger with lower adaptation is not getting food, so at this time the scavengers need to fly to other places to find food. When the sparrows realize the danger, they will abandon the immediate food and enter the warning state; the specific expression is shown in Equation (12) below.

$$X_{i,j}^{t+1} = \begin{cases} Q \cdot \exp(\frac{X_{worst}^{t} - X_{i,j}^{t}}{i^2}), i > n/2; \\ X_{P}^{t+1} + \left| X_{i,j}^{t} - X_{P}^{t+1} \right| \cdot A^{+}, otherwise; \end{cases} \tag{11}$$

$$X_{i,j}^{t+1} = \begin{cases} X_{best}^{t} + \beta \cdot \left| X_{i,j}^{t} - X_{best}^{t} \right|, f_i > f_g \\ X_{i,j}^{t} + K \cdot (\frac{\left| X_{i,j}^{t} - X_{worst}^{t} \right|}{(f_i - f_w) + \varepsilon}), f_i = f_g; \end{cases} \tag{12}$$

where, X_{worst} is the current global worst position; X_P^{t+1} is the optimal position of the producer at the t + 1th iteration; A is a matrix of $1 \times d$ with elements in the range $[-1, 1]$ and which has to satisfy $A^{+} = A^{T}\left(AA^{T}\right)^{-1}$; K is a randomized measure and belongs to the range $[-1, 1]$; β is a step control parameter with a random normal distribution; ε is a minimal constant which avoids the error of having a zero denominator in the division; X_{best} is the current globally optimal position; f_i is the fitness value of an individual sparrow at this point in time; f_w and f_g are the current global worst and best fitness, respectively.

If $f_i > f_g$, it means that at this time the sparrow is at the edge of the population, and there is a great possibility of being attacked by predators. If $f_i = f_g$, it means that the sparrows located in the center of the population have found the danger; in order to avoid being preyed upon, they need to approach the other sparrows.

LSTM [21] is a kind of recurrent neural network with complex and powerful asymptotic processing ability. Due to the existence of before and after correlation of logging data, the model can extract the logging sequence feature data along the depth, respectively, forward and backward, and the change rule of formation pressure before and after depth extraction. x_t and x_{t-1} represent the input states at the current time and before time respectively; H_t and H_{t-1} denote the current time and the previous time; C_t and C_{t-1} represent the cell states that send the current time and the previous time, respectively. σ denotes the sigmoid activation function with the range of [0, 1] and Tanh is the hyperbolic tangent function with the range of $[-1, 1]$.

First, the LSTM decides what information to discard from the cell state through the forgetting gate. The formula is as in Equation (13):

$$f_t = \sigma\left(W_f \cdot [h_{t-1}, x_t] + b_f\right) \tag{13}$$

f_t is the output of the oblivious gate; W_f is the weight matrix of the oblivious gate; b_f is the bias term of the oblivious gate. Next, the LSTM determines the information that needs to be updated through the input gate according to Equation (13):

$$i_t = \sigma(W_i \cdot [h_{t-1}, x_t] + b_i) \tag{14}$$

i_t is the output of the input gate, W_i is the weight matrix of the input gate, and b_i is the bias term of the input gate. Then we update the cytosolic state C_t. The candidate cytosolic state $\hat{C}_t$ and the current cell state C_t are expressed as Equations (15) and (16).

$$\hat{C}_t = tanh(W_c \cdot [h_{t-1}, x_t] + b_c) \tag{15}$$

$$C_t = f_t C_{t-1} + i_i \hat{C}_t \tag{16}$$

where, W_c is the weight matrix of the candidate vector, and its deviation is represented by b_c. Finally, the LSTM determines the output information state of the unit through the output gate. The output of the output gate is shown in Equation (17):

$$O_t = \sigma(W_o \cdot [h_{t-1}, x_t] + b_o) \tag{17}$$

O_t is the output of the output gate; W_o is the weight matrix output gate; b_o is the bias term of the output gate. The hidden state output of the LSTM unit is shown in Equation (18):

$$h_t = O_t tanh(C_t) \tag{18}$$

However, the hyperparameters of this model have a significant impact on the model, and during the drilling process, different parameters should be used at different intervals or lithological layering. Considering the effectiveness of the drilling process and the addition of new knowledge in the drilled section, the practical sparrow algorithm is a good way to optimize the hyperparameters of the model. The optimization process is shown in Figure 4:

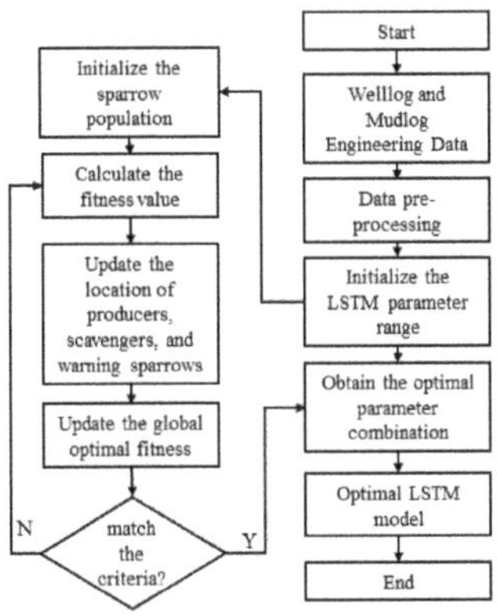

Figure 4. SSA-LSTM algorithm flowchart.

2.4.2. Data Preparation and Evaluation Indicators

The burrs or abrupt changes in logging and logging curves may be abnormal jumps, or they may indeed be sudden changes in the rock environment. In order to eliminate the influence of abnormal data on correlation and model, data smoothing is also required before this, and a fast Fourier transform smoothing method is used. Normalization processing takes into account the dimensional impact between data at different scales and normalizes the logging data to a range of 0~1. After normalization, the logging data y_i is shown in Equation (19).

$$y_i = \frac{x_i - \min(x_i)}{\max(x_i) - \min(x_i)}, 1 < i < n \tag{19}$$

where, y_i is the normalized logging data at the corresponding depth H_i, and x_i represents the original logging data at the measured depth H_i.

The maximum information coefficient (MIC) is used to measure the degree of correlation between two variables x and y, as well as the strength of linearity or nonlinearity. Compared to conventional Pearson correlation analysis, MIC is more suitable for complex nonlinear relationships and has the advantages of low computational complexity and higher robustness. The range of values for the MIC correlation coefficient is [0, 1], and the closer it is to 1, the stronger the correlation.

The mean absolute percentage error (*MAPE*) [13] and root mean square error (*RMSE*) of the indicator are shown in Equations (20) and (21), respectively.

$$RMSE = \sqrt{\frac{1}{N}\sum_{i=1}^{N}(\hat{y}_i - y_i)^2} \tag{20}$$

$$MAPE = \frac{100\%}{N}\sum_{i=1}^{N}\left|\frac{\hat{y}_i - y_i}{y_i}\right| \tag{21}$$

3. Example Analysis and Comparison

3.1. Examples of Lost Pressure in Drilled Wells

A total of twelve wells have been drilled in this block. This article uses two wells as validation to simulate the drilling process, while the rest are adjacent wells. Through the calculation of formation pore pressure and leakage pressure in the previous section, three adjacent well-pressure profiles were constructed and compared based on the actual leakage situation in the well history. The following is an example of the L4 well calculation, as shown in Figure 5. The well has experienced a total of 27 losses, with the types of losses being expansionary and differential pressure losses:

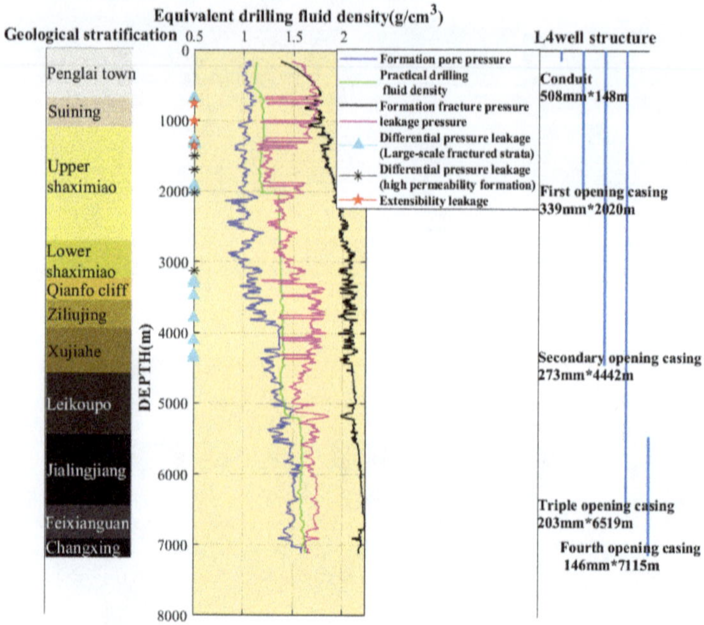

Figure 5. Pressure profile of L4 well.

(1) At 674 m–1397 m (from Penglaizhen Formation to Suining Formation), microfractures coexist with large-scale fractures. For example, at 1008 m, extended leakage occurs. When the wellbore pressure is equal to the formation pressure, the crack width is 0.5 mm, the critical crack width is 2.5 mm, the formation pore pressure is 10.56 MPa, the minimum horizontal principal stress is 16.85 MPa, A is 0.15, a is 6, and the leakage pressure at this depth is calculated to be 11.61 MPa;

(2) Intermittent permeability leakage occurs in a large section from 1398 m to 1890 m (Shang Shaximiao Formation). The leakage well section is long, the leakage volume is large, and the leakage rate is low, but the success rate of plugging is high. At 1398 m, the average leakage flow is 0.56 L/s, the plastic viscosity is 0.035 Pa·s, the permeability is

0.25 D, the hole radius is 0.22 m, the average leakage half diameter is 5 m, the porosity is 3%, the drilling-fluid yield stress is 7 Pa·s, the formation pore pressure is 12.56 MPa, and the leakage pressure is 16.55 MPa;

(3) There are many large-scale fractures in the 3475–4361 m (artesian well group) formation, resulting in lateral differential pressure leakage. The leakage rate is relatively high, and the leakage amount is relatively large. Based on the actual differential pressure leakage situation of adjacent wells (C5, C6, C7), the relationship between pressure difference and leakage rate is obtained as follows: $\Delta P = 1.47Q^{0.562}$.

3.2. Data Sample Instance

In order to demonstrate the improvement effect of the previous drilling data on the drilling formation model, this article constructed adjacent well data sample 1 and sample 2 integrated into the drilled formation, respectively, to predict the leakage pressure while drilling. MIC analysis included a total of 19 feature parameters, excluding parameters with a correlation less than 0.4 and retaining 12 parameters with a strong correlation with leakage pressure such as DEPTH, MwIN, MwOUT, and Inlet Resilience for model training, as shown in Figure 6. Similarly, the L4 well was used as validation, and training sets were constructed using L5, L6, and L9. The constructed sample dataset 1 (8294 × 11) and the data from the first two wells of L4 were integrated into the drilled dataset (12,197 × 11).

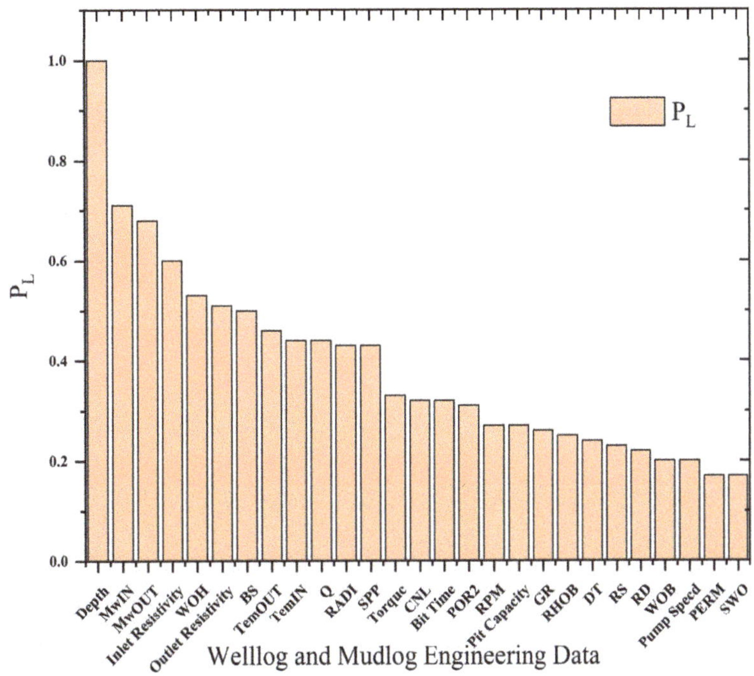

Figure 6. MIC correlation coefficient.

3.3. SAA-LSTM Model Parameter Settings

The first layer neuron range is [5–150], the second layer neuron range is [5–150], and the dropout ratio is [0.05–0.8]. The batch size is [8–32], the sparrow algorithm searches for the optimal combination are (10, 20, 0.6788, 8), the sparrow population size is set to 20, and the maximum number of iterations is 100. The proportion of producers in the population is 20%, and its safety threshold is 0.8.

3.4. Comparison of Positive Drilling Model Predictions

Through the data samples and model construction constructed in the previous section, the comparison results of Figure 7 were obtained, and the two predicted results were compared with the actual leakage pressure. Using only drilled data samples for training, testing was conducted on a positive drilling well (with a depth range of 4450–7200 m). Sample 1 showed an RMSE of 0.053 and a MAPE of 2.8%. However, for a positive drilling well, incorporating the upper drilling data can reduce the RMSE to 0.0405 and the MAPE to 2.4%. And this data sample can better reflect the leakage pressure situation of vulnerable formations (pressure profile fluctuations). Furthermore, it has been proven that accurate prediction of leakage pressure is difficult, and using adjacent well data as samples alone cannot effectively characterize the situation of leakage pressure in normal drilling, especially in areas with uneven lateral distribution of the formation. After adding the upper strata data, this not only proves the applicability of the SSA-LSTM model but also demonstrates the impact of data quality on its results.

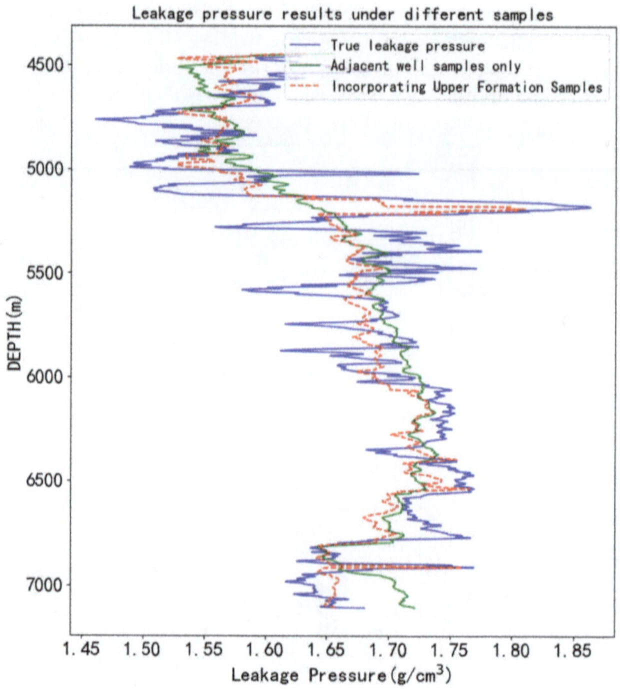

Figure 7. Prediction results of drilling leakage pressure under different samples.

4. Conclusions

This article proposes a method for predicting the leakage pressure during the drilling process, which utilizes various traditional mechanism models to calculate the leakage profile of the drilled well. The constructed leakage samples are used as the output of the machine-learning model to achieve the inversion of the leakage pressure in the drilling formation while drilling. In order to highlight the importance of upper formation information in drilling, we also compared two different datasets and obtained the following summary:

1. The leakage situation of the block was analyzed, and a leakage pressure mechanism model suitable for the entire well section of the block was obtained. Instead of using a single model to calculate the leakage pressure, different models are used for different leakage formations to build the leakage pressure of the whole well section. And the actual

leakage situation will be checked, and the accurate calculation of the leakage pressure of the drilled well will provide a basis for the training of the forward drilling model;

2. The constructed well leakage profile data can be used as the output of the model to effectively predict the leakage pressure of the drilling formation. Different data samples have been constructed to prove that incorporating the upper formation data of the drilling into the samples can effectively improve the accuracy of the model, with an RMSE of only 0.0405 and a MAPE of only 2.4%. And it can better reflect the leakage pressure of vulnerable formations (pressure-profile fluctuations);

3. Future work will start with software integration of input features and combined models, integrating more logging while drilling data into the model, and achieving leakage pressure prediction by combining geological and engineering data. We will always keep up with onsite requirements and integrate this method into the software to achieve risk prediction of drilling leakage and assist in adjusting the drilling construction plan.

Author Contributions: Conceptualization, D.C. and B.H.; methodology, Y.X.; software, C.H.; validation, Y.W. (Yanshu Wang), Y.W. (Yucong Wang) and Y.X.; formal analysis, Y.W. (Yucong Wang); investigation, D.C.; resources, D.C.; data curation, B.H.; writing—original draft preparation, B.H.; writing—review and editing, B.H.; visualization, C.H.; supervision, Y.W. (Yanshu Wang); project administration, D.C.; funding acquisition, D.C. All authors have read and agreed to the published version of the manuscript.

Funding: This research received no external funding.

Institutional Review Board Statement: Not applicable.

Informed Consent Statement: Not applicable.

Data Availability Statement: Not applicable.

Conflicts of Interest: The authors declare no conflict of interest.

References

1. Sun, J.; Liu, F.; Cheng, R.; Feng, J.; Hao, H.; Wang, R.; Bai, Y.; Liu, Q. Research progress and prospects of machine learning in lost circulation control. *Acta Pet. Sin.* **2022**, *43*, 10.
2. Wang, C.; Liu, H.; Liu, Y.; Xiao, F.; Zheng, N. Managed pressure drilling technology: A research on the formation adaptability. *Fluid Dyn. Mater. Process.* **2022**, *18*, 1865–1875. [CrossRef]
3. Xie, R.; Zhang, X.; He, B.; Zheng, N.; Yu, Y. Analysis of the applicability of a risk quantitative evaluation method to high temperature-pressure drilling engineering. *Fluid Dyn. Mater. Process.* **2023**, *19*, 1385–1395. [CrossRef]
4. Wang, S. Judgment of drilling fluid leakage and overflow during the drilling process. *West-China Explor. Eng.* **2020**, *32*, 49–52.
5. Yang, J.; Sun, J.; Bai, Y.; Lv, K.; Zhang, G.; Li, Y. Status and prospect of drilling fluid loss and lost circulation control technology in fractured formation. *Gels* **2022**, *8*, 260. [CrossRef] [PubMed]
6. Zhang, X.; Xie, R.; Liu, K.; Li, Y.; Xu, Y. Analysis of the lost circulation problem. *Fluid Dyn. Mater. Process.* **2023**, *19*, 1721–1733. [CrossRef]
7. Zhang, L.; Wang, X.; Xie, T.; Lin, H.; Yuan, W. A New leakage pressure prediction model for natural fractured formation. *Drill. Prod. Technol.* **2018**, *41*, 5.
8. Fang, J.; Yang, F.; Jia, X.; Deng, M. Method of calculating formation leakage pressure based on mud logging data. *Mud Logging Eng.* **2020**, *31*, 5.
9. Zhai, X.; Chen, H.; Lou, Y.; Wu, H. Prediction and control model of shale induced fracture leakage pressure. *J. Pet. Sci. Eng.* **2020**, *198*, 108186. [CrossRef]
10. Sabah, M.; Talebkeikhah, M.; Agin, F.; Talebkeikhah, F.; Hasheminasab, E. Application of decision tree, artificial neural networks, and adaptive neuro-fuzzy inference system on predicting lost circulation: A case study from Marun oil field. *J. Pet. Sci. Eng.* **2019**, *177*, 236–249.
11. Pang, H.; Meng, H.; Wang, H.; Fan, Y.; Nie, Z.; Jin, Y. Lost circulation prediction based on machine learning. *J. Pet. Sci. Eng.* **2022**, *208*, 109364. [CrossRef]
12. Matinkia, M.; Amraeiniya, A.; Behboud, M.M.; Mehrad, M.; Bajolvand, M.; Gandomgoun, M.H.; Gandomgoun, M. A novel approach to pore pressure modeling based on conventional well logs using convolutional neural network. *J. Pet. Sci. Eng.* **2022**, *211*, 110156. [CrossRef]
13. Song, X.; Yao, X.; Li, G.; Xiao, L.; Zhu, Z. A novel method to calculate formation pressure based on the LSTM-BP neural network. *Pet. Sci. Bull.* **2022**, *7*, 12–23.

14. Pu, J. Research on Comprehensive Control of Lost Circulation Technology in Northeast Sichuan Region. Ph.D. Thesis, Southwest Petroleum University, Chengdu, China, 2015.
15. Lietard, O.; Unwin, T.; Guillot, D.; Hodder, M. Fracture width LWD and drilling mud/LCM selection guidelines in naturally fractured reservoirs. In Proceedings of the European Petroleum Conference, Milan, Italy, 22–24 October 1996.
16. Fuh, G.F.; Morita, N.; Boyd, P.A.; McGoffin, S.J. A new approach to preventing lost circulation while drilling. In Proceedings of the SPE Annual Technical Conference and Exhibition, Washington, DC, USA, 4–7 October 1992.
17. Li, D.; Kang, Y.; Liu, X.; Zeng, Y.; Du, C. The lost circulation pressure of carbonate formations on the basis of leakage mechanisms. *Acta Pet. Sin.* **2011**, *32*, 900–904.
18. Zhang, J.; Yue, H.; Zhang, D.; Hou, R.; Liu, K. New analysis and calculation method for permeable leakage layer depth and pressure. *Oil Drill. Prod. Technol.* **2013**, *35*, 12–15.
19. Le, M.; Wang, D.; Xu, H.; Wang, J. Analysis of Leakage Pressure in Different Formation. *Liaoning Chem. Ind.* **2016**, *45*, 195–197.
20. Zhang, L.; Xu, J.; Xie, T.; Lin, H.; Liu, H. Comparison of Several Calculation Models for Loss Pressure of Fractured Formation. *China Pet. Mach.* **2018**, *46*, 13–17.
21. Chen, L. *Ultra-Short-Term Wind Speed Forecast Based on the LSTM Optimized by the Sparrow Search Algorithm*; Jiangxi University of Finance and Economics: Nanchang, China, 2022.

processes MDPI

Article

Prediction of Lost Circulation in Southwest Chinese Oil Fields Applying Improved WOA-BiLSTM

Xianming Liu [1], Wen Jia [1], Zhilin Li [2], Chao Wang [1,*], Feng Guan [1], Kexu Chen [2] and Lichun Jia [2]

[1] School of Mechanical Engineering, Yangtze University, Jinzhou 434000, China; jiawen0855@163.com (W.J.)
[2] Research Institute of Drilling and Production Engineering Technology, Chuanqing Drilling Engineering Co., Ltd., Guanghan 618300, China
* Correspondence: cw_paper@163.com

Abstract: Drilling hazards can be significantly decreased by anticipating potential mud loss and then putting the right well control measures in place. Therefore, it is critical to provide early estimates of mud loss. To solve this problem, an enhanced WOA (Whale Optimization Algorithm) and a BiLSTM (Bidirectional Long Short Term Memory) optimization based prediction model of lost circulation prior to drilling has been created. In order to minimize the noise in the historical comprehensive logging data, a wavelet filtering technique was first used. Then, according to the nonlinear Spearman rank correlation coefficient between mud loss and logging parameter values from large to small, seven characteristic parameters were preferred, and the sliding window was used to extract the relevant data. Secondly, the number of neurons in the first and second hidden layers, the maximum training time, and the initial learning rate of the BiLSTM model were optimized using the enhanced WOA method. The BiLSTM network was given the acquired superparameters in order to improve the model's ability to predict occurrences. Finally, the model was trained and tested using the processed data. In comparison to the LSTM model, BiLSTM model, and WOA-BiLSTM model, respectively, the improved WOA-BiLSTM early mud loss prediction in southwest Chinese oil fields suggested in this study beat the others, receiving 22.3%, 18.7%, and 4.9% higher prediction accuracy, respectively.

Keywords: lost circulation prior to drilling; prediction model; correlation analysis; improved whale optimization algorithm; Bidirectional Long Short Term Memory

Citation: Liu, X.; Jia, W.; Li, Z.; Wang, C.; Guan, F.; Chen, K.; Jia, L. Prediction of Lost Circulation in Southwest Chinese Oil Fields Applying Improved WOA-BiLSTM. *Processes* **2023**, *11*, 2763. https://doi.org/10.3390/pr11092763

Academic Editor: Jie Zhang

Received: 23 August 2023
Revised: 8 September 2023
Accepted: 10 September 2023
Published: 15 September 2023

1. Introduction

Mud circulates from the annulus back to the earth as it travels through the drill pipe during the drilling operation. Mud is essential for maintaining hydrostatic pressure, wellbore stability, and bit temperature, in addition to being utilized for suspending cuttings. As a consequence, the wellbore's mud circulation does not want to cause mud loss [1]. However, deep drilling in complex geological environments is becoming more common with increased exploration and development, particularly in the process of drilling "three high" oil and gas wells, which involves drilling into the cracks of carbonate rocks or other abnormal pressure geospheres with complex and variable pressure systems [2]. It is challenging in this situation to precisely assess the ground pressure and geological conditions, and it is easy to introduce the risk of mud loss. In addition, the amount of mud loss is different, and the corresponding plugging methods are different. If the mud loss is detected in advance, it can be solved by adding plugging agents within 48 h to prevent it from further developing into more serious mud loss [3,4]. Therefore, the purpose of this study is to train neural networks using historical drilling data and establish a predictive model capable of forecasting early mud circulation loss. By predicting these early mud losses, appropriate measures can be taken in advance to prevent or, at the very least, significantly mitigate early mud loss. This approach aims to address downhole crossflow, blowout, and wellbore instability incidents in their infancy, thereby achieving safe and efficient drilling.

In recent years, several researchers both domestically and internationally have begun to apply neural network approaches for lost circulation prior prediction. Moazzenii et al. (2010) built a multilayer feed-forward network learned by backpropagation to forecast lost circulation events in the Maroun oil field [5]. Jahanbakhshi et al. (2014) created a multilayer perceptron model to estimate mud loss and demonstrate the impact of geomechanical factors [6,7]. Aljubran et al. (2017) created many ML and DL models to predict circulation loss, including RF (Random Forest), ANN (Artificial Neural Network), CNN (Convolutional Neural Network), and LSTM (Long Short-Term Memory). The CNN model was shown to be the best [8]. Sabah et al. (2019) created many smart systems to anticipate circulation loss in the Maroun oil field such as MLP (Multi-Layer Perceptron), RBF (Radial Basis Function), GA-MLP (Genetic Algorithm Multi-Layer Perceptron, DP (Decision Tree), and ANFIS (Adaptive Neuro-Fuzzy Inference System). The findings revealed that DT is the best prediction mode [9–15]. Ahmed et al. (2020) employed artificial neural network models to foresee lost circulation in both naturally occurring and artificially produced fractures [16,17]. Mardanirad et al. (2021) used a comparison between different DL (deep learning) algorithms, CNN (Convolutional Neural Network), GRU (Gated Recurrent Unit), and LSTM (Long Short-Term Memory) for the classification of mud loss intensity in the Azadegan oil field, which showed the superior accuracy of the LSTM compared to other DL algorithms [18–20]. Jafarizadeh et al. (2022) used a fusion of an optimization algorithm and a modular neural network to address the problem of mud loss. The topology, threshold, and weight of the neural network were optimized to effectively solve the shortcomings of the traditional neural network, such as improper setting of hyperparameters and easy to fall into local optimization [21]. SiamiNamini et al. (2022) carried out depth analysis using traditional RNN, LSTM, and BILSTM network algorithms in deep water drilling condition identification. The results showed that the BILSTM network has good performance [22]. Xiang et al. (2022) predicted horizontal in situ stresses by using a CNN-BiLSTM-Attention hybrid neural network. The verification showed that compared with convolutional neural networks, LSTM and BiLSTM can extract the autocorrelation characteristics of the dynamic changes of the comprehensive logging curve and can better predict [23]. Li et al. (2022) proposed a deep learning method for early mud loss prediction based on the CNN-LSTM fusion network. They verified that the prediction accuracy of the network structure fused by the optimization algorithm is better than that of the CNN or LSTM structure alone [24].

BiLSTM has a distinct advantage in dealing with the complex mapping relationship of high-dimensional nonlinear long time series and can fully account for the time effect and parameter influence of the drilling process, which has good potential in mud loss prediction, according to research of existing early mud loss prediction models. At the same time, the network's prediction accuracy varies substantially according to the effect of structural characteristics. If the network parameters are not appropriately configured, the trained model will struggle to obtain the desired result. Furthermore, at the moment, early mud loss prediction is generally used to forecast whether or not mud loss would occur, whereas there is little research on mud loss volume prediction. Therefore, it is necessary to select a suitable neural network method to realize the early prediction of mud loss during the drilling process so as to guide the drilling operation more effectively.

In order to solve the problems in the existing methods, this paper chooses a two-layer BiLSTM as the basic neural network. The improved WOA is used to optimize the number of neurons in the input layer, the number of neurons in the hidden layer, the maximum training period, and the initial learning rate in the BiLSTM structure [25]. Based on the comprehensive consideration of measurement while drilling parameters, logging parameters, and fine pressure control drilling parameters, an improved early prediction model of WOA-BiLSTM of mud loss is constructed. First, textual data are converted into numerical values. Characteristic parameters are then selected through Spearman rank correlation analysis. Subsequently, wavelet filtering is applied to mitigate the impact of noise on the data. The selected characteristic parameters are used. Finally, the collected data are partitioned into training sets, test sets, and validation sets for training and verification purposes.

2. Relevant Theories

Data preprocessing is an important step in building a real model. A good data preprocessing process includes important steps such as data denoising, data conversion, and data dimensionality reduction [26].

2.1. Data Denoising

When we talk about real data, noise is an inevitable component, with at least 5% even under the strictest controls [27]. In this study, the term "data denoising" refers to the use of filtering to lessen the effect of noise on the data. Data conversion requires the use of data in multiple units, and the distribution of hyperparameters may be impacted by the scale disparity, homogenizing the processing. The Spearman rank correlation coefficient's correlation analysis is used to determine the order of the influencing factors during data dimensionality reduction. The main factors are then chosen from a list of parameters affecting drilling mud loss.

2.2. Data Normalization

At present, the main methods of dimensionless data processing are standardization, averaging, and standard deviation [28]. Considering that the covariance matrix composed of the original data after averaging processing can not only reflect the difference in the degree of variation of each index in the data but also contain information on the degree of mutual influence of each index, the data in this paper chose the averaging method to normalize it to the scale range of $[-1, 1]$. The equation for the data normalization is given below:

$$X = 2(\frac{X - X_{min}}{X_{max} - X_{min}}) - 1 \tag{1}$$

where X_{max} represents the data maximum; X_{min} represents the data minimum.

2.3. Feature Selection

When using the neural network model to train the sample data, we need to consider the high dimension, which will cause the neural network model to run slowly and consume hardware. In addition, in the case of large data dimensions, there is the problem of "dimension disaster" [29]. Therefore, it is necessary to select features of the data to achieve the purpose of dimensionality reduction. The Spearman rank correlation coefficient, also called the rank correlation coefficient, is a nonparametric statistic whose value is unrelated to two groups of variables related to the specific value but only the size of the relationship between its values. Therefore, it is very suitable for studying the correlation between nonlinear relations.

2.4. LSTM Principle of Neural Network

LSTM neural network is another neural network algorithm improved on the basis of RNN neural network to solve time series problems [30,31]. The structure is shown in Figure 1. The shaded areas represent the previous and next moments, while the non-shaded area represents the current moment. By adding three control units, forgetting gate, memory gate, and output door, the network can alleviate the problem of RNN being prone to gradient explosion and gradient disappearance with a special way of storing "memory" and setting gradient range threshold. The main operation flow of LSTM is as follows:

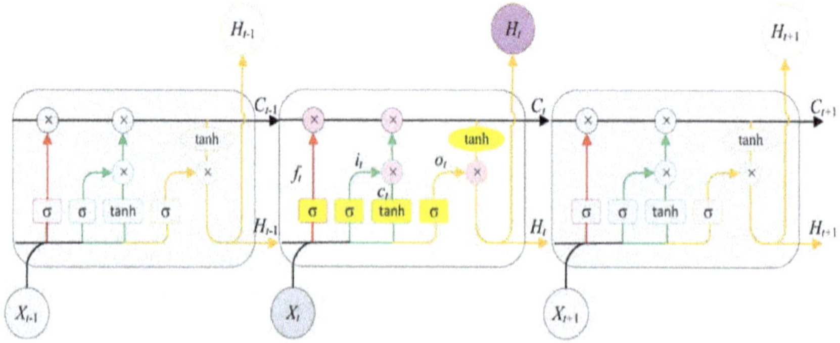

Figure 1. LSTM structure.

(1) Use the forgetting door, as shown by the red arrow in Figure 1, combined with the output of the previous moment H_{t-1}, to cellular information C_{t-1} filtering; purposefully screen out the cell information that has influence on this moment. The mathematical expression is as follows:

$$f_t = \sigma(W_f \cdot [H_{t-1}, x_t] + b_f) \tag{2}$$

where H_{t-1} represents the last moment output; f_t represents the output of the forgetting door; x_t represents the input at the current moment; σ represents the sigmoid function; W_f and b_f, respectively, represent the weight coefficient and offset of the linear relationship.

(2) The memory gate, as shown by the green arrow in Figure 1, is used to retain the effective information of the cell information C_{t-1} combined with the output H_{t-1} of the previous moment. The mathematical expression is as follows:

$$i_t = \sigma(W_i \cdot [h_{t-1}, x_t] + b_i) \tag{3}$$

$$c_t = \tanh(W_C \cdot [h_{t-1}, x_t] + b_c) \tag{4}$$

$$C_t = f_t \times C_{t-1} + i_t \times c_t \tag{5}$$

where i_t represents the put of the first part; c_t represents the put of the second part; W_i, b_i, and W_c represent the corresponding weight coefficient and offset, respectively; C_t represents the state of the updated cells.

(3) The output door, as shown by the orange arrow section in Figure 1, combined with the output of the previous moment H_{t-1} and current cell information C_t after the calculation, input to the neural network for operation. The mathematical expressions are as follows:

$$o_t = \sigma(W_O[h_{t-1}, x_t] + b_o) \tag{6}$$

$$h_t = o_t \times \tanh(C_t) \tag{7}$$

where o_t represents the state of the upper hidden layer; the value of h_{t-1} and x_t is computed by the sigmod function.

2.5. BiLSTM Principle of Neural Network

The basic principle of BiLSTM neural network is composed of two LSTM neural networks [32], and the training sequence can be transmitted forward and backward. It can achieve a more complete analysis of the characteristics and laws of the data. Below, Figure 2 shows the structure of a single-layer BiLSTM neural network expanded over time. The shaded areas in the diagram represent the previous and next moments, while the non-shaded areas represent the current moment. At the same time, the blue arrows represent forward propagation, and the yellow arrows represent backward propagation, thus achieving bidirectional propagation. x is the input value of the neuron. The hidden layer of the bidirectional convolutional neural network needs to save two values, "A" participates in forward calculation, and "A*" participates in backward calculation. The final output value y depends on both "A" and "A*".

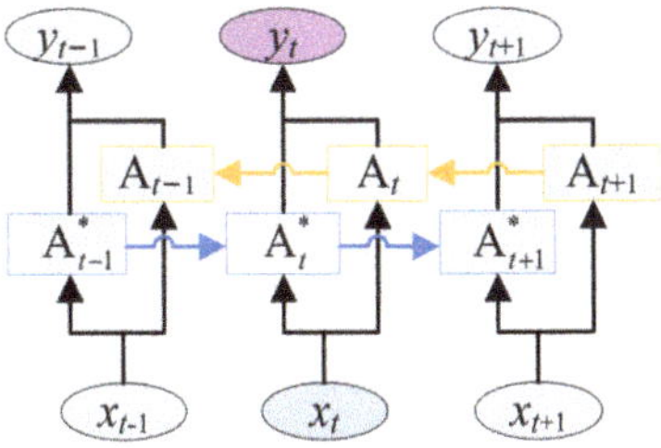

Figure 2. Single layer BiLSTM neural network.

2.6. *Principle of Optimization Algorithm*

2.6.1. Whale Optimization Algorithm (WOA)

WOA is a metaheuristic swarm intelligence optimization algorithm which mainly includes three steps: randomly searching for prey, encircling target prey, and preying on selected prey [33].

(1) Finding the solution to a problem is the process of finding prey by a herd of whales, choosing a prey at random, and the process can be translated into the following:

$$D = |X(t) - CX_{rand}(t)| \tag{8}$$

$$X(t+1) = X_{rand}(t) - AD \tag{9}$$

where $X_{rand}(t)$ represents randomly selected position vectors from the current population of whales; $X(t)$ represents the position vector of the individual; t represents the current iterations; A and C represent the coefficient and are calculated as follows:

$$A = 2a{\cdot}r_1 - a \tag{10}$$

$$C = 2r_2 \tag{11}$$

$$a = 2 - 2t/T_{max} \tag{12}$$

where r_1 and r_2 are random vectors belonging to the interval [0, 1]; a is linear reduction from 2 to 0 during iteration; T_{max} is the maximum number of iterations.

(2) The best candidate solution is the target prey or a near optimal solution. After the optimal solution is found, the other candidate positions will move closer to the target prey, surround the prey, and update its position. The mathematical model is as follows:

$$D = |X(t) - CX^*(t)| \tag{13}$$

$$X(t+1) = X^*(t) - AD \tag{14}$$

where t represents the current iterations; A represents the coefficient; $X^*(t)$ represents the current best position; $X(t)$ represents the current location.

(3) Humpback whales update their positions by spiraling up to hunt selected prey. The mathematical model is as follows:

$$X(t+1) = D'{\cdot}e^{bl}{\cdot}\cos(2\pi l) + X^*(t) \tag{15}$$

$$D' = |X^* - X(t)| \tag{16}$$

where b represents the constant; l represents a random number in the interval $[-1, 1]$. In the Formula (16), D' represents the distance between the best whale individual in the t iteration and the current whale.

In order to achieve this simultaneous behavior, it is assumed that there is a 50% chance to choose whether to shrink the encirclement mechanism or the spiral model to update the position of the whale in the optimization process. The mathematical model is as follows:

$$X(t+1) = \left\{ \begin{array}{ll} X^*(t) - A \cdot D & p < 0.5 \\ X^*(t) + D' \cdot e^{bl} \cdot \cos(2\pi l) & p \geq 0.5 \end{array} \right. \tag{17}$$

where p represents a random number in the interval $[0, 1]$.

2.6.2. Improved Whale Optimization Algorithm (WOA)

The original WOA still has some disadvantages similar to other swarm intelligence optimization algorithms [34], such as low solution accuracy, slow convergence speed, and easy to fall into local optimization. In order to overcome these shortcomings, this paper will improve WOA from two aspects: location update strategy and prevention of falling into local optimization [35].

(1) Nonlinear convergence factor

As the WOA Formula (9) knows, the global and local exploration abilities of the algorithm mainly depend on the parameters A as setting a larger A in the early stages of the iteration can speed up the algorithm. The algorithm's ability to perform local searches is enhanced by lowering parameter A in later iterations. By Formulas (10) and (12), it is known that the value of the parameter A mainly depends on the convergence factor a. In this research, a nonlinear convergence factor a is proposed because the linear variation of convergence factor a cannot demonstrate its searching ability. The mathematical model is as follows:

$$a = 2 - 2\sin(\mu \frac{t}{T_{max}} \pi + \varphi) \tag{18}$$

where t is current iterations; T_{max} is maximum iterations.

(2) Adaptive weight strategy and random difference variation strategy

In order to keep the diversity of the population and jump out of the local optimization in time, Yao Ning proposes an adaptive weight strategy and a random difference mutation strategy [31]. The mathematical expression of the adaptive weight strategy is as follows:

$$w(t)_i = \left\{ \begin{array}{ll} w_1 - \frac{w_2 - w_1}{T_{max}} \cdot \frac{f(t)_i - f(t)_{min}}{f(t)_{max} - f(t)_{min}} & f(t)_i < f(t)_{avg} \\ w_1 - \frac{w_2 - w_1}{T_{max}} \cdot \frac{f(t)_i - f(t)_{min}}{f(t)_{max} - f(t)_{min}} & f(t)_i > f(t)_{avg} \end{array} \right. \tag{19}$$

where $w(t)_i$ represents the weight of the search i in the iteration t; T_{max} represents the maximum number of iterations; w_1 represents the initial minimum weight; w_2 represents the initial maximum weight; $f(t)_{avg}$ represents the average adaptation value of the population after the current iteration of t times; $f(t)_{min}$ and $f(t)_{max}$ represent the minimum and maximum fitness values after the iteration of t times, respectively. The position update strategy expression occurs when the improved prey is formed by substituting the Formula (19) into (17) is as follows:

$$X(t+1) = \left\{ \begin{array}{ll} w(t) g X^*(t) - A g D & p < 0.5 \\ w(t) g X^*(t) + D' g e^{bl} g \cos(2\pi l) & p \geq 0.5 \end{array} \right. \tag{20}$$

The mathematical expression of the random difference variation strategy is as follows:

$$X(t+1) = r_1 \times (X^*(t) - X(t)) + r_2 \times (X'(t) - X(t)) \quad (21)$$

where r_1 and r_2 are random numbers within the range of [0, 1]; $X'(t)$ is randomly selected individuals from a population.

2.7. Improved WOA-BiLSTM Prediction Model of Early Mud Loss

The selection of structural parameters of the BiLSTM prediction model has a great influence on the final prediction ability of the model. In order to find the optimal superparameters of the early mud loss prediction model, the WOA is used to optimize the number of units in the hidden layer (L_1, L_2), the maximum cycle period (T), and initial learning rate of the cycle (I_r). Taking these four key hyperparameters as the characteristics of optimization, the WOA algorithm is used to adjust and optimize the LSTM model to make the network structure model more compatible with the characteristics of comprehensive logging data. The main implementation steps are shown in Figure 3. Firstly, the yellow flowchart represents the computational process of BiLSTM. The initial steps involve inputting historical well data and conducting correlation analysis on the data. Subsequently, training is performed using a predictive model with conventional parameter settings to select highly correlated predictive parameters. Then, the selected parameters are subjected to wavelet filtering for noise reduction. After identifying the four hyperparameters that need optimization, the blue flowchart illustrates the process of optimizing these parameters using the training error of the BiLSTM model as the fitness value. Finally, an ideal predictive model is obtained through training.

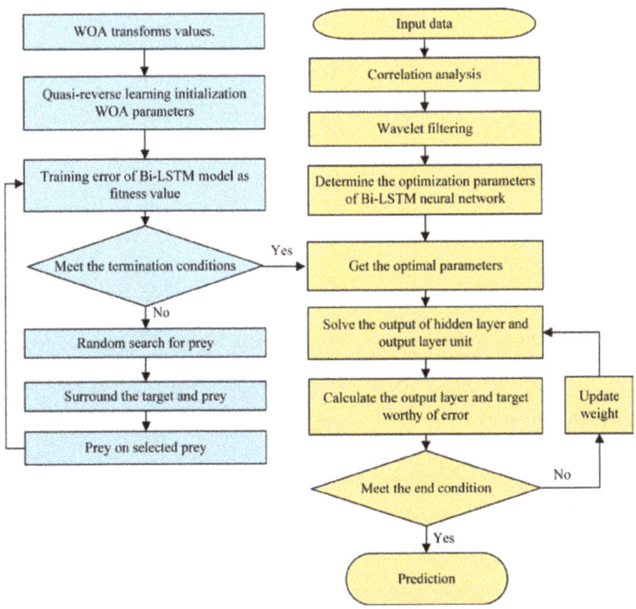

Figure 3. WOA-BiLSTM predictive model flow chart.

3. Experimental Study

3.1. Input Data

The data used for the comprehensive recording of wells came from three oil and gas wells located in the Sichuan Basin in the southwestern region of China, named respectively as A, B, and C. A full set of drilling data was collected and integrated in the drilling conditions of the three wells within 5 h prior to the loss of sludge, including PWD (Pressure While Drilling) underground pressure and surface microflow monitoring data during drilling. The extracted data were stacked together to form a sample data set. The interval between each set of data collected was 20 s, the data within 10 min were selected as a time sequence and the length of the sequence was 30 time steps [34]. The integrated data consisted of 144,000 sets of data. The 115,200 (80%) groups of data were modeled as data A and the remaining 28,800 (20%) groups were modeled for data B validation. Some of the primary historical data collected include drilling parameters as shown in Figure 4. Figure 4A–H offers an in-depth depiction of the drilling process, showcasing the fluctuations in crucial parameters such as well depth, casing pressure, hook weight, torsion moment, standpipe pressure, drilling pressure, inlet flow, and total pool volume. The substantial variations in the hook weight's load and torque are particularly striking, underscoring the dynamic nature of the drilling operation.

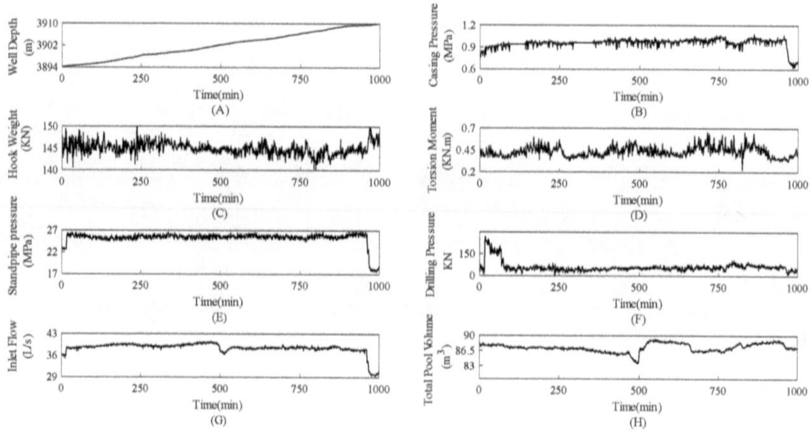

Figure 4. Partial logging data.

From Figure 4A,H, it can be clearly seen that at approximately 3902 m, a significant turning point is observed in the total volume of the mud pool. This decrease suggests a possible mud loss incident prior to this depth. This critical observation underscores the importance of continuous monitoring and analysis in identifying potential issues during drilling operations. In addition to these findings, it's also crucial to consider other factors that may influence these parameters. Therefore, we conducted a correlation analysis on multiple drilling parameters to identify the most sensitive key parameters, in order to better establish a predictive model.

3.2. Correlation Analysis

The occurrence of mud loss will be demonstrated by comprehensive drilling recording parameters. The characteristic parameters for extracting the state of mud loss from large amounts of data are a key step in predicting well loss using the BiLSTM neural algorithm. The actual drilling process at the site is mostly used to observe the changes in the total pool volume as a judgment criterion for the mud loss. Therefore, in this article, the amount of change in the total pool volume is used as a "reference value", by analyzing the Spearman rank correlation coefficient to evaluate the depth of the well, the steering pressure, the measurement of the pipe pressure and set the piping pressure, the position of the flow valve, suspension, drilling pressure, pumping, input flow, input density, meter drilling time, working current, height, speed, torque, and the degree of association of mud loss. This article provides an analysis based on relevance. From Figure 5 it can be seen that the surface pressure pump correlation is the highest at 0.865, followed by torque correlations of

0.657. According to the correlation values of the feature parameters, the feature parameters are sorted from high to low. The sorted parameters are shown in Table 1.

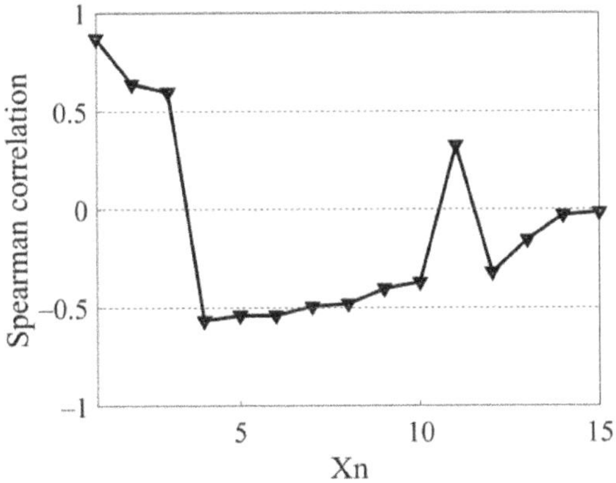

Figure 5. Grey relational degree comprehensive evaluation.

Table 1. Values of the gathered parameters drilled in Southwest Chinese Oil fields.

X_n	Parameter	Unit	Min	Max	Mean
X_1	Surface Pressure Pump	MPa	10.02	29.98	20.00
X_2	Torque	KN.m	0	1.73	0.86
X_3	Casing Pressure	MPa	0	8.77	4.39
X_4	Hook Weight	KN	0	774.4	387.2
X_5	Inlet Flow	L/min	0	49.03	24.51
X_6	Inlet Density	g/cm^3	0	2.52	1.26
X_7	Measured Depth	m	810.00	5540.00	3175.00
X_8	Overpull	KN	210.80	1000.00	605.40
X_9	Tripping in	h	0.89	2.98	1.93
X_{10}	Choke position	m	13.23	95.03	54.13
X_{11}	Rotational speed	rpm	12.00	70.00	41.00
X_{12}	Traveling block height	m	10.00	30.00	20.00
X_{13}	Pore pressure	MPa	8.65	72.45	40.55
X_{14}	Fracture pressure	MPa	13.58	93.83	53.71
X_{15}	Drilling time	min/m	0.94	10.00	5.47

Based on extracted feature parameters, it is observed that within the same block, mud loss events in oil and gas wells typically occur within a specific range of measurement depths. This phenomenon arises due to variations in reservoir pressure, differing rock properties at various depths, and discrepancies in inlet flow rates and densities. These factors reflect distinct drilling conditions, thereby influencing the incidence of mud loss.

The recorded changes in casing pressure reflect fluctuations in annular pressure, while variations in mud density affect the pressure differential between the wellbore and the formation. During mud loss incidents, drilling pressure, annular casing pressure, and torque also undergo changes in response to subsurface conditions. When mud loss occurs, drilling pressure decreases, annular casing pressure diminishes, and concurrently, drill bit speed increases, resulting in an upsurge in inlet flow rates. In contrast, during wellbore overflow events, mud density decreases, and with the overall increase in mud pit volume, annular casing pressure reduction occurs gradually. This, in turn, leads to a decrease in the impact of buoyancy on downhole fluids as mud density decreases, ultimately resulting in an increase in hook load.

Considering the on-site testing accuracy of engineering parameters and the computational efficiency of the prediction model, we determined the order of sorted drilling parameters based on correlation analysis to increase the number of input neurons in the BiLSTM model, denoted as 'n'. The criterion for selecting the number of input parameters is based on the prediction model's error. From Figure 6, it is evident that when the number of input neurons reaches 7, the error reaches its minimum. With 8 neurons, there is a slight increase in error, after which the error remains relatively stable. Therefore, to conserve computational resources and optimize the efficiency of the prediction model, we chose 7 drilling parameters as input neurons.

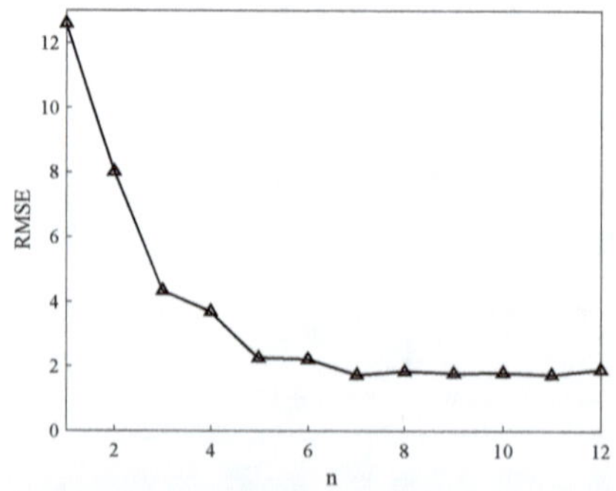

Figure 6. Feature parameter extraction.

In the process of oil drilling, the movement of the previous period of time affects the next movement. Therefore, in the tag construction, the 8 drilling parameters selected by the feature over the length of the time series are used as the multidimensional variable X. The change in total pool volume after the time series length is used as a regression prediction of mud loss (label Y). In this paper, a time series matrix with time series length of 30 is constructed by using the time window sample structure method. Select the time series matrix using the form of window slide, and the time step of each slide is 1. The drilling data within the first 10 min are used to predict the mud loss at the next moment, achieving the purpose of predicting 10 min in advance, as shown in Figure 7. The red box represents the matrix of the first sample data input, while the green and blue boxes represent the input of sample data for the next and the following moments, respectively, proceeding in an orderly manner.

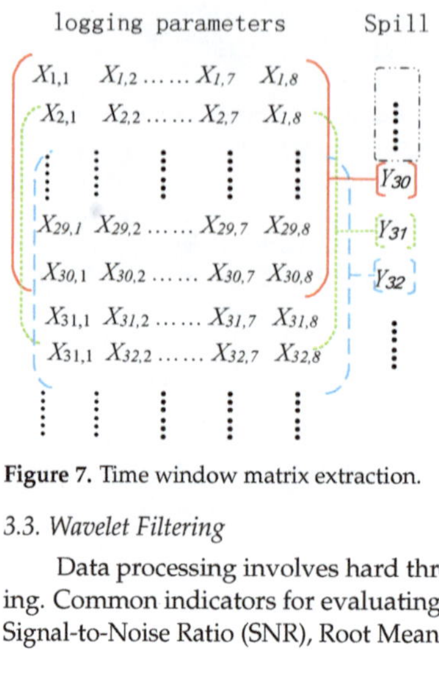

Figure 7. Time window matrix extraction.

3.3. Wavelet Filtering

Data processing involves hard thresholding, soft thresholding, and fixed thresholding. Common indicators for evaluating the effect of wavelet threshold denoising include Signal-to-Noise Ratio (SNR), Root Mean Square Error (RMSE), smoothness, and correlation coefficient. In this study, we use SNR and RMSE to evaluate the effect of wavelet threshold denoising. SNR is defined as the ratio of the energy of the original signal to the noise signal. The higher the SNR, the better the denoising effect. RMSE is the square root of the variance between the original signal and the denoised signal. The smaller the RMSE value, the better the denoising effect.

As shown in Figure 8, the green line, blue line, and red line represent denoising methods using hard thresholding, soft thresholding, and fixed thresholding respectively. From Figure 8A, it can be seen that when using hard thresholding and soft thresholding, the SNR values are similar, but overall lower than the noise reduction effect of using a fixed threshold. This indicates that filtering with a fixed threshold is better. Similarly, from Figure 8B, it can be seen that when using a fixed threshold, the RMSE values are overall lower than those obtained using hard thresholding and soft thresholding methods. This again proves the superiority of filtering with a fixed threshold. Therefore, in our research, we chose to filter with a fixed threshold.

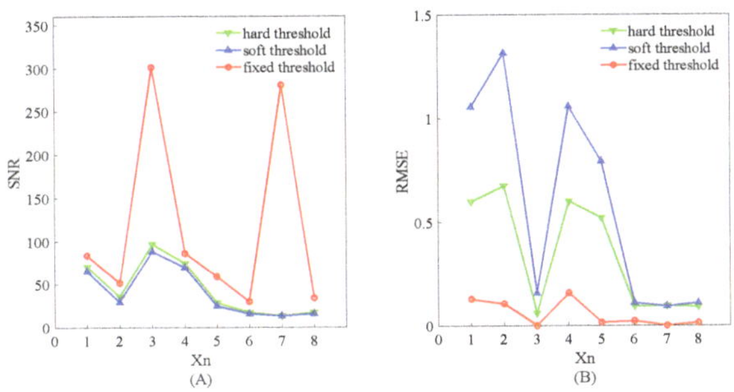

Figure 8. Wavelet noise reduction ratio and square error.

The 8 data after feature selection are displayed by wavelet filtering. Figure 9A–H represent measured depth (MD), surface pressure pump (SPP), torque (TQ), casing pressure (CP), weight on bit (WOB), inlet flow rate, outlet flow rate, and total pool volume (TPV) respectively. the black line represents the original data before filtering, and the red line represents the data after wavelet filter denoising.

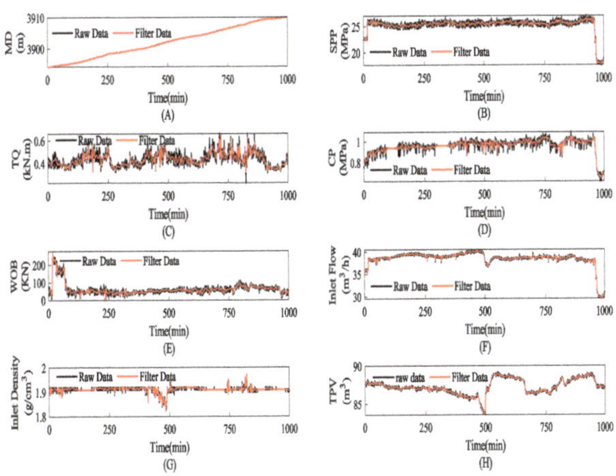

Figure 9. Wavelet Filtering.

Figure 9H illustrates two significant events in the change of total pit volume. At approximately 500 min, there is a decrease in total pit volume, indicating mud loss occurring around a drilling depth of approximately 3902 m. Around 890 min, the total pit volume first increases and then decreases, signifying an occurrence of mud overflow followed by mud loss around a drilling depth of approximately 3910 m. We can observe noticeable fluctuations in inlet density, inlet flow rate, torque, drilling pressure, and casing pressure before and after these events, indicating their correlation with changes in total pit volume. This suggests that these parameter variables can be used to establish a predictive model.

3.4. Improved WOA Optimization Algorithm Parameter Design

Selecting the appropriate number of hidden layer neurons and the right initial learning rate is a crucial decision in neural network design. Typically, this requires experimentation and validation to determine the optimal configuration that meets the specific task requirements, while ensuring the network possesses excellent performance and generalization ability. Increasing the number of hidden layer neurons can expand the network's capacity and generalization capabilities, enabling it to better fit complex data patterns and functional relationships. However, if there are too many neurons, it may lead to overfitting the training data, thereby reducing generalization performance.

The choice of learning rate is equally critical. Smaller learning rates usually demand more training epochs because each weight update has a smaller magnitude, while larger learning rates can result in rapid model changes, requiring fewer training epochs. Inappropriate learning rate settings can lead to the model converging quickly to suboptimal solutions. Due to the lack of mature theoretical guidance, this study relied on existing research results to determine these key parameters [36,37].

The parameter settings for the WOA optimization algorithm are as follows: a population size of 100, with a maximum weight of 0.9 and a minimum weight of 0.2. Additionally, the whale algorithm is configured with a population size of 50 and a maximum iteration count of 30. Considering the need to optimize four parameters, each corresponding to a dimension, constraints were applied within a limited parameter search space, as outlined in Table 2. Simultaneously, the number of neurons in the first and second layers of the neural network was set within a range of 10 to 50, the initial learning rate was within the range of 0.001 to 0.01, and the maximum cycle count was within the range of 50 to 200.

Table 2. Bi-LSTM optimizing parameter range table.

Parameter	L_1	L_2	T	I_r
scope	10~50	10~50	0.001~0.01	50~200
step size	2	2	0.0005	5

According to the parameters and constraints set above, the genetic algorithm, the particle swarm optimization algorithm, and the improved whale optimization setting parameters are shown in Table 2 (WOA maximum weight, 0.9; minimum weight, 0.5; population size, 50; the maximum number of iterations, 100).

The weight of WOA-BiLSTM tends to be stable at about 63 generations and the best fitness value is 0.022. Compared with the genetic algorithm and the particle swarm optimization algorithm, the convergence speed is faster and the error value is the smallest. Select the population optimal solution when the number of iterations is 100, $[L_1, L_2, T, I_r] = [9, 12, 16, 0.03, 100]$, as the combined value of the parameters to be optimized for the Bi-LSTM structure.

4. Model Evaluation

Figure 10A,B illustrate the comparison between the predicted results of four models and the actual values from randomly selected test data sets. The black line represents our prediction target, which is the total pool volume change (i.e., mud circulation loss). The green line shows the prediction results obtained using the LSTM model, the dark blue line shows the prediction results obtained using the BiLSTM model, the purple line shows the prediction results obtained using the WOA-BiLSTM model, and the red line shows the prediction results obtained using the improved WOA-BiLSTM model.

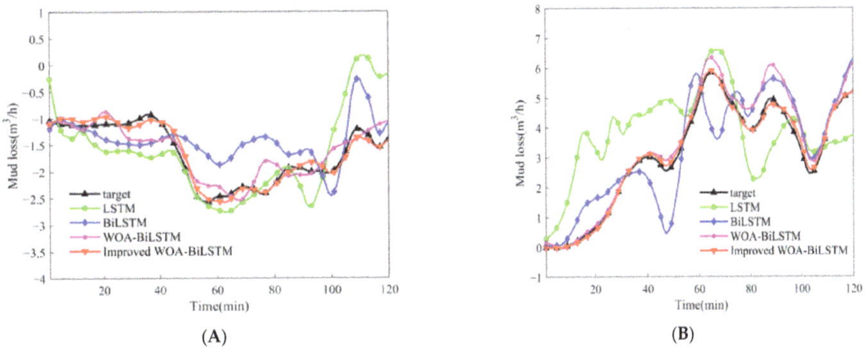

Figure 10. Regression graph.

In Figure 10A, real drilling data from a measurement depth range of 3900 m to 3901.5 m are used, while in Figure 10B, data from a measurement depth range of 4710 m to 4712 m are employed. The objective is to predict the change in total pool volume 10 min in advance to assess the occurrence of mud circulation loss, as indicated in Figure 10A for potential mud circulation loss and in Figure 10B for potential drilling fluid overflow, in the respective depth ranges.

Through comparative analysis, it was observed that LSTM and Bi-LSTM models with random parameters exhibited significant disparities between their training and testing sets. These models displayed fluctuations and deviations from the target values throughout the training process, indicating overfitting and resulting in suboptimal test results. Conversely, the WOA-BiLSTM and the improved WOA-BiLSTM models did not exhibit overfitting and demonstrated a more accurate performance. The research findings underscore the substantial impact of hyperparameter configurations on BiLSTM neural network models.

The statistic that measures goodness of fit is the coefficient of determination, also known as the coefficient of certainty (R^2), with a maximum value of 1. The closer the value of R^2 is to 1, the better the regression line fits the observed value. In addition, RMSE is also an evaluation index to measure the fitting performance with the target value, and the smaller the RMSE, the smaller the error. After 4320 groups of data verification, the comparison results are shown in Table 3. It can be found that the BiLSTM neural network prediction model shows better prediction performance than LSTM. At the same time, the realization of LSTM and BiLSTM in the training set and test set is very different, and the overfitting phenomenon appears. The prediction model optimized for BiLSTM neural network parameters by the WOA optimization algorithm shows relatively stable regression fitting performance, which also indicates that for different regression problems, the superparameter setting of LSTM and BiLSTM has a great impact on the performance of the neural network. Improper setting makes it easy to fall into the local optimal solution and has poor generalization ability. The swarm intelligent optimization algorithm can solve the problem of improper setting of superparameters. At the same time, prediction accuracy can be improved by improving the WOA optimization algorithm.

Table 3. R^2 and RMSE values.

		LSTM	BiLSTM	WOA-BiLSTM	Improved-WOA-BiLSTM
Training	RMSE	0.365	0.315	0.254	0.206
Training	R^2	0.821	0.908	0.942	0.996
Test	RMSE	0.448	0.412	0.271	0.225
Test	R^2	0.783	0.823	0.932	0.984

Furthermore, the improved WOA-BiLSTM model achieved the best performance in predicting mud loss on the test dataset, with an RMSE (Root Mean Square Error) of 0.225 and an R^2 (Coefficient of Determination) of 0.984. When compared to the three models mentioned above, this model demonstrated significantly closer alignment with actual values in both trends and accuracy. As shown in Figure 11, the graph illustrates the absolute error between the predicted and actual values. Figure 11A presents the error when predicting mud loss, with the absolute error primarily concentrated around ±0.2. When the absolute error of predicting mud loss is less than 0.2, the prediction accuracy can reach up to 90.8%. On the other hand, Figure 11B displays the error when predicting mud overflow, where the prediction accuracy can reach 88.3% if the absolute error of predicting mud overflow is less than 0.2. These results indicate that this model performs excellently in terms of fitting accuracy, stability, and predictive performance, and can effectively predict mud circulation loss. The output during the training phase shows that in the early stages of mud loss prediction for all models, the predicted values lag behind the actual values, which may be due to the time window function we used. At the same time, we found that the accuracy of predicting mud circulation loss is higher than that of predicting mud overflow. The reason for this may be that there are more sample data sets where mud loss occurs in the training sample data, so the training effect is better.

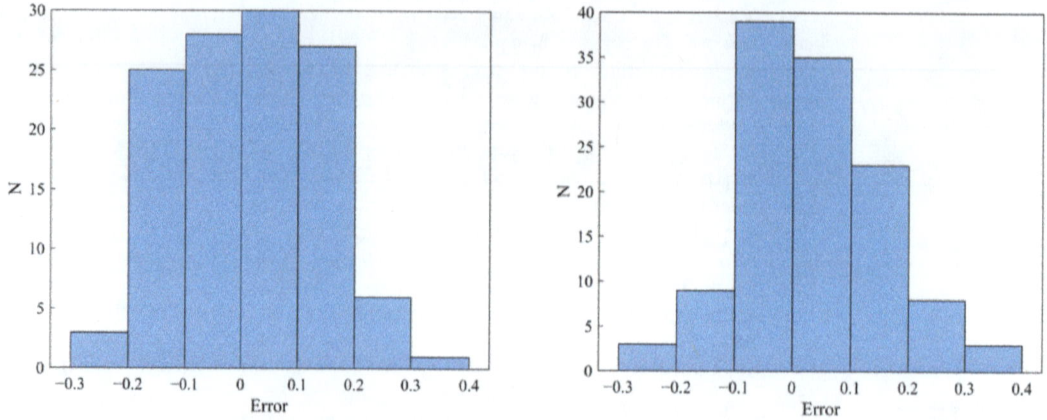

(A) **(B)**

Figure 11. Error distribution.

5. Discussion

The primary innovation of this study lies in the amalgamation of historical drilling data with an improved WOA-BiLSTM. This fusion resulted in the development of a mud loss circulation prediction model. This model, by utilizing a time window matrix, is capable of forecasting changes in the total pool volume ten minutes in advance, thus replacing the conventional manual recording of the total pool volume. The application of this predictive model indirectly accomplishes the prediction of mud circulation loss, with the potential to assist in early risk identification and the implementation of corresponding well control measures, ultimately mitigating the risk of blowouts at an early stage. Key aspects highlighted by our research include:

(1) The research underscores the critical role of the time window matrix in data processing. Through the incorporation of the time window matrix during the training process, we have successfully achieved early predictions of total pool volume changes. This method provides a new avenue for applying artificial neural networks to predict other drilling data and highlights the critical role of the time window matrix in data processing. It enables early predictions of total pool volume changes, opening possibilities for applying neural networks to other drilling data.

(2) Using the improved WOA for hyperparameter selection enhances predictive accuracy, offering valuable guidance for handling complex drilling datasets.

(3) While our study has made significant strides, limitations remain. It focuses on specific block wells, necessitating validation across different blocks for mud circulation loss prediction efficacy. Limited dataset size and quality may constrain model performance. Future research should consider expanding data diversity and quantity to enhance capabilities.

6. Conclusions

The upgraded WOA-BiLSTM neural network is the foundation for the early mud loss prediction model used in this study. The model primarily makes use of the upgraded WOA to overcome the challenge of configuring the parameters of the conventional BiLSTM neural network and raise the model's prediction accuracy. The following are the main findings:

(1) According to the size of the linked coefficients, extracted characteristic parameters are sorted using Spearman rank-related coefficients. The outcomes of WOA optimization reveal that seven criteria can be used to obtain extremely good accuracy. Therefore, these seven traits were tested in the modeling of the BiLSTM neural network algorithm: total pool volume (TPV), inlet flow rate, inlet iensity rate, weight on bit (WOB), surface pressure pump (SPP), torque (TQ), casing pressure (CP), and measured depth (MD).

(2) In this study, the maximum cycle, the initial learning rate, and the number of units in hidden layers 1 and 2 of the Bi-LSTM neural network structure are all optimized using the enhanced WOA. Based on this, the three prediction models LSTM, BiLSTM, and WOA-BiLSTM are compared with the early prediction model of WOA-BiLSTM of mud loss. There has been an increase in prediction accuracy of 22.3%, 18.7%, and 4.9%, respectively. The findings demonstrate that the enhanced WOA-BiLSTM model is more accurate in estimating early mud loss.

(3) The model can estimate changes in the total pit volume 10 min in advance, thereby predicting loss circulation with a high degree of accuracy. This precise forecasting contributes significantly to taking timely countermeasures, reducing the adverse effects of mud loss on drilling operations. For on site operators, this functionality is crucial as it allows for better work planning and management, resulting in increased production efficiency and reduced environmental risks.

(4) The research addresses a critical research gap in the field of petroleum drilling. To date, there have been relatively limited methods for predicting and managing mud loss events, and the model offers an efficient approach to addressing this issue. This is of great significance in ensuring the smooth progress of drilling operations and minimizing unnecessary downtime. Additionally, it underscores the potential of machine learning and deep learning in the petroleum engineering field y introducing advanced computational methods into traditional drilling processes. This will contribute to accelerating the digital transformation of petroleum engineering, improving industry efficiency, and sustainability.

Author Contributions: Methodology, F.G. and L.J.; Software, C.W.; Validation, X.L., W.J., C.W. and F.G.; Resources, Z.L.; Data curation, Z.L. and K.C.; Writing—original draft, W.J.; Writing—review & editing, X.L. and W.J.; Visualization, W.J. All authors have read and agreed to the published version of the manuscript.

Funding: This research was funded by Oil and Gas Reservoir Geology, Development Engineering National Key Laboratory Open Fund Project (Grant No. PLN2020-12) and Hubei Provincial Department of Education Scientific Research Program Funded Project (Grant No. D20201305).

Data Availability Statement: The data presented in this study are available on request from the corresponding author. The data are not publicly available due to privacy.

Acknowledgments: We express our gratitude to the anonymous reviewers for their invaluable insights and constructive feedback. This research was supported by Oil and Gas Reservoir Geology, Development Engineering National Key Laboratory Open Fund Project (Grant No. PLN2020-12) and Hubei Provincial Department of Education Scientific Research Program Funded Project (Grant No. D20201305).

Conflicts of Interest: The authors declare no conflict of interest.

References

1. Peng, Q.; Fan, H.; Xu, S.; Zhou, H.; Lai, M.; Ma, G.; Fu, S. *A Real-Time Warning System for Identifying Drilling Accidents*; SPE: Astana, Kazakhstan, 2014; p. SPE-172303-MS. [CrossRef]
2. Yi, H.; Ming, L.; Wentuo, L.; Wenbiao, D.; Chao, Z.; Zhenxiang, Z.; Tianwei, Z.; Xin, Z.; Yuqun, H.; Can, Z.; et al. *Application of Formation Pressure While Drilling Prediction Technology in Offshore HTHP Wells*; OnePetro: Richardson, TX, USA, 2019.
3. Arshad, U.; Jain, B.; Ramzan, M.; Alward, W.; Diaz, L.; Hasan, I.; Aliyev, A.; Riji, C. *Engineered Solution to Reduce the Impact of Lost Circulation During Drilling and Cementing in Rumaila Field, Iraq*; IPTC: Doha, Qatar, 2015; p. IPTC-18245-MS. [CrossRef]
4. Liu, Z.; Luo, R.; Yang, Z.; Wang, L.; Wang, L. Research and Practice of Risk Early Warning Technology for Lost Circulation with Drilling under the Conditions of Geological Engineering Information Fusion: The Example of the Yuanba Area. *Processes* **2022**, *10*, 2516. [CrossRef]
5. Moazzeni, A.R.; Nabaei, M.; Shahbazi, K.; Shadravan, A. *Mechanical Earth Modeling Improves Drilling Efficiency and Reduces Non-Productive Time (NPT)*; SPE: Manama, Bahrain, 2010; p. SPE-131718-MS. [CrossRef]
6. Alkinani, H.H.; Al-Hameedi, A.T.T.; Dunn-Norman, S. Artificial neural network models to predict lost circulation in natural and induced fractures. *SN Appl. Sci.* **2020**, *2*, 1980. [CrossRef]
7. Carpenter, C. Liner-Drilling Technology Mitigates Lost Circulation Offshore Mexico. *J. Pet. Technol.* **2014**, *66*, 104–107. [CrossRef]
8. Aljubran, M.; Ramasamy, J.; Albassam, M.; Magana-Mora, A. Deep Learning and Time-Series Analysis for the Early Detection of Lost Circulation Incidents During Drilling Operations. *IEEE Access* **2021**, *9*, 76833–76846. [CrossRef]
9. Sabah, M.; Talebkeikhah, M.; Agin, F.; Talebkeikhah, F.; Hasheminasab, E. Application of decision tree, artificial neural networks, and adaptive neuro-fuzzy inference system on predicting lost circulation: A case study from Marun oil field. *J. Pet. Sci. Eng.* **2019**, *177*, 236–249. [CrossRef]
10. Al-Hameedi, A.T.T.; Alkinani, H.H.; Dunn-Norman, S.; Flori, R.E.; Hilgedick, S.A.; Amer, A.S.; Alsaba, M. Mud loss estimation using machine learning approach. *J. Pet. Explor. Prod. Technol.* **2018**, *9*, 1339–1354. [CrossRef]
11. Tang, H.; Zhang, S.; Zhang, F.; Venugopal, S. Time series data analysis for automatic flow influx detection during drilling. *J. Pet. Sci. Eng.* **2019**, *172*, 1103–1111. [CrossRef]
12. Alkinani, H.H.; Al-Hameedi, A.T.T.; Dunn-Norman, S.; Flori, R.E.; Alsaba, M.T.; Amer, A.S.; Hilgedick, S.A. Using data mining to stop or mitigate lost circulation. *J. Pet. Sci. Eng.* **2019**, *173*, 1097–1108. [CrossRef]
13. Geng, Z.; Wang, H.; Fan, M.; Lu, Y.; Nie, Z.; Ding, Y.; Chen, M. Predicting seismic-based risk of lost circulation using machine learning. *J. Pet. Sci. Eng.* **2019**, *176*, 679–688. [CrossRef]
14. Abbas, A.K.; Bashikh, A.A.; Abbas, H.; Mohammed, H.Q. Intelligent decisions to stop or mitigate lost circulation based on machine learning. *Energy* **2019**, *183*, 1104–1113. [CrossRef]
15. Abbas, A.K.; Al-Haideri, N.A.; Bashikh, A.A. Implementing artificial neural networks and support vector machines to predict lost circulation. *Egypt J. Pet.* **2019**, *28*, 339–347. [CrossRef]
16. Ahmed, A.; Elkatatny, S.; Abdulraheem, A.; Abughaban, M. *Prediction of Lost Circulation Zones Using Support Vector Machine and Radial Basis Function*; IPTC: Dhahran, Saudi Arabia, 2020; p. IPTC-19628-MS. [CrossRef]
17. Hou, X.; Yang, J.; Yin, Q.; Liu, H.; Chen, H.; Zheng, J.; Wang, J.; Cao, B.; Zhao, X.; Hao, M.; et al. *Lost Circulation Prediction in South China Sea Using Machine Learning and Big Data Technology*; OTC: Houston, Texas, USA, 2020; p. OTC-30653-MS. [CrossRef]
18. Mardanirad, S.; Wood, D.A.; Zakeri, H. The application of deep learning algorithms to classify subsurface drilling lost circulation severity in large oil field datasets. *SN Appl. Sci.* **2021**, *3*, 785. [CrossRef]
19. Sabah, M.; Mehrad, M.; Ashrafi, S.B.; Wood, D.A.; Fathi, S. Hybrid machine learning algorithms to enhance lost-circulation prediction and management in the Marun oil field. *J. Pet. Sci. Eng.* **2021**, *198*, 108125. [CrossRef]
20. Yingzhuo, X.; Liupeng, W.; Yang, Z.; MingZhen, L. Real-time monitoring and early warning of well leakage based on big data analysis. *J. Phys. Conf. Ser.* **2021**, *1894*, 012084. [CrossRef]
21. Jafarizadeh, F.; Larki, B.; Kazemi, B.; Mehrad, M.; Rashidi, S.; Neycharan, J.G.; Gandomgoun, M.; Gandomgoun, M.H. A new robust predictive model for lost circulation rate using convolutional neural network: A case study from Marun Oilfield. *Petroleum* **2022**, 468–485. [CrossRef]
22. Siami-Namini, S.; Tavakoli, N.; Namin, A.S. The Performance of LSTM and BiLSTM in Forecasting Time Series. In Proceedings of the IEEE International Conference on Big Data (Big Data), Los Angeles, CA, USA, 9–12 December 2019; IEEE: Piscataway, NJ, USA, 2019; pp. 3285–3292. [CrossRef]
23. Ma, T.; Xiang, G.; Shi, Y.; Liu, Y. Horizontal in situ stresses prediction using a CNN-BiLSTM-attention hybrid neural network. *Géoméch. Geophys. Geo-Energy Geo-Resour.* **2022**, *8*, 152. [CrossRef]
24. Li, P.; Zhang, J.; Krebs, P. Prediction of Flow Based on a CNN-LSTM Combined Deep Learning Approach. *Water* **2022**, *14*, 993. [CrossRef]
25. Shen, Y.; Zhang, C.; Gharehchopogh, F.S.; Mirjalili, S. An improved whale optimization algorithm based on multi-population evolution for global optimization and engineering design problems. *Expert Syst. Appl.* **2023**, *215*, 119269. [CrossRef]
26. Fukuda, S.; Uchida, H.; Fujii, H.; Yamada, T. Short-term prediction of traffic flow under incident conditions using graph convolutional recurrent neural network and traffic simulation. *IET Intell. Transp. Syst.* **2020**, *14*, 936–946. [CrossRef]

27. Ulman, H.; Gütter, J.; Niebling, J. Uncertainty is not sufficient for identifying noisy labels in training data for binary segmentation of building footprints. *Front. Remote Sens.* **2023**, *3*, 1100012. [CrossRef]
28. Liu, Z.; Ma, Q.; Cai, B.; Liu, Y.; Zheng, C. Risk assessment on deepwater drilling well control based on dynamic Bayesian network. *Process Saf. Environ. Prot.* **2021**, *149*, 643–654. [CrossRef]
29. Gooneratne, C.P.; Gomez Gonzalez, E.S.; Al-Musa, A.S.; Osorio, H.F. *Thirsty Reservoirs—Challenges in Drilling Through Severe Lost Circulation Zones*; SPE: Abu Dhabi, United Arab Emirates, 2017; p. SPE-188461-MS. [CrossRef]
30. Chaweewanchon, A.; Chaysiri, R. Markowitz Mean-Variance Portfolio Optimization with Predictive Stock Selection Using Machine Learning. *Int. J. Financ. Stud.* **2022**, *10*, 64. [CrossRef]
31. Chatterjee, S.; Mahapatra, S.S.; Bharadwaj, V.; Upadhyay, B.N.; Bindra, K.S. Prediction of quality characteristics of laser drilled holes using artificial intelligence techniques. *Eng. Comput.* **2021**, *37*, 1181–1204. [CrossRef]
32. Liu, P.-L.; Du, Z.-C.; Li, H.-M.; Deng, M.; Feng, X.-B.; Yang, J.-G. Thermal error modeling based on BiLSTM deep learning for CNC machine tool. *Adv. Manuf.* **2021**, *9*, 235–249. [CrossRef]
33. Deng, H.; Liu, L.; Fang, J.; Qu, B.; Huang, Q. A novel improved whale optimization algorithm for optimization problems with multi-strategy and hybrid algorithm. *Math. Comput. Simul.* **2023**, *205*, 794–817. [CrossRef]
34. Mirjalili, S.; Lewis, A. The Whale Optimization Algorithm. *Adv. Eng. Softw.* **2016**, *95*, 51–67. [CrossRef]
35. Aljarah, I.; Faris, H.; Mirjalili, S. Optimizing connection weights in neural networks using the whale optimization algorithm. *Soft Comput.* **2018**, *22*, 1–15. [CrossRef]
36. Wood, D.A.; Mardanirad, S.; Zakeri, H. Effective prediction of lost circulation from multiple drilling variables: A class imbalance problem for machine and deep learning algorithms. *J. Pet. Explor. Prod. Technol.* **2022**, *12*, 83–98. [CrossRef]
37. Sirisha, B.; Goud, K.K.C.; Rohit, B.T.V.S. A Deep Stacked Bidirectional LSTM (SBiLSTM) Model for Petroleum Production Forecasting. *Procedia Comput. Sci.* **2023**, *218*, 2767–2775. [CrossRef]

MDPI
St. Alban-Anlage 66
4052 Basel
Switzerland
www.mdpi.com

Processes Editorial Office
E-mail: processes@mdpi.com
www.mdpi.com/journal/processes